SUPERFLUID STATES
OF MATTER

SUPERFLUID STATES
OF MATTER

SUPERFLUID STATES OF MATTER

Boris Svistunov
Egor Babaev
Nikolay Prokof'ev

CRC Press
Taylor & Francis Group
Boca Raton London New York

CRC Press is an imprint of the
Taylor & Francis Group, an **informa** business

Chapter 10 is written in coauthorship with Evgeny Kozik. Manuscript editing by Elizabeth Earl Phillips.

CRC Press
Taylor & Francis Group
6000 Broken Sound Parkway NW, Suite 300
Boca Raton, FL 33487-2742

First issued in paperback 2021

© 2015 by Boris Svistunov, Egor Babaev, and Nikolay Prokof'ev
CRC Press is an imprint of Taylor & Francis Group, an Informa business

No claim to original U.S. Government works

Version Date: 20141202

ISBN 13: 978-0-367-78352-5 (pbk)
ISBN 13: 978-1-4398-0275-5 (hbk)

Library of Congress Cataloging-in-Publication Data

Svistunov, Boris.
    Superfluid states of matter / Boris Svistunov, Egor Babaev, and Nikolay Prokof'ev.
        pages cm
    "A CRC title."
    Includes bibliographical references and index.
        ISBN 978-1-4398-0275-5 (hardcover : alk. paper) 1. Superfluidity. 2. Fluid dynamics. 3.
    Bose-Einstein condensation. I. Babaev, Egor, 1973- II. Prokof'ev, Nikolay. III. Title.

QC175.4.S85 2015
530.4'2--dc23
                                                                                                2014043091

Visit the Taylor & Francis Web site at
http://www.taylorandfrancis.com

and the CRC Press Web site at
http://www.crcpress.com

Contents

Preface

Superfluidity is the central topic across many fields of physics, including condensed matter, critical phenomena, quantum field theory, nuclear matter, and quantum chromodynamics, with a dozen Nobel Prizes awarded for related discoveries. In the last two decades the field has undergone an important transformation by a synergistic combination of theoretical progress and the vastly expanding scope of available experimental systems and potential applications. New superfluid materials include high-temperature and multicomponent superconductors, ultracold atomic bosons and fermions, exciton and polariton condensates, magnets, Josephson junction arrays, and helium supersolids, to name just a few. Certain theoretical models, which, until recently, were envisaged only theoretically, can now be implemented experimentally with great precision, bringing novel topological states of matter into focus.

Traditionally, the phenomenon of superfluidity is introduced and interpreted within the framework of quantum-field theory. Starting with the Bose–Einstein condensation/condensate (BEC) of ideal bosons, one proceeds with Bogoliubov's microscopic theory of normal modes in a weakly interacting Bose gas (WIBG), often employing the Landau criterion to argue that the system is superfluid. Similarly, a discussion of superfluidity (superconductivity) of fermions normally starts—once again resorting to the Landau criterion—with the Bardeen–Cooper–Schrieffer (BCS) solution for the elementary excitations of weakly interacting systems. Meanwhile, with the notable exception of ^{4}He, nearly all superfluid/superconducting systems discovered before 1995 consisted of fermionic particles, and ^{4}He is as far from being in the weakly interacting limit as one can imagine. Measured critical velocities turn out to be orders of magnitude smaller than predicted by the Landau criterion. It is also important to keep in mind that the Landau criterion is limited to Galilean systems only, and only in the zero-temperature limit—when one can speak of well-defined elementary excitations. Similarly, most of the key aspects of BCS theory are simply not generic for superfluidity. Finally, in the 1970s, Berezinskii, Kosterlitz, and Thouless proved that BEC is not required for superfluidity.

To explain superfluidity in the most general case, that is, under the conditions of (a) essentially finite temperature, including the vicinity of the normal-liquid transition, (b) possible absence of BEC, and (c) possible absence of Galilean invariance, one must invoke the notion of the topological order responsible for the emergence of an extra constant of motion. Since superfluid topological order is a natural property of classical complex-valued fields,* superfluidity of a bosonic quantum field is nothing but the property inherited by the system from its classical-field counterpart.

Hence, a generic discussion of fundamental aspects of superfluidity and superconductivity should start with the theory of classical complex-valued matter fields and their topological properties, while the quantum-field origin of the classical-field description for interacting bosons and fermions can be discussed later separately. This is the approach we adopt in this book. We start with a general reasoning of why one must separate superfluid properties of the field from its underlying quantum nature, if any, and introduce all essential concepts through the classical complex-valued matter field described by nonlinear wave equations. Their role is similar to that of Maxwell equations in the theory of electromagnetism.† The topological constant of motion is naturally introduced in terms of the phase of the classical matter field, revealing the fundamental origin of superfluidity. Next, we introduce topological defects and argue that the appearance of free topological defects is the only mechanism for destroying superfluidity. This brings about the idea of the superfluid-to-normal-fluid phase transition through proliferation of topological defects. We emphasize the difference between the genuine long-range order (BEC) and topological order taking place even in the absence of BEC.‡ It is this aspect that makes the theory of superfluidity prototypical with respect to more sophisticated cases of emergent topological orders.

In the first six chapters of the book (split into two parts), we establish fundamental macroscopic properties of superfluids and superconductors—including the finite-temperature superfluid-/superconductor-to-normal phase transitions—within the paradigm of the classical matter field. Part I (Chapters 1 through 4) deals with a single-component neutral matter field. Part II (Chapters 5 and 6) discusses charged matter fields, fundamentals of superconductivity, and multicomponent systems (both charged and neutral), including, in particular, the discussion of topological defects in systems that break various symmetries.

Part III (Chapters 7 through 10) deals with quantum-mechanical aspects and macrodynamics. In Chapter 7, we present the quantum-field perspective on the conditions under which classical-field description is relevant in bosonic and fermionic systems. An important aspect of our book is the use of Feynman path integrals to discuss subtle issues about supersolid and insulating ground states,

---

* All by itself, the classical-field description implies neither quantum mechanics nor superfluidity. It is a widespread misconception that the very presence of a classical-field component in a quantum system (often incorrectly identified with BEC) automatically implies superfluidity.

† Curiously, the idea of classical-field limit for quantum-particle systems goes back to Einstein's famous papers (1924–1925) on the application of Bose statistics to quantum gases. Unfortunately, Einstein's conjecture concerning the existence of the scalar matter-wave equation—an analog of Maxwell equations—was not immediately appreciated by the community. Decades later, and rather gradually, the understanding was built that the Gross–Pitaevskii (aka nonlinear Schrödinger) equation describes not only BECs but the normal state as well, thus forming a natural framework for a full-scale discussion of superfluid and normal states of matter.

‡ The condensate is known to be destroyed by temperature or quantum fluctuations of the phase of the order parameter in 2D and 1D, respectively.

the relationship between superfluidity and BEC, disordered systems, quantum phase transitions, etc. The path integral formalism is introduced in Chapter 8. In this chapter, we also show how Feynman path integrals can be efficiently simulated with the worm algorithm. The language of path integrals is heavily used in Chapter 9 to explain why nonsuperfluid (insulating) ground states of regular and disordered bosons occur under appropriate conditions. We also discuss in this chapter superfluid solids (supersolids). Chapter 10, written in coauthorship with Evgeny Kozik, discusses the rich dynamics of superfluids and various aspects of superfluid turbulence at $T \to 0$.

Our skepticism about starting a book on superfluidity with the theory of the weakly interacting Bose gas (WIBG) should not be interpreted as lack of appreciation for the importance of that system. In fact, we feel that the theory presented in Part IV (Chapters 11 through 13) is unique in its completeness. We present (a) thermodynamic functions derived from the Beliaev diagrammatic technique, adopted to provide a controlled unified approach to interacting gases in all physical dimensions, (b) an exhaustive description of the fluctuation region based on the universal—to all weakly interacting systems of the U(1) universality class—parameters and scaling functions obtained by first-principle simulations, and (c) the theory of the BEC formation kinetics. We also present a brief account of BCS theory for the weakly interacting Fermi gas.

In the historical overview (Chapter 14), we do not attempt to provide a comprehensive account of events, ideas, human relationships, or progress milestones—this information is available elsewhere. Rather, we highlight and critically analyze the most crucial developments that led us to the current understanding of superfluidity and superconductivity. We lay out the necessary arguments first, followed by our judgment of the events. It is only against the background of a fully developed theory providing an accurate scientific language and perspective that one can properly evaluate a particular advance in a dramatic and instructive evolution of physics knowledge. That is why our historical overview is the penultimate chapter in the book.

In the concluding chapter (Chapter 15), we briefly review the variety of superfluid and superconducting systems available today in nature and the laboratory, as well as the states that experimental realization is currently actively pursuing. As opposed to the generic theoretical discussion in Chapters 1 through 13, here we emphasize the specificity of each particular superfluid system.

The prime readership of our book is graduate students interested in systems and topics where superfluid states of matter manifest themselves: from quantum fluids, superconductors, and ultracold atomic gases to neutron matter statistical models and semiconducting structures. We believe that the format of the book will be appropriate for use as a main text for a graduate course on the theory of superfluidity and superconductivity. We hope that the book will be valuable to researchers in the field as well. Those who used to think about superfluidity in terms of the Landau criterion, the two-fluid model, or genuine long-range order might appreciate the power and simplicity of the approach based on the emergent topological constant of motion. Many in the community working on quantum gases are still not used to thinking of bosons in terms of Feynman's paths. Our book is intended as an introduction to this language. Experimentalists, as well as theorists, might find it useful to learn the basics of the state-of-the-art first-principle numeric approaches. As was mentioned earlier, our narration is based on the idea of separating key, topology-determined superfluid properties from the system's "quanta," allowing one to formulate all the relevant notions in terms of dynamics and statistics of the classical complex-valued field. This approach renders the material of the first few chapters understandable at the undergraduate level. The notion and significance of the emergent constant of motion is suitable for teaching the essence of superfluidity even in introductory physics courses. A continuously updated list of typos will be available at https://sites.google.com/site/superfluidstatesofmatter/. We encourage readers to submit typos they discover in the book to superfluidstates@gmail.com.

We are grateful to Elizabeth Earl Phillips, our manuscript editor, for significantly improving the readability of the book. We are also indebted to our colleagues Robert Hallock and Jonathan Machta for numerous valuable comments and suggestions. We acknowledge the support from the National Science Foundation.

Authors

Boris Vladimirovich Svistunov earned an MSc in physics in 1983 at Moscow Engineering Physics Institute, Moscow, Russia. In 1990, he earned a PhD in theoretical physics at Kurchatov Institute (Moscow), where he worked from 1986 to 2003 (and is still affiliated with). In 2003, he joined the Physics Department of the University of Massachusetts, Amherst.

Egor Sergeevich Babaev earned an MSc in physics in 1996 at St. Petersburg State Polytechnical University and A. F. Ioffe Physical Technical Institute, St. Petersburg, Russia. In 2001, he earned a PhD in theoretical physics at Uppsala University (Sweden). In 2007, after several years as a postdoctoral research associate at Cornell University, he joined the faculty of the Physics Department of the University of Massachusetts, Amherst. He is currently a faculty member at the Royal Institute of Technology, Sweden.

Nikolay Victorovich Prokof'ev earned an MSc in physics in 1982 at Moscow Engineering Physics Institute, Moscow, Russia. In 1987, he earned a PhD in theoretical physics from Kurchatov Institute (Moscow), where he worked from 1984 to 1999. In 1999, he joined the Physics Department of the University of Massachusetts, Amherst.

Abbreviations

bcc	Body-centered cubic
BCS	Bardeen–Cooper–Schrieffer
BEC	Bose–Einstein condensation/condensate
BG	Bose glass
BKT	Berezinskii–Kosterlitz–Thouless
CLT	Central limit theorem
1D	One dimension; one-dimensional
2D	Two dimensions; two-dimensional, etc.
GL	Ginzburg–Landau
GPE	Gross–Pitaevskii equation
hcp	Hexagonal close-packed
KT	Kosterlitz–Thouless
KW	Kelvin wave
LIA	Local induction approximation
LMH	Liquid metallic hydrogen
MC	Monte Carlo
MG	Mott glass
MI	Mott insulator
NK	Nelson–Kosterlitz
NLSE	Nonlinear Schrödinger equation
NMR	Nuclear magnetic resonance
n.n.	Nearest neighbor
ODLRO	Off-diagonal long-range order
OLE	Optical-lattice emulator
PIMC	Path integral Monte Carlo
RG	Renormalization group
SF	Superfluid
SG	Superglass
SQUID	Superconducting quantum interferometric device
ST	Superfluid turbulence
TLRO	Topological long-range order
UV	Ultraviolet
WIBG	Weakly interacting Bose gas

Part I

Superfluidity from a Classical-Field Perspective

CHAPTER 1

Neutral Matter Field

Superfluidity is the phenomenon of matter flow—referred to as a *superflow*—without any resistance from the walls or other obstacles. It is common to refer to superfluidity as a *macroscopic quantum* phenomenon. The word "quantum" here is only conditionally correct. Furthermore, it is somewhat confusing since it creates an impression that superfluidity cannot exist within the domain of classical physics. It is unquestionably true that the system of classical *particles* cannot support a superflow, and it is from that classical-particle perspective that superfluidity is associated with the quantum world. However, it turns out that for certain complex-valued classical *fields*, the superfluid phenomenon is absolutely natural. Hence, superfluidity looks quantum only from either the classical-particle or quantum-field point of view, while being classical from the complementary perspective. In general, this is reflective of the dual nature of quantum fields, which under certain conditions can mimic the behavior of purely classical particles, while under other conditions can also behave as classical fields.

One might think naively that it is easy to immediately tell the essentially quantum phenomena from conditionally quantum ones by just looking at the equations describing them: If the formulae explicitly contain the Planck's constant $\hbar$, then one is dealing with a quantum case; otherwise, it is classical. The loophole here is what is meant by "explicitly." Let us look at the central relation in the theory of superfluidity (the origin and precise meaning of which will be discussed later),

$$\mathbf{v}_s(\mathbf{r}) = \frac{\hbar}{m}\nabla\tilde{\Phi}(\mathbf{r}), \tag{1.1}$$

where $\mathbf{v}_s(\mathbf{r})$ is the velocity of superflow at point $\mathbf{r}$, m is a *single* particle mass, and $\tilde{\Phi}(\mathbf{r})$ is the phase of a certain classical complex-valued field, $\tilde{\Psi} = |\tilde{\Psi}|e^{i\tilde{\Phi}}$, characterizing the collective behavior of an ensemble of particles. Apart from $\hbar$, all other entities in Equation (1.1) seem classical, so that there is no way that $\hbar$ will simply cancel with yet another purely quantum factor. Meanwhile, the superfluidity is fundamentally of the *classical-field* microscopic origin. The subtlety about $\hbar$ in Equation (1.1) comes from the fact that it enters this and other expressions relevant to superfluidity exclusively as the ratio $\hbar/m$. Hence, as long as measurements do not go beyond the domain of superfluid responses, we have no tools to "observe" $\hbar$ and m separately; in other words, studies of superfluid properties do not allow one to reveal that the underlying field is quantized, that is, consists of discrete particles of the mass m. It is thus instructive to write Equation (1.1) in the form

$$\mathbf{v}_s(\mathbf{r}) = \gamma\nabla\tilde{\Phi}(\mathbf{r}), \tag{1.2}$$

with some coefficient γ. Then the only reminder of the quantum field theory—remarkable on its own but not at all necessary for introducing and understanding the phenomenon of superfluidity—will be the relation

$$\gamma = \frac{\hbar}{m}. \tag{1.3}$$

In essence, the previous two equations define our strategy in this book. We first develop the theory of superfluidity and, in particular, reveal the origin of Equation (1.2) on a purely classical-field basis. Only after that, substantially later, we trace the quantum-field-theoretical aspect of the parameter γ, expressed by Equation (1.3). Correspondingly, the parameter γ—or parameters γ_i ($=\hbar/m_i$ in the quantum case) for a multicomponent system with i labeling the components—will be playing the role of fundamental constant(s) throughout the first few chapters. We believe that this approach is best for newcomers to the field because it avoids unnecessary allusions to quantum mechanics. On the other hand, readers with a background in quantum fluids, who enjoy working with Equation (1.1) featuring $\hbar$ and m, may find our γ-parameterization somewhat annoying. For those advanced readers, we suggest the following way out: setting $\hbar = 1$ allows one to think of γ as an inverse mass, and vice versa, $m \equiv 1/\gamma$. The trick also allows the beginners to read comfortably the "quantum" superfluid literature, by purging all the relations from $\hbar$ and m.

In the rest of this chapter, we will build the theory of the superfluid state starting from the classical complex-field description. Since this is not the traditional way of looking at things and since most texts on the dynamics of classical fields are limited to electricity and magnetism, to establish the proper mathematical foundation for the rest of our discussion, we start with the formal introduction to the theory of Hamiltonian dynamics of complex-valued fields. Readers familiar with the subject may skip the next section.

1.1   Classical Hamiltonian Formalism

1.1.1   Hamiltonian Mechanics

In classical Hamiltonian formalism, the state of a mechanical system is specified by a set of pairs of conjugated variables, $\{q_j, p_j\}$, $(j = 1, 2, 3, \ldots)$ referred to as generalized coordinates and momenta, respectively. The equations of motion are generated with the Hamiltonian function, $H(\{q_j, p_j\})$, by the following axiom:

$$\dot{q}_j = \frac{\partial H}{\partial p_j}, \tag{1.4}$$

$$\dot{p}_j = -\frac{\partial H}{\partial q_j}. \tag{1.5}$$

From the form of Equations (1.4) and (1.5), one can see that the difference between generalized coordinates and corresponding generalized momenta is merely terminological. Indeed, either of the two transformations, (1) $q_j \to -q_j$ or (2) $p_j \to -p_j$, changes the signs in Equations (1.4) and (1.5), thus turning the corresponding "coordinates" into "momenta" and "momenta" into "coordinates." In this sense, more relevant is the term *pair of canonical*, or, equivalently, *canonically conjugated* variables, which stands for the whole pair (q_j, p_j), treating both variables on equal footing. Moreover, sometimes it is reasonable—we will see it soon (!)—to unify explicitly the two canonically conjugated variables (q_j, p_j) by introducing a single *complex canonical variable* a_j, such that

$$a_j = (\alpha q_j + i\beta p_j), \tag{1.6}$$

where α and β are complex numbers (that can also depend on j, but we are not interested here in such a generalization). The equations of motion then acquire the form

$$i\lambda \dot{a}_j = \frac{\partial H}{\partial a_j^*}, \tag{1.7}$$

with $\lambda = 1/(\alpha\beta^* + \alpha^*\beta)$. That is, each pair of real Equations (1.4) and (1.5) is replaced with one complex Equation (1.7). Equivalently, one may use the complex-conjugated equation $-i\lambda \dot{a}_j^* = \partial H/\partial a_j$ (by definition, H is a real-valued function).

A mathematical remark is in order here. In Equation (1.7), we use the notion of a partial derivative with respect to complex variables. For a function $F(z, z^*)$ of the complex variable $z = z' + iz''$, the partial derivatives $\partial/\partial z$ and $\partial/\partial z^*$ are defined as

$$\partial/\partial z = (\partial/\partial z' - i\partial/\partial z'')/2, \tag{1.8}$$

$$\partial/\partial z^* = (\partial/\partial z' + i\partial/\partial z'')/2. \tag{1.9}$$

The reason for introducing the definitions (1.8) and (1.9) is that with respect to these operations, the variables z and z^* may be formally considered as independent.

Problem 1.1 *Prove the previous statement. Show that Equation (1.7) is equivalent to Equations (1.4) and (1.5).*

Problem 1.2 *Derive the following useful formula for taking derivatives with respect to the phase φ of the complex number z $(z \equiv |z|e^{i\varphi})$:*

$$\partial/\partial\varphi = z' \partial/\partial z'' - z'' \partial/\partial z' = i(z \partial/\partial z - z^* \partial/\partial z^*). \tag{1.10}$$

The complex form of Hamiltonian mechanics is especially convenient in the case when the Hamiltonian function $H(\{a_j, a_j^*\})$ is invariant under the transformation $a_j \to a_j e^{i\varphi}$ (the phase φ is one and the same for all j's)—such a property is called a *global U(1) symmetry*. In this case, the following quadratic form of the complex variables,

$$N = \sum_j |a_j|^2, \tag{1.11}$$

is a constant of motion:

$$\dot{N} = 0. \tag{1.12}$$

Problem 1.3 *Prove Equation (1.12). Hint: A convenient starting point is the observation that by the global U(1) symmetry, $\partial H(\{a_j e^{i\varphi}, a_j^* e^{-i\varphi}\})/\partial \varphi \equiv 0$.*

We note in passing that, when going from a classical model to its quantum counterpart, complex variables $\{a_j\}$ prove very convenient. The variables a and a^* are then simply replaced with annihilation and creation operators, respectively; and the quantity N acquires the meaning of the total number of quanta, conserved by U(1)-symmetric Hamiltonians.

Poisson bracket, $\{A, B\}$, is a function of coordinates and momenta constructed from the two other functions, $A(\{q_j, p_j\})$ and $B(\{q_j, p_j\})$, by the following prescription:

$$\{A, B\} = \sum_j \left[\frac{\partial A}{\partial p_j} \frac{\partial B}{\partial q_j} - \frac{\partial A}{\partial q_j} \frac{\partial B}{\partial p_j} \right]. \tag{1.13}$$

As is seen from (1.4) and (1.5)—and this is the most important property of Poisson brackets—for any function of coordinates and momenta, $A(\{q_j, p_j\})$, its time derivative during the evolution can be expressed as

$$\dot{A} = \{H, A\}. \tag{1.14}$$

In particular, if $\{H, A\} \equiv 0$, then the quantity A is a constant of motion.

As is readily checked, in terms of complex variables (1.6), the definition (1.13) reads

$$\{A, B\} = (i/\lambda) \sum_j \left[\frac{\partial A}{\partial a_j} \frac{\partial B}{\partial a_j^*} - \frac{\partial A}{\partial a_j^*} \frac{\partial B}{\partial a_j} \right]. \tag{1.15}$$

1.1.2 Complex-Valued Scalar Field as a Canonical Variable

Very important for the purposes of this book is a special subclass of the Hamiltonian classical-field theories in which a classical complex-valued scalar field $\psi(\mathbf{r})$ is a *complex canonical variable*. We emphasize that the mere fact that a dynamical equation for a complex-valued classical field can be obtained within Hamiltonian formalism does not yet imply that the field is a complex canonical variable in the sense of definition (1.6), since the latter requires that the real and imaginary parts of the field be canonically conjugated variables—which is not generally true.

In a direct analogy with (1.7)—with the parameter λ set equal to unity by appropriately rescaling the canonical variables—the equation of motion reads

$$i\dot{\psi}(\mathbf{r}) = \frac{\delta H[\psi, \psi^*]}{\delta \psi^*(\mathbf{r})}. \tag{1.16}$$

The Hamiltonian H is a real-valued functional of the field ψ. Note that from the very form of Equation (1.16), it is clear that we are dealing with a special case of Hamiltonian field theory, since the equation of motion for $\psi(\mathbf{r})$ is always first order in terms of the time derivatives.

Similar to (1.8) and (1.9), the variational derivatives with respect to a complex-valued function are defined through the variational derivatives with respect to its real and imaginary parts. Accordingly, with respect to the operations $\delta/\delta\psi(\mathbf{r})$ and $\delta/\delta\psi^*(\mathbf{r})$, the fields $\psi(\mathbf{r})$ and $\psi^*(\mathbf{r})$ formally behave as independent ones.

Problem 1.4 *Find the variational derivatives with respect to $\psi(\mathbf{r})$ and $\psi^*(\mathbf{r})$ for the functionals $\int |\psi|^2 d^d r$, $\int |\psi|^4 d^d r$, and $\int |\nabla\psi|^2 d^d r$.*

Problem 1.5 *Let the field ψ depend on some parameter α: $\psi \equiv \psi(\mathbf{r},\alpha)$. Show that for any functional $F[\psi,\psi^*]$, the following formula takes place:*

$$\frac{\partial F[\psi,\psi^*]}{\partial \alpha} = \int \left(\frac{\delta F}{\delta\psi} \frac{\partial\psi}{\partial\alpha} + \frac{\delta F}{\delta\psi^*} \frac{\partial\psi^*}{\partial\alpha} \right) d^d r, \tag{1.17}$$

where the variable α can, in particular, be complex, in which case the partial derivative is understood in the sense of Equation (1.8). For a complex α, replace $\alpha \to \alpha^*$ and make sure that (1.17) still applies. Prove that the energy $E \equiv H[\psi,\psi^*]$ is conserved by the equation of motion (1.16): Use (1.17) with $\alpha \equiv t$.

In the case of the classical-field theory, the global U(1) symmetry of the Hamiltonian is defined as invariance with respect to the transformation $\psi(\mathbf{r}) \to \psi(\mathbf{r}) e^{i\varphi}$; the word "global" here means that the phase φ is independent of $\mathbf{r}$. If global U(1) symmetry takes place, then the *amount of matter*,

$$N = \int |\psi|^2 d^d r, \tag{1.18}$$

is conserved:

$$\dot{N} = 0. \tag{1.19}$$

Problem 1.6 *Prove this statement by using (1.17) with $\alpha \equiv \varphi$.*

The Poisson bracket, $\{A,B\}$, is defined as a functional formed out of the functionals $A[\psi]$ and $B[\psi]$ by the rule

$$\{A,B\} = i \int \left[\frac{\delta A}{\delta\psi(\mathbf{r})} \frac{\delta B}{\delta\psi^*(\mathbf{r})} - \frac{\delta A}{\delta\psi^*(\mathbf{r})} \frac{\delta B}{\delta\psi(\mathbf{r})} \right] d^d r. \tag{1.20}$$

Similar to the Hamiltonian mechanics of discrete variables, for any functional $A[\psi]$ of the complex field $\psi(\mathbf{r})$ obeying the Hamiltonian equation of motion (1.16), its time derivative is given by

$$\dot{A} = \{H,A\}. \tag{1.21}$$

The validity of (1.21) is readily seen from (1.16) and (1.17).

1.1.3   Canonical Transformation

Along with a set of complex *canonical* variables,[*] $\{a_j\}$, one can work with a set of *arbitrary* variables, $\{b_s\}$, $b_s \equiv b_s(\{a_j\})$, employing the Poisson brackets to produce the corresponding equations of motion:

$$i\dot{b}_s = i\{H, b_s\} = \sum_j \left[\frac{\partial b_s}{\partial a_j} \frac{\partial H}{\partial a_j^*} - \frac{\partial b_s}{\partial a_j^*} \frac{\partial H}{\partial a_j} \right]. \tag{1.22}$$

Remarkably, there exist such sets of new variables that—independent of the particular form of H—are also canonical; that is, Equation (1.22) is equivalent to

$$i\dot{b}_s = \frac{\partial H}{\partial b_s^*}, \tag{1.23}$$

for *any* H. To arrive at the necessary and sufficient conditions for the new variables to be canonical, we note that what we require amounts to

$$\sum_j \left[\frac{\partial b_s}{\partial a_j} \frac{\partial}{\partial a_j^*} - \frac{\partial b_s}{\partial a_j^*} \frac{\partial}{\partial a_j} \right] \equiv \frac{\partial}{\partial b_s^*} \equiv \sum_j \left[\frac{\partial a_j^*}{\partial b_s^*} \frac{\partial}{\partial a_j^*} + \frac{\partial a_j}{\partial b_s^*} \frac{\partial}{\partial a_j} \right]. \tag{1.24}$$

By linear independence of the differential operators $\{\partial/\partial a_j, \partial/\partial a_j^*\}$, Equation (1.24) leads to

$$\frac{\partial b_s}{\partial a_j} = \frac{\partial a_j^*}{\partial b_s^*}, \qquad \frac{\partial b_s}{\partial a_j^*} = -\frac{\partial a_j}{\partial b_s^*}. \tag{1.25}$$

These are the necessary and sufficient conditions for $\{b_s\}$ to be canonical.

Problem 1.7 *Show that the following two elementary transformations dealing with only one variable, a_j, are canonical:*

(a) *Shift of the variable: $a_j \rightarrow a_j + c_j$, where c_j is a complex constant.*

(b) *Shift of the variable's phase: $a_j \rightarrow e^{i\varphi_j} a_j$, where φ_j is a real constant.*

Show that the linear transformation $b_s = \sum_j u_{sj} a_j$ is canonical, if and only if the matrix u is unitary.

Using Equation (1.25), one can show that the Poisson brackets are invariant with respect to the canonical transformation. That is, the result for $\{A, B\}$ is *the same* for *any* set of canonical variables.

---

[*] For the sake of definiteness and simplicity of notation, we treat our canonical variables as discrete. The translation into the complex-field language is straightforward (see, e.g., the end of Section 1.1.4).

Problem 1.8 *Prove this.*

The invariance of the Poisson brackets under some transformation of variables is a *sufficient* condition for that transformation to be canonical. Indeed, noticing that, for any A,

$$\frac{\partial A}{\partial a_j^*} = i\{A, a_j\}, \qquad \frac{\partial A}{\partial a_j} = i\{a_j^*, A\}, \qquad (1.26)$$

and requiring the invariance of the Poisson brackets, we arrive at (1.25):

$$\frac{\partial b_s}{\partial a_j} = i\{a_j^*, b_s\} = \frac{\partial a_j^*}{\partial b_s^*}, \qquad \frac{\partial b_s}{\partial a_j^*} = i\{b_s, a_j\} = -i\{a_j, b_s\} = -\frac{\partial a_j}{\partial b_s^*}. \qquad (1.27)$$

1.1.4 Normal Modes by Bogoliubov Transformation

An equilibrium solution of the equation of motion is the solution that does not evolve in time. By the Hamiltonian equations of motion, the equilibrium solution corresponds to such a point, $\{a_s^{(0)}\}$, in the space of variables $\{a_s\}$, where for any s

$$\frac{\partial H}{\partial a_s} = \frac{\partial H}{\partial a_s^*} = 0 \qquad \text{(at the equilibrium point).} \qquad (1.28)$$

We see that the equilibrium points are the points of *extremal* behavior of the Hamiltonian.

At the equilibrium point $\{a_s^{(0)}\}$, a natural question arises of what happens if the state of the system is slightly perturbed. The general way of answering this question is to expand the Hamiltonian in the vicinity of the equilibrium in terms of the shifted variables (recall that the shifted variables are canonical), $a_s \rightarrow a_s - a_s^{(0)}$, assuming that their absolute values are small enough. Omitting the dynamically irrelevant constant term, the resulting Hamiltonian reads

$$H = \sum_{ij} \left(A_{ij} a_i^* a_j + \frac{1}{2} B_{ij} a_i a_j + \frac{1}{2} B_{ij}^* a_i^* a_j^* \right), \qquad (1.29)$$

$$A_{ij} = \frac{\partial^2 H}{\partial a_i \, \partial a_j^*}, \qquad B_{ij} = \frac{\partial^2 H}{\partial a_i \, \partial a_j} \qquad \text{(at the equilibrium point).} \quad (1.30)$$

[Linear terms of the expansion are zero because of Equation (1.28).] Given (a) that the Hamiltonian has to be real and (b) that we can always symmetrize the matrix B in the case when it is not symmetric automatically, from now on, we will assume, without loss of generality, that

$$A_{ij} = A_{ji}^*, \qquad B_{ij} = B_{ji}. \qquad (1.31)$$

For all equilibrium points that are (local) minima of the function H, the Hamiltonian (1.29) can be *diagonalized* by the Bogoliubov transformation

$$b_s = \sum_j (u_{sj} a_j + v_{sj} a_j^*), \qquad (1.32)$$

which is a linear canonical transformation with certain coefficients u and v. By definition, a Hamiltonian is diagonal with respect to canonical variables $\{b_s\}$ if it has the form

$$H = \sum_s \omega_s b_s^* b_s, \tag{1.33}$$

equivalent to a set of noninteracting harmonic oscillators (called the normal modes). We will show how to relate the u and v coefficients to the matrices A and B, but first we establish their general properties. From the requirement that the transformation (1.32) be canonical, we get [differentiate both sides with respect to a_j, a_j^*, b_r, and b_r^* and use (1.25)]

$$\sum_j [u_{sj} u_{rj}^* - v_{sj} v_{rj}^*] = \delta_{sr}, \qquad \sum_j [u_{sj} v_{rj} - v_{sj} u_{rj}] = 0. \tag{1.34}$$

Now we make sure that the conditions (1.34) are not only necessary, but also *sufficient* for the transformation to be canonical. To this end, we observe that if coefficients u and v satisfy (1.34), then the inverse transformation is given by

$$a_j = \sum_s (\tilde{u}_{js} b_s + \tilde{v}_{js} b_s^*), \qquad \tilde{u}_{js} = u_{sj}^*, \qquad \tilde{v}_{js} = -v_{sj}. \tag{1.35}$$

Problem 1.9 *Derive the observation by substituting a_j from (1.35) into the right-hand side of (1.32).*

Now when we know how to express a's in terms of b's, we need to make sure that Equations (1.25) are satisfied; and this is easily seen from (1.35).

The relations (1.34) for the transformation (1.35) read

$$\sum_s [\tilde{u}_{js} \tilde{u}_{ks}^* - \tilde{v}_{js} \tilde{v}_{ks}^*] = \delta_{jk}, \qquad \sum_s [\tilde{u}_{js} \tilde{v}_{ks} - \tilde{v}_{js} \tilde{u}_{ks}] = 0. \tag{1.36}$$

With (1.35), these can be rewritten as

$$\sum_s [u_{sj}^* u_{sk} - v_{sj} v_{sk}^*] = \delta_{jk}, \qquad \sum_s [u_{sj}^* v_{sk} - v_{sj} u_{sk}^*] = 0. \tag{1.37}$$

To obtain the equations for the u and v coefficients, we start with

$$\omega_s b_s = \frac{\partial H}{\partial b_s^*}, \tag{1.38}$$

and then treat b's as functions of a's. Differentiating both sides with respect to a_j and a_j^* and taking into account

$$\frac{\partial b_s}{\partial a_j} \equiv u_{sj}, \qquad \frac{\partial b_s}{\partial a_j^*} \equiv v_{sj}, \tag{1.39}$$

we get

$$\omega_s u_{sj} = \frac{\partial}{\partial a_j}\frac{\partial H}{\partial b_s^*} = \frac{\partial}{\partial b_s^*}\frac{\partial H}{\partial a_j}, \qquad\qquad \omega_s v_{sj} = \frac{\partial}{\partial a_j^*}\frac{\partial H}{\partial b_s^*} = \frac{\partial}{\partial b_s^*}\frac{\partial H}{\partial a_j^*}. \qquad (1.40)$$

Now we note that here we can change the order of operators $(\partial/\partial b)$'s and $(\partial/\partial a)$'s, because our transformation is linear and, say, $(\partial/\partial b)$'s are just linear combinations of $(\partial/\partial a)$'s:

$$\frac{\partial}{\partial b_s^*} = \sum_j \left[\frac{\partial a_j}{\partial b_s^*}\frac{\partial}{\partial a_j} + \frac{\partial a_j^*}{\partial b_s^*}\frac{\partial}{\partial a_j^*} \right] = \sum_j \left[\tilde{v}_{js}\frac{\partial}{\partial a_j} + \tilde{u}_{js}^*\frac{\partial}{\partial a_j^*} \right]. \qquad (1.41)$$

Calculating the derivatives

$$\frac{\partial H}{\partial a_j} = \sum_i [A_{ij}\, a_i^* + B_{ij}\, a_i], \qquad\qquad \frac{\partial H}{\partial a_j^*} = \sum_i [A_{ji}\, a_i + B_{ij}^*\, a_i^*], \qquad (1.42)$$

and observing that

$$\frac{\partial a_i}{\partial b_s^*} \equiv \tilde{v}_{is} = -v_{si}, \qquad\qquad \frac{\partial a_i^*}{\partial b_s^*} \equiv \tilde{u}_{is}^* = u_{si}, \qquad (1.43)$$

we finally arrive at the system of equations:

$$\sum_i [A_{ij}\, u_{si} - B_{ji}\, v_{si}] = \omega_s\, u_{sj}, \qquad\qquad \sum_i [B_{ji}^*\, u_{si} - A_{ji}\, v_{si}] = \omega_s\, v_{sj}. \qquad (1.44)$$

Using the vector notation

$$|u_s\rangle = (u_{s,1}, u_{s,2}, u_{s,3}, \ldots), \qquad |v_s\rangle = (v_{s,1}, v_{s,2}, v_{s,3}\ldots), \qquad (1.45)$$

and taking into account (1.31), we write this system as

$$\begin{cases} A^*|u_s\rangle - B|v_s\rangle = \omega_s|u_s\rangle, \\ B^*|u_s\rangle - A|v_s\rangle = \omega_s|v_s\rangle. \end{cases} \qquad (1.46)$$

We can also write this system in terms of $\tilde{u}$'s and $\tilde{v}$'s. Introducing the vectors

$$|\tilde{u}_s\rangle = (\tilde{u}_{1,s}, \tilde{u}_{2,s}, \tilde{u}_{3,s}, \ldots), \qquad\qquad |\tilde{v}_s\rangle = (\tilde{v}_{1,s}, \tilde{v}_{2,s}, \tilde{v}_{3,s}, \ldots), \qquad (1.47)$$

we have

$$\begin{cases} A|\tilde{u}_s\rangle + B^*|\tilde{v}_s^*\rangle = \omega_s|\tilde{u}_s\rangle, \\ B|\tilde{u}_s\rangle + A^*|\tilde{v}_s^*\rangle = -\omega_s|\tilde{v}_s^*\rangle. \end{cases} \qquad (1.48)$$

The frequencies of the normal modes, ω_s's, arise as the eigenvalues of the problem. The relations (1.34) and (1.36), which in the vector notation read

$$\langle u_r|u_s\rangle - \langle v_r|v_s\rangle = \delta_{sr}, \qquad\qquad \langle v_r^*|u_s\rangle - \langle u_r^*|v_s\rangle = 0, \qquad (1.49)$$

$$\langle \tilde{u}_s|\tilde{u}_r\rangle - \langle \tilde{v}_r|\tilde{v}_s\rangle = \delta_{sr}, \qquad\qquad \langle \tilde{u}_s|\tilde{v}_r\rangle - \langle \tilde{u}_r|\tilde{v}_s\rangle = 0, \qquad (1.50)$$

are automatically guaranteed at $\omega_r \neq \omega_s$.

Problem 1.10 *Prove that this is the case.*

The system (1.46) is invariant with respect to the transformation

$$|u_s\rangle \rightarrow |v_s^*\rangle, \qquad |v_s\rangle \rightarrow |u_s^*\rangle, \qquad \omega_s \rightarrow -\omega_s. \tag{1.51}$$

This means that each solution has its counterpart of the opposite frequency. However, only one solution of the pair is physically relevant, namely, the one with $\langle u_s|u_s\rangle - \langle v_s|v_s\rangle > 0$. For its counterpart, we then have $\langle u_s|u_s\rangle - \langle v_s|v_s\rangle < 0$, which means that it is impossible to normalize it to unity, in contradiction with (1.49).

For a complex-valued field, the transformation (1.35) reads

$$\psi(\mathbf{r}) = \sum_s [\, \tilde{u}_s(\mathbf{r})\, b_s + \tilde{v}_s(\mathbf{r})\, b_s^* \,]. \tag{1.52}$$

The functions $\tilde{u}_s(\mathbf{r})$ and $\tilde{v}_s(\mathbf{r})$ can be found by solving Equations (1.48), with A and B understood as operators (Hermitian and symmetric, respectively). In this case, Equations (1.48) acquire the form of a system of two partial-derivative equations, referred to as *Bogoliubov–de Gennes* (BdG) equations.

If the classical-field system is spatially uniform, the normal modes are naturally labeled with their wavevectors, $s \rightarrow \mathbf{k}$, the functions $\tilde{u}_{\mathbf{k}}(\mathbf{r})$ and $\tilde{v}_{\mathbf{k}}(\mathbf{r})$ being the plane waves:

$$\tilde{u}_{\mathbf{k}}(\mathbf{r}) = \alpha_{\mathbf{k}}\, e^{i\mathbf{k}\cdot\mathbf{r}}, \qquad\qquad \tilde{v}_{\mathbf{k}}(\mathbf{r}) = \beta_{\mathbf{k}}\, e^{-i\mathbf{k}\cdot\mathbf{r}}. \tag{1.53}$$

The form of the plane-wave exponentials follows from the requirement that the modes $\{b_{\mathbf{k}}\}$ (not necessarily normal) are the eigenmodes of translations by an arbitrary vector $\mathbf{r}_0$:

$$\psi(\mathbf{r}) \rightarrow \psi(\mathbf{r} + \mathbf{r}_0) \qquad \Leftrightarrow \qquad b_{\mathbf{k}} \rightarrow e^{i\mathbf{k}\cdot\mathbf{r}_0} b_{\mathbf{k}}. \tag{1.54}$$

Consistent with (1.50), the transformation (1.52)–(1.53) diagonalizes the momentum:

$$\mathbf{P} = \sum_{\mathbf{k}} \mathbf{k}\,|b_{\mathbf{k}}|^2. \tag{1.55}$$

The uniformity of the system also guarantees that plane waves are the eigenfunctions of the operators A and B. As a result, BdG equations reduce to a system of algebraic equations:

$$\left[\begin{array}{c|c} A_{\mathbf{k}} - \omega_{\mathbf{k}} & B_{\mathbf{k}}^* \\ \hline B_{\mathbf{k}} & A_{\mathbf{k}} + \omega_{\mathbf{k}} \end{array} \right] \left[\begin{array}{c} \alpha_{\mathbf{k}} \\ \beta_{-\mathbf{k}}^* \end{array} \right] = \left[\begin{array}{c} 0 \\ 0 \end{array} \right], \tag{1.56}$$

where $A_{\mathbf{k}}$ and $B_{\mathbf{k}}$ are the eigenvalues of the operators A and B for the plane wave $e^{i\mathbf{k}\cdot\mathbf{r}}$. (Note that $A^\dagger = A$ implies $A_{\mathbf{k}}^* = A_{\mathbf{k}}$.) The characteristic equation for the problem then yields the dispersion of the modes:

$$\omega_{\mathbf{k}} = \pm\sqrt{A_{\mathbf{k}}^2 - |B_{\mathbf{k}}|^2}, \tag{1.57}$$

the right choice of the sign corresponding to the solution with $|\alpha_{\mathbf{k}}|^2 - |\beta_{-\mathbf{k}}^*|^2 > 0$ [see the comment after Equation (1.51)].

1.1.5   Statistical and Dynamical Instabilities

Consider the case when the bilinear Hamiltonian, while diagonalizable by the Bogoliubov transformation, features ω's that are real but not all of them are positive. Clearly, the equilibrium point of such a Hamiltonian is not the point of local energy minimum, but the weakly perturbed dynamics is still a superposition of oscillating normal modes. However, in an open system when there are terms in the Hamiltonian that weakly couple it to a heat bath, the amplitudes of the modes with negative ω will gradually drift to higher and higher values. Such a situation is called *statistical instability*. Moreover, even in the microcanonical system, the amplitudes of some (or all) modes may increase slowly with time due to nonlinear couplings between them since this leads to the entropy increase at a constant energy: the positive contribution to the total energy from the positive-ω modes is compensated by negative contribution from the negative-ω modes.

A point of equilibrium is called *dynamically unstable* if the amplitude of the perturbed solution—no matter how small—grows exponentially, taking the system away from the equilibrium. In a close analogy to the problem of normal modes, the question of dynamical instability can be addressed within the bilinear Hamiltonian (1.29). The difference comes from the fact that at the point of dynamical instability, the bilinear Hamiltonian cannot be (fully) diagonalized,* and, instead of applying the Bogoliubov transformation, one has to start with the equations of motion:

$$i\frac{\partial}{\partial t}|a\rangle = A|a\rangle + B^*|a^*\rangle. \tag{1.58}$$

Since this is a linear equation, its general solution can be written as a sum of elementary solutions (the necessity of the term with complex-conjugated ω comes from the term with complex-conjugated a):

$$|a\rangle = e^{-i\omega t}|f\rangle + e^{i\omega^* t}|g\rangle. \tag{1.59}$$

Substituting this into (1.58) and separating terms with $e^{-i\omega t}$ and $e^{i\omega^* t}$, which are linear independent at $\mathrm{Re}\,\omega \neq 0$, we get

$$\begin{cases} A|f\rangle + B^*|g^*\rangle = \omega|f\rangle, \\ B|f\rangle + A^*|g^*\rangle = -\omega|g^*\rangle. \end{cases} \tag{1.60}$$

By continuity with respect to Hamiltonian parameters (i.e., by setting $\mathrm{Re}\,\omega$ to be infinitesimally small, but finite), this system of equations applies also to the $\mathrm{Re}\,\omega = 0$ case.

It is not surprising that (1.60) takes the same form as (1.48), since in a situation with real ω's, the elementary solutions (1.59) correspond to the normal modes found by the Bogoliubov transformation. Complex-conjugating Equation (1.60), we readily reveal the symmetry

---

* As an instructive example, consider Equations (1.56) and (1.57) with $A_{\mathbf{k}}^2 < |B_{\mathbf{k}}|^2$.

$$|f\rangle \rightarrow |g\rangle, \qquad |g\rangle \rightarrow |f\rangle, \qquad \omega \rightarrow -\omega^*, \tag{1.61}$$

which generalizes (1.51) to the case of complex frequencies.

In the case when at least one of the frequencies is not real, there always exists a mode with $\mathrm{Im}\,\omega > 0$, leading to dynamical instability. The fact that the existence of a complex solution with $\mathrm{Im}\,\omega < 0$ implies the existence of yet another one, with $\mathrm{Im}\,\omega > 0$, and vice versa, follows from the symmetries of the matrices A and B, as expressed by Equation (1.31). [Note that the symmetry (1.61) is not helpful for our purposes, since it does not change the sign of the imaginary part of ω.] To reveal the relevant symmetry of the eigenvalues, consider the characteristic equation for the problem (1.60), written in terms of the determinant of a block (2×2) matrix:

$$\det \begin{bmatrix} A - \omega & B^* \\ B & A^* + \omega \end{bmatrix} = 0. \tag{1.62}$$

Now transposing the block (2×2) matrix, swapping its rows and columns, and finally utilizing the symmetries (1.31), we arrive at the same characteristic equation, but with $\omega \rightarrow -\omega$. Hence, if ω is an eigenvalue of the problem, then $-\omega$ is also an eigenvalue. For real ω's, this statement is equivalent to (1.61), while for complex ω's, it yields the desired change of the sign of the imaginary part.

Problem 1.11 *Consider the one-mode Hamiltonian*

$$H = a^* a + \frac{1}{2}[\gamma\, a\, a + \gamma^* a^* a^*]. \tag{1.63}$$

(a) *Show that without loss of generality (up to a simple canonical transformation), one may choose γ to be real and positive.*

(b) *Explore the possibility of solving for the dynamics of this model by the Bogoliubov transformation.*

(c) *In the region of γ where the Bogoliubov transformation is not helpful, directly solve the equation of motion.*

1.2   Basic Dynamic and Static Properties

1.2.1   Nonrelativistic Limit for a Complex-Valued Scalar Field

At this point, readers must suppress their natural desire to think of matter as an ensemble of particles and get used to the idea that—at least under certain conditions—matter can be perfectly described by a classical complex-valued relativistic field $\Psi(\mathbf{r})$. At first glance, it might seem irrelevant to deal with the relativistic field while being interested in the essentially nonrelativistic case: One could simply take the corresponding Hamiltonian equations of motion as an axiom. Our seemingly exotic choice is motivated, in the first place, by an amazing efficiency of the special relativity principle in narrowing down the class of possible equations of motion. It turns out that what we need is just the simplest equation among the ones featuring the nonrelativistic limit! Last but not least, the scalar relativistic field has trivial transformation properties when changing inertial frames: The field at a given spacetime point does not change at all. In the nonrelativistic limit, the Galilean transformation of the field is less trivial, and the easiest way to establish the rules of the Galilean transformation would be to take the nonrelativistic limit of the relativistic description.

The simplest equation consistent with special relativity—that is, invariant with respect to the Lorentz transformation—is the wave equation

$$\frac{\partial^2 \Psi}{\partial t^2} = c^2 \Delta \Psi, \tag{1.64}$$

with c the velocity of light. This equation is too restrictive, since the waves it describes always propagate with the velocity of light; we are interested in the nonrelativistic limit when characteristic velocities are much smaller than c. The simplest equation featuring the nonrelativistic limit is the Klein–Gordon equation

$$\frac{\partial^2 \Psi}{\partial t^2} = c^2 \Delta \Psi - \lambda^2 \Psi, \tag{1.65}$$

where λ is a real (and positive definite, without loss of generality) parameter of the dimensionality of inverse time. We also assume that the field Ψ is complex-valued.*

To arrive at the nonrelativistic limit, we first represent Ψ as

$$\Psi = \psi e^{\mp i\lambda t}. \tag{1.66}$$

The two signs of the exponential correspond to matter and antimatter solutions, respectively. They are related to each other by complex conjugation—with respect to which Equation (1.65) is invariant—thus, it is sufficient to consider just one of them. Picking the parameterization with the minus sign, we obtain

$$\frac{\partial^2 \Psi}{\partial t^2} = -\lambda^2 \Psi - e^{-i\lambda t} \left[2i\lambda \frac{\partial \psi}{\partial t} - \frac{\partial^2 \psi}{\partial t^2} \right], \tag{1.67}$$

which brings us to the conclusion that under the condition

$$\left| \frac{\partial^2 \psi}{\partial t^2} \right| \ll \lambda \left| \frac{\partial \psi}{\partial t} \right|, \tag{1.68}$$

Equation (1.65) reduces to its nonrelativistic limit

$$i \frac{\partial \psi}{\partial t} = -\frac{\gamma}{2} \Delta \psi, \tag{1.69}$$

---

* The assumption of complex-valuedness of the field Ψ is mostly a matter of taste/definiteness, because we are interested only in the nonrelativistic limit of the Klein–Gordon equation, in which case even the real-field equation leads, by the substitution $\Psi = \psi e^{-i\lambda t} + \psi^* e^{i\lambda t}$, to the same Equation (1.69) in terms of the complex-vaued field ψ.

with

$$\gamma = c^2/\lambda. \tag{1.70}$$

It should be emphasized that while Equation (1.69) has the form of the Schrödinger equation, the field ψ should not be interpreted as a wavefunction. This is yet another opportunity to reiterate that at this stage of developing the theory, an allusion to quantum mechanics is more confusing than helpful. The only exception can be made for purely *mathematical* aspects of the intuition developed over the years for the Schrödinger equation.

Now we make an observation that Equation (1.69) can be obtained from (1.16) with the Hamiltonian functional

$$H = \frac{\gamma}{2} \int |\nabla\psi|^2 d^d r. \tag{1.71}$$

It is worth noting that it is only in the nonrelativistic limit of Equation (1.65) and only with the parameterization (1.66) that there arises the classical field ψ being a canonical complex variable, in the sense of the definition (1.6), of a complex-valued Hamiltonian formalism. This is immediately clear from the fact that the original equation, Equation (1.65), is a second-order differential equation with respect to time derivatives, while in the complex-valued Hamiltonian formalism, the equation of motion is—by definition (1.16)—a first-order differential equation with respect to time derivatives.

In accordance with the Hamiltonian mechanics, the right-hand side of Equation (1.71) is a constant of motion called energy. The translational invariance of the Hamiltonian implies conservation of momentum

$$\mathbf{P} = (i/2) \int (\psi\nabla\psi^* - \psi^*\nabla\psi)d^d r = \int |\psi|^2 (\nabla\Phi)d^d r, \tag{1.72}$$

and the rotational invariance guarantees conservation of the angular momentum

$$\mathbf{M} = (i/2) \int \mathbf{r} \times (\psi\nabla\psi^* - \psi^*\nabla\psi)d^d r = \int |\psi|^2 (\mathbf{r} \times \nabla\Phi)d^d r. \tag{1.73}$$

(Here, Φ is the phase of the field ψ.) From Equations (1.72) and (1.73), it is seen that the quantity $i(\psi\nabla\psi^* - \psi^*\nabla\psi)/2 = |\psi|^2\nabla\Phi$ has the meaning of momentum density.

Since the Hamiltonian (1.71) is globally U(1)-invariant, the equation of motion conserves the amount of matter N (1.18). [A reader familiar with the quantum field theory might find it tempting to refer to N as the total number of particles (up to a dimensional factor of $1/\hbar$). We deliberately avoid this correspondence because at the classical-field level the notion of individual particles is irrelevant, if not misleading.]

The law of the conservation of the amount of matter can be written in the differential form. To this end, we first introduce the notion of matter density,

$$n(\mathbf{r}) = |\psi(\mathbf{r})|^2, \tag{1.74}$$

and then observe that Equation (1.69) implies

$$\dot{n} + \nabla \cdot \mathbf{j} = 0, \tag{1.75}$$

where

$$\mathbf{j} = i(\gamma/2)[\psi \nabla \psi^* - \psi^* \nabla \psi] = \gamma |\psi|^2 \nabla \Phi. \tag{1.76}$$

By direct analogy with hydrodynamics, Equation (1.75) has the meaning of the continuity equation, and thus the vector $\mathbf{j}(\mathbf{r})$ is the density of matter flux at point $\mathbf{r}$.

Problem 1.12 *Derive Equation (1.75) from Equation (1.69).*

Note a simple relation between $\mathbf{j}(\mathbf{r})$ and momentum density: The former differs from the latter by a factor of γ. Hence,

$$\int \mathbf{j}(\mathbf{r}) d^d r = \gamma \mathbf{P}. \tag{1.77}$$

Consider now the Galilean transformation, which naturally leads to the notion of the microscopic flow velocity. An instructive and transparent way of introducing the Galilean transformation of the field ψ is through the Lorentz invariance of the field Ψ. Since for relativistic fields in the moving reference frame (primed variables)

$$\Psi'(\mathbf{r}', t') = \Psi(\mathbf{r}, t), \tag{1.78}$$

we have

$$\psi'(\mathbf{r}', t') = \psi(\mathbf{r}, t) e^{i\lambda(t'-t)}. \tag{1.79}$$

By taking the $v_0/c \ll 1$ limit in the Lorentz transformation, one readily gets

$$\psi'(\mathbf{r}, t) = \psi(\mathbf{r}' - \mathbf{v}_0 t, t) \exp\left[\frac{i}{\gamma}\left(\mathbf{v}_0 \cdot \mathbf{r}' - \frac{v_0^2 t}{2}\right)\right], \tag{1.80}$$

where the velocity of the moving frame with respect to the original one is $-\mathbf{v}_0$:

$$\mathbf{r}' = \mathbf{r} + \mathbf{v}_0 t \qquad (t' = t). \tag{1.81}$$

Problem 1.13 *Derive Equation (1.80) from Equation (1.79).*

With the expression (1.80) for the Galilean transformation of the field ψ, one straightforwardly obtains the following important laws for Galilean transformation of the flux density, momentum, and energy:

$$\mathbf{j}' = \mathbf{j} + n\mathbf{v}_0, \tag{1.82}$$

$$\mathbf{P}' = \mathbf{P} + N\mathbf{v}_0/\gamma, \tag{1.83}$$

$$E' = E + \mathbf{P} \cdot \mathbf{v}_0 + Nv_0^2/2\gamma. \tag{1.84}$$

Let us define now the notion of local velocity of the flow. We start with the ground state $\psi = \sqrt{n}$ = const. that does not evolve in time in a given reference frame, and look at it from the frame moving with respect to the original one with the velocity $-\mathbf{v}_0$. By Galilean relativity, what we see from the moving frame is supposed to be a uniform flow of matter with the velocity $\mathbf{v}_0$. Thus, such a flow is described by

$$\psi(\mathbf{r}, t) = \sqrt{n} \exp\left[\frac{i}{\gamma}\left(\mathbf{v}_0 \cdot \mathbf{r} - \frac{v_0^2 t}{2}\right)\right]. \tag{1.85}$$

We see that Equation (1.85) implies Equation (1.2), with $\Phi(\mathbf{r}) = (1/\gamma)/(\mathbf{v}_0 \cdot \mathbf{r} - v_0^2 t/2)$ being understood as the phase of ψ and $\mathbf{v}_s$ being identified with the velocity of the uniform flow, $\mathbf{v}_0$. It is useful then to generalize the notion of velocity to the case of nonuniform flow by *defining* the microscopic velocity in terms of the phase $\Phi(\mathbf{r})$ of the field $\psi(\mathbf{r})$:

$$\mathbf{v}(\mathbf{r}) = \gamma \nabla \Phi(\mathbf{r}). \tag{1.86}$$

One may find this definition somewhat strange since it appears to be identical to Equation (1.2), which was declared to be the central relation in the theory of superfluidity. Can a key relation be a mere definition? The subtlety here is that the coincidence between Equations (1.2) and (1.86) takes place only for the previously-discussed case of the uniform ground-state flow (1.85). In a general case, both $\mathbf{v}_s(\mathbf{r})$ and the phase in Equation (1.2) are coarse-grained mesoscopic entities ($\tilde{\Phi} \neq \Phi$) that fundamentally cannot be reduced to the local properties of the field ψ. Also, let us not forget that Equation (1.86) is defined for the classical field description, which itself has to be derived as an emergent phenomenon for a system of particles.

With our definition of local velocity and the previously introduced notion of local density, the expression for the flux becomes intuitive:

$$\mathbf{j} = n\mathbf{v}. \tag{1.87}$$

Yet another definition, naturally suggested by the notion of microscopic velocity, is the *mass density*, $\rho(\mathbf{r})$, and the *total mass*, M:

$$\rho = n/\gamma, \qquad M = N/\gamma. \tag{1.88}$$

In terms of ρ and $\mathbf{v}$, the expressions for energy, momentum, and angular momentum acquire very intuitive quasihydrodynamic form. Using M in the Galilean transformations (1.83) and (1.84) renders them most intuitive as well:

$$\mathbf{P}' = \mathbf{P} + M\mathbf{v}_0, \qquad E' = E + \mathbf{P} \cdot \mathbf{v}_0 + Mv_0^2/2. \tag{1.89}$$

At this point, it is also instructive to mention that the interpretation of ρ as the mass density (with M the total mass) is perfectly consistent with quantum-mechanical perspective, where we have the relation $\gamma = \hbar/m$ together with the interpretation of $n/\hbar$ as the number density of (quantum) particles.

In accordance with Noether's theorem, Galilean symmetry implies a certain conservation law, which proves to be nothing but the law of uniform motion of the "center-of-mass" position vector defined as

$$\mathbf{R} = \frac{1}{N} \int \mathbf{r}|\psi|^2 d^d r. \tag{1.90}$$

With Equation (1.69), we readily prove that

$$\dot{\mathbf{R}} = \frac{\mathbf{P}}{M} \quad \Rightarrow \quad \mathbf{R}(t) = \mathbf{R}(0) + \frac{\mathbf{P}t}{M}. \tag{1.91}$$

(So that the earlier mentioned conservation law is $M\mathbf{R} - \mathbf{P}t = \text{const.}$)

1.2.2 Gross–Pitaevskii Equation, aka Nonlinear Schrödinger Equation

To demonstrate the phenomenon of superfluidity, the linear model (1.69) must be generalized to become nonlinear, the particular type of nonlinearity (i.e., self-interaction) being not very important, provided the U(1) invariance is preserved. The simplest self-interaction consistent with U(1) invariance of the Hamiltonian is introduced by the $|\psi|^4$-term,

$$H \to H + (g/2) \int |\psi|^4 d^d r, \tag{1.92}$$

with positive interaction constant g (at $g < 0$, the homogeneous ground state is unstable). Finally, to distinguish superfluidity from trivial conservation of momentum (1.72), guaranteed by translational invariance of the Hamiltonian, we introduce a term with an inhomogeneous external potential $\mathcal{V}(\mathbf{r})$:

$$H \to H + \int \mathcal{V}(\mathbf{r})|\psi(\mathbf{r})|^2 d^d r. \tag{1.93}$$

With the new Hamiltonian

$$H = \int \left[(\gamma/2)|\nabla\psi|^2 + \mathcal{V}(\mathbf{r})|\psi|^2 + (g/2)|\psi|^4 \right] d^d r, \tag{1.94}$$

the equation of motion (1.16) reads

$$i\dot{\psi} = -\frac{\gamma}{2}\Delta\psi + g|\psi|^2\psi + \mathcal{V}(\mathbf{r})\psi. \tag{1.95}$$

In various subfields of physics, this equation is known under the generic name of *nonlinear Schrödinger* equation (NLSE). In the theory of quantum gases—closest to the subject of this book—this equation is called the *time-dependent Gross–Pitaevskii* equation (GPE), with the reservation that GPE is traditionally associated with the Bose-Einstein condensate (BEC), and thus superfluidity. In the next sections, we will see that while certain solutions of Equation (1.95) do feature superfluidity, there is a normal-fluid regime as well.

Since the Hamiltonian (1.94) is U(1) symmetric, the quantity N (1.11) is conserved by GPE. For a given N, there is a ground state, ψ_0, corresponding to the minimal possible energy. To find ψ_0, one can use the technique of Lagrange multipliers. This amounts to finding the global minimum, $\psi_0(\mathbf{r}, \mu)$, of the functional

$$H' = H - \mu N, \tag{1.96}$$

with the Lagrange multiplier μ and demanding that μ satisfies the condition

$$\int |\psi_0(\mathbf{r}, \mu)|^2 d^d r = N. \tag{1.97}$$

The global minimum of the functional H' is found from the requirement of vanishing variational derivative

$$\frac{\delta H'}{\delta \psi^*} = 0, \tag{1.98}$$

which leads to the equation

$$-(\gamma/2)\Delta\psi_0 + \mathcal{V}(\mathbf{r})\psi_0 + g|\psi_0|^2\psi_0 - \mu\psi_0 = 0, \tag{1.99}$$

known as the *stationary GPE*.

Problem 1.14 *Check that (1.98) means (1.99). Argue that the ground-state solution of Equation (1.99) is real, up to a global phase factor.*

The parameter μ in the stationary GPE has the meaning of the chemical potential (note, however, an unusual dimensionality $[\mu] = [\mathcal{V}] = [1/t]$, as well as the dimensionality $[N] = [E][t]$).

Problem 1.15 *Prove this statement by showing at zero temperature μ equals $dE(N)/dN$, where $E(N)$ is the ground-state energy for the given amount of matter.*

Consistent with the global U(1) invariance, Equation (1.99) defines the function ψ_0 only up to a global phase. At $\mu \neq 0$, this phase has to evolve in time to satisfy both the stationary equation (1.99) and the dynamical equation (1.95):

$$\psi_0(t) = \psi_0(0)e^{-i\mu t}. \tag{1.100}$$

Uniform phase evolution has no physical implication and is not even observable unless we introduce interactions changing N because it is indistinguishable from the phase evolution resulting from placing the system in a uniform external potential $\mathcal{V}(\mathbf{r}) \equiv \mathcal{V}_0$, which does not change the physics: All the solutions of Equation (1.95) simply acquire a phase factor $\exp(-i\mathcal{V}_0 t)$. It makes then perfect sense to introduce $\mathcal{V}_0 = -\mu$ to render the phase of the ground state time-independent.

1.2.3   Healing Length

To appreciate the crucial role of the nonlinear term in GPE, consider a case when the field is trapped in a large box with nonpenetrable walls: $V(\mathbf{r}) = 0$, if $\mathbf{r}$ is inside the box, and $V(\mathbf{r}) = \infty$, otherwise. For $g = 0$, the density profile is nonuniform at length scales of the box size. With the nonlinear term added, the nonuniform profile becomes energetically unfavorable. To prove that, introduce the mean density $\bar{n} = N/V$, with V the volume of the box, and consider the quantity

$$\int [n(\mathbf{r}) - \bar{n}]^2 d^d r \geq 0. \tag{1.101}$$

The potential energy associated with the nonlinear term in the Hamiltonian (1.94) is

$$E_{\text{pot}} = (g/2) \int n^2 d^d r, \tag{1.102}$$

and according to (1.101) satisfies the inequality

$$E_{\text{pot}} \geq \frac{g}{2} \frac{N^2}{V}. \tag{1.103}$$

For the homogeneous density profile, $n(\mathbf{r}) \equiv \bar{n}$, the energy reaches its minimal possible value $E_{\text{pot}} = (g/2)N^2/V$. The kinetic energy term,

$$E_{\text{kin}} = (\gamma/2) \int |\nabla\psi|^2 d^d r, \tag{1.104}$$

also favors flat density distribution. This means that inside a large enough box, where the effect of boundaries on the density profile in the bulk is negligible, the density profile is flat: $n(\mathbf{r}) = n_0$ (up to macroscopically small corrections due to boundary effects, $n_0 = \bar{n}$). Substituting $\sqrt{n_0}$ for ψ_0 in the stationary GPE (1.99), we see that

$$n_0 = \mu/g. \tag{1.105}$$

The density profile of the *noninteracting* field in a box solves the linear Schrödinger equation. The smooth nonuniform distribution naturally follows from the kinetic energy minimization, which results in small gradients inversely proportional to the system size so that the kinetic energy is macroscopically small; making it smaller is impossible without violating the condition that ψ_0 vanishes at the box walls!

Having established that the density is uniform in the bulk, one is interested in determining the typical length, l_0, at which the density saturates (away from the box boundaries) to its bulk value n_0. This length, referred to as *healing length*, can be estimated from a dimensional argument. By introducing the dimensionless field variable $\tilde{\psi}_0$ by $\psi_0 = \sqrt{n_0}\tilde{\psi}_0$ (so that $\tilde{\psi}_0 = 1$ in the bulk) and using it in

Equation (1.99) along with the (1.105) relation, we find that the resulting equation contains no dimensional variables if distances are measured in units of

$$l_0 = \sqrt{\gamma/gn_0} = \sqrt{\gamma/\mu}. \tag{1.106}$$

With $\tilde{\mathbf{r}} = \mathbf{r}/l_0$, the GPE takes the form

$$-(1/2)\Delta_{\tilde{\mathbf{r}}}\tilde{\psi}_0 + |\tilde{\psi}_0|^2\tilde{\psi}_0 - \tilde{\psi}_0 = 0. \tag{1.107}$$

In terms of new coordinates, the characteristic scale of the field variation is of order unity, meaning that l_0 is indeed a typical variation length in terms of the original coordinates. In agreement with what was said earlier about the noninteracting case, l_0 diverges as $g \to 0$, explicitly demonstrating the role of interactions in healing the effects of boundary conditions.

Problem 1.16 *Find the density profile $n(\mathbf{r})$ in the vicinity of a flat nonpenetrable wall. Hint: This problem is 1D, and partial derivatives can be replaced with ordinary ones. Take advantage of relations $2 f'(x) \cdot f''(x) \equiv [(f'(x))^2]'$ and $2 f(x) \cdot f'(x) \equiv [f(x)^2]'$.*

In a slowly varying external potential, one can introduce a local healing length

$$l_0(\mathbf{r}) = \sqrt{\gamma/gn_0(\mathbf{r})}, \tag{1.108}$$

where $n_0(\mathbf{r}) = |\psi_0(\mathbf{r})|^2$. The notion of the local healing length is appropriate when

$$l_0(\mathbf{r}) \ll L, \tag{1.109}$$

where L is the typical scale of the density variation. Under the condition (1.109), the differential term $(\gamma/2)\Delta\psi_0$ in the stationary GPE is much smaller than the self-interaction term $g|\psi_0|^2\psi_0$, and thus can be omitted, rendering this equation purely algebraic and trivial to solve:

$$n_0(\mathbf{r}) = \begin{cases} [\mu - \mathcal{V}(\mathbf{r})]/g, & \text{if positive,} \\ 0, & \text{otherwise.} \end{cases} \tag{1.110}$$

Equation (1.110) is a particular case of the so-called *quasihomogeneous* distribution—often referred to as the *local density approximation*—taking place in *any* equilibrium macroscopic system in a smooth enough external potential, provided the characteristic length, L, of the spatial variation of thermodynamic functions is much larger than all appropriate microscopic length scales (l_0 in our case). In the quasi-homogeneous regime, one can view the system as a set of cells of the size l, where $l_0 \ll l \ll L$, so that a cell with its center at point $\mathbf{r}$ approximately behaves as a *homogeneous* system in a *uniform* external potential of the magnitude $\mathcal{V}(\mathbf{r})$. Under these conditions, for any thermodynamic quantity A, the form of its spatial distribution in the external potential is approximately determined by the $A(\mu)$ function in the homogeneous system transformed as $A(\mu) \to A(\mu - \mathcal{V}(\mathbf{r}))$. As it readily follows from general thermodynamic relations, a quasihomogeneous density distribution always implies the *hydrostatic equilibrium*

$$\nabla p(\mathbf{r}) + n(\mathbf{r})\nabla\mathcal{V}(\mathbf{r}) = 0, \tag{1.111}$$

where p is pressure, n is density, and $\mathcal{V}$ is external potential.

Problem 1.17 *Explicitly check the validity of the relation (1.111) for our classical field* ψ *in the quasihomogeneous regime. Hint: Recall that in a homogeneous system,* $n = (\partial p / \partial \mu)_{T,V}$.

1.2.4 Vortices

The field ψ is a solution of the second-order (with respect to spatial coordinates) differential Equation (1.95) and thus has to be a smooth function of $\mathbf{r}$, implying that the gradient of the phase Φ of the field ψ is a well-defined function at any point where $|\psi| \neq 0$. As opposed to its gradient, the field $\Phi(\mathbf{r})$ itself is defined only up to a multiple of 2π. As a result, the field $\Phi(\mathbf{r})$ can feature *vortex topological defects* (vortices). Vortices are characterized by a set of points—at which $\Phi(\mathbf{r})$ is fundamentally ill defined—such that for any contour C going *around* these points, there is a nonzero *phase winding number*

$$M_{\text{def}} = (2\pi)^{-1} \oint_C \nabla \Phi \cdot d\mathbf{l} \neq 0. \tag{1.112}$$

For vortex defects to have finite energy, the modulus of the field ψ must go to zero at the point of $\nabla \Phi$ singularity. The integer number M_{def} is a topological characteristic (topological charge) of the vortex defect. By continuity, it is independent of the shape of the contour C: Integers cannot change "a little" when the contour is transformed continuously. By the same argument, vortices in 3D are lines with no free ends in the bulk: They either form closed loops or terminate at the box boundaries. Vortex defects in 3D are often referred to as vortex lines or filaments. In 2D, elementary vortex defects have point singularities.

Consider a solution of the stationary GPE (1.99) describing a straight, infinitely long vortex line in three dimensions.[*] By cylindrical symmetry, one looks for the solution ψ_0 in the form

$$\psi_0(\rho, \varphi) = \sqrt{n(\rho)} e^{\pm i \varphi}, \tag{1.113}$$

where φ is the azimuthal angle and ρ is the distance from the vortex line. The solution is not diverging at $\rho = 0$, implying

$$n(\rho) \propto \rho^2 \quad \text{at} \quad \rho \to 0. \tag{1.114}$$

By dimensional argument, it is clear that the size of the core—the region where the density n is significantly different from its asymptotic uniform value n_0—is on the order of the healing length l_0.

Quantitatively, the form of the density profile, $n(\rho)$, can be found by explicitly solving the GPE. Using the dimensionless Equation (1.107) in cylindrical

---

[*] By uniformity of the solution along the direction of the line, it automatically describes a point vortex in 2D, in a plane perpendicular to the line.

coordinates with the parameterization $\tilde{\psi}_0(\tilde{\rho}, \varphi) = f(\tilde{\rho})e^{\pm i\varphi}$ (here $\tilde{\rho} = \rho/l_0$), we get a second-order ordinary differential equation for the function $f(\tilde{\rho})$:

$$f'' + \frac{f'}{\tilde{\rho}} - 2\left(f^2 - 1 + \frac{1}{2\tilde{\rho}^2}\right)f = 0, \tag{1.115}$$

to be solved with the boundary conditions $f(0) = 0$ and $f(\infty) = 1$.

To reveal the asymptotic behavior of the solution in the $\tilde{\rho} \ll 1$ limit, we keep only the leading terms in Equation (1.115):

$$\tilde{\rho}^2 f'' + \tilde{\rho} f' - f = 0 \qquad (\tilde{\rho} \ll 1). \tag{1.116}$$

The solution of (1.116) consistent with $f(0) = 0$ is $f(\tilde{\rho}) \propto \tilde{\rho}$, as stated in (1.114).

To find the asymptotic behavior in the $\tilde{\rho} \gg 1$ limit, we represent the solution as $f = 1 + g$ ($g \to 0$ at $\tilde{\rho} \to \infty$). Then, keeping only the leading terms in (1.115)—note that all differential terms are subleading in the limit of interest—we find $g = -1/4\tilde{\rho}^2$. The constant density is recovered following a power rather than exponential law:

$$f(\tilde{\rho}) \approx 1 - \frac{1}{4\tilde{\rho}^2} \qquad (\tilde{\rho} \gg 1). \tag{1.117}$$

This effect is due to long-range phase gradients. The law (1.117) tells us that the flow around the vortex is compressible even far outside the core, since the $1/\rho^2$ correction implies *logarithmic divergence* of such additive quantity as the vortex-induced depletion of the amount of matter N (1.18).

Despite compressibility of the flow around the vortex, a delicate situation takes place with respect to the vortex energy $H' = H - \mu N$, for which the logarithmic divergence comes entirely from the kinetic energy term (R is the radial size of the cylindrical system):

$$\mathcal{E}_{\text{vort}} = \pi \gamma n_0 \ln(R/l_0) + \text{subleading terms} \qquad (\text{per unit length in 3D}). \tag{1.118}$$

This happens due to exact cancelation of leading-order corrections to the potential energy functional since n_0, by definition, is the solution minimizing H' in a homogeneous system and thus the variational derivative $\delta H'/\delta n$ is zero at $n = n_0$.

In a general case of a vortex line with topological charge M, the exponential in Equation (1.113) is $e^{iM\varphi}$, and the leading term in the energy (1.118) acquires an extra factor of M^2, meaning, in particular, that a vortex line with a charge $|M| > 1$ is energetically unstable against splitting into two lines with smaller charges: $|M_1|, |M_2| < |M|$ (such that $M_1 + M_2 = M$ and $M_1^2 + M_2^2 < M^2$). By this argument, it ultimately splits into $|M|$ lines with the same charge $M/|M|$.

Due to slow logarithmic dependence on the system size, formula (1.118) applies also to a large vortex loop of radius $R \gg l_0$ (in 2D, the same is true for the vortex–antivortex pair at distance $R \gg l_0$), where, by the energy of a vortex loop/pair, we mean the increase of the minimal value of E' due to the presence of the defect, the spatial configuration of the latter being fixed.

Furthermore, if the energy functional H' is minimized for an arbitrary system of fixed-shape vortex loops (and radii $R \gg l_0$), then the minimal energy is expressed by the general formula

$$E_{\text{vort}} = \frac{\pi}{2}\gamma n_0 \sum_{i,j} M_i M_j \int_{|\mathbf{r}_i - \mathbf{r}_j| > l_*} \frac{d\mathbf{l}_i \cdot d\mathbf{l}_j}{|\mathbf{r}_i - \mathbf{r}_j|}. \tag{1.119}$$

Here, the (double) line integration is over all the oriented vortex loops. The positive integer M_i is the winding number of the ith loop, and the direction on the loop is associated with the sense of the circulation of velocity (the sign is a matter of convention). If not properly regularized at $|\mathbf{r}_i - \mathbf{r}_j| \to 0$, the integral logarithmically diverges; that is why the cutoff parameter $l_* \sim l_0$ is necessary. With a logarithmic accuracy, one can simply set $l_* = l_0$ [cf. Equation (1.118)]. For even better result, the value of l_* can be fine-tuned in such a way that (1.119), as written, is accurate up to higher-order corrections in the parameter $l_0/R \ll 1$. Indeed, both the energy contributions from the vortex cores and the cutoff-dependent part of the integral (1.119) are linearly proportional to the total length of the vortex lines, so that the two can be rendered equal by fine-tuning.

The origin of Equation (1.119) can be traced by *magnetostatic analogy*. Observe that, away from the vortex cores, the functional being minimized depends only on the phase Φ of the field ψ,

$$H'_{\text{vort}}[\Phi] = \frac{1}{2}\gamma n_0 \int d^d r \, (\nabla\Phi)^2. \tag{1.120}$$

(In a direct analogy with the case of a single straight vortex line, this pseudo-incompressible form of the energy functional is justified by the fact that the variational derivative $\delta H'/\delta n = 0$ at $n = n_0$.) Minimization of the functional (1.120) amounts then to solving the Laplace equation

$$\Delta\Phi = 0, \tag{1.121}$$

for which the fixed vortex loops play the role of boundary conditions. For the velocity field $\mathbf{v} = \gamma\nabla\Phi$, Equation (1.121) is nothing but the condition of identically zero divergence, while the boundary conditions prescribe the law of circulation of the field $\mathbf{v}(\mathbf{r})$ around the vortex lines. The analogy with the magnetostatics is now obvious. Think of the i-th oriented vortex filament as if it is an infinitely thin wire carrying the current M_i. Then, up to a proportionality coefficient, the field $\mathbf{v}(\mathbf{r})$ is nothing but the static magnetic field created by the system of fixed loop currents, and the energy (1.119) is the corresponding magnetostatic energy.[*]

Along with the relation (1.119), the magnetostatic analogy readily yields an expression for the field $\mathbf{v}(\mathbf{r})$ in the form of the Biot–Savart law

$$\mathbf{v}(\mathbf{r}) = \frac{\gamma}{2}\sum_i M_i \int \frac{d\mathbf{l}_i \times (\mathbf{r} - \mathbf{r}_i)}{|\mathbf{r} - \mathbf{r}_i|^3}. \tag{1.122}$$

---

[*] It is worth reminding that the expression for the magnetostatic energy in terms of currents is a direct analog of the electrostatic energy in terms of charges.

The magnetostatic analogy can be put in a broader *hydrodynamic* context. The zero-divergence velocity field

$$\nabla \cdot \mathbf{v} \equiv 0 \tag{1.123}$$

and the associated energy functional (1.120) correspond to the flow of incompressible liquid, the condition (1.123) implied by the general hydrodynamic continuity equation [cf. Equation (1.75)]. In our case, the vorticity of the flow is concentrated along the vortex lines. In a generic hydrodynamical case, the vorticity field

$$\mathbf{w} = \nabla \times \mathbf{v} \tag{1.124}$$

is a regular function, the regime of vortex filaments being the limit when $\mathbf{w}(\mathbf{r})$ becomes a generalized function. The magnetostatic counterpart of $\mathbf{w}(\mathbf{r})$ is the density of currents. Hence, the problem of obtaining the magnetic field from the known pattern of current density is mathematically identical to restoring the field $\mathbf{v}(\mathbf{r})$ from the regular field $\mathbf{w}(\mathbf{r})$. The condition (1.123) allows one to conveniently parameterize the velocity field as

$$\mathbf{v} = \nabla \times \mathbf{a}, \tag{1.125}$$

where—same as in magnetostatics—the form of the "vector potential" $\mathbf{a}$ is fixed by an extra condition

$$\nabla \cdot \mathbf{a} \equiv 0, \tag{1.126}$$

consistent with (1.125) by the gauge freedom.*

In terms of the field $\mathbf{a}$, the vorticity (1.124) is expressed as

$$\mathbf{w} = -\Delta \mathbf{a}, \tag{1.127}$$

and, using the known solution of the 3D Poisson equation, we get

$$\mathbf{a}(\mathbf{r}) = \frac{1}{4\pi} \int \frac{\mathbf{w}(\mathbf{r}_0)}{|\mathbf{r} - \mathbf{r}_0|} d^3 r_0, \tag{1.128}$$

$$\mathbf{v}(\mathbf{r}) = \frac{1}{4\pi} \int \frac{\mathbf{w}(\mathbf{r}_0) \times (\mathbf{r} - \mathbf{r}_0)}{|\mathbf{r} - \mathbf{r}_0|^3} d^3 r_0. \tag{1.129}$$

Taking the limit of infinitely thin (quantized) vortex filaments amounts to transforming the bulk integrals into the line ones, in accordance with the rather obvious (componentwise) rule

$$\int w^{(\alpha)}(\mathbf{r}_0) d^3 r_0 (\ldots) \rightarrow 2\pi\gamma \sum_i M_i \int dl_i^{(\alpha)} (\ldots)|_{\mathbf{r}_0 \rightarrow \mathbf{r}_i}. \tag{1.130}$$

Applied to (1.129), the rule yields the Biot–Savart law (1.122).

---

* That is, the freedom of replacing $\mathbf{a} \rightarrow \mathbf{a} + \nabla f$, where f is an arbitrary smooth scalar field; for $\nabla \cdot \mathbf{a} \neq 0$, choose f such that $\Delta f = -\nabla \cdot \mathbf{a}$.

The continuous-vorticity prototype of (1.119) is[*]

$$E_{\text{vort}} = \frac{n_0}{8\pi\gamma} \int \frac{\mathbf{w}(\mathbf{r}_1) \cdot \mathbf{w}(\mathbf{r}_2)}{|\mathbf{r}_1 - \mathbf{r}_2|} d^3 r_1 \, d^3 r_2. \tag{1.131}$$

The rule (1.130) applied to (1.131) converts it into (1.119).

Along with the energy of a (spatially localized) vortex configuration, one might be also interested in the momentum, $\mathbf{P} = n_0 \int d^d r \nabla\Phi = (n_0/\gamma) \int \mathbf{v} \, d^d r$ [see (1.72)], associated with the flow pattern. Starting with the relationship between the momentum and the nonsingular vorticity field,[†]

$$\mathbf{P}_{\text{vort}} = \frac{n_0}{2\gamma} \int d^3 r \, \mathbf{r} \times \mathbf{w}(\mathbf{r}), \tag{1.132}$$

we take the vortex-line limit by applying the rule (1.130). This brings us to

$$\mathbf{P}_{\text{vort}} = \pi n_0 \sum_i M_i \int \mathbf{r}_i \times d\mathbf{l}_i. \tag{1.133}$$

The parameter γ is absent in (1.133), which is not surprising since this parameter is not a part of the definition (1.72).

In two dimensions, an extra simplification arises due to the analogy with 2D *electrostatics*. The mapping onto the 2D electrostatics is performed by parameterizing

$$\nabla\Phi = \hat{z} \times \nabla u, \tag{1.134}$$

where $u(\mathbf{r})$ is a scalar potential satisfying, in accordance with (1.121), the Laplace equation $\Delta u = 0$, everywhere except for the vortex centers, at which u is singular and $\hat{z}$ is a unit 3D vector perpendicular to the xy-plane in which the vector $\mathbf{r}$ lives. The boundary condition at vortex centers for the field Φ translates into the condition defining the flux of the field $u(\mathbf{r})$ through a closed contour enclosing the vortex. Hence, the field $u(\mathbf{r})$ can be interpreted as the 2D electrostatic potential of the system of point charges located at the vortex centers, the "electric" charge of the ith vortex being equal to its vorticity charge M_i; we adopt the sign convention that $M_i > 0$ ($M_i < 0$) if the ith vortex is anticlockwise (clockwise). The solution for the potential then is

$$u(\mathbf{r}) = \sum_i M_i \ln|\mathbf{r} - \mathbf{r}_i|. \tag{1.135}$$

This equation can be viewed as the limiting case of continuous (nonsingular) charge density $\rho(\mathbf{r})$, in which case $u(\mathbf{r})$ obeys the Poisson equation

$$\Delta u = 2\pi\rho, \tag{1.136}$$

---

[*] Integrate by parts, $\int d^3 r v^2 = \int d^3 r \mathbf{v} \cdot (\nabla \times \mathbf{a}) = \int d^3 r \mathbf{a} \cdot (\nabla \times \mathbf{v}) = \int d^3 r \mathbf{a} \cdot \mathbf{w}$, and then use (1.128).
[†] The relation is nothing but a mathematical identity $\int d^3 r \mathbf{r} \times (\nabla \times \mathbf{v}) \equiv 2 \int d^3 r \mathbf{v}$, readily checked by integrating by parts.

the solution to which reads

$$u(\mathbf{r}) = \int d^2r_1 \, \rho(\mathbf{r}_1) \ln|\mathbf{r} - \mathbf{r}_1|. \tag{1.137}$$

Here, we are dealing with the equivalence between electrostatics and hydrodynamics of incompressible 2D liquid with continuous rather than singular vorticity, the latter being described by the scalar field $\rho(\mathbf{r})$. This, in particular, leads to the 2D analog of the relation (1.129),

$$\mathbf{v} = \gamma \nabla \Phi = \gamma \hat{z} \times \int \frac{\mathbf{r} - \mathbf{r}_1}{|\mathbf{r} - \mathbf{r}_1|^2} \, \rho(\mathbf{r}_1) \, d^2r_1. \tag{1.138}$$

The continuous-vorticity representation is convenient for deriving* the expressions for the energy and momentum:

$$E_{\text{vort}} = \frac{n_0 \gamma}{2} \int d^2r \, (\nabla \Phi)^2 = -\pi n_0 \gamma \int d^2r_1 \, d^2r_2 \, \rho(\mathbf{r}_1)\rho(\mathbf{r}_2) \ln|\mathbf{r}_1 - \mathbf{r}_2|, \tag{1.139}$$

$$\mathbf{P}_{\text{vort}} = n_0 \int d^2r \, \nabla \Phi = n_0 \hat{z} \times \int d^2r \, \nabla u = -2\pi n_0 \hat{z} \times \int d^2r \, \mathbf{r} \cdot \rho(\mathbf{r}). \tag{1.140}$$

Note that the two expressions are meaningful only if the system is "neutral," $\int d^2r \rho(\mathbf{r}) = 0$.

The quantized-vorticity counterparts of Equations (1.139) and (1.140) are obtained by taking the limit of point "electric" charges in the function $\rho(\mathbf{r})$,

$$E_{\text{vort}} = \frac{n_0 \gamma}{2} \int d^2r \, (\nabla \Phi)^2 = -\pi n_0 \gamma \sum_{i \neq j} M_i M_j \ln|\mathbf{r}_i - \mathbf{r}_j| + \text{const.}, \tag{1.141}$$

$$\mathbf{P}_{\text{vort}} = n_0 \int d^2r \, \nabla \Phi = n_0 \hat{z} \times \int d^2r \, \nabla u = -2\pi n_0 \hat{z} \times \sum_j M_j \mathbf{r}_j, \tag{1.142}$$

where $\mathbf{r}_j$ is the position of the jth vortex. The neutrality condition necessary for Equations (1.141) and (1.142) to be meaningful is $\sum_i M_i = 0$. The constant term in (1.141) originates from the vortex cores.

1.2.5 Topological Invariant, Persistent Current, and Role of Vortices

Suppose the field ψ is confined within a torus and, for simplicity, its modulus is nonzero everywhere in the bulk. Then, for any closed contour Γ inside the torus, there is a well-defined integral, a direct analog of Equation (1.112) for a vortex,

$$I = \oint_\Gamma \nabla \Phi \cdot d\mathbf{l} = 2\pi M, \tag{1.143}$$

---

* Using, respectively, $\int d^2r (\nabla \Phi)^2 = \int d^2r (\nabla u)^2 = -\int d^2r \, u \Delta u = -2\pi \int d^2r \, u \rho$ and the identity $\int d^2r \nabla u \equiv -\int d^2r \, \mathbf{r} \cdot \Delta u.$

with an integer winding number M. There are two cases for Γ and M to consider. For any *trivial* contour, that is, a contour that is continuously shrinkable to a zero radius, we have $M = 0$ and thus $I = 0$. In what follows, we focus on nontrivial contours Γ (see Figure 1.1) that go around the central hole. The winding number can take nonzero values on Γ.

In our special case of $|\psi| \neq 0$, the two relations, Equations (1.143) and (1.86), reveal the topological origin of superfluidity at the microscopic level. The integral I (1.143) is *discrete* and thus *conserved* under time evolution, because as long as $|\psi| \neq 0$ on any nontrivial Γ, the function $I(t)$ should be a continuous function of time. In other words, as long as the absolute value of the field ψ remains nonzero, the time evolution of the field ψ can be viewed as a continuous transformation with fixed topological invariant I. Now if $I \neq 0$, then its conservation implies, in accordance with (1.143), a systematic gradient of phase and, by (1.86), a systematic *superflow* (also called a *persistent current*) along the contour Γ.

We come to an instructive conclusion that for a classical complex-valued field, superfluidity is a natural phenomenon, and, in fact, it is the *destruction* of superfluidity that requires conceptual understanding. The way to change the winding number M for a given contour Γ with the least expensive energy increase is to introduce a vortex topological defect in the field of phase $\Phi(\mathbf{r})$ and to have this defect cross the contour. With the presence of topological defects, the integral I (1.143) is, strictly speaking, no longer a constant of motion. There are, however, two qualitatively different situations. If all topological defects are microscopic in size (small vortex loops in 3D or small vortex–antivortex pairs in 2D), the non-conservation of I does not yet mean the absence of superfluidity. As long as the

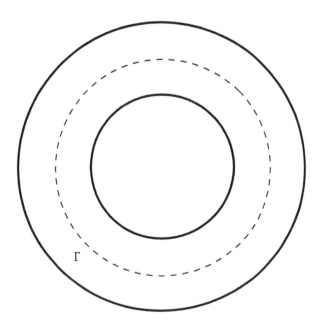

Figure 1.1 Toroidal system: the contour Γ cannot be continuously reduced to a zero radius and thus can have a finite winding number M.

defects are microscopically small, they introduce only short-lived fluctuations of I about a certain nonzero average. Moreover, in the case of microscopic defects, it is always possible to coarse-grain the field $\psi(\mathbf{r})$—that is, to introduce a new field $\tilde{\psi}(\mathbf{r}) = Q([\psi], \mathbf{r})$, where Q is some local functional—in such a way that, on one hand, the field $\tilde{\psi}(\mathbf{r})$ inherits the long-range topological structure of the field $\psi(\mathbf{r})$ and, on the other hand, is free from topological defects. One of the simplest ways of regularizing ψ this way is as follows: Take some local vortex loop in 3D (a vortex–antivortex pair in 2D) and introduce a supplementary complex field $F(\mathbf{r})$ satisfying the requirements that (a) this field has exactly one vortex loop (a vortex–antivortex pair) and at exactly the same location as in the field ψ but with *opposite* orientation of the phase gradients and (b) $F \to 1$ away from the defect. Then the regularization procedure $\psi(\mathbf{r}) \to \psi(\mathbf{r})F(\mathbf{r})$ removes the isolated defect. In view of the requirement (b), the procedure is local and modifies the field ψ only in the region of the size on the order of the size of the topological defect. By applying it to each defect, we ultimately obtain the field $\tilde{\psi}$ that has no vortices. The described procedure possesses one convenient property, which we will be utilizing from now on. Namely, for any regular (free of topological defects) complex function $f(\mathbf{r})$, we have

$$Q([f(\mathbf{r})\psi(\mathbf{r})], \mathbf{r}) = f(\mathbf{r}) Q([\psi(\mathbf{r})], \mathbf{r}). \tag{1.144}$$

The regularization allows us to accurately define the topological invariant in terms of the phase $\tilde{\Phi}(\mathbf{r})$ of the field $\tilde{\psi}(\mathbf{r})$,

$$I = \oint_{\Gamma} \nabla \tilde{\Phi} \cdot d\mathbf{l}. \tag{1.145}$$

1.2.6 Normal Modes, Landau Criterion, and Critical Velocity

In this section, we review the Landau theory of stability of the superflow at $T \to 0$ with homogeneous $n = n_0$ (or local-density-approximation type) density profile in the bulk. In its original formulation, Landau theory deals with the spectrum of normal modes. This will be our starting point as well. Next, we will address the crucial role of vortices, readily revealed by Landau's type of argument.

First, we establish the spectrum of normal modes in a uniform system with no superflow; the spectrum with a superflow will then follow by the Galilean transformation. To this end, we obtain the bilinear Hamiltonian in the vicinity of the ground state ψ_0. For the theory of Section 1.1.4 to be directly applicable, the ground state must be a genuine point of equilibrium, which is not exactly the case if $\mu \neq 0$, because of the time-dependent phase factor [Equation (1.100)]. This formal complication is easily circumvented by introducing a global uniform potential rendering $\mu = 0$, which amounts to working with $H' = H - \mu N$ instead of H. The bilinear Hamiltonian can be derived by substituting $\psi = \psi_0 + \psi_1$ into $H'[\psi]$ and keeping only the leading terms in powers of the small perturbation ψ_1.

Linear in ψ_1 terms are absent since ψ_0 satisfies Equation (1.98). By choosing ψ_0 to be real, $\psi_0 = \sqrt{n_0}$, we obtain

$$H_1 = \frac{\gamma}{2} \int |\nabla \psi_1|^2 d^d r + g n_0 \int |\psi_1|^2 d^d r + \frac{g n_0}{2} \int (\psi_1 \psi_1 + \psi_1^* \psi_1^*) d^d r, \qquad (1.146)$$

where we used $\mu = g n_0$ [see Equation (1.105)]. The Hamiltonian (1.146) is *not* U(1) invariant. This is a manifestation of the *spontaneous symmetry breaking*, when the symmetry of the ground-state solution is lower than that of the Hamiltonian. Here, the phase of the ground-state solution ψ_0 dictates the "preferable" global phase of the field ψ_1 (the term in the parenthesis can be written as $2|\psi_1|^2 \cos 2\Phi_1$). Note that, in the absence of the uniform potential shift, this global phase would evolve in time as $-\mu t$.

Following the theory outlined in Section 1.1.4, we look for the normal modes, $\{b_\mathbf{k}\}$:

$$\psi_1(\mathbf{r}) = \sum_\mathbf{k} [\alpha_\mathbf{k} b_\mathbf{k} + \beta_{-\mathbf{k}}^* b_{-\mathbf{k}}^*] e^{i\mathbf{k}\cdot\mathbf{r}}, \qquad (1.147)$$

which are found from the system of linear equations

$$\left[\begin{array}{c|c} \gamma k^2/2 + g n_0 - \omega_\mathbf{k} & g n_0 \\ \hline g n_0 & \gamma k^2/2 + g n_0 + \omega_\mathbf{k} \end{array}\right] \left[\begin{array}{c} \alpha_\mathbf{k} \\ \beta_{-\mathbf{k}}^* \end{array}\right] = \left[\begin{array}{c} 0 \\ 0 \end{array}\right]. \qquad (1.148)$$

The characteristic equation of the problem (1.148) yields frequencies of the normal modes

$$\omega_\mathbf{k} = \sqrt{\frac{\gamma k^2}{2}\left(\frac{\gamma k^2}{2} + 2 g n_0\right)}. \qquad (1.149)$$

There are two limits (the inverse healing length plays the role of the characteristic momentum):

$$\omega_\mathbf{k} = \begin{cases} ck & (\text{with } c = \sqrt{\gamma g n_0}), & k l_0 \ll 1, \\ \gamma k^2/2, & k l_0 \gg 1. \end{cases} \qquad (1.150)$$

The second limit shows that, at $k l_0 \gg 1$, the nonlinearity becomes irrelevant, and the first limit reveals the crucial importance of self-interaction in the long-wave limit. At $k l_0 \ll 1$, the dispersion is *linear*, meaning that the normal modes are the sound waves (phonons) with the sound velocity c equal to $\sqrt{\gamma g n_0}$.

In terms of the normal modes, the Hamiltonian H_1 and the momentum functional read

$$H_1 = \sum_\mathbf{k} \omega_\mathbf{k} |b_\mathbf{k}|^2, \qquad \mathbf{P} = \sum_\mathbf{k} \mathbf{k} |b_\mathbf{k}|^2. \qquad (1.151)$$

In the state with a superflow at the velocity $\mathbf{v}$, we take advantage of the Galilean transformation for momentum and energy [Equations (1.83) and (1.84)] and find (by replacing $\mathbf{v}_0 \to -\mathbf{v}$ and omitting constant terms)

$$H_1^{(\mathrm{flow})} = \sum_{\mathbf{k}} (\omega_{\mathbf{k}} + \mathbf{v} \cdot \mathbf{k}) |b_{\mathbf{k}}|^2. \tag{1.152}$$

As discussed in Section 1.1.5, the statistical stability of the system requires that for any $\mathbf{k}$,

$$\omega_{\mathbf{k}} + \mathbf{v} \cdot \mathbf{k} > 0 \qquad (\forall \mathbf{k}) \qquad \text{(Landau criterion)}. \tag{1.153}$$

Otherwise, the interaction with the walls, even if infinitesimally weak, will result in the steady growth of the amplitudes of the modes with $\omega_{\mathbf{k}} + \mathbf{v} \cdot \mathbf{k} \leq 0$. The physical meaning of the Landau criterion is identical to that of the absence of the Cherenkov radiation. If the criterion is satisfied, there exists a setup when the walls do not radiate and the uniform flow is stable. It should be kept in mind, however, that at large superflow velocities (on approach to the violation of the Landau criterion), the requirement on the shape of the walls gets more restrictive.*

Formally, the Landau criterion and the generation of normal modes *per se* does not yet imply the slowing down of the superflow since the topological invariant (1.145) remains intact as long as the macroscopically large vortex loops† are absent. In this sense, the Landau criterion is *not* a criterion for superfluidity. The mechanism of generation and proliferation of macroscopically large topological defects can be understood by the same Landau-type argument, now applied to a vortex ring of radius R. Formula (1.118) and the general expression (1.72) for the momentum allow one to calculate both the energy and momentum (E_{vort} and P_{vort}, respectively) of the vortex ring to obtain

$$E_{\mathrm{vort}} \approx 2\pi^2 \gamma n_0 R \ln(R/l_0), \qquad\qquad P_{\mathrm{vort}} \approx 2\pi^2 n_0 R^2. \tag{1.154}$$

We assume that the ring is large $R \gg l_0$ and keep only the leading logarithmic term in the expression for E_{vort}. Equation (1.154) holds for the generic loop of size R except for the numerical values of prefactors. The resulting energy-momentum dispersion relation $E_{\mathrm{vort}} \propto \sqrt{P_{\mathrm{vort}}} \ln(P_{\mathrm{vort}}/n_0 l_0^2)$ is such that if it is used in the Landau criterion (1.153) instead of the normal modes

$$E_{\mathrm{vort}}^{(\mathrm{flow})} = E_{\mathrm{vort}} + \mathbf{P}_{\mathrm{vort}} \cdot \mathbf{v}, \tag{1.155}$$

one would immediately find that the criterion is always violated for large enough vortex loops! For the loop with $\mathbf{P}_{\mathrm{vort}}$ oriented against the flow, we get

$$E_{\mathrm{vort}}^{(\mathrm{flow})}(R) \sim n_0 \left[\gamma R \ln(R/l_0) - vR^2 \right], \tag{1.156}$$

---

* The reason for that is the mechanism of shedding the vortices by macroscopically nonuniform walls, which is discussed later in this section.

† For definiteness, we consider the 3D case. In 2D, the role of vortex loops is played by vortex–antivortex pairs.

and see that, for any v, there exists a critical size R_v such that $E_{vort}^{(flow)}(R_v) = 0$. The corresponding critical vortex loop is unstable with respect to increasing its size. Microscopically, the increase is due to the interaction of the loop with the walls, either direct, if the loop is just pinned by the walls, or via the normal modes, if the latter are excited.

Problem 1.18 *Find 2D analogs of Equations (1.154) through (1.156).*

Thinking of realistic superfluids like ^{4}He flowing through a channel of diameter D, one should remember that typically there will be some vortex lines of length $\sim D$ pinned by the channel walls. This history-dependent state of a realistic superfluid is characterized by the term *remanent vorticity*. When the considerations mentioned earlier are directly applied to the remanent vorticity state, one obtains the *critical velocity* as (with $R \to D$)

$$v_{crit} \sim \frac{\gamma}{D \ln(D/l_0)} \qquad \text{(remanent vorticity)}. \qquad (1.157)$$

At faster velocities, the superflow is unstable with respect to blowing up the pinned vortices. This result was first derived by R. Feynman.

In the absence of remanent vorticity (which is a typical case with trapped BECs of ultracold gases), the superflow remains *metastable* up to velocities comparable to the sound velocity c. At $T = 0$, the destabilization of superflow must be analyzed by looking at normal modes, taking into account the specific geometry of the walls. The true critical velocity corresponds to the situation when unstable normal modes lead to seeding and blowing up a vortex loop—after all, without vortices, the flow will never stop!

Let us go back to the Landau criterion (1.153) and address the question of what exactly happens when it gets violated. In turns out that the answer depends dramatically on the particular shape of the dispersion ω_k. Here, we will confine ourselves to discussing the case of Equation (1.149) or a similar one—the crucial qualitative feature being that the violation of the criterion first occurs at arbitrarily small wavevectors.[*] Initially, this leads to generation of long-wave density waves with large amplitudes transforming into macroscopically large domains with different densities. This process cannot be stabilized at any finite minimal density because, in domains with lower density, the Landau criterion is violated even more strongly. At some point of the evolution, the density will reach zero and a vortex line seed will be created. The vortex will then be blown up to a macroscopic size. If the distance between the walls has substantial nonuniformity (e.g., a narrowing of a channel), then the Landau criterion will be violated locally in the place where the velocity of the flow is the fastest. As a result, the nullifying of the density with subsequent creation and blowing up of a vortex will happen locally *before* the criterion is violated everywhere in the bulk.

---

[*] An example of a very different behavior will be given in Section 1.2.7.

1.2.7   Flow-Induced Density Wave: Supersolid

An instructive generalization of the energy functional deals with nonlocal interaction:

$$(g/2) \int |\psi|^4 d^d r \rightarrow (1/2) \int U(|\mathbf{r}_1 - \mathbf{r}_2|) |\psi(\mathbf{r}_1)|^2 |\psi(\mathbf{r}_2)|^2 d^d r_1 d^d r_2. \tag{1.158}$$

In this case, the only change in Equation (1.148) is the replacement $g \rightarrow U_k$, where
U_k is the Fourier harmonic of the potential $U(r)$. The spectrum of normal modes
then becomes

$$\omega_{\mathbf{k}} = \sqrt{\frac{\gamma k^2}{2} \left(\frac{\gamma k^2}{2} + 2 U_k n_0 \right)}. \tag{1.159}$$

The condition $U(k = 0) \geq 0$ (necessary to prevent the system from collapse to the
state with infinitely high density) is not incompatible with the possibility of having $U_k < 0$ for *certain* k's, and it is the latter option that distinguishes the model
with nonlocal potential from the original one. With the potential $U(r)$ having
negative Fourier harmonics,[*] the dispersion (1.159) becomes nonmonotonic[†] at
large enough values of $n_0 |U(k)|$. As a result, the dynamics of unstable superflow
violating the Landau criterion becomes different. With the dispersion (1.159) having a minimum at finite $k = k_{\text{rot}}$, an instability may first occur at some finite
wavevector $\mathbf{k}_*$ (with $|k_*|$ close to k_{rot}) such that $\omega_{\mathbf{k}_*} = k_* v$. Initially, it will lead to the
growth of amplitudes of harmonics with momenta close to $\mathbf{k}_*$. However, this can-
not change the velocity circulation despite reducing the current density. Further
analysis then shows—see the end of this section—that when the amplitude of the
perturbation becomes large enough, the *negative feedback* of the perturbation on
the spectrum of normal modes stabilizes the system. The superflow persists with
the same velocity circulation, but the density of the system is no longer uniform:
There appears a *density wave* with a period defined by the wavevector $\mathbf{k}_*$. This consideration shows that the Landau criterion is best characterized as the criterion for
stability of superflows with uniform density profile.

   At large enough amplitude of $U_k < 0$, even the ground-state solution (obviously it is stable!) is nonuniform. Indeed, in this case, there appear dynamically
unstable modes corresponding to imaginary values of $\omega_{\mathbf{k}}$ (1.159), meaning that
the uniform solution is not even metastable. The genuine ground state is now a
solid, that is, a state with broken translational symmetry. Such solids are called
supersolids because they can support a superflow of the matter they are made from.
In this context, they were first introduced by E. Gross. A systematic description of
realistic supersolids requires quantum mechanics. Corresponding discussion will
be presented in Chapter 9.

---

[*] For example, a stepwise potential $U(r) = U(0)$ at $r < r_0$, $U(r) = 0$ at $r \geq r_0$.

[†] Qualitatively, this resembles the dispersion of elementary excitations in liquid ^{4}He. This analogy, however, cannot be pushed to arbitrarily high momenta: in ^{4}He, the curve features an *end
   point* of essentially quantum-mechanical origin.

It is easy to show that the transition between a homogeneous ground state and a supersolid in two and three dimensions is always of the first order; that is, by the time the homogeneous ground state becomes dynamically unstable, there already exists a better solid state with finite-amplitude density modulation. Consider the cubic terms of the Hamiltonian. The general form of these terms is as follows:

$$H^{(\text{cubic})} = \sum_{\mathbf{k}_1 + \mathbf{k}_2 + \mathbf{k}_3 = 0} \left[C_{\mathbf{k}_1 \mathbf{k}_2 \mathbf{k}_3} \, b_{\mathbf{k}_1} b_{\mathbf{k}_2} b_{\mathbf{k}_3} + D_{\mathbf{k}_1 \mathbf{k}_2 \mathbf{k}_3} \, b^*_{-\mathbf{k}_1} b_{\mathbf{k}_2} b_{\mathbf{k}_3} \right] + \text{c.c.}, \tag{1.160}$$

where $C_{\mathbf{k}_1 \mathbf{k}_2 \mathbf{k}_3}$ and $D_{\mathbf{k}_1 \mathbf{k}_2 \mathbf{k}_3}$ are certain complex-valued functions of $(\mathbf{k}_1, \mathbf{k}_2, \mathbf{k}_3)$. The constraint $\mathbf{k}_1 + \mathbf{k}_2 + \mathbf{k}_3 = 0$ follows from the translation symmetry of the Hamiltonian, which is invariant with respect to the transformation $b_{\mathbf{k}} \to b_{\mathbf{k}} \exp(i\mathbf{k} \cdot \mathbf{r}_0)$ with arbitrary $\mathbf{r}_0$. By rotation symmetry, the instability, $\omega_k = 0$, is simultaneously satisfied on a sphere (circle) $|\mathbf{k}| = k_{\text{rot}}$ in 3D (2D). Thus, it is possible to select three vectors, $\mathbf{k}_1, \mathbf{k}_2, \mathbf{k}_3$, such that $k_1 = k_2 = k_3 = k_{\text{rot}}$ and $\mathbf{k}_1 + \mathbf{k}_2 + \mathbf{k}_3 = 0$. For those three vectors, the quadratic terms are zero while the cubic terms are not. The sign of the cubic terms is not definite since it changes by, say, $b_{\mathbf{k}_2} \to -b_{\mathbf{k}_2}$. Hence, it is energetically favorable to have *finite* amplitudes for some b's (the amplitudes are limited by the quartic terms in the Hamiltonian). This proves that the homogeneous solution is not the ground state.

The situation with the flow-induced density wave is different in that the violation of Landau criterion takes place at a single, nonzero momentum $\mathbf{k}_*$, so that the relevant cubic terms are absent. By the same requirement of the zero sum of all the momenta, there is only one quartic term, $\propto |b_{\mathbf{k}_*}|^4$, which controls the growth of the $\mathbf{k}_*$-harmonic amplitude. The proportionality coefficient has to be positive for a well-defined Hamiltonian; otherwise, the system does not have a ground state as the energy tends to $-\infty$ as $|b_{\mathbf{k}_*}| \to \infty$. The quartic term stabilizes the growth of the density wave amplitude and, in particular, guarantees (in the absence of cubic terms) that the transition between the uniform ground state and the flow-induced density wave is continuous.

1.3   Matter Field under Rotation

The external potential $\mathcal{V}(\mathbf{r})$ in the GPE (1.95) can also depend on time. For obvious reasons, a general case of time-dependent $\mathcal{V}(\mathbf{r})$ goes beyond the equilibrium statistics. There is, however, a very important special case—*rotation of the system*—where the time dependence of $\mathcal{V}(\mathbf{r})$ does not lead to energy nonconservation and can be removed by an appropriate transformation of coordinates, namely, by going to the *rotating reference frame*. In this section, we develop a theory of the field ψ in the rotating frame. It is important to realize that for cases in which the rotation is enforced by the walls, the rotating external potential should not necessarily be present in the GPE, since the direct effect of the walls vanishes at microscopic distances on the order of the healing length. In such a case, the rotating $\mathcal{V}(\mathbf{r})$ is merely a tool that allows one to arrive at the notion of the rotating frame. We will see that the hallmark of the rotating state is the appearance of topological defects (vortex lines in 3D and point vortices in 2D). For definiteness, we consider only the 3D case, since adapting 3D results to the 2D case is straightforward.

1.3.1 Rotating Frame

The rotation of the external potential means that its coordinate-time dependence is of the form

$$\mathcal{V}(\mathbf{r}, t) \equiv \mathcal{V}(\tilde{\mathbf{r}}), \tag{1.161}$$

with $\tilde{\mathbf{r}} = \tilde{\mathbf{r}}(\mathbf{r}, t)$ being the time-dependent radius vector related to $\mathbf{r}$ and t by the rotation transformation

$$\tilde{r}_\alpha = \sum_\beta T_{\alpha\beta}(t) r_\beta, \tag{1.162}$$

where $T_{\alpha\beta}$ is an orthogonal time-dependent matrix, whose explicit form (not necessary for our purposes) follows from the differential relation

$$\frac{\partial \tilde{\mathbf{r}}}{\partial t} = -\mathcal{W} \left[\hat{\mathbf{n}} \times \tilde{\mathbf{r}} \right], \tag{1.163}$$

where $\mathcal{W}$ is the rotation frequency and $\hat{\mathbf{n}}$ is the unit vector along the rotation axis. Equation (1.161) suggests changing the variables in the GPE as

$$(\mathbf{r}, t) \rightarrow (\tilde{\mathbf{r}}, \tilde{t}), \tag{1.164}$$

with $\tilde{t} = t$. The spatial differential operators then transform as follows ($\partial_\alpha \equiv \partial/\partial r_\alpha$, $\tilde{\partial}_\alpha \equiv \partial/\partial \tilde{r}_\alpha$):

$$\partial_\alpha = \sum_\beta T_{\beta\alpha} \tilde{\partial}_\beta. \tag{1.165}$$

From (1.165) and the fact that $T_{\alpha\beta}$ is real and unitary, we have

$$\Delta \equiv \sum_\alpha \partial_\alpha \partial_\alpha = \sum_{\alpha,\beta,\gamma} T_{\beta\alpha} T_{\gamma\alpha} \tilde{\partial}_\beta \tilde{\partial}_\gamma = \sum_\beta \tilde{\partial}_\beta \tilde{\partial}_\beta \equiv \tilde{\Delta} \tag{1.166}$$

and conclude that the only additional term generated in the rotating frame comes from the time derivative

$$\frac{\partial}{\partial t} = \frac{\partial}{\partial \tilde{t}} + \frac{\partial \tilde{\mathbf{r}}}{\partial t} \cdot \frac{\partial}{\partial \tilde{\mathbf{r}}} = \frac{\partial}{\partial \tilde{t}} - \mathcal{W} \left[\hat{\mathbf{n}} \times \tilde{\mathbf{r}} \right] \cdot \frac{\partial}{\partial \tilde{\mathbf{r}}}. \tag{1.167}$$

The resulting equation reads (from now on, we work exclusively in the rotating frame and thus omit the tilde notation)

$$i\dot{\psi} = -\frac{\gamma}{2}\Delta\psi + g|\psi|^2\psi + \mathcal{V}(\mathbf{r})\psi + i\mathcal{W} \left[\hat{\mathbf{n}} \times \mathbf{r} \right] \cdot \nabla\psi. \tag{1.168}$$

As can be checked directly, Equation (1.168) is Hamiltonian, with the following *time-independent* Hamiltonian functional (use $[\mathbf{n} \times \mathbf{r}] \cdot \nabla = \mathbf{n} \cdot [\mathbf{r} \times \nabla]$)

$$H = \int \left[(\gamma/2)|\nabla\psi|^2 + \mathcal{V}(\mathbf{r})|\psi|^2 + (g/2)|\psi|^4 \right] d^3r - \mathcal{W}\hat{\mathbf{n}} \cdot \mathbf{M}, \tag{1.169}$$

where $\mathbf{M}$ is the angular momentum (1.73).

It is instructive to note that for a system with conserved angular momentum, the Hamiltonian (1.169) is nothing but the "primed" Hamiltonian of the grand canonical—with respect to the additive constant of motion $\mathbf{M}$—Gibbs distribution. In a general case of rotating external potential to which the Hamiltonian (1.169) applies, for both equilibrium and nonequilibrium problems, the angular momentum is not conserved, and the last term in (1.169) is exclusively due to changing the reference frame.

1.3.2 Vector Potential, Gauge Equivalence

Let us rewrite the Hamiltonian (1.169) identically as

$$H = \int \left[(\gamma/2)|(\nabla - i\mathbf{A})\psi|^2 + \mathcal{V}_{\text{eff}}(\mathbf{r})|\psi|^2 + (g/2)|\psi|^4 \right] d^3r, \tag{1.170}$$

with

$$\mathbf{A}(\mathbf{r}) = (\mathcal{W}/\gamma)(\hat{\mathbf{n}} \times \mathbf{r}) \tag{1.171}$$

and the effective (=bare + centrifugal) external potential

$$\mathcal{V}_{\text{eff}}(\mathbf{r}) = \mathcal{V}(\mathbf{r}) - \frac{\mathcal{W}^2\rho^2}{2\gamma}, \tag{1.172}$$

where ρ is the distance from the point $\mathbf{r}$ to the rotation axis. The crucial difference between (1.170) and the nonrotating Hamiltonian (1.94) is in the structure of gradient term. To explore the new physics that comes with it, consider a generalization of (1.170) to the case when the vector field $\mathbf{A}$ is an *arbitrary* function of coordinate referred to as *vector potential*. Moreover, allow vector potential $\mathbf{A}$ and the external potential to be time-dependent as well. The resulting Hamiltonian equation of motion reads

$$i\dot{\psi} = \frac{\gamma}{2}[i\nabla + \mathbf{A}(\mathbf{r}, t)]^2\psi + g|\psi|^2\psi + \mathcal{V}(\mathbf{r}, t)\psi. \tag{1.173}$$

Here, we use the notation $\mathcal{V}$ rather than $\mathcal{V}_{\text{eff}}$ to emphasize that in this general case we do not assume any special relationship between $\mathcal{V}$ and $\mathbf{A}$.

With time-dependent $\mathbf{A}$ and $\mathcal{V}$, the only additive quantity conserved generically by Equation (1.173) is the amount of field N (1.18). The conservation of N can still be written in the form of continuity Equation (1.75), but now the expression for the matter flux density reads

$$\mathbf{j} = i(\gamma/2)[\psi\nabla\psi^* - \psi^*\nabla\psi] - \gamma\mathbf{A}|\psi|^2 = \gamma|\psi|^2(\nabla\Phi - \mathbf{A}). \tag{1.174}$$

This suggests a generalization of the definition of microscopic velocity [Equation (1.86)], to

$$\mathbf{v} = \gamma(\nabla\Phi - \mathbf{A}), \tag{1.175}$$

to preserve the formula $\mathbf{j} = n\mathbf{v}$ [see (1.87)]. In addition, there emerges a useful relation between $\mathbf{j}$, H, and vector potential:

$$\mathbf{j}(\mathbf{r}) = -\frac{\delta H}{\delta \mathbf{A}(\mathbf{r})}. \tag{1.176}$$

Equation (1.173) features an important property of *gauge equivalence*. Introduce a new field ψ' by

$$\psi = \psi' e^{i\phi}, \tag{1.177}$$

with an arbitrary real phase field $\phi(\mathbf{r}, t)$. The phase Φ' of the field ψ' is related to the phase Φ of the field ψ by

$$\Phi' = \Phi - \phi. \tag{1.178}$$

A direct check shows that the equation of motion for ψ' has exactly the same form as (1.173) when vector and external potentials are simultaneously transformed as

$$\mathbf{A}' = \mathbf{A} - \nabla\phi, \tag{1.179}$$

$$\mathcal{V}' = \mathcal{V} + \dot{\phi}. \tag{1.180}$$

Equations (1.178) through (1.180) are known as the *gauge transformation*.* The gauge equivalence can be used to simplify Equation (1.173) since certain terms in $\mathbf{A}$ and $\mathcal{V}$ can be removed. Under gauge transformation, the values of $\mathbf{j}(\mathbf{r})$ (1.174) and $\mathbf{v}(\mathbf{r})$ (1.175) are preserved. In the case of time-independent ϕ, the gauge transformation reduces to (1.178) and (1.179) and preserves the value of the Hamiltonian functional.

1.3.3 Vortex Array

In a macroscopically large system and at finite angular velocity, the state of rotating field ψ is quite nontrivial, even at $T = 0$. The general features of the ground state are understood readily by revealing the necessary conditions to be met to minimize the Hamiltonian in the rotating frame. The representation (1.170) in terms of the vector and effective potentials proves very handy when combined with the density-phase parameterization of the field

$$\psi = \sqrt{n} e^{i\Phi}. \tag{1.181}$$

---

* It is worth emphasizing that here the vector and scalar potentials are the external parameters of the Hamiltonian rather than dynamic degrees of freedom (cf. Chapter 5). This is the reason for using the "gauge equivalence" term (for Hamiltonians in different external fields) rather than "gauge invariance"—the latter referring to the invariance of H when the vector and scalar potentials are dynamic variables.

By splitting energy into the kinetic and potential parts, $E = E_{kin} + E_{pot}$ as

$$E_{kin} = (\gamma/2) \int |(\nabla - i\mathbf{A})\psi|^2 d^3r = (\gamma/2) \int \left[n(\nabla\Phi - \mathbf{A})^2 + (\nabla\sqrt{n})^2 \right] d^3r, \qquad (1.182)$$

$$E_{pot} = \int \left[\mathcal{V}_{eff}(\mathbf{r})n + (g/2)n^2 \right] d^3r, \qquad (1.183)$$

we arrive at the crucial conclusion that, in the ground state of a macroscopic system, the presence of the gauge field induces a macroscopic number of vortices in the field $\Phi(\mathbf{r})$; moreover, this number is essentially independent of the details of the potential energy.

Indeed, the energy of the ground state is not supposed to diverge with the system size faster than L^d; that is, the energy density has to remain finite. Were the contribution from $\nabla\Phi$ of subleading character, the kinetic energy would inevitably scale as L^{d+2} due to terms containing $\mathbf{A}^2$. Kinetic energy density can be made finite only if the field $\Phi(\mathbf{r})$ obeys the relation

$$\nabla\Phi = \mathbf{A} + \text{intensive terms.} \qquad (1.184)$$

There is no way to satisfy Equation (1.184) without introducing topological defects. To demonstrate this, take a closed contour C in the plane perpendicular to the rotation axis, and integrate both sides of (1.184) along C:

$$\oint_C \nabla\Phi \cdot d\mathbf{l} = \oint_C \mathbf{A} \cdot d\mathbf{l} + \mathcal{O}(\text{contour length}). \qquad (1.185)$$

Assume that C is a circle of radius R. Then, with (1.171) taken into account, we find

$$2\pi M_{top}(R) = 2\pi(\mathcal{W}/\gamma)R^2 + \mathcal{O}(R), \qquad (1.186)$$

where the integer $M_{top}(R)$ stands for the net charge of topological defects inside the circle. We conclude that, irrespective of the details of the external potential and interaction term, the rotating field ψ features a *homogeneous* coarse-grained density of the net topological charge, n_{top}, that depends only on the parameter γ and the angular velocity of rotation:

$$n_{top} = \frac{\mathcal{W}}{\pi\gamma}. \qquad (1.187)$$

The Onsager–Feynman relation (1.187) applies not only to the ground state. Indeed, our analysis was based exclusively on the absence of superextensive scaling of energy. Clearly, the argument applies to any equilibrium state and, generally speaking, certain nonequilibrium states as well.

Given the finite concentration of vorticity in the rotating field ψ, the question arises of whether such a state can be superfluid and, if yes, then what is the difference between a superfluid and a normal state? Recalling that the phenomenon of superfluidity is based on the existence of topological invariant, we realize that the answer to this question depends on the dynamics of the vortices induced by rotation. If these vortices are mobile, they form a relaxation channel for the topological invariant I and thus destroy superfluidity. When the vortex array is in a solid state pinned by the walls and/or impurities, the positions of individual vortices, while being subject to microscopic fluctuations, cannot perform the macroscopic drift necessary to relax I. We conclude then that, in a rotating system, the phase transition to the normal state at high enough temperature can be viewed as melting of the vortex lattice (Figure 1.2).

Only at a macroscopically small $\mathcal{W}$, it may turn out that it is energetically favorable to have no vortices at all. To find the critical angular frequency $\mathcal{W}_c$ at which the energy minimum corresponds to having exactly one vortex in the system, we resort to Equation (1.118) for the energy of the vortex line. Using this result with the Hamiltonian (1.169), we find (one needs to compute the angular momentum of a vortex line)

$$\mathcal{W}_c = (\gamma/R^2)\ln(R/l_0) + \text{subleading terms}. \tag{1.188}$$

Problem 1.19 *Perform the calculation leading to Equation (1.188).*

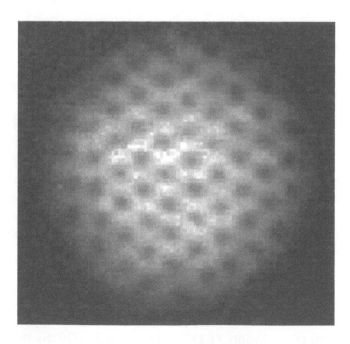

Figure 1.2   Top view of the vortex array in a rotating BEC. As we will learn later, BECs are typical examples of the classical-field superfluids being considered in this chapter. (The image is courtesy of D. S. Hall.)

1.3.4   Twisted Boundary Condition

In a simply connected system, a time-independent vector potential with identically zero curl,

$$\nabla \times \mathbf{A} \equiv 0, \tag{1.189}$$

is trivial: it can be "gauged out" immediately by the transformation (1.178) and (1.179) with

$$\phi(\mathbf{r}) = \int\limits_{C(\mathbf{r},\mathbf{r}_0)} \mathbf{A} \cdot d\mathbf{l}, \tag{1.190}$$

where $C(\mathbf{r}, \mathbf{r}_0)$ is a path from a certain fixed point $\mathbf{r}_0$ to the point $\mathbf{r}$; independence of the integral on the actual path is guaranteed by the condition (1.189). In cases of multiple connectivity—the simplest and most important being the case of periodic boundary conditions—the situation gets less trivial when for a certain class of closed paths (e.g., the one in Figure 1.1), we have

$$\oint\limits_{\Gamma} \mathbf{A} \cdot d\mathbf{l} = \varphi_0 + 2\pi M, \qquad 0 < |\varphi_0| < \pi, \tag{1.191}$$

where M is an integer. In this case, Equation (1.190) is inconsistent with the single-valuedness of the fields ψ and ψ', requiring

$$(2\pi)^{-1} \oint\limits_{\Gamma} \nabla \phi \cdot d\mathbf{l} = \text{integer}. \tag{1.192}$$

Hence, in this case, the vector potential brings about some physical effect characterized by the irreducible *imposed phase* φ_0.

Consider a homogeneous system with periodic boundary conditions along the x-axis and a vector potential

$$\mathbf{A} = -\hat{\mathbf{e}}_x \varphi_0 / L_x, \qquad 0 < |\varphi_0| < \pi, \tag{1.193}$$

where L_x is the system size in the x-direction. We observe that for $|\varphi_0| < \pi$ the ground state of the system (we ignore here the global time-dependent phase) is just $\psi = \sqrt{n}$. From Equation (1.174), we then find that the ground state features a uniform flux directly proportional to the imposed phase φ_0 and inversely proportional to the system size in the direction of the flow:

$$\mathbf{j} = \hat{\mathbf{e}}_x \gamma n \varphi_0 / L_x. \tag{1.194}$$

We conclude that the ground state is the same as the state (1.85) in the moving reference frame corresponding to the flow with velocity

$$\mathbf{v}_0 = \hat{\mathbf{e}}_x \gamma \varphi_0 / L_x. \tag{1.195}$$

The imposed phase φ_0 allows one to continuously extrapolate between the quantized velocity circulation values, $\gamma 2\pi M/L_x$; with finite φ_0, the allowed values become $\gamma(2\pi M + \varphi_0)/L_x$. Often $\varphi_0 \neq 0$ is referred to as the twist phase in a system with *twisted boundary conditions*. Indeed, if the vector potential (1.193) is gauged out using relation (1.190), then the field obeys a relation $\psi(x + L_x) = e^{-i\varphi_0}\psi(x)$.

It is instructive to see how the persistent current arising in response to the twist is related in a very simple and universal way to the twist phase dependence of the free energy (or the grand canonical potential, depending on the ensemble being used). For definiteness, we choose the canonical ensemble with a fixed amount of matter N and work with the Helmholtz free energy, F. As a result of the calculus of variations—a vector-field analog of the scalar-field expression (1.17)—we can express the partial derivative of F with respect to φ_0 as follows:

$$\frac{\partial F}{\partial \varphi_0} = \int \frac{\delta F}{\delta \mathbf{A}} \cdot \frac{\partial \mathbf{A}}{\partial \varphi_0}\, d^d r. \tag{1.196}$$

Now we utilize a known theorem of statistical mechanics, which provides thermodynamic relations for the expectation value of a partial derivative (with respect to a certain external parameter) of the system's Hamiltonian. The theorem states that such an expectation value equals to the partial derivative—with respect to the parameter in question, the rest of thermodynamic variables being fixed—of the thermodynamic potential associated with the fixed variables. With this theorem applied to the variational derivative of interest and using relation (1.176), we have

$$\frac{\delta F}{\delta \mathbf{A(r)}} = \left\langle \frac{\delta H}{\delta \mathbf{A(r)}} \right\rangle = -\langle \mathbf{j(r)} \rangle. \tag{1.197}$$

Since the vector potential $\mathbf{A}$ responsible for the twisted boundary condition has zero curl, we can write it as the gradient of the multiple-valued scalar potential u (the contour Γ is like that in Figure 1.1):

$$\mathbf{A} = -\varphi_0 \nabla u, \qquad \oint_\Gamma \nabla u \cdot d\mathbf{l} = 1, \tag{1.198}$$

so that

$$\frac{\partial \mathbf{A}}{\partial \varphi_0} = -\nabla u. \tag{1.199}$$

With this parameterization, and also taking into account that, in accordance with (1.75), the vector field $\mathbf{j}$ has zero divergence, we get (for brevity, we omit the statistical average sign for current $\mathbf{j}$)

$$\frac{\partial F}{\partial \varphi_0} = \int \mathbf{j} \cdot \nabla u\, d^d r = \int \nabla \cdot (u\mathbf{j})\, d^d r. \tag{1.200}$$

The bulk integral from $\nabla \cdot (u\mathbf{j})$ can be converted into a surface integral from $u\mathbf{j}$, with special attention paid to the multiple-valued nature of u. The latter amounts to introducing a cross-sectional cut, $\mathcal{S}$, rendering torus a simply connected region in which u is a single-valued function and a unit step across the cut: $(u_+ - u_-) = 1$. Then,

$$\int \mathbf{j} \cdot \nabla u \, d^d r = \int_{\mathcal{S}} u_+ \mathbf{j} \cdot d\mathbf{s} - \int_{\mathcal{S}} u_- \mathbf{j} \cdot d\mathbf{s} = \int_{\mathcal{S}} (u_+ - u_-)\mathbf{j} \cdot d\mathbf{s} = \int_{\mathcal{S}} \mathbf{j} \cdot d\mathbf{s} = J, \quad (1.201)$$

with J being the (average) current through the torus cross section. This leads to an elegant generic relation

$$J = \frac{\partial F}{\partial \varphi_0}. \tag{1.202}$$

If a rotating toroidal system of radius R has a typical cross-section diameter much smaller than R and if the angular velocity of rotation is small enough

$$\mathcal{W} = -\frac{\gamma \varphi_0}{2\pi R^2}, \qquad 0 < |\varphi_0| < \pi, \tag{1.203}$$

then the effect of such a macroscopically slow rotation reduces to the twisted boundary condition with the twist φ_0 along the toroidal channel. This induces persistent current in the rotating frame with velocity $v = -\mathcal{W}R$, which corresponds to the situation when the field remains static in the original frame.

Apart from its relevance to the macroscopically slow rotation, the twisted boundary condition plays an important role in first-principle simulations of classical-field and quantum superfluids. Later, we will see that an elegant way of accessing superfluid properties of a system is through its static response to the twist phase. With this goal in mind, we generalize the definition of the topological number (1.145) to the gauge invariant form

$$I = \oint_\Gamma (\nabla \tilde{\Phi} - \mathbf{A}) \cdot d\mathbf{l} = 2\pi M + \varphi_0, \tag{1.204}$$

suitable for working with twisted boundary conditions. It also makes sense to introduce the gauge-invariant momentum

$$\mathbf{P} = \int |\psi|^2 (\nabla \Phi - \mathbf{A}) d^d r. \tag{1.205}$$

The conservation of $\mathbf{P}$ in a translationally invariant system—in which there exists a gauge such that $\mathbf{A}$ is just a constant vector [see Equation (1.193)]—immediately follows from the fact that the gauge-invariant momentum (1.205) differs from the originally defined $\mathbf{P}$ (1.72) by a constant shift equal to $N\mathbf{A}$. The definition (1.205) renders the flux-momentum relation (1.77) gauge invariant.

1.3.5 Josephson Junction, SQUID

Suppose we are dealing with a superfluid in a toroidal container with a very weak link, a situation that occurs, for example, when there is a strong potential barrier across the toroidal channel that results in the exponential suppression of the density n in a certain cross-sectional area. We will be interested in the case when the link is so weak that the topological invariant is no longer conserved. Such weak links are called *Josephson junctions*. Because of the nonconservation of I, the system does not support persistent currents associated with different winding numbers, and thus is qualitatively different from a toroidal superfluid. Nevertheless, one characteristic feature of a toroidal superfluid (namely, the response to the twist phase φ_0) partially survives. The word "partially" reflects the fact that now the value of $J(\varphi_0)$ is defined exclusively by the properties of Josephson junction, with the rest of the system merely ensuring the toroidal topology (or boundary conditions away from the junction). In this sense, it is said that φ_0 is applied to the Josephson junction, and the current $J(\varphi_0)$ is referred to as the Josephson current. The reason for this physical picture is that it is energetically favorable to have the entire contribution to the gauge-invariant integral

$$\left\langle \oint_{\Gamma} (\nabla \Phi - \mathbf{A}) \cdot d\mathbf{l} \right\rangle = \varphi_0 \tag{1.206}$$

coming from the Josephson junction region and, correspondingly, to have $\langle (\nabla \Phi - \mathbf{A}) \rangle \approx 0$ away from that region. The Josephson junction is completely described by the function $J(\varphi_0)$, a certain 2π-periodic, odd [due to the reflection symmetry of the Hamiltonian $H(\varphi_0)$] function of φ_0. These properties hint at the qualitative behavior of the function $J(\varphi_0)$: It starts from zero at $\varphi_0 = 0$, increases until it reaches a maximum at some $\varphi_0 \in (0, \pi)$, and then decreases, vanishing at $\varphi_0 = \pi$. One can show that for extremely weak links, the function $J(\varphi_0)$ takes on a universal form

$$J(\varphi_0) = J_0 \sin \varphi_0 \tag{1.207}$$

(in which case, the parameter J_0 is called the Josephson constant). We will derive Equation (1.207) later, using the general relation (1.202) and a special loop representation for the partition function [see Chapter 3, Equation (3.37)]. Meanwhile, we want to mention a certain subtlety of the limiting regime (1.207). The simplest case is when the limit is reached already at $T = 0$ (say, by using the height of the external potential to control the potential energy barrier). Less trivial is the situation when the regime (1.207) takes place only at high enough temperature, being driven by proliferation of thermally activated topological defects.

A toroidal superfluid with two Josephson junctions and two superleads can be used as a gyrometric device sensitive to the vector potential produced by rotation. Originally, this device (see Figure 1.3) was invented in the context of superconductivity (and, correspondingly, the electromagnetic, rather than rotation-induced, vector potential). This context explains the abbreviation SQUID (superconducting

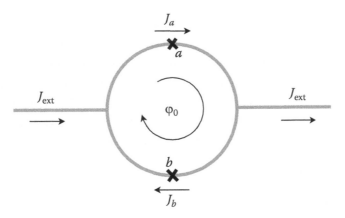

Figure 1.3   Schematics of the SQUID. Josephson junctions, a and b, are shown with crosses; $J_a \equiv J_a(\varphi_a)$ and $J_b \equiv J_b(\varphi_b)$ are the corresponding Josephson currents, while J_{ext} is the external current applied to the SQUID. Arrows indicate positive directions of currents and phases. The phase φ_0 is the twist phase being measured.

quantum interference device) used to name the devices with this type of structure. Incidentally, in connection with the dual meaning of the term "quantum," see the beginning of this chapter.

The circulation of the vector potential in SQUID is measured through the phase φ_0 it induces in the loop. The superfluid channels are supposed to be narrow enough, so that all the effects of the vector potential are reduced to the twist phase. The recipe of measuring φ_0 is as follows: One applies to the device an external current, J_{ext}, and ramps it until the critical value $J_{\text{ext}}^{(\text{crit})}$ is reached, at which point the response becomes dissipative. The quantity $J_{\text{ext}}^{(\text{crit})} \equiv J_{\text{ext}}^{(\text{crit})}(\varphi_0)$ is then used to extract the value of φ_0.

To reveal the underlying physics, we need to understand (a) why at small enough J_{ext} there is no dissipation, (b) where the critical value of J_{ext} comes from, and (c) why this value is sensitive to φ_0. The answer to the first question is quite simple. The Josephson junctions, a and b, support equilibrium (and thus nondissipative) currents, as long as the relation

$$J_{\text{ext}} = J_a - J_b \qquad (1.208)$$

can be satisfied with *Josephson currents* $J_a \equiv J_a(\varphi_a)$ and $J_b \equiv J_b(\varphi_b)$, where the phases φ_a and φ_b are the phases applied across the junctions a and b, respectively. The phases φ_a and φ_b are subject to the obvious constraint (see Figure 1.3)

$$\varphi_a + \varphi_b = \varphi_0. \qquad (1.209)$$

Hence, there is only one degree of freedom, which defines the thermodynamic equilibrium condition (1.208). Without loss of generality, we choose this degree of freedom to be the phase φ_a. Since the left-hand side of

$$J_a(\varphi_a) + J_b(\varphi_a - \varphi_0) = J_{\text{ext}} \qquad (1.210)$$

is a bounded function of φ_a, its upper bound defines the critical value of $J_{\text{ext}}^{(\text{crit})}$. The dependence of the left-hand side on φ_0 readily translates into $J_{\text{ext}}^{(\text{crit})} \equiv J_{\text{ext}}^{(\text{crit})}(\varphi_0)$.

An elegant experimental way of scanning the left-hand side of (1.210) as a function of φ_a, with automatically obtaining $J_{\text{ext}}^{(\text{crit})}(\varphi_0)$, is to attach the external leads to two reservoirs with slightly different chemical potentials. In this case, the phase φ_a will be a (slow) linear function of time, with the slope given by the chemical potential difference between the reservoirs:

$$\varphi_a(t) = \varphi_a(0) + \Delta\mu t. \tag{1.211}$$

At zero temperature, this formula directly follows from Equation (1.100). As we will show in Chapter 4, the same result holds true at finite temperature as well.

CHAPTER 2

Superfluidity at Finite Temperatures and Hydrodynamics

2.1 Equilibrium Statistics of a Superfluid

In this section, we derive the Gibbs distribution for a classical-field superfluid and utilize it to establish general (independent of microscopic details) thermodynamic relations and fundamental physical properties implied by them. In fact, the relations that we obtain for the classical field apply to quantum fields as well. We will see it when generalizing the theory to the quantum case. The physical reason for universality is that the relations in question are derived for the long-wave limit. The fact that we are dealing with long-wave physics is also crucial for understanding that the so-called ultraviolet (UV) catastrophe (the UV divergence of the integral for N at any finite temperature) inherent to most classical-field models, including the one considered here, does not affect these universal results. The particular form of the UV regularization is relevant only to the explicit form of the equations of state, playing no role in this section.

2.1.1 Gibbs Distribution in the Presence of a Superflow

The constant of motion (1.145), since it is not additive, cannot be directly incorporated into the Gibbs distribution. However, one can introduce an additive topological invariant and constant of motion. For the sake of simplicity, in what follows, we confine ourselves to a rectangular geometry with periodic boundary conditions. The additive topological invariant is nothing but the volume integral of the gradient of the phase $\tilde{\Phi}$:

$$\mathbf{I} = \int \nabla \tilde{\Phi} d^d r. \tag{2.1}$$

The conservation of (2.1) immediately follows from the conservation of (1.145). Consider, for definiteness, the x-component of the vector $\mathbf{I}$:

$$I_x = \int dy \, dz \int \nabla \tilde{\Phi} \cdot \hat{\mathbf{e}}_x \, dx. \tag{2.2}$$

For any fixed y and z, the integral over x in (2.2) is nothing but the invariant (1.145) along a particular closed contour going parallel to the x-axis. Hence, I_x reduces to the (y, z)-integral of a conserved quantity and thus is a conserved quantity itself.

With the additive constant of motion $\mathbf{I}$, we can introduce the superfluid grand canonical partition function as

$$Z = \int \mathcal{D} \psi \, e^{-H'[\psi]/T}, \tag{2.3}$$

where $H'[\psi]$ is the effective Hamiltonian, obtained from the original Hamiltonian $H[\psi]$ in the standard way, by subtracting terms proportional to additive constants of motion—the amount of field N, the total momentum $\mathbf{P}$, and the topological invariant $\mathbf{I}$ (in this section, we are not interested in the rotating case, and thus do not introduce the terms with the angular momentum):

$$H' = H - \mu N - \mathbf{v}_n \cdot \mathbf{P} - \mathbf{s} \cdot \mathbf{I}. \tag{2.4}$$

Here, μ is the chemical potential, $\mathbf{v}_n$ is the so-called *normal velocity*, which can be viewed as the velocity of the walls (which the system is supposed to be in equilibrium with). The parameter $\mathbf{s}$ is introduced as an intensive variable, thermodynamically conjugated to $\mathbf{I}$; its physical meaning will be revealed later. All thermodynamic relations follow then from the grand canonical potential

$$\Omega(T, \mu, \mathbf{v}_n, \mathbf{s}) = -T \ln Z. \tag{2.5}$$

In particular, there is a specific to the superfluid system relation:

$$\langle \mathbf{I} \rangle = -\frac{\partial \Omega}{\partial \mathbf{s}}, \tag{2.6}$$

immediately following (2.3) and (2.4). Similarly, the momentum, averaged over the Gibbs distribution, is given by

$$\langle \mathbf{P} \rangle = -\frac{\partial \Omega}{\partial \mathbf{v}_n}. \tag{2.7}$$

They are direct analogs of the more familiar relation $\langle N \rangle = -\partial \Omega / \partial \mu$. Given the general relation between the momentum and the spatially integrated current density, the aforementioned result can also be viewed as the thermodynamic expression for the spatially averaged current density:

$$\langle\!\langle \mathbf{j} \rangle\!\rangle \equiv \int \langle \mathbf{j} \rangle \, d^d r / V = \gamma \langle \mathbf{P} \rangle / V. \tag{2.8}$$

At this point, we make an observation concerning the conceptually important case of a system with a random external potential $\mathcal{V}(\mathbf{r})$ or, more generally, any system without Galilean invariance. In this case, despite the nonconservation of momentum, the partition function (2.3) with the Hamiltonian (2.4) featuring the momentum term with an infinitesimally small $\mathbf{v}_n$ remains meaningful. Indeed, the function $\Omega(T, \mu, \mathbf{v}_n, \mathbf{s})$ with an infinitesimally small $\mathbf{v}_n$ can be viewed as a generating function that, while producing the grand canonical potential at $\mathbf{v}_n = 0$, is handy for calculating average momentum and the (spatially averaged) density of matter flux by Equations (2.7) and (2.8).

The function $\Omega(T, \mu, \mathbf{v}_n, \mathbf{s})$ features the following general property. For any $\mathbf{v}_*$,

$$\Omega(T, \mu, \mathbf{v}_n, \mathbf{s}) = \Omega\left(T, \mu + (2\mathbf{v}_n \cdot \mathbf{v}_* - v_*^2)/2\gamma, \mathbf{v}_n - \mathbf{v}_*, \mathbf{s}\right) - V\mathbf{v}_* \cdot \mathbf{s}/\gamma, \tag{2.9}$$

where V is the system volume. This relation is insensitive to the specific form of the interaction term of the Hamiltonian, provided the interaction is independent of the phase of the field ψ. In the case of a Galilean invariant system, Equation (2.9) is nothing but the Galilean transformation of the grand canonical potential. In this sense, Equation (2.9) is really nontrivial only in the case of a system *without* Galilean invariance.

Problem 2.1 *Prove Equation (2.9). Hint: In the functional integral (2.3), make the substitution of the integration variable, $\psi(\mathbf{r}) = \psi'(\mathbf{r}) \exp(i\mathbf{v}_* \cdot \mathbf{r}/\gamma)$. Do not forget to take into account (1.144).*

Setting $\mathbf{v}_* = \mathbf{v}_n$, we get a useful specific case of Equation (2.9):

$$\Omega(T, \mu, \mathbf{v}_n, \mathbf{s}) = \Omega(T, \mu + v_n^2/2\gamma, \mathbf{v}_n = 0, \mathbf{s}) - V\mathbf{v}_n \cdot \mathbf{s}/\gamma. \tag{2.10}$$

Considering Equation (2.9) in the limit $\mathbf{v}_* \to 0$, we get its differential form:

$$\frac{\mathbf{v}_n}{\gamma} \frac{\partial \Omega}{\partial \mu} - \frac{\partial \Omega}{\partial \mathbf{v}_n} - V\mathbf{s}/\gamma = 0. \tag{2.11}$$

It is worth emphasizing that Equations (2.9) (at $\mathbf{v}_* = 0$) and (2.11) are equivalent. Mathematically, Equation (2.9) is a solution of the partial differential Equation (2.11) with respect to variables $(\mu, \mathbf{v}_n)$ at fixed $(T, \mathbf{s})$.

Recalling that $\partial \Omega/\partial \mu = -\langle N \rangle$ and $\partial \Omega/\partial \mathbf{v}_n = -\langle \mathbf{P} \rangle$, Equation (2.11) means

$$\langle \mathbf{P} \rangle = \langle N \rangle \mathbf{v}_n/\gamma + V\mathbf{s}/\gamma. \tag{2.12}$$

To clearly see the implications of this formula, rewrite it in terms of the spatially averaged density

$$\langle\langle n \rangle\rangle \equiv \int \langle n \rangle \, d^d r / V = \langle N \rangle / V \tag{2.13}$$

and current density (2.8) to arrive at

$$\langle\langle \mathbf{j} \rangle\rangle = \langle\langle n \rangle\rangle \mathbf{v}_n + \mathbf{s}. \tag{2.14}$$

In the absence of the second term, Equation (2.14) would be trivial, because the first term is just what one expects from the Galilean invariance (and in the absence of Galilean invariance, both $\langle\langle \mathbf{j} \rangle\rangle$ and $\mathbf{v}_n$ would be zero). The nonzero second term means that we are dealing with the situation of an *incomplete drag*, when, in effect, not all of the fluid follows the motion of the walls, despite the fact that we are dealing with an equilibrium state.

2.1.2 Two-Fluid Picture: Superfluid and Normal Components

Let us further elaborate on the phenomenon of incomplete drag revealed by Equation (2.12) and show that the physical picture that emerges from it is identical to that of Tisza's two-fluid model, when two fluids—so-called superfluid and normal components—can exchange the amount of matter, but not the momentum.

We start with *defining* the superfluid velocity, $\mathbf{v}_s$, as an essentially *macroscopic* quantity:

$$\mathbf{v}_s = \gamma \langle \mathbf{I} \rangle / V, \tag{2.15}$$

which, in accordance with Equation (2.6), can be calculated from $\Omega(T, \mu, \mathbf{v}_n, \mathbf{s})$ as

$$\mathbf{v}_s = -\frac{\gamma}{V} \frac{\partial \Omega}{\partial \mathbf{s}}. \tag{2.16}$$

The term "velocity" is justified by the fact that $\mathbf{v}_s$ transforms as velocity under the Galilean transformation. This property immediately follows from Equations (1.144) and (1.145).

Similarly to $\langle \mathbf{I} \rangle$, the superfluid velocity remains meaningful even if the Galilean invariance is broken. In particular, the crucial property of $\mathbf{v}_s$ implied by the definition (2.15), that is, its relation to the gradient of $\tilde{\Phi}$ averaged over the Gibbs ensemble and system volume,

$$\mathbf{v}_s = \gamma \int \langle \nabla \tilde{\Phi} \rangle d^d r / V, \tag{2.17}$$

is by no means affected by breaking of the Galilean invariance. We already mentioned that, even in the absence of translation symmetry, the potential $\Omega(T, \mu, \mathbf{v}_n, \mathbf{s})$ plays the role of a generating function for calculating the average momentum and the (spatially averaged) matter flux density:

$$\langle \mathbf{P} \rangle = -\frac{\partial \Omega}{\partial \mathbf{v}_n}\bigg|_{\mathbf{v}_n=0} = V\mathbf{s}/\gamma \qquad (\mathbf{v}_n = 0), \tag{2.18}$$

$$\langle\langle \mathbf{j} \rangle\rangle \equiv \int \langle \mathbf{j} \rangle d^d r / V = \gamma \langle \mathbf{P} \rangle / V = \mathbf{s} \qquad (\mathbf{v}_n = 0). \tag{2.19}$$

The latter quantity, characteristic of superfluid systems, will be referred to as the supercurrent density. Equation (2.19) reveals the physical meaning of $\mathbf{s}$ as the supercurrent density in the reference frame where $v_n = 0$.

We are in a position now to introduce the notion of *superfluid density*, n_s; we define n_s as a coefficient relating the superfluid velocity and the supercurrent density (in the reference frame where $v_n = 0$):

$$\langle\langle \mathbf{j} \rangle\rangle = n_s \mathbf{v}_s \qquad (\mathbf{v}_n = 0). \tag{2.20}$$

The only implicit assumption behind this definition is that the vector $\langle\langle \mathbf{j} \rangle\rangle$ is parallel to $\mathbf{v}_s$, which is always true when the system is isotropic at the macroscopic level; otherwise, the superfluid density is a tensor quantity. The definition (2.20) *does not* imply direct proportionality between the two vectors: speaking generally, n_s depends on v_s (saturating to a finite limit when $v_s \to 0$, as discussed later).

A few more general steps can be made for the Galilean invariant systems. Here, we note that the microscopic property (1.82) is inherited by the spatially averaged current,

$$\langle\langle \mathbf{j} \rangle\rangle' = \langle\langle \mathbf{j} \rangle\rangle + \langle\langle n \rangle\rangle \mathbf{v}_0, \tag{2.21}$$

in terms of spatially averaged density, $\langle\langle n \rangle\rangle = N/V$, the property being readily seen from (1.83) and (2.19). For the sake of brevity in what follows, we omit the sign $\langle\langle \ldots \rangle\rangle$. We will reserve the superscript "(0)" for the special reference frame in which $v_n = 0$, while the absence of a superscript implies a generic reference frame. With our choice of the special reference frame, the velocity of the Galilean transformation to this frame is $\mathbf{v}_0 \equiv \mathbf{v}_n$. Since $\mathbf{v}_s = \mathbf{v}_s^{(0)} + \mathbf{v}_0 \equiv \mathbf{v}_s^{(0)} + \mathbf{v}_n$, as the result of the Galilean transformation (2.21) and the definition (2.20), we have a relation

$$\mathbf{j} = n_s \mathbf{v}_s + (n - n_s)\mathbf{v}_n. \tag{2.22}$$

Equation (2.22) suggests yet another definition,

$$n_n = n - n_s, \tag{2.23}$$

rendering (2.22) symmetric with respect to superfluid and normal velocities,

$$\mathbf{j} = n_s \mathbf{v}_s + n_n \mathbf{v}_n, \tag{2.24}$$

and suggesting a remarkable physical *interpretation*. Equation (2.24) can be viewed as having two distinctive components of the flow: the superfluid one, characterized by the density n_s and moving with the velocity $\mathbf{v}_s$, and the normal component, characterized by the density n_n and moving with the velocity $\mathbf{v}_n$. It is also natural to define the corresponding superfluid and normal currents, thus representing Equation (2.24) as

$$\mathbf{j} = \mathbf{j}_s + \mathbf{j}_n, \qquad \mathbf{j}_s = n_s \mathbf{v}_s, \qquad \mathbf{j}_n = n_n \mathbf{v}_n. \tag{2.25}$$

By comparison with (2.14), Equation (2.22) yields the following generic relation for $\mathbf{s}$:

$$\mathbf{s} = n_s(\mathbf{v}_s - \mathbf{v}_n). \tag{2.26}$$

The right-hand side of (2.26) is Galilean invariant, leading us to the important conclusion that going from one inertial frame to another one does not involve changing $\mathbf{s}$. This, in particular, is consistent with the fact that, in a translation invariant system, the property (2.10) of the grand canonical potential Ω is equivalent to changing the reference frame.

2.1.3 Superfluid and Normal Densities as Static Response Functions

In this section, we consider superfluid density n_s in the important limit $|\mathbf{v}_s - \mathbf{v}_n| \to 0$, corresponding to an infinitesimally weak superflow, since the difference $\mathbf{v}_s - \mathbf{v}_n$ is nothing but the superfluid velocity in the reference frame of the walls. We work in this reference frame by setting $v_n = 0$, or taking the limit $v_n \to 0$, if we need to differentiate with respect to $\mathbf{v}_n$. With this choice of the reference frame, our results apply also to systems that lack microscopic translation invariance.

In this limiting case, n_s can be immediately related to the static response of the system with respect to $\mathbf{s}$:

$$\frac{1}{n_s} = -\frac{\gamma}{Vd}\Delta_s\Omega \qquad (v_n = 0, \quad \mathbf{s} \to 0), \tag{2.27}$$

where d is the dimension of space and $\Delta_s = \sum_{\alpha=1}^{d} \partial^2/\partial s_\alpha^2$. Indeed, combining the definition of n_s (2.20) with the relations (2.16) and (2.19), we find $-(\gamma/V)(\partial\Omega/\partial\mathbf{s}) = \mathbf{s}/n_s$, implying Equation (2.27). Recalling that $\Omega = -pV$, where $p \equiv p(\mu, T, \mathbf{v}_n, \mathbf{s})$ is the pressure expressed as a function of the grand canonical variables $(\mu, T, \mathbf{v}_n, \mathbf{s})$, we can write Equation (2.27) in the intensive form

$$\frac{1}{n_s} = \frac{\gamma}{d}\Delta_s p(\mu, T, \mathbf{v}_n = 0, \mathbf{s}) \qquad (\mathbf{s} \to 0). \tag{2.28}$$

We have been dealing so far with the grand canonical ensemble in which the particular average value of the superfluid velocity, $\mathbf{v}_s$, occurs as a function of control parameter $\mathbf{s}$. Sometimes, it is desirable to work in the canonical ensemble, where $\mathbf{v}_s$ is the main thermodynamic variable. The standard thermodynamic procedure of changing from grand canonical to canonical description implies introducing the free-energy thermodynamic potential. In our case, the free energy is [for clarity, we omit the variables $(\mu, T, \mathbf{v}_n)$]

$$\mathcal{F}(\mathbf{v}_s) = \Omega(\mathbf{s}) + \mathbf{s}\cdot\mathbf{I} = \Omega(\mathbf{s}) + (V/\gamma)\mathbf{s}\cdot\mathbf{v}_s. \tag{2.29}$$

Do not confuse $\mathcal{F}$ with the Helmholtz free energy because, with respect to the total number of particles and momentum, it is still a grand canonical potential. We use the calligraphic letter $\mathcal{F}$ as a formal reminder of this circumstance.

Confining ourselves to the case—most important for further application—of small values of $\mathbf{v}_s$, we expand Ω in powers of s:

$$\Omega(\mathbf{s}) = \Omega(0) - V\frac{s^2}{2\gamma n_s} \qquad (v_n = 0, \quad s \to 0). \tag{2.30}$$

Taking into account that $\mathbf{s} = n_s\mathbf{v}_s$ [see Equations (2.19) and (2.20)], we get

$$\mathcal{F}(\mathbf{v}_s) = \mathcal{F}(0) + V\frac{n_s v_s^2}{2\gamma} \qquad (v_n = 0, \quad v_s \to 0). \tag{2.31}$$

This leads to a useful expression for n_s in terms of $\mathcal{F}$,

$$n_s = \frac{\gamma}{Vd}\Delta_{v_s}\mathcal{F} \qquad (v_n = 0, \quad v_s \to 0), \tag{2.32}$$

that directly follows from (2.31). The intensive form of this relation is

$$n_s = -\frac{\gamma}{d}\Delta_{v_s}p(\mu, T, \mathbf{v}_n = 0, \mathbf{v}_s) \qquad (v_s \to 0). \qquad (2.33)$$

Note that, in contrast to (2.28), the pressure here is expressed as a function of $\mathbf{v}_s$, not $\mathbf{s}$, because $\mathcal{F}(\mu, T, \mathbf{v}_n, \mathbf{v}_s) = -Vp(\mu, T, \mathbf{v}_n, \mathbf{v}_s)$.

We can also express n_n as a static response to $\mathbf{v}_n$ at $\mathbf{v}_s = 0$. The corresponding formula is valid (and useful in practice) in the absence of translation invariance, with $n_n \equiv n - n_s$. Consider Equation (2.29), in which we now set $\mathbf{v}_s = 0$ and restore the variable $\mathbf{v}_n$:

$$\mathcal{F}(\mathbf{v}_s = 0, \mathbf{v}_n) = \Omega(\mathbf{s} = -n_s\mathbf{v}_n, \mathbf{v}_n). \qquad (2.34)$$

Differentiating twice with respect to $\mathbf{v}_n$, we then find

$$-\frac{\gamma}{d}\Delta_{v_n}\mathcal{F}(\mathbf{v}_s = 0, \mathbf{v}_n) = V(n - n_s) = Vn_n \qquad (v_n \to 0). \qquad (2.35)$$

The intensive form of this relation is

$$n_n = \frac{\gamma}{d}\Delta_{v_n}p(\mu, T, \mathbf{v}_s = 0, \mathbf{v}_n) \qquad (v_n \to 0). \qquad (2.36)$$

Now, let us find the equivalent of the relation (2.9) in terms of $\mathcal{F}$. To this end, we consider the differential of $\mathcal{F}$ (keeping T and V fixed and thus having $dT \equiv dV \equiv 0$):

$$d\mathcal{F} = -N\,d\mu - \mathbf{P}\cdot d\mathbf{v}_n + (V/\gamma)\mathbf{s}\cdot d\mathbf{v}_s \qquad (dT \equiv dV \equiv 0), \qquad (2.37)$$

and note that Equation (2.12) allows us to replace $(V/\gamma)\mathbf{s}$ in (2.37) with $\mathbf{P} - N\mathbf{v}_n/\gamma$. As a result, we get

$$d\mathcal{F} = -N\,(d\mu + \mathbf{v}_n\cdot d\mathbf{v}_s/\gamma) - \mathbf{P}\cdot d(\mathbf{v}_n - \mathbf{v}_s), \qquad (2.38)$$

from which one can readily see that $\mathcal{F}$ is constant on a certain manifold parameterized with a running vector $\mathbf{v}_*$; namely,

$$\mathcal{F}(\mu, \mathbf{v}_n, \mathbf{v}_s) = \mathcal{F}\left(\mu + (2\mathbf{v}_n\cdot\mathbf{v}_* - v_*^2)/2\gamma, \mathbf{v}_n - \mathbf{v}_*, \mathbf{v}_s - \mathbf{v}_*\right). \qquad (2.39)$$

Yet another way of establishing Equation (2.39) is to start directly from the canonical (with respect to the topological invariant) counterpart of the partition function (2.3), in which the integration over the field ψ is restricted to a given topological sector. The derivation is based on the same substitution $\psi(\mathbf{r}) = \psi'(\mathbf{r})\exp(i\mathbf{v}_*\cdot\mathbf{r}/\gamma)$ as in Problem 2.1.

2.1.4 Fluctuations of the Superfluid Velocity Field, Topological versus Genuine Long-Range Order

Superfluid velocity, $\mathbf{v}_s$, can be ascribed not only to the macroscopic system as a whole but also to its significantly large part—a macroscopic subsystem. This way, one can introduce a fluctuating *superfluid velocity field*, $\mathbf{v}_s(\mathbf{r})$, and study its correlation properties. These properties are directly relevant to the long-range behavior of the so-called off-diagonal correlators. By off-diagonal correlator, we mean a correlation function that depends on the phase of the field ψ. Correspondingly, the diagonal correlator is the correlation function that depends only on the absolute value of the field ψ, such as the density–density correlation function $\langle n(\mathbf{r}_1)n(\mathbf{r}_2)\rangle$. The simplest off-diagonal correlator is

$$\rho(\mathbf{r}_1,\mathbf{r}_2) = \langle\psi^*(\mathbf{r}_2)\psi(\mathbf{r}_1)\rangle. \tag{2.40}$$

It is also known as the single-particle density matrix in the quantum context. As long as we are dealing with the classical field, the term "single particle" should be considered as jargon that becomes meaningful only at the quantum-field level. In a superfluid system, the off-diagonal correlators have interesting asymptotic behavior. In dimensionality larger than 2, the off-diagonal correlators *do not vanish* at large distances. For example,

$$\rho(\mathbf{r}_1,\mathbf{r}_2) \to \psi_0^*(\mathbf{r}_2)\psi_0(\mathbf{r}_1) \quad \text{at} \quad |\mathbf{r}_1 - \mathbf{r}_2| \to \infty \qquad (d \ge 3). \tag{2.41}$$

This behavior of off-diagonal correlators is called the *off-diagonal long-range order* (ODLRO) and is referred to as *spontaneous symmetry breaking*. In a macroscopic system, one normally expects that correlators factorize in the long-range limit since fluctuations of the corresponding quantities become independent. For the single-particle density, matrix such a factorization is

$$\rho(\mathbf{r}_1,\mathbf{r}_2) \to \langle\psi^*(\mathbf{r}_2)\rangle\langle\psi(\mathbf{r}_1)\rangle \quad \text{at} \quad |\mathbf{r}_1 - \mathbf{r}_2| \to \infty. \tag{2.42}$$

By U(1) symmetry, the average $\langle\psi(\mathbf{r})\rangle$ is supposed to be identically equal to zero since there is no preferable phase for this complex quantity. Thus, there are two essentially equivalent ways of interpreting the property (2.41): (a) saying that the U(1) symmetry is spontaneously broken in such a way that now

$$\langle\psi(\mathbf{r})\rangle = \psi_0(\mathbf{r}) \tag{2.43}$$

or (b) declaring factorization (2.42) inadequate in view of the presence of the long-range order. The language of spontaneous symmetry breaking has certain technical advantages, especially if it is introduced in a mathematically rigorous way through Bogoliubov's *quasi* averages. The technique of quasi averages implies adding to the Hamiltonian infinitesimally small terms, explicitly breaking the symmetry in question. In the case of U(1) symmetry, the procedure is as follows ($h \to 0$ is a positive-definite real number; φ_0 is a certain fixed phase):

$$H' \to H' + h \int (e^{-i\varphi_0}\psi + \text{c.c.})\,d^d r. \tag{2.44}$$

Then, the quasi average, also known as *anomalous* average, is defined by

$$\langle \psi(\mathbf{r}) \rangle_{\text{quasi}} \equiv \lim_{h \to 0} \langle \psi(\mathbf{r}) \rangle = \psi_0(\mathbf{r}), \tag{2.45}$$

so that the factorization of off-diagonal correlators does take place, but in terms of quasi-averages rather than genuine averages. The latter differs from the former by an additional average over the global phase φ_0, which nullifies the genuine average but is irrelevant to the thermodynamic properties of the system. The idea behind adding the infinitesimal terms (2.44) is to suppress this physically irrelevant averaging over the global phase of ψ by simply forcing it to take the value of φ_0. In the presence of ODLRO—that is, at $\psi_0(\mathbf{r}) \neq 0$—the integral on the right of (2.44) is macroscopically large. Hence, the value of h sufficient to fix the phase is macroscopically small, in which case the extra term in the Hamiltonian does not distort thermodynamic properties of a macroscopic system.

The existence of ODLRO (2.41) is a sufficient, though not a necessary, condition for superfluidity to occur. The theory of the previous sections was based exclusively on the conservation of the topological invariant. For finite anomalous average $\psi_0(\mathbf{r})$, the existence of the topological invariant (1.145) is guaranteed since one can associate the phase $\tilde{\Phi}(\mathbf{r})$ with the phase of $\psi_0(\mathbf{r})$ to obtain the phase field without defects. The opposite, however, is not true. The topological order, by which one means the existence of a well-defined conserved topological invariant (1.145), does not necessarily require ODLRO (2.45). The regime of topological order without ODLRO (TLRO) takes place in a 2D superfluid at any finite temperature. The physical reason for that is quite instructive: ODLRO is destroyed by long-range fluctuations of the field $\mathbf{v}_s(\mathbf{r})$.

Let us derive the effective free energy that controls the long-wavelength properties of $\mathbf{v}_s(\mathbf{r})$ and apply it to the problem of asymptotic behavior of the off-diagonal correlators. Our starting point is the general theory of Gaussian fluctuations of additive constants of motion in the grand canonical ensemble. For any additive constant of motion X, by the central limit theorem, fluctuations in a volume with linear size much larger than all correlation scales are described by the Gaussian probability density $w(X)$:

$$w(X) \propto e^{-(X-\bar{X})^2/2C^2}, \tag{2.46}$$

where $\bar{X} \equiv \langle X \rangle$ is the average value of X and C is its dispersion:

$$C^2 = \langle (X - \bar{X})^2 \rangle \equiv \int (X - \bar{X})^2 w(X) \, dX. \tag{2.47}$$

The parameters $\bar{X}$ and C of the distribution (2.46) can be found directly from the grand canonical potential $\Omega(\lambda)$, where λ is the variable thermodynamically conjugated to X [utilizing $C^2 = \langle X^2 \rangle - \bar{X}^2$]:

$$\bar{X} = -\frac{\partial \Omega}{\partial \lambda}, \qquad C^2 = -T \frac{\partial^2 \Omega}{\partial \lambda^2}. \tag{2.48}$$

Recall Equation (2.5) and the fact that grand canonical and canonical with respect to X partition functions are related by $Z_{GC}(\lambda) = \int dX Z_C(X) \exp\{-\lambda X/T\}$.

In our case, X and λ are the topological invariant $\mathbf{I}$ and the vector $\mathbf{s}$, respectively. In accordance with (2.1) and (2.15), the ratio $\gamma \mathbf{I}/V$ can be interpreted as the instant value of superfluid velocity, $\mathbf{v}_s$. Previously, we used the symbol $\mathbf{v}_s$ for the ensemble average—see Equation (2.15); in this section, for the ensemble averaged $\mathbf{v}_s$, we will be using $\bar{\mathbf{v}}_s$. With (2.27) and (2.46) through (2.48), we write the distribution of $\mathbf{v}_s$ as

$$w(\mathbf{v}_s) \propto \exp[-(n_s V/2\gamma T)(\mathbf{v}_s - \bar{\mathbf{v}}_s)^2]. \tag{2.49}$$

We can view (2.49) as the distribution function for the values of the superfluid velocity field averaged over a macroscopic subvolume V of an even larger macroscopic system. This interpretation allows us to employ (2.49) for formulating the effective free-energy functional for fluctuations of $\mathbf{v}_s(\mathbf{r})$. We formally divide our system into macroscopically large—and thus statistically independent—blocks and write the partition function Z as a product of partition functions for each individual block. For a given block ν, we first perform a summation over all the microscopic variables except for the instant value of the superfluid velocity in that block, $\mathbf{v}_s^{(\nu)}$. The final summation is then over the configurations of block variables, $\{\mathbf{v}_s^{(\nu)}\}$. By construction, the partially summed partition function in the block ν is the distribution function $w(\mathbf{v}_s^{(\nu)})$ defined earlier, so that the global partition function is given by

$$Z = \sum_{\{\mathbf{v}_s^{(\nu)}\}} \prod_\nu w\left(\mathbf{v}_s^{(\nu)}\right). \tag{2.50}$$

The discrete form of Equation (2.50) is rather inconvenient. For practical applications, it is easier to work with the smooth field $\mathbf{v}_s(\mathbf{r})$, by writing $\mathbf{v}_s^{(\nu)} \equiv \mathbf{v}_s(\mathbf{r}_\nu)$, where $\mathbf{r}_\nu$ points to the block ν. The blocks are assumed to be large enough for the typical fluctuations of $\mathbf{v}_s(\mathbf{r})$ within a single block to be negligibly small.

With (2.49) taken into account (here, we confine ourselves to the case $\bar{\mathbf{v}}_s = 0$), we arrive at the partition function in the form of the functional integral over the field $\mathbf{v}_s(\mathbf{r})$:

$$Z = \int \mathcal{D}\mathbf{v}_s \exp\left\{-\frac{n_s}{2\gamma T} \int \mathbf{v}_s^2 \, d^d r\right\}. \tag{2.51}$$

It is essential that the integral (2.51) is over configurations with zero curl

$$\nabla \times \mathbf{v}_s = 0, \tag{2.52}$$

since $\mathbf{v}_s$ is proportional to a gradient of the regularized phase $\nabla\tilde{\Phi}$. This immediately suggests the parameterization

$$\mathbf{v}_s(\mathbf{r}) = \gamma \nabla \varphi(\mathbf{r}), \tag{2.53}$$

where $\varphi(\mathbf{r})$ is the long-wave part of the field $\nabla\tilde{\Phi}$ (or another similarly coarse-grained field). In terms of the phase field $\varphi(\mathbf{r})$, the partition function (2.51) acquires the form

$$Z = \int e^{-A[\varphi]}\mathcal{D}\varphi, \tag{2.54}$$

$$A[\varphi] = \frac{\gamma n_s}{2} \int (\nabla\varphi)^2 d^d r. \tag{2.55}$$

Problem 2.2 *Explain why replacing the integration over $\mathcal{D}\mathbf{v}_s$ with the integration over $\mathcal{D}\varphi$ does not involve a nontrivial (i.e., φ-dependent) Jacobian.*

By construction, the field φ varies significantly only over large enough distances. Mathematically, this fact can be expressed by introducing a UV cutoff, k_*, for the Fourier harmonics of φ:

$$\varphi(\mathbf{r}) = \frac{1}{\sqrt{V}} \sum_{k<k_*} \varphi_\mathbf{k}\, e^{i\mathbf{k}\cdot\mathbf{r}}. \tag{2.56}$$

The cutoff momentum is set equal to the inverse block size used to coarse-grain the field (ultimately, it can be increased up to the inverse of the largest correlation length in the system). The free-energy functional (2.55) is bilinear with respect to φ, and the Fourier expansion (2.56) allows one to write it as the sum of independent terms because, apart from the constraint

$$\varphi_{-\mathbf{k}} = \varphi_\mathbf{k}^*, \tag{2.57}$$

necessary for φ to be real, all the harmonics fluctuate independently:

$$Z \propto \prod_\mathbf{k}^{\sim} Z_\mathbf{k}, \tag{2.58}$$

$$Z_\mathbf{k} = \int e^{-(\gamma n_s k^2/T)|\varphi_\mathbf{k}|^2} d\varphi_\mathbf{k}. \tag{2.59}$$

The tilde sign on the product means that, in accordance with the constraint (2.57), the subscript $\mathbf{k}$ simultaneously stands for a given wavevector and its counterpart of the opposite direction, so that each factor in the product is statistically independent. This is also the reason for why there is no factor of 2 in the denominator of the exponent in (2.59).

Due to the bilinearity of (2.55), the correlation functions of the field φ are expressed, by Wick's theorem, in terms of sums of products of the pair correlator (here, we also take into account translation invariance of the problem)

$$G(\mathbf{r}) = \langle\varphi(\mathbf{r}')\varphi(\mathbf{r}'+\mathbf{r})\rangle = \langle\varphi(0)\varphi(\mathbf{r})\rangle. \tag{2.60}$$

Taking Fourier transform and utilizing (2.59), we obtain

$$G(\mathbf{r}) = \frac{1}{V}\sum_{k<k_*} G_\mathbf{k}\, e^{i\mathbf{k}\cdot\mathbf{r}}, \qquad G_\mathbf{k} = \langle|\varphi_\mathbf{k}|^2\rangle = \frac{T}{\gamma n_s k^2}. \tag{2.61}$$

We are now in a position to study the asymptotic long-range behavior of off-diagonal correlators. As a typical case, consider single-particle density matrix (2.40). In the limit of $r = |\mathbf{r}_1 - \mathbf{r}_2| \to \infty$, the leading contribution to the spatial dependence of $\rho(r)$ comes from the long-range fluctuations of the phase. The free energy of these fluctuations is associated only with the gradients (rather than with the absolute values) of $\varphi(\mathbf{r})$ and thus becomes vanishingly small in the long-wave limit. Hence, we have

$$\rho(\mathbf{r}) \propto \langle e^{i\varphi(\mathbf{r})-i\varphi(0)} \rangle, \qquad r \to \infty. \tag{2.62}$$

To evaluate the average (2.62), we use the formula

$$\langle e^{iA} \rangle = e^{-\langle A^2 \rangle/2}, \tag{2.63}$$

valid for any Gaussian quantity A, by Wick's theorem.

Problem 2.3 *Derive Equation (2.63). Hint: Expand the exponential into the power series and apply Wick's theorem.*

Substituting $[\varphi(\mathbf{r}) - \varphi(0)]$ for A in Equation (2.63), we find

$$\rho(\mathbf{r}) = n_{k_*} \exp\left[-\int_{k<k_*} (1 - \cos\mathbf{k} \cdot \mathbf{r}) \, G_{\mathbf{k}} \, d^d k/(2\pi)^d \right], \qquad r \to \infty. \tag{2.64}$$

We replace summation over momenta with integration using the standard Fourier analysis rule $\sum_k \to V \int d^d k/(2\pi)^d$ and introduce prefactor n_{k_*}, which depends on the choice of the cutoff momentum k_* in such a way that the product of n_{k_*} and the exponential are k_* independent.

In 3D, $n_{k_*} \to n_0$ at $k_* \to 0$, where $n_0 = |\psi_0|^2$ is a finite constant called condensate density. Correspondingly, the integral in the exponent (2.64) becomes arbitrarily small as $k_* \to 0$, so that we can expand the exponential and arrive at Bogoliubov's formula for the Fourier transform of $\rho(r)$:

$$\rho_{\mathbf{k}} \to (2\pi)^3 n_0 \, \delta(\mathbf{k}) + \frac{n_0 T}{\gamma n_s k^2}, \qquad k \to 0 \qquad (d = 3). \tag{2.65}$$

In the real space, we thus have

$$\rho(r) \to n_0 \left(1 + \frac{T}{4\pi\gamma n_s r} \right), \qquad r \to \infty \qquad (d = 3). \tag{2.66}$$

We see that at finite temperature the saturation of $\rho(r)$ to its asymptotic value n_0 with $r \to \infty$ is rather slow because of the thermodynamic fluctuations of the phase field. Since the gradients of the phase imply currents, these long-range fluctuations are often referred to as hydrodynamic fluctuations.

Problem 2.4 *Complete the calculation leading to (2.65) and (2.66).*

In 2D, the integral in (2.64) behaves like $(T/2\pi\gamma n_s)\ln(k_* r)$, at $r \gg 1/k_*$. To see this, note that at $kr \ll 1$, we have $1 - \cos \mathbf{k} \cdot \mathbf{r} \approx (\mathbf{k} \cdot \mathbf{r})^2/2$, while at $kr \gg 1$, the cosine term oscillates so strongly that the integral involving it can be neglected. As a result, the main contribution to the integral comes from the interval between two cutoffs $1/r < k < k_*$ where the function under the integral is $\propto 1/k^2$. This leads to the logarithmic behavior mentioned earlier. Hence, in a 2D superfluid, the hydrodynamic fluctuations lead to the power-law decay of the single-particle density matrix:

$$\rho(r) \propto 1/r^\alpha, \qquad r \to \infty \qquad (d = 2), \qquad (2.67)$$

with the exponent α given by

$$\alpha = \frac{T}{2\pi\gamma n_s}. \qquad (2.68)$$

From this expression, we see that n_{k_*} in Equation (2.64) should vanish as $n_{k_*} \propto k_*^\alpha$ when k_* is decreased to ensure that the physical answer remains cutoff independent.

In a normal state, the decay of a single-particle density matrix is exponential. The power-law decay of the off-diagonal correlator (2.67) is a sign of the topological order in the system and is often referred to as *algebraic order*.

2.1.5 Generic Off-Diagonal Correlations: Long-Range Asymptotics, Mean-Field and Non-Mean-Field Regimes, and Elongated Systems

The theory of long-range asymptotic behavior of the single-particle density matrix (developed in the previous section) is readily generalized to the case of an arbitrary off-diagonal correlator. The key circumstance remains the same: at large distances, only phase fluctuations are important and their effect can be factored out as

$$\psi = \tilde{\psi}\, e^{i\varphi}. \qquad (2.69)$$

The cutoff momentum k_* in Equation (2.56) is supposed to be small enough to guarantee (a) the statistical independence of the fluctuations of φ from the fluctuations of $\tilde{\psi}$ and (b) the Gaussian character of the fluctuations of φ. (A particular value of k_* is not important as it drops from final expressions.)

In view of (a) and (b), for an "*m*-particle"* correlator, we have

$$\mathcal{K} = \langle \psi_1 \psi_2^* \psi_3 \psi_4^* \cdots \psi_{2m-1} \psi_{2m}^* \rangle, \qquad (2.70)$$

where the subscripts label the position variables. Using (2.63)—under the condition of statistical independence of φ and $\tilde{\psi}$—we then get

$$\mathcal{K} = \tilde{\mathcal{K}} e^{-\Lambda}, \qquad (2.71)$$

---

* We remind that invoking the notion of particles, while dealing with the classical fields, is nothing but a jargon that becomes meaningful only at the quantum-field level.

with

$$\Lambda = \frac{1}{2}\left\langle\left(\sum_{j=1}^{2m}(-1)^j \varphi_j\right)^2\right\rangle, \tag{2.72}$$

where $\tilde{\mathcal{K}}$ is obtained from $\mathcal{K}$ by substituting $\psi_j \to \tilde{\psi}_j$, for all j's.

At appropriately large distances between *any* two coordinates, the correlator $\tilde{\mathcal{K}}$ saturates to constant, and the remaining coordinate dependence is exclusively due to the fluctuations of φ. It is important, however, that the correlator $\tilde{\mathcal{K}}$ does depend on the cutoff momentum k_*. This dependence is crucial for compensating the k_* dependence of the function Λ.

The independence of $\tilde{\mathcal{K}}$ on coordinates (in the asymptotic limit) leads to the relation

$$\mathcal{K}(\vec{X}) = \mathcal{K}(\vec{X}') e^{\Lambda(\vec{X}') - \Lambda(\vec{X})} \tag{2.73}$$

between correlators at different sets of variables, $\vec{X}$ and $\vec{X}'$, provided both are in the asymptotic region. Relation (2.73) explains how $\mathcal{K}(\vec{X}')$ extrapolates to $\mathcal{K}(\vec{X})$ at *arbitrary* $\vec{X}$ in the asymptotic domain. By the structure of (2.73), the difference $\Lambda(\vec{X}') - \Lambda(\vec{X})$ is independent of k_*, as opposed to individual Λ's. The Gaussian character of the field Φ implies [see (2.72)] that

$$\Lambda(\vec{X}) - \Lambda(\vec{X}') = \sum_{s<j}(-1)^{s+j}\,\Xi(\mathbf{r}_{sj}, \mathbf{r}'_{sj}), \tag{2.74}$$

where $\mathbf{r}_{sj} = \mathbf{r}_s - \mathbf{r}_j$ (the same for primed variables) and

$$\Xi(\mathbf{r}, \mathbf{r}') = \langle\varphi(\mathbf{r})\,\varphi(0) - \varphi(\mathbf{r}')\,\varphi(0)\rangle = \int\left(e^{i\mathbf{k}\cdot\mathbf{r}} - e^{i\mathbf{k}\cdot\mathbf{r}'}\right)G_{\mathbf{k}}\,d^d k/(2\pi)^d. \tag{2.75}$$

The integral in (2.75) is free of infrared singularity, consistently with the fact that the cutoff momentum k_* drops out from the final answer. The integration is readily performed by the observation—see Equation (2.61)—that $\Delta_{\mathbf{r}}\Xi(\mathbf{r},\mathbf{r}') = (T/\gamma n_s)\delta(\mathbf{r})$ and $\Delta_{\mathbf{r}'}\Xi(\mathbf{r},\mathbf{r}') = -(T/\gamma n_s)\delta(\mathbf{r}')$: combined with $\Xi(\mathbf{r},\mathbf{r}) \equiv 0$, this leads to the solution $\Xi(\mathbf{r},\mathbf{r}') = (T/\gamma n_s)[f(r) - f(r')]$, with $f(r)$ the Green's function of the d-dimensional Laplace operator.[*] By Gauss theorem, $df/dr = -\Gamma(d/2)/2\pi^{d/2}r^{d-1}$, so that a generic answer for $\Xi(\mathbf{r},\mathbf{r}')$ is conveniently written in the integral form:

$$\Xi(\mathbf{r},\mathbf{r}') = -\frac{T\Gamma(d/2)}{2\pi^{d/2}\gamma n_s}\int_{r'}^{r}\frac{d\tilde{r}}{\tilde{r}^{d-1}}. \tag{2.76}$$

---

[*] Note that $f(r)$ does not coincide with $G(r)$ since the latter depends on the cutoff parameter k_*; see Equation (2.60).

For the first four dimensions,[*] the explicit answers are

$$\Xi(\mathbf{r},\mathbf{r}') = \frac{T}{\gamma n_s} \times \begin{cases} (1/2)(r'-r), & d = 1, \\ (1/2\pi)\ln(r'/r), & d = 2, \\ (1/4\pi)(1/r - 1/r'), & d = 3, \\ (1/4\pi^2)\left[r^{-2} - (r')^{-2}\right], & d = 4. \end{cases} \qquad (2.77)$$

In three and two dimensions, this result reproduces, in particular, Equations (2.66) through (2.68), respectively.

The 1D case is special. Strictly speaking, here, the superfluidity is absent at any finite temperature due to thermally activated phase slippages. Hence, Equation (2.77) can be used in 1D only at low enough temperatures *and* short enough distances, at which one can neglect extra suppression of correlations caused by phase slippages (not taken into account by the harmonic hydrodynamic action). Nevertheless, it is instructive to note that even in the absence of phase slippages, that is, without breaking the topological order, the harmonic thermal fluctuations of the phase field lead to an exponential decay of the off-diagonal correlations.[†]

In the above treatment, we were assuming that the distances between each pair of coordinates were large enough. The same analysis is readily adapted for cases when only some of the coordinates in the m-particle correlation function are in the asymptotic regime; only these coordinates have to be present in (2.73).

From Equation (2.76), we see that in any dimensions, the long-range off-diagonal correlations are controlled by a single parameter, $T/\gamma n_s$, a remarkable similarity. The parameter itself has an important physical meaning. By dimensional analysis, one can associate it (in $d \neq 2$) with a certain characteristic distance:

$$r_\varphi = \left(\frac{T}{\gamma n_s}\right)^{\frac{1}{d-2}} \qquad (d \neq 2). \qquad (2.78)$$

In 1D, the radius r_φ is nothing but the correlation radius of the phase field, beyond which the phase fluctuations exponentially suppress the off-diagonal correlations. The 2D case is special. Here, the parameter $T/\gamma n_s$ is dimensionless. In accordance with (2.77) [see also (2.67) and (2.68)], it controls the powers of the power-law decay of the off-diagonal correlators.

In $d \geq 3$, thermal fluctuations of the phase field do not destroy the genuine long-range order, and r_φ cannot be associated directly with a correlation radius. Rather, it plays a role of scaling parameter nondimensionalizing the distance in the asymptotic regime of the off-diagonal correlators. Correspondingly, an important role is played by the ratio r_φ/r_c, where r_c is the typical radius below which

---

[*] The practical importance of the 4D case will become clear later, when we show (in Chapter 7) that zero-point off-diagonal correlations in a d-dimensional quantum system correspond to an effective finite-temperature $(d + 1)$-dimensional classical system.

[†] In accordance with (2.77), the correlation length of this exponential decay scales only as $1/T$ with decreasing temperature, while the phase slippage correlation length explodes exponentially with $1/T$.

the thermodynamic notion of the phase field becomes meaningless. At $r_\varphi/r_c \ll 1$, we are in the so-called *mean-field* regime, when, to a good approximation, one can completely neglect the fluctuations of the phase field, representing the ψ-field as

$$\psi(\mathbf{r}) = \psi_0 + \psi'(\mathbf{r}), \tag{2.79}$$

where the noncondensate part $\psi'(\mathbf{r})$ is essentially independent of the condensate part ψ_0, so that the notion of condensate applies not only to the long-range correlations, but to the local correlators as well. The necessary and sufficient condition for the mean-field regime to take place is $r_\varphi/l_0 \ll 1$, where l_0 is the healing length (1.106) with $n_0 = |\psi_0|^2$. The condition $r_\varphi/l_0 \ll 1$ guarantees that the finite-temperature behavior of the system is almost the same as in the ground state of the density n_0, the noncondensate field $\psi'(\mathbf{r})$ playing the role of an external heat bath/field reservoir. For that reason, in the mean-field regime, the quantities n_0 and n_s are close to each other, up to higher-order corrections in parameter r_φ/l_0.

In any dimension $d \geq 3$, a rather trivial mean-field regime—of essentially vanishing $\psi'(\mathbf{r})$—takes place at an appropriately low temperature, since $r_\varphi \to 0$ at $T \to 0$, while l_0 saturates to a constant. Less trivial mean-field regime occurs in $d > 4$, and especially in 4D, on the approach to the critical point. In this limit, both r_φ and l_0 diverge, because of vanishing n_0, but in $d \geq 4$, the healing length l_0 diverges faster than r_φ. In $d > 4$, this is immediately seen by comparing powers of n_0 and n_s in Equations (1.106) and (2.78), respectively. The 4D case is more subtle. Here, the powers are equal, meaning that at the microscopic level, the mean-field regime does not have to take place. However, careful renormalization group analysis shows that, close enough to the critical point, the mean-field regime develops because of fluctuational renormalization (asymptotic vanishing) of the effective coupling constant replacing g in Equation (1.106). We conclude that nontrivial regimes of superfluid criticality take place only in 2D and 3D, while in higher dimensions, the criticality is of the mean-field character.[*]

Since the character of long-range fluctuations of the phase crucially depends on the dimension of space, especially if we are talking of 1D and 2D, it is clear that substantially elongated—in one, two, or more directions—systems have to demonstrate a mixed character of long-range off-diagonal correlations. Consider the case when one of the sizes of the system, say, along the x-direction, is much smaller than the others, setting the latter equal to infinity.[†] Let L_x be the system size in the x-direction. As long as $r, r' \ll L_x$, the situation is the same as at $L_x = \infty$. At $r, r' \sim L_x$, there takes place a crossover regime. Quantitatively, it is described by replacing, in Equation (2.75), the integration over k_x with summation over discrete Fourier harmonics. At $r, r' \gg L_x$, only the $k_x = 0$ harmonic survives, the contributions of the other harmonics being exponentially small in the parameter r/L_x.

---

[*] In the general theory of continuous phase transition, the mean-field (aka Landau) criticality is described by an order parameter, ψ_0 in our case, whose spatial fluctuations are negligibly small on approach to the critical point.

[†] Apart from being an instructive example on its own, this case (with periodic boundary conditions along the x-axis) corresponds to the low-temperature regime of a $(d-1)$-dimensional quantum system; see Chapter 7.

Correspondingly, the system behaves as a $(d-1)$-dimensional one, with the effective $(d-1)$-dimensional parameter $T/\gamma n_s$ obtained by replacing $T/\gamma n_s \rightarrow T/\gamma L_x n_s$, which can be naturally interpreted as replacing the original d-dimensional n_s with the $(d-1)$-dimensional superfluid density equal to $L_x n_s$.

2.1.6 Statistics of Supercurrents and Mesoscopic Effects

In the previous section, the equilibrium fluctuations of superfluid velocity were discussed in the context of large enough subsystems rather than a system as a whole. One might think that studying an equilibrium distribution of $\mathbf{v}_s = \gamma \mathbf{I}/V$ in a macroscopic system (under the assumption that the system has enough time to equilibrate through exponentially rare events leading to configurations with macroscopically large topological defects) is senseless, since the equilibrium probability of finding any $\mathbf{I} \neq 0$ appears to be macroscopically small. It turns out, however, that the aforementioned intuitive guess is true only in $d > 2$, while in 2D at a finite temperature, there is a finite probability of finding a state with nonzero $\mathbf{I}$. Indeed, rewriting (2.49) in terms of $\mathbf{I}$, we get the distribution

$$w_{\mathbf{I}} \propto \exp\left(-\frac{\gamma n_s \mathbf{I}^2}{2TV}\right). \tag{2.80}$$

For winding numbers $\mathbf{M} = (M_x, M_y, M_z)$ (1.112) of order unity, the ratio $\mathbf{I}^2/V$ scales as L^{d-2} with the linear system size L. We see that the exponent is macroscopically large (for a nonzero winding number) only at $d > 2$, while in 2D the macroscopic factor is absent.

Let us study Equation (2.80) in a 2D system with dimensions $L_x \times L_y$ and periodic boundary conditions. We have

$$\mathbf{I} = 2\pi L_y M_x \hat{\mathbf{e}}_x + 2\pi L_x M_y \hat{\mathbf{e}}_y. \tag{2.81}$$

Since $\mathbf{I}^2 = I_x^2 + I_y^2$, the fluctuations of M_x and M_y are statistically independent, and without loss of generality, we can consider fluctuations of M_x described by the distribution

$$w_{M_x} \propto \exp\left[-\frac{2\pi^2 \gamma n_s}{T}\left(\frac{L_y}{L_x}\right)M_x^2\right]. \tag{2.82}$$

The very essence of the phenomenon of superfluidity is that switching times, τ_M, between different values of M_x exceed the age of the Universe. Under what conditions then could the distribution (2.82) be observed? Clearly, the switching between M_x's becomes reasonably fast at temperatures close enough to the critical temperature T_c corresponding to the onset of superfluidity, since in the normal phase, τ_M is much smaller than the experimental time scale, $\tau_{\exp}$, by definition. Thus, as the physical system is slowly cooled down across the transition point, there necessarily will be a point where τ_M is short enough to ensure equilibrium statistics of winding numbers in the form of Equation (2.82). The condition

$\tau_M \sim \tau_{\exp}$ defines the ratio n_s/T, which must be used in the distribution function when looking at the ensemble of systems that are no longer equilibrating their winding numbers. The other way to reveal (2.82) is to radically increase τ_M by adding and then removing a weak link.

In Chapter 4, we will see that in 2D the critical temperature satisfies the universal Nelson–Kosterlitz relation:

$$T_c = \frac{\pi \gamma n_s}{2} \qquad (d = 2) \tag{2.83}$$

(implying, in particular, that n_s remains finite at $T = T_c$). The Nelson–Kosterlitz formula then leads to the universal critical distribution:

$$w_{M_x} \propto e^{-4\pi(L_y/L_x)M_x^2} \qquad (T = T_c). \tag{2.84}$$

The frozen distribution (2.84) is what one should expect to observe at $T < T_c$ (in an *ensemble* of macroscopic systems of the same aspect ratio L_y/L_x), since at temperatures below T_c relaxation of winding numbers will be essentially stopped.

Consider now a static response of the supercurrent to the twist phase φ_0. With Equation (1.202) applied to a uniform (at the macroscopic level) rectangular system, we have (the twist is applied along the x-axis)

$$\langle j_x \rangle = \frac{L_x}{V} \frac{\partial \Omega}{\partial \varphi_0}, \tag{2.85}$$

where the free energy should be evaluated for the full statistical ensemble, including all possible winding numbers. In the limit of $\varphi_0 \to 0$, the average current density is proportional to the twist phase

$$\langle j_x \rangle = n_s^{(\varphi)} \gamma \frac{\varphi_0}{L_x}, \tag{2.86}$$

and one can introduce the coefficient of proportionality $n_s^{(\varphi)}$. From (2.85), we see that it can be expressed as a linear response coefficient:

$$n_s^{(\varphi)} = \frac{L_x^2}{\gamma V} \left. \frac{\partial^2 \Omega}{\partial \varphi_0^2} \right|_{\varphi_0=0} \equiv \frac{\gamma}{V} \left. \frac{\partial^2 \Omega}{\partial (\gamma \varphi_0/L_x)^2} \right|_{\varphi_0=0}. \tag{2.87}$$

It is very tempting to identify $n_s^{(\varphi)}$ with superfluid density, and, in fact, such identification is a widespread mistake. For example, the identification is at the origin of the famous winding number formula for superfluid density used in numerical simulations. One may appeal to the similarity of Equations (2.20) and (2.86) as well as Equations (2.32) and (2.87) by arguing that in the state of thermal equilibrium, the gauge invariant expression for superfluid velocity,

$$v_s = \gamma \left(\frac{\langle I_x \rangle}{V} + \frac{\varphi_0}{L_x} \right), \tag{2.88}$$

coincides with $\gamma\varphi_0/L_x$ because $\langle I_x \rangle$ is zero. However, this argument is valid only in $d > 2$ but is incorrect in 2D and 1D. [It turns out that $n_s^{(\varphi)}$ is extremely close to n_s in 2D systems with square geometry $L_x = L_y$ so that the difference between the two is virtually impossible to detect.]

The dependence of the free energy on φ_0 is rather simple. By its topological nature, it reduces to the statistics of winding numbers, Equation (2.80), where the topological invariant is written in the form (1.204). The explicit expression for the free energy,

$$\Omega(\varphi_0) = -T\ln Z, \qquad Z \propto \sum_{M_x} \exp\left(-\frac{\gamma n_s V(2\pi M_x + \varphi_0)^2}{2L_x^2 T}\right), \qquad (2.89)$$

then readily leads to the result

$$\langle j_x \rangle = \frac{\gamma n_s}{L_x}\left(\varphi_0 + 2\pi\overline{M_x}\right), \qquad (2.90)$$

which perfectly agrees with (2.88). The expectation value of the winding number M_x, corresponding to the distribution

$$w_{M_x} \propto \exp\left[-\frac{\gamma n_s V(2\pi M_x + \varphi_0)^2}{2TL_x^2}\right], \qquad (2.91)$$

can be easily analyzed in various dimensions and geometries.

In 3D, $\overline{M_x}$ is exponentially small because the exponent in (2.91) is proportional to the linear system size for nonzero values of M_x. Thus,

$$n_s^{(\varphi)} = n_s \qquad (d \geq 3), \qquad (2.92)$$

which is reminiscent of the ground-state formula (1.194). In 2D and 1D, however, the value of $\overline{M_x}$ is finite because the probability of finding nonzero values of M_x is not vanishingly small. This underlines the important fact that in lower dimensions, the superfluid velocity field is strongly fluctuating and thus response to the twist phase is *not* reduced to $\gamma\varphi_0/L_x$.

Problem 2.5 *Show that*

$$n_s^{(\varphi)} = n_s\left[1 - \frac{4\pi^2\gamma n_s V\langle M_x^2 \rangle}{TL_x^2}\right], \qquad (2.93)$$

where $\langle M_x^2 \rangle$ is the dispersion of the winding number M_x at $\varphi_0 = 0$, and show that the value of the second term in the square brackets for the 2D system with $L_x = L_y$ at the critical point when $\gamma n_s/T_c = 2/\pi$ is approximately given by $16\pi e^{-4\pi} \approx 1.75 \times 10^{-4}$.

Equation (2.93) is what one should keep in mind when doing numeric simulations in low-dimensional superfluids at finite temperatures. The second derivative of the free energy with respect to the twist phase is the most typical observable used in simulations for extracting $n_s^{(\varphi)}$, but not n_s. Neglecting the second term in the right-hand side of (2.93) might lead to a wrong result for n_s.

2.2 Basics of Superfluid Hydrodynamics, Thermomechanical Effects, Hydrodynamic Hamiltonian and Action

The two-fluid equilibrium picture at $T > 0$ naturally leads to two-fluid hydrodynamics, with the extra constraint that the flow of the superfluid component is potential (apart from vortices, if any) and carries no entropy. Detailed consideration of the two-fluid hydrodynamics[*] goes beyond the scope of this book. Here, we will confine ourselves to basic relations of which the central one is the Beliaev–Josephson–Anderson formula[†] for the time derivative of $\tilde{\Phi}$ averaged over the Gibbs ensemble:

$$-\langle \dot{\tilde{\Phi}} \rangle = \mu + \mathbf{v}_n \cdot \mathbf{v}_s / \gamma \equiv \tilde{\mu}. \tag{2.94}$$

Equation (2.94)—to be derived in the next section—applies to systems without translation invariance as well (with setting $v_n \equiv 0$), and it is also valid for quantum systems (with $\gamma = \hbar/m$, where m is the particle mass).

Due to U(1) symmetry, the global uniform change of the phase of an isolated system described by Equation (2.94) is not observable. In the presence of (weak) gradients in thermodynamic quantities, the rate of changing $\tilde{\Phi}$ becomes spatially inhomogeneous and results in the evolution of $\tilde{\Phi}$ gradients. Translating gradients of $\tilde{\Phi}$ into superflow velocity, by Equation (2.17), we see that Equation (2.94) describes a nontrivial superfluid response to the gradients of thermodynamical characteristics:

$$\dot{\mathbf{v}}_s = -\nabla(\gamma\mu + \mathbf{v}_n \cdot \mathbf{v}_s) \equiv -\nabla\tilde{\mu}. \tag{2.95}$$

Hence, the gradient in $\tilde{\mu}$ (in the absence of translation symmetry, $v_n = 0$ and $\tilde{\mu} \equiv \mu$) implies *acceleration* of superflow, meaning in particular that, under conditions of temperature gradient, a steady state is possible only if (a) $\tilde{\mu}$ is *uniform* along the whole system, that is, $\tilde{\mu}(\mathbf{r}) = \mu_0$, or (b) the genuine superfluidity—conservation of the topological invariant—is destroyed. In the former case, the temperature gradient must be accompanied by a nonrelaxing gradient of pressure: since all points in the system with $v_n = 0$ are supposed to have one and the same chemical potential μ_0, *independently of the temperature* they must have different pressures $p \equiv p(\mu_0, T)$. This leads to nontrivial mechanical regimes. The steady states of type (b) are characterized by self-sustained structures of topological defects (called *superfluid turbulence*), which are responsible for the nonexistence of the topological invariant. As a result, a number of the so-called thermomechanical effects take place. Those include superconductivity of heat, counterflow, fountain, and Josephson effects. Along with persistent currents and vortex arrays, the thermomechanical effects are the hallmarks of superfluidity.

It is worth noting that Equation (2.95) can be derived purely from the Galilean transformation in a translation invariant system (Landau, 1941 [1]) without invoking the phase $\tilde{\Phi}$. We emphasize, however, that Equations (2.94) and (2.95) apply

---

[*] Which can be found, for example, in Reference [2].

[†] See also a discussion in Chapter 14.

equally well to the case when translation symmetry is absent (with $v_n \equiv 0$). Furthermore, Equation (2.94) plays the key role in the Josephson effect.

2.2.1 Time Evolution of the Phase

To arrive at Equation (2.94), we recall that for any quantity $A \equiv A[\psi]$, its dynamical time derivative can be expressed as a Poisson bracket; see Equations (1.20) and (1.21). The next useful observation concerns averaging this type of Poisson bracket with the corresponding Gibbs distribution: For any functional $A[\psi]$, the identity

$$\langle \{H', A\} \rangle \equiv 0 \tag{2.96}$$

takes place, where

$$\langle ... \rangle = \frac{\int (...) e^{-H'[\psi]/T} \mathcal{D}\psi}{\int e^{-H'[\psi]/T} \mathcal{D}\psi}, \tag{2.97}$$

and H' is the grand canonical Hamiltonian. This identity is of a purely mathematical nature. It has nothing to do with the specific form of H', or the physical meaning A, and amounts to showing that the two terms in the Poisson bracket exactly cancel each other under averaging. The proof is rather formal, but straightforward.[*] The first term is

$$\int \mathcal{D}\psi \, e^{-H'/T} \int \frac{\delta H'}{\delta\psi(\mathbf{r})} \frac{\delta A}{\delta\psi^*(\mathbf{r})} d^d r = -T \int \mathcal{D}\psi \int \frac{\delta e^{-H'/T}}{\delta\psi(\mathbf{r})} \frac{\delta A}{\delta\psi^*(\mathbf{r})} d^d r, \tag{2.98}$$

while the second term is

$$-\int \mathcal{D}\psi \, e^{-H'/T} \int \frac{\delta H'}{\delta\psi^*(\mathbf{r})} \frac{\delta A}{\delta\psi(\mathbf{r})} d^d r = T \int \mathcal{D}\psi \int \frac{\delta e^{-H'/T}}{\delta\psi^*(\mathbf{r})} \frac{\delta A}{\delta\psi(\mathbf{r})} d^d r. \tag{2.99}$$

To show that (2.99) equals minus (2.98), we take the prosaic approach of discretizing the field variable: $\psi(\mathbf{r}) \to \psi_j$, with j labeling discrete spatial points. Discretization allows one to do integration by parts in the standard way (which, in principle, is also possible with variational derivatives, but requires the extra work of developing the corresponding mathematical machinery for readers not familiar with it; the result is the same). The discrete version of (2.98) can be rewritten as (using $\psi_j = \psi_j' + i\psi_j''$)

$$-T \int \prod_k d\psi_k' d\psi_k'' \sum_j \frac{\partial e^{-H'/T}}{\partial\psi_j} \frac{\partial A}{\partial\psi_j^*} = T \int \prod_k d\psi_k' d\psi_k'' \sum_j e^{-H'/T} \frac{\partial^2 A}{\partial\psi_j \partial\psi_j^*}. \tag{2.100}$$

---

[*] Curiously, the quantum mechanical analog of (2.96), written in terms of the commutator of the operators H' and A, $\langle [H', A] \rangle \equiv 0$, is immediately proven by the cyclic symmetry of the trace operation: $\text{Tr}[H', A]\exp(-H'/T) = 0$. There is, however, one subtlety concerning both quantum and classical cases. The two terms formally canceling each other can prove ill defined, rendering illegal the mathematical transformations done with them. A way out is to replace H' with some well-behaved $\tilde{H}'$ such that $\tilde{H}' \to H'$ as a function of some vanishing regularization parameter ϵ, then establish the identity (2.96) for $\tilde{H}'$, and then argue that, by continuity, it remains valid as $\epsilon \to 0$, that is, for $\tilde{H}' \to H'$.

To complete the proof, we similarly process the second term and then take into account that

$$\frac{\partial^2 A}{\partial \psi_j \, \partial \psi_j^*} = \frac{\partial^2 A}{\partial \psi_j^* \, \partial \psi_j}. \tag{2.101}$$

With (1.21) and (2.96), we have

$$\langle \dot{\tilde{\Phi}} \rangle = \langle \{H, \tilde{\Phi}\} \rangle \equiv \langle \{H - H', \tilde{\Phi}\} \rangle. \tag{2.102}$$

With Equation (2.4) taken into account, we get

$$\langle \dot{\tilde{\Phi}} \rangle = \mu \langle \{N, \tilde{\Phi}\} \rangle + \mathbf{v}_n \cdot \langle \{\mathbf{P}, \tilde{\Phi}\} \rangle + \mathbf{s} \cdot \langle \{\mathbf{I}, \tilde{\Phi}\} \rangle. \tag{2.103}$$

[Clearly, in the absence of translation invariance, the second term in the right-hand side should be omitted.]

The structure of the result (2.103) is quite instructive. The Poisson brackets deal exclusively with the constants of motion rather than with the system-specific Hamiltonian, meaning that the final expression in terms of thermodynamic functions will be universal.

The first two Poisson brackets in the right-hand side of (2.103) are

$$\{N, \tilde{\Phi}(\mathbf{r})\} = -1, \tag{2.104}$$

$$\{\mathbf{P}, \tilde{\Phi}(\mathbf{r})\} = -\nabla \tilde{\Phi}(\mathbf{r}). \tag{2.105}$$

Problem 2.6 *Derive (2.104). Hint: In accordance with (1.144), variation of $\tilde{\Phi}(\mathbf{r})$ is equal to the variation of the phase $\Phi(\mathbf{r})$ of the field ψ, which means that $\delta\tilde{\Phi}(\mathbf{r}_1)/\delta\Phi(\mathbf{r}) = \delta(\mathbf{r}_1 - \mathbf{r})$. Pay attention to Problem 1.2 and relation (1.10).*

Problem 2.7 *Derive (2.105). Hint: Take into account the translation invariance of the regularization procedure: if the field ψ is translated by some vector $\mathbf{r}_0$, $\psi(\mathbf{r}) \rightarrow \psi(\mathbf{r} + \mathbf{r}_0)$, then the field $\tilde{\psi}$ and, correspondingly, its phase $\tilde{\Phi}$ are also shifted by the same vector, $\tilde{\Phi}(\mathbf{r}) \rightarrow \tilde{\Phi}(\mathbf{r} + \mathbf{r}_0)$. Express this property in terms of the functional derivatives for $\tilde{\Phi}([\psi, \psi*], \mathbf{r})$.*

Since Equations (2.104) and (2.105) already yield the right-hand side of Equation (2.94), all we need is to prove that the third term in the right-hand side of (2.103) is zero. This follows immediately from the discrete nature of the conserved topological invariant in the superfluid state: $\delta \mathbf{I}/\delta\psi \equiv 0$, $\delta \mathbf{I}/\delta\psi^* \equiv 0$.

2.2.2 Super Thermal Conductivity, Counterflow, and the Fountain Effect

One striking implication of Equation (2.95) is the phenomenon of *supertransport of heat*. It turns out that the standard mechanism of heat conductance by diffusion can take place in a superfluid *only* if the normal flow is completely suppressed, for example, by placing the system in a porous medium. Otherwise, the heat is transported *mechanically* by the normal flow with the conservation of matter being preserved by having the superflow in the opposite direction. Indeed, in a Galilean system, a purely diffusive heat transport (i.e., heat transport at $v_n \equiv 0$) can take place only in the absence of pressure gradients (no force field on the normal component). But constant pressure in combination with temperature gradients leads to the gradients of μ and, by Equation (2.95), to the accelerated superflow in the direction of the temperature gradient (toward regions with higher temperature). This can be seen from Equation (2.95) and the general thermodynamic relation

$$dp = sdT + nd\mu, \tag{2.106}$$

where s is the entropy density. Since s and n are positive definite, the two equations imply

$$\dot{\mathbf{v}}_s \uparrow\uparrow -\nabla\mu \uparrow\uparrow \nabla T \qquad (v_n = 0, \quad P = \text{const.}). \tag{2.107}$$

By conservation of matter in a steady state, the superflow should be compensated with the normal flow in the opposite direction. We thus arrive at the contradiction with the assumption $v_n \equiv 0$.

Consider the setup shown in Figure 2.1. A smaller reservoir with a heater is connected by a channel to the main reservoir with a superfluid. In accordance with the previous analysis, the only possible steady state corresponds to a counterflow—normal and superflows in opposite directions. The normal flow transports heat from the smaller reservoir *mechanically* with the value of j_n (and thus j_s, since $j_s = -j_n$ by conservation of matter) being controlled by the rate of heat production in the smaller reservoir. At large enough heat production rate, the supercounterflow becomes unstable with respect to proliferation of topological defects, and a self-sustained vortex tangle, referred to as *superfluid turbulence*, develops. At the macroscopic level, the vortex tangle leads to coupling between the normal and superfluid components (the topological invariant is not conserved any longer). It is important, however, that even in the presence of the vortex tangle, the behavior of the system is radically different from that of a normal fluid with finite or even vanishing viscosity. While bringing about friction between the normal and superfluid components, the superfluid turbulence is fundamentally incapable of stopping the counterflow completely because it is sustained by the balance between production of vorticity by the counterflow and natural relaxation of the vortex tangle.

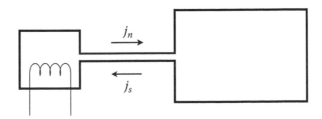

Figure 2.1 The counterflow setup.

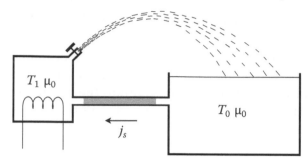

Figure 2.2   The fountain effect.

Now, we turn to the fountain effect setup shown in Figure 2.2. The setup is similar to that of counterflow (Figure 2.1) but with two essential modifications: (a) the channel between the main and heated reservoir is filled with porous material, which completely suppresses the normal flow, and (b) there is a second channel with a valve connected to the heated reservoir. In a static state with closed valve, the chemical potentials in both reservoirs must share the same value, μ_0, thus creating a pressure difference, $P_1 \equiv P(\mu_0, T_1) > P_0 \equiv P(\mu_0, T_0) \approx 0$, in response to the temperature difference $T_1 > T_0$, the so-called *fountain pressure* (we assume that the low-temperature liquid has open surface). For small ΔT, we have $\Delta P \approx s\Delta T$, according to (2.106). Opening the valve creates a *persistent jet*, supported by the *superflow* through the porous channel and energy input from the heater.

Problem 2.8 *Describe the transient regime of the fountain effect starting with opening the valve in the setup of Figure 2.2. In particular, argue that there is no alternative to the fountain steady state. Consider the situation when the channel between the reservoirs has no porous filling but is extremely narrow.*

There is a certain analogy between the fountain and *osmotic* effects. In both cases, the pressure difference takes place in the two-reservoir setup with a partially permeable partition allowing equilibration of chemical potentials,[*] while some other thermodynamic parameters remain different—different temperatures, in the case of fountain effect, and different concentrations of the solute, in the case of osmotic effect.

Curiously enough, realizing the fountain effect setup of Figure 2.2 with superfluid ^{4}He would require covering the walls of the main reservoir with a material that is *not wetted* by ^{4}He; an example of such a material is Cs. If walls are wetted by ^{4}He (typical case in practice), then one faces the problem of helium escaping from the reservoir by developing the critical flow in the surface film. This phenomenon, illustrated in Figure 2.3, is yet another hallmark of superfluidity. The underlying physics is very close to that of the counterflow and fountain effects.

---

[*] For the solvent but not the solute, in the case of osmosis.

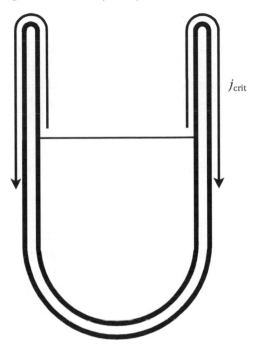

j_{crit}

Figure 2.3 Superfluid ^{4}He escaping from a Dewar due to a critical superflow within the surface film.

Once again, the origin is Equation (2.95), now applied to the superfluid film at the surface. The gradient of the chemical potential between the inside of the film and its edge accelerates the superflow up to some critical value. Further acceleration is prevented by the formation of a self-sustaining system of vortices (the situation is qualitatively similar to the self-sustained superfluid turbulence in thermal counterflow). When compared to a viscous flow of a normal film, the critical flow of the superfluid film is orders-of-magnitude faster.

The fountain effect and Equation (2.107) can create the wrong impression that there is some law that always directs the superflow toward the higher-temperature region. As an instructive counterexample, consider the setup of Figure 2.2, with closed valve, and analyze the transient regime taking place after switching on the heater in the initially equilibrium ($T_1 = T_0$) system. What can be stated on the basis of Equations (2.95) and (2.106) is that (a) after the transient, we have $P_1 > P_0$, because $T_1 > T_0$, while the chemical potential is the same and (b) if, after the transient, we reduce the pressure (by opening the tap), while keeping the temperature T_1 constant, the drop of the chemical potential will result in the superflow *toward* the heated chamber. However, Equations (2.95) and (2.106) do not fix the direction of the superflow during the transient. This direction can easily be *away* from the heated reservoir, a perfect example being the superfluid made out of the weakly interacting Bose gas in which the number density *decreases* with increasing temperature (by a factor of 2 when the temperature goes from zero to the critical value).

2.2.3 AC Josephson Effect

Let two superfluids, with slightly different chemical potentials, μ_1 and μ_2, be connected via a Josephson link (see Section 1.3.5). In accordance with Equation (2.94), the phase difference between two superfluids will grow linearly with time ($t = 0$ corresponding to a moment when the two phases were equal):

$$\langle \tilde{\Phi}_2 - \tilde{\Phi}_1 \rangle = (\mu_1 - \mu_2)t. \tag{2.108}$$

Equation (1.207) then implies that there will be an *alternating flow* across the link:

$$J(t) = J_0 \sin(\mu_1 - \mu_2)t. \tag{2.109}$$

This phenomenon is known as the *AC Josephson effect*.[*]

With increasing the value of $|\mu_1 - \mu_2|$, the result (2.109) gets less accurate, and, in particular, the dissipation becomes nonnegligible. Indeed, increasing $|\mu_1 - \mu_2|$ means increasing the frequency of flow oscillations. Correspondingly, the response of superfluids to the flow between them cannot be treated as purely adiabatic. Meanwhile, adiabaticity is implicit in Equation (2.108) based on the equilibrium relation (2.94).

2.2.4 Hydrodynamic Hamiltonian and Action

Let us concentrate now on the zero temperature case when Equation (2.94) becomes (in the absence of normal density or in the presence of disorder, the parameter $\mathbf{v}_n$ can be set equal to zero)

$$\dot{\Phi} = -\mu, \qquad \mu = \frac{\partial \mathcal{E}(n, \mathbf{v}_s)}{\partial n} \qquad (T = 0, \ V = \text{const.}), \tag{2.110}$$

with $\mathcal{E}(n, \mathbf{v}_s) = E/V$, the ground-state energy density. According to standard thermodynamic relations, $\mathcal{E}$ is the canonical thermodynamic potential with respect to the thermodynamic variables n and $\mathbf{v}_s$:

$$d\mathcal{E} = \mu dn + (1/\gamma)\mathbf{j} \cdot d\mathbf{v}_s \qquad (T = 0, \ V = \text{const.}). \tag{2.111}$$

It is important not to confuse $\mathcal{E}$, n, and $\mathbf{j}$ with the microscopic local values of the corresponding densities: in these relations, $\mathcal{E} \equiv \langle\langle \mathcal{E} \rangle\rangle$, $n \equiv \langle\langle n \rangle\rangle$, and $\mathbf{j} \equiv \langle\langle \mathbf{j} \rangle\rangle$, which is crucial when the system is not spatially uniform at the microscopic level.

Equation (2.110) leads to the zero-temperature hydrodynamics in the case of macroscopically slow spatial variations of the fields n and Φ (from now on, for briefness, we replace $\tilde{\Phi} \rightarrow \Phi$). The adiabaticity of the flow[†] is guaranteed by the slow evolution of vanishingly small gradients when locally the system is at $T = 0$.

---

[*] In superconductors—charged superfluids—the AC Josephson effect is readily detected through radiation produced by the time-dependent electric current.

[†] Here, it is crucial that vorticity is quantized because otherwise one would face the inevitable breakdown of the adiabatic flow in a spatially nonuniform system by a threshold-less generation of turbulence.

The complete set of equations describing the zero-temperature hydrodynamics in terms of the fields n and Φ is obtained by adding the continuity equation $\dot{n} = -\mathrm{div}\,\mathbf{j}$. Taking into account that $\mathbf{v}_s = \gamma\nabla\Phi$ and Equation (2.111), the pair of equations can be written in the canonical form

$$\dot{\Phi} = -\frac{\delta H_{\mathrm{hd}}}{\delta n}, \qquad \dot{n} = \frac{\delta H_{\mathrm{hd}}}{\delta\Phi}, \tag{2.112}$$

with the *hydrodynamic Hamiltonian*

$$H_{\mathrm{hd}}[n,\Phi] = \int \mathcal{E}(n, \mathbf{v}_s = \gamma\nabla\Phi)d\mathbf{r}. \tag{2.113}$$

One can also introduce the hydrodynamic action, either in the canonical form,

$$S_{\mathrm{hd}}[n,\Phi] = -\int dt\,d^d r\left[n\dot{\Phi} + \mathcal{E}(n, \mathbf{v}_s = \gamma\nabla\Phi)\right], \tag{2.114}$$

or in terms of the field Φ only

$$S_{\mathrm{hd}}[\Phi] = \int p_0(\mu = -\dot{\Phi}, \mathbf{v}_s = \gamma\nabla\Phi)\,dt\,d\mathbf{r}, \tag{2.115}$$

where $p_0(\mu, \mathbf{v}_s) = -\mathcal{F}/V$ is the ground-state pressure. In the latter case, the action leads to a closed equation of motion for the phase field Φ. Expressions (2.114) and (2.115) correspond to what is known as Popov's hydrodynamic action.[*]

Speaking generally, actions (2.114) and (2.115) are strongly anharmonic. If the density variations are small compared to the density itself, and the superfluid velocities are also appropriately small, it makes sense to expand the action in the vicinity of the equilibrium solution with $v_s = 0$ and $\Phi = -\mu t$, where μ is the uniform equilibrium potential. To this end, we make replacements $\Phi \to \Phi - \mu t$ and $n \to n+\eta$ (meaning that now we reckon Φ from its equilibrium value $-\mu t$ and that we work with the density deviation η from the uniform equilibrium density n) leading to $n\dot{\Phi} + \mathcal{E}(n) \to (n+\eta)\dot{\Phi} + \mathcal{E}(n+\eta) - \mu(n+\eta)$ and $p_0(-\dot{\Phi}) \to p_0(\mu - \dot{\Phi})$ in the actions. Then, we Taylor-expand the actions in terms of $\dot{\Phi}$, $\nabla\Phi$, and η.

Expanding p_0 at the point $(\mu, \mathbf{v}_s = 0)$ [below $p_0^{(0)} = p_0(\mu, \mathbf{v}_s = 0)$], we have

$$p_0\left(\mu - \dot{\Phi}, \mathbf{v}_s = \frac{\nabla\Phi}{m}\right) = p_0^{(0)} - \frac{\partial p_0}{\partial\mu}\dot{\Phi} + \frac{1}{2}\frac{\partial^2 p_0}{\partial\mu^2}\dot{\Phi}^2 + \frac{\gamma^2}{2d}\Delta_{v_s}p_0\,(\nabla\Phi)^2$$
$$- \frac{\gamma^2}{2d}\frac{\partial}{\partial\mu}\Delta_{v_s}p_0\,\dot{\Phi}(\nabla\Phi)^2 - \frac{1}{6}\frac{\partial^3 p_0}{\partial\mu^3}\dot{\Phi}^3 + \dots. \tag{2.116}$$

[We show all terms up to the cubic ones, inclusively. By reflectional symmetry, odd powers of $\nabla\Phi$ are absent.] The first three partial derivatives in the expansion (2.116) are conveniently expressed in terms of basic thermodynamic quantities by

---

[*] We must remark, though, that the original (quantum-mechanical) derivation by Popov [3] contained an error because his p_0 was simultaneously a function of μ and n while μ and n are not independent thermodynamic variables.

utilizing general $T = 0$ relations: $\partial p_0/\partial \mu = n$ and $\partial^2 p_0/\partial \mu^2 = \partial n/\partial \mu = \varkappa$, where $\varkappa$ is compressibility. Also, $(\gamma/d)\Delta_{v_s} p_0 = -n_s$ (see Section 2.1.3). For cubic derivatives, we have

$$\frac{\gamma}{d}\frac{\partial}{\partial\mu}\Delta_{v_s} p_0 = -\varkappa_s, \qquad \varkappa_s = \frac{\partial n_s}{\partial\mu} = \varkappa\frac{\partial n_s}{\partial n} = \frac{\varkappa\gamma}{d}\frac{\partial}{\partial n}\Delta_{v_s}\mathcal{E}, \qquad (2.117)$$

$$\frac{\partial^3 p_0}{\partial\mu^3} = \frac{\partial\varkappa}{\partial\mu} = -\varkappa^3\lambda, \qquad \lambda = \frac{\partial^3\mathcal{E}}{\partial n^3}. \qquad (2.118)$$

By analogy with compressibility $\varkappa$, the quantity $\varkappa_s$ can be referred to as superfluid compressibility. In a system with Galilean invariance, $n_s \equiv n$ and $\varkappa_s \equiv \varkappa$.

We thus arrive at the following structure of the hydrodynamic action:

$$S_{\mathrm{hd}}[\Phi] = S_{\mathrm{hd}}^{(\mathrm{top})}[\Phi] + S_{\mathrm{hd}}^{(\mathrm{harm})}[\Phi] + S_{\mathrm{hd}}^{(\mathrm{n.h.})}[\Phi], \qquad (2.119)$$

where

$$S_{\mathrm{hd}}^{(\mathrm{top})}[\Phi] = -n\int d\mathbf{r}\int \dot{\Phi}\,dt \qquad (2.120)$$

is a very specific topological term that is relevant only when the field Φ features topological defects and

$$S_{\mathrm{hd}}^{(\mathrm{harm})}[\Phi] = \int dt\,d\mathbf{r}\left[\frac{\varkappa}{2}\dot{\Phi}^2 - \frac{\gamma n_s}{2}(\nabla\Phi)^2\right] \qquad (2.121)$$

is the harmonic part of the action describing noninteracting phonons. The rest of the action is the hierarchy of nonharmonic terms:

$$S_{\mathrm{hd}}^{(\mathrm{n.h.})}[\Phi] = \int dt\,d\mathbf{r}\left[\frac{\gamma\varkappa_s}{2}\dot{\Phi}(\nabla\Phi)^2 + \frac{\varkappa^3\lambda}{6}\dot{\Phi}^3 + \ldots\right]. \qquad (2.122)$$

Similarly, the Taylor expansion of $\mathcal{E}$ leads to the canonical action in the form (up to cubic terms, inclusively)

$$S_{\mathrm{hd}}[\eta,\Phi] = -\int dt\,d\mathbf{r}\left[(n+\eta)\dot{\Phi} + \frac{\gamma n_s}{2}(\nabla\Phi)^2 + \frac{1}{2\varkappa}\eta^2 + \frac{\gamma\varkappa_s}{2\varkappa}\eta(\nabla\Phi)^2 + \frac{\lambda}{6}\eta^3\right]. \qquad (2.123)$$

Actions (2.119) and (2.123) capture the dynamics of phonons and, with certain reservations, the dynamics of vortices and vortex–phonon interaction. The reservations in the case of vortex dynamics come from the fact that the hydrodynamic action captures only the long-wave part of the vortex action, ignoring microscopic contributions associated with the vortex cores. In disordered systems, the latter can qualitatively change the vortex dynamics, to the extent that vortices get pinned. In a translation-invariant system, the effect of the cores is less dramatic, allowing one to proceed within the paradigm of hydrodynamics while taking care of the vortex core physics with an appropriate UV cutoff.

A detailed account of vortex dynamics and vortex–phonon interaction will be given in Chapter 10. Here, we will make a few crucial steps toward this goal by introducing an appropriate parameterization of the field Φ:

$$\Phi = \Phi_0 + \varphi, \tag{2.124}$$

where the singular (non-single valued) field Φ_0 contains all the vortices, so that the field φ is regular (single valued) and $\nabla\varphi$ is circulation-free. This decomposition is not yet unique, and the standard way to proceed is by introducing the constraint:

$$\Delta\Phi_0(\mathbf{r}, t) = 0 \qquad \text{(apart from vortex cores).} \tag{2.125}$$

The convenience of this constraint—based on the properties of the velocity potential of an incompressible fluid—is that it eliminates the coupling between Φ_0 and φ in the leading order. With Equations (2.124) and (2.125), the action becomes

$$S_{\text{hd}}^{(\text{n.F.})}[\Phi_0, \varphi] = S_{\text{vor}}[\Phi_0] + S_{\text{ph}}[\varphi] + S_{\text{vor–ph}}[\Phi_0, \varphi], \tag{2.126}$$

where

$$S_{\text{vor}}[\Phi_0] = -\int dt\, d\mathbf{r} \left[n\dot{\Phi}_0 + \frac{\gamma n_s}{2} (\nabla\Phi_0)^2 \right] \tag{2.127}$$

is the vortex part, in which the terms $(\varkappa/2)\dot{\Phi}_0^2$, $(\gamma\varkappa_s/2)\dot{\Phi}_0(\nabla\Phi_0)^2$, and $(\varkappa^3\lambda/6)\dot{\Phi}_0^3$ are omitted because they are much smaller than the term $(\gamma n_s/2)(\nabla\Phi_0)^2$ by the parameter

$$v_{\text{vor}}/c \ll 1, \tag{2.128}$$

where v_{vor} is the characteristic velocity of vortex motion, which is much smaller than the sound velocity in the hydrodynamic limit, implying that the distances considered are much larger than the vortex core radius. Action (2.127) describes the dynamics of vortices in incompressible-liquid approximation. Corresponding theory will be developed in Chapter 10, where the vortex action will be expressed directly in terms of the positions of vortex lines, naturally yielding the least-action equation of motion for the lines.

Next,

$$S_{\text{ph}}[\varphi] = \int dt\, d\mathbf{r} \left[\frac{\varkappa}{2}\dot{\varphi}^2 - \frac{\gamma n_s}{2}(\nabla\varphi)^2 + \frac{\gamma\varkappa_s}{2}\dot{\varphi}(\nabla\varphi)^2 + \frac{\varkappa^3\lambda}{6}\dot{\varphi}^3 \right] \tag{2.129}$$

is the phonon part. The remaining terms,

$$S_{\text{vor-ph}}[\Phi_0, \varphi] = \int dt \, d\mathbf{r} \left[\varkappa \, \dot{\varphi} \, \dot{\Phi}_0 + \frac{\gamma \varkappa_s}{2} \, \dot{\varphi} \, (\nabla \Phi_0)^2 + \gamma \varkappa_s \, \dot{\varphi} \, (\nabla \varphi \cdot \nabla \Phi_0) \right], \qquad (2.130)$$

describe the vortex–phonon coupling; here, we retain only the leading-in-parameter (2.128) terms.

The semi-canonical counterparts of Equations (2.129) and (2.130) are

$$\tilde{S}_{\text{ph}}[\eta, \varphi] = -\int dt \, d\mathbf{r} \left[\frac{1}{2\varkappa} \eta^2 + \frac{\gamma n_s}{2} (\nabla \varphi)^2 + \frac{\gamma \varkappa_s}{2\varkappa} \eta \, (\nabla \varphi)^2 + \frac{\lambda}{6} \eta^3 \right], \qquad (2.131)$$

$$\tilde{S}_{\text{vor-ph}}[\Phi_0, \eta, \varphi] = -\int dt \, d\mathbf{r} \left[\eta \, \dot{\Phi}_0 + \frac{\gamma \varkappa_s}{2\varkappa} \eta \, (\nabla \Phi_0)^2 + \frac{\gamma \varkappa_s}{\varkappa} \eta \, (\nabla \varphi \cdot \nabla \Phi_0) \right]. \qquad (2.132)$$

Obviously, the hydrodynamic expansion we developed in this section is not limited to the explicitly presented cubic terms. If necessary, one can continue expanding the hydrodynamic action to an arbitrary high order.

In conclusion of this section, consider the relationship between the aforementioned hydrodynamic action and hydrodynamics of classical ideal liquid. The former deals with the notions of (superfluid) matter density and phase, while the latter is naturally expressed in terms of mass density and velocity. The connection between the two representations is straightforward [cf. Equation (1.88)]: the mass (superfluid) density is defined by dividing its matter-density counterpart by γ, while the scalar velocity potential is defined as $\gamma \Phi$. The definition of mass density implies that the (superfluid) mass compressibility corresponds to the (superfluid) matter compressibility divided by γ^2. Parameter γ then drops out from the hydrodynamic action—as long as the latter is expressed in terms of velocity potential and mass densities/compressibilities—meaning that the only distinction of the zero-point superfluid hydrodynamics from its classical ideal-liquid counterpart is the quantization of vorticity. Hydrodynamic role of a particular value of the parameter γ is merely that of a scaling coefficient. It is thus very convenient to adopt units such that $\gamma = 1$, in which case one does not need to distinguish between mass and matter densities/compressibilities as well as between the fields of phase and the scalar velocity potential. The proper power of γ (in practice, that of the ratio $\gamma = \hbar/m$), is readily restored in final expressions by dimensional analysis.

References

1. L. Landau, Theory of superfluidity of helium II, *Phys. Rev.* **60**, 356 (1941).

2. L. D. Landau and E. M. Lifshitz, *Fluid Mechanics*, Course of Theoretical Physics, Vol. 6, Pergamon Press, London, U.K., 1987.

3. V. N. Popov, *Functional Integrals in Quantum Field Theory and Statistical Physics*, Reidel, Dordrecht, the Netherlands, 1983.

CHAPTER 3

Superfluid Phase Transition

In the general framework of critical phenomena, all systems that share the same relevant symmetries, dimension of space, and interaction range are supposed to exhibit a certain degree of universality at the critical point. Though the transition point itself and contributions of critical (long-wavelength) fluctuations to various physical properties are system specific, the dependence on detuning from the critical point, the decay of correlation functions at criticality, and the amplitude ratios are system independent. We do not attempt here a comprehensive discussion of critical phenomena that can be found elsewhere.* We rather apply these ideas to the superfluid–normal transition, mention the main results, and discuss their significance.

The concept of the universality class is extremely powerful for continuous phase transitions since it is sufficient to study in detail just one representative model to be able to make solid predictions about all other systems featuring a similar broken-symmetry transition. This is so no matter whether the system is quantum or classical, continuous or on a lattice, and regardless of the microscopic details of interactions. In this chapter, we concentrate on the standard XY and $|\psi|^4$ models. Later in the book (see Chapter 9) we will explore the effects of disorder and particle localization (in quantum models) on the normal–superfluid transition.

Finally, it is often the case that quantum phase transitions in d-dimensional systems can be mapped precisely to the classical finite-temperature transition in $(d + z)$-dimensions, where z is the dynamic critical exponent. We will discuss the relevant mapping later in Chapters 8 and 9. Here, it is assumed that we are always dealing with the classical finite-temperature superfluid transition.

3.1 *XY* Universality Class

The simplest model one can imagine that undergoes the U(1) symmetry-breaking transition is given by the system of planar unit-vector lattice spins, $\mathbf{n_i}$, with the nearest neighbor (n.n.) ferromagnetic interaction J:

$$H_{XY} = -J \sum_{\langle \mathbf{ik} \rangle} \mathbf{n_i} \cdot \mathbf{n_k} = -J \sum_{\langle \mathbf{ik} \rangle} \cos(\varphi_i - \varphi_k). \tag{3.1}$$

To be specific, we consider a simple hypercubic lattice in d-dimensions and denote pairs of n.n. sites by $\langle \mathbf{ik} \rangle$. In the last equality, we have parameterized planar vectors of unit length using the angle variable $\mathbf{n_i} \equiv (\cos \varphi_i, \sin \varphi_i)$. The U(1) symmetry

---

* See, for example, Reference [1].

immediately follows from the fact that H_{XY} is invariant under rotation of all spins by the same angle $\varphi_i \to \varphi_i + \Phi$. The connection between the XY model (3.1) and the classical complex-field model on a lattice

$$H_\psi = -t \sum_{\langle ik \rangle} \left(\psi_i^* \psi_k + \text{c.c.} \right) + \frac{U}{2} \sum_i |\psi_i|^4 - \mu \sum_i |\psi_i|^2 \tag{3.2}$$

is straightforward. Indeed, the last two terms can be combined into one expression $(U/2)(|\psi_i|^2 - \mu/U)^2$, which, in the limit of $U \to \infty$, $\mu/U \to 1$, becomes equivalent to the "fixed modulus" constraint $|\psi_i|^2 = 1$. In this limit, $\psi_i = e^{i\varphi_i}$ and the two models coincide at the microscopic level, $H_\psi = H_{XY}(J = 2t)$.

Formally, nothing is "flowing" in the purely statistical XY model. The transition point is associated with the development of long-range correlations and order in the directions of unit vectors, or phases φ_i. As mentioned previously, superfluidity is about topological order in the phase field; whether the phase gradient is related to the current flow or not is a secondary consideration that does not change the essence of the transition point and emergence of order at low temperature. One can apply the same notions of topological protection and quantization of closed-contour integrals $\oint \nabla \varphi \cdot d\mathbf{l}$, introduce vortices, and consider the free energy increase in response to phase gradients [see Equation (2.55)]:

$$F(\Phi) - F(0) = \frac{\Lambda_s}{2} \int d^d r (\nabla \varphi)^2, \tag{3.3}$$

though often one finds a "name change" for Λ_s in the literature—instead of superfluid stiffness $\Lambda_s = \gamma n_s$, it is called the helicity modulus. To avoid confusion, we choose to keep the same language for all models. The advantage of using statistical models (3.1) and (3.2) is their conceptual simplicity and the possibilities they offer for obtaining high-accuracy numerical results for very large systems (up to 10^7 lattice points).

We start with the superfluid transition in $d = 3$ dimensions. On the approach to the critical point $T \to T_c$ from the normal phase, one observes the emergence of the diverging correlation length ξ, which separates the power-law decay from the exponential decay of the correlation function (see Figure 3.1)

$$g(\mathbf{i} - \mathbf{k}) = \langle \mathbf{n}_i \cdot \mathbf{n}_k \rangle \equiv \langle e^{i\varphi_i} e^{-i\varphi_k} \rangle, \tag{3.4}$$

which is a direct analog of (2.40). We assume here periodic boundary conditions that allow us to write the argument of the correlation function in the translation invariant form. The power-law decay $g(r) \propto r^{2-d-\eta} = r^{-1-\eta}$ is characterized by the critical exponent $\eta = 0.0380(2)$ [2,3], while the divergence of the correlation length for small reduced temperatures, $t = |T - T_c|/T_c \to 0$, is given by the power law $\xi \propto t^{-\nu}$ with $\nu = 0.6717(1)$ [3–6]. We can immediately read from the plot in Figure 3.1 that, in the ordered phase, the correlation function saturates at a plateau corresponding to the condensate fraction [see Equations (2.42) and (2.43) and the text after these equations], $n_0 \sim g(\xi) \propto \xi^{-(1+\eta)} \propto t^{\nu(1+\eta)}$; that is, the order parameter critical exponent is $\beta = \nu(1 + \eta)/2 = 0.3486(1)$.

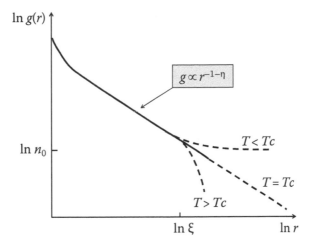

Figure 3.1   Schematic behavior of the correlation function across the transition point in 3D superfluid.

One may argue, using the *hyperscaling* hypothesis—stating that there is about one critical mode per correlation volume*—that the superfluid stiffness should scale as $\Lambda_s \propto \xi^{-1} \propto t^\nu$ for $T < T_c$ (Josephson relation [7]), and critical exponents for the correlation length and specific heat are related by $\alpha = 2 - d\nu = -0.0151(3)$. Indeed, the free energy of phase fluctuations with $\nabla\varphi \sim 1/\xi$ is given by $\Lambda_s(\nabla\varphi)^2\xi^3 \sim \Lambda_s\xi$ and this ought to be of order temperature according to the hyperscaling hypothesis. Negative α implies that the celebrated specific heat divergence in ^{4}He, which gave the transition its name, "the λ-point," is actually a finite-amplitude spike with divergent derivative $C_\pm \approx C(T_c) - C_\pm t^{-\alpha}$. As early as 1953, Feynman noticed that the superfluid transition can be reduced to the statistical problem of proliferation of closed oriented loops [8]. His consideration was based on path integrals and macroscopic exchange cycles. In the next section, we will show how the same statistical problem emerges for the XY and $|\psi|^4$ models. In Chapter 8, we will derive a rigorous relation between the superfluid stiffness and macroscopic exchange cycles discussed by Feynman. In this language, the hyperscaling hypothesis states that at $T > T_c$, there is about one exchange cycle of diameter ξ in the correlation volume.

It is also worth mentioning how properties of the XY universality class are used in the finite-size scaling analysis of the superfluid critical points [9]. For sufficiently large system size L and small t, the dependence of physical properties on L can only happen in the form of a dimensionless ratio L/ξ, implying

$$n_0 = t^{2\beta} f_0(L/\xi) \equiv L^{-(1+\eta)} \bar{f}_0(tL^{1/\nu}), \tag{3.5}$$

---

* The hyperscaling hypothesis/property is equivalent to the hypothesis/property of *scale invariance* of critical fluctuations.

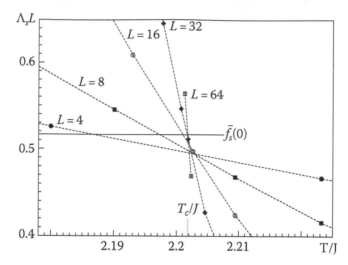

Figure 3.2   Finite-size scaling curves for $\Lambda_s L$ for the 3D XY model (linear system sizes are mentioned next to the dashed lines). The solid horizontal line marks the value of the universal function $\bar{f}_s(0) = 0.516(2)$ in the thermodynamic limit.

where f_0 and $\bar{f}_0$ are universal functions for a given universality class. Similar expression can be written for the superfluid stiffness,

$$\Lambda_s = L^{-1}\bar{f}_s(tL^{1/\nu}),\tag{3.6}$$

and other singular quantities.

Note that $\bar{f}_0$ and $\bar{f}_s$ are analytic functions at the origin since, in a finite system, all thermodynamic quantities are analytic functions of temperature. This last property is very convenient for extracting critical points from the intersection of $\bar{f}$-curves calculated for different system sizes: at $t = 0$, the argument of the $\bar{f}$-function does not depend on L. In Figure 3.2, we show data for the scaled superfluid stiffness, $L\Lambda_s$, of the XY model in the vicinity of the critical point $T/J = 2.20184$. Remarkably, nearly three-digit accuracy is obtained already from simulations of the smallest systems. The crossing is not perfect due to corrections to scaling laws that vanish as $L^{-\omega}$ with $\omega = 0.785(20)$ [3].

Problem 3.1 *Let $R(t, L)$ be some measured property that is supposed to be scale invariant at the critical point. Calculate convergence of the crossing points for finite-size systems to the critical point $t = 0$ and the universal value $\bar{f}(0)$ by taking into account corrections to scaling using the following formula:*

$$R(t, L) = \bar{f}_R(tL^{1/\nu})\left[1 + CL^{-\omega}\right].$$

In two dimensions, the superfluid transition is of the Berezinskii–Kosterlitz–Thouless (BKT) type; that is, it does not fit the standard picture of the scale-invariant second-order criticality. In Chapter 4, we discuss the BKT theory at length and provide derivations of all basic relations. Here we summarize them shortly to have a complete review of transition points in one place. The behavior of the correlation function in the normal state and at $T = T_c$ is qualitatively similar to that shown in Figure 3.1. What is different is the absence of a plateau in $g(r)$ at large distances in the superfluid state at any finite temperature. At $T = T_c$, the asymptotic power-law decay is given by $g(r) \propto r^{-1/4}$. In the vicinity of the transition point, this law is approximately observed up to the exponentially large correlation length $\xi \propto a e^{1/\sqrt{C|t|}}$ where a is a lattice constant and C is a dimensionless, system-specific, constant. In the superfluid phase, one observes a slow power-law decay $g(r) \propto r^{-s}$ with the temperature-dependent exponent $s = T/2\pi\Lambda_s$. All thermodynamic functions remain continuous and differentiable at the transition point with the notable exception of the superfluid stiffness (or superfluid density), which has a universal jump at $T = T_c$ from

$$\Lambda_s(T_c) = \frac{2}{\pi} T_c \qquad (T = T_c - 0) \tag{3.7}$$

to zero. In the vicinity of the transition point, on the superfluid side, the superfluid stiffness has a square-root singularity:

$$\Lambda_s(T) = \frac{2}{\pi} T \left(1 + \frac{1}{2} \sqrt{C|t|} \right). \tag{3.8}$$

The superfluid transition in 2D is special because it is governed by the unbinding of vortex–antivortex pairs as the temperature is increased. Moreover, the theory is asymptotically exact at large scales, because the relative (with respect to their size) concentration of macroscopic vortex pairs at the transition point is vanishingly small, and one may neglect interactions between the pairs. This also brings about violations of scale invariance at the BKT transition—one can use the average size squared of the pairs in an $L \times L$ box to determine how close the system is to the transition point. Still, the previously mentioned laws for the superfluid stiffness and the correlation length are universal for all systems featuring the BKT transition. We defer the discussion of finite-size effects to Chapter 4.

Formally, there is no superfluidity at finite temperature in 1D. We may stop short here if the discussion is about systems with fixed temperature in the thermodynamic limit. In Chapter 2, we explained that, in 1D (and quasi-1D) systems at low enough temperature, one has to distinguish between the two definitions of superfluid density [see Equations (2.93) and (2.86)]. More specifically, an exponential suppression of Λ_s^φ with temperature is linked to the broad statistical distribution of states with nonzero phase windings (in $d > 2$, these states have energies much larger than temperature, and their contribution to statistics is negligible; the 2D case is marginal). At the same time, the typical time scale for changing the phase winding in a realistic dynamic system gets extremely long at low temperature and can easily exceed an experimental time scale by orders of magnitude. Likewise, at low enough temperature, the probability of having a phase slip in the phase field *in a finite-size system* will be extremely small in the statistical system. Under these conditions, one can introduce Λ_s as the superfluid stiffness *within a given phase-winding sector*. In this sense, 1D superfluidity at low temperature has the same origin as in higher-dimensional systems, with the only difference being that it is rather a dynamic crossover revealed in a particular experimental setup rather than a phase transition and the temperature needed for this to occur scales to zero with the system size. Classically, the crossover temperature to the regime where the notion of superfluid density Λ_s starts making sense can be estimated from the probability of having just one phase-slip configuration in the entire system: $(L/l_0)\exp\{-E_0/T\} \ll 1$, where l_0 is the healing length and E_0 is the energy penalty for making a zero in the density profile. This leads to $T \ll E_0/\ln(L/l_0)$. Note that Λ_s^φ gets exponentially small already at temperatures $T \gg \Lambda_s/L$, which is much smaller than the crossover estimate in the large L limit.

Finally, $d = 4$ is the upper critical dimension separating scale-invariant critical phenomena from the mean-field transitions in $d > 4$; that is, critical exponents for the correlation length and order parameter have the mean-field values $\nu = 1/2$, $\beta = 1/2$, but scaling laws have to be corrected by additional logarithmic terms (see, e.g., Reference [1]). More importantly, an effective theory of order-parameter fluctuations at large distances is asymptotically free, and this fact is often used to set up a perturbative calculation of critical exponents in $4 - \epsilon$ dimensions using ϵ as a small parameter.

3.2   High-Temperature Expansion for the XY Model

There is an alternative view on normal and superfluid phases and the transition between them in terms of loop models. Loop representation is particularly useful for showing similarities between the classical and quantum systems, as well as establishing the universality class of the superfluid transition. In what follows we derive an exact mathematical mapping for the XY model, which has its quantum counterpart in the form of Feynman's exchange cycles (see Chapter 8) within the path-integral representation of quantum statistics [10].

We start by noting that periodic functions of angle φ can be written in terms of a discrete Fourier sum. Thus,

$$e^{K\cos(\varphi)} \equiv \sum_{j=-\infty}^{\infty} e^{ij\varphi} F(j), \tag{3.9}$$

where $K = J/T$ and $F(j) = \int_0^{2\pi} e^{K\cos(\varphi)} \cos(j\varphi)\,(d\varphi/2\pi) > 0$. Since the Gibbs factor in the partition function factorizes into a product of exponentials defined on lattice bonds, we have $\left[\int \mathcal{D}\varphi\,(\ldots) \equiv \prod_i \int_0^{2\pi}(d\varphi_i/2\pi)\,(\ldots) \right]$

$$Z = \int \mathcal{D}\varphi \prod_b e^{K\cos(\varphi_{i'}-\varphi_i)} = \int \mathcal{D}\varphi \prod_b \sum_{j_b=-\infty}^{\infty} e^{ij_b(\varphi_{i'}-\varphi_i)} F(j_b). \tag{3.10}$$

Here $b = (\mathbf{i}, \alpha)$ is a shorthand notation for lattice bonds with $\alpha = 1, 2, \ldots, d$ enumerating bonds pointing in the positive axis directions; bonds pointing in the negative direction will be denoted as $(\mathbf{i}, -\alpha)$, or, identically, $(\mathbf{i} - \hat{\alpha}, \alpha)$ with the understanding that $\mathbf{k} = \mathbf{i} - \hat{\alpha}$ is the n.n. site of $\mathbf{i}$ in the direction $-\hat{\alpha}$.

By changing places the summation over bond numbers $\{j_b\}$ with the integration over phases $\{\varphi_i\}$, we notice that the phase integrals can have only two values, 0 or 1. Indeed, on every site, we are dealing with

$$\int_0^{2\pi} e^{-iN_i\varphi_i}\,\frac{d\varphi_i}{2\pi} = \delta_{N_i,0},\tag{3.11}$$

where $N_i = \sum_\alpha (j_{i,\alpha} - j_{i-\hat\alpha,\alpha})$. Thus, the phase integral is nonzero and equal to unity if, and only if, all $N_i = 0$. There is a convenient graphical picture of meaningful configurations of bond numbers in which positive (negative) $j_{i,\alpha}$ are represented by arrows, or "currents," going out of (into) the site i with the corresponding rule that $j_{i,-\alpha} = -j_{i-\hat\alpha,\alpha}$. The condition of zero $\{N_i\}$ is then nothing, but the zero-divergence current constraint

$$\sum_{\mu=\pm 1,\pm 2,\dots,\pm d} j_{i,\mu} = 0,\tag{3.12}$$

stating that incoming and outgoing current fluxes are equal on all sites. A typical example of allowed (i.e., making a nonzero contribution to Z) configuration of bond currents is shown in Figure 3.3. With this constraint in mind (for brevity, we denote it as $\nabla\cdot\mathbf{j} = 0$)

$$Z = \sum_{\{j_b\}}^{\nabla\cdot\mathbf{j}=0} \prod_b F(j_b) = \sum_{\{j_b\}}^{\nabla\cdot\mathbf{j}=0} e^{-F_{\text{loop}}[\{j_b\}]/T},\tag{3.13}$$

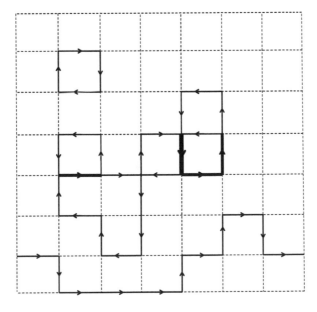

Figure 3.3   Graphical representation of allowed bond configurations in terms of oriented currents. The thickness of the line is proportional to the bond number.

we arrive at the loop formulation of the *XY* model. The loop language is nothing but a simple alternative way to express the zero-divergence constraint since it is automatically satisfied for loops. In other words, all allowed $\{j_b\}$ states are accounted for by considering configurations constructed from arbitrary loops with overlaps and self-intersections. [The decomposition of $\{j_b\}$ into the loop state is not unique for the *XY* model.] When loops overlap, the corresponding bond current is an algebraic sum of individual loop currents.

Before we proceed with the discussion of normal and superfluid phases in the loop language, let us generalize the mapping to the correlation function $g(\mathbf{i}-\mathbf{j})$, Equation (3.4), which we express as $g(\mathbf{i}-\mathbf{j}) = G(\mathbf{i}-\mathbf{j})/Z$. Upon expansion of bond Gibbs factors into Fourier series, we find that the only difference between Z, Equation (3.13), and $G(\mathbf{i}-\mathbf{j})$ is that the zero-divergence constraint has to be satisfied everywhere except at sites $\mathbf{i}$ and $\mathbf{j}$ where $N_{\mathbf{i}} = 1$ and $N_{\mathbf{j}} = -1$. This is happening because we have an extra exponential $e^{i\varphi_{\mathbf{i}}}$ on site $\mathbf{i}$ coming from the correlation function, *not* the expansion of partition function into Fourier series; a similar situation occurs on site $\mathbf{j}$. The corresponding configuration space is thus composed of oriented loops plus one line originating at $\mathbf{i}$ and terminating at $\mathbf{j}$ (see Figure 3.4). The configuration weight is still given by the product of $F(j_b)$ factors.

At high temperature, when $K \ll 1$ and the $F(j)$-function is rapidly decaying with j, the statistics are best viewed as those of a dilute gas of small, one lattice plaquette in size, loops. In this regime, the probability of finding a large loop, or having $|\mathbf{i}-\mathbf{j}| \gg 1$, is exponentially small, as expected for the normal state.

As the temperature is decreased, the concentration of loops is increasing until they start to self-intersect, overlap, and form larger and larger structures with

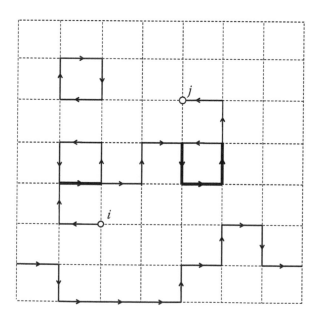

Figure 3.4   Current configuration contributing to the two-point correlation function $G(\mathbf{i}-\mathbf{j})$, Equation (3.4). The thickness of the line is proportional to the bond number.

fractal properties at intermediate distances. The power-law decay of $g(\mathbf{i} - \mathbf{j})$ with divergent normalization sum at the superfluid transition is thus linked to the pro-liferation of loops with macroscopic sizes. In the opposite limit of $K \gg 1$, we are dealing with large bond currents, and in a typical configuration, one can construct a macroscopic loop with probability close to unity. This point of view is further supported by the winding number formula for superfluid stiffness, $\Lambda_s^{(\varphi)}$, that is derived below.

Consider a state with a constant gauge phase φ_0/L applied in the $\hat{\alpha} = 1$ direc-tion and changing each term in Equation (3.1) from $\cos(\varphi_i - \varphi_j)$ to $\cos(\varphi_i - \varphi_j - \varphi_0/L)$ for n.n. sites in the $\hat{\alpha}$ direction. Repeating all steps in the derivation of the loop representation for the partition function, we get

$$Z = \prod_{\mathbf{i}} \int_0^{2\pi} \frac{d\varphi_{\mathbf{i}}}{2\pi} \prod_{b=\mathbf{i},\alpha} \sum_{j_b=-\infty}^{\infty} e^{ij_b(\varphi_{\mathbf{i}'}-\varphi_{\mathbf{i}}+\delta_{\alpha,1}\,\varphi_0/L)} F(j_b)$$

$$= \sum_{\{j_b\}}^{\nabla j=0} \prod_b F(j_b) e^{i\varphi_0 W} \equiv \sum_{W=-\infty}^{\infty} Z_W e^{iW\varphi_0}, \tag{3.14}$$

where an integer $W = L^{-1} \sum_{\mathbf{i}} j_{\mathbf{i},1}$ is called a winding number. Indeed, loops that do not wind around the system with periodic boundary condition make no con-tribution to W since the net sum of bond currents for such loops in the $\hat{\alpha} = 1$ direction is zero (when going along the loop, one makes as many steps forward as backward; this is why one returns to the same point). On the other hand, loops that wind around the system necessarily contribute a multiple of L to the sum of bond currents (see a winding loop at the bottom of Figure 3.3).

We are in a position to calculate the superfluid stiffness, $\Lambda_s^{(\varphi)} = \gamma n_s^{(\varphi)}$, straight-forwardly from the thermodynamic relation $\Lambda_s^{(\varphi)} = L^{2-d}[d^2 F(\varphi_0)/d\varphi_0^2]_{\varphi_0=0}$ [the relation between Λ_s and $\Lambda_s^{(\varphi)}$ is the same as between n_s and $n_s^{(\varphi)}$ in Equation (2.93)]. The result is

$$\Lambda_s^{(\varphi)} = \frac{T}{L^{d-2}} \langle W^2 \rangle. \tag{3.15}$$

By symmetry between the positive and negative winding numbers, $\langle W \rangle = 0$. Equa-tion (3.15) establishes a direct connection between the superfluid transition to the ordered state and the picture of macroscopic loops. From the results for the super-fluid stiffness mentioned in the previous section, we deduce that winding number fluctuations are described by the universal function $\bar{f}_s((L/\xi)^{1/\nu})$ and are of order unity at the transition point. Numerically, one finds that at the 3D critical point, $\langle W^2 \rangle = 0.516(2)$. In 2D, the Kosterlitz–Nelson formula (3.7) leads to an accurate relation $\langle W^2 \rangle \approx 2/\pi$ [see Equation (2.93) and Problem 4.4 in the next chapter].

It is clear from our discussion that statistics of loops and decay of the correla-tion function are closely related. To characterize this relation at the scale-invariant critical point of the 3D system quantitatively, we introduce the probability dis-tribution $P(r, M)$ of finding the two ends of the correlation function $g(r)$ being connected by a line of length M. One should not be disturbed by the fact that there are multiple ways of decomposing the same bond configuration for G into loops and a discontinuous line: any decomposition is equally representative. Also, at criticality, the short-distance structure of the theory is irrelevant for scaling arguments. One may thus simply consider a different statistical model such that self-intersections and overlaps are prohibited at the microscopic level (this leads to the unique decomposition of bond currents into loops). In the asymptotic limit $M \gg 1$ and $r \gg 1$, a typical length of lines contributing to $g(r)$ scales with the separation between the end points as a power law

$$\langle M \rangle_r \propto r^{D_H}, \tag{3.16}$$

where D_H is the Hausdorff dimension of critical loops. Assuming further a power-law decay of the probability distribution for values of $M \gg \langle M \rangle_r$, one arrives at the following generic form:

$$P(r, M) = M^{-\rho} f(r^{D_H}/M) \quad \text{for} \quad r \gg 1, \ M \gg 1. \tag{3.17}$$

By definition of $P(r, M)$, its integral over M has to reproduce the correlation function $g(r)$. Thus

$$g(r) \propto \int dM M^{-\rho} f(r^{D_H}/M) \propto r^{(1-\rho)D_H} \int dx f(x), \tag{3.18}$$

leading to

$$\rho = 1 + \frac{d - 2 + \eta}{D_H}. \tag{3.19}$$

The other important result for $P(r = 0, M)$ follows directly from the Josephson relation $\Lambda_s \propto L^{2-d}$, meaning that there is about one loop of size L in a volume L^d. One way of arriving at this conclusion is by employing the winding number formula Equation (3.15) and noticing that only loops of size L can make nonzero winding numbers. The same outcome follows from scale invariance at criticality stating that, under coarse graining up to length scale L, one should observe the same structure of loop configurations after defining L as the unit of length. The probability of hitting a loop of length M when selecting a lattice site in the box of linear size M^{1/D_H} at random is proportional to $M/M^{d/D_H}$; an extra factor of M is necessary to account for equivalence of all points on the loop. It can also be expressed as $MP(0, M)$, since any site on the loop can be considered as the end point of the correlation function. It is a mistake, however, to think that

$$P(0, M) \propto M^{-d/D_H} \tag{3.20}$$

can be used to establish an immediate relation between η and D_H from Equation (3.19) by assuming that $f(0) = const.$ in (3.17). Finite $P(0, M)$ is perfectly consistent with the function $f(x)$ in Equation (3.17) approaching zero as x^{θ/D_H} with $\theta > 0$, or, equivalently, with $P(r) \propto r^\theta$, which is increasing for $1 \ll r \ll M^{1/D_H}$ starting from a nonzero constant at short distances. This is because, for r of the order of unity, we are no longer in the asymptotic limit and Equation (3.17) is not valid. In other words,

$$P(0,M) \propto M^{-\rho-\theta/D_H},\qquad(3.21)$$

with the following relation between η, D_H, and θ

$$D_H = 2 - \eta - \theta.\qquad(3.22)$$

Numerical simulations for the XY universality class predict that $D_H = 1.7655(20)$ and $\theta = 0.1965(20)$ [11]. In Section 8.5, we will discuss further connections between the properties of loop configurations and phase correlations in the superfluid system.

3.3   J-Current Models

Formally, Equation (3.13) with arbitrary energy functional E_{loop} describes a generic set of loop, or j-current, models, all in the same universality class as long as the superfluid transition remains continuous and couplings between the currents remain short range. One choice for the energy functional is motivated by the Gaussian form of $F(j) = e^{-Uj^2}$:

$$E_{\text{loop}}[\{j_b\}] = U \sum_b j_b^2,\qquad(3.23)$$

which is also known as the Villain model [12,13]. This model has remarkable properties (to be derived in the next section) because it allows an exact mapping to the dual representation that works directly with the vortex degrees of freedom. Furthermore, the same mapping can be used to establish duality between charged and neutral superfluids (this is done in Chapter 5) and is relevant for the discussion of properties of two-component systems (see Chapter 6). In both cases, one finds that the transition point is in the XY universality class.

J-current models are extremely convenient for numerical studies. Aiming at universal properties of a continuous phase transition, it makes sense to consider a model with the smallest configuration space by demanding that $|j_b| \le 1$ and using $F(1)/F(0) = e^{-U}$ as a control parameter. There are several efficient Monte Carlo methods for simulating the superfluid transition. Cluster algorithms (see Reference [14]) work directly with the original spin variables. On the other hand, worm algorithm [15] works directly with the loop representation for $G(i-j)$ and thus is a method of choice for simulations of j-current models. It is as efficient as cluster algorithms and has several additional advantages such as (a) simple implementation of local updates*—just one (!) update for all models specified in this

---

* In Monte Carlo simulations, an "update" is a procedure of modifying the configuration space parameters. For j-current models, the configuration space is based on the collection of $\{j_b\}$ numbers and locations of the end points, i and j, of the correlation function.

section—(b) winding number estimator for $\Lambda_s^{(\varphi)}$, and (c) automatic computation of the correlation function G. The quantum version of the worm algorithm will be described in Section 8.7. A complete description of the classical worm algorithm for the loop model (3.13) is as follows (note the special role played by the end points **i** and **j** of the correlation function):

- If **i** = **j**, select the new value of **i** = **j** out of the L^d lattice sites at random; otherwise skip this step. Select any of the bonds attached to the site **j** at random and propose to move **j** along this bond to the nearest neighbor site **j**'. Moving in the positive/negative axis direction will increase/decrease the bond current $j_b' = j_b \pm 1$. Accept the proposed move with probability $F(j_b')/F(j_b)$. The distribution function of relative distances between the end points **i** and **j** is proportional to $G(\mathbf{i} - \mathbf{j})$.

3.4   Dual Descriptions of Normal and Superfluid Phases

Consider a generalized 3D Villain model with nonlocal current–current interactions:

$$E_j = \sum_{nm} U_m\, \mathbf{j_n} \cdot \mathbf{j_{n+m}}. \tag{3.24}$$

It turns out that after a number of identity transformations, the partition function of this model can be mapped onto the partition function of a dual model that has exactly the same structure of the energy functional as (3.24) in terms of divergence-free integer currents $\mathbf{w_n}$ coupled by a dual interaction potential $U_m^{(d)}$:

$$Z = \sum_{\{\mathbf{j_n}\}}^{\nabla \cdot \mathbf{j} = 0} \exp\left\{ -\sum_{nm} U_m\, \mathbf{j_n} \cdot \mathbf{j_{n+m}} \right\} \longleftrightarrow \sum_{\{\mathbf{w_n}\}}^{\nabla \cdot \mathbf{w} = 0} \exp\left\{ -\pi^2 \sum_{nm} U_m^{(d)}\, \mathbf{w_n} \cdot \mathbf{w_{n+m}} \right\}. \tag{3.25}$$

The relation between the direct and dual potentials is most straightforward in Fourier space:

$$U_q^{(d)} = \frac{1}{Q_q U_q}, \qquad Q_q = \sum_\alpha 2\,(1 - \cos q_\alpha). \tag{3.26}$$

We immediately see that if the direct potential is short range, $U_{q \to 0} \to C$, then the dual potential is of the long-range Coulomb type, $U_{q \to 0}^{(d)} \to (1/C)q^{-2}$. This leads to the following coarse-grained (long-wave) expression for contribution of large $\mathbf{w}$-loops to the energy functional

$$E_w = \pi^2 \sum_{nm} U_m^{(d)}\, \mathbf{w_n} \cdot \mathbf{w_{n+m}} \qquad \longrightarrow \qquad \frac{\pi}{4C} \sum_{nm} \frac{\mathbf{w_n} \cdot \mathbf{w_{n+m}}}{|\mathbf{m}|}. \tag{3.27}$$

One cannot miss the one-to-one correspondence between this expression and the vortex energy functional, Equations (1.119) and (1.131). It proves that the superfluid–normal transition, if viewed from the ordered phase, is driven by pro-liferation of large vortex excitations, and the dual variables $\mathbf{w}$ should be identi-fied with the vortex lines (in the 2D case, $\mathbf{w}$ variables are reduced to point vortex charges). The $\mathbf{j} \to \mathbf{w}$ mapping establishes the dual description of the XY transitions and introduces the so-called *inverted XY* universality class (it has the same crit-ical exponents and inverse universal amplitude ratios as the standard XY point, except the anomalous dimension).

Apart from explicitly revealing the nature of criticality of the superfluid-to-normal transition, the duality mapping proves crucial for establishing a direct link between critical properties of charged and neutral systems: in Chapter 5, we will employ the $\mathbf{j} \to \mathbf{w}$ mapping to show that the superconducting transition in a charged system also belongs to the inverted XY universality class.

In a broader context of dualities in 3D j-current models, we observe that the model with the potential $U_{\mathbf{q}} = U/\sqrt{Q_{\mathbf{q}}}$ (it corresponds to $1/r^2$ long-range interac-tions) is self-dual under the $\mathbf{j} \leftrightarrow \mathbf{w}$ mapping, with the critical point $U_c = \pi$. Here it is also worth noting that the $1/r^2$ interaction in 3D is rather special. A simple scal-ing analysis shows that this interaction is a marginal operator. In general, under duality mapping, $U_{\mathbf{q}}$ and $U_{\mathbf{q}}^{(d)}$ both belong to the $1/r^2$ class but have different strengths of the $1/r^2$ term; that is, the criticality is not self-dual.

Let us now outline the derivation leading to (3.25). One starts with going one step back and implementing the $\nabla \cdot \mathbf{j} = 0$ constraint using site phase integrals [see (3.11)]:

$$Z = \int \mathcal{D}\varphi \sum_{\{j_n\}} \exp\left\{ -\sum_{nm} U_m \mathbf{j_n} \cdot \mathbf{j}_{n+m} + i \sum_n \nabla \varphi_n \cdot \mathbf{j_n} \right\}, \qquad (3.28)$$

where $(\nabla\varphi)_{n\alpha} = \varphi_{n+\hat{\alpha}} - \varphi_n$ is the standard continuous-type notation for lattice gradients (in Fourier space, it takes the form $(\nabla\varphi)_{\mathbf{q}} = \mathbf{s_q}\varphi_{\mathbf{q}}$, with $s_{\mathbf{q}\alpha} = 1 - e^{iq_\alpha}$). Next, the summation over integer bond variables j_b is replaced with integration over j_b in combination with summation over integer variables m_b and integration over phases θ_b as follows:

$$\sum_{j_b} \cdots \equiv \int_{-\infty}^{\infty} dj_b \sum_{m_b} \delta(j_b - m_b)\cdots \equiv \int_{-\infty}^{\infty} dj_b \sum_{m_b} \int_{-\infty}^{\infty} \frac{d\theta_b}{2\pi} e^{i\theta_b(j_b - m_b)}\cdots, \qquad (3.29)$$

At this junction, one changes the order of integrals/sums and performs the Gaussian integration over $\{j_b\}$ variables first (Gaussian integrals decouple in the Fourier representation), followed by summation over $\{m_b\}$ and integration over $\{\theta_b\}$ variables. The result is

$$\sum_{\{j_n\}} e^{-\sum_{nm} U_m \mathbf{j_n} \cdot \mathbf{j}_{n+m} + i\sum_n \nabla\varphi_n \cdot \mathbf{j_n}} = f \sum_{\{v_n\}} e^{-(1/4)\sum_{nm} U_m^{(v)} (\nabla\varphi - 2\pi\mathbf{v})_n \cdot (\nabla\varphi - 2\pi\mathbf{v})_{n+m}}, \qquad (3.30)$$

where $U_q^{(v)} = 1/U_q$, $\mathbf{v_n}$ is the dual integer current field, and f is an irrelevant, for our purposes, prefactor originating from the Gaussian integration over the original bond variables.

Problem 3.2 *Perform all steps leading to Equation (3.30).*

Now we conveniently parameterize the field $\mathbf{v}$ in terms of the integer divergence-free vorticity field $\mathbf{w} = \nabla \times \mathbf{v}$ [in lattice notations, $w_{n\alpha} = \sum_{\nu\mu} \epsilon_{\alpha\nu\mu}(v_{n\mu} - v_{n+\hat{\nu}\mu})$] and integer scalar field $N \equiv N_n$. To this end, we observe that if two fields, $\mathbf{v}$ and $\mathbf{v}'$, share one and the same vorticity field, then their difference has identically zero curl and can be represented as a lattice gradient of an integer scalar field N. Hence, $\mathbf{v}' = \mathbf{v} + \nabla N$. Specifically, for any given configuration of $\mathbf{w}$, we pick up one reference configuration of $\mathbf{v}$ and obtain other $\mathbf{v}$-configurations with the same $\mathbf{w}$ by adding ∇N and summing over all N-configurations.[*]

Problem 3.3 *Show that if parameters $A_{i\alpha}$ form a smooth (i.e., slowly varying) function of lattice index $\mathbf{i}$, then the continuous version of*

$$\sum_{\alpha<\beta}\left[(A_{i+\hat{\alpha}\beta} - A_{i\beta}) - (A_{i+\hat{\beta}\alpha} - A_{i\alpha})\right]^2$$

is given by

$$a^2 \left[\nabla \times \mathbf{A}(\mathbf{r})\right]^2,$$

where a is the lattice constant.

Given the structure of the exponent in (3.30), the $2\pi N_n$ field is naturally combined with the φ_n field, amounting to the (convenient) extension of the integration domain for on-site phase variables from $-\infty$ to $+\infty$. Now the Gaussian integrals over the phase field φ can be performed explicitly in Fourier representation (the prefactor originating from Gaussian integrations can be omitted; when combined with f in Equation (3.30), it does not depend on the Hamiltonian parameters):

$$E_v = \pi^2 \sum_q U_q^{(v)} \left| \mathbf{v_q} - \frac{\mathbf{v_q} \cdot \mathbf{s_q}}{|\mathbf{s_q}|^2} \right|^2 \equiv \pi^2 \sum_q U_q^{(d)} |\mathbf{s_q} \times \mathbf{v_q}|^2. \qquad (3.31)$$

The derivation of the duality mapping is complete because, in Fourier representation, $\mathbf{w_q} = \mathbf{s_q} \times \mathbf{v_q}$, and Equation (3.31) immediately leads to the duality transformation described by Equations (3.25) and (3.26).

---

[*] The particular choice of the reference $\mathbf{v}$-configuration corresponding to a given $\mathbf{w}$ proves irrelevant, since the statistical weight of the configuration depends on $\mathbf{w}$ only.

In 2D systems, the **w** field has only one nonzero component $w_3 \equiv w$ that is identified with the point vortex charge. In this case, the energy functional takes the form

$$\pi^2 \sum_{\mathbf{nm}} U_{\mathbf{m}}^{(d)} w_{\mathbf{n}} w_{\mathbf{n+m}}, \tag{3.32}$$

with $U^{(d)}$ displaying logarithmic behavior at large distances, as expected. Using special properties of the so-called sine–Gordon model (this discussion goes beyond the scope of our book), one can show that the 2D XY critical point is self-dual (see Reference [16]).

Even if we knew nothing about the nature of the superfluid state and the crucial role played by vortex excitations, the duality mapping all by itself would be instructive in clarifying it (cf. Reference [4]).

3.5   High-Temperature Expansion for the $|\psi|^4$ Model

Since the canonical approach to superfluid systems is based on the Ginzburg–Landau approach and statistical properties of the minimal classical-field Hamiltonian, we will briefly discuss below the oriented-loop representation for the Hamiltonian (3.2). This mapping is slightly more general than that for the XY model in the following respect: When individual loops are allowed to overlap and intersect, the Gibbs factor for them depends, in general, on the total number of currents entering each site and on the current numbers in the positive and negative directions along the bonds—Equations (3.13) and (3.23) depend only on the algebraic sum of all bond currents. The mapping starts with expanding exponentials $e^{(t/T)\psi_i^*\psi_{i'}}$ and $e^{(t/T)\psi_i\psi_{i'}^*}$ into Taylor series for each bond:

$$Z = \prod_i \int d\psi_i \, e^{-(U/2T)|\psi_i|^4 + (\mu/T)|\psi_i|^2} \prod_b \left\{ \sum_{j_{b+}=0}^{\infty} \frac{[(t/T)\psi_{i'}^*\psi_i]^{j_{b+}}}{(j_{b+})!} \sum_{j_{b-}=0}^{\infty} \frac{[(t/T)\psi_{i'}\psi_i^*]^{j_{b-}}}{(j_{b-})!} \right\}. \tag{3.33}$$

As before, we will call j_{b+} and j_{b-} bond currents propagating in the positive and negative directions along the bond b.

For each term in the multidimensional sum over the set $\{j_{b+}, j_{b-}\}$, one has to evaluate integrals

$$2^{-1} \int_0^{\infty} d|\psi_i|^2 \int_0^{2\pi} d\varphi_i \, e^{-(U/2T)|\psi_i|^4 + (\mu/T)|\psi_i|^2} (\psi_i^*)^{N_i} (\psi_i)^{M_i} \tag{3.34}$$

for every lattice site, where N_i and M_i represent the total current entering and exiting the site i, respectively. Similarly to the XY model, integrals over the phase are zero unless the number of incoming and outgoing currents are equal, $N_i = M_i$, leading to the familiar constraint that allowed current configurations must be divergence-free (i.e., must consist of closed loops). The remaining integral over the modulus squared defines on-site interactions between the intersecting and overlapping loops:

$$Q(N_i) = \pi \int_0^\infty dx\, e^{-(U/2T)x^2 + (\mu/T)x}\, x^{N_i}.$$

(3.35)

The final expression for the partition function reads

$$Z = \sum_{\{j_{b+}, j_{b-}\}}^{\nabla(j_{b+} - j_{b-}) = 0} \prod_i Q(N_i) \prod_b \frac{(t/T)^{j_{b+} + j_{b-}}}{(j_{b+})!(j_{b-})!}.$$

(3.36)

Though microscopically this is quite different from Equations (3.13) and (3.23), the most important aspect is the configuration space of oriented closed loop characteristic of the normal-to-superfluid transition. Of course, microscopic details may change the nature of the transition from second to first order, in which case the universality considerations no longer apply.

In exact analogy with our discussion of the XY model [see Equation (3.4)], the correlation function $g(\mathbf{i} - \mathbf{j}) = \langle \psi_i \psi_j^* \rangle \equiv G(\mathbf{i} - \mathbf{j})/Z$ generates the configuration space that contains closed loops superimposed on the disconnected line, which originates at site $\mathbf{i}$ and terminates at site $\mathbf{j}$ with the same configuration weight as in Equation (3.36). Finally, long-range order in the correlation function g implies that typical Z-configurations contain macroscopically large loops.

We conclude the discussion of the loop representation with derivation of the Josephson current relation (1.207) in Chapter 1 starting from the general formula (3.14) for the system with the phase twist φ_0. We assume that the geometry is such that loops with nonzero W_x numbers necessarily cross the weak link. For extremely weak links, the probability of having a nonzero winding number W_x is very small, meaning that the partition function sum over W in (3.14) is dominated by $W_x = 0$ terms with the leading correction coming from $W_x = \pm 1$ contributions; that is,

$$Z = \sum_{W = -\infty}^{\infty} Z_W e^{iW\varphi_0} \approx Z_0 + 2Z_1 \cos \varphi_0 \quad \Rightarrow \quad F = -T \ln Z \approx F_0 + 2\frac{Z_1}{Z_0} \cos \varphi_0.$$

(3.37)

The cosine term in the free energy leads to the sinusoidal dependence of the Josephson current in response to the twist phase through relation (1.202).

The derivation of (1.207) is based entirely on the suppressed probability of having nonzero winding numbers in the $\hat{x}$-direction. This may happen for a number of reasons from (a) the most straightforward situation when the link is created because there is a large potential energy barrier across the channel to (b) a moderate geometrical constriction on approach to the phase transition point out of the superfluid phase.

References

1. J. Zinn-Justin, *Quantum Field Theory and Critical Phenomena*, Oxford University Press, Oxford, U.K., The International Series of Monographs on Physics, Vol. 113 (2003).

2. M. Campostrini, M. Hasenbusch, A. Pelissetto, P. Rossi, and E. Vicari, Critical behavior of the three-dimensional XY universality class, *Phys. Rev. B* **63**, 214503 (2001).

3. M. Campostrini, M. Hasenbusch, A. Pelissetto, and E. Vicari, Theoretical estimates of the critical exponents of the superfluid transition in ^{4}He by lattice methods, *Phys. Rev. B* **74**, 144506 (2006).

4. G. A. Williams, Vortex-loop phase transitions in liquid helium, cosmic strings, and high-Tc superconductors, *Phys. Rev. Lett.* **82**, 1201 (1999).

5. E. Burovski, J. Machta, N. Prokofev, and B. Svistunov, High-precision measurement of the thermal exponent for the three-dimensional XY universality class, *Phys. Rev. B* **74**, 132502 (2006).

6. J. A. Lipa, J. A. Nissen, D. A. Stricker, D. R. Swanson, and T. C. P. Chui, Specific heat of liquid helium in zero gravity very near the lambda point, *Phys. Rev. B* **68**, 174518 (2003).

7. B. D. Josephson, Relation between the superfluid density and order parameter for superfluid He near T_c, *Phys. Lett.* **21**, 608 (1966).

8. R. P. Feynman, The λ-transition in liquid helium, *Phys. Rev.* **90**, 1116 (1953); Atomic theory of the λ-transition in helium, *Phys. Rev.* **91**, 1291 (1953).

9. K. Binder, Critical properties from Monte Carlo coarse graining and renormalization, *Phys. Rev. Lett.* **47**, 693 (1981).

10. R. Feynman, Space-time approach to non-relativistic quantum mechanics, *Rev. Mod. Phys.* **20**, 367 (1948).

11. N. Prokof'ev and B. Svistunov, Comment on "Hausdorff dimension of critical fluctuations in Abelian Gauge Theories", *Phys. Rev. Lett.* **96**, 219701 (2006).

12. J. Villain, Theory of one- and two-dimensional magnets with an easy magnetization plane. II. The planar, classical, two-dimensional magnet, *J. Phys.* **36**, 581 (1975).

13. M. Wallin, E. S. Sørensen, S. M. Girvin, and A. P. Young, Superconductor-insulator transition in two-dimensional dirty boson systems, *Phys. Rev.* **49**, 12115 (1994).

14. M. E. J. Newman and G. T. Barkema, *Monte Carlo Methods in Statistical Physics*, Clarendon Press, Oxford, U.K. (1999).

15. N. Prokofiev and B. Svistunov, Worm algorithms for classical statistical models, *Phys. Rev. Lett.* **87**, 160601 (2001).

16. P. Lecheminant, A. O. Gogolin, and A. A. Nersesyan, Criticality in self-dual sine-Gordon models, *Nucl. Phys. B* **639**, 502 (2002).

CHAPTER 4

Berezinskii–Kosterlitz–Thouless Phase Transition

The Berezinskii–Kosterlitz–Thouless (BKT) transition [1–3] is the finite-temperature transition between superfluid and normal states in two dimensions.* A crucial feature of the BKT transition is that it is driven by the statistics of an *asymptotically dilute* gas of large-size vortex pairs: Up to the very critical point from the superfluid side, the distribution of vortex–antivortex pairs is such that the probability of finding at least one pair with the distance between the two vortices on the order of the linear system size is macroscopically small.† The asymptotically vanishing concentration of the vortex pairs renders BKT transition distinctively different from other second-order superfluid–normal phase transitions, the most specific feature being the fact that the superfluid density remains finite up to the transition point and then jumps to zero (in the thermodynamic limit). This jump in the superfluid density should not be interpreted as a first-order phase transition, since it comes exclusively from an infinitesimal concentration of free vortices nullifying the static response to the twisted boundary condition (or gauge phase in annulus geometry). The vanishing density of the vortex–antivortex pairs implies that other thermodynamic quantities remain continuous (including all derivatives, as we will see later) across the BKT transition point.

As revealed by Kosterlitz and Thouless [2,3], the dilute gas of vortex pairs leads to a relatively simple and accurate renormalization-group (RG) description of the phase transition in terms of the statistics of large vortex pairs. The Kosterlitz–Thouless (KT) theory predicts rather specific behavior of the superfluid density, n_s, and correlation radius, ξ, in the vicinity of the critical point T_c:

$$n_s(T) - n_s(T_c) \propto \sqrt{T_c - T} \qquad (T \leq T_c), \tag{4.1}$$

$$\ln \xi(T) \propto |T_c - T|^{-1/2}. \tag{4.2}$$

The square-root cusp in $n_s(T)$ reflects a nontrivial fact that the dilute gas of large vortex pairs plays a significant role in renormalizing the macroscopic superfluid density. The exponential divergence of the correlation radius at the critical point implies that all standard thermodynamic quantities are smooth functions across the phase transition.

---

* It is also important that continuous quantum phase transitions in 1D systems, if associated with suppression of superfluidity, normally belong to the same or very similar universality class as the BKT transition (see Chapter 9).

† In the BKT-type quantum phase transitions in 1D systems, the analogs of vortex pairs are instanton pairs. An effective action for a 1D superfluid can be written in terms of the phase field in (1 + 1) dimensions; the instantons are the vortices in this field (see Chapter 9).

At the critical point of the BKT transition, there takes place the universal Nelson–Kosterlitz (NK) [4] relation between $n_s(T_c)$ and T_c:

$$\frac{\gamma n_s(T_c)}{T_c} = \frac{2}{\pi}. \tag{4.3}$$

Historically, the NK relation was derived from the KT theory. However, this relation is a simple consequence of the vanishingly small concentration of the largest vortex pairs at the critical point. It does not depend on the details of statistics of vortex pairs with sizes significantly smaller than the system size. The only relevant question is whether it is thermodynamically favorable to have at least one free vortex per system—and the answer is determined by the *sign* of the free energy of a single vortex. As long as n_s is finite, the free energy scales as $\propto \ln L$ with the system size L. The two competing terms, both $\propto \ln L$, are the effective energy (e.g., the free energy of the system with the vortex core fixed at a certain position) and the entropy associated with selecting different positions for the vortex core. The BKT transition takes place when the positional entropy of a single free vortex compensates exactly the effective energy—the NK relation (4.3) being a necessary and sufficient condition for this compensation to happen.[*]

In what follows, we give a detailed account of the KT theory. Along with deriving relations (4.1) and (4.2), the theory establishes an important result that the BKT transition is characterized by logarithmically slow renormalization of n_s to its macroscopic value satisfying the NK relation. In view of this circumstance, the finite-size effects are very important for the BKT transition. Speaking practically, the BKT transition in a realistic system looks like a moderately sharp crossover, for which the logarithm of the system size behaves as a control parameter defining the function $n_s(T - T_c, \ln L)$.

4.1   Statistics of Vortices at Large Scales

4.1.1   Effective Action for Vortices

We work in the canonical—with respect to superflow—ensemble. We use expressions (2.54) and (2.55) for the partition function, in which the vortex-free field φ is replaced with the field Φ that can have both (large-scale) vortices, as well as a winding number associated with a global superflow:

$$Z = \int e^{-A[\Phi]} \mathcal{D}\Phi, \tag{4.4}$$

$$A[\Phi] = K(l_0) \int (\nabla \Phi)^2 \, d^d r, \tag{4.5}$$

$$K(l_0) = \frac{\gamma n_s(l_0)}{2T}, \tag{4.6}$$

---

[*] Similarly, the energy–entropy argument applies to BKT-type quantum phase transitions in 1D, leading to the NK-type relations for the critical value of n_s that is made dimensionless using other relevant quantities (see Chapter 9).

where $n_s(l_0)$ is the "bare" superfluid number density that is obtained by averaging out microscopic fluctuations up to some mesoscopic scale l_0. The structure of the field $\Phi(\mathbf{r})$ is as follows:

$$\Phi(\mathbf{r}) = \mathbf{k}_0 \cdot \mathbf{r} + \Phi_0(\mathbf{r}) + \varphi(\mathbf{r}). \tag{4.7}$$

It includes a nonzero global equilibrium superflow with velocity $\mathbf{v}_s = \gamma \mathbf{k}_0$ (parameterized using the vector $\mathbf{k}_0$) and large-scale (separated by a distance larger than l_0) vortices. The latter are taken into account by the field $\Phi_0(\mathbf{r})$ that satisfies the nonzero phase winding condition:

$$\oint_{C_j} \nabla \Phi_0(\mathbf{r}) \cdot d\mathbf{l} = 2\pi M_j, \tag{4.8}$$

where C_j is a contour enclosing only the jth vortex and M_j is an integer. Otherwise, $\Delta \Phi_0 = 0$ [see Equations (2.124) and (2.125)]. The field $\varphi(\mathbf{r})$ is the regular, single-valued part of the phase field.

The structure of the field Φ_0 corresponds to the quantized-vortex pattern in an incompressible fluid and is fixed by the positions of vortices (see Section 1.2.4). With the condition $\Delta \Phi_0 = 0$, the cross-term between the fields Φ_0 and φ vanishes:

$$\int (\nabla \varphi \cdot \nabla \Phi_0) d^d r = -\int (\varphi \cdot \Delta \Phi_0) d^d r = 0. \tag{4.9}$$

Noting also that $\int \mathbf{k}_0 \cdot \nabla \varphi \, d^d r = 0$ since φ is a regular field, we arrive at a peculiar, if not counterintuitive, result:

$$A[\Phi] = \text{const.} + A_{\text{reg}}[\varphi] + A_{\text{def}}[\Phi_0], \tag{4.10}$$

$$A_{\text{reg}}[\varphi] = K(l_0) \int (\nabla \varphi)^2 d^d r, \tag{4.11}$$

$$A_{\text{def}}[\Phi_0] = K(l_0) \int \left[(\nabla \Phi_0)^2 + 2\mathbf{k}_0 \cdot \nabla \Phi_0 \right] d^d r. \tag{4.12}$$

Equations (4.10) through (4.12) imply that statistics of the field φ and that of the vortices[*] are *independent*.[†] This fact appears surprising because it creates an illusion that the field φ describes phonon modes, which is indeed the case in the absence of vortices. In the presence of vortices, however, the field φ does not have a distinct physical meaning, so that the decomposition (4.7) with $\Delta \Phi_0 = 0$ should be viewed mostly as a mathematical trick rather than physical circumstance. We will see that while being really useful for addressing the statistics of vortices *locally* in

---

[*] Incidentally, note that $\Delta \Phi_0 = 0$ allows one to explicitly express A_{def} as a function of the positions of the vortices [see Equations (1.141) and (1.142)].

[†] In addition to the separability of the action, statistical independence of variables also requires that the Jacobian of the transformation from Φ to φ and Φ_0 also does not contain cross-terms leading to the coupling φ and Φ_0. In our case, φ is related to Φ by a simple shift, Equation (4.7), implying that $\mathcal{D}\varphi = \mathcal{D}\Phi$.

the space of scales of distance, the decomposition trick becomes rather inconvenient if applied globally to all vortices.

Since the field φ is decoupled from $\mathbf{k}_0$, its fluctuations do not contribute to the dependence of Z on $\mathbf{k}_0$ and, correspondingly, to corrections to the superfluid density. The field of vortices does couple to $\mathbf{k}_0$, and this results, in accordance with Equation (2.32), in the renormalization of the superfluid density. Hence, as long as l_0 in (4.6) is large enough to guarantee the mesoscopic description (4.4) through (4.6), the dependence of $n_s(l_0)$ on l_0 is solely due to the contributions from vortices.

4.1.2 Nelson–Kosterlitz Relation

Now we are in a position to derive the Nelson–Kosterlitz relation (4.3). We assume that the critical point is the point at which the formation of at least one isolated vortex becomes favorable, while n_s remains constant at $T_c - 0$. That is, T_c is the point at which the single-vortex partition function

$$Z_{\text{def}} \propto \int e^{-A_{\text{def}}[\Phi_0]} \, d^2 r_0, \tag{4.13}$$

where $\mathbf{r}_0$ is the position of the vortex, does not vanish in the thermodynamic limit. With only one vortex in a system,

$$\int (\nabla \Phi_0)^2 \, d^2 r = 2\pi \ln (L/l_0) + o\,(\ln L). \tag{4.14}$$

The mesoscopic cutoff parameter $l_0 \ll L$ is supposed to be large enough to justify the single-vortex consideration by neglecting vanishingly small contributions from the vortex pairs of the size larger than l_0. With this reservation about l_0, we can set $K(l_0) \approx \gamma n_s / 2T$, where n_s is the macroscopic superfluid density. Estimate (4.14) for the action translates into the following estimate for the partition function:

$$Z_{\text{def}} \propto 1/L^{2\pi K - 2}, \tag{4.15}$$

consistent with the abrupt change of n_s at the critical point. As long as the parameter K is larger than the critical value

$$K_c = \pi^{-1}, \tag{4.16}$$

the single-vortex partition function remains macroscopically small. At $K > K_c$, the macroscopic scaling of Z_{def} radically changes. Condition (4.16) is nothing but the Nelson–Kosterlitz relation (4.3).

4.1.3 Dilute Gas of Vortex Pairs

Let us select the mesoscopic cutoff parameter l_0 large enough to guarantee that the system of vortex pairs of size $R > l_0$ is dilute in the following sense (strong criterion): For any pair of size R, the typical distance to the closest pair of *any* size is much larger than R. The smallest possible size of the vortex pair in this picture is $\sim l_0$, by definition. In what follows, we will see that the KT RG treatment allows one to relax this criterion by considering only pairs with the same order-of-magnitude size R; for the purposes of the present section, we do not need this generalization.

The dilute gas criterion ensures statistical independence of pairs—modulo permutation symmetry. By standard statistical mechanics, the latter is automatically taken into account by computing the number of pairs through the free energy of one pair:

$$F_{\text{pair}} = -T \ln Z_{\text{pair}}, \tag{4.17}$$

$$Z_{\text{pair}} = \int e^{-A_{\text{pair}}} d^2 R. \tag{4.18}$$

Here $\mathbf{R} = \mathbf{r}_+ - \mathbf{r}_-$ is the vector of distance between the antivortex located at $\mathbf{r}_-$ and vortex located at $\mathbf{r}_+$, and

$$A_{\text{pair}} = 4\pi K \left[\ln R - \mathbf{k}_0 \cdot (\hat{z} \times \mathbf{R}) \right] \tag{4.19}$$

is the effective energy of the $\mathbf{R}$-pair; its explicit form is readily found from (4.12) using (1.141) and (1.142).

Problem 4.1 *Given partition function $Z_N = \exp\{-N F_1/T + N \ln(Ve/N)\}$, find the most probable value $N = N_{\text{opt}}$ and show that at this point, $Z_{N_{\text{opt}}} = \exp\{N_{\text{opt}}\}$.*

Define

$$W_{\mathbf{k}_0}(\mathbf{R}) \propto e^{-A_{\text{pair}}} = \frac{e^{4\pi K \, \mathbf{k}_0 \cdot (\hat{z} \times \mathbf{R})}}{R^{4\pi K}} \tag{4.20}$$

as the distribution function of a vortex pair over $\mathbf{R}$ normalized to the concentration of pairs. In this case, the differential $W_{\mathbf{k}_0}(\mathbf{R}) 2\pi R\, dR$ is equal to the concentration of pairs with $R \in [R, R + dR]$ and the integral

$$n_{\text{p}}(l_0) = \int_{R > l_0} W_{\mathbf{k}_0}(\mathbf{R})\, d^2 R \tag{4.21}$$

yields the concentration of vortex pairs with $R > l_0$.

The total free energy, $F(L)$, of a square system $L \times L$ is then written as

$$F(L) = (L/l_0)^2 F(l_0) - TV \int_{l_0 \le R \le L} W_{\mathbf{k}_0}(\mathbf{R})\, d^2 R. \tag{4.22}$$

The first term in the right-hand side of (4.22) is the free energy in the absence of vortex pairs of sizes $R > l_0$. We see that the integral in the second term depends on orientation between $\mathbf{R}$ and $\mathbf{k}_0$, meaning that the superflow *polarizes* vortex pairs.

This polarization, in turn, suppresses the net current density, thereby renormalizing the macroscopic superfluid density. In accordance with the general formula (2.32), the superfluid density can be obtained by doubly differentiating the free energy with respect to $\mathbf{k}_0$:

$$\gamma n_s = \frac{1}{2V} \Delta_{k_0} F. \tag{4.23}$$

By applying this rule to Equation (4.22), we obtain

$$\gamma n_s(L) = \gamma n_s(l_0) - 16\pi^3 T K^2(l_0) \int_{l_0}^{L} R^3 W_0(R) dR, \tag{4.24}$$

where

$$W_0(R) \equiv W_{\mathbf{k}_0=0}(\mathbf{R}) = \left(2K - \pi^{-1}\right) \frac{n_p(l_0)}{l_0^2} \left(\frac{l_0}{R}\right)^{4\pi K}. \tag{4.25}$$

To justify (and quantify) the dilute vortex-pair gas approximation, we estimate the radius, $R_{typ}(L)$, of the largest pair that can be found in the system with probability of order unity. We require that, with macroscopic accuracy, $R_{typ}(L) \ll L$. With the normalization (4.21), from the definition of $R_{typ}(L)$, we have $W_0(R_{max}) R_{max}^2 L^2 \sim 1$, which yields the estimate

$$\frac{R_{max}(L)}{L} \sim \left[n_p(l_0) l_0^2\right]^{\frac{1}{4\pi K - 2}} \left(\frac{l_0}{L}\right)^{\frac{2\pi K - 2}{2\pi K - 1}}. \tag{4.26}$$

This brings us back to the Nelson–Kosterlitz critical condition (4.16). Indeed, as long as $K > K_c = \pi^{-1}$, the value of R_{max} is smaller than L by a macroscopic factor and the theory is consistent. In the thermodynamic limit of $L \to \infty$, the theory becomes inconsistent for $K < K_c$. The $K = K_c$ point is marginal and requires a special RG analysis, which we will present in the next section. Its final outcome is that the ratio $R_{max}(L)/L$ still vanishes in the thermodynamic limit.

Apart from the dilute gas condition, we also need to justify the use of $n_s(l_0)$ [or $K(l_0)$] on the right-hand side of Equation (4.24). This is possible provided the difference between $n_s(L)$ and $n_s(l_0)$ caused by the polarization of pairs with sizes $R > l_0$ is small. This condition turns out to be more restrictive than $R_{max}/L \ll 1$ since, in accordance with (4.24), the contributions to n_s from large pairs contain an extra factor of R^2, as compared to their contributions to n_p. Let us first rewrite (4.24) identically by replacing n_s with the dimensionless quantity K:

$$K(L) = K(l_0) - 8\pi^3 K^2(l_0) \int_{l_0}^{L} R^3 W_0(R) dR. \tag{4.27}$$

With the explicit expression for W_0, Equation (4.25), we arrive at the second condition in the form (we substitute $L \to \infty$)

$$\frac{K(l_0) - K(\infty)}{K(l_0)} \sim \frac{n_p(l_0)}{l_0^2} \int\limits_{l_0}^{\infty} \left(\frac{l_0}{R}\right)^{4\pi K(l_0)} R^3 \, dR \sim \frac{n_p(l_0)l_0^2}{\pi K(l_0) - 1} \ll 1. \tag{4.28}$$

As expected, on approach to the critical value, $K(l_0) \to K_c = \pi^{-1}$, the condition (4.28) becomes progressively more restrictive than the condition of having a dilute vortex-pair gas at the mesoscopic length scale: $n_p(l_0)l_0^2 \ll 1$. Thus, to justify our calculation on approach to the critical point, apart from the necessary condition $K(l_0) > K_c$, we must keep *increasing* the mesoscopic scale l_0. The physical picture behind this behavior is as follows: While the macroscopic critical value of K is fixed by the Nelson–Kosterlitz relation, the mesoscopic values $K(l_0)$ turn out to be significantly larger than K_c. The logarithmic divergence of the integral (4.27) at $K(l_0) \to K_c$ means that the renormalization of K from $K(l_0)$ to K_c comes from exponentially large scales of distance, implying a peculiar dependence of $K(\infty)$ on $T - T_c$ and calling for the RG description of the flow of $K(l_0)$ with l_0. The corresponding theory, developed by Kosterlitz and Thouless, is presented in the next section.

4.2 Kosterlitz–Thouless Renormalization-Group Theory

4.2.1 Flow Equations

The divergence of the integral in (4.27) at $K \to K_c$ is logarithmic. This allows one to employ the RG approach, which is based on the idea of utilizing (4.27) *iteratively*. In the RG treatment, l_0 becomes a running cutoff distance; to emphasize this fact, in the final RG relations, we use the letter l instead of l_0. Rather than computing corrections to superfluid density from pairs of arbitrarily large sizes, we confine ourselves to accounting for the "next" scale of distance $l' > ul$, where $u > 1$, such that the renormalization of K remains small. The accuracy of this approach is controlled by the small parameter

$$\frac{K(l) - K(l')}{K(l)} \ll 1. \tag{4.29}$$

As we will see shortly, this parameter becomes progressively smaller at $l \to \infty$ even at the critical point, rendering the RG treatment asymptotically exact.

Given $K(l)$ and $W_0(R \sim l)$ for a certain scale l, we utilize (4.27) and (4.25) to obtain $K(l')$ and $W_0(R \sim l')$ at a larger scale l'. The procedure is then repeated iteratively. In practice, this RG theory has the form of two coupled differential equations that describe the evolution (referred to as "flow") of $K(l)$ and $W_0(l)$ with growing the scale of distance

$$\lambda = \ln(l/l_0), \tag{4.30}$$

where l_0 is the microscopic cutoff introduced previously. The first equation is readily obtained by simply differentiating Equation (4.27) written as

$$K(ul) - K(l) = -8\pi^3 K^2(l) \int_l^{ul} R^3 W_0(R,l) \, dR, \tag{4.31}$$

with respect to λ (or u) and then (but not before!) setting $u = 1$. The replacement $u = 1$ simply means that l is the running distance. The other implicit assumption is that $K(l)$ changes so slowly that it can be pulled out of the integral over vortex pairs; that is, we rely on the condition (4.29) and slow change of K with scale of distance. The result is

$$\frac{df}{d\lambda} = -(g/2)f^2, \tag{4.32}$$

where we introduced two convenient dimensionless functions:

$$f(\lambda) = \pi K(l), \tag{4.33}$$

$$g(\lambda) = 16\pi^2 l^4 W_0(l). \tag{4.34}$$

The function f is self-explanatory. Apart from the relatively large numeric prefactor, the function g is the typical number of vortex pairs of the size $\sim l$ within a box $l \times l$. From the analysis presented in the previous section, we know that this number gets progressively smaller with $l \to \infty$. Note also that the condition

$$g \ll 1 \tag{4.35}$$

is crucial to guarantee the slow dependence of f (and thus K) on λ.

To get the second RG equation, we observe that formula (4.25) can be rewritten as

$$W_0(ul)(ul)^4 = W_0(l)l^4 \left(\frac{l}{ul}\right)^{4\pi K(l)-4} \qquad \text{or} \qquad g(ul) = g(l)u^{-4(f-1)}, \tag{4.36}$$

where $K(l)$ on the right-hand side is assumed to be nearly constant when going from l to l'. For the RG treatment, we differentiate (4.36) with respect to u and set $u = 1$, as before. This leads to the differential equation

$$\frac{d \ln g}{d\lambda} = -4(f - 1). \tag{4.37}$$

Further simplification of the flow equations comes from the observation that pronounced renormalization effects take place only in close vicinity of the critical point, that is, at $f \approx 1$. This allows one to set $f^2 = 1$ in the right-hand side of (4.32) and introduce yet another small variable

$$w = 2(f - 1) \ll 1 \tag{4.38}$$

to simplify the algebra. In terms of two small quantities, w and g, the KT RG theory reads

$$\frac{dg}{d\lambda} = -2wg, \tag{4.39}$$

$$\frac{dw}{d\lambda} = -g. \tag{4.40}$$

Equations (4.39) and (4.40) form a system of two first-order differential equations and have the general solution containing two free constants. Since the right-hand sides are independent of λ, one constant is trivial. It is associated with an arbitrary shift of λ, which, given the definition (4.30), corresponds to a particular choice of l_0. The second, and the only nontrivial, free constant must then control how close to the BKT point the system is.

The first integral of the system (4.39) and (4.40) is readily obtained by dividing (4.39) by (4.40). A straightforward integration then yields

$$w^2(\lambda) - g(\lambda) = C. \tag{4.41}$$

Here C is the free constant that depends on $T - T_c$. It is crucial that C should be an analytic function of $T - T_c$, since Equation (4.41) holds true at large but *finite* λ, where $w(\lambda)$ and $g(\lambda)$ cannot have singularities at $T - T_c$. At $T \le T_c$, we have $g(\infty) = 0$; that is,

$$w(\infty) = \sqrt{C(T)} \qquad (T \le T_c). \tag{4.42}$$

By the definition of w, this means that $C(T_c) = 0$. Then, taking into account the smoothness of the function $C(T)$ and the fact that we are working in close vicinity of the critical point, we write

$$C = \zeta(T_c - T) \qquad (|T - T_c| \ll T_c), \tag{4.43}$$

where ζ is some nonuniversal constant. Relations (4.42) and (4.43) then yield the law (4.1) for the temperature dependence of the macroscopic superfluid density on approach to the critical point.

To find the complete solution of the system (4.39) and (4.40), we use (4.41) to exclude function g from (4.40) and then find $w(\lambda)$ by integrating the differential equation

$$\frac{dw}{d\lambda} = C - w^2. \tag{4.44}$$

With known $w(\lambda)$, we then get $g(\lambda)$ from (4.41). At the critical point, $C = 0$, we obtain:

$$w(l) = \frac{1}{\ln(l/l_0)}, \qquad g(l) = \frac{1}{\ln^2(l/l_0)} \qquad (T = T_c). \tag{4.45}$$

We see that the renormalization of the superfluid density with the scale of distance is logarithmically slow, justifying the central assumption of the RG theory. We also see that the function $g(l)$—the parameter characterizing the dilute gas approximation for vortex pairs of size l—vanishes in the $l \to \infty$ limit. Hence, our treatment is self-consistent at $T \to T_c$.

When $C \neq 0$, we rescale our variables as $y = w/\sqrt{|C|}$ and $x = \sqrt{|C|}\lambda$ so that the resulting differential equation does not explicitly contain C:

$$y'(x) = \pm 1 - y^2, \tag{4.46}$$

with the sign corresponding to the sign of C. At $C > 0$, the relevant solution* is $y = 1/\tanh(x)$. We thus have

$$w(\lambda) = \frac{\sqrt{C}}{\tanh(\sqrt{C}\lambda)}, \qquad g(\lambda) = \frac{C}{\sinh^2(\sqrt{C}\lambda)} \qquad (C > 0). \tag{4.47}$$

At $\lambda \ll 1/\sqrt{C}$, the behavior of w and g is C-independent and corresponds to the critical behavior (4.45). In the opposite limit, we find

$$w(l) \approx \sqrt{C}[1 + 2(l_0/l)^{2\sqrt{C}}], \qquad g(l) \approx 4C\,(l_0/l)^{2\sqrt{C}}, \qquad l/l_0 \gg e^{1/\sqrt{C}} \qquad (C > 0). \tag{4.48}$$

Hence, the logarithmic decay of w and g with l is replaced at large distances with a power-law decay, the exponent of which is vanishingly small in the limit of $T \to T_c$. The requirement of having g and w small is met only when (the same will be true for negative values of C)

$$|C| \ll 1. \tag{4.49}$$

From (4.48), we see that the criterion for the value of $w(l)$ to saturate to its macroscopic value $w(\infty) = \sqrt{C}$ is

$$l \gg \xi^{(+)} = l_0\, e^{\frac{1}{2\sqrt{C}}}. \tag{4.50}$$

It defines the correlation length on the superfluid side, $\xi^{(+)}$.

At $C < 0$, the proper solution is $y = 1/\tan(x)$. Indeed, by continuity, at $x \ll 1$, our solution has to be C-independent and follow the critical law (4.45). Hence,

$$w(\lambda) = \frac{\sqrt{|C|}}{\tan(\sqrt{|C|}\lambda)}, \qquad g(\lambda) = \frac{|C|}{\sin^2(\sqrt{|C|}\lambda)} \qquad (C < 0). \tag{4.51}$$

In contrast to the $C > 0$ case, the solution (4.51) is not supposed to be physically meaningful at arbitrarily large distances. At $\lambda = \pi/2\sqrt{|C|}$, the formal solution

---

* There are two other solutions: $y \equiv 1$ and $y = \tanh(x)$. These are nonphysical, since y is supposed to decrease with x.

diverges, which is inconsistent with the underlying assumptions. This defines the correlation length on the normal side of the transition:

$$\xi^{(-)} = l_0\, e^{\pi/\sqrt{|C|}}. \tag{4.52}$$

Physically, it means that pairs of size $R \sim \xi^{(-)}$ are no longer well isolated, and at length scales $l > \xi^{(-)}$, we are instead dealing with a plasma of *single* vortices. The density of single vortices is finite, $\sim [\xi^{(-)}]^{-2}$, but, in accordance with (4.52), exponentially small in the parameter $\propto 1/\sqrt{T - T_c}$.

Given that nonanalytic critical behavior of thermodynamic functions develops only in the limit of infinite length scale, we understand that, at any fixed λ, the pair of solutions (4.47) and (4.51) for w, as well as the pair of solutions (4.47) and (4.51) for g, correspond to one and the same analytic function of the microscopic parameter C.

Problem 4.2 *Check the above statement by expanding the hyperbolic and trigonometric functions into Taylor series. Make sure that the expansion is in powers of $C\lambda^2$ and, in particular, show that*

$$w(\lambda) \approx \frac{1}{\lambda}\left(1 + \frac{C\lambda^2}{3}\right), \qquad g(\lambda) \approx \frac{1}{\lambda^2}\left(1 - \frac{C\lambda^2}{3}\right) \qquad (|C|\lambda^2 \ll 1). \tag{4.53}$$

4.2.2 Off-Diagonal Correlations in the Vicinity of the BKT Transition

Let us use solutions (4.47) and (4.51) for w to find an explicit expression for the behavior of the single-particle density matrix $\rho(r)$ in the vicinity of the BKT point. Apart from Equations (4.47) and (4.51), we employ the RG ideas to write the asymptotic expression (1.197) of Chapter 1 in the form

$$\rho(r) = \rho(l)\left(\frac{l}{r}\right)^{\frac{1}{4\pi K}}, \tag{4.54}$$

with $K \equiv K(l)$. Note the similarity with (4.36): We relate $\rho(r)$ at a certain distance r (large enough to guarantee that the gas of vortex pairs is dilute) to $\rho(l)$, such that the scale of distance l is not dramatically different from r. By doing so, we also assume that the renormalization of K is slow enough to allow us to treat $K(l)$ as a constant within the range $l \in [l, r]$. By rewriting (4.54) in the form of a differential equation and using the smallness of w, we find

$$\frac{d\ln\rho}{d\lambda} = -\frac{1}{4\pi K} \approx -\frac{1}{4} + \frac{w(\lambda)}{8}, \qquad \lambda = \ln(r/l_0). \tag{4.55}$$

Explicit integration of this equation with $w(\lambda)$ given by Equations (4.47) and (4.51) then yields

$$\rho(r) = \frac{B}{r^{1/4}}\left[\frac{1}{\sqrt{C}}\sinh\left(\sqrt{C}\ln\frac{r}{l_0}\right)\right]^{1/8} \qquad (C > 0), \tag{4.56}$$

$$\rho(r) = \frac{B}{r^{1/4}}\left[\frac{1}{\sqrt{|C|}}\sin\left(\sqrt{|C|}\ln\frac{r}{l_0}\right)\right]^{1/8} \qquad (C < 0), \tag{4.57}$$

where B is a certain C-independent constant. The C-dependence is fixed by the requirement that Equations (4.56) and (4.57) correspond to one and the same analytic function of C. At the critical point, we have

$$\rho(r) = \frac{B}{r^{1/4}} \left[\ln(r/l_0)\right]^{1/8} \qquad (C = 0). \qquad (4.58)$$

We see that the $\rho(r) \propto r^{-1/4}$ law turns out to be rather robust. The effects of renormalization of K with r are reduced to the extremely slow varying prefactor that, for almost all practical purposes, behaves as a constant. It is also useful to note that

$$\rho(r) \approx \frac{B}{r^{1/4}} \left[\ln(r/l_0)\right]^{1/8} \left[1 + \frac{C}{48} \ln^2(r/l_0)\right] \qquad (r \ll \xi^{(\pm)}). \qquad (4.59)$$

4.2.3 Alternative Form of the Effective Energy for Vortices

Finally, we would like to address* an important theoretical issue of the physical meaning of parameterization (4.7), with $\Delta\Phi_0 = 0$ leading to the statistical independence of vortices and the regular field $\varphi(\mathbf{r})$, Equations (4.10) through (4.12). If this parameterization is used within a restricted interval $[l, ul]$ of scales of distance, such that the difference between $K(l)$ and $K(ul)$ is small, the field $\varphi(\mathbf{r})$ can be naturally interpreted as the phonon field with the wavelength restricted to be in the interval $\lambda_{\text{ph}} \in [l, ul]$. We will keep track of this restriction using $\varphi_l(\mathbf{r})$ notation. The fluctuations of $\varphi(\mathbf{r})$ are responsible for the suppression of the off-diagonal correlations in accordance with

$$\rho(r) = \rho(l) \langle e^{i\varphi_l(\mathbf{r}) - i\varphi_l(0)} \rangle, \qquad (4.60)$$

leading to the result (4.54). In this picture, the field $\varphi_l(\mathbf{r})$ has nothing to do with the renormalization of K and vice versa. As long as we confine ourselves to the local application of Equation (4.7) with $\Delta\Phi_0 = 0$ and employ the iterative RG treatment to go from microscopic to arbitrarily large scales of distance, we are not making any mistake because the $\varphi_l(\mathbf{r})$ field is appropriately redefined at the next scale. An apparent paradox arises, however, with the observation that the parameterization (4.7) with $\Delta\Phi_0 = 0$ *per se* implies neither locality in the space of distances, nor even that the gas of vortex pairs is dilute! It is a mathematically accurate procedure, and the representation of the effective energy of the system as a sum of the regular part (4.11) and the vortex part (4.12) is valid as soon as the cutoff length l_0 is significantly larger than the vortex core radius to justify the mesoscopic description (4.5). Given *global* separation of variables into the Coulomb system of vortices and the regular field $\varphi(\mathbf{r})$, it is tempting to interpret $\varphi(\mathbf{r})$ as the phonon field that contains all harmonics with $\lambda_{\text{ph}} > l_0$. Nevertheless, such an interpretation would be definitely misleading. Indeed, if φ stood for phonons, its longest-wave harmonics would be controlled by the macroscopic value of $K \equiv K(\infty)$. Meanwhile, the field φ is decoupled from the rest of the degrees of freedom, and thus, *all of*

---

* Following closely Reference [5].

its harmonics fluctuate in accordance with the energy functional (4.11), where, instead of $K(\infty)$, we have $K(l_0)$.

The complementary paradox is based on interpreting Φ_0 as a *purely* vortex field. Since, in the superfluid phase, all vortices are bound in microscopic pairs, one would not expect them to be *directly* observable in long-range correlation properties. In particular, this is exactly why the vortex pairs of sizes $R < l_0$ are absent in the field $\Phi_0(\mathbf{r})$. The only physical way for vortex pairs to manifest themselves at the macroscopic scale is to renormalize the superfluid density. However, the condition $\Delta\Phi_0 = 0$ forces small vortex pairs to contribute to the long-range correlations! The way they do this reveals a "conspiracy" between mathematically independent fields Φ_0 and φ. The statistical independence of the two fields implies factorization of the single-particle density matrix (at large distances):

$$\rho(\mathbf{r}) \propto \langle\exp[i\Phi(\mathbf{r}) - i\Phi(0)]\rangle \propto \Gamma_{\Phi_0}(\mathbf{r})\Gamma_\varphi(\mathbf{r}), \tag{4.61}$$

$$\Gamma_{\Phi_0}(\mathbf{r}) = \langle\exp[i\Phi_0(\mathbf{r}) - i\Phi_0(0)]\rangle, \qquad \Gamma_\varphi(\mathbf{r}) = \langle\exp[i\varphi(\mathbf{r}) - i\varphi(0)]\rangle. \tag{4.62}$$

If taken separately, the correlation functions Γ_{Φ_0} and Γ_φ make little physical sense, since they both depend on $K(l_0)$ and thus are sensitive to the particular choice of an arbitrary cutoff scale l_0. It is only when the two are combined in $\rho(\mathbf{r})$ that the dependence on l_0 disappears. A rational explanation of this peculiar agreement between Φ_0 and φ is that, while being mathematically independent, the two fields are deeply connected physically. The previously mentioned structure of the correlation function even suggests the qualitative form of the connection: The long-wave part of the vortex-pair contribution to field Φ_0 actually belongs to phonons, not vortices. Correspondingly, there should exist a more physical set of global variables, $(\tilde{\varphi}, \tilde{\Phi}_0)$, so that the regular field $\tilde{\varphi}$ describes real phonon fluctuations and the field $\tilde{\Phi}_0$ is free from long-range contributions of short-range vortex pairs. A crucial observation that allows one to construct the variables $(\tilde{\varphi}, \tilde{\Phi}_0)$ is that, for a neutral system of vortices, the asymptotic form of the field $\Phi_0(\mathbf{r})$ is regular (see the following problem).

Problem 4.3 *Make sure that for a neutral system of vortices* $(\sum_j M_j = 0)$, *centered at the point*

$$\mathbf{r}_* = \frac{\sum_j |M_j|\mathbf{r}_j}{\sum_j |M_j|}, \tag{4.63}$$

the asymptotic form of the function $\Phi_0(\mathbf{r})$ is governed by the dipole moment

$$\vec{\mathcal{P}} = \hat{z} \times \sum_j M_j\mathbf{r}_j. \tag{4.64}$$

Namely,

$$\Phi_0(\mathbf{r}) \to \frac{\vec{\mathcal{P}}\cdot(\mathbf{r} - \mathbf{r}_*)}{|\mathbf{r} - \mathbf{r}_*|^2} \qquad at \qquad |\mathbf{r} - \mathbf{r}_*| \to \infty. \tag{4.65}$$

Hence, for a vortex pair of size R, centered at $\mathbf{r}_p = (\mathbf{r}_+ + \mathbf{r}_-)/2$, we can introduce a regular field $\varphi_{R,\mathbf{r}_p}(\mathbf{r})$ with the same asymptotic behavior as the contribution of the pair to the field $\Phi_0(\mathbf{r})$:

$$\varphi_{R,\mathbf{r}_p}(\mathbf{r}) \to \frac{(\hat{z} \times \mathbf{R}) \cdot (\mathbf{r} - \mathbf{r}_p)}{|\mathbf{r} - \mathbf{r}_p|^2} \qquad \text{at} \qquad |\mathbf{r} - \mathbf{r}_p| \gg R. \tag{4.66}$$

The details of the short-range behavior of $\varphi_{R,\mathbf{r}_p}(\mathbf{r})$ are not important. Generally speaking, it can be any regular function centered at the vortex pair.

Problem 4.4 *Show that for a regular field $\varphi_{R,\mathbf{r}_p}(\mathbf{r})$ to satisfy condition (4.66) it is necessary and sufficient to have*

$$\int \mathbf{r} \, \Delta\varphi_{R,\mathbf{r}_p}(\mathbf{r}) d^2 r = 2\pi (\mathbf{R} \times \hat{z}). \tag{4.67}$$

For our purposes, it is also important to require that $\varphi_{R,\mathbf{r}_p}(\mathbf{r})$ goes to the asymptotic solution at length scales $\sim R$ and does not contain short-wave harmonics with wavelength smaller than R. This requirement allows one to define the transformation of the regular field locally in the space of distance scales and to ensure that contributions from different vortex pairs do not interfere.

We define alternative fields as

$$\tilde{\varphi}(\mathbf{r}) = \varphi(\mathbf{r}) + \sum_p \varphi_{R,\mathbf{r}_p}(\mathbf{r}), \tag{4.68}$$

$$\tilde{\Phi}_0(\mathbf{r}) = \Phi_0(\mathbf{r}) - \sum_p \varphi_{R,\mathbf{r}_p}(\mathbf{r}), \tag{4.69}$$

where the sum is over all vortex pairs. [Note that $\mathcal{D}\tilde{\varphi} = \mathcal{D}\varphi$ since the two fields are related by a shift.] After this transformation, the long-range behavior of the density matrix is described in terms of the field $\tilde{\varphi}$ only:

$$\rho(\mathbf{r}) \propto \langle \exp[i\tilde{\varphi}(\mathbf{r}) - i\tilde{\varphi}(0)] \rangle, \tag{4.70}$$

meaning that the field $\tilde{\varphi}$ is the field of phonon fluctuations. As far as the effective energy is concerned, we get the following changes: (a) Up to irrelevant higher-order corrections, vortex pairs do not interact with each other *directly*, as long as they form a dilute gas and Equation (4.35) is satisfied. (b) A vortex pair of size R is now coupled to the long-wave ($\lambda_{ph} \gg R$) harmonics of the phonon field $\tilde{\varphi}$ by the following term:

$$A_{int} = 4\pi K (\mathbf{R} \times \hat{z}) \cdot \nabla\tilde{\varphi}\Big|_{\mathbf{r}_p}. \tag{4.71}$$

As expected, the vortex pair now interacts with the long-wave part of $\tilde{\varphi}$ exactly the same way it interacts with a homogeneous velocity flow $\gamma \nabla \tilde{\varphi}\big|_{r_p}$ [see Equation (4.19)].* The requirement that $\varphi_{R,r_p}(\mathbf{r})$ is smooth on the length scale $\sim R$ guarantees that pairs do not interact with the short-wave phonon harmonics.

Problem 4.5 *Derive Equation (4.71). Also, make an estimate showing that the coupling of a vortex pair to a phonon field, including harmonics $\lambda_{ph} \sim R$, is negligibly small as compared to the logarithmic vortex–antivortex interaction.*

For the RG treatment of the previous sections, the new and the old effective energies are equivalent, since the phonon–vortex coupling (4.71) becomes relevant only for $\lambda_{ph} \gg R$. What we actually gain in the new formulation is direct physical insight into the way that vortex pairs renormalize properties of phonon fluctuations. In the original formulation, the physics of the vortex–phonon interaction was hidden and its effects were obtained iteratively, in the RG sense, through response of the vortex subsystem to the homogeneous flow. With new fields, this physics manifests itself explicitly. It is also worth noting that in terms of the (rather formal, as we understand it now!) global mapping onto the 2D Coulomb gas, the alternative effective energy formulated in this section corresponds to the standard trick of eliminating direct interactions between charge–anticharge pairs by introducing the dielectric function.

References

1. V. L. Berezinskii, Destruction of long-range order in one-dimensional and two-dimensional systems possessing a continuous symmetry group. II. Quantum systems, *Sov. Phys. JETP* **34**, 610 (1972).

2. J. M. Kosterlitz and D. J. Thouless, Long range order and metastability in two dimensional solids and superfluids, *J. Phys. C Solid State Phys.* **5**, L124 (1972); Ordering, metastability and phase transitions in two-dimensional systems, *J. Phys. C Solid State Phys.* **6**, 1181 (1973).

3. J. M. Kosterlitz, The critical properties of the two-dimensional XY model, *J. Phys. C Solid State Phys.* **7**, 1046 (1974).

4. D. R. Nelson and J. M. Kosterlitz, Universal jump in the superfluid density of two-dimensional superfluids, *Phys. Rev. Lett.* **39**, 1201 (1977).

5. E. Kozik, N. Prokof'ev, and B. Svistunov, Vortex–phonon interaction in the Kosterlitz–Thouless theory, *Phys. Rev. B* **73**, 092501 (2006).

---

* Incidentally, note that the absence of such an interaction in the original formulation was very counterintuitive and obviously paradoxical. Indeed, how can vortex pairs interact with a homogeneous global flow but not with a local one, despite the fact that, at the scale of distance $\sim R$, the latter cannot be physically distinguished from the former?

Part II

Superconducting and Multicomponent Systems

CHAPTER 5

Charged Matter Fields

In this chapter, we will generalize our discussion of classical-field superfluids to charged matter fields. We will be referring to them as superconductors. This does not imply any particular microscopic realizations such as the presence of a crystal lattice. We will describe generic qualitative macroscopic properties for which microscopic details of the system demonstrating supertransport phenomena are irrelevant.

Superconductors are characterized by the dissipationless charge transfer, along with the matter transfer. We already know that superfluids are not just ideal fluids. The quantization of vorticity and Josephson effect in superfluids reflect the underlying (topological) order characterized by the field of phase (of a complex-valued field/order parameter), the supertransport being a consequence of the emerging constant of motion, the topological invariant. The same is true with respect to superconductors. Furthermore, there are dramatic U(1)-order-based effects that fundamentally distinguish superconductors from both ideal conductors and neutral superfluids. First is the Meissner effect of expulsion of magnetic field from the superconductor bulk and then its twin effect, the expulsion of supercurrents from the bulk. Another effect directly related to the Meissner effect is the Anderson effect of absence—despite the existence of long-range order—of gapless (Goldstone) normal mode. These three effects are due to a specific gauge coupling of the phase of the order parameter to the vector potential. Another effect, closely related in its origin to the previously mentioned ones, is the effect of magnetic flux quantization: Gauge coupling of the vector potential to the phase of the order parameter implies a direct relationship between the (quantized) circulation of the gradient of the phase and (isolated by a massive toroidal superconductor) magnetic flux. Finally, the rotational response of superconductors proves to be radically different from that of neutral superfluids. Instead of forming a vortex lattice, rotating superconductors can mimic a solid-body rotation in the bulk, at the expense of having surface currents and, correspondingly, uniform magnetic field in the bulk. This amazing *London effect*, while seemingly "contradicting" the Meissner effect, is actually a close relative of the latter: Mathematically, the two phenomena are isomorphous to each other, both involving perfect diamagnetic response (with respect to either a real or a fictitious magnetic field).

The state of a superconductor with expelled magnetic field and supercurrents is called the Meissner state. Interestingly, the hydromagnetostatics of the Meissner state turn out to be almost trivial. Apart from exponentially small (as functions of distance from the surface or a vortex core) tails of magnetic and supercurrent fields, the Meissner state is a state with no magnetic field or bulk current.

Detailed description of vortex cores, as well as essentially nonuniform states—such as vortex lattices—requires going beyond the hydrodynamic approach.

Except for the presence of uniform magnetic field and rigid-body supercurrent pattern, the London state of rotating superconductor is analogous to the Meissner state of nonrotating superconductor in the sense that *extra* magnetic fields and supercurrents will be expelled. This illustrates the fact that expelling either the magnetic field or supercurrent or both is not the most fundamental property of a superconductor. The Meissner–London physics is characteristic of a generic gauge theory with the coupling of the form $\propto [(\nabla\theta + q\mathbf{A})^2 + (\text{curl}\,\mathbf{A})^2]$, where $\mathbf{A}$ is the vector potential, q is a coupling constant, and θ is a certain real scalar field. As a result, the supercurrent density is proportional to $(\nabla\theta + q\mathbf{A})$, implying, in particular, that a necessary condition for penetration of magnetic field into the bulk is finite bulk supercurrent density. In many cases, it is thermodynamically favorable to have the Meissner state, that is, to nullify both the magnetic field and supercurrent density. Magnetostatically, it is also clear that having finite bulk density of supercurrent and zero net magnetic field is only possible if the supercurrent density is exactly compensated by a background density of normal currents (or supercurrents of other components, in a multicomponent system).

Nontrivial aspects arise if the field θ is not single valued. In what follows, θ will be the field of the phase[*] of a certain complex field. It is then the possibility of having $\oint d\mathbf{l} \cdot \nabla\theta = 2\pi \times \text{integer} \neq 0$ that brings about quantized magnetic flux and vortices, the latter in some cases allowing the system to become normal at high enough temperature.

5.1   U(1) Gauge Theory

5.1.1   Nonrelativistic $|\psi|^4$ Gauge Model

A classical-field model featuring the phenomenon of superconductivity is very well known in field theory. It is nothing but the nonrelativistic limit of the U(1) gauge theory for the scalar complex field $\psi = |\psi|e^{i\theta}$ with $|\psi|^4$-type self-interaction. For a charged finite-size system, the condition of electrostatic neutrality is necessary to guarantee spatial uniformity. The two simplest ways of meeting this condition are (a) to have two components of the matter field with opposite electric charges and (b) to introduce a uniform neutralizing background. In this chapter, we consider option (b), which, on one hand, is minimalistic and, on the other hand, is directly relevant to superconductivity of metals, where the supertransport is due to electrons and the neutralizing background is due to ions. Given that only the nonrelativistic regime is relevant for our proposes, we can formulate the corresponding U(1) gauge model by upgrading the Gross–Pitaevskii theory to the case of charged field ψ. The latter must be coupled to the electromagnetic field described by the 4-vector potential $A = (A_0, \mathbf{A})$, the strength of the coupling

---

[*] In the context of charged matter fields, we use θ for the phase.

being characterized by a real[*] constant q. The generalization is readily performed in Lagrangian formalism.

If decoupled from the electromagnetic field ($q = 0$), the Lagrangian for the Gross–Pitaevskii field reads

$$\mathcal{L} = -\int d\mathbf{r}\, \mathrm{Im}\psi^*\dot{\psi} - H[\psi], \tag{5.1}$$

where H is the Hamiltonian of the Gross–Pitaevskii theory. We split H into kinetic and potential parts, $H = H_{\mathrm{kin}} + H_{\mathrm{pot}}$, $H_{\mathrm{kin}} = (\gamma/2)\int d\mathbf{r}\,|\nabla\psi|^2$, $H_{\mathrm{pot}} = \int d\mathbf{r}\,\mathcal{V}|\psi|^2 + (g/2)\int d\mathbf{r}\,|\psi|^4$. The gauge-invariant coupling to the 4-vector potential is achieved by replacing $\nabla\psi \to \nabla\psi + iq\mathbf{A}$ in the term H_{kin} and by replacing $\mathcal{V} \to \mathcal{V} - qA_0$ in the external potential term, $\mathcal{V} \equiv \mathcal{V}(\mathbf{r})$. (In the minimalistic model, the external potential reduces to a uniform neutralizing background.) The corresponding Euler–Lagrange equation of motion for the field ψ is

$$i\dot{\psi} = \frac{\gamma}{2}[i\nabla - q\mathbf{A}]^2\psi + (\mathcal{V} - qA_0)\psi + g|\psi|^2\psi. \tag{5.2}$$

Up to rescaling the vector potential, Equation (5.2) is similar to Equation (1.173), discussed in Chapter 1 in the context of rotating matter field. The *crucial* difference, however, comes from the fact that Equation (5.2) does not yet form a closed theory, since $A_0 \equiv A_0(\mathbf{r}, t)$ and $\mathbf{A} \equiv \mathbf{A}(\mathbf{r}, t)$ are not independent external fields. Rather, the 4-vector potential is a degree of freedom on its own and as such is supposed to obey its own Euler–Lagrange equation of motion. The equation of motion has the general 4-vector form of Maxwell equations, with the system-specific expression for the electric current density following from varying H_{kin} with respect to $\mathbf{A}$ [cf. Equations (1.174) and (1.176)]:

$$\mathbf{J} = \frac{i\gamma q}{2}(\psi^*\nabla\psi - \psi\nabla\psi^*) - \gamma q^2|\psi|^2\mathbf{A} = -\gamma q|\psi|^2(\nabla\theta + q\mathbf{A}), \tag{5.3}$$

and the electric charge density equal to $-q|\psi|^2$. The expression for the electric charge density follows from varying H_{pot} with respect to A_0.

The gauge invariance of the resulting theory is the invariance of both Equation (5.2) and the Maxwell equations[†] with respect to the transformation [cf. Equations (1.177) through (1.180)]

$$\theta' = \theta + \phi, \tag{5.4}$$

$$\mathbf{A}' = \mathbf{A} - q^{-1}\nabla\phi, \tag{5.5}$$

$$A_0' = A_0 + q^{-1}\dot{\phi}, \tag{5.6}$$

---

[*] As long as the theory contains only one field ψ, the sign of q is merely a matter of convention; with more than one component, only relative signs of corresponding q's matter. In view of the standard convention that the electron charge is negative, we find it convenient to relate the absolute sign of q of a given matter field to the sign of electron charge by the following rule: The sign of q is positive if the electric charge density associated with the field, $-q|\psi|^2$ (see below), is negative; that is, it has the same sign as the electric charge density of electrons.

[†] Including, in particular, gauge invariance of the expression (5.3) for the electric current density and the expression $q|\psi|^2$ for the electric charge density.

in which $\phi(\mathbf{r}, t)$ is an arbitrary *single-valued* real function. [The requirement of $\phi(\mathbf{r})$ to be single valued is necessary to exclude cases with nonzero values of counter integrals $\oint_C d\mathbf{l} \cdot \nabla\phi$, which would otherwise change physical values of the magnetic field fluxes, $\oint_C d\mathbf{l} \cdot \mathbf{A}$.] This type of gauge invariance is also called local $U(1)$ invariance, since the phase of the field ψ is shifted locally in spacetime.

A generic property of gauge theories is degeneracy of the equations of motion: One of them has to be an automatic implication of the others. This property becomes obvious if we select the scalar function ϕ of the transformation (5.4) through (5.6) to be one of the degrees of freedom of the system. The fact that this degree of freedom is absolutely unconstrained by the equations of motion means the degeneracy of the latter. Equivalently, we can say that the degeneracy of the equations of motion is necessary to guarantee the *ambiguity* of the solutions expressed by Equations (5.4) through (5.6).

Physically, the gauge ambiguity of the solutions is removed by postulating that all these solutions are equivalent; that is, they describe one and the same physical state. Hence, the degeneracy of the equations of motion is compensated by *redundancy* of degrees of freedom. As a result, the number of physical degrees of freedom is reduced by two. For the nonrelativistic $|\psi|^4$ gauge model, the number of physical degrees of freedom is $(4 + 1) - 2 = 3$—four components of the 4-vector potential plus one degree of freedom of the scalar field* minus two gauge-redundant degrees of freedom.

Further insight into the structure of the equations of motion of a Lagrangian gauge theory can be achieved by looking at the gauge symmetry in the light of Noether's theorem. Local continuous symmetry implies local constant of motion—the *field* of constants of motion. This conserving field is nothing but the generalized momentum canonically conjugated to the field ϕ. Indeed, the insensitivity of the physical state to any time-independent shift of the field ϕ implies that the variational derivative of the Lagrangian with respect to ϕ is identically equal to zero.[†] The Euler–Lagrange equation of motion for ϕ then reads

$$\frac{d}{dt} \frac{\delta \mathcal{L}}{\delta \dot{\phi}} = \frac{\delta \mathcal{L}}{\delta \phi} = 0. \tag{5.7}$$

We conclude that the system of equations of motion is not only degenerate but can also be cast in such a form that one of the equations has a form of conservation of a certain local combination of the fields.

As in the case of neutral superfluids, understanding the quantum-mechanical origin of the classical local gauge $U(1)$ theory allows one to relate parameters γ and q to the properties of underlying quantum particles. If the field ψ corresponded to charged bosons of the mass m and electric charge q_0, we would have $\gamma = \hbar/m$ and $q = -q_0/(\hbar c)$, where c is the velocity of light, which, for briefness,

---

* We remind that, in the nonrelativistic theory, the real and imaginary parts of ψ are dynamically conjugated rather than independent degrees of freedom.

† Or to a total derivative of some functional. The latter only slightly modifies the argument we present in the following—see Equation (5.7).

we normally set equal to unity. In the bosonic case, we would also have to interpret $|\psi|^2/\hbar$ as the number density of bosons, so that $|\psi|^2/\gamma$ and $-q|\psi|^2$ ($c = 1$) are the mass and charge densities, respectively. Electrons, however, are not bosons—they are fermions—and to demonstrate boson-type classical behavior, they must be *paired*. Correspondingly, for electrons (paired fermions), we have $\gamma = \hbar/(2m_e)$ and $q = 2e/\hbar$ ($c = 1$), with m_e being the electron mass and $e > 0$ the absolute value of electron charge [1,2]; similarly, $|\psi|^2/\hbar$ is the number density of electronic pairs. (A discussion of quantum-mechanical aspects will be presented in Chapter 7.)

5.1.2 Superfluid Velocity and Topological Invariant

Identically to the case of a neutral matter field, one can utilize the Galilean transformation* to introduce the notion of microscopic velocity

$$\mathbf{v} = \gamma(\nabla\theta + q\mathbf{A}). \tag{5.8}$$

Similarly, if ψ features the topological long-range order, as defined in Chapter 1, one can introduce the notions of superfluid velocity

$$\mathbf{v}_s = \gamma(\nabla\tilde{\theta} + q\mathbf{A}), \tag{5.9}$$

and topological invariant (responsible for the persistent current)

$$I = \oint_C d\mathbf{l} \cdot \nabla\tilde{\theta} = 2\pi \times \text{integer}, \tag{5.10}$$

where $\tilde{\theta}$ is the phase of the coarse-grained field $\tilde{\psi}$; see Chapter 1. All three quantities are gauge invariant. The gauge invariance of $\mathbf{v}$ and $\mathbf{v}_s$ is seen by inspection, while I is invariant because ϕ is necessarily single valued.

A slightly more subtle situation takes place in a toroidal superconductor subject to a magnetic flux penetrating the torus without *direct* contact with the superconducting matter field. In this case, simultaneously changing the magnetic flux by $\Phi_0 M_I$, with $\Phi_0 = 2\pi/q$ and integer M_I, and applying the *pseudo*†-gauge transformation with ϕ such that $\oint_C d\mathbf{l} \cdot \nabla\phi = 2\pi M_I$ on the contour C going around the torus leaves the state of the superconductor unchanged.‡ It would be deeply wrong, however, to interpret this equivalence of superconducting states as an indication that the topological invariant (5.10) is irrelevant to the physics of superconductors. As with superfluids, the crucial circumstance here is the *discrete* nature of I, implying that, in the absence of proliferated topological defects, the states with different values of I cannot be continuously connected to each other by the equations of motion and that I is an emergent constant of motion.

---

* Being the $c \to \infty$ limit of the Lorentz transformation, Galilean transformation does not change $\mathbf{A}$.

† Because the magnetic flux does change.

‡ This effect is closely related to the phenomenon of magnetic flux quantization in units of Φ_0 (in massive toroidal superconductors, as well as in vortices), to be discussed in Section 5.1.3.

As we see from (5.9), the circulation of superfluid velocity is not quantized. Furthermore, as we will show later, the Meissner physics implies that the circulation of $\mathbf{v}_s$ simply vanishes far away from the center of topological defects, so that there is only finite energy cost (per unit vortex-line length) for creating a vortex.*

In the case of superconductors, the concept of topological—as opposed to genuine—long-range order is even more important than in the case of neutral superfluids, where the notion is indispensable only in low-dimensional cases, while 3D superfluids can be discussed in terms of the genuine long-range order associated with the quasi-average (condensate) $\psi_0 = \langle \psi \rangle_{\text{quasi}} \neq 0$. In a gauge theory, the field ψ is not uniquely defined. In particular, a gauge-invariant analog of the single-particle density matrix of a neutral superfluid reads

$$\rho(\mathbf{r}_1, \mathbf{r}_2) = \left\langle \psi^*(\mathbf{r}_2) \psi(\mathbf{r}_1) \exp\left[-\frac{i}{q} \int_{\mathbf{r}_1}^{\mathbf{r}_2} d\mathbf{l} \cdot \mathbf{A} \right] \right\rangle. \tag{5.11}$$

Based on (5.11), one can, in principle, introduce the notion of the off-diagonal long-range order, associating it with the existence of the finite "condensate density"

$$n_0 = \lim_{r \to \infty} \rho(\mathbf{0}, \mathbf{r}). \tag{5.12}$$

However, it is problematic to introduce the notion of quasi-average (and thus spontaneously broken symmetry) such that $\sqrt{n_0} = |\langle \psi \rangle_{\text{quasi}}|$, because the averaged quantity does not reduce to a product of two local operators.

An important ground-state property, distinguishing superconductors from neutral superfluids, is the structure of long-wave elementary excitations. Recalling that long-wave normal modes in neutral superfluids are gapless sound waves, consistent with the general Goldstone argument for existence of gapless modes in systems with broken continuous symmetry, one might naively expect the existence of a Goldstone mode in a superconductor. However, this expectation is fundamentally misleading since the superconducting ground state features the so-called *Anderson effect* of mutual "cancellation" of gapless phonons and gapless photons. At the microscopic level, the Anderson effect, along with all possible normal modes, is revealed by considering the linearized system of dynamic equations for the matter waves, Equation (5.2), and Maxwell equations for the vector potential $\mathbf{A}$ with the current density (5.3). Given that the effect is concerned with the long-wave properties that are insensitive to microscopic details of the system, it is more instructive to reveal it at the general hydrodynamic level, as we shall do in the next section.

---

* A crucial immediate implication of this fact is the absence of superconductivity in 2D at any finite temperature: No matter how high the energy price is for creating a vortex, it is finite, so that a finite concentration of free vortices is guaranteed entropically.

5.1.3   Magnetic Flux Quantization

Consider a superconductor with a hole as shown in Figure 5.1. The geometry of the hole and the rest of the superconductor does not matter, provided the superconductor is macroscopically large. For example, the hole can be either macroscopically large or microscopically small. Furthermore, in an important particular case, the "hole" can be nothing but the center of a topological defect—a vortex. We are interested in the most general situation, with a finite winding of the superfluid phase $\tilde{\theta}$ around the hole, and nonzero magnetic field inside and in the vicinity of the hole.

Now take a large enough contour σ (see Figure 5.1) and combine Equation (5.9) with the standard relationship (Stokes' theorem) between magnetic flux, Φ, through a surface S_σ bounded by the contour σ, and the vector potential circulation along σ:

$$\Phi = \int_{S_\sigma} d\mathbf{S} \cdot \mathbf{B} = \int_{S_\sigma} d\mathbf{S} \cdot \operatorname{curl} \mathbf{A} = \oint_\sigma d\mathbf{l} \cdot \mathbf{A} = \frac{1}{q} \oint_\sigma d\mathbf{l} \cdot (\mathbf{v}_s/\gamma - \nabla\tilde{\theta}). \tag{5.13}$$

As will be shown later, in the Meissner state, both the magnetic field and the field of superfluid velocity decay rapidly in the bulk of a superconductor. Hence, we can always choose the contour σ far enough from the hole/vortex core, so that, on one hand, we can neglect the term with $\mathbf{v}_s$ in the right-hand side of (5.13), and, on the other hand, the left-hand side of (5.13) yields the total magnetic flux, Φ_{tot}, associated with the hole. This brings us to a remarkable observation that the total magnetic flux through a hole/vortex line in a massive superconductor is determined entirely by the circulation of the gradient of the phase around the hole, and vice versa.

The circulation of phase $\tilde{\theta}$ can only take values $2\pi \times N$, with N an integer. Hence, the total magnetic flux associated with the hole/vortex can only take a discrete set of values:

$$\Phi_{\text{tot}} = -\frac{1}{q} \oint_\sigma d\mathbf{l} \cdot \nabla\theta = -\frac{2\pi N}{q} \equiv -\Phi_0 N \qquad (N = 0, \pm 1, \pm 2, \ldots), \tag{5.14}$$

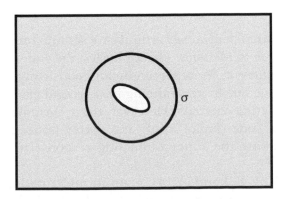

Figure 5.1   Integration path σ around flux-carrying hole/vortex core.

the multiples of the *magnetic flux quantum*

$$\Phi_0 = \frac{2\pi}{q}. \tag{5.15}$$

Recalling that the quantum-mechanical origin of the classical field ψ in realistic superconducting metals implies $q = 2e/\hbar$, we conclude that for all those systems $\Phi_0 = \pi\hbar/e \approx 2.07 \times 10^{-7}$ G·cm^2.

We have derived the quantization of magnetic flux as an immediate implication of a strict relation between the magnetic flux through a contour σ and the phase winding (the circulation of the gradient of the phase) along σ. Crucial for the relation is the fact that $\mathbf{v}_s$ vanishes at σ. Actually, in a broader context involving a macroscopic number of vortices and a macroscopic magnetic flux through a system—here we are talking about states other than the Meissner state that will be introduced later—the flux-to-phase-winding relation can still hold true, but now only in a macroscopic (coarse-grained) sense. A sufficient condition is that the velocity v_s be an intensive (i.e., system-size-independent) variable. In this case, for the magnetic flux, Φ_C, through a macroscopic contour C, we have

$$\Phi_C = -\Phi_0 N_C + \text{macroscopically irrelevant term}, \tag{5.16}$$

where $N_C = (1/2\pi)\oint_C d\mathbf{l} \cdot \nabla\theta$ is the winding number of the phase along C. The "macroscopically irrelevant term" is proportional to the integral $\oint_C d\mathbf{l} \cdot \mathbf{v}_s$ [similarly to (5.13)]. With an intensive v_s, this term scales only as a diameter of C, while both Φ_C and N_C scale as an area of C. This explains the macroscopic irrelevance of the term, which will be omitted in what follows, so that the magnetic flux in the system is directly related to the vortex number. In that sense, one still uses the term flux "quantization," meaning that each vortex "carries" one magnetic flux quantum. Note that Equation (5.16) does not rely on magnetic field screening: it is equivalent to the Onsager–Feynman relation (1.187) [3] between vortex density and rotation frequency in neutral superfluids.

5.1.4 London Law

In Chapter 1, we saw that neutral superfluids are set into (quasi)rotation by forming vortex lattices. Due to Meissner effect physics, the superconducting response to rotation is very different. In superconductors consisting of charged particles (but neutral on average due to neutralizing background charge), vortices are also surrounded by circulating currents. However, these currents decay rapidly away from the cores. Thus, one cannot apply the vortex-lattice-formation argument as in Chapter 1, making the superconducting response to rotation even more intriguing.

As first realized by F. London [4], a uniformly rotating single-component superconductor generates a magnetic field (London field) that perfectly matches the pseudo-vector potential of the rotating frame. As a result, in the rotating frame, there is no bulk flow whatsoever. In the rotating frame, we immediately understand the necessity of surface currents needed to create the uniform magnetic field. This uniform magnetic field does penetrate into the bulk, but, energetically, this is precisely what is needed to minimize the (free) energy in the rotating frame, or, equivalently, to set the superconducting matter field into a rigid-body rotation, as observed from the laboratory frame. From both the rotating and the laboratory frames, there is no net bulk electric current, provided the neutralizing background—ions in superconducting metals—rotates with the same angular velocity as the charged matter field. In the London theory, a mismatch between the rotation of the superfluid matter field and that of the neutralizing background takes place only near the surface. The mismatch is the source of net surface current that creates the uniform magnetic field in the bulk.

The London law relates the self-induced uniform magnetic field to the rotation frequency Ω. Assuming that the rotation axis is along the z-direction and introducing vector notation $\boldsymbol{\Omega} = \hat{z}\Omega$, we write the rotation-induced velocity as

$$\mathbf{v}_0 = \boldsymbol{\Omega} \times \mathbf{r}. \tag{5.17}$$

For the rigid-body rotation, the field of superfluid velocity (5.9) should be identically the same (in the bulk). Hence,

$$\gamma(\nabla\tilde{\theta} + q\mathbf{A}) = \boldsymbol{\Omega} \times \mathbf{r}. \tag{5.18}$$

Taking curl from both sides, and noticing that $\operatorname{curl} \nabla\tilde{\theta} \equiv 0$ in view of the absence of vortices, we arrive at the London law

$$\mathbf{B} = -\frac{2}{\gamma q}\boldsymbol{\Omega}. \tag{5.19}$$

Identically rewriting $\gamma q \equiv q|\psi|^2/(|\psi|^2/\gamma)$, we realize that the parameter γq is nothing but the (constant) charge-to-mass ratio. Correspondingly, for electrons, it will be equal to e/m_e, the fact of pairing being irrelevant. Hence, for the electronic superconductors of all possible types, the London law reads (restoring c): $\mathbf{B} = -2c(m_e/e)\boldsymbol{\Omega}$.

5.1.5 Hydroelectrodynamic Action, Plasmons, and Anderson Effect

The universal long-wave dynamics of a U(1) charged field at $T \to 0$ is described by a hydroelectrodynamic action. The latter is a straightforward generalization of Popov's hydrodynamic action of Section 2.2.4): gauge-invariant coupling to the vector potential corresponds to replacing $\nabla\theta \to \nabla\theta + q\mathbf{A}$, $\dot{\theta} \to \dot{\theta} - qA_0$. For the harmonic part of the hydroelectrodynamic action—sufficient for our purposes here—we get

$$S_{\mathrm{hd}}^{(\mathrm{harm})}[\theta, A] = \int dt d\mathbf{r}\left[\frac{\varkappa}{2}(\dot{\theta} - qA_0)^2 - \frac{\gamma n_s}{2}(\nabla\theta + q\mathbf{A})^2\right] + S_{\mathrm{EM}}[A], \tag{5.20}$$

with $S_{EM}[A]$ being the action of the noninteracting electromagnetic field. The least-action principle then leads to a system of coupled equations: the Euler–Lagrange equation for θ,

$$\ddot{\theta} = \frac{\gamma n_s}{\varkappa}(\Delta\theta + q\nabla \cdot \mathbf{A}) + q\dot{A}_0, \tag{5.21}$$

and Maxwell equations for the 4-vector potential $A = (A_0, \mathbf{A})$

$$\Delta A_0 + \nabla \cdot \dot{\mathbf{A}} = -\rho, \qquad \Box\mathbf{A} + \nabla(\nabla \cdot \mathbf{A} + \dot{A}_0) = \mathbf{J}, \tag{5.22}$$

with the charge and current densities given, respectively, by

$$\rho = q\varkappa(\dot{\theta} - qA_0), \qquad \mathbf{J} = -\gamma q n_s(\nabla\theta + q\mathbf{A}). \tag{5.23}$$

[We use units in which the velocity of light is equal to unity.]

Upon removing the gauge redundancy, the solution to the linear system of Equations (5.21) through (5.23) can be represented as a linear superposition of three branches of normal modes. To this end, we first decompose the vector field $\mathbf{A}$ into longitudinal and transverse parts:

$$\mathbf{A} = \mathbf{A}_\| + \mathbf{A}_\perp, \qquad \text{curl}\,\mathbf{A}_\| \equiv 0, \qquad \nabla \cdot \mathbf{A}_\perp \equiv 0. \tag{5.24}$$

Doing the same for the current density and observing that

$$\mathbf{J}_\| = -\gamma q n_s(\nabla\theta + q\mathbf{A}_\|), \qquad \mathbf{J}_\perp = -\gamma q^2 n_s\mathbf{A}_\perp, \tag{5.25}$$

we arrive at two decoupled systems of equations: Klein–Gordon equation for the transverse part,

$$(\Box + \mathcal{M}^2)\mathbf{A}_\perp = 0, \qquad \mathcal{M}^2 = \gamma q^2 n_s, \tag{5.26}$$

and three coupled equations for the longitudinal components

$$\ddot{\theta} = \frac{\gamma n_s}{\varkappa}(\Delta\theta + q\nabla \cdot \mathbf{A}_\|) + q\dot{A}_0, \tag{5.27}$$

$$\Delta A_0 + \nabla \cdot \dot{\mathbf{A}}_\| + q\varkappa(\dot{\theta} - qA_0) = 0, \tag{5.28}$$

$$\Box\mathbf{A}_\| + \nabla(\nabla \cdot \mathbf{A}_\| + \dot{A}_0) + \gamma q n_s(\nabla\theta + q\mathbf{A}_\|) = 0. \tag{5.29}$$

Since the transverse part $\mathbf{A}_\perp$ is a gauge-invariant quantity, Equation (5.26) has no gauge redundancy. It describes two "massive" transverse modes.[*]

---

[*] In the field theory, the term "massive" stands for modes having finite energies at zero wavevector; the condensed-matter equivalent is "gapped mode."

The third, longitudinal, normal mode must be extracted from Equations (5.27) and (5.28). To this end, we make a standard observation that gauge freedom allows one to set $\mathbf{A}_\parallel \equiv 0$. Indeed, the condition $\mathrm{curl}\,\mathbf{A}_\parallel \equiv 0$ implies that $\mathbf{A}_\parallel$ can be parameterized as a gradient of a certain scalar function, $\mathbf{A}_\parallel = \nabla f$, and then eliminated by the gauge transformation with $\theta = qf$. Setting $\mathbf{A}_\parallel = 0$ yields

$$\ddot{\theta} = \frac{\gamma n_s}{\varkappa} \Delta\theta + q\dot{A}_0, \tag{5.30}$$

$$\Delta A_0 + q\varkappa(\dot{\theta} - qA_0) = 0, \tag{5.31}$$

$$\nabla(\dot{A}_0 + \gamma q n_s \theta) = 0. \tag{5.32}$$

The last equation states that θ and A_0 are not two independent degrees of freedom. The two are related to each other by the constraint[*]

$$\theta = -\frac{1}{\gamma q n_s}\dot{A}_0, \tag{5.33}$$

which reduces Equations (5.30) and (5.31) to the same scaled Klein–Gordon equation

$$(\tilde{\Box} + \mathcal{M}^2)A_0 = 0, \qquad \tilde{\Box} = \frac{\partial^2}{\partial t^2} - \frac{\gamma n_s}{\varkappa}\Delta. \tag{5.34}$$

We conclude that all three normal modes are massive and have the same frequency $\mathcal{M}$ in the zero-wavenumber limit. The dispersion of transverse modes is universal and is controlled by the speed of light. The dispersion of the longitudinal mode is sensitive to material properties. Namely, it is dictated by the speed of sound of the neutral ($q = 0$) counterpart of the theory. Note also that the gap $\mathcal{M}$ depends on $|q|$, γ, and n_s but is insensitive to $\varkappa$.

The normal modes we obtained are nothing else but the familiar (in electrodynamics of ideal conductors) transverse and longitudinal branches of *plasmons*. In particular—up to a minor and natural distinction of replacing the total density of matter, n, with the superfluid density n_s (recall that $n_s \neq n$ in non-Galilean systems)—the expression for $\mathcal{M}$ in Equation (5.26) is the standard expression for plasma frequency. There is, however, an important broader context for plasmons in superconductors associated with the previously mentioned Anderson effect.

The Anderson effect profoundly reveals an implicit assumption made in the Goldstone theorem. The argument for existence of a gapless normal mode on top of a ground state with broken continuous global symmetry rests on the simple observation that the energy of the state obtained by applying the symmetry transformation to the ground state is zero. The underlying assumption then is that this transformation leads to a different state, which is a limiting, $k \to 0$, case of a

---

[*] After setting $\mathbf{A}_\parallel \equiv 0$, we still have a freedom of gauge transformations with any coordinate independent function $\phi \equiv \phi(t)$. We use this freedom to suppress an arbitrary function of time, which arises when going from Equations (5.32) through (5.33).

normal mode with a finite wavevector k, the Goldstone mode. Anderson's counterexample of plasmons in superconductors demonstrates that long-range order does not necessarily imply the existence of the Goldstone mode. The transformation $\theta \to \theta + \theta_0$ with a small constant θ_0 does not correspond to the $k \to 0$ limit of any normal mode, nor does it lead to a physically distinguishable state. In other words, an implicit assumption behind Goldstone theorem is the absence of redundant degrees of freedom. When all degrees of freedom are physical, any small perturbation must be a linear superposition of normal modes, thus guaranteeing the existence of a Goldstone mode. In relativistic quantum field theory, the Anderson effect is a standard way of obtaining massive gauge bosons in Yang–Mills-type theories.[*]

5.2 U(1) Mean-Field Gauge Theories. London and Ginzburg–Landau Models

The previously formulated U(1) gauge $|\psi|^4$ model features all qualitative properties of a generic single-component superconductor. Nevertheless, at the quantitative level, its predictions are quite different from what is observed in realistic superconductors. In the first place, this quantitative difference concerns the structure of the vortex cores. Speaking of the classical matter-field description of superconductivity for electrons in metals, it would be incorrect to directly associate the field ψ with *all* conducting electrons and treat $|\psi|^2$ as the electron density. The crucial observation here is that the classical field $\psi(\mathbf{r})$ represents coarse-grained pairing correlations in the electronic subsystem that are only indirectly related to the local electron density. The field $\psi(\mathbf{r})$ is usually called the *superconducting order parameter*[†] and is quite nontrivial, being based on the effect of *electron pairing* in an interacting fermionic quantum field.[‡] As a consequence, the depletion of $\psi(\mathbf{r})$ has little to do with depletion of the electron density but rather reflects local

---

[*] Brout, Englert, and Higgs (see Chapter 14) extended Anderson's analysis to the case of the relativistic matter field. In the relativistic case, as pointed out earlier by Anderson, the dispersions of transverse and longitudinal "plasmons" are the same, consistent with the Lorentz invariance of the theory. An additional circumstance—revealed already in Goldstone's analysis of relativistic $|\psi|^4$ theory—is the existence, at least at the classical-field level, of yet another massive mode, associated with the fact that relativistic complex scalar field ψ, as opposed to its nonrelativistic counterpart, has two, rather than one, independent degrees of freedom (complex and real parts of ψ).

[†] For historical reasons, it is termed "superconducting order parameter" even when, strictly speaking, it has no relationship to the notion of "order parameter" from Landau theory of the second-order phase transitions. We adopt here this standard terminology, but it should be kept in mind that, in a broader context, appearance of the field $\psi(\mathbf{r})$ does not necessarily signal spontaneous symmetry breakdown. One example that will be discussed in Chapter 6 is the multiband U(1) superconductors that are described by multicomponent classical-field theories.

[‡] The quantum-mechanical relations $\gamma = \hbar/(2m_e)$ and $q = 2e/\hbar$—mentioned earlier in the context of dynamical classical-field model—take place also in the finite-temperature classical-field description. Similarly, if electrons were (weakly interacting) bosons, rather than fermions, we would have $\gamma = \hbar/m_e$ and $q = e$.

changes in the microscopic state of the electronic subsystem. Moreover, in realistic materials, many-body effects are very important in screening the Coulomb forces.

At the coarse-grained level, however, the quantum-field-theoretic details play a minor (essentially quantitative) role, provided the effective classical-field theory is properly formulated to avoid the spurious effects of dramatic charge depletion in large vortex cores. A reasonable way to achieve the latter goal is to absorb the effects of coupling to the scalar part A_0 of the 4-vector potential into the local effective $|\psi|^4$ term. The resulting model is simplified by removing the electrostatic effects, including the necessity of having the neutralizing background. This procedure implies that (a) A_0 disappears from Equation (5.2), (b) the term $q|\psi|^2$ disappears from the right-hand side of the Maxwell equation for A_0. Even though the gap in the spectrum of longitudinal plasmons in the long-wave limit is removed by omitting A_0, this deficiency is not relevant for thermodynamics. The theory still features local U(1) gauge invariance but now with a restriction that the field ϕ be time-independent. At the level of *gauge equivalence* [see Equations (1.177) through (1.180) and the corresponding discussion], one can still apply the gauge transformation with time-dependent ϕ. The only violation of invariance that will result is the generation of the A_0 term in Equation (5.2).

For many realistic superconductors, there is an important simplifying aspect of the classical-field description. The finite-temperature mean-field description in terms of the nonfluctuating order parameter ψ applies virtually up to the very critical point (i.e., the critical region is extremely narrow for many materials, such as low-temperature superconductors). Mathematically, the finite-temperature mean-field description is equivalent to finding the ground state of a classical-field model in the grand canonical formalism. That is, one deals with a theory similar to the previously described dynamic theory, in which the Hamiltonian is replaced with a free-energy functional of the same form but with temperature-dependent parameters. At $T > 0$, the free-energy functional has nothing to do with dynamics. Its only physical meaning is to define the equilibrium structure of the field $\psi(\mathbf{r})$ at a given temperature.

5.2.1 London Model. Hydromagnetostatic Regime

The structure of the self-interaction term in either the field-theoretical dynamical model or the free-energy functional should not necessarily be of the $|\psi|^4$ form. In particular, the self-interaction can be replaced with the constraint $|\psi|^2 = n = \text{const}$. The corresponding theory is known as the London model of superconductivity. In this model, the state of the matter field is described by the phase $\theta(\mathbf{r})$ of the order parameter* $\psi = \sqrt{n}\, e^{i\theta}$. The kinetic part of the free-energy density then reads

$$F_k = \frac{n}{2} |(\nabla\theta + q\mathbf{A})|^2 . \tag{5.35}$$

---

* The original formulation of the London theory (F. London and H. London [5]) was not explicitly based on the notion of superconducting order parameter ψ and, correspondingly, did not explicitly involve the phase $\theta(\mathbf{r})$ in its modern sense; see discussion in Chapter 14.

Adding the magnetic field energy density

$$F_m = \frac{1}{2}(\operatorname{curl}\mathbf{A})^2, \tag{5.36}$$

we get the free-energy density of the London model:

$$F_L = \frac{n}{2}|(\nabla\theta + q\mathbf{A})|^2 + \frac{1}{2}(\operatorname{curl}\mathbf{A})^2. \tag{5.37}$$

The Meissner effect and other interesting phenomena follow from minimization of the free-energy functional* $F_L[\theta,\mathbf{A}] = \int d\mathbf{r}\, F_L$ to appropriate boundary conditions.

Comparing the functional $F_L[\theta,\mathbf{A}]$ to the hydrodynamic action (5.20) reveals a very important broader context of the London model. Upon replacing $n \to n_s$, the model (5.37) describes the universal hydromagnetostatic regime of a superconductor. The only requirement for the hydromagnetostatic regime to take place is the existence of a spatially uniform limit away from boundaries/vortex cores.

5.2.2 Behavior of a Magnetic Field in a Superconductor. The Meissner Effect and London Penetration Length

We are ready to derive the Meissner effect. The effect is a direct consequence of coupling between the phase gradient and vector potential. It is readily established in the London/hydromagnetostatic limit $|\psi|^2 \to n_s$. Combining Maxwell equation $\operatorname{curl}\mathbf{B} = \mathbf{J}$ with the definition of the electric current (5.3), we find that $\nabla\theta + q\mathbf{A} = -\operatorname{curl}\mathbf{B}/qn_s$. Substituting this relation back to the London/hydromagnetostatic free energy, we obtain

$$F = \frac{1}{2}\left[\lambda^2(\operatorname{curl}\mathbf{B})^2 + \mathbf{B}^2\right], \tag{5.38}$$

where

$$\lambda = \frac{1}{q\sqrt{n_s}} \tag{5.39}$$

is called the magnetic field penetration length.[†]

The theory (5.38) is a theory of a single massive[‡] vector field $\mathbf{B}$, with the mass $\mu_A = \lambda^{-1}$. The mass μ_A nullifies at the critical point of superconducting phase transition.

---

* Here and in the following, we use one and the same letter for the free-energy density and free-energy functional. Whenever we need to avoid potential ambiguity of this notation, we specify functional arguments in square brackets to distinguish the free-energy functional from the free-energy density.

[†] With restored γ, we have $\lambda = [q\sqrt{\gamma n_s}]^{-1}$.

[‡] A field—stationary as well as time-dependent—is called massive if its perturbation decays exponentially in space away from local infinitesimal nonuniformity, the decrement of the exponential decay being referred to as the mass.

The equations governing variations of the magnetic field in space immediately follow from (5.38), after taking a variational derivative with respect to the magnetic field and taking integrals by parts. The result is

$$\mathbf{B} + \lambda^2 \operatorname{curl}\operatorname{curl}\mathbf{B} = 0. \tag{5.40}$$

To see that the externally applied field indeed decays exponentially inside the superconductor with the decay length λ, consider a flat superconductor surface perpendicular to the x-axis (with $x > 0$ corresponding to the superconductor bulk), with the magnetic field parallel to the z-axis; that is, $\mathbf{B}(x) = (0, 0, B(x))$. Then (5.40) with boundary condition $B(0) = B_0$ reduces to

$$B - \lambda^2 \frac{d^2 B}{dx^2} = 0 \quad \Rightarrow \quad B = B_0 e^{-x/\lambda} \quad (x > 0). \tag{5.41}$$

This characteristic solution gives λ its name—the London penetration length. According to Maxwell equation $\operatorname{curl}\mathbf{B} = \mathbf{J}$, the screening of the magnetic field is a result of the nondissipative surface current

$$\mathbf{J} = \left(0, J_y(x), 0\right), \quad J_y(x) = -\frac{dB}{dx} = \frac{B_0}{\lambda} e^{-x/\lambda} \quad (x > 0). \tag{5.42}$$

The Meissner effect has a twin: the effect of expulsion of the supercurrent from the bulk of a superconductor. Indeed, if it penetrated the bulk, the supercurrent would create a bulk magnetic field, which would be inconsistent with the Meissner effect. Mathematically, the twin effects are described by one and the same set of equations; see the previous example, Equations (5.41) and (5.42). The London penetration length thus equally characterizes the exponential decay of the magnetic field and the current density away from the surface of a superconductor. Apart from the boundary conditions reflecting the physical statement of the problem, the two effects are simply two aspects of one and the same phenomenon.

Drawing a parallel between the previously developed theory of plasmons and the hydromagnetostatic theory of this section, one can see the direct relationship between the Anderson and Meissner effects. Both effects originate from a gauge coupling between the phase of the order parameter and the vector potential. The coupling yields mass to the gauge field.

5.2.3 Ginzburg–Landau Model

The London theory was the first phenomenological theory of superconductivity introduced well before the discovery of pairing mechanisms between fermions. One controversy that surrounded it in the early days of superconductivity was that it predicted negative energy of the superconductor–normal boundary in an external magnetic field, which was, erroneously, considered as a sign of unphysical instability. Ginzburg and Landau extended the London theory by introducing the order parameter ψ with spatially varying absolute value and $|\psi|^4$-type self-interaction [6]. The form of the free-energy density of the GL model,

$$F = \frac{\gamma}{2}|(\nabla + iq\mathbf{A})\psi|^2 - a|\psi|^2 + \frac{b}{2}|\psi|^4 + \frac{(\text{curl}\,\mathbf{A})^2}{2}, \tag{5.43}$$

coincides with the energy density of the previously discussed $|\psi|^4$-type classical matter-field gauge theory (with temperature-dependent parameters a and b), in which the field ψ is coupled to the vector potential $\mathbf{A}$ but decoupled from the scalar potential A_0. Since the temperature-independent parameter γ can be absorbed into the definition of ψ by rescaling the fields, from now on, we set $\gamma = 1$.

While originally introduced at a phenomenological level, the GL model was later shown to emerge from the Bardeen–Cooper–Schrieffer (BCS) theory of pairing between weakly attracting electrons [2]. It emerges in this theory under the condition that the system is close enough to the critical temperature T_c, where a superconductor undergoes a phase transition between normal and superconducting states. In that case, the model is microscopic; that is, coefficients a and b are expressed through microscopic parameters of the system. Physically, the weakness of the interaction is necessary for the mean-field description to apply, and the closeness to the critical temperature is needed to Taylor-expand the free-energy density in powers of $|\psi|^2$ up to the leading terms (keeping the $|\psi|^4$ term crucial because the coefficient a nullifies at the critical point). However, despite the fact that the GL model, within BCS theory, is microscopically justified only on the approach to T_c, in many respects, it is a more general theory than BCS since it is not necessary that it stems from that particular microscopic physics. Moreover, it yields a qualitatively accurate description for many superconducting materials, even in the low-temperature regime.

Equilibrium configurations of the fields $\psi(\mathbf{r})$ and $\mathbf{A}(\mathbf{r})$ are obtained by minimizing the free-energy functional $F[\psi, \mathbf{A}] = \int d\mathbf{r}\, F$. For homogeneous states $\psi = $ const., in the absence of vector potential, the equilibrium value, $\bar{\psi}$, of the order parameter is given by

$$|\bar{\psi}| = \sqrt{a/b}. \tag{5.44}$$

The so-called condensation energy, defined as the difference between free energies of the normal, $\psi = 0$, and superconducting states, is often expressed in terms of the magnetic energy density:

$$F(\psi = 0) - F(\psi = |\bar{\psi}|) = \frac{a^2}{2b} = \frac{H_c^2}{2}. \tag{5.45}$$

The value of the magnetic field H_c defined by this equation is called the thermodynamic critical magnetic field. Indeed, while the external magnetic field penetrates the volume occupied by the normal state without change (magnetic susceptibility of most materials is extremely small and can safely be neglected here), it is expelled from the superconductor. Thus, free energy of the normal state in magnetic field $H > H_c$ is lower than that of the uniform superconductor with the expelled field. In the simplest case, discussed in the following, H_c corresponds to the value of the external field, which destroys superconductivity.

5.2.4   Ginzburg–Landau Equations

The two coupled equations for the equilibrium configuration of the fields ψ and $\mathbf{A}$ are obtained by minimizing the GL free-energy functional by the standard technique of variational derivatives: $\delta F[\psi, \mathbf{A}]/\delta\psi^*(\mathbf{r}) = 0$, $\delta F[\psi, \mathbf{A}]/\delta\mathbf{A}(\mathbf{r}) = 0$. These equations—called, respectively, the first and the second GL equations—describe how ψ and $\mathbf{A}$ vary in space, under some specific boundary conditions.

For the variation of $F[\psi, \mathbf{A}]$ with respect to $\psi^*(\mathbf{r})$, we have

$$\int d\mathbf{r}\left[-a\psi\delta\psi^* + b\psi|\psi|^2\delta\psi^* + \frac{1}{2}(i\nabla\delta\psi^* + q\mathbf{A}\delta\psi^*)\cdot(-i\nabla\psi + q\mathbf{A}\psi)\right] = 0. \qquad (5.46)$$

Denoting $\mathbf{v} = -i\nabla\psi + q\mathbf{A}\psi$ and using identity $(\nabla\delta\psi^*)\cdot\mathbf{v} \equiv \nabla\cdot(\delta\psi^*\mathbf{v}) - \delta\psi^*\nabla\cdot\mathbf{v}$ to integrate the term involving the gradient of $\delta\psi^*$ by parts,

$$\int d\mathbf{r}\,(\nabla\delta\psi^*)\cdot\mathbf{v} = \oint_\sigma \delta\psi^*\mathbf{v}\cdot d\mathbf{s} - \int d\mathbf{r}\delta\psi^*\nabla\cdot\mathbf{v} \qquad (5.47)$$

(σ stands for the surface of the superconductor), we get

$$\int d\mathbf{r}\left[-a\psi + b\psi|\psi|^2 + \frac{1}{2}(-i\nabla + q\mathbf{A})^2\psi\right]\delta\psi^* + \oint_\sigma[-i\nabla\psi + q\mathbf{A}\psi]\delta\psi^*\cdot d\mathbf{s} = 0. \quad (5.48)$$

The surface term is identically equal to zero under the boundary condition

$$[-i\nabla\psi + q\mathbf{A}\psi]\cdot\mathbf{n} = 0, \qquad (5.49)$$

where $\mathbf{n}$ is the unit vector normal to the superconductor's surface. As we will see shortly, Equation (5.49) is the condition of zero density of normal current. The first term in (5.48) is zero when the field satisfies the first GL equation

$$\frac{1}{2}(-i\nabla + q\mathbf{A})^2\psi - a\psi + b\psi|\psi|^2 = 0. \qquad (5.50)$$

Varying $F[\psi, \mathbf{A}]$ with respect to $\mathbf{A}$

$$\int d\mathbf{r}\left[\frac{iq}{2}(\psi\nabla\psi^* - \psi^*\nabla\psi) + q^2|\psi|^2\mathbf{A}\right]\cdot\delta\mathbf{A} + \operatorname{curl}\mathbf{A}\cdot\operatorname{curl}\delta\mathbf{A} = 0, \qquad (5.51)$$

we use identity

$$\mathbf{V}\cdot\operatorname{curl}\mathbf{W} = \mathbf{W}\cdot\operatorname{curl}\mathbf{V} - \nabla\cdot(\mathbf{W}\times\mathbf{V}), \qquad (5.52)$$

to integrate the second term by parts

$$\int d\mathbf{r}\operatorname{curl}\mathbf{A}\cdot\operatorname{curl}\delta\mathbf{A} = \int d\mathbf{r}\delta\mathbf{A}\cdot(\operatorname{curl}\operatorname{curl}\mathbf{A}) - \oint_{\sigma'}d\mathbf{s}\cdot(\delta\mathbf{A}\times\operatorname{curl}\mathbf{A}). \qquad (5.53)$$

We assume in the following that σ' is a certain distant surface—not to be confused with the surface of the superconductor σ—satisfying the requirement that the distortion of the external vector potential on this surface due to the presence of the superconductor is negligible. This means that $\delta \mathbf{A}|_{\sigma'} = 0$ and the surface integral is automatically zero, while the configuration of $\mathbf{A}$ on the surface σ in the absence of a superconductor forms the boundary condition for the problem.

The requirement that the bulk integral be zero leads to the second GL equation

$$\frac{iq}{2}(\psi\nabla\psi^* - \psi^*\nabla\psi) + q^2|\psi|^2\mathbf{A} + \operatorname{curl}\operatorname{curl}\mathbf{A} = 0. \tag{5.54}$$

With Equation (5.3) for the electric current and $\mathbf{B} = \operatorname{curl}\mathbf{A}$ for the magnetic field, the second GL equation can be interpreted as Ampere's law

$$\operatorname{curl}\mathbf{B} = \mathbf{J}. \tag{5.55}$$

5.3   Type-1 and Type-2 Superconductors

In the fixed-modulus description, there is only one characteristic length scale in the superconductor: the London penetration length λ. This picture, however, is too simplified for description of key properties of superconductors. In particular, it fails to deal with the internal structure of topological defects. Vortices cause variations of $|\psi|$ in space, introducing the second important length scale into the problem.

5.3.1   Coherence Length

As in the previous example, consider a superconductor occupying the $x > 0$ half-space. Also, assume ψ to be real and fix the gauge $\mathbf{A} = 0$ (i.e., there are no currents and no magnetic field). Now consider a configuration where inhomogeneous behavior of $\psi(x)$ is imposed by the boundary condition $\psi(x = 0) = 0$. With this boundary condition, the field $\psi(x)$ should gradually recover its ground-state value away from the boundary. By rescaling the field value so that it approaches unity at $x \to \infty$, that is, $\psi \to \sqrt{a/b}\,\tilde{\psi}$, we arrive at the GL equation

$$-2\xi^2\frac{d^2\tilde{\psi}}{dx^2} - \tilde{\psi} + \tilde{\psi}^3 = 0, \tag{5.56}$$

with

$$\xi = \frac{1}{2\sqrt{a}}, \tag{5.57}$$

the so-called coherence length. The solution of (5.56) is given by $\tilde{\psi} = \tanh(x/2\xi)$. By looking at the limit of large x when the amplitude variation caused by the boundary condition recovers its equilibrium bulk value, we find

$$1 - \tilde{\psi} = 2e^{-\frac{x}{\xi}}. \tag{5.58}$$

We see that the asymptotic decay is exponential and the weakly perturbed field $|\psi|$ recovers its equilibrium value on the characteristic length scale ξ. According to definition of the "mass of the field" as the inverse decay length of small local perturbations, we find the mass μ of the field $|\psi|$ equal to

$$\mu = \xi^{-1}. \tag{5.59}$$

For historic reasons, one can often find in the literature a definition of coherence length that differs from that adopted here by a factor of $1/\sqrt{2}$ (and the GL parameter larger by a factor of $\sqrt{2}$). The traditional definition of ξ does *not* represent a characteristic length scale of the GL theory. In contrast, our definition is directly linked to the exponential decay length and leads to the more compact expressions presented in the following.

In Chapter 1, we introduced the healing length as a characteristic length scale of the overall shape over which a perturbation of the complex field $|\psi|$ recovers its ground-state value. The difference between the coherence and healing length should be emphasized: the coherence length is a characteristic exponent of the recovery of a *tail* of the perturbation. It is the fundamental length scale of a linearized theory. By contrast, the healing length is defined as a characteristic length scale of the field variation in the full nonlinear problem. Although in this simplest single-component example these length scales are similar, in most cases, they are very different.* In general, the healing length depends on the boundary conditions and is not a fundamental length scale. For example, we will see in the following that in some circumstances, the presence of currents can dramatically affect the characteristic length scales of the variation of $|\psi|$.

Finally, note that the thermodynamic magnetic field (and thus condensation-energy density) of a superconductor in the GL theory can be expressed entirely in terms of the fundamental length scales λ and ξ, and the magnetic flux quantum as

$$H_c = \frac{\Phi_0}{4\pi\xi\lambda}. \tag{5.60}$$

5.3.2 Ginzburg–Landau Parameter and Surface Tension between the Superconducting and Normal Phases

The Meissner state of a superconductor in a low magnetic field is structurally simple. Our goal now is to study the more complicated inhomogeneous states that appear in higher fields. The key characteristic of an inhomogeneous state (essentially, a mixture of normal and superconducting domains) is the energy of interface between the normal and superconducting states, the so-called surface

---

* For example, in the case of multiband superconductors, which we will consider in the next chapter, the coherence lengths (exponents of the asymptotics of the fields) will, in general, have nothing to do with the *overall* length scale at which the nonlinear theory recovers from a perturbation.

tension. As we will see, surface tension can be both positive and negative; the different signs imply very different structures of inhomogeneous states.

Having only two characteristic length scales in the single-component case, GL theory is entirely characterized by the dimensionless ratio of the two,

$$\kappa = \frac{\lambda}{\xi}, \tag{5.61}$$

called the GL parameter.

A remark on temperature dependence of the GL quantities is in order. In the mean-field theory of second-order phase transitions, parameters of the effective free-energy functional are analytic functions of $T - T_c$, so that one can Taylor-expand them in powers of $(1 - T/T_c)$ in the vicinity of T_c. Within the mean-field regime/approximation, the coefficient $a(T)$ changes sign at the transition point, from negative at $T > T_c$ to positive at $T < T_c$, while the temperature dependence of b is neglected. Hence, in the vicinity of the critical point,

$$a(T) \approx a_0(1 - T/T_c). \tag{5.62}$$

Then, from Equation (5.44), the field ψ has a nonzero ground-state value only at $T < T_c$, with the temperature dependence

$$|\psi(T)| \propto \sqrt{1 - T/T_c}. \tag{5.63}$$

Equation (5.57) then implies that, at the level of mean-field theory, the coherence length diverges on the approach to T_c as

$$\xi \propto 1/\sqrt{T_c - T}. \tag{5.64}$$

The London penetration length has the same temperature scaling at $T \to T_c - 0$,

$$\lambda \propto |\psi|^{-1} \propto a^{-1/2} \propto 1/\sqrt{T_c - T}, \tag{5.65}$$

so that the GL parameter remains constant.

Consider now a flat yz-boundary between normal and superconducting states in an external field $\mathbf{H}$ oriented in the $\hat{z}$-direction. Let the normal state be to the left of the interface. The appropriate thermodynamic potential to be minimized for the system in a fixed external field ($\mathbf{H}$) and at fixed temperature is Gibbs free-energy density, G, which is related to Helmhotz free energy, F, as $G = F - \mathbf{H} \cdot \mathbf{B}$.* Let us consider the situation where the value of the external magnetic field is set equal to the thermodynamical critical field H_c. On the right of the boundary, in the superconductor bulk, the Gibbs free energy is given by $G(x \to \infty) = F_s^{(0)}$, where $F_s^{(0)}$ is the free energy of a superconductor without magnetic field. On the left, we have the normal state in external field H_c with $F = F_n + H_c^2/2$, where F_n is the free energy of the normal metal. Thus, $G(x \to -\infty) = F_n - H_c^2/2$, which is the same as $G(x \to \infty)$

---

* A detailed discussion of thermodynamic aspects can be found in Reference [7].

by definition of $H_c^2/2$ as the condensation energy. The boundary surface energy is defined then as the difference

$$
\sigma_{ns} = \int\limits_{-\infty}^{+\infty} dx[G(x) - F_n + H_c^2/2]
$$

$$
= \int\limits_{-\infty}^{+\infty} dx\left[-a|\psi|^2 + \frac{b}{2}|\psi|^4 + \frac{1}{2}|(\nabla + iq\mathbf{A})\psi|^2 + \frac{1}{2}\mathbf{B}^2 - \mathbf{B}\cdot\mathbf{H}_c + \frac{1}{2}\mathbf{H}_c^2\right]. \tag{5.66}
$$

In a general case, the minimization of this expression should be done numerically, but the limiting cases $\kappa \ll 1$ and $\kappa \gg 1$, shown in Figure 5.2, are easy to analyze.

First, consider $\kappa \gg 1$. Here $|\psi|$ recovers its equilibrium bulk value very quickly. Also, in this case, we can neglect energy costs associated with gradients of $|\psi|$. The potential-energy terms on the superconducting side compensate the $H_c^2/2$ term. The problem reduces to dealing with magnetic field screening in a layer of width $\sim\lambda$ with the surface energy defined by the integral

$$
\sigma_{ns}^{\kappa\gg 1} \approx \int\limits_{0}^{\infty} dx\left[\frac{\lambda^2}{2}\left(\frac{dB}{dx}\right)^2 + \frac{1}{2}B^2 - BH_c\right], \tag{5.67}
$$

where $B = B_c e^{-x/\lambda}$ [see Equation (5.41)]. Elementary integration yields a negative value

$$
\sigma_{ns}^{\kappa\gg 1} \approx -\frac{1}{2}H_c^2\lambda = -\frac{1}{2q}\frac{a^{3/2}}{b^{1/2}}. \tag{5.68}
$$

In the $\kappa \ll 1$ limit, the magnetic field is screened at a length scale much smaller than the coherence length ξ, and the dominant energy contribution is coming from the variation of $|\psi|$. This results in the energetically expensive boundary of width ξ, where the system does not fully benefit from having a full-fledged

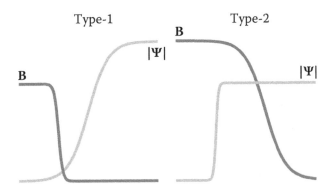

Figure 5.2   A schematic picture of the interface between normal and superconducting phases in type-1 and type-2 regimes, in the cases $\kappa \ll 1$ and $\kappa \gg 1$, respectively.

superconducting state. Now the surface tension is approximately defined by the integral [using the same notations as in Equation (5.56)]

$$\sigma_{ns}^{\kappa \ll 1} \approx \frac{a^2}{2b} \int_0^\infty dx \left[\left(1 - |\tilde{\psi}|^2 \right)^2 + 4\xi^2 \left(\frac{d|\tilde{\psi}|}{dx} \right)^2 \right]. \tag{5.69}$$

This energy is positive and can be roughly estimated as $\xi H_c^2/2$. When the solution of Equation (5.56) is substituted into the integral, one finds

$$\sigma_{ns}^{\kappa \ll 1} \approx \frac{4a^2}{3b} \xi = \frac{2}{3} \frac{a^{3/2}}{b}. \tag{5.70}$$

The solution of the full problem shows that the crossover from positive to negative interface energy happens at the critical value $\kappa_c = 1$. We will demonstrate it in Section 5.6 in the more general context of vortex interaction.

Historically, the sign of the interface energy (where the interface was supposed to be translationally invariant and without phase gradients along it) was determining the division line between type-1 ($\kappa < 1$, positive σ_{ns}) and type-2 ($\kappa > 1$, negative σ_{ns}) superconductors. At first glance, this division is quite natural because there are only two phases and the interface energy between them can be either positive or negative. In the following chapters, we will see, however, that the division does not have a straightforward generalization in multicomponent systems and thus cannot furnish a general classification scheme for superconductors. In multicomponent systems (see Chapter 6), one finds simultaneously different kinds of boundaries and domain shapes depending on phase gradients of the order-parameter fields and other factors. In particular, there are boundaries that are not translationally invariant and thus are very different from the previous 1D solution. In addition, some of these boundaries can have positive energy and some negative energy, leading to a distinct "type-1.5" superconductivity specific to multicomponent systems with several coherence lengths ξ_i, such that $\xi_j < \lambda < \xi_k$.

For the purposes of this chapter, it is sufficient to work with the traditional classification scheme. For positive σ_{ns} (i.e., type-1 superconductors), we are dealing with the standard situation found in first-order phase transitions where the energy penalty for forming interfaces at the coexistence line ensures that, in external fields, the system forms macroscopically large normal and superconducting domains. This is no longer the case for negative surface tension, since free energy is reduced by creating large numbers of smaller and smaller domains until their sizes become comparable to certain minimal characteristic length scale (to be determined in Section 5.4). Clearly, such domain structure would radically modify the microscopic properties of the phase, invalidating all basic assumptions behind previous calculations. In particular, one must consider the possibility of having states with nonzero phase windings, which were excluded from the previous derivation.

The negative surface tension for $\kappa > 1$ is nothing but an indication that the system enters a new phase of matter distinct from both the normal state and the superconductor with the expelled magnetic field; moreover, the new phase is supposed to be stable in a finite interval of fields near H_c. Indeed, if an external field is slightly below H_c, the free energy of the superconductor with expelled field is lowered by creating domains of the normal phase as long as $\sigma_{ns}S + \left(G_n - F_s^{(0)}\right)V_n < 0$, where S and V_n are the total interface area and normal state volume, respectively. This suggests the existence of the lower critical field $H_{c1} < H_c$, such that the Meissner state in type-2 superconductors survives only in fields $H < H_{c1}$. By the same token, if an external field is slightly above H_c, the free energy of the normal state is lowered by creating superconducting domains. Correspondingly, there should exist an upper critical field $H_{c2} > H_c$, such that for $H > H_{c2}$ the true normal state becomes stable. In the following, we will discuss the upper and lower critical fields in more detail. Note that type-1 superconductors become normal in fields $H > H_c$.

5.3.3   Intermediate Landau State

Positive interface energy in type-1 superconductors implies that the phase transition in external field is firstorder; that is, in the thermodynamic limit, the system abruptly changes from one uniform state to another through nucleation and growth of domains of the competing phase. In finite samples, however, one may observe macroscopic phase separation between normal and superconducting domains known as the Landau state. At first sight, for $\sigma_{ns} > 0$, the optimal domain shape in an external magnetic field $H\hat{z}$ should be circular in the xy-plane for the best surface-to-volume ratio. This assertion, as shown by Landau [8], is not valid for finite samples because it neglects boundary effects and does not take into account magnetic energy associated with stray (demagnetization) fields emitted *outside* of the superconducting surface. These effects lead to formation of macroscopically large multidomains, or laminar structures, which are frequently observed experimentally. A schematic picture of Landau domains is shown in Figure 5.3.

Similar to other first-order transitions, slow kinetics of the nucleation process (the process of formation of seeds of the competing phase) means that the superconducting phase in type-1 materials can survive in fields exceeding H_c (an analog of the "superheating" phenomenon). Also, when domains of the competing phase form, they are system and history dependent.

5.4   Vortices in a Superconductor

We already know—from our discussion of the Meissner effect and magnetic flux quantization—that coupling to the vector potential changes vortex properties quite dramatically compared to neutral superfluids. In addition to the previously mentioned general aspects, it is important to analyze the structure of the vortex core (see Figure 5.4), as well as to derive the interaction between the vortices. These questions will be addressed in this section.

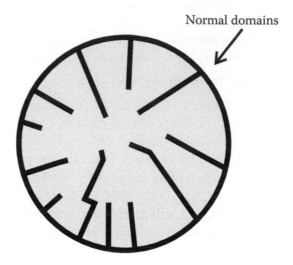

Normal domains

Figure 5.3 A schematic picture of normal phase domains in a disk-shaped superconductor of finite thickness (the Landau state). In the immediate vicinity of the boundary of the sample, the domains split and form smaller-scale structures that have more detail in the x–y plane (not shown here).

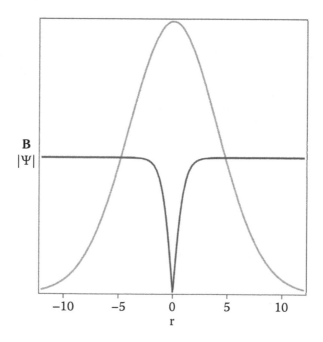

Figure 5.4 Schematic picture of a cross section of a vortex showing the magnetic field and the absolute value of the order parameter varying with distance from the vortex core.

We consider an infinite straight vortex line with the phase winding $2\pi N$, thus carrying a magnetic flux $-\Phi_0 N$ [see (5.14) and (5.15)]. We use cylindric coordinates, the position of the line corresponding to the z-axis. We will derive in the following that in the London model, the elementary vortex solution, $N = 1$, corresponds to the minimal free energy per unit phase winding number. However, for generality, we consider an arbitrary integer phase winding N.

5.4.1 Vortices and Their Interaction in the London Model

In the regime $\xi \ll \lambda$, the absolute value of the order parameter quickly recovers its bulk value, relative to the scale of magnetic field localization. The structure of the magnetic field can then be obtained by setting $|\psi| = $ const. with appropriate regularization at the $r \sim \xi$ scale (in the London limit, this approximation is simply a sharp cutoff: $|\psi|$ is set to zero at $r < \xi$). And the superconducting current density around the vortex is given by

$$\mathbf{J} = -\lambda^{-2}\left(\frac{1}{q}\nabla\theta + \mathbf{A}\right). \tag{5.71}$$

For the simplest single-component superconductor, a vortex solution is axially symmetric. In cylindric coordinates, the vector potential around the core has the form

$$\mathbf{A} = \frac{\mathbf{r} \times \hat{z}}{r} A(r), \tag{5.72}$$

where $\mathbf{r}$ is the radius vector in the xy-plane. Since magnetic fields and currents decay exponentially in the bulk, asymptotically, the vector potential must compensate exactly the phase gradients in (5.71); that is,

$$A(r \to \infty) \longrightarrow -\frac{N}{qr}, \tag{5.73}$$

consistently with the value of $-\Phi_0 N$ for the total magnetic flux carried by the vortex.

Substituting (5.71) into Maxwell equation (5.55) and applying curl operation to both sides, we obtain a closed equation for the magnetic field around the vortex

$$\mathbf{B} + \lambda^2 \text{curl}\,\text{curl}\,\mathbf{B} = (\Phi_0/2\pi)\text{curl}\,\nabla\theta. \tag{5.74}$$

The right-hand side of (5.74) is a generalized function because $\text{curl}\,\nabla\theta$ for $\nabla\theta = N\mathbf{r} \times \hat{z}/r^2$ is zero everywhere except at the origin, where it is infinite. Integrating this function over an area inside the contour σ, which includes the origin, and performing the Stokes area-to-contour transformation, we find that the generalized function is identical to

$$\text{curl}\,\nabla\theta = 2\pi N\hat{z}\,\delta(\mathbf{r}). \tag{5.75}$$

Since the magnetic field is also oriented along the z-direction, we have to deal with the 2D cylindrically symmetric problem (in the following, Δ_r is a radial Laplace operator)

$$B - \lambda^2 \Delta_r B = \Phi \delta(\mathbf{r}), \tag{5.76}$$

which must be solved with the boundary condition

$$B(r \to \infty) \longrightarrow 0. \tag{5.77}$$

This is a standard equation that has the following solution:

$$B = \frac{\Phi}{2\pi\lambda^2} K_0(r/\lambda). \tag{5.78}$$

Here, K_0 is the modified Bessel function with asymptotic properties

$$K_0(x \to 0) \longrightarrow -\ln(x), \qquad K_0(x \to \infty) \longrightarrow e^{-x}/\sqrt{x}. \tag{5.79}$$

One can check, by explicit differentiation, that $(-\Delta_r + \lambda^{-2})K_0(r/\lambda)$ is a generalized function $2\pi\delta(\mathbf{r})$, ensuring that the right-hand side of Equation (5.76) is satisfied.

This solution predicts an unphysical divergence of the magnetic field at $r \to 0$. Recall, however, that the vortex has a finite core and thus our solution cannot be used at distances $r < \xi$. By imposing a cutoff at finite $r \sim \xi$, we estimate (with logarithmic accuracy) the magnitude of the magnetic field at the origin as

$$B(0) \approx \frac{\Phi}{2\pi\lambda^2} \ln(\lambda/\xi) \qquad (\lambda \gg \xi). \tag{5.80}$$

We are now ready to calculate the vortex energy per unit length. It is given by the sum of kinetic and magnetic terms (again, neglecting the smaller energy of the core)

$$E_v = \frac{1}{2} \int d^3r \left[\lambda^2 (\operatorname{curl} \mathbf{B})^2 + \mathbf{B}^2 \right]. \tag{5.81}$$

The integration volume is a cylinder of unit length in z-direction and infinite radius. Using identity $(\operatorname{curl} \mathbf{B})^2 = \mathbf{B} \cdot \operatorname{curl} \operatorname{curl} \mathbf{B} + \operatorname{div}(\mathbf{B} \times \operatorname{curl} \mathbf{B})$, we express energy as

$$E_v = \frac{1}{2} \int d^3r \left[\mathbf{B} \cdot (\mathbf{B} + \lambda^2 \operatorname{curl} \operatorname{curl} \mathbf{B}) \right] + \frac{\lambda^2}{2} \int d^3r \operatorname{div}(\mathbf{B} \times \operatorname{curl} \mathbf{B}), \tag{5.82}$$

and employ the Gauss theorem to transform the second integral

$$\int d^3r \operatorname{div}(\operatorname{curl} \mathbf{B} \times \mathbf{B}) = \oint (\operatorname{curl} \mathbf{B} \times \mathbf{B}) \cdot d\mathbf{S} = 0. \tag{5.83}$$

Indeed, the integral over infinitely remote cylindrical surface is zero because B decays exponentially in the bulk. Thus,

$$E_v = \frac{1}{2} \int d^3r \left[\mathbf{B} \cdot (\mathbf{B} + \lambda^2 \operatorname{curl} \operatorname{curl} \mathbf{B}) \right], \tag{5.84}$$

or, taking into account that $\mathbf{B}$ is the solution of Equation (5.76),

$$E_v = \frac{\Phi}{2} B(0) \approx \frac{1}{4\pi} \left(\frac{\Phi}{\lambda} \right)^2 \ln \left(\frac{\lambda}{\xi} \right). \tag{5.85}$$

As expected, the vortex energy increases quadratically with $\Phi = N\Phi_0$, meaning that, in the London model, vortices with $N \neq 1$ are unstable against breakup into N elementary vortices. In what follows, we will frequently assume that $N = 1$.

To avoid divergence of magnetic and kinetic energy, we introduced a cutoff at the length scale ξ. An estimate for the core energy of one quantum vortex is given by the condensation-energy density times $\pi\xi^2$

$$E_c \sim \frac{\pi\xi^2}{2} H_c^2 = \frac{1}{32\pi} \left(\frac{\Phi_0}{\lambda} \right)^2. \tag{5.86}$$

Here we used definitions (5.60), (5.39), (5.57), and Φ_0. As stated earlier, the core energy can be neglected in the London limit as long as $\ln(\lambda/\xi) \gg 1$. In the $q \to 0$ limit, the London penetration length diverges, and so does the logarithm in Equation (5.85). Since the ratio Φ_0/λ is independent of q, we recover the standard expression for the vortex energy in a neutral superfluid.

Consider now interaction between two vortices each having 2π phase winding. Our calculations apply equally to a pair of vortices in 2D and two straight parallel vortex lines in 3D (in the latter case, we deal with interaction energy per unit length). In the London model, we neglect core energies. Then, the energy of a vortex pair can be obtained from

$$E_v^{(1+1)} = \frac{1}{2} \int d^3r \left[\lambda^2 (\operatorname{curl} \mathbf{B})^2 + \mathbf{B}^2 \right], \tag{5.87}$$

where the magnetic field is now a superposition of two solutions (5.78), one centered at point $\mathbf{r}_1$ and the other at point $\mathbf{r}_2$; that is,

$$\mathbf{B} + \lambda^2 \operatorname{curl} \operatorname{curl} \mathbf{B} = \Phi_0 \left[\delta(\mathbf{r} - \mathbf{r}_1) + \delta(\mathbf{r} - \mathbf{r}_2) \right] \hat{z}. \tag{5.88}$$

Therefore, for two vortices in the London limit,

$$E_v^{(1+1)} = (\Phi_0/2)[B(\mathbf{r}_1) + B(\mathbf{r}_2)]. \tag{5.89}$$

Since the fields $B(\mathbf{r}_1)$ and $B(\mathbf{r}_2)$ have contributions from both vortices, $E_v^{(1+1)}$ depends on the separation $R = |\mathbf{r}_1 - \mathbf{r}_2|$ between them:

$$E_v^{(1+1)} = 2E_v + E_{\text{int}} = 2E_v + \Phi_0 B(R), \tag{5.90}$$

where $B(R)$ is a field of a single vortex at the distance R from the core. This yields

$$E_{\text{int}} = \frac{\Phi_0^2}{2\pi\lambda^2} K_0(R/\lambda), \tag{5.91}$$

decaying exponentially with R.

For a pair of vortices with opposite phase windings, the interaction potential is attractive. In the $q \to 0$ limit, we recover the familiar long-range logarithmic interaction between vortices in a neutral superfluid. The divergence of (5.91) at $R = 0$ is unphysical and is cut off at $R \sim \xi$, when one accounts for vanishing $|\psi|$ at the center of the vortex core.

The force between vortices is

$$F = -\frac{dE_{\text{int}}}{dR} = -\Phi_0 \frac{dB}{dR} = J(R)\Phi_0, \tag{5.92}$$

where $J(R)$ is the current density induced by the first vortex at the center of the second vortex. This formula also describes the force experienced by the vortex subject to the generic current flow. Indeed, there is no physical mechanism to differentiate the origin of the local current density, whether it is due to other vortices or external sources. Equation (5.92) is also known as the Lorentz force.

5.4.2   Vortices and Their Interaction in the Ginzburg–Landau Model

So far, we have considered only the interaction between vortices due to magnetic and kinetic energies. There is also important core–core intervortex interaction. In order to calculate it, density variations should be treated on equal footing with magnetic fields.

Let us consider now a vortex solution in the GL model. For a cylindrically symmetric vortex line, the GL equations take the form (note that $\nabla\theta \cdot \nabla|\psi| = 0$ and $\mathbf{A} \cdot \nabla|\psi| = 0$ by symmetry)

$$-\frac{1}{r}\frac{d}{dr}\left(r\frac{d}{dr}|\psi|\right) + \left[\left(\frac{N}{r} - qA\right)^2 + \left(2b|\psi|^2 - 2a\right)\right]|\psi| = 0, \tag{5.93}$$

$$-\frac{d}{dr}\left[\frac{1}{r}\frac{d}{dr}(rA)\right] + \left(q^2A - \frac{qN}{r}\right)|\psi|^2 = 0. \tag{5.94}$$

The boundary conditions are

$$|\psi|(r = 0) = 0, \qquad |\psi|(r \to \infty) \longrightarrow \sqrt{\frac{a}{b}} \equiv u, \tag{5.95}$$

and

$$A(r \to \infty) \longrightarrow \frac{N}{qr}. \tag{5.96}$$

Before analyzing the solution in detail, let us make general remarks on the magnetic field penetration length in GL theory. The GL theory supports the simplest vortex solutions of the form

$$\psi = f(r)e^{i\theta}, \qquad\qquad (A_1, A_2) = \frac{a(r)}{r}(-\sin\theta, \cos\theta), \qquad (5.97)$$

where $f = |\psi|/u$ are real vortex-profile functions with boundary behavior $f(0) = a(0) = 0$, $f(\infty) = u$, $a(\infty) = 1/q$. Very far from a vortex, all the fields are close to their ground-state values. We thus expand, up to quadratic order, in powers of small quantities $s = |\psi| - u$ and A. This linearized theory reads

$$F_{\text{lin}} = \frac{1}{2}\left(|\nabla s|^2 + \mu s^2\right) + \frac{1}{2}(\partial_1 A_2 - \partial_2 A_1)^2 + \frac{1}{2\lambda^2}|A|^2. \qquad (5.98)$$

The key aspect here is that the fields decouple: We have a Klein–Gordon functional for modulus field s, whose mass is the inverse coherence length, and independent Proca functional for vector potential. The latter is a massive field with the mass being the inverse penetration length λ^{-1}, just like in our previous discussion of the magnetic field in the London model, Equation (5.38).

At the origin of a vortex, the fields behave as

$$|\psi|(r \to 0) \propto r^N, \qquad\qquad A(r \to 0) \propto r. \qquad (5.99)$$

Indeed, $A \propto r$ ensures that magnetic flux inside the circle of radius r is proportional to r^2 and thus the magnetic solution is regular at the origin. The behavior of $|\psi|$ follows then directly from (5.93) with only the most singular $1/r^2$ term retained in the square brackets. For higher phase windings N, the density recovers slower and the core size increases. We will focus on the $N = 1$ solution. Neglecting density variation at large distances, we use identity $-(A/r + A') = H$, the solution (5.78), and identity $K_1(x)/x + K_1'(x) = -K_0(x)$ to suggest the asymptotic behavior of the vector potential, the Nielsen–Olesen ansatz [9]:

$$A(r \to \infty) \approx \frac{1}{qr} - \frac{C}{q\lambda}K_1(q|\psi|r) = \frac{1}{qr} - \frac{C}{q\lambda}\sqrt{\frac{\pi\lambda}{2r}}e^{-r/\lambda}, \qquad (5.100)$$

where constant C is close to unity in the strongly type-2 limit. This follows from the requirement that the flux $\Phi = 2\pi R A(R)$ is small for the area inside the circle of radius R such that $\xi \ll R \ll \lambda$ and observing that, for $q|\psi|r \ll 1$, one has $K_1(q|\psi|r) \approx 1/(q|\psi|r)$.

Let us now look more closely at the structure of the core. In the "extreme" type-2 limit $q \to 0$, one might think of ignoring the vector potential in the first GL equation. This would lead to the same equation and solution as in the neutral superfluid; see Equation (1.115) in Chapter 1. This solution is *not* saturating to $|\psi| = \sqrt{a/b}$ exponentially, contrary to the prediction of Equation (5.56), which does not take into account phase gradients. Instead, $|\psi|$ approaches its bulk value in a power law fashion ($1/r^2$ in the neutral case). This observation alone tells us that taking into account phase gradients and vector potential is important for obtaining a correct solution.

To simplify expressions, we will use $f(r) = |\psi|\sqrt{b/a}$ and $\alpha(r) = Aqr$. Then GL equations take the form

$$f''(r) + \frac{1}{r}f'(r) - \frac{1}{r^2}[1 - \alpha(r)]^2 f(r) - \frac{1}{2\xi^2}[f^2(r) - 1]f(r) = 0, \tag{5.101}$$

$$\alpha'' - \frac{1}{r}\alpha' + \frac{1}{\lambda^2}(1 - \alpha)f^2 = 0. \tag{5.102}$$

When looking for the asymptotic large-distance behavior, we can set $f = 1$ in the second equation. The solution of linear equation $\alpha'' - \alpha'/r + (1 - \alpha)/\lambda^2 = 0$ is already known to us, $\alpha(r) \approx 1 - C\sqrt{\pi r/2\lambda}\,e^{-r/\lambda}$: see Equation (5.100).

It is clear that, under certain conditions, it is possible to omit from Equation (5.101) the term containing $(1 - \alpha)^2 \propto r\,e^{-2r/\lambda}$. Without this term, the asymptotic solution saturates exponentially at the length scale of coherence length

$$1 - f \propto e^{-r/\xi}. \tag{5.103}$$

Therefore, omitting the term $(1 - \alpha)^2 \propto r\,e^{-2r/\lambda}$ is consistent for type-1 and moderately type-2 superconductors when $\lambda < 2\xi$, or $\kappa < 2$. Further analysis of Equation (5.101) [10] indeed shows that for $\kappa > 2$, the order parameter modulus in a vortex has exponential dependence with the characteristic decay length being half that of the magnetic field

$$1 - f \propto \frac{\xi}{(\kappa - 2)r}\,e^{-2r/\lambda}, \qquad \kappa > 2. \tag{5.104}$$

Problem 5.1 *Assuming that $\kappa < 2$, consider the equation*

$$f''(r) + \frac{1}{r}f'(r) - \frac{1}{2\xi^2}[f^2(r) - 1]f(r) = 0$$

and find the leading and the first subleading terms in the asymptotic decay of $s = 1 - f$.

A quite common misconception in literature is that the asymptotic behavior of the density in a vortex is determined by coherence length and is given by (5.103). In fact, as we discussed earlier, in strongly type-2 superconductors, the asymptotic behavior of density field is different and is determined by $\lambda/2$. The reason behind this is that the linearized theory (5.98) fails for the problem of this asymptotical behavior, and the asymptotical behavior of density is determined by nonlinearities. At the same time, it should be emphasized that another common misconception is that overall vortex core size is determined by the same length scale as the one in the density asymptotics. In contrast for type-2 superconductors with $\lambda \gg \xi$, the decay length $\lambda/2$ should not be confused with the core size. As mentioned earlier, when screening effects are neglected, $1 - f$ decays as ξ^2/r^2, meaning that f is very close to unity at distances $r > \lambda/2$ and ξ is still a more accurate core size scale, despite it not being associated with the density field asymptotic.

For type-1 and type-2 superconductors, the intervortex interaction mediated by magnetic and current–current couplings, $E_{int} = E_v^{(1+1)} - 2E_v$, at large distances is well approximated by Equation (5.91) multiplied by some function of κ. Generically, at short distances, there is also core–core interaction mediated by the density field. For large vortex separations, it can be calculated by linearizing the GL theory [11] (when linearization holds). Similar calculation can be conducted in a more compact form by linearizing the theory first and adding point sources such that at a large distance the asymptotic vortex solution is reproduced [12]. Technically, this formalism is similar to what we did for the magnetic and current–current interaction. Finally, even though the full interaction potential for vortices in the GL model is not known analytically, it can be computed numerically.

Assuming that the field f deviates only slightly from the ground-state value (which is the case for tails of the density modulation produced by vortices), we introduce $s = 1 - f$ and linearize GL functional by assuming that s is small. The resulting expression for energy density has the form of the Klein–Gordon theory

$$F = \frac{1}{2}(\nabla s)^2 + \frac{1}{2\xi^2}s^2, \tag{5.105}$$

and leads to the following equation for density modulation s:

$$s'' + \frac{1}{r}s' = \frac{1}{\xi^2}s. \tag{5.106}$$

The solution is

$$s(r) = C_2 K_0(r/\xi), \tag{5.107}$$

with C_2 determined by the strength of the source term, $k(\mathbf{r})s = -2\pi C_2 \delta(\mathbf{r})s$, added to the right-hand side of energy density (5.105). It can be interpreted as a point particle at $\mathbf{r} = 0$, which produces the long-range density modulation identical to that of a vortex (5.107).

For a well-separated pair of vortices, the linearized solution is a superposition of two single-vortex solutions centered at $(\mathbf{r}_1)$ and $(\mathbf{r}_2)$; that is, $(s_1 + s_2)$. Substituting $s_1 + s_2$ into the linear theory (5.105) with two-point sources $k(\mathbf{r}_1)s_1$ and $k(\mathbf{r}_2)s_2$, and subtracting energies of two isolated vortices, we obtain the core–core interaction energy

$$E_{int}^{(c)} \propto \int d^2r\, k(\mathbf{r}_1)s_2 = -2\pi C_2^2 \int d^2r\, \delta(\mathbf{r} - \mathbf{r}_1) K_0(\xi^{-1}|\mathbf{r} - \mathbf{r}_2|)$$

$$= -2\pi C_2^2 K_0(\xi^{-1}|\mathbf{r}_1 - \mathbf{r}_2|) = -2\pi C_2^2 K_0(R/\xi), \tag{5.108}$$

where $R = |\mathbf{r}_1 - \mathbf{r}_2|$.

Since phase gradients and vector potential were ignored, this contribution to the interaction potential originates purely from the density suppression at the cores. In contrast to current–current and magnetic interactions, it is always attractive; that is, its sign does not depend on the sign of the current circulation.

An interaction potential between co-directed vortices that originates from current and magnetic energies has the same form as before [see Equation (5.91)]. Thus, for two vortices with similar vorticity and $\kappa < 2$, one has the following interaction energy consisting of repulsive current–current electromagnetic part and attractive core–core interaction:

$$E_{\text{int}} = 2\pi[C_1^2 K_0(r/\lambda) - C_2^2 K_0(r/\xi)], \tag{5.109}$$

where C_1 and C_2 are certain functions of GL parameters. In what follows, we will perform the analysis of the full nonlinear GL theory to show that, at exactly the critical value $\kappa = 1$, the coefficients C_1 and C_2 become equal and thus vortices do not interact at any distance; that is, the attraction between cores exactly compensates the current–current and magnetic repulsion [11,13,14].*

Note that, given that linearization fails for superconductors with $\kappa > 2$, the form (5.108) of core–core interaction is accurate only for type-1 and weakly type-2 superconductors with $\kappa < 2$. On the other hand, in superconductors $\kappa \gg 2$, the core–core intervortex interaction is negligible compared to current–current and magnetic interaction. At $\kappa < 1$ (type-1 superconductors), the core–core interaction dominates over magnetic and current–current interactions at all intervortex separations. This implies instability against clustering with subsequent formation of one "megavortex," or a domain of the normal phase. This result is consistent with previously established positive surface tension for the normal–superconductor interface: The best configuration corresponds to the smallest possible interface area. At $\kappa > 1$ (type-2 superconductors), the electromagnetic and current–current repulsion given by the first term in (5.109) dominates over core–core attraction at all distances and, for finite flux density, leads to the onset of a new state of matter—a stable vortex structure.

Inverting the phase winding of one of the two vortices, one gets a vortex–antivortex pair, in which both core–core and current/magnetic interactions are attractive:

$$E_{\text{int}} = -2\pi[C_1^2 K_0(r/\lambda) + C_2^2 K_0(r/\xi)]. \tag{5.110}$$

Apart from being functions of κ, coefficients C_1 and C_2 depend on winding numbers.

5.5 Upper and Lower Critical Magnetic Fields (H_{c1} and H_{c2})

Negative surface tension for the normal–superconductor interface found for $\kappa > 1$ in the external magnetic field $H = H_c$ is nothing but a signature of the system's instability and emergence of a new phase with magnetic flux penetrating into the

---

* Obviously, in real materials, the truly noninteracting regime cannot be realized even at $\kappa = 1$, because the interaction vanishes only at the level of classical field theory. There will always be tiny nonuniversal (i.e., microscopic physics-specific) corrections to the interaction potential. They result in various complicated (including nonmonotonic)—but tiny—material-specific intervortex potentials that go beyond the GL theory.

superconductor bulk. In the following, we will discuss the structure of the new *vortex-lattice* phase and critical magnetic fields required to stabilize it in type-2 superconductors [15].

5.5.1 Lower Critical Field H_{c1}

Similarly to rotation that induces vortices in a superfluid [3], an external magnetic field can induce vortices in a superconductor. The problems are in fact asymptotically equivalent in the limit $\lambda \to 0$ (which is the case when density gets depleted in high magnetic fields). At low fields, however, the two cases are different, because screening effects are absent in the neutral case. In a direct analogy with the effective vector potential for a neutral superfluid in the rotating frame, a magnetic field cannot penetrate into the superconductor bulk without causing singular vortex-like flow pattern: In the absence of vortices the field of phase can be gauged out, making it obvious that the state with the bulk magnetic field will cost an enormous (superextensive) energy $\propto \int d^3 r \mathbf{A}^2$. The requirement that the energy per unit volume be intensive implies a certain finite concentration of vortices, strictly related to the value of magnetic flux inside the sample—one magnetic flux quantum per elementary vortex [see Equation (5.16) and corresponding discussion]. The problem is analogous to the description of a neutral superfluid in rotating frame, where the concentration of vortices is dictated by the angular frequency of the rotation. In both cases, vortices naturally form a lattice to further minimize the (free) energy.

In an external field, $\mathbf{H}$, the Gibbs free energy of a superconductor with a vortex (we assume that the vortex-line direction is parallel to $\mathbf{H}$) is given by

$$G - F_s^{(0)} = E_v - \int d^3 r \, \mathbf{B} \cdot \mathbf{H} = E_v - \Phi_0 H, \tag{5.111}$$

where $\mathbf{B}$ is the magnetic field of a vortex. For positive E_v, there is a threshold magnetic field strength for which the free-energy difference goes negative. At the critical field value, it becomes energetically beneficial to introduce a vortex into the superconductor. This defines the lower, or the first, critical magnetic field

$$H_{c1} = \frac{E_v}{\Phi_0}. \tag{5.112}$$

In the London limit, we can use (5.85) to determine the critical value with logarithmic accuracy

$$H_{c1} = \frac{1}{4\pi} \left(\frac{\Phi_0}{\lambda^2} \right) \ln \left(\frac{\lambda}{\xi} \right). \tag{5.113}$$

The thermodynamic field H_c can be expressed as $H_c = \Phi_0/(4\pi\xi\lambda)$. We immediately see that, for strongly type-2 superconductors, H_{c1} is much smaller than H_c:

$$\frac{H_{c1}}{H_c} = \frac{\xi}{\lambda} \ln(\lambda/\xi) \qquad (\lambda \gg \xi). \tag{5.114}$$

Thus, type-1 and type-2 superconductors, while behaving similarly in weak magnetic fields, demonstrate a remarkably different behavior in stronger fields. In the type-1 case, the supercritical field simply destroys superconductivity in favor of the normal state. In the type-2 case, the magnetic field starts penetrating the bulk in the form of quantized vortices when H exceeds H_{c1}. The effect of gradual penetration of magnetic field in the bulk of superconductors was experimentally discovered by Shubnikov et al. [16].

Note that type-1 superconductors also have vortex solutions, but these solutions describe excited states, not an equilibrium thermodynamic phase.

5.5.2 Vortex-Lattice State

Let the external magnetic field be slightly above H_{c1}. To determine the (2D) density of vortices, $n_v = N_v/S$, where N_v is the number of vortices and S is the sample cross-sectional area perpendicular to the applied field, one should account for vortex–vortex interaction E_{int}. When the interaction can be approximated by pairwise forces, the Gibbs free energy reads

$$G_v = N_v E_v + \sum_{\alpha > \beta} E_{int}(R_{\alpha\beta}) - S\,\mathbf{B}\cdot\mathbf{H}, \tag{5.115}$$

with $\mathbf{B}$ the *average* magnetic field in the sample. In accordance with (5.16), we have $n_v = B/\Phi_0$. Then

$$G_v = S(B/\Phi_0)E_v + \sum_{\alpha > \beta} E_{int}(R_{\alpha\beta}) - SBH. \tag{5.116}$$

Repulsive interactions between vortices limit their number in the sample. Minimizing free energy with respect to the magnetic field B, we obtain an equation

$$\frac{\partial G_v}{\partial B} = \frac{\partial}{\partial B}\sum_{\alpha > \beta} E_{int}(R_{\alpha\beta}) + S(H_{c1} - H) = 0. \tag{5.117}$$

We immediately see that for $H \to H_{c1} + 0$, the density of vortices goes to zero. As the field increases above H_{c1}, the vortex density rises continuously and so does magnetization. Penetration of magnetic field in a type-2 superconductor at H_{c1} is a continuous phase transition.

An estimate for $n_v = B/\Phi_0$ in the vicinity of H_{c1} in the $\kappa \gg 1$ case is obtained by using expression (5.91) for the interaction energy with $R^2(H) = 1/n_v$ and assuming that each vortex couples only to six nearest neighbors (recall that we discuss low vortex density with exponentially decaying interaction potential). An elementary algebra yields

$$\frac{3\Phi_0}{2\pi\lambda^2}\frac{\partial}{\partial x}\left[x K_0(x^{-1/2})\right] = H - H_{c1}, \tag{5.118}$$

where $x = B\lambda^2/\Phi_0 = n_v\lambda^2$. With logarithmic accuracy, the solution is

$$n_v = \lambda^{-2}\ln^{-2}\left(\frac{\Phi_0/\lambda^2}{H - H_{c1}}\right), \tag{5.119}$$

predicting infinitely sharp phase transition features.

In the ground state, it is straightforward to show numerically that, for the pairwise interaction potential (5.91), the lowest energy configuration corresponds to the triangular vortex lattice. With this knowledge, one has to compute the sum in Equation (5.117) with lattice spacing $R^2 = 2/\sqrt{3}n_v$ and solve the corresponding equation.

5.5.3   Upper Critical Field H_{c2}

Further increasing magnetic field, one reaches the state when vortex cores start to overlap. At a certain field H_{c2}, called the upper, or second, critical field, the superconductivity is destroyed in favor of the normal state. When vortex cores overlap, the absolute value of the field ψ is suppressed everywhere. Correspondingly, the effective magnetic field penetration length increases and the magnetic field becomes more uniform. In the vicinity of H_{c2}, we are dealing with the state that has nearly homogeneous magnetic field in the bulk and vanishing $|\psi|$. This circumstance holds the key to determining the exact value of H_{c2}.

For small $|\psi|$, we can linearize the GL equation

$$(-i\nabla + q\mathbf{A})^2\psi = 2a\psi, \tag{5.120}$$

and solve it under the assumption that the magnetic field is homogeneous. Note that (5.120) is identical to the Schrödinger equation for a charged particle. The spectrum of this equation is highly degenerate and admits solutions that are either entirely localized around some point in space or nearly homogeneous over an arbitrary large area; the latter case can be represented as a superposition of localized solutions. Since the lowest possible eigenvalue is qH, we must require that $a > qH/2$. This condition defines the second critical magnetic field

$$H_{c2} = \frac{2a}{q} = \frac{\Phi_0}{4\pi\xi^2}. \tag{5.121}$$

Comparing the value of H_{c2} to the thermodynamic field, we find

$$\frac{H_{c2}}{H_c} = \frac{\lambda}{\xi} > 1, \tag{5.122}$$

as expected.

The value of $|\psi|^2$ near H_{c2} can be estimated by adding nonlinear terms to the equation and balancing them against the linear contribution $b|\psi|^2 = a - qH/2$; that is, at the level of this mean-field argument, we have a continuous phase transition with $|\psi|^2 = q(H_{c2} - H)/2b$.

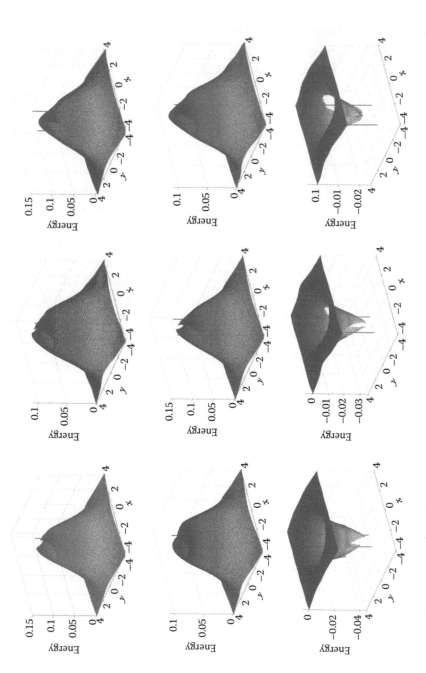

Figure 5.5 Intervortex interactions (total, pairwise, and three-body contributions are shown separately in the first, second, and third rows, respectively) in a system of three vortices in a type-2 superconductor computed numerically using GL functional with $a = 1$, $b = 1$, and $q = 1.5$. Interaction energy is shown as a function of position of the third vortex relative to a vortex pair fixed at $(\pm R_1/2, 0) = 1.2, 1.6, 2.0$ (shown by vertical lines). Each column corresponds to a certain value of the intervortex separation in the first pair, R_1. No data are shown for extremely short intervortex distance where the numerical scheme becomes ill-defined. (From Reference [17].)

When magnetic field is expelled from the superconductor, we have the largest possible diamagnetic response (perfect diamagnetism), with magnetization $M = -H$ (neglecting boundary effects). As more and more vortices penetrate into the bulk, the magnetization of the system increases until it reaches zero at H_{c2} (ignoring other mechanisms of magnetic response).

Note that, at H_{c2}, the magnetic field is essentially uniform and thus there is little difference between a superconductor in $H \approx H_{c2}$ and a superfluid under rotation. Similarly, at the mean-field level at $T > 0$, there exists a counterpart of H_{c2} in the superfluid state, which is the critical angular rotation velocity Ω_c at which the system becomes normal.

5.5.4 Nonpairwise Intervortex Forces

Since GL equations are nonlinear, a straightforward superposition of single-vortex solutions does not necessarily yield a solution. Therefore, in general, the interaction between vortices is nonpairwise.

The crossover from low- to high-field regime is associated with the growing importance of multivortex forces that, in a dense lattice, compete with pairwise interactions. In this regime, we cannot use linearized GL theory to calculate forces between vortices, and this is the reason the interaction potential is no longer given by the sum of pairwise contributions. In Figure 5.5, we show results of the numerical solution for three-body intervortex forces in a type-2 superconductor. These forces are indeed attractive and thus compete with the repulsive two-body terms. By contrast, the three-body intervortex forces in type-1 superconductors are typically repulsive (while two-body forces are attractive). As will be shown in the following, when $\kappa = 1$, the multibody forces are zero.

5.6   Critical Coupling $\kappa = 1$: Bogomolny Bound and Equations

At the critical point, $\kappa = 1$, separating type-1 and type-2 regimes, superconductors have special properties as revealed in detail by Bogomolny [14]. This regime is often called the Bogomolny point. To discuss this regime in universal terms, we rescale the variables: We measure length in units of λ, normalize ψ and $\mathbf{A}$ to the ground-state value of $|\bar{\psi}|$, and measure free energy in units of $q^2|\bar{\psi}|^4$. As a result, we get

$$2F = |(\nabla + i\mathbf{A})\psi|^2 + \frac{\kappa^2}{4}\left(|\psi|^2 - 1\right)^2 + (\text{curl}\mathbf{A})^2. \tag{5.123}$$

For simplicity, in what follows, we deal with a 2D system. The significance of the $\kappa = 1$ point transpires if, instead of the complex field ψ, we work with its real and imaginary parts

$$\psi = \phi_1 + i\phi_2 \tag{5.124}$$

and introduce the operator

$$\mathbb{D}\phi_a = \nabla\phi_a - \epsilon_{ab}\mathbf{A}\phi_b, \tag{5.125}$$

where $a, b = 1, 2$ and ϵ_{ab} is the Levi-Civita symbol. Here and in the following, the summation over repeating subscripts is implied. Then

$$2F = \mathcal{F}_{ij}^2 + \mathbb{D}_i\phi_a\mathbb{D}_i\phi_a + (\kappa^2/4)(\phi_a\phi_a - 1)^2, \tag{5.126}$$

with $\mathcal{F}_{ij} = \partial_i A_j - \partial_j A_i$. Our analysis will be based on the so-called Bogomolny inequality; to arrive at which, we first rewrite (5.126) as

$$2F = \left(\mathcal{F}_{ij} - \frac{\epsilon_{ij}}{2\sqrt{2}}[1 - \phi_a\phi_a]\right)^2 + \frac{1}{2}(\epsilon_{ik}\mathbb{D}_k\phi_a + \epsilon_{ab}\mathbb{D}_i\phi_b)^2 \tag{5.127}$$

$$+ \frac{\kappa^2 - 1}{4}(\phi_a\phi_a - 1)^2 + \frac{\epsilon_{ij}}{\sqrt{2}}\mathcal{F}_{ij}(\phi_a\phi_a - 1) - \epsilon_{ik}\epsilon_{ab}\mathbb{D}_k\phi_a\mathbb{D}_i\phi_b. \tag{5.128}$$

The last two terms in this expression have the form, $\partial_k S_k$, where

$$S_k = -\epsilon_{ik}(\epsilon_{ab}\phi_a\mathbb{D}_i\phi_b + A_i). \tag{5.129}$$

Terms based on divergence of a vector can be integrated using the Gauss theorem.

Consider a group of N co-directed vortices. The first term in the right-hand side of (5.129) is the supercurrent density. Then, integrating along a path that lies at a distance much larger than λ away from the group of the vortices, we get

$$\int \partial_k S_k \, dx dy = \oint \mathbf{S} \cdot d\mathbf{l} = \oint \mathbf{A} \cdot \mathbf{l} = \Phi_0 N. \tag{5.130}$$

In our units, $\Phi_0 = 2\pi$; thus, after integration over the system volume, for the total free energy of the system, we get

$$2F(\Delta\theta = 2\pi N) = 2\pi N + \int dx dy \left\{ \left(\mathcal{F}_{ij} - \frac{\epsilon_{ij}}{2\sqrt{2}}[1 - \phi_a\phi_a] \right)^2 \right.$$

$$\left. + \frac{1}{2}(\epsilon_{ik}\mathbb{D}_k\phi_a + \epsilon_{ab}\mathbb{D}_i\phi_b)^2 + \frac{\kappa^2 - 1}{4}(\phi_a\phi_a - 1)^2 \right\}. \tag{5.131}$$

Given that all terms under the integral are based on squares, we arrive at the so-called Bogomolny bound on the energy of N vortices for $\kappa > 1$*:

$$2F(\Delta\theta = 2\pi N) \geq 2\pi N. \tag{5.132}$$

---

* This kind of inequality is not specific to vortices in superconductors but is also widely used for various theories supporting topological solitons; see Reference [14].

At $\kappa = 1$, the last term vanishes. Furthermore, one can satisfy the bound if the vortex solution simultaneously satisfies two *first-order* Bogomolny partial-differential equations:

$$\mathcal{F}_{ij} - \frac{\epsilon_{ij}}{2\sqrt{2}}[1 - \phi_a \phi_a] = 0, \qquad \epsilon_{ik}\mathbb{D}_k \phi_a + \epsilon_{ab}\mathbb{D}_i \phi_b = 0. \qquad (5.133)$$

It can be shown that these equations indeed have solutions $\kappa = 1$, making it a special critical point on the phase diagram where Bogomolny inequality is saturated.

When deriving (5.132), we specified the total number of vortices (or the total winding number), but the argument did not require us to know the vortex positions. This immediately proves that vortices have neither pairwise nor multivortex interactions at $\kappa = 1$ because the total energy is degenerate with respect to vortex positions.* Furthermore, the Gibbs free energy of a critical superconductor in external magnetic field with N vortices also has remarkably simple properties:

$$G - F_s^{(0)} = N E_v - \int d^2 r \, \mathbf{B} \cdot \mathbf{H} = \pi N - N \Phi_0 H \equiv \pi N (1 - 2H). \qquad (5.134)$$

This free energy goes negative at $H > 1/2$, irrespective of the number of vortices. Thus,

$$H_{c1} = H_{c2} = H_c = 1/2 \qquad (\kappa = 1). \qquad (5.135)$$

Independence on N allows one to consider a giant vortex that can be viewed, in some sense, as a macroscopic normal domain. The degeneracy of free energy on the cluster size implies that the surface tension between the normal and superconducting states is zero at $\kappa = 1$.

5.7   Summary of the Magnetic Response and Vortex Liquid

Let us now summarize the magnetic response of superconductors. The schematic picture of magnetization curves in type-1 and type-2 superconductors is shown in Figure 5.6. The schematic phase diagram is shown in Figure 5.7.

As long as thermal fluctuations of the vortex lattice are negligible, or when vortex lines are pinned by crystalline defects, the system retains superconductivity. This is similar to pinning of the vortex lattice and remanent vorticity in neutral superfluids. Also, the order-parameter field remains ordered in the direction parallel to the oriented vortex structure, and thus there exists a superfluid response in that direction even for unpinned vortices.

On the approach to $H_{c1} + 0$, when the vortex lattice gets extremely dilute, the interaction between the vortices vanishes. As a result, thermal fluctuations melt the vortex lattice. A similar argument applies when the magnetic field is close to H_{c2}: then the intervortex distance is small. Also the phase stiffness is small near H_{c2}. Thus, again, the vortex lattice can melt very easily (see, e.g., [18]). The

---

* This was earlier noticed by Sarma [13] and Kramer [11].

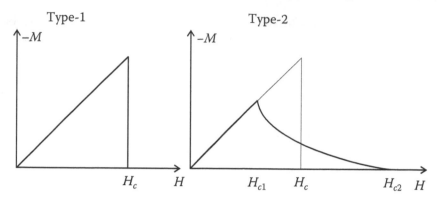

Figure 5.6   Sketch of magnetization curves in a superconductor. The magnetic moment of the type-1 superconductor is defined by perfect diamagnetic response in all fields below H_c. At $H = H_c$, the type-1 superconductor undergoes the first-order phase transition to the normal state. In the type-2 superconductor, the magnetization undergoes a gradual change in fields $H_{c1} < H < H_{c2}$ (i.e., within the Shubnikov state), due to the formation of the Abrikosov vortex lattice.

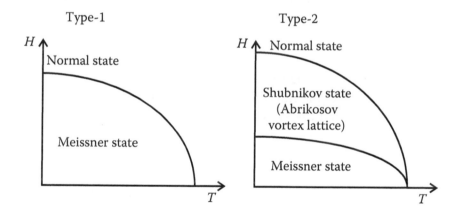

Figure 5.7   Schematic mean-field phase diagram of type-1 and type-2 superconductors.

resulting phase diagram is shown schematically in Figure 5.8. As discussed in the previous chapters, disordered vortex structures destroy order on large scales and, if not pinned, signal the onset of the normal state. In a typical *vortex liquid* state, vortex lines tend to entangle (see detailed discussion in [18]) and destroy phase coherence even in the direction of their preferential orientation. Indeed, consider a system with very large spatial extension in the z-direction, $z \in [0, L]$, with entangled vortices. Such vortices start at one position in xy-plane at $z = 0$ and terminate at entirely different xy position at $z = L$. Now, even if we draw a large area contour residing in a plane with one axis being parallel to $\hat{z}$, we will nevertheless have vortices going through this contour. A current in either of the three spatial dimensions will then agitate and move the vortex liquid, which in turn will create relaxation. Thus, a vortex liquid represents a normal, not a superconducting, state.

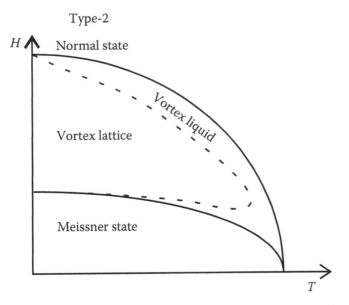

Figure 5.8 Schematic phase diagram of vortex states in the type-2 superconductor. A dashed line represents a first-order vortex-lattice-melting transition. Melting of the vortex-lattice disorders the phase field and represents superconductor–to-normal-state phase transition.

The specific aspect of the normal state represented by a vortex liquid is that it originates in the restoration of broken spatial symmetry of the vortex lattice and does not require proliferation of vortex loops. This is because vortex lines are thermodynamically stable when the system is subjected to a sufficiently strong magnetic field.

Basic characteristics of type-1 and type-2 superconductors are summarized in Table 5.1.

We conclude this discussion by noting that multicomponent superconductors can have several coherence lengths and several types of coexisting boundaries between different phases. This calls for a more sophisticated classification of superconducting states, which will be discussed in Chapter 6.

Problem 5.2 *Consider a thin superconducting cylinder with inner radius R_1 and outer radius R_2. Calculate magnetic flux that this cylinder encloses if it has a 2π phase winding.*

Problem 5.3 *Consider a vortex near a boundary of superconductor in a zero applied field. Calculate the interaction of the vortex with the boundary and the flux carried by such a vortex. (Hint: One can use a method of "image charge" from electrostatics and calculate interaction with "image antivortex.")*

Table 5.1 Comparative Summary of Basic Properties of Type-1 and Type-2 Superconductors

	Type-1	Type-2
Characteristic length scales	Penetration length λ and coherence length $\xi > \lambda$.	Penetration length λ and coherence length $\xi < \lambda$.
Surface tension for flat superconducting/ normal boundary	Positive.	Negative.
Intervortex interaction	Attractive at all distances.	Repulsive at all distances.
Magnetic field required to form a vortex	Larger than the thermodynamic critical magnetic field.	Smaller than the thermodynamic critical magnetic field.
Thermodynamic stability of vortices	Vortices are thermodynamically unstable	Vortices are thermodynamically stable within a finite range of magnetic fields
Phases in external magnetic field	(a) Meissner state at low fields; (b) macroscopically large normal domains near $H = H_c$.	(a) Meissner state at low fields; (b) vortex lattices at $H_{c1} < H < H_{c2}$.
Energy per winding, $E(N)/N$, for winding-N axially symmetric vortex solution	$E(N)/N$ decreases with N for all N, implying that $N \gg 1$ vortices tend to collapse into a single "megavortex." A normal domain may form without a phase winding if it is in contact with the sample boundary.	$E(N)/N$ increases with N for all N, meaning thermodynamic favorability of splitting into N elementary vortices, with formation of vortex lattices at $N \gg 1$.

5.8   Magneto-Rotational Isomorphism

We now return to the problem of the rotating superconductor to reveal behavior beyond the London law. The general solution to the problem is obtained by observing that there is an isomorphism between a rotating superconductor and a nonrotating superconductor in a uniform external magnetic field [19].

Assuming electroneutrality as a natural physical condition and confining ourselves, for simplicity, with the constant-density regime, we put the superconducting matter field onto a uniformly rotating, uniformly charged background.* In Chapter 1, we saw that, in a neutral superfluid, rotation is equivalent to introduction of a fictitious vector potential (1.170) (and also the centrifugal potential, which we ignore here). Thanks to electroneutrality, the equivalence directly

---

* In superconducting metals, the background charge is associated with the crystal lattice of positively charged ions. In the case of protonic superconductivity in a neutron star, the background charge comes from normal electrons.

applies to our case as well. The key circumstance enforced by electroneutrality is the invariance of the net electric current $\mathbf{J}_{net} = \mathbf{J} + \mathbf{J}_n$, where $\mathbf{J}$ is the supercurrent and $\mathbf{J}_n$ is the electric current of the normal background. When going to the rotating frame, $\mathbf{J}_{net}$ remains the same, thus implying the same field $\mathbf{A}$.

The free-energy functional of a superconductor in the rotating frame thus reads[*]

$$F[\phi, \mathbf{A}] = \frac{\gamma n}{2}\left(\nabla\phi + q\mathbf{A} - \gamma^{-1}\mathbf{W}\right)^2 + \frac{1}{2}(\text{curl}\,\mathbf{A})^2, \tag{5.136}$$

where $\mathbf{W} = \mathbf{\Omega} \times \mathbf{r}$ is the fictitious vector potential (1.171).

For a type-1 superconductor, there will be a critical rotation frequency when the London magnetic field energy density $B^2/2$, coming from (5.19), becomes equal to the superconducting condensation energy. At this rotation frequency, the type-1 superconductor experiences a first-order phase transition from the London regime to a normal state. Within the GL model for a type-1 superconductor, the value of this magnetic field coincides with the thermodynamical critical magnetic field given by $H_c = \Phi_0/(4\pi\xi\lambda)$, where ξ and λ are coherence and magnetic field penetration lengths and Φ_0 is the magnetic flux quantum. Thus,

$$\Omega_c = \frac{\gamma q \Phi_0}{8\pi\xi\lambda}. \tag{5.137}$$

For a type-2 superconductor, Equation (5.136) allows the superconducting state to persist at higher rotation frequencies by forming a vortex lattice. The picture becomes immediately clear by the following mapping: introduce *shifted* vector potential

$$\tilde{\mathbf{A}} = \mathbf{A} - (q\gamma)^{-1}\mathbf{W} \tag{5.138}$$

and observe that, in terms of $\tilde{\mathbf{A}}$, Equation (5.136) becomes isomorphic to free energy of a superconductor in external field $\tilde{\mathbf{H}} = -(q\gamma)^{-1}\text{curl}\,\mathbf{W} \equiv -\mathbf{B}_L$ [with $\mathbf{B}_L$ the London field of Equation (5.19)]:

$$F[\phi, \tilde{\mathbf{A}}] = \frac{\gamma n}{2}\left(\nabla\phi + q\tilde{\mathbf{A}}\right)^2 + \frac{1}{2}\left[\text{curl}\left(\tilde{\mathbf{A}} + \frac{\mathbf{W}}{q\gamma}\right)\right]^2. \tag{5.139}$$

Hence, the type-2 superconductor will have first and second critical rotation frequencies:

$$\Omega_{c1} = \frac{\gamma q}{8\pi}\left(\frac{\Phi_0}{\lambda^2}\right)\ln\left(\frac{\lambda}{\xi}\right), \qquad \Omega_{c2} = \frac{\gamma q \Phi_0}{8\pi\xi^2}. \tag{5.140}$$

At Ω_{c1}, the rotating vortex lattice will appear, and at Ω_{c2}, the system will become normal. The London state—characterized by a perfect diamagnetic response to the fictitious magnetic field $\tilde{H}$—emerges as a counterpart of the Meissner state. The magnetization diagram, Figure 5.6, applies, with the horizontal axis understood as $\tilde{H}$.

---

[*] Note that, while the uniformly rotating, uniformly charged background does not explicitly enter the free-energy functional, the expression is senseless in the absence of the background.

Observe that, in the limit $\Omega \to \Omega_{c2}$, the vortices of the lattice carry no magnetic flux. In this limit, the fictitious vector potential $(q\gamma)^{-1}\mathbf{W}$ is compensated by the vortex phase windings in $\nabla\phi$, rather than by $\mathbf{A}$. The only relevant vector potential here is $\mathbf{W}$, and the standard Feynman relationship between the flux of corresponding fictitious field and the vortex density holds. Physically, the fluxless vortex lattice in this state mimics the solid-body rotation of the superfluid matter field, corotating with the oppositely charged normal component. Apart from possible short-length-scale effects, there is no net transfer of electric charge, $\mathbf{J}_{net} = 0$, and thus $\mathbf{A} = 0$.

In a general case, the rotational response will be a combination of London response and vortex lattice, the number of vortices (antivortices, if negative) inside the system satisfying the relation [the counterpart of Equation (5.13)]

$$N = (\Phi - \tilde{\Phi})/\Phi_0, \tag{5.141}$$

where $\Phi_0 = -2\pi/q$ is the magnetic flux quantum, Φ is the magnetic flux though the system, and $\tilde{\Phi} = (q\gamma)^{-1} \oint_{\text{syst.}} \mathbf{W}\cdot d\mathbf{l} = -2(q\gamma)^{-1}\Omega S$ (with S the system area) is the flux of the London field $\mathbf{B}_L$ through the system. The total magnetic flux per vortex is not quantized due to the existence of the London background field. Nevertheless, an addition of a vortex (antivortex) *at fixed rotational frequency* amounts to addition (subtraction) of exactly one flux quantum Φ_0 to (from) the total magnetic flux.

The rotational response of a superconductor is summarized as follows: Slow rotation results in creation of a uniform magnetic field, in accordance with the London picture. The uniformity of the field sets in at the lengths scale λ from the boundary of the system. At the first critical rotation frequency of type-2 superconductor, the vortex lattice appears. Each vortex *reduces* the total magnetic flux through the system by one flux quantum—in contrast to the case of Abrikosov lattice, where a vortex does the opposite. Close to the second critical rotation frequency, the vortex lattice is essentially free of magnetic flux.

The previously discussed picture is illustrative of one fundamental aspect: In a superconducting state, the magnetic field is not necessary expelled. The presence of magnetic field in a rotating superconductor does not contradict the Anderson effect. Rather, the isomorphism between the rotating and nonrotating superconductor shows that the Meissner and London effects are counterparts of each other and are closely related to the Anderson effect. The massiveness of the rotation-induced magnetic field explains why the uniformity of the field sets in at the length scale λ from the sample's boundary. Similarly, the decay of any perturbation of magnetic field in the bulk—relative to the London field—is described by a massive vector field theory.

5.9   Superconducting Phase Transition

It is fundamentally important to understand how fluctuations affect the superconducting phase transition. For most of the early-discovered low-temperature superconductors, the mean-field theory description is very accurate, even very close to

T_c. In general, however, this is not the case, and one needs to go beyond mean-field approximation to describe superconducting phase transition. The mechanism that drives the phase transition to the normal state is proliferation of thermally excited topological defects [except for strongly first-order phase transitions in type-1 superconductors] [20]. We discussed this mechanism previously in connection with superfluid-to-normal phase transitions. In the case of a superconductor, the crucial difference is that topological defects have finite energy (per unit length in 3D).

We will start with 2D systems. In that case, finiteness of the energy of topological excitations has a radical consequence: Fluctuations destroy superconductivity at any finite temperature. Indeed, the energy of a superconducting vortex is given by (5.85) up to a finite core correction. Then, in a 2D system of size $L \times L$, the probability of such a vortex to be thermally excited can be estimated as

$$P = \left(\frac{L}{\xi}\right)^2 e^{-\frac{E_v}{T}}, \tag{5.142}$$

where ξ is a characteristic size of vortex core. The prefactor $(L/\xi)^2$ estimates how many possible vortex placement positions are there in the sample. In the limit $L \to \infty$, there will be thermally excited free vortices at any, no matter how low, temperature. The free energy of a vortex is given by

$$F = E_v - TS. \tag{5.143}$$

It is negative at any finite temperature, due to logarithmic divergence of the entropy $S = 2\log(L/\xi)$ in the limit $(L/\xi) \to \infty$. Thus, at any finite temperature, there exist free topological defects, and the phase transition to the superconducting state happens only at[*]

$$T_c = 0^+ \qquad \text{(in 2D)}. \tag{5.144}$$

In 3D, as discussed in Chapter 3, a neutral system with the global U(1) symmetry has a continuous phase transition in the 3D XY universality class. Coupling to the vector potential changes the symmetry type from global to local U(1). In the absence of applied magnetic field, topological defects are vortex loops, so that the normal state sets in when macroscopically large vortex loops proliferate and fill the system. Given that the symmetry of the system (and correspondingly the nature of topological defects) is fundamental for the universal critical properties, one can naturally expect that the universality class of the superconducting transition would be different from that of a normal superfluid. In this context, quite instructive is the prediction by Halperin et al. [20], who considered the GL model in the type-1 superconductor regime and found that the model features the

---

[*] Real superconducting films have finite thickness d. Also, stray magnetic fields outside the sample renormalize penetration length effectively to $\tilde{\lambda} = 2\lambda^2/d$ [21]. Finally, real superconductors are finite in the xy-plane, while low-temperature concentration of vortices is exponentially suppressed; see Equation (5.142). This means that, for all practical purposes, 2D superconductivity is possible at low enough temperatures.

first-order normal-to-superconductor phase transition. However, it turns out that, in certain cases in 3D, the normal-to-superfluid and superconductor–to-normal second-order criticalities can be directly mapped onto each other, meaning that both transitions are in the same 3D XY universality class. A nontrivial aspect of the mapping is that the direction of the transition becomes inverted: Superconducting phase is mapped onto the neutral *normal* phase, and the neutral superfluid phase is mapped onto the *normal* phase of the superconductor. That is, the superconducting transition is in the *inverted* 3D XY universality class, or, equivalently, the superconducting and superfluid transitions in 3D are *dual* to each other.

Thanks to the universality of the second-order phase transitions, the previously mentioned mapping can be established in a mathematically rigorous way by addressing a particular model of superconductor [22]. A class of models that readily allow us to establish mapping are lattice (London-type) models, in which the state of discrete matter field is specified by on-site phase variables, similar to the XY-type models discussed in Chapter 3. In the absence of the coupling to the vector potential, the partition function of those models reads [we use the same lattice notation as before; see Equation (3.10)]

$$Z_{XY} = \int \mathcal{D}\varphi \prod_{n\alpha} e^{f(\varphi_{n+\hat{\alpha}} - \varphi_n)}, \tag{5.145}$$

with $f(\varphi)$ some 2π-periodic function of the phase argument. To formulate a lattice model for a superconductor, we must introduce a lattice version of vector potential. On a hypercubic lattice, this is done by specifying scalar bond variables* $A_{n\alpha}$. The coupling to the lattice vector potential is through the gauge-invariant combination. In lattice gauge theories, it is customary to scale the vector potential with electric charge: $\varphi_{n+\hat{\alpha}} - \varphi_n - A_{n\alpha}$. The partition function then becomes

$$Z_A = \int \mathcal{D}A \mathcal{D}\varphi \, e^{-S[A]} \prod_{n\alpha} e^{f(\varphi_{n+\hat{\alpha}} - \varphi_n - A_{n\alpha})}, \tag{5.146}$$

where (e is the electric charge)

$$S[\mathbf{A}] = \frac{1}{2e^2} (\nabla \times \mathbf{A})^2, \tag{5.147}$$

and the lattice curl is defined as $(\nabla \times \mathbf{A})_{n\alpha} = \sum_{\nu\mu} \epsilon_{\alpha\nu\mu} (A_{n\mu} - A_{n+\hat{\nu}\mu})$. The model (5.146) and (5.147) is invariant with respect to the lattice gauge transformation (modulo 2π for the phase variables φ_n)

$$\begin{cases} \varphi_n \to \varphi_n + \phi_n, \\ A_{n\alpha} \to A_{n\alpha} + \phi_{n+\hat{\alpha}} - \phi_n, \end{cases} \tag{5.148}$$

---

* In the continuum limit, $A_{n\alpha}$ corresponds to the α-component of the vector potential at the position of the **n**-th lattice site.

where ϕ_n is an arbitrary lattice scalar field. It is seen that the long-wave part of the transformation (5.148) corresponds to the continuous-space gauge transformation.

A somewhat unpleasant implication of the gauge invariance of the theory (5.146) and (5.147) is that the integral over A in (5.146) is not only redundant—counts infinitely many physically identical states—but also ill-defined.* There are two ways to fix the problem. One way is to fix the gauge—for instance, by requiring that the longitudinal part of A be identically equal to zero. Another way—proving more convenient for our purposes since it does not restrict the space of configurations of A—is to introduce an infinitesimal massive term in $S[A]$ (with setting the mass equal to zero in final answers). The second way still involves multiple counting of physically identical configurations, but (with convergent integral) this amounts only to a physically irrelevant global factor in front of the partition function.

Our goal now is to show that the model (5.146) and (5.147) belongs to the inverted XY universality class. To this end, we cast it in the j-current form. The 2π-periodicity of $f(\varphi)$ allows us to represent the exponential as a Fourier series $e^{f(\varphi)} = \sum_{j=-\infty}^{\infty} e^{ij\varphi} F(j)$ and write

$$Z_A = \int \mathcal{D}A\mathcal{D}\varphi \, e^{-S[A]} \prod_{n\alpha} \sum_{j_{n\alpha}=-\infty}^{\infty} e^{ij_{n\alpha}(\varphi_{n+\hat{\alpha}}-\varphi_n-A_{n\alpha})} F(j_{n\alpha}). \tag{5.149}$$

As with the XY model, we then integrate over the phase variables φ_n and arrive at the J-current representation

$$Z_A = \int \mathcal{D}A \, e^{-S[A]} \sum_{\{j_n\}}^{\nabla\cdot j=0} \prod_{n\alpha} e^{-ij_{n\alpha}A_{n\alpha}} F(j_{n\alpha}), \tag{5.150}$$

in which the currents are coupled to the vector potential. The next step is to observe that the integral over A is Gaussian and thus can be performed explicitly, resulting in a J-current model with pairwise current–current interaction:

$$\sum_{\{j_n\}}^{\nabla\cdot j=0} \exp\left\{-\sum_{nm} U_m j_n \cdot j_{n+m}\right\} \prod_{n'\alpha} F(j_{n'\alpha}). \tag{5.151}$$

Here we omitted the irrelevant global prefactor. The integration over A is readily done in the Fourier representation, where we have

$$S[A] = \frac{1}{2e^2} \sum_q |s_q \times A_q|^2, \qquad s_{q\alpha} = 1 - e^{iq_\alpha}. \tag{5.152}$$

As explained earlier, we regularize the integral by adding a massive term, $S[A] \to S[A] + (\Lambda/2e^2)\sum_q |A_q|^2$.

---

* The magnitude of the longitudinal part of the vector potential turns out to be unrestricted, because the action $S[A]$ depends only on the transverse part of A.

The integral we need to evaluate is

$$\mathcal{I} = \int \mathcal{D}A \prod_{\mathbf{q}} e^{-\sum_{\alpha\beta} M_{\alpha\beta}(\mathbf{q}) A^*_{\mathbf{q}\alpha} A_{\mathbf{q}\beta} + \lambda (j^*_{\mathbf{q}} \cdot A_{\mathbf{q}} + j_{\mathbf{q}} \cdot A^*_{\mathbf{q}})}, \tag{5.153}$$

$$M_{\alpha\beta}(\mathbf{q}) = \frac{1}{2e^2} \left[(\Lambda + |s_{\mathbf{q}}|^2) \delta_{\alpha\beta} - s^*_{\mathbf{q}\alpha} s_{\mathbf{q}\beta} \right], \tag{5.154}$$

at $\lambda = i/2$. In view of the constraints

$$\mathbf{A}_{-\mathbf{q}} = \mathbf{A}^*_{\mathbf{q}}, \qquad \mathbf{j}_{-\mathbf{q}} = \mathbf{j}^*_{\mathbf{q}}, \qquad M_{\alpha\beta}(-\mathbf{q}) = M^*_{\alpha\beta}(\mathbf{q}), \tag{5.155}$$

we use a trick of proceeding with real λ: As long as λ is real, we can shift the integration variables

$$\mathbf{A}_{\mathbf{q}} \to \mathbf{A}_{\mathbf{q}} + \bar{\mathbf{A}}_{\mathbf{q}}, \qquad \bar{A}_{\mathbf{q}\alpha} = \lambda \sum_{\beta} M^{-1}_{\alpha\beta}(\mathbf{q}) j_{\mathbf{q}\beta}, \tag{5.156}$$

remaining consistent with conditions (5.155). With the shift (5.156), the dependence on j's factors out,

$$\mathcal{I} \propto \prod_{\mathbf{q}} e^{\lambda^2 \sum_{\alpha\beta} M^{-1}_{\alpha\beta}(\mathbf{q}) j^*_{\mathbf{q}\alpha} j_{\mathbf{q}\beta}}, \tag{5.157}$$

and this is all we need thanks to the irrelevance of the global j-independent factor. The exponent is analytic in λ, allowing us to restore $\lambda = -i/2$.

The observation that the matrix $M^{-1}_{\alpha\beta}$ has to be a superposition of matrices $\delta_{\alpha\beta}$ and $s^*_{\mathbf{q}\alpha} s_{\mathbf{q}\beta}$ readily leads to the expression

$$M^{-1}_{\alpha\beta}(\mathbf{q}) = \frac{2e^2}{\Lambda + |s_{\mathbf{q}}|^2} \left[\delta_{\alpha\beta} + \Lambda^{-1} s^*_{\mathbf{q}\alpha} s_{\mathbf{q}\beta} \right]. \tag{5.158}$$

The second term in the right-hand side is singular in the physical limit of $\Lambda \to 0$. This, however, is compensated by the fact that, at any finite Λ, the contribution of this term to the exponent in (5.157) is zero—due to the condition $\nabla \cdot \mathbf{j} = 0$ having the form $s_{\mathbf{q}} \cdot j_{\mathbf{q}} = 0$ in the Fourier space. After omitting this term, we safely set $\Lambda = 0$. The exponent in (5.157) becomes equal to $-(e^2/2) |j_{\mathbf{q}}|^2 / |s_{\mathbf{q}}|^2$, meaning that the Fourier component of the current–current potential in (5.151) is

$$U_{\mathbf{q}} = \frac{e^2}{2Q_{\mathbf{q}}}, \qquad Q_{\mathbf{q}} = |s_{\mathbf{q}}|^2 = \sum_{\alpha} 2(1 - \cos q_{\alpha}). \tag{5.159}$$

Notice that $Q_{\mathbf{q}}$ is precisely the function arising in the context of duality transformation for 3D XY universality class—see Chapter 3. This completes the proof.[*]

---

[*] We do not need to discuss the details of the form of the function $F(j)$ since those are irrelevant for the long-range current–current interaction.

References

1. J. Bardeen, L. N. Cooper, and J. R. Schrieffer, Theory of superconductivity, *Phys. Rev.* **108**, 1175 (1957).

2. L. P. Gor'kov, Microscopic derivation of the Ginzburg–Landau equations in the theory of superconductivity, *Sov. Phys. JETP* **36**, 1364 (1959).

3. R. P. Feynman, Application of quantum mechanics to liquid helium. In: *Progress in Low Temperature Physics*, Vol. 1 (ed. C. J. Gorter), Elsevier, p. 17 (1955).

4. F. London, *Superfluids*, Wiley, New York (1950).

5. F. London and H. London, The electromagnetic equations of the supraconductor, *Proc. Roy. Soc. (London)* **A149**, 71 (1935).

6. V. L. Ginzburg and L. D. Landau, To the theory of superconductivity, *Zh. Eksp. Teor. Fiz.* **20**, 1064 (1950).

7. L. D. Landau and E. M. Lifshitz, *Electrodynamics of Continuous Media*, Butterworth-Heinemann, Burlington, MA (1984).

8. L. Landau, The intermediate state of supraconductors, *Nature* **141**, 688 (1938).

9. H. B. Nielsen and P. Olesen, Vortex-line models for dual strings, *Nucl. Phys. B* **61**, 45 (1973).

10. B. Plohr, The behavior at infinity of isotropic vortices and monopoles, *J. Math. Phys.* **22**, 2184 (1981).

11. L. Kramer, Thermodynamic behavior of type-ii superconductors with small κ near the lower critical field, *Phys. Rev. B* **3**, 3821 (1971).

12. J. M. Speight, Static intervortex forces, *Phys. Rev. D* **55**, 3830 (1997).

13. D. Saint-James, G. Sarma, and E. J. Thomas, *Type II Superconductivity*, Pergamon Press, Oxford, U.K. (1969).

14. E. B. Bogomolny, Stability of classical solutions, *Sov. J. Nucl. Phys.* **24**, 449 (1976).

15. A. A. Abrikosov, On the magnetic properties of superconductors of the second group, *Sov. Phys. JETP* **5**, 1174 (1957).

16. L. W. Schubnikow, W. I. Chotkewitsch, J. D. Schepelew, and J. N. Rjabinin, Magnetische Eigenschaften Supraleitender Metalle und Legierungen, *Sondernummer Phys. Z. Sowjet. Arbeiten auf dem Gebiete tiefer Temperaturen*, Juni 1936, 39 (1936); L. W. Schubnikow, W. I. Chotkewitsch, J. D. Schepelew, and J. N. Rjabinin, Magnetische Eigenschaften Supraleitender Metalle und Legierungen, *Phys. Z. Sowjet.* Bd.10, H.2, 165 (1936); L. V. Shubnikov,

V. I. Khotkevich, Yu. D. Shepelev, and Yu. N. Riabinin Magnetic properties of superconducting metals and alloys, *Zh. Exper. Teor. Fiz. (USSR)* **7**, 221 (1937).

17. A. Edstrom, Three- and four-body intervortex forces in the Ginzburg–Landau models of single- and multicomponent superconductivity, *Physica C* **487**, 19 (2013).

18. D. Nelson, *Defects and Geometry in Condensed Matter Physics*, Cambridge University Press, Cambridge, U.K. (2002).

19. E. Babaev and B. Svistunov, Rotational response of superconductors: Magneto-rotational isomorphism and rotation-induced vortex lattice, *Phys. Rev. B* **89**, 104501 (2014).

20. B. I. Halperin, T. C. Lubensky, and S. K. Ma, First-order phase transitions in superconductors and smectic-A liquid crystals, *Phys. Rev. Lett.* **32**, 292 (1974).

21. J. Pearl, Current distribution in superconducting films carrying quantized fluxoids, *Appl. Phys. Lett.* **5**, 65 (1964).

22. C. Dasgupta and B. I. Halperin, Phase transition in a lattice model of superconductivity, *Phys. Rev. Lett.* **47**, 1556 (1981).

CHAPTER 6

Multicomponent Superconductors and Superfluids. Superconducting and Metallic Superfluids

In previous chapters, we saw that the most quintessential properties of superfluids and superconductors are described by the classical complex scalar field. These properties are universal and robust in the sense that they are independent of the microscopic origin of the classical field and insensitive to the presence of disorder. In single-component systems, several effects are considered to be the hallmarks of superfluidity. One of them is the quantization of superfluid-velocity circulation with the circulation quantum depending only on the microscopic parameter γ (recall that, in a quantum system, γ is directly related to the bare particle mass: $\gamma = \hbar/m$). Another example is the Onsager–Feynman relation (1.187) between the density of rotation-induced vortices and angular velocity, with γ being the only system-specific parameter entering the expression. In a single-component super-conductor, an even more impressive universality takes place in the context of magnetic flux quantization: With quantum field theory taken into account, the magnetic flux quantum turns out to be a ratio of fundamental constants only—the particle mass drops out: $\Phi_0 = \pi\hbar c/e$. The extraordinary robustness of those phenomena is dictated by topology—quantization of the phase gradient circulation for the (coarse-grained) complex field.

Multicomponent superfluids and superconductors are described by several complex fields.* Clearly, topological considerations associated with the individual phases of complex fields take place in multicomponent systems as well. However, as we will establish in this chapter, this does not necessarily imply the other universal phenomena characteristic of single-component systems. A crucially new aspect that arises in a multicomponent superfluid or superconductor is that, in general, such a system can have different broken symmetries. As a result, many symmetry- and topology-specific properties found in single-component

---

* The microscopic origin of multiple components—although not important for understanding this chapter—is worth a brief comment. Examples of superfluid mixtures are abundant and quite diverse. Historically, the first experimentally studied mixture was the mixture of super-fluid Cooper pairs of ^{3}He atoms and superfluid ^{4}He (see Reference [1]). A neutron star interior is also argued to be a mixture of coexisting superfluids based on neutron and proton Cooper pairs; some models add superconducting Cooper pairs of Σ^- hyperons. In the context of metallic hydrogen, the superfluid mixture may consist of electronic and protonic pairs and even deuterons [2]. A special class of multicomponent systems is formed by multiband superconductors, where electrons belonging to different bands give rise to two or more coupled superconducting components; see, for example, Reference [3]. In the field of ultracold gases, one can load several species of bosons/fermions into the trap.

systems get "undermined" in multicomponent systems. For example, we will see that, in a superfluid mixture of several electrically charged superconducting components, it is possible to not obey Onsager–Feynman quantization of super-fluid velocity. Likewise, the quantization of magnetic flux may be nonuniversal. The London law, which universally describes rotational response of a single-component superconductor below first critical rotation frequency, does not always hold in superconducting mixtures. In contrast to the London–Anderson picture, the electrodynamics of a multicomponent superconductor do not reduce to a theory of a massive vector field only. Instead, it involves (a) a massive vector field arising from the London–Anderson mechanism and (b) self-generated Skyrme-like terms.

Superconducting mixtures are, in general, characterized by several coherence lengths ξ_i, meaning that they cannot be exhaustively classified by the single GL parameter κ. This leads to breakdown of the standard type-1/type-2 dichotomy in the classification of superconductors and emergence of a new class, the so-called "type-1.5" superconductors.

Besides that, new topological invariants exist in systems with higher numbers of components.

6.1   Mixtures of Superfluids and Entrainment Effect

Consider a mixture of two superfluids with independently conserved amounts of matter (total numbers of particles). The broken symmetry is thus U(1)×U(1), and the order parameter consists of two complex fields with two independent phases, $\psi_\alpha = |\psi_\alpha| e^{i\phi_\alpha}$ ($\alpha = 1, 2$). Each component is described by its own parameter γ_α (1.3), which can be related to the particle mass in the quantum case:

$$\gamma_\alpha = \hbar/m_\alpha \qquad \text{(in a quantum system)}. \tag{6.1}$$

Quite generically, the superflows of two components interact [1].* The corresponding terms in the free-energy functional should respect the U(1)×U(1) symmetry. In the constant-density (hydrodynamic) approximation, the free energy of a mixture of interacting superfluids has the form

$$
\begin{aligned}
F &= \frac{1}{2} \int d\mathbf{r} \left[(\rho_1 - \rho_d)\, \mathbf{v}_1^2 + (\rho_2 - \rho_d)\, \mathbf{v}_2^2 + 2\rho_d\, \mathbf{v}_1 \cdot \mathbf{v}_2 \right] \\
&= \frac{1}{2} \int d\mathbf{r} \left[\rho_1 \mathbf{v}_1^2 + \rho_2 \mathbf{v}_2^2 - \rho_d\, (\mathbf{v}_1 - \mathbf{v}_2)^2 \right],
\end{aligned}
\tag{6.2}
$$

where

$$\mathbf{v}_\alpha = \gamma_\alpha \nabla \theta_\alpha \qquad (\alpha = 1, 2) \tag{6.3}$$

---

* The microscopic origin of this interaction is not important for this chapter. It can arise, for example, from the intercomponent scattering [1,4]. In the case of BECs in optical lattices, interactions between the currents are substantially modified by the underlying periodic potential [5,6].

is the velocity of the component α. The current–current interaction is represented by the $\rho_d \mathbf{v}_1 \cdot \mathbf{v}_2$ term; consistent with the U(1)×U(1) symmetry, it depends only on the phase gradients. Alternatively, one can work with $\bar{\rho}_\alpha \equiv (\rho_\alpha - \rho_d)$ and write the current–current interaction as $\rho_d \mathbf{v}_1 \cdot \mathbf{v}_2$. Variational derivatives of the free energy yield relations between the currents and velocities:

$$\mathbf{j}_1 = (\rho_1 - \rho_d)\mathbf{v}_1 + \rho_d \mathbf{v}_2, \tag{6.4}$$

$$\mathbf{j}_2 = (\rho_2 - \rho_d)\mathbf{v}_2 + \rho_d \mathbf{v}_1. \tag{6.5}$$

They clarify the physical meaning of the parameter ρ_d: It plays the role of superfluid density of one component dragged by the motion of the other. An immediate consequence of these relations is the *entrainment effect* predicting a superflow in one of the components at zero superfluid velocity (zero phase gradient). Indeed, at $\nabla\theta_2 \equiv 0$, $\nabla\theta_1 \neq 0$, we have

$$\mathbf{j}_1 = (\rho_1 - \rho_d)\gamma_1 \nabla\theta_1, \qquad \mathbf{j}_2 = \rho_d \gamma_1 \nabla\theta_1. \tag{6.6}$$

Applying Equation (6.6) to a vortex defect in $\theta_1(\mathbf{r})$, we observe that component 1 has nonzero circulations of both current and superfluid velocities (the latter characterized by the standard Onsager circulation quantum $\gamma_1/2\pi$), while component 2 has nonzero entrainment-induced current circulation at identically zero superfluid velocity.

6.2 Multicomponent Superconductors

Consider now a mixture of coexisting superconducting components. The crucial difference between the multicomponent superconductor and the previously considered superfluid mixture is an unavoidable electromagnetic coupling between all phases, leading to numerous counter-intuitive consequences.

6.2.1 Liquid Metallic Hydrogen–Type System

Consider a U(1)×U(1) superconducting system, with two components having opposite electric charges of the same magnitude. For example, it can be the projected liquid metallic hydrogen (LMH) state[*] or the neutron star interior (more details can be found in Chapter 15). At the macroscopic level, all these *LMH-type systems* are similar and are described by the following minimal (constant-density/hydrodynamic) model for the free-energy density:

$$F = \frac{1}{2}|\psi_1|^2(\nabla\theta_1 + q\mathbf{A})^2 + \frac{1}{2}|\psi_2|^2(\nabla\theta_2 - q\mathbf{A})^2 + \frac{(\text{curl } \mathbf{A})^2}{2}. \tag{6.7}$$

(In view of the symmetry associated with changing the sign of electric charge with simultaneously complex-conjugating the ψ-fields, the sign of q is a matter of taste.)

---

[*] Condensates of electronic and protonic Cooper pairs conserve their particle numbers independently.

The densities of the two components are not supposed to be equal, with the overall electroneutrality being provided by some background. To simplify the notation, we rescale ψ_α to absorb γ_α in the constant field magnitude. Focusing exclusively on the physics brought by the coupling to the vector potential $\mathbf{A}$, first we do not consider here the current–current coupling terms $\rho_d (\nabla\theta_1 + q\mathbf{A}) \cdot (\nabla\theta_2 - q\mathbf{A})$, which otherwise can also be present.

Allowing for spatial variations of the $|\psi_\alpha|$ fields yields the "minimal" GL U(1)×U(1)-symmetric free-energy functional*

$$F = \frac{1}{2}|(\nabla + iq\mathbf{A})\psi_1|^2 + \frac{1}{2}|(\nabla - iq\mathbf{A})\psi_2|^2 + \frac{(\mathrm{curl}\,\mathbf{A})^2}{2}$$
$$- a_1|\psi_1|^2 + \frac{b_1}{2}|\psi_1|^4 - a_2|\psi_2|^2 + \frac{b_2}{2}|\psi_2|^4. \tag{6.8}$$

Variation with respect to the fields ψ_α—analogous to the procedure discussed in Chapter 5—yields the boundary conditions

$$[i\nabla\psi_1 - q\mathbf{A}\psi_1] \cdot \mathbf{n} = 0, \qquad\qquad [i\nabla\psi_2 + q\mathbf{A}\psi_2] \cdot \mathbf{n} = 0 \tag{6.9}$$

(here $\mathbf{n}$ is the unit vector normal to the superconductor's surface), and a system of GL equations

$$\frac{1}{2}(i\nabla - q\mathbf{A})^2\psi_1 - a\psi_1 + b\psi_1|\psi_1|^2 = 0,$$
$$\frac{1}{2}(i\nabla + q\mathbf{A})^2\psi_2 - a\psi_2 + b\psi_2|\psi_2|^2 = 0. \tag{6.10}$$

The expression for the net electric supercurrent is obtained by varying the vector potential; cf. Equation (5.3):

$$\mathbf{J} = \frac{iq}{2}(\psi_1^*\nabla\psi_1 - \psi_1\nabla\psi_1^*) - \frac{iq}{2}(\psi_2^*\nabla\psi_2 - \psi_2\nabla\psi_2^*) - q^2\rho^2\mathbf{A}, \tag{6.11}$$

$$\rho^2 \equiv |\psi_1|^2 + |\psi_2|^2. \tag{6.12}$$

The magnetic penetration length now is given by

$$\lambda = \frac{1}{|q|\rho}. \tag{6.13}$$

Problem 6.1 *Derive Equation (6.13).*

6.2.2   Other Multicomponent Ginzburg–Landau Models

A different class of multicomponent systems is formed by condensates that are not independently conserved, so that the system has only U(1) broken symmetry while

---

* In general, this model can have other terms consistent with U(1)×U(1) symmetry, such as $|\psi_1|^2|\psi_2|^2$, or the aforementioned current–current terms. Again, we omit these terms here since our goal is to start by investigating the physics that originates from the electromagnetic inter-component coupling.

being described by several classical complex fields. The most common example is a multiband superconductor.* Interband transitions of electrons lead to the non-conservation of the particle number in each individual component. At the macroscopic (classical-field) level, this results in the intercomponent Josephson terms of the following structure: $\psi_1\psi_2^* + \psi_1^*\psi_2$.

In the following, we will discuss many other physical examples where the multicomponent GL model appears. In contrast to the LMH-type system, the model in general can have similar signs of electric charges q, more than two components, and more complicated effective potential V

$$F = \frac{1}{2}\sum_{\alpha,\beta}|(\nabla + iq_\alpha\mathbf{A})\psi_\alpha|^2 + \frac{(\operatorname{curl}\mathbf{A})^2}{2} + V\left(|\psi_\alpha|, \theta_\alpha - \theta_\beta\right), \tag{6.14}$$

where α, β are component indices. In electronic superconductors, the effective potential often contains terms that depend on the phase difference such as the aforementioned terms $\psi_\alpha\psi_\beta^* + \psi_\alpha^*\psi_\beta$, $|\psi_\alpha|^2|\psi_\beta|^2$ as well as other similar-type terms such as $\psi_\alpha^2\left(\psi_\beta^2\right)^* + \left(\psi_\alpha^2\right)^*\psi_\beta^2$. Besides that, many multicomponent superconductors can have various mixed gradient terms.

6.3 Fractional Vortices

Superconducting and superfluid properties of a multicomponent system are controlled by the physics of topological defects in the system. In a two-component superconductor, described by (6.8), there are two kinds of vortex excitations: fractional vortices [9] and their integer-flux bound states. First, we observe that since there are now two phases, $\theta_{1,2} \in [0, 2\pi)$, it is possible to have two different phase windings and two different species of "elementary" vortices. In what follows, we denote a general vortex as (M_1, M_2), where M_α ($\alpha = 1, 2$) is the phase winding: $M_\alpha = (2\pi)^{-1}\oint_\sigma d\mathbf{l} \cdot \nabla\theta_\alpha$ [the closed contour σ is going around the vortex core]. Consider the simplest $(1, 0)$ vortex. In the absence of vector potential as an independent degree of freedom, such a vortex would have logarithmically divergent energy, while in a single-component superconductor (see Chapter 5), its energy per unit length remains finite, because the phase gradient is (asymptotically) compensated by the vector potential field. In the two-component model (6.8), the situation changes radically. An attempt to compensate the gradient of θ_1 in the first term by the corresponding configuration of $\mathbf{A}$ will result in the logarithmic divergence of energy originating from the second term due to the same vector potential $\mathbf{A}$. Hence, one can alleviate the divergent behavior of energy, but one cannot remove it completely.

The model (6.8) also has vortex solutions with finite energy per unit length, namely, $(M, -M)$ vortices. In this case, the system generates magnetic flux that

---

* Multiband superconductivity—Cooper pairing in different bands—is found in materials with nontrivial band structure [3,7,8].

compensates both phase gradients. To see that, rewrite (6.7)—the hydrodynamic limit of (6.8)—as

$$F = \frac{1}{2\rho^2} |\psi_1|^2 |\psi_2|^2 [\nabla(\theta_1 + \theta_2)]^2$$
$$+ \frac{1}{2\rho^2} \left(|\psi_1|^2 \nabla\theta_1 - |\psi_2|^2 \nabla\theta_2 + q\rho^2 \mathbf{A} \right)^2 + \frac{(\text{curl } \mathbf{A})^2}{2}. \tag{6.15}$$

For $(M, -M)$ vortex, we have $\nabla\theta_2 = -\nabla\theta_1$, so that the first term in (6.15) is identically equal to zero, while the second term acquires the form equivalent to a single-component superconductor, thus proving the statement.

For the model (6.8) and similar systems, rewriting it in the form (6.15), we decompose the flow into the *neutral* (or *superfluid*) sector associated with coflow of oppositely charged fields and the *charged* (or *superconducting*) sector associated with electric currents. Diverging vortex energy is associated with the neutral mode only, and for $(M, -M)$ vortices, there are no currents in the neutral sector. The last circumstance is directly related to the equality of the electric charge modules. If, instead, we had a GL functional with two condensates having the same (rather than opposite) electric charges, then the neutral sector would be associated with gradients of the phase difference, and the finite-energy-vortex solutions would correspond to equal phase windings (M, M). One can also generalize this terminology to the case of commensurate charges when the ratio q_1/q_2 is a rational number. It is easy to see that the (M_1, M_2) vortex features zero topological charge in the neutral sector if (and only if) $M_2/M_1 = q_1/q_2$. The fact that all other vortices inevitably carry the neutral superflow is readily seen (at $q_1 = -q_2$) from the structure of (6.15), where all three gradient terms are nonnegative and the first term does not involve the vector potential. Hence, the only way to get rid of the divergent contribution coming from the first term is to have $\nabla(\theta_1 + \theta_2) = 0$, which corresponds to the $(M, -M)$ vortex. If the model (6.15) is an (infinite) type-2 superconductor, then only the composite $(1, -1)$ vortices are induced by applying an external magnetic field.

Vortices that have phase winding in only one component carry a fraction of the flux quantum. As an illustration, let us calculate the magnetic flux carried by the $(1, 0)$ vortex. The supercurrent density around the core is

$$\mathbf{J} = -q\rho^2 \left[\sin^2(\beta/2)\nabla\theta_1 + q\mathbf{A} \right], \tag{6.16}$$

where we introduced a convenient parameter β such that

$$\sin(\beta/2) = |\psi_1|/\rho, \qquad \cos(\beta/2) = |\psi_2|/\rho. \tag{6.17}$$

Integrating both sides of (6.16) over a large closed contour σ placed at a distance much larger than λ from the core where the magnitude of $\mathbf{J}$ on σ is vanishingly small, we arrive at the following flux-quantization condition [9]:

$$\Phi_{(1,0)} = -\sin^2(\beta/2)\Phi_0, \tag{6.18}$$

where $\Phi_{(1,0)} = \oint_\sigma \mathbf{A} \cdot d\mathbf{l}$ is the magnetic flux carried by the $(1,0)$ vortex and $\Phi_0 = 2\pi/q$ is the magnetic flux quantum. Similar calculation for the $(0,-1)$ vortex yields

$$\Phi_{(0,-1)} = -\cos^2(\beta/2)\Phi_0. \tag{6.19}$$

In a general case of the (M_1, M_2) vortex, the flux is given by

$$\Phi_{(M_1, M_2)} = -\Phi_0\left[\sin^2(\beta/2)M_1 - \cos^2(\beta/2)M_2\right]. \tag{6.20}$$

Thus, in a two-component superconductor, a vortex can carry an arbitrary fraction of the flux quantum controlled by the continuous parameter β (6.17). For the finite-energy $(1,-1)$ "composite vortex," the magnetic flux adds up to one flux quantum.

Problem 6.2 *Derive (6.20).*

Let us now discuss in detail the structure of the $(1,0)$ vortex in the framework of the two-component London model (in the following, r is the radial distance from the core):

- Within London approximation, the vortex core is modeled as a sharp cutoff (boundary condition) at the distance $r = \xi$.

- The region $\xi < r < \lambda$ features both the neutral superflow associated with the gradients of the variable $(\theta_1 + \theta_2)$ and the charged supercurrent $\mathbf{J}$.

- The region $r \gg \lambda$ features only neutral superflow; similarly to a neutral system, it decays as $1/r$ away from the core. In this region, $\mathbf{J}$ and the magnetic field vanish more rapidly.

Equation (6.15) allows one to calculate the energy of the fractional-flux vortex in the London limit [10]. The first term leads to the logarithmically divergent contribution to vortex energy, albeit with a different prefactor. The contribution coming from the second and third terms can be treated the same way as in the single-component case, Equation (5.85). In a system of the linear size R, we get

$$E_{(1,0)} = \frac{1}{4\pi}\sin^4(\beta/2)\left[\frac{\Phi_0}{\lambda}\right]^2\log\frac{\lambda}{\xi} + \frac{\pi}{2}\rho^2\sin^2\beta\log\frac{R}{\xi},$$

$$E_{(0,1)} = \frac{1}{4\pi}\cos^4(\beta/2)\left[\frac{\Phi_0}{\lambda}\right]^2\log\frac{\lambda}{\xi} + \frac{\pi}{2}\rho^2\sin^2\beta\log\frac{R}{\xi}. \tag{6.21}$$

The first term in (6.21) is the energy of the charged current and magnetic field [note that $\sin^2(\beta/2)\Phi_0 = \Phi_1$ and $\cos^2(\beta/2)\Phi_0 = \Phi_2$ are fractions of flux quantum carried by $(1,0)$ and $(0,1)$ vortices]. The second term in (6.21) is the kinetic energy of the neutral flow. The energy per unit length of the composite vortex $(1,-1)$ is finite even in an infinite system:

$$E_{(1,-1)} = \frac{1}{4\pi}\left[\frac{\Phi_0}{\lambda}\right]^2\log\frac{\lambda}{\xi}. \tag{6.22}$$

Representation (6.15) is also convenient for deriving the interaction energy between fractional vortices. The contribution coming from the first term can be calculated analogously to Equation (1.141), while the contribution coming from the second and third terms can be derived similarly to Equation (5.91). For two $(1,0)$ vortices at a distance r, the interaction energy has a repulsive logarithmic contribution [coming from the first term in Equation (6.15)], as well as a repulsive exponentially screened contribution [coming from the second term and third terms in Equation (6.15)]:

$$
\begin{aligned}
E_{(1,0)+(1,0)} &= \frac{\pi}{2}\rho^2\sin^2\beta\log\frac{R}{r} + 2\pi\rho^2\sin^4(\beta/2)K_0(r/\lambda) \\
&\equiv \frac{\pi}{2}\rho^2\sin^2\beta\log\frac{R}{r} + \frac{\Phi_1^2}{2\pi\lambda^2}K_0(r/\lambda).
\end{aligned}
\tag{6.23}
$$

The interaction between $(1,0)$ and $(0,-1)$ vortices contains an attractive logarithmic part and a repulsive exponentially screened part:

$$
\begin{aligned}
E_{(1,0)+(0,-1)} &= -\frac{\pi}{2}\rho^2\sin^2\beta\log\frac{R}{r} + \frac{\pi}{2}\rho^2\sin^2\beta K_0(r/\lambda) \\
&\equiv -\frac{\pi}{2}\rho^2\sin^2\beta\log\frac{R}{r} + \frac{\Phi_1\Phi_2}{2\pi\lambda^2}K_0(r/\lambda).
\end{aligned}
\tag{6.24}
$$

For the $(1,0)+(0,-1)$ pair of vortices, the attractive and repulsive contributions cancel each other at separations $r\ll\lambda$. This is because, at length scales much smaller than the magnetic field penetration length, the system becomes effectively neutral, which implies effective decoupling between the components and thus vanishing of the intercomponent interaction between topological defects.

6.4   Superfluid Sector of Liquid Metallic Hydrogen–Type System

In the previous section, we saw that in a U(1)×U(1) superconductor, it is possible to have a neutral supertransport of matter without charge supertransport. In this section, we will discuss the superfluid properties of a mixture of two charged condensates that are quite different from those of a neutral condensate and also from the neutral mixture (this was considered in Reference [11]). In fact, the neutral sector *alone* allows one to arrive at important qualitative and quantitative conclusions concerning the rotational response of the system, such as a finite density of (fractional) vortices at any rotational frequency.

Keeping LMH in mind as a prototypical example, we continue working with two matter fields of exactly opposite charges but with a disparity in the other parameters: $\gamma_1 \neq \gamma_2$ and $|\psi_1| \neq |\psi_2|$. For the same reason, we will often speak of the electronic ($\alpha = 1$) and protonic ($\alpha = 2$) components.

6.4.1   Nonuniversal Quantization of Superfluid Velocity

The simplest vortex having nontrivial winding in the phase sum $(\theta_1 + \theta_2)$ is the one with the winding in only one of the phases: $(\pm1,0)$ or $(0,\pm1)$. Since the first

Table 6.1 Fractional Quantization of Magnetic Flux and Superfluid-Velocity Circulation in a Two-Component Superconductor (LMH-Type System)

	$(1,0)$ **Vortex**	$(0,-1)$ **Vortex**
Magnetic flux quantum	$\sin^2(\beta/2)\Phi_0$	$\cos^2(\beta/2)\Phi_0$
Electronic superfluid-velocity circulation quantum	$\cos^2(\beta/2)K_1^{(0)}$	$\cos^2(\beta/2)K_1^{(0)}$
Protonic superfluid-velocity circulation quantum	$\sin^2(\beta/2)K_2^{(0)}$	$\sin^2(\beta/2)K_2^{(0)}$

Source: Reference [11].

term in (6.15) is symmetric with respect to electronic and protonic condensates, both vortices have identical energy associated with the neutral superflow. As seen in Equation (6.21), the energy difference between the two cases is due to the second term in (6.15) dealing with the charged sector. Consider the $(0,1)$ vortex. The solution for the vector potential $\mathbf{A}$, at distances much larger than the penetration length (since now parameters $\gamma_{1,2}$ cannot be completely absorbed into rescaled fields, we prefer to restore them everywhere), is given by

$$|\mathbf{A}|_{r\to\infty} = \frac{\gamma_2|\psi_2|^2}{q\rho^2 r}, \qquad \rho^2 = \gamma_1|\psi_1|^2 + \gamma_2|\psi_2|^2, \tag{6.25}$$

where r is the radial distance from the core center. This solution follows from the condition that $\mathbf{A}$ should compensate the gradient of θ_2 in the second term in (6.15). At $r \gg \lambda$, the superfluid velocities of electronic and protonic components are

$$\mathbf{v}_1 = \gamma_1 q\mathbf{A}, \qquad \mathbf{v}_2 = \gamma_2(\nabla\theta_2 - q\mathbf{A}). \tag{6.26}$$

With Equations (6.25) and (6.26), we readily see that, at $r \gg \lambda$, the superfluid-velocity circulation quanta are

$$K_1 = \oint_\sigma \mathbf{v}_1 \cdot d\mathbf{l} = 2\pi\gamma_1\cos^2(\beta/2), \qquad K_2 = \oint_\sigma \mathbf{v}_2 \cdot d\mathbf{l} = 2\pi\gamma_2\sin^2(\beta/2), \tag{6.27}$$

$$\sin^2(\beta/2) = \gamma_1|\psi_1|^2/\rho^2, \qquad \cos^2(\beta/2) = \gamma_2|\psi_2|^2/\rho^2. \tag{6.28}$$

[Definitions (6.28) are identical to (6.17); here we simply restore $\gamma_{1,2}$.] It can be explicitly checked that the same quantization conditions hold also for the $(1,0)$ vortex. This is a radical deviation from the Onsager–Feynman quantization rule. The circulation quanta $K_{1,2}$ differ from their neutral counterparts $K_{1,2}^{(0)} = 2\pi\gamma_{1,2}$ by *non-universal* factors (the same as in the case of magnetic flux quantization). All the results so far are summarized in Table 6.1.

6.4.2 Rotational Response in the Neutral Sector

Similarly to Equation (6.15), let us represent London free energy of the two-component superconductor in a rotating coordinate system as a sum of three non-negative terms:

$$F = \frac{1}{2\rho^2}\gamma_1|\psi_1|^2\gamma_2|\psi_2|^2\left[\nabla(\theta_1+\theta_2)-\gamma_*^{-1}\mathbf{W}\right]^2$$

$$+ \frac{1}{2\rho^2}\left[\gamma_1|\psi_1|^2\nabla\theta_1-\gamma_2|\psi_2|^2\nabla\theta_2+q\rho^2\mathbf{A}-\left(|\psi_1|^2-|\psi_2|^2\right)\mathbf{W}\right]^2+\frac{(\mathrm{curl}\,\mathbf{A})^2}{2},$$

$$(6.29)$$

where

$$\gamma_* = \frac{\gamma_1\gamma_2}{\gamma_1+\gamma_2}. \tag{6.30}$$

Consider the first term. To ensure that energy scaling with the system size is not hyperextensive,[*] the density of vortex topological charges, $n_{\mathrm{top}}^{(\mathrm{total})}$, must satisfy the condition

$$\oint_\sigma d\mathbf{l}\cdot\nabla(\theta_1+\theta_2) = \gamma_*^{-1}\oint d\mathbf{l}\cdot\mathbf{W} \quad\Rightarrow\quad n_{\mathrm{top}}^{(\mathrm{total})}\equiv n_{\mathrm{top}}^{(1)}+n_{\mathrm{top}}^{(2)} = \frac{\Omega}{\pi\gamma_*}. \tag{6.31}$$

Here, $n_{\mathrm{top}}^{(\alpha)}$ is the density of topological charges in the component α. For $(0,1)$ vortices, for example, we have $n_{\mathrm{top}}^{(1)} = 0$ and $n_{\mathrm{top}}^{(2)} = n_{\mathrm{vort}}^{(0,1)}$, where $n_{\mathrm{vort}}^{(0,1)}$ is the number density of $(0,1)$ vortices. For composite vortices, both $n_{\mathrm{top}}^{(1)}$ and $n_{\mathrm{top}}^{(2)}$ are nonzero.

First, consider the rotational response of a finite system at zero temperature when there is no normal component. The formation of a vortex is energetically favorable if

$$E_v - \mathbf{M}\cdot\Omega < 0, \tag{6.32}$$

where E_v is the (free) energy of the vortex and $\mathbf{M}$ is the angular momentum of the vortex. Observe that, for a $(0,1)$ [or $(1,0)$] vortex, both protonic and electronic components contribute to the angular momentum:

$$|\mathbf{M}| = |\mathbf{M}_2+\mathbf{M}_1| = \int\left[\frac{|\psi_2|^2}{\gamma_2}v_2(r)+\frac{|\psi_1|^2}{\gamma_1}v_1(r)\right]rd^3r. \tag{6.33}$$

In a cylindrical system of radius R, we have (note that $\sin^2\beta$ is $1\leftrightarrow2$ symmetric)

$$|\mathbf{M}_{(0,1)}| = |\mathbf{M}_{(1,0)}| = \frac{\pi}{4}R^2\rho^2\gamma_*^{-1}\sin^2\beta \qquad\text{(per unit vortex-line length).}$$

$$(6.34)$$

First, we consider a single vortex in a finite system with $R\gg\lambda$ at slow rotation, thus neglecting the angular momentum associated with the charged sector, because charged currents are screened. We also neglect now magnetic field that can be generated by surface currents. Then, according to Equation (6.21), the energetically preferred vortices forming in response to rotation are the $(0,1)$ vortices with the phase winding in the protonic condensate and carrying a smaller fraction of Φ_0.

---

[*] The same argument as in the case of rotating single-component superfluid.

Under these conditions, we do not consider the contribution of composite $(\pm1, \mp1)$-type vortices to the rotational response, since they do not have phase winding in the sum of the phases. Also, it is straightforward to see that here the $(\pm1, \pm1)$ composite vortices are energetically unstable. The argument is the same as in ordinary superfluids for vortices with winding number magnitude larger than unity; see Chapter 1.

Assuming the condition $R \gg \lambda$, we neglect the first term in (6.21) to arrive at the first critical velocity, where such a system can create a vortex:

$$\Omega_c \approx \frac{\gamma_*}{R^2} \log \frac{R}{a}. \tag{6.35}$$

Equation (6.35) for the critical velocity is the same as for the single-component superfluid matter field with $\gamma \to \gamma_*$. In a quantum system, γ_* is directly related to the "total mass": $\gamma_* = \hbar/(m_1 + m_2)$, naively suggesting an idea of a "composite neutral particle." Nevertheless, in our case, there is no real-space pairing between protonic and electronic Cooper pairs, not to mention that the superfluid densities and velocities [see Equation (6.26)] of these components are different. Apart from fractional (nonuniversal) quantization of velocity circulation, the additional difference from Feynman's vortex-array picture is that (at $\Omega \approx \Omega_c$) topological defects are induced only in the phase field of one of the two condensates [these are vortices carrying the smaller fraction of flux quantum]. Also, in contrast to single-component systems, only a (small) fraction of total density participates in the neutral superflow, even at zero temperature: The bare superfluid stiffness in the neutral sector is only $(1/4)\rho^2 \sin^2 \beta$.

If system components have the *same* electric charge, the neutral sector is associated with the gradients of the phase difference, that is, with counterflow of two components. Correspondingly, the superfluid sector responds to rotation with countercirculation of components rather than by mimicking a solid-body rotation. For $\gamma_1 > \gamma_2$, we have [11]

$$\Omega_c \approx \frac{1}{R^2} \frac{\gamma_1 \gamma_2}{\gamma_1 - \gamma_2} \log \frac{R}{a}, \tag{6.36}$$

$$n_{\text{vort}} = \frac{\gamma_1 - \gamma_2}{\gamma_1 \gamma_2} \frac{\Omega}{\pi}. \tag{6.37}$$

At $\gamma_1 = \gamma_2$, there is no *net* matter (mass) transfer in the neutral mode, and vortices do not carry angular momentum associated with the superfluid mode. Thus, this system does not form a vortex array and represents a peculiar superfluid state where vortices cannot be induced by rotation. This case applies, for example, to electronic superconductors with multicomponent-order parameters that break U(1)×U(1) symmetry.

Finally, we briefly consider the thermodynamic limit $R \to \infty$. Now, we can no longer neglect the magnetic field and cannot rule out the formation of other type of vortices. Let us first apply the condition of absence of hyperextensive scaling of energy, Equation (5.141), for each of the two components. Dividing both sides of Equation (5.141) by system area, we get

$$2\pi n_{\text{top}}^{(1,2)} = \frac{2\Omega}{\gamma_{1,2}} \pm q\bar{B}. \tag{6.38}$$

Here, $\bar{B}$ is the average magnetic field. We should then minimize the second and third terms in Equation (6.29), obeying the relations (6.38). In the second term in Equation (6.29), one can compensate phase gradients by generating magnetic flux, as in the London effect at the cost of magnetic field $\bar{B}$ in the third term. One can see that the magnetic flux can be nullified if the density of topological defects is given by

$$n_{\text{top}}^{(1,2)} = \frac{\Omega}{\pi\gamma_{(1,2)}}. \tag{6.39}$$

However, in general, there should be deviations from (6.39), for example, if one goes beyond the London limit and allows for different core energies of the vortices.

6.5 Rotational Response of a Charged Sector in the Multicomponent Superconductor and Violation of the London Law

We will now analyze the full rotational response of the LMH-type system (i.e., the detailed pattern of supercurrents and the magnetic field) rather than the structure of the vortex array only. It is instructive to start with the finite-size system and consider macroscopically small Ω when there are no vortices. This allows one to readily appreciate the generic importance and some implications of the fact that both superconducting components are coupled to one and the same vector potential. Consider $\mathbf{A} = q_\alpha^{-1}\left(\nabla\theta_\alpha - \gamma_\alpha^{-1}\mathbf{v}_\alpha\right)$ [with $q_{(1,2)} = \pm q$ for the LMH-type system], take the curl of both sides of this expression, in the absence of vortices (i.e., at $\Omega < \Omega_c$), and then equate the right-hand sides to arrive at

$$\frac{\text{curl}\mathbf{v}_2}{q_2\gamma_2} = \frac{\text{curl}\mathbf{v}_1}{q_1\gamma_1}. \tag{6.40}$$

This relation establishes a strict correspondence between the curls of the two superfluid velocities in an equilibrium flow pattern. The simplest is the zero-temperature case when the two charged condensates can remain irrotational in the laboratory frame, $\mathbf{v}_2 = \mathbf{v}_1 = 0$, or, equivalently, perform solid-type corotation with the angular velocity $-\Omega$ in the rotational frame. Thanks to the electroneutrality and the absence of a normal component ($q_1 = -q_2$, $|\psi_1| = |\psi_2|$), this solution involves no net electric (normal or super-) currents, neither in the laboratory nor in the rotational frame, and thus is consistent. The constraint (6.40) is trivially satisfied.

In the presence of a normal component with a net electric charge, the rotation of the charged background produces an electric current, so that the superconducting component necessarily has to respond and $\mathbf{v}_2 = \mathbf{v}_1 = 0$ can no longer be a stationary solution. From (6.40), it is directly seen that, in contrast to the London picture in single-component superconductors, the two matter fields with

$q_1 = -q_2$ cannot follow the rotation of normal component, since that would imply $\mathrm{curl}\,\mathbf{v}_2 = \mathrm{curl}\,\mathbf{v}_1 \neq 0$, violating the constraint.

For a finite-size vortex-free system, Equation (6.40) suggests a solution of the generalized rigid-body form: $\mathbf{v}_\alpha = c_\alpha \Omega \times \mathbf{r}$, with constant coefficients such that $c_1/\gamma_1 = -c_2/\gamma_2$. To find c_α, we utilize the requirement of stationarity. The rotation-induced electric current of the normal component, $\mathbf{J}_n = -\left(q_1|\psi_1|^2 + q_2|\psi_2|^2\right)\Omega \times \mathbf{r}$, must be equal to the rotation-induced current response of the charged sector, $\mathbf{J}_s = \left(q_1 c_1 |\psi_1|^2 + q_2 c_2 |\psi_2|^2\right)\Omega \times \mathbf{r}$, with the opposite sign. We thus find

$$\mathbf{v}_1 = \gamma_1 \rho^{-2}\left(|\psi_1|^2 - |\psi_2|^2\right)\Omega \times \mathbf{r}, \qquad \mathbf{v}_2 = \gamma_2 \rho^{-2}\left(|\psi_2|^2 - |\psi_1|^2\right)\Omega \times \mathbf{r}.$$

To sustain these countercurrents, a rotating superconductor generates a bulk vector potential and an associated magnetic field [11], which is different from that in the London law (5.19):

$$\mathbf{B}_{\mathrm{rot}} = \frac{2}{q\rho^2}\left(|\psi_2|^2 - |\psi_1|^2\right)\Omega \times \mathbf{r}. \tag{6.41}$$

Equation (6.41) demonstrates that, even in the absence of the vortex lattice, the London law takes place only under special circumstances (e.g., in a single-component system). In the two-component system, the London law is obeyed if components have the same γ's (masses) and values of electric charges.

While the superfluid velocities $\mathbf{v}_{1,2}$ are observed in the bulk, the magnetic field $\mathbf{B}_{\mathrm{rot}}$ is generated by currents in the surface layer of the penetration length $\lambda = (q\rho)^{-1}$ width. This follows from the familiar equation $-\lambda^2 \nabla^2 \mathbf{B} + \mathbf{B} = \mathbf{B}_{\mathrm{rot}}$.

We considered several characteristic examples of rotational response of a two-component charged system that illustrate that a substantial part of the physics of a rotating single-component system is not generic and does not hold in the more general example of superfluid states such as superconducting superfluids. To describe the response of a two-component superconductor in general requires introducing vortex cores and coherence lengths and is beyond the scope of this section. Finally, we remark on the magnetorotational isomorphism. For single-component superconductors, as discussed in Chapter 5, rotational and magnetic responses are isomorphic. For a multicomponent system, if, in the rotating frame, all these components have similar coupling to the fictitious gauge field, as well as to real magnetic field, the rotational response of such system is also isomorphic to its magnetic response. As is clear from the earlier discussion, in the case of a mixture of components with different masses and U(1)×U(1) or higher broken symmetry, a rotational response of the system may be different from the magnetic response because of different couplings to the fictitious vector potential.

6.6   Violation of London Electrodynamics

In previous sections, we focused on the constant-density London approximation. Remarkably, when absolute values of the fields are allowed to vary, the magnetic field behavior in a multicomponent superconductor becomes fundamentally different from single-component behavior [for brevity, the latter case will be termed London electrodynamics]. To discuss the magnetic field behavior, we will start with a concrete example—vortex solutions in the two-component GL model—and then follow with the general theory of magnetic field behavior in a multicomponent system.

6.6.1 Fractional Vortices in the Two-Component Ginzburg–Landau Model and Magnetic Flux Delocalization and Field Inversion

Consider fractional flux vortices in the 2D* U(1)×U(1)-symmetric GL model (6.14) with effective potential (without loss of generality, we set $q_1 = q_2 = q$):

$$V = \frac{v_1}{2}\left[u_1^2 - |\psi_1|^2\right]^2 + \frac{v_2}{2}\left[u_2^2 - |\psi_2|^2\right]^2. \tag{6.42}$$

Here, $u_{1,2}$ are the equilibrium values of the two fields. We will discuss that, in contrast to the London model, magnetic field in these vortex solutions is not exponentially localized and can also invert its direction [12].

Vortices are solutions of the GL equations [see (6.10) and (6.11)]

$$\left(\partial_k + iqA_k\right)^2 \psi_1 + 2v_1\left(u_1^2 - |\psi_1|^2\right)\psi_1 = 0, \tag{6.43}$$

$$\left(\partial_k + iqA_k\right)^2 \psi_2 + 2v_2\left(u_2^2 - |\psi_2|^2\right)\psi_2 = 0, \tag{6.44}$$

$$-\epsilon_{kj}\partial_j B = J_k, \tag{6.45}$$

with appropriate asymptotic behavior at large distances. We look for a (M_1, M_2) vortex solution in polar coordinates $x + iy = re^{i\theta}$, using axially symmetric ansatz

$$(A_1, A_2) = \frac{a(r)}{r}(-\sin\theta, \cos\theta), \qquad \psi_\alpha = \sigma_\alpha(r)e^{-iM_\alpha\theta}. \tag{6.46}$$

Since, away from the vortex, supercurrent vanishes, the boundary conditions are $|\mathbf{J}| \to 0$ and $\sigma_\alpha \to u_\alpha$ as $r \to \infty$. The sum of phase gradients should be compensated by the vector potential in the asymptotic limit, implying

$$a(r \to \infty) = a_\infty = \varphi/q, \qquad \varphi = \frac{M_1 u_1^2 + M_2 u_2^2}{u_1^2 + u_2^2}, \tag{6.47}$$

and leading to the expression $\Phi = \Phi_0 \varphi$ for the enclosed magnetic flux [same as derived in the London model (6.20)].

Substituting ansatz (6.46) into Equations (6.43) through (6.45), we get a system of differential equations:

$$\sigma_1'' + \frac{\sigma_1'}{r} - \frac{(M_1 - qa)^2}{r^2}\sigma_1 + 2v_1\left(u_1^2 - \sigma_1^2\right)\sigma_1 = 0, \tag{6.48}$$

$$\sigma_2'' + \frac{\sigma_2'}{r} - \frac{(M_2 - qa)^2}{r^2}\sigma_2 + 2v_2\left(u_2^2 - \sigma_2^2\right)\sigma_2 = 0, \tag{6.49}$$

$$a'' - \frac{a'}{r} - q\left[aq\left(\sigma_1^2 + \sigma_2^2\right) - M_1\sigma_1^2 - M_2\sigma_2^2\right] = 0. \tag{6.50}$$

---

* Obviously, our analysis applies also to a 3D case with translation invariance in the third direction.

The boundary conditions are $a \to a_\infty$, $\sigma_1 \to u_1$, $\sigma_2 \to u_2$ as $r \to \infty$. The vortices with $M_1 = M_2$, carrying integer number of flux quanta, are special in that, similarly to their counterpart in single-component systems, they have modulations of $|\psi_\alpha|$ and $|B|$ exponentially localized in space (cf. discussion in Chapter 5). By contrast, if $M_1 \neq M_2$, both $(M_1 - qa)$ and $(M_2 - qa)$ do not vanish at $r \to \infty$. As a result, neither σ_1 nor σ_2 can approach its boundary value (u_1, u_2, respectively) exponentially fast because vorticity in the neutral sector implies slowly decaying gradients of the phase difference, which, in turn, affect the density fields at large distance from the core. Hence, at $M_1 \neq M_2$, the densities $|\psi_\alpha|$ recover their asymptotic values according to a certain power law. Given that the third terms in (6.48) and (6.49) decay as $1/r^2$, one can guess—with subsequent verification—that the equilibrium density is recovered by the same power law as in a neutral system:

$$\sigma_\alpha(r) \underset{r \to \infty}{\sim} u_\alpha - w_\alpha/r^2 \qquad (\alpha = 1,2), \tag{6.51}$$

with certain real coefficients w_1, w_2. Given that σ_α'' and σ_α'/r are $O\left(r^{-4}\right)$ and demanding that the leading (order r^{-2}) term vanishes lead to

$$w_\alpha = \frac{(M_\alpha - \varphi)^2}{4v_\alpha u_\alpha} \qquad (\alpha = 1,2). \tag{6.52}$$

Note that $w_{1,2} > 0$, so that $\sigma_{1,2}$ approach their boundary values from below.

Equation (6.50) then suggests—the assumption is verified in the following—that

$$a(r) \underset{r \to \infty}{\sim} \varphi/q - \Upsilon/r^2, \tag{6.53}$$

with a certain real coefficient Υ. Again, given that a'' and a'/r are order r^{-4} and demanding that the leading term in (6.50) vanishes lead to

$$\Upsilon = \frac{1}{2q\left(u_1^2 + u_2^2\right)} \left[\frac{(M_1 - \varphi)^3}{v_1} + \frac{(M_2 - \varphi)^3}{v_2} \right]. \tag{6.54}$$

Now, $B = a'(r)/r$. If $\varphi > 0$ (e.g., for $M_1, M_2 \geq 0$ case), then $a(r)$ interpolates between $a(0) = 0$ and $a_\infty > 0$, suggesting that $a'(r) > 0$ uniformly and thus $B(r) > 0$. In particular, one expects $a(r)$ to approach its boundary value a_∞ from below; that is, $\Upsilon > 0$. However, according to Equation (6.54), it is possible to have negative Υ as well. In the latter case, since $B(r) \sim 2\Upsilon/r^4$ at large r, the magnetic field must flip its sign from positive to negative as one goes away from the vortex core.

It is convenient to introduce complex/polar representation for the pairs of parameters (u_1, u_2) and (v_1, v_2): $u_1 + iu_2 = ue^{i\zeta}$ and $v_1 + iv_2 = \eta e^{i\phi}$, with $0 < \zeta$, $\phi < \pi/2$. With this parameterization,

$$\Upsilon = \frac{(M_1 - M_2)^3}{2qu^2\eta} \left[\frac{\sin^6 \zeta}{\cos \phi} - \frac{\cos^6 \zeta}{\sin \phi} \right], \tag{6.55}$$

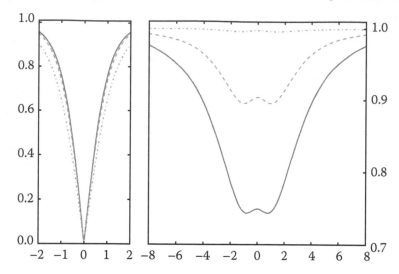

Figure 6.1   The behavior of the field magnitudes (shown on the vertical axis) as a function of the distance from the center of the vortex core (shown on the horizontal axis) of a fractional vortex with the 2π phase winding in the component ψ_1: $|\psi_1|$ (left) and $|\psi_2|$ (right) with flux fractions 5/6 (solid), 5/7 (dashed), and 1/3 (dash-dotted). The component with the phase winding, $|\psi_1|$, features a singularity. The other component always has a nonsingular W-shaped suppression of density due to the current induced in that component by the vector potential coupling. (From Reference [12].)

and we see that $\Upsilon < 0$ if, and only if, $\tan\phi < \cot^6\zeta$, which holds on precisely one-half of the $\zeta\phi$ square. Hence, the peculiar magnetic flux reversal proves to be a *generic* effect taking place on one-half of the parameter space of the model. This effect is absent in the London model. Since flux quantization is the same in London and GL models, the inversion phenomenon implies that, in the latter case, there is an additional flux located near the vortex center to offset the flux of the inverted tail. The London model solution is approximating the overall structure of the fractional vortex solution for extremely large $v_{1,2}$ and short coherence lengths relative to λ. In that case, the delocalized tail of magnetic field has small magnitude. Numerical solutions for fractional vortices are presented in Figures 6.1 and 6.2.

Nonexponential and sign-alternating behavior of magnetic field in fractional vortices signals that the electrodynamics of multicomponent superconductors are fundamentally different from that of single-component superconductors. Next, we will address this question in more detail and show that, indeed, the magnetic field behavior of multicomponent systems cannot be reduced to the theory of a massive vector field that we discussed in Chapter 5.

6.6.2 Magnetic Field Behavior and Higher Order Derivative Terms in the Free-Energy Functional

Let us now consider the magnetic field behavior in a general GL free-energy functional for a two-component superconductor (6.14), including the possibility

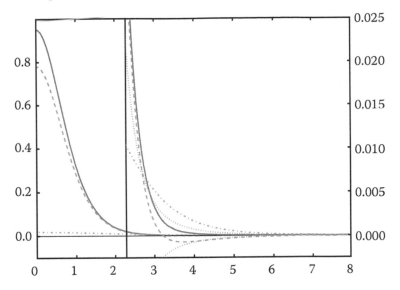

Figure 6.2   The behavior of B_z (shown on the vertical axis) as a function of the distance from the center of the vortex. Left panel shows the field near the vortex center, and right panel shows the field at a large distance for flux fractions 1 (solid), 5/6 (dashed), and 1/6 (dash-dotted). It is strikingly different from that of the Abrikosov vortex: In the 1/6 case, the magnetic field is extremely delocalized, without a pronounced maximum at the origin; despite small flux, at $r \approx 3.5$, the field value is already larger than the field value of the single-flux vortex. In the 5/6 case, the flux is accumulated near the core, closely mimicking the behavior of Abrikosov vortex. However, the field rapidly goes to zero at $r = 3.275 \pm 0.0125$, flips its direction, and follows a slowly decaying power-law tail with inverse flux. The delocalized magnetic flux in the outer region subtracts from the strongly localized flux near the origin to produce net flux $(5/6)\Phi_0$. Dotted lines in the right panel depict the asymptotic behavior predicted by Equation (6.53). (From Reference [12].)

of symmetry-breaking interband coupling terms, with the general effective potential

$$V = V\left[|\psi_\alpha|, (\theta_1 - \theta_2)\right]. \tag{6.56}$$

Without loss of generality, we assume here that all components have the same mass and charge and set $\gamma_\alpha = 1$. The following arguments apply to systems with various broken symmetries, as long as the components are coupled to the same vector potential. For systems with U(1) broken symmetry, the potential V is a function of $|\psi_\alpha|$ and the phase difference $(\theta_1 - \theta_2)$. In a number of physical contexts— apart from superconductivity—the SU(2)-symmetric GL model is highly relevant. In that special case, the potential term, $V = V_{SU(2)}$, sets a preferential value for the sum $|\psi_1|^2 + |\psi_2|^2$, but not for individual $|\psi_{1,2}|$; the most typical choice is

$$V_{SU(2)} = v\left(|\psi_1|^2 + |\psi_2|^2 - K^2\right)^2. \tag{6.57}$$

We employ the following additional notation:

$$\chi_1 = \cos(\beta/2)e^{i\theta_1}, \qquad\qquad \chi_2 = \sin(\beta/2)e^{i\theta_2},$$

$$\mathbf{j} = i \sum_{\alpha=1,2} [\chi_\alpha \nabla \chi_\alpha^* - \chi_\alpha^* \nabla \chi_\alpha], \qquad\qquad \mathbf{C} = \frac{\mathbf{J}}{q\rho^2}. \tag{6.58}$$

Next, we define the vector field

$$\vec{n} = \frac{\psi^\dagger \hat{\sigma} \psi}{\psi^\dagger \psi}, \tag{6.59}$$

where $\psi = \begin{pmatrix} \psi_1 \\ \psi_2 \end{pmatrix}$, $\psi^\dagger = (\psi_1^*, \psi_2^*)$, and $\hat{\sigma}$ the vector of Pauli matrices $(\sigma_1, \sigma_2, \sigma_3)$. In components,

$$\vec{n} = [\cos(\theta_1 - \theta_2)\sin\beta, \sin(\theta_1 - \theta_2)\sin\beta, \cos\beta]. \tag{6.60}$$

The following identity holds true:

$$\frac{\rho^2}{2}\left[|\nabla\chi_1|^2 + |\nabla\chi_2|^2 - \frac{\mathbf{j}^2}{2}\right] = \frac{\rho^2}{8}(\nabla\vec{n})^2. \tag{6.61}$$

In this parameterization, the model (6.56) can be identically represented as [13]

$$F = \frac{\rho^2}{8}(\nabla\vec{n})^2 + \frac{\rho^2}{2}\mathbf{C}^2 + \frac{1}{2q^2}\left[\epsilon_{kij}\left(\partial_i C_j + \frac{1}{4}\vec{n}\cdot\partial_i\vec{n}\times\partial_j\vec{n}\right)\right]^2 + \frac{1}{2}(\nabla\rho)^2 + V(\rho,\vec{n}). \tag{6.62}$$

This reveals that the model is mapped onto a theory that contains a massive scalar field ρ and a massive vector field $\mathbf{C}$ (the mass of the latter is the inverse magnetic penetration length). The fact that it acquires a mass is nothing but a straightforward generalization of London–Anderson physics to the two-component case. The crucial difference with the single-component case is associated with the variable $\vec{n}$, which requires special attention. First, for simplicity, consider the case where effective GL potential V is SU(2) symmetric. Then the ρ-field is massive, while the $\vec{n}$-field spontaneously breaks the global O(3) symmetry. On the other hand, in the case where V breaks the symmetry down to U(1)×U(1), then each of the fields $|\psi_1|$ and $|\psi_2|$ has a preferred value in the ground state. In terms of $\vec{n}$, the effective potential in (6.62) acquires an explicit dependence on n_3 of the form $An_3^2 + Bn_3$, breaking the O(3) symmetry for $\vec{n}$ down to U(1) symmetry. Under the constraint of constant $|\psi_{1,2}|$, we recover the representation (6.15), with the term $\left(\rho^2/8\right)(\nabla\vec{n})^2$ describing the phase gradients associated with the neutral mode. In the (6.62) representation, intercomponent Josephson coupling, $\eta\left[\psi_1\psi_2^* + \psi_1^*\psi_2\right] = 2|\psi_1||\psi_2|\cos(\theta_1 - \theta_2)$, generates a term $\propto n_1$ that breaks the remaining global U(1) symmetry associated with the field $\vec{n}$. This is how the new representation accommodates previously discussed cases by breaking various symmetries with anisotropic potential for the field $\vec{n}$.

Allowing relative density fluctuations in the multicomponent system leads to new *electrodynamic* effects associated with the form of the third term in (6.62) representing the magnetic field energy density. It acquires dependence on the second-order gradients of $\vec{n}$; namely, the contribution $\propto (\vec{n} \cdot \partial_i \vec{n} \times \partial_j \vec{n})^2$. Even though the original GL energy density is second order in field derivatives, the new parameterization reveals that the gradients of phases and densities result in self-generated contribution to the magnetic energy density, which is fourth order in derivatives. Thus, in a multicomponent superconductor, the electrodynamics do not reduce to a theory for a massive vector field: It contains the aforementioned fundamentally different contributions [13,14].

The model Equation (6.62) has the embedded nonlinear σ-model

$$F = \frac{1}{2}(\nabla \vec{n})^2 + \frac{c}{2}\left(\epsilon_{kij} \vec{n} \cdot \partial_i \vec{n} \times \partial_j \vec{n}\right)^2. \tag{6.63}$$

However, the critical difference between the functionals (6.62) and (6.63) is in the coupling between the $\vec{n} \cdot \partial_i \vec{n} \times \partial_j \vec{n}$ terms and gradients of the massive vector field in the third term.

Let us comment on the physical origin of the $\vec{n} \cdot \partial_i \vec{n} \times \partial_j \vec{n}$ terms. To this end, rewrite Equation (6.11) as

$$\mathbf{A} = \frac{1}{q^2 \rho^2}\left[-\mathbf{J} + \frac{iq}{2}(\psi_1^* \nabla \psi_1 - \psi_1 \nabla \psi_1^*) + \frac{iq}{2}(\psi_2^* \nabla \psi_2 - \psi_2 \nabla \psi_2^*)\right], \tag{6.64}$$

and use it to compute the magnetic field

$$\mathbf{B} = \mathrm{curl}\,\mathbf{A} = \mathrm{curl}\,\frac{i}{2q\rho^2}\left[(\psi_1 \nabla \psi_1^* - \psi_1^* \nabla \psi_1) + (\psi_2 \nabla \psi_2^* - \psi_2^* \nabla \psi_2)\right] - \mathrm{curl}\,\frac{1}{q^2 \rho^2}\mathbf{J}. \tag{6.65}$$

We see that $\mathbf{B}$ necessarily contains products of density gradients and gradients of the phases of the form

$$\partial_i \left(\frac{|\psi_\nu|^2}{|\psi_\nu|^2 + |\psi_\mu|^2}\right)\partial_j \theta_\nu. \tag{6.66}$$

Since such gradients involve derivatives of the relative condensate densities, they arise exclusively in superconductors with two and more components. This implies that the magnetic energy density has contributions that are fourth order in derivatives. These terms render electrodynamics of the mixture dramatically different from the London electrodynamics considered in Chapter 5—in particular leading to the field-inversion effects in the fractional vortex considered in the previous subsection. Indeed, we saw that in single-component systems, for magnetic field to penetrate superconductor deeper than λ, it is required to create phase singularities, as implied by Equation (5.74). By contrast, from the Equation (6.66), it follows that in multicomponent systems, the mixed phase-difference and relative density gradients can be a source of magnetic field.

The aforementioned considerations are generic for multicomponent systems. The magnetic field depends on the curl of total current **J** and mixed gradients of phase differences and densities. In the case of a superconductor with N-components, it can be expressed in the form

$$B_k = -\epsilon_{kij}\partial_i\left(\frac{J_j}{q|\Psi|^2}\right) - \frac{i\epsilon_{kij}}{q^2|\Psi|^4}\left[|\Psi|^2\partial_i\Psi^\dagger\partial_j\Psi + \Psi^\dagger\partial_i\Psi\partial_j\Psi^\dagger\Psi\right], \tag{6.67}$$

where **J** is the total current and $\Psi^\dagger = \left(\psi_1^*, \psi_2^*, \psi_3^*, \ldots, \psi_N^*\right)$ and $|\Psi|^2 = \Psi^\dagger\Psi$.

6.7   Skyrmion and Hopfion Topological Defects

6.7.1   Skyrmions in Two-Component Superconductors

In accordance with Equation (6.62), in the SU(2)-symmetric two-component superconductor, the field $\vec{n}$ spontaneously breaks the global O(3) symmetry. If one considers the O(3) nonlinear sigma model that contains vector field $\vec{n}$ only, the model does not admit stable topological defects in the form of ordinary vortices, but it does allow other types of topological excitations. In 2D, topologically nontrivial textures of $\vec{n}$ are skyrmions. They correspond to a situation where a planar texture of the vector $\vec{n}$ contains all the possible directions a vector $\vec{n}$ can have. That is, for any direction on the unit sphere characterized by the angles $\theta_1 - \theta_2 = \tilde{\theta}_1 - \tilde{\theta}_2$ and $\beta = \tilde{\beta}$, there will be at least one point in the $x - y$ plane such that the vector $\vec{n}$ located there will be pointing in that direction (i.e., having $\tilde{\theta}_1 - \tilde{\theta}_2$ and $\tilde{\beta}$). This generalizes the notion of vortex as a topological defect. Indeed, in the case of a vortex, one can select a circle around it in the $x - y$ plane. Then, for any phase value, one can identify a point on that circle in the physical space where the phase will have such value (for N-quantum vortex, there will be N points on the circle where that specific phase value will occur). The simplest skyrmion in the $\vec{n}$ field is shown in Figure 6.3. As a vortex is characterized by the topological invariant (the phase winding), in the nonlinear sigma model, the skyrmion texture is similarly characterized by a conserved integer-valued topological invariant. However, in the case of the skyrmion, this topological invariant is given by an integral over the xy-plane of some 2D topological charge density:

$$Q(\mathbf{n}) = \frac{1}{4\pi}\int \vec{n}\cdot\partial_x\vec{n}\times\partial_y\vec{n}\,dxdy. \tag{6.68}$$

In 3D, one can have skyrmion lines (e.g., a skyrmion in the xy-plane, translated in the z-direction) and loops (closed skyrmion lines).

Consider now a skyrmion in the two-component U(1)×U(1) superconductor. In this case, the n_3 component of $\vec{n}$ has the energetically preferred value, which corresponds to the ground-state values of $|\psi_{1,2}|$. However, because the skyrmion texture is localized in 2D, and there is no winding of $\vec{n}$ at infinity, the anisotropy results in only a finite-energy penalty. The skyrmion texture in $\vec{n}$ should retain the same topology that is shown in Figure 6.3. According to definitions (6.58) and

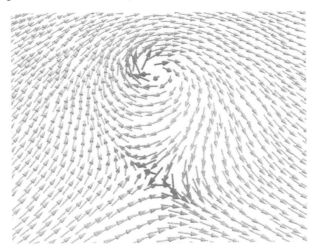

Figure 6.3 Closeup view of a skyrmion texture of the three-component unit vector field $\vec{n}$. This texture has no core and no winding at infinity: All vectors $\vec{n}$ point in the same direction $\vec{n}_0$ far away from the center. At the skyrmion center, $\vec{n}$ points in the direction opposite to $\vec{n}_0$. Extending this texture in the third dimension—perpendicular to the plane—gives a skyrmion line.

(6.60), the positions in space where $\vec{n}$ vector points to the north and the south poles of the unit sphere correspond to zeros in $\psi_{1,2}$. In skyrmions, these zeros are spatially separated and represent fractional-flux vortices. Their combined flux adds up to one flux quantum. Similarly, a skyrmion with higher topological charge Q is a bound state of $2Q$ spatially separated fractional vortices, the fluxes of which add up to Q flux quanta.

6.7.2 CP^{N-1} Skyrmions

N-component superconductors can have not only integer and fractional vortex excitations but also CP^{N-1} skyrmions.* These are defined as topological defects characterized by the CP^{N-1} topological invariant. In a theory involving multiple complex fields, the skyrmion can be viewed as a defect that consists of QN spatially separated fractional vortices with the combined flux adding up to Q flux quanta. The topological invariant associated with such excitations is given as an integral over the xy-plane:

$$Q[\Psi] = \int \frac{i\epsilon_{ji}}{2\pi|\Psi|^4}\left[|\Psi|^2\partial_i\Psi^\dagger\partial_j\Psi + \Psi^\dagger\partial_i\Psi\partial_j\Psi^\dagger\Psi\right]dxdy, \qquad (6.69)$$

with $\Psi^\dagger = \left(\psi_1^*,\psi_2^*,\psi_3^*,\ldots,\psi_N^*\right)$. In the case of an ordinary vortex with a core where all superconducting condensates vanish simultaneously, we obviously have $Q=0$. By contrast, in the case of skyrmion, zeros occur at different locations. Then, $Q \neq 0$ and the quantization condition $Q = B/\Phi_0 = N$ hold. It can be checked, for

---

* The abbreviation CP originates from complex projective.

example, numerically, that (6.69) does not change if the fractional vortex positions are shifted, as long as the cores are separated in space.

Finally, let us comment on the stability of skyrmions. These objects are made of fractional vortices that, as discussed above, attract each other logarithmically.* There are multiple mechanisms, however, that stabilize skyrmions by preventing fractional vortices from collapsing onto an axially symmetric integer vortex. For example, either density–density interaction $|\psi_\alpha|^2|\psi_\beta|^2$ or mixed gradient terms stabilize the skyrmions under certain conditions [15]. Also, skyrmions are rather generically (meta)stable in N-component models where potential breaks the U(1)^N symmetry down to a discrete symmetry such as U(1)×Z_2 [16].

6.7.3   Hopfions

In 3D, there are different topological defects in the form of $\vec{n}$-field texture: hopfions, also called knotted solitons. These defects are characterized by the Hopf topological invariant [17]. One can visualize the simplest of hopfion topological defects as follows: Take the skyrmion line and twist it around the axis corresponding to the center of the line and glue the ends. One can also twist the skyrmion line multiple times around the center of the texture. In addition, one can knot it before closing the ends. All these operations produce hopfions of different topological charges. Schematically, a toroidal hopfion is shown in Figure 6.4.

The conserved topological invariant of the hopfion is defined as follows: First, introduce $F_{ij} \equiv \vec{n} \cdot \partial_i \vec{n} \times \partial_j \vec{n}$. Next, define $\vec{A}$ such that $F_{ij} = \partial_i \mathcal{A}_j - \partial_j \mathcal{A}_i$.

Figure 6.4   Schematic picture of the toroidal hopfion. Any cross section of this topological defect produces the skyrmion texture of $\vec{n}$, like the one shown in Figure 6.3. The skyrmion line is twisted: The helical lines schematically show where, in physical space, the vector $\vec{n}$ points in a certain direction on the unit sphere S^2 (i.e., everywhere on a helix, $\vec{n}$ points in the same direction). Here, the skyrmion line was twisted twice before its ends were glued to produce the hopfion.

---

* In the presence of intercomponent Josephson coupling, the attraction is linear, as discussed in the following.

The integer-valued hopfion topological invariant is then given by the following volume integral:

$$Q = \frac{1}{32\pi^2} \int \varepsilon^{ijk} A_i F_{jk} \, dx dy dz. \tag{6.70}$$

Even though hopfions can formally be introduced in any two-component system of this kind, they can be unstable against shrinkage. Naively, for a hopfion texture of a characteristic size L, the term $(\partial \vec{n})^2$ integrated over 3D scales as L. It was demonstrated by Faddeev and Niemi [17,18] that hopfions are stable in the nonlinear σ- model (6.63) where the fourth-order derivative term $\left(\epsilon_{kij} \vec{n} \cdot \partial_i \vec{n} \times \partial_j \vec{n} \right)^2$ integrated over 3D space scales as $1/L$. As we discussed earlier, such terms exist also in superconductors, but in contrast to the model (6.63), their stability against shrinkage is not guaranteed in a general case because of the coupling to the field **C**; however, under certain conditions, these defects should be energetically stable [14].

Since a hopfion is a twisted skyrmionic line, in a superconductor, it can be viewed as a bound state of twisted non-overlapping loops of fractional vortices, the total flux of which sums up to an integer number of flux quanta [14].

6.8   Magnetic Responses and Type-1.5 Superconductivity

Here, we will discuss that, for multicomponent superconductors, the classification that divides them into type-1 and type-2 classes (and a Bogomolny point in between) is insufficient and that there is an additional class, which was recently termed *type-1.5*.*

For simplicity, we will begin the discussion with the U(1)×U(1) case; after that, we will discuss generic multicomponent GL models, including those for two-band U(1) superconductors, as well as models with more than two components. Observe first that, as was shown previously, only the composite (M, M) vortices are thermodynamically stable in a bulk sample and should be considered when discussing the phase diagram of the system in an external magnetic field. By contrast, the fractional vortices $(M, 0)$ and $(0, M)$ have divergent energy [see the discussion around Equations (6.21) and (6.22)]. We will first focus on the minimal free-energy functional (6.8) with $q_\alpha = q$, in which case the coherence lengths of the two components are

$$\xi_1 = \frac{1}{2\sqrt{a_1}}, \qquad \xi_2 = \frac{1}{2\sqrt{a_2}}, \tag{6.71}$$

and London magnetic field penetration length is

$$\lambda = \frac{1}{|q|\sqrt{u_2^2 + u_1^2}}, \tag{6.72}$$

---

* This regime was discussed first theoretically in References [19,20]. The term "type-1.5 superconductivity" was coined in the experimental work by Moshchalkov et al. [21]. Currently, this is a subject of active experimental research [22].

where, again, $u_{1,2}$ stand for the ground-state values of $|\psi_{1,2}|$. In contrast to the single-component case, we must deal with three fundamental length scales, rendering it impossible to characterize a superconductor by a single dimensionless GL parameter κ.

Type-1, Type-2 Regimes, and the Bogomolny Point in Multicomponent Systems: Type-1 superconductors expel weak external magnetic fields by generating surface currents at the length scale of λ, while strong fields give rise to formation of macroscopic normal domains. Clearly, this regime is realized when a superconductor consists of two components such that $\xi_\alpha > \lambda_\alpha$ and $\xi_1, \xi_2 > \lambda$ (where the characteristic constant λ_α is the penetration length that the component α would have if it were decoupled from the other component). Indeed, in that case, the composite vortex has two cocentered cores, each being much larger than λ. Similar to the single-component type-1 case, such a vortex is unstable thermodynamically (irrespective of the number of carried flux quanta): Formally, for the vortex to become a part of an equilibrium state in magnetic field, the value of the field must be larger than the thermodynamic critical magnetic field.

Type-2 superconductors expel external magnetic field if it is weaker than the first critical field H_{c1}. At $H = H_{c1}$, the type-2 superconductor has a second-order phase transition to a vortex state. The vortices interact repulsively with each other and form either lattices or liquids. Superconductivity is destroyed only by fields $H > H_{c2}$. It is obvious that in the model (6.8), the type-2 regime is realized at $\xi_1, \xi_2 \ll \lambda$, when a composite vortex $(1,1)$ has two cocentered cores that are much smaller than λ.

In contrast to the single-component case, in multicomponent systems, type-1 and type-2 regimes are, in general, not separated by a Bogomolny point. This is due to the fact that here the parameter space is characterized, besides λ, by several coherence lengths. The saturation of the Bogomolny bound (5.132) in two-component systems requires fine-tuning of three parameters $\xi_1 = \xi_2 = \lambda$.

Type-1.5 Superconductivity: Due to multiple fundamental length scales, in multicomponent systems, there is a distinctively new "type-1.5" superconducting regime that takes place when $\xi_1 < \lambda < \xi_2$. Here, for a composite $(1,1)$ vortex, the core of the second component is larger than the flux-carrying area, while the core of the first component is smaller (Figure 6.5). We know from Chapter 5 that, for a pair of co directed vortices, the force mediated by magnetic flux and current–current interaction is repulsive. On the other hand, the interaction mediated by overlaps of the cores is attractive. Now we have two cores capable of providing attractive forces at different length scales. The long-range interaction forces can be calculated in the linearized theory by the methods of Chapter 5; see Equation (5.109). We can use those results to write down the interaction potential between widely separated composite vortices as a sum of three terms:

$$E_{\text{int}} = 2\pi \left[C_B^2 K_0(r/\lambda) - C_1^2 K_0(r/\xi_1) - C_2^2 K_0(r/\xi_2) \right]. \tag{6.73}$$

Here, the first term is the current–current and electromagnetic repulsion, while the second and third terms represent attractive interaction coming from overlaps

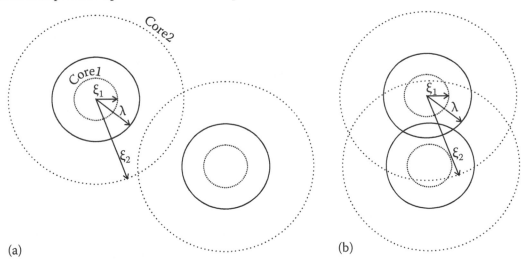

Figure 6.5   A pair of composite vortices in the regime $\xi_1 < \lambda < \xi_2$. (a) Two vortices inter-act primarily by the outer-core overlaps. (b) Two vortices interact by both the outer-core overlaps and the current-carrying-area overlaps.

of the cores. The prefactors C_B, C_1, and C_2 are determined by nonlinearities in GL equations. Note that if $C_2 \gg C_1$ [in a U(1)×U(1) system, C_2 is guaranteed to be large if the ground-state value of $|\psi_2|$ is large], then a single vortex of that type is not thermodynamically stable; that is, it cannot be produced by an applied external magnetic field because of the large energy cost of the vortex core in the second component.

Consider now a different limiting case where $\xi_1 < \lambda < \xi_2$, but the constant C_2 is small in comparison with C_1. Such a case is realized if $u_1 \gg u_2$ and $\xi_2 \gg \xi_1$. The relative smallness of u_2 guarantees that its contribution to the penetration length is small:

$$\lambda \approx \frac{1}{|qu_1|}. \tag{6.74}$$

Furthermore, in this limit, the vortex energy is determined mostly by the first component and thus cannot be very different from the vortex energy in the single-component case. In this regime, a single composite vortex is thermodynamically stable if the first component is such that it would be in the substantially type-2 regime were it uncoupled from the second component. However, the coupling to the second component, even with small u_2, changes the magnetic response qual-itatively: Since, in our example, ξ_2 is the largest length scale in the problem, the long-range interaction forces between composite vortices are dominated by the core–core interactions in ψ_2, according to (6.73), and are attractive. However, since, in this example, $\xi_2 \gg \lambda \gg \xi_1$ and $u_2 \ll u_1$, the interaction at short range is guaranteed to be dominated by the repulsive first term in (6.73). Thus, in the type-1.5 regime, vortex excitations attract each other at large distances but repel at short range. Other physical properties of this regime are discussed in the following.

Even in the case where a single vortex is not thermodynamically stable, in the regime $\xi_1 < \lambda < \xi_2$, a vortex cluster can nonetheless be thermodynamically stable, thus resulting in type-1.5 behavior. Indeed, for a vortex cluster, which carries N flux quanta and has a characteristic size $R(N)$, the positive free energy associated with its boundary is proportional to $R(N)$, while the negative energy associated with the vortex lattice inside the cluster is proportional to R^2. Thus, for a range of parameters, there will be a critical number N at which a vortex cluster becomes thermodynamically stable [cf. Equation (5.111)].*

6.8.1   Phase Diagrams and Magnetization Curves

Nonmonotonic intervortex interactions in the type-1.5 regime convert the second-order transition at H_{c1} to a first-order transition. Indeed, when the type-1.5 system forms a vortex lattice at H_{c1}, the intervortex distance—dictated by minimization of the interaction energy—must be finite, as opposed to the infinite intervortex separation in type-2 systems (cf. discussion in Chapter 5). This follows from the Gibbs free energy:

$$G_v = N_v E_v + \sum_{\alpha > \beta} E_{\text{int}}\left(R_{\alpha\beta}\right) - \Phi H, \qquad (6.75)$$

where N_v is the number of vortices, E_v is the energy of a single vortex, and Φ is the magnetic flux. In an experimentally relevant situation of a finite system in low external field, one should expect, in general, a phase separation picture: domains of vortex clusters immersed in domains of two-component Meissner state. To a certain extent, this is reminiscent of phase separation into domains of normal and Meissner states in type-1 superconductors, with the crucial difference, though, that, instead of normal domains, we now have vortex clusters.

With increasing magnetic field, the total number of vortices increases, and at a certain stage, the average intervortex distance gets smaller than the separation corresponding to the minimum of the interaction potential. Thus, in higher fields, the structure of vortex arrays transitions to either a vortex lattice or a vortex liquid. Finally, at the mean-field level, there is a continuous phase transition at H_{c2} to a normal state similar to that in type-2 superconductors. A schematic picture of magnetization curves and the phase diagram for type-1.5 superconductors are shown in Figures 6.6 and 6.7.

6.8.2   Vortex Clusters and Phase Separation in Vortex and Meissner Domains

If the minimum of the intervortex interaction potential is very shallow, then, with respect to most properties, a vortex cluster in type-1.5 superconductor can be well characterized by a superposition of single-vortex solutions. In a more general case, the structure of vortex clusters depends also on non pairwise intervortex forces [23], in addition to the sum of pairwise forces (6.73). Vortices in a cluster can

---

* We will discuss the interfaces in type-1.5 regime in more detail below.

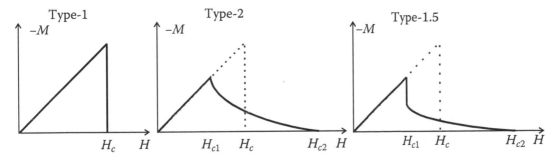

Figure 6.6   Sketch of magnetization curves in type-1.5 superconductors in comparison with those of type-1 and type-2 systems.

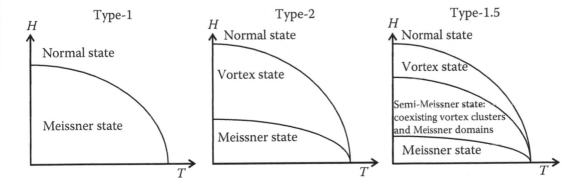

Figure 6.7   Schematic phase diagram of type-1.5 superconductors in comparison with phase diagrams of type-1 and type-2 systems.

strongly deform each other, making the vortices that are located deep inside the cluster structurally different from the vortices outside.

Figures 6.8 and 6.9 show examples of numerical solutions for vortex clusters in the GL model of type-1.5 superconducting state exhibiting a variety of patterns influenced by both pairwise and non-pairwise interactions. They feature coexistence and competition between the type-1 and type-2 tendencies in the two constituent superconducting components. One can see from these solutions that, despite being coupled to one and the same vector potential, the two components respond very differently to the external magnetic field. The interior of the vortex cluster is dominated by the first component, which prefers to form a hexagonal vortex lattice. The density of the second component is rather uniformly depleted in the core, and its supercurrent is concentrated at the cluster surface, in a way resembling surface currents on normal domains in type-1 superconductors.*

---

* In various multicomponent type-1.5 theories, the phase separation into vortex and Meissner domains also results in phase separation into states of different broken symmetries. That is, a state inside vortex clusters can have broken symmetry different from a state in the Meissner domain (e.g., U(1) vs. U(1)×U(1), or U(1) vs. U(1)×Z$_2$ [24]).

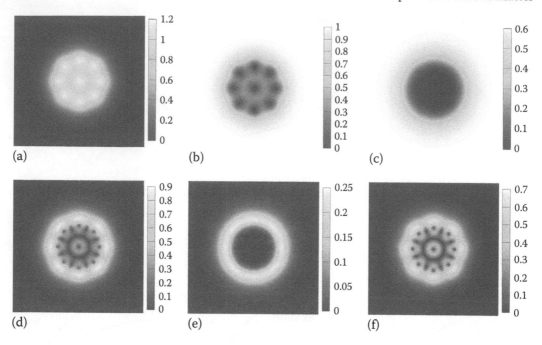

Figure 6.8   Numerical solution for a cluster of $N = 9$ flux quanta in type-1.5 superconductor. The parameters of the GL model are $(a_1, b_1) = (-1.00, 1.00)$, $(a_2, b_2) = (-0.60, 1.00)$ and $q = 1.55$. The physical quantities displayed are (a) the magnetic flux density; (b) [*resp.* (c)] the density of the first (*resp.* second) condensate $|\psi_{1,2}|^2$; (e) the magnitude of the total supercurrent; and (e) [*resp.* (f)] the magnitude of the supercurrent in the first (*resp.* second) component. A hexagonal vortex-lattice pattern formed predominantly by the first component coexists with quasi-type-1 nearly axially symmetric behavior of the second component: its density is depleted inside the cluster, and its current is concentrated at the cluster boundary. (From Reference [23].)

Let us remark on the original argument that, among other aspects, used interface energy for the classification of superconductors into type-1 and type-2 (Ginzburg and Landau [25]; see also Chapter 5). In the single-component system, one can assume that there are no phase gradients along the boundary and then reduce the calculation to the 1D problem. However, in general, the interface problem in GL equations is *not* 1D in the presence of phase winding. This circumstance becomes especially important in the multicomponent case. In the type-1.5 regime, there appears an additional characteristic interface energy associated with the boundary of a vortex cluster that necessarily involves phase gradients along the boundary. The situation shown in Figures 6.8 and 6.9 clearly illustrates the following: Inside the cluster, the system behavior is dominated by one of the components and has a negative interface energy and thus tries to maximize the superconducting-to-normal-state interface by forming a vortex lattice. However, due to the existence of multiple coherence lengths, the cluster's boundary has excess current of the second component, rendering the corresponding

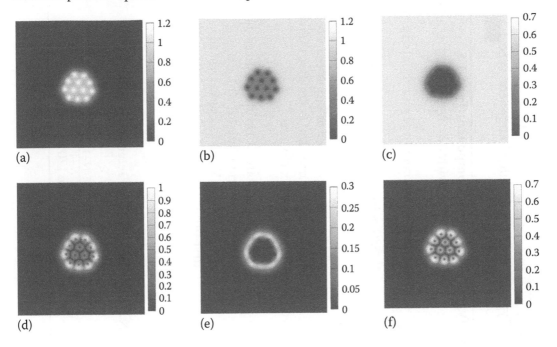

Figure 6.9 Numerical solution for a cluster with $N = 12$ flux quanta in type-1.5 U(1)×U(1) superconductor. The parameters of the potential are $(a_1, b_1) = (-1.00, 1.00)$, $(a_2, b_2) = (-0.60, 1.00)$, and $q = 1.55$. The physical quantities displayed here are (a) the magnetic flux density; (b) [resp. (c)] the density of the first (resp. second) condensate $|\psi_{1,2}|^2$; (d) the magnitude of the total supercurrent; and (e) [resp. (f)] the magnitude of the supercurrent in the first (resp. second) component. This pattern demonstrates the same coexistence and competition of type-1-like and type-2-like tendencies, as in Figure 6.9. (From Reference [23].)

interface energy positive and forcing the system to minimize the vortex-cluster-to-Meissner-domain interface.

In the general case of N-component superconductor ($N \geq 2$), type-1.5 regime occurs when some of the coherence lengths are smaller than the magnetic field penetration length while some are larger: $\xi_1, \ldots, \xi_k < \lambda < \xi_{k+1}, \ldots, \xi_N$, and there are vortices with nonmonotonic interaction. Concluding this subsection, properties of type-1, type-2, and type-1.5 regimes are summarized in Table 6.2.

In Chapter 5, we discussed that, in the single-component GL theory, the ration ξ/λ is temperature independent [see (5.65) and (5.64)]. By contrast, a multicomponent superconducting system in general has multiple critical temperatures, that is, second-order phase transitions between different superconducting states [such as between U(1) and U(1)×U(1) or between U(1) and U(1)×Z_2 states]. When a superconductor breaks an additional symmetry, there appears additional coherence length, which diverges at the transition simultaneously with vanishing amplitude of the corresponding component. On the other hand, the magnetic field penetration length stays finite at such a transition. Thus, in such cases, appearance of type-1.5 superconductivity should be rather generic.

Table 6.2 Basic Characterization of Superconductors into Single-Component Type-1/Type-2 and Multi Component Type-1.5 Regimes, Based on GL Model

	Single-Component Type-1	Single-Component Type-2	Two-Component Type-1.5
Characteristic length scales	λ and ξ with $\lambda/\xi < 1$	λ and ξ with $\lambda/\xi > 1$	λ, ξ_1, and ξ_2 with $\xi_1 < \lambda < \xi_2$.
Inter vortex interaction	Attractive	Repulsive	Attractive at long range and repulsive at short range.
Energy of the superconducting/normal state interface	Positive	Negative	Negative inside the vortex cluster and positive at cluster's boundary.
Magnetic field stabilizing a vortex	Larger than the thermodynamical critical magnetic field H_c	Smaller than H_c	Can be either smaller or larger than H_c for a single vortex but is smaller than H_c for a vortex cluster.
Phases in external magnetic field	(a) Meissner state, (b) normal state, (c) phase separation into (a)+(b). First-order (a)–(b) phase transition	(a) Meissner state, (b) vortex lattices, (c) normal state. Second-order (a)–(b) and (b)–(c) phase transitions (at the mean-field level)	(a) Meissner state, (b) vortex lattices, (c) phase separation into (a)+(b), (d) normal state. Vortices appear via the first-order (a)–(b) phase transition. The (b)–(d) transition is the second order.
Energy $E(N)$ of N-quantum axially symmetric vortex solutions	$\frac{E(N)}{N} < \frac{E(N-1)}{N-1}$ for all N.	$\frac{E(N)}{N} > \frac{E(N-1)}{N-1}$ for all N.	There is a characteristic number N_c such that $\frac{E(N)}{N} < \frac{E(N-1)}{N-1}$ for $N < N_c$, while $\frac{E(N)}{N} > \frac{E(N-1)}{N-1}$ for $N > N_c$ (see Reference [20]).

6.9 Effects of Intercomponent Interactions in Multiband Systems

Multicomponent superconductivity is not limited to systems with higher broken symmetries; it arises in many materials due to features of electronic band structures [3,7,8]. In this section, we will consider such systems.

6.9.1 Intercomponent Josephson Coupling

If there are multiple bands in a material, there can be several superconducting components originating from these bands. The superconducting components in different bands are not independently conserved, and in two-band superconductors, for example, one should consider the GL free energy (6.8) with additional intercomponent coupling of the form*

$$V \to V + \frac{\eta}{2}[\psi_1\psi_2^* + \psi_1^*\psi_2] = |\psi_1||\psi_2|\cos(\theta_1 - \theta_2). \tag{6.76}$$

Here, $\alpha = 1,2$ labels two electronic bands that give rise to two superconducting components. The bilinear term is nothing but interband Josephson coupling. It is similar to Josephson coupling between two spatially separated superconductors, but now it describes intrinsic coupling between (not individually conserved) superconducting components in different bands. Note that in this case $q_\alpha = q$, and the neutral sector is associated with the phase difference.[†] This term sets the preferential value for the phase difference: $\theta_1 - \theta_2 = 0$ if $\eta < 0$, and $\theta_1 - \theta_2 = \pi$ if $\eta > 0$.

Consider the London regime $|\psi_\alpha| = $ const. The part of the free energy that depends on the phase difference,

$$F_{\theta_1-\theta_2} = \frac{|\psi_1|^2|\psi_2|^2}{2\rho^2}|\nabla(\theta_1 - \theta_2)|^2 + \eta|\psi_1||\psi_2|\cos(\theta_1 - \theta_2), \tag{6.77}$$

features a massive mode in the presence of interband Josephson coupling, with the mass

$$m_J \equiv \frac{1}{\lambda_J} = \sqrt{\frac{\eta\rho^2}{|\psi_1||\psi_2|}}. \tag{6.78}$$

The parameter λ_J—often referred to as the Josephson length—sets the characteristic length scale at which the perturbed phase difference recovers its ground-state value. This phase locking, originating from the fact that components are not independently conserved, explicitly breaks the U(1) symmetry—thus removing

---

* Microscopically, this model was discussed first by Tilley [3]; for formal justification, see Reference [26].

† The symmetry also allows the system to have mixed gradient terms of the form $\{[(\nabla + iq\mathbf{A})\psi_1](\nabla - iq\mathbf{A})\psi_2^* + \text{c.c.}\}$. At the microscopic level, such terms are induced by interband impurity scattering; see, for example, Reference [27].

superfluidity—in the neutral sector.* The system retains the second U(1) symmetry and thus remains superconducting. Obviously, the Josephson coupling cannot dramatically affect composite $(1,1)$ integer-flux vortices, because they do not have gradients of the phase difference. However, this coupling leads to linear confinement of fractional $(1,0)$ and $(0,1)$ vortices, which can be viewed as constituents of the $(1,1)$ composite topological defect. The linear confinement can be seen as follows: In U(1)×U(1) systems, the gradients of the phase difference result in logarithmic interaction between $(1,0)$ and $(0,1)$ vortices; see Equations (6.23) and (6.24). In the presence of the $\eta |\psi_1||\psi_2| \cos(\theta_1 - \theta_2)$ coupling, the gradients in $\theta_1 - \theta_2$ must be contained in a stripe of the width λ_J (i.e., forming a string that connects two vortices). For a string of the length $L \gg \lambda_J$ the behavior of the phase difference in the cross section of the string can be approximated by a solution of the 1D sine–Gordon equation that derives from Equation (6.77). Thus, the interaction energy should asymptotically be a linear function of the separation between the vortices. We will present in the following the solution for the Josephson string in the more general three-band case.

6.9.2 Hybridization of Coherence Lengths in Multiband Systems

In the presence of Josephson or other couplings between the components (such as density–density interaction $\propto |\psi_1|^2 |\psi_2|^2$), the definitions (6.71) for coherence lengths are no longer valid. This kind of interaction leads to the hybridization of coherence lengths [28,30]. This effect is crucial for deriving the vortex interaction and classification of the magnetic response of the system.

Consider the model with the most general effective potential (6.56) that depends on $|\psi_1|$, $|\psi_2|$, and $(\theta_1 - \theta_2)$. The allowed terms include (but are not restricted to) Josephson terms and the biquadratic (density–density) interaction. For definiteness, assume that the phase difference is locked at 0. The coherence length is defined as the characteristic length controlling an asymptotic exponential decay of density variations. By definition, it is characteristic of a linear regime, and its calculation is based on GL equations linearized with respect to small deviations of $|\psi_{1,2}|$ from their equilibrium values, $\sigma_1 = |\psi_1| - u_1$, $\sigma_2 = |\psi_2| - u_2$. [Decay of strong density perturbations, $|\psi_\alpha| \ll u_\alpha$, may initially involve different scales.] We can use the following axially symmetric ansatz for the solution:

$$\psi_\alpha = f_\alpha(r)e^{i\theta_\alpha}, \qquad (A_1, A_2) = \frac{a(r)}{r}(-\sin\theta, \cos\theta), \qquad (6.79)$$

where f_1, and f_2, and a are profile functions subject to the boundary conditions $f_\alpha(0) = a(0) = 0$, $f_\alpha(\infty) = u_\alpha$, $a(\infty) = -1/q$, and $\theta_1 = \theta_2 = \theta$. Being interested in the coherence lengths only, we neglect variations in the phase difference.† Once we

---

* A reservation should be made here: Nonconservation of matter (number of particles) *per se* does not generically suppress superfluidity. In particular, in Chapter 15, we will consider examples of realistic superfluid systems with nonconserved numbers of particles.

† In general, intervortex forces in a vortex cluster involve interactions via gradients of the phase difference [23]; however, such interactions are coming from nonlinear effects that we do not consider here.

identify the coherence lengths and magnetic penetration length, we will use them to obtain the long-range intervortex forces.[*]

The linearized theory for GL functional with the most general effective potential (6.56) is given by

$$F_{\text{lin}} = \frac{1}{2} \sum_{\alpha=1,2} |\nabla \sigma_\alpha|^2 + \frac{1}{2} \begin{bmatrix} \sigma_1 \\ \sigma_2 \end{bmatrix} \cdot \mathcal{H} \begin{bmatrix} \sigma_1 \\ \sigma_2 \end{bmatrix} + \frac{1}{2} (\partial_1 A_2 - \partial_2 A_1)^2 + \frac{1}{2q^2} \left(u_1^2 + u_2^2 \right) |A|^2,$$

(6.80)

where $\mathcal{H}$ is the Hessian matrix of $V(|\psi_1|, |\psi_2|, 0)$ about the ground state:

$$\mathcal{H}_{\alpha\beta} = \frac{\partial^2 V}{\partial |\psi_\alpha| \partial |\psi_\beta|} \bigg|_{|\psi_1|=u_1, |\psi_2|=u_2, \theta_1=\theta_2}.$$

(6.81)

The vector potential A is decoupled from $\sigma_{1,2}$ and has a mass based on contributions from both fields:

$$\mu_A = q\sqrt{u_1^2 + u_2^2} = \lambda^{-1}.$$

(6.82)

Thus, λ dependence on various interband couplings, including the Josephson coupling, is only through the field amplitudes.

By contrast, the density fields $\sigma_{1,2}$ are, in general, coupled in (6.80), and their asymptotic behavior follows the same exponential law. This, however, does not mean that there is only one coherence length in the system. Let us express (σ_1, σ_2) in terms of eigenvectors $v_{1,2}$ of matrix $\mathcal{H}$:

$$(\sigma_1, \sigma_2)^T = s_1 v_1 + s_2 v_2.$$

(6.83)

This results in the decoupled free energy for all fields, thereby yielding two coherence lengths ξ_α:

$$F_{\text{lin}} = \frac{1}{2} \sum_{\alpha=1,2} \left(|\nabla \varsigma_\alpha|^2 + \frac{1}{\xi_\alpha^2} \varsigma_\alpha^2 \right) + \frac{1}{2} (\text{curl} A)^2 + \frac{\lambda^2}{2} |A|^2.$$

(6.84)

Thus, the effect of the interaction, such as the intercomponent Josephson coupling, is the following: The coherence lengths ξ_1 and ξ_2 could no longer be directly associated with the individual condensates ψ_1, ψ_2. Instead, the interband interactions of this kind produce a *hybridization*. The coherence lengths can be associated only with their linear combinations ς_1, ς_2, defined as follows:

$$\varsigma_1 = (|\psi_1| - u_1) \cos \Theta - (|\psi_2| - u_2) \sin \Theta,$$

(6.85)

$$\varsigma_2 = -(|\psi_1| - u_1) \sin \Theta - (|\psi_2| - u_2) \cos \Theta.$$

(6.86)

---

[*] A reservation should be made that (as discussed in Chapter 5) one cannot always rely on linearized theory even for the long-range asymptotic behavior of the fields in a vortex solution. For example, in a strongly type-2 superconductor, the asymptotic behavior of $|\psi|$ is controlled by nonlinearities. This circumstance, however, is not relevant to the main topic of interest in the present chapter: The type-1.5 regime where the fundamental length scales are $\xi_1 < \lambda < \xi_2$ and the long-range forces are determined by the competition between the largest coherence length ξ_2 and λ.

These may be interpreted as rotated (in the fields space) versions of $\sigma_1 = |\psi_1| - u_1$, $\sigma_2 = |\psi_2| - u_2$. We can define a *mixing angle* as the angle between the ς and σ axes, which is Θ, where the eigenvector v_1 of $\mathcal{H}$ is $(\cos\Theta, \sin\Theta)^T$. The mixing angle can be determined from $\mathcal{H}$. Unfortunately, in the case of a general system, there are no analytically known general expressions for coherence lengths and mixing angles in terms of parameters of the GL model. However, they can easily be obtained numerically, and also, when interband interactions are weak, one can obtain analytical expressions perturbatively [28].*

With the fields ς_1 and ς_2, one can apply the methods of Chapter 5 to derive the intervortex interaction as

$$E_{int} = 2\pi\left[C_B^2 K_0(r/\lambda) - C_1^2 K_0(r/\xi_1) - C_2^2 K_0(r/\xi_2)\right]. \tag{6.87}$$

It has the same functional form as (6.73) in U(1)×U(1) theory. Correspondingly, the magnetic phase diagram of the U(1) two-band system is similar to that of U(1) × U(1) theory (summarized in Table 6.2). In particular, it features the type-1.5 regime at $\xi_1 < \lambda < \xi_2$ also in the presence of various interband interactions.†
Figure 6.10 shows the structure of magnetic and matter fields in the cross section taken through the midpoint between two vortices in a two-band superconductor, with and without weak interband Josephson coupling. We see that small symmetry-breaking terms do not affect dramatically the overall profile of the fields. At the same time, even weak coupling significantly modifies the asymptotic decay of the fields. Quantitative effects of various interband couplings on the intervortex interaction potential are shown in Figure 6.11.

6.9.3   Passive-Band Superconductivity

Let us now consider a concrete example of length scales and intervortex interactions in the presence of interband coupling. Notice first that, since the interband interaction term $(\eta/2)\left(\psi_1\psi_2^* + \psi_1^*\psi_2\right)$ is negative, one can have superconductivity in both bands even when one or both prefactors of quadratic terms are positive. A rather generic situation in two-band superconductivity is when $a_1 > 0$ and $a_2 < 0$. In such a case, the superconductivity in the second band appears exclusively due to interband Josephson coupling to the second band. Often, in such an example, the first band is called "active" and the second band "passive." This physical situation can be illustrated by the minimal GL model with the potential

$$V = -a_1|\psi_1|^2 + \frac{b_1}{2}|\psi_1|^4 + a_2|\psi_2|^2 + (\eta/2)(\psi_1\psi_2^* + \psi_1^*\psi_2), \tag{6.88}$$

---

* A similar mixing effect is indeed also present in multicomponent microscopic theories including the regimes where the GL approach is inapplicable [29].

† In two-component GL theory, one can always define two coherence lengths; note, however, that having multiple bands is not a sufficient condition for having multiple coherence lengths. In many multiband materials, the interband interaction is strong, and the system does not have a description in terms of two-component GL theory with two coherence lengths [26].

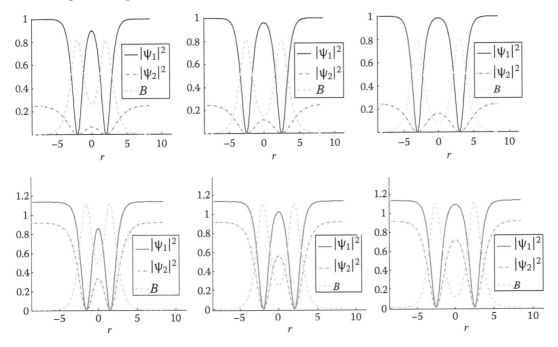

Figure 6.10   Cross-sectional view of field amplitudes and magnetic field between two vortices in a type-1.5 superconductor with and without interband Josephson coupling (upper and lower rows, respectively). Images on the left represent short intervortex separation when the interaction potential is repulsive. Images in the middle represent the intervortex distance corresponding to the minimum of the interaction potential. Images on the right represent large intervortex separation when the interaction potential is dominated by core overlaps and is attractive. The parameters are $a_1 = -1$, $b_1 = 1$, $a_2 = -0.0625$, $b_2 = 0.25$, and $q = 1$. The value of the interband Josephson coupling is $\eta = 0.3$. (From Reference [28].)

where $a_\alpha, b_\alpha > 0$. Then V is minimized when

$$u_1 = \sqrt{\frac{a_1}{b_1}\left(1 + \frac{\eta^2}{4a_1 a_2}\right)}, \qquad u_2 = \frac{\eta}{2a_2}u_1. \tag{6.89}$$

Note that, in this example, the ground-state value u_2 of the passive band field is linearly proportional to the coupling coeffcient and the ground-state value of the density in the active band. Since Josephson coupling increases ground-state values of the complex fields in both bands, it decreases the value of the penetration length:

$$\lambda = 1/q\sqrt{u_1^2 + u_2^2} = \left(qu_1\sqrt{1 + \frac{\eta^2}{4a_2^2}}\right)^{-1}. \tag{6.90}$$

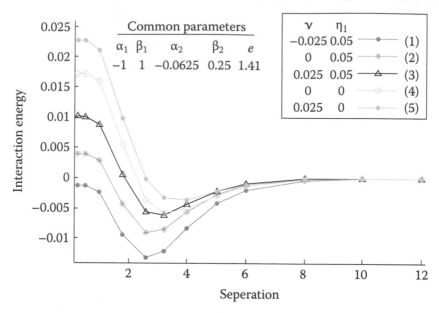

Figure 6.11 Numerical results for intervortex interaction potentials for the type-1.5 regime in GL models with interband couplings. The plot is adapted from Reference [28], with the original notation for the parameters of the model: $F = \sum_{j=1,2}$ $\left[(1/2)\left|(\nabla + ie\mathbf{A})\psi_j\right|^2 + \alpha_j\left|\psi_j\right|^2 + (\beta_j/2)\left|\psi_j\right|^4 \right] + \frac{1}{2}(\text{curl } \mathbf{A})^2 + \eta_1\left|\psi_1\right|\left|\psi_2\right|\cos(\theta_1 - \theta_2) + (\nu/2)$ $\left\{ \left[(\nabla - ie\mathbf{A})\psi_1^*\right](\nabla + ie\mathbf{A})\psi_2 + \text{c.c.} \right\}$. The "common parameters" are shared by curves. Curve (4) is for the case when the bands are coupled by the vector potential only. In this case, the ratio of the coherence lengths is $\xi_2/\xi_1 = 4$. Curve (2) shows the effect of adding the Josephson term, while curve (5) shows the effect of adding the mixed gradient term. Curves (1) and (3) show the combined effect of both the mixed gradient and Josephson terms, with similar and opposite signs. The unit for the interaction energy is chosen to be the energy of an isolated vortex. (From Reference [28].)

The Hessian matrix of the effective potential V about the field's ground-state values (u_1, u_2) is

$$\mathcal{H} = \begin{pmatrix} 4a_1 + \frac{3\eta^2}{2a_2} & \eta \\ \eta & 2a_2 \end{pmatrix}. \tag{6.91}$$

In this case, one can straightforwardly compute analytically the values of coherence lengths and the mixing angle. It is instructive, however, to take the small-η limit. Then we find that Josephson coupling gives a correction to the coherence lengths that is second order in η:

$$\xi_1^{-1} = 2\sqrt{a_1} + O\left(\gamma^2\right), \qquad \xi_2^{-1} = \sqrt{2a_2} + O\left(\gamma^2\right). \tag{6.92}$$

Now consider the mixing effect: The normalized eigenvector associated with eigenvalue ξ_1^{-2} is

$$
v_1 = \begin{pmatrix} 1 \\ (2a_1 - a_2)^{-1}(\eta/2) \end{pmatrix} + O(\eta^2),
\tag{6.93}
$$

so the normal modes associated with the fluctuations around the ground state (which defines the coherence lengths) are characterized by the mixing angle:

$$
\Theta = (2a_1 - a_2)^{-1}\eta/2 + O\left(\eta^2\right).
\tag{6.94}
$$

So, in this example, the main effect of weak Josephson coupling is associated with the mixing of the normal modes. By contrast, the weak Josephson coupling affects the coherence lengths less. Therefore, the model with the passive band can easily have distinct coherence lengths and supports type-1.5 superconductivity when $\xi_1 < \lambda < \xi_2$ [30].

6.9.4   Spontaneous Breaking of Time-Reversal Symmetry

Superconductors and superfluids can break not only continuous but also discrete symmetries. As a characteristic example, we will consider spontaneous breaking of time-reversal symmetry in three-band superconductors. The minimal GL free energy to model a three-band superconductor is identical to (6.8), trivially generalized to $\alpha = 1, 2, 3$ and supplemented with the interband coupling:

$$
V_J = \sum_{\alpha=1,2,3} \sum_{\beta > \alpha} \eta_{\alpha\beta} |\psi_\alpha| |\psi_\beta| \cos\left(\theta_\alpha - \theta_\beta\right).
\tag{6.95}
$$

In the London limit, it can be rewritten as

$$
F = \frac{1}{2\rho^2} \left(\sum_{\alpha=1,2,3} |\psi_\alpha|^2 \nabla\theta_\alpha + q\rho^2 \mathbf{A} \right)^2
$$

$$
+ \frac{1}{4\rho^2} \left[|\psi_1|^2 |\psi_2|^2 [\nabla(\theta_1 - \theta_2)]^2 \right.
$$

$$
\left. + |\psi_2|^2 |\psi_3|^2 [\nabla(\theta_2 - \theta_3)]^2 + |\psi_1|^2 |\psi_3|^2 [\nabla(\theta_1 - \theta_3)]^2 \right]
$$

$$
+ \sum_{\alpha=1,2,3} \sum_{\beta > \alpha} \eta_{\alpha\beta} |\psi_\alpha| |\psi_\beta| \cos\left(\theta_\alpha - \theta_\beta\right)
$$

$$
+ \frac{1}{2}(\mathrm{curl}\mathbf{A})^2,
\tag{6.96}
$$

where $\rho^2 = \sum_{\alpha=1,2,3} |\psi_\alpha|^2$. In the absence of Josephson couplings, the system features $U(1) \times U(1) \times U(1)$ symmetry. There are two neutral modes, but these cannot be fully characterized using only two phase differences; see the second line in Equation (6.96).

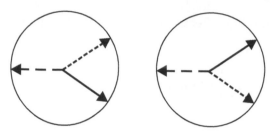

Figure 6.12   Phase-locking patterns in a three-band superconductor with spontaneously broken time-reversal symmetry. Arrows represent phase values. The states are related to each other by complex conjugation of the fields.

When all Josephson couplings are negative $\eta_{\alpha\beta} < 0$, the phase differences are locked to zero, and the symmetry is reduced to U(1). If, however, all couplings are positive, $\eta_{\alpha\beta} > 0$, the model features *frustration* in the intercomponent coupling: Each individual term is minimized at $\theta_\alpha - \theta_\beta = \pi$, which cannot be simultaneously true for all three phase differences. The resulting phase locking is based on a compromise. There are two degenerate states, with inequivalent phase-locking patterns, which minimize the energy. They are related to each other by complex conjugation of the fields (i.e., time-reversal transformation). Hence, when assuming one of the two phase-locked states, the system breaks the time-reversal symmetry. To illustrate this state, consider the simplest case where all three components have similar coupling. Then two possible ground states are given by (for brevity, we define $\varphi_{\alpha\beta} = \theta_\alpha - \theta_\beta$) $\varphi_{12} = 2\pi/3$, $\varphi_{13} = -2\pi/3$ or $\varphi_{12} = -2\pi/3$, $\varphi_{13} = 2\pi/3$; see Figure 6.12.

Hence, in this case, the remaining symmetry is U(1)×Z_2, as opposed to U(1). Like any other system with broken Z_2 symmetry, such superconductors allow an additional set of topologically nontrivial configurations. Z_2 topological defects are domain walls that interpolate between the ground states that are different with respect to Z_2 transformation. Under certain conditions, the system also allows composite topological excitations, which are bound states of closed domain walls and vortices and are characterized by CP^2 topological invariants [16].

6.9.5   Fractional Vortices and Domain Walls in the Presence of Symmetry-Breaking Interactions

Fractional-flux vortices in three-band systems provide an important example of how Josephson terms affect topological excitations in multiband systems. Consider the case of one fractional vortex in which the phase of ψ_1 winds by 2π, while ψ_2 and ψ_3 do not have windings.

A calculation analogous to that in Equation (6.18) gives that, in this model, a vortex with phase winding in only one phase carries flux

$$\Phi_{(1,0,0)} = -\frac{|\psi_1|^2}{\rho^2}\Phi_0. \tag{6.97}$$

In the London limit, it follows from Equation (6.96) that the total free energy of any configuration satisfies the lower bound (L is the system size):

$$F - F_{GS} \geq F_{SG}, \qquad F_{SG} = \sum_{\alpha < \beta} v_{\alpha\beta} \int_{r \leq L} \left[|\nabla \varphi_{\alpha\beta}|^2 + \frac{1}{2} m_{\alpha\beta}^2 \left(1 - \cos \varphi_{\alpha\beta}\right) \right] d^2 r, \quad (6.98)$$

where $v_{\alpha\beta} = |\psi_\alpha|^2 |\psi_\beta|^2 / \rho^2$, $m_{\alpha\beta}^2 = 2\eta_{\alpha\beta} \rho^2 / |\psi_\alpha| |\psi_\beta|$ and F_{GS} denotes the energy of the vortex-less ground state. Further progress can be obtained by noting (a) that F_{SG} simplifies to a sum of sine–Gordon energies and (b) that $|\nabla \varphi_{\alpha\beta}|^2 \geq r^{-2} (\partial \varphi_{\alpha\beta}/\partial \theta)^2$, with r and θ, are the polar coordinates around the vortex core. Hence, we have

$$F_{SG} \geq \sum_{\alpha < \beta} v_{\alpha\beta} \int_{r \leq L} d^2 r \left[\frac{1}{r^2} \left(\frac{\partial \varphi_{\alpha\beta}}{\partial \theta} \right)^2 + m_{\alpha\beta}^2 \sin^2 \frac{\varphi_{\alpha\beta}}{2} \right] \qquad (6.99)$$

$$= \sum_{\alpha < \beta} v_{\alpha\beta} \int_{r \leq L} d^2 r \left[\left(\frac{1}{r} \frac{\partial \varphi_{\alpha\beta}}{\partial \theta} - m_{\alpha\beta} \sin \frac{\varphi_{\alpha\beta}}{2} \right)^2 + \frac{2 m_{\alpha\beta}}{r} \frac{\partial \varphi_{\alpha\beta}}{\partial \theta} \sin \frac{\varphi_{\alpha\beta}}{2} \right] \qquad (6.100)$$

$$\geq \sum_{\alpha < \beta} 2 m_{\alpha\beta} v_{\alpha\beta} \int_{r_0}^{L} dr\, r \int_0^{2\pi} d\varphi \frac{1}{r} \frac{\partial \varphi_{\alpha\beta}}{\partial \theta} \sin \frac{\varphi_{\alpha\beta}}{2} \qquad (6.101)$$

$$= 8 \left(m_{12} v_{12} + m_{13} v_{13} \right) \left(L - r_0 \right), \qquad (6.102)$$

where r_0 is the short-distance cutoff [31].* In the last expression, we explicitly used the boundary condition that φ_{12} and φ_{13} wind once, while φ_{23} does not wind. We see that F_{SG}, and thus the total free energy $F - F_{GS}$, grows (at least) linearly with the system size L. This illustrates the aforementioned fact that, in the presence of Josephson coupling, a fractional vortex emits strings, which asymptotically behave as sine–Gordon solitons or kinks (in the phase difference). In the three-component case, a fractional vortex emits two such strings, which should terminate either on an antivortex or at the system boundary or on other fractional vortices with phase windings in θ_2 and θ_3. This is why, in the presence of Josephson coupling, a composite vortex is a state of linearly confined fractional vortices.

6.10 Composite U(1) Order and Coflow/Counterflow Superfluidity

Under a number of microscopically different circumstances, a multicomponent (classical-field, bosonic, or fermionic) system can feature U(1) order that does

---

* The lower bound on F_{SG} cannot be attained [31]: To achieve this, one would need $\varphi_{\alpha\beta}$ to satisfy $\partial \varphi_{\alpha\beta}/\partial \theta = r m_{\alpha\beta} \sin(\varphi_{\alpha\beta}/2)$. Meanwhile, no solution to this partial differential equation is consistent with the correct boundary behavior $\varphi_{12}(r, 2\pi) - \varphi_{12}(r, 0) = 2\pi$ for all r.

not reduce to broken U(1) symmetry of individual components. The (irreducible) anomalous averages* characterizing this order have the form

$$\left\langle \psi_{\alpha_1}^{S_1} \psi_{\alpha_2}^{S_2} \psi_{\alpha_3}^{S_3} \cdot \ldots \cdot \psi_{\alpha_P}^{S_P} \right\rangle \neq 0, \tag{6.103}$$

where ψ_α is the αth component of the classical field or the quantum field operator, $P \leq N$ (with N the total number of components), and S_j ($j = 1, 2, \ldots, P$) are some integers with the convention that negative S_j implies complex/Hermitian conjugation of the field taken to power $|S_j|$. By irreducibility of (6.103), we mean that it has no nonzero contributions from the product of two or more anomalous averages based on its constituents; in particular, $\left\langle \psi_{\alpha_j} \right\rangle = 0$ for each $j = 1, 2, \ldots, P$.

From the quantum-mechanical perspective, some classes of multicomponent orders (6.103) are trivial, reflecting nothing but the microscopic structure of quantum particles of which a superfluid is made. For example, U(1) order in superfluid ^{4}He, when expressed in terms of the field operators of (spin-up and spin-down) electrons, $\psi_\uparrow$, $\psi_\downarrow$, and the nuclei, ψ_n, will read $\left\langle \psi_n \psi_\uparrow \psi_\downarrow \right\rangle$. From the long-range classical-field perspective, the triviality of such cases is that the composite nature of the order parameter does not manifest itself at the macroscopic (or even mesoscopic) scale. In this sense, one can also consider trivial—within the context of multicomponent superfluidity, not on its own—the case of Cooper pairing, $\left\langle \psi_\uparrow \psi_\downarrow \right\rangle \neq 0$.

In this section, we will instead discuss nontrivial cases of multicomponent U(1) orders (6.103). These are not connected with the direct real-space particle pairing. In this category, one of the most instructive —and, as we will see later, not at all physically unfeasible—cases is when $\left\langle \psi_{\alpha_j} \right\rangle$ orders are destroyed by *increasing* temperature. Here, the interpretation in terms of "particle binding" becomes especially physically misleading. This should be kept in mind, for example, when the $\left\langle \psi_1 \psi_2 \right\rangle \neq 0$ order is termed (just for brevity) a *paired state*. We term these states "composite U(1)" order. Such composite states were discussed to appear under certain conditions in 2D two-component superconductors [10], 3D superconductors in external field [2], 3D superconductors in the absence of external field [32], and two-component bosonic systems in a (nearly) commensurate external potential—promoting dissipationless drag through Mott physics [5,6].

The nontrivial composite U(1) states could be understood through the physics of vortices: When vortices of a certain type are present at finite concentration and are free to move, they destroy superfluidity in the channel where they carry a topological charge. For example, a superconductor in external magnetic field or a superfluid under rotation loses its ability to sustain supercurrent or superflow if the field-induced vortex lattice melts. In the absence of rotation or external field, superfluid/superconducting phase transitions are linked to proliferation of thermally excited topological defects. What all the single-component cases have in common is that there is only one topological defect responsible for the transition.

---

* For brevity, in this section, we will often use the terminology of genuine long-range order and broken symmetries. The discussion straightforwardly applies to charged systems as well as to lower dimensional systems with (algebraic) long-range order.

By contrast, in a multicomponent system, a different topology originating from a higher broken symmetry along with certain types of intercomponent interactions leads to a set of composite topological excitations that can have a dramatic impact on the nature of phase transitions and the structure of the phase diagram.[*] We will begin the discussion with a mixture of two charged condensates interacting electromagnetically.

6.10.1 Metallic Superfluid

In a charged mixture, the proliferation of composite vortices leads to a new state of matter, *metallic superfluid*. We begin by considering the 2D LMH-type model (6.8) in the London limit (subsequently we will cover the 3D case as well). At $T = 0$, the system is a two-component superconducting superfluid, discussed earlier in this chapter. Since the condensates are oppositely charged, $(1, -1)$ vortices have the lowest energy. At finite temperature, in a 2D system of size $R \times R$, the free energy of a composite vortex is given by

$$F_v = E_{(1,-1)} - TS \approx E_{(1,-1)} - 2T \ln(R/a), \qquad (6.104)$$

where $E_{(1,-1)}$ is the vortex energy (6.22), and a is the characteristic size of the vortex core. Given that $E_{(1,-1)}$ is finite, while the entropy term diverges as $R \to \infty$, we conclude that $F < 0$ in the thermodynamic limit. Thus, there will be a finite density of (free) $(1, -1)$ and $(-1, 1)$ vortices at finite T destroying the order in each of the components. When these vortices proliferate and are free to move, the phase difference $(\theta_1 - \theta_2)$ is disordered, implying $\langle \psi_{1,2} \rangle = 0$, and there is no superconductivity. However, the $(1, -1)$ and $(-1, 1)$ vortices do not carry topological charge in the phase sum $(\theta_1 + \theta_2)$, and thus, their proliferation cannot destroy superfluidity. On the other hand, recall that the simplest topological defects in the superfluid sector are fractional vortices $(1, 0)$ and $(0, 1)$. Those have logarithmically divergent energy (6.21) in the $R \to \infty$ limit. Correspondingly, from Equation (4.3), it follows that the BKT transition for fractional vortices happens at a finite temperature:

$$T_{BKT} = \frac{\pi}{2} \Lambda_p, \qquad (6.105)$$

where Λ_p is the superfluid stiffness in the "phase sum" channel. Thus, at temperatures $0 < T < T_{BKT}$, the composite "paired" field $\psi_1 \psi_2$ will still possess the topological (algebraic) order in the $(\theta_1 + \theta_2)$ field, and the system will be a metallic superfluid in the sense that only mass transport that does not involve net charge transport will be dissipationless. Lack of real-space pairing in the metallic superfluid in particular implies that the charged sector is electrically conducting and has finite ohmic resistance, but it is not superconducting. Phase configurations in these states are schematically illustrated in Figure 6.13.

---

[*] An alternative description is based on the observation that in the vicinity of the second-order transition to the nontrivial U(1) state, the "pair" field remains ordered on both sides of the transition; thus, order in the single-particle field is the first one to be destroyed by thermal fluctuations.

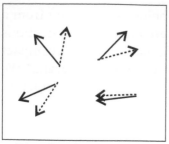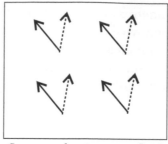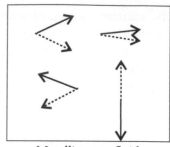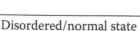

Disordered/normal state   Superconducting superfluid   Metallic superfluid

Figure 6.13   Schematic picture of phase configurations at four randomly selected points in space. The arrows represent phases at those points with convention $\theta_\alpha \in (-\pi, \pi]$. In the normal state, all phases are disordered. In the superconducting superfluid, both phases are ordered. In the metallic superfluid, the phases are disordered individually, but the phase sum is ordered (here, $\theta_1 + \theta_2 = 0$).

From this example, we see a general principle for how nontrivial composite orders emerge: *Composite* topological defects are energetically cheaper than *elementary* ones and proliferate first.

Another example where composite vortices are lower in energy than elementary ones is a 3D system described by Equation (6.8) in external magnetic field that induces composite integer-flux vortices $(1, -1)$. At finite temperature, the vortex lattice can melt. In high fields, where intervortex distance is short, or in low fields, where vortices are far apart, the lattice can melt even at relatively low temperature. If, after melting, the composite vortices do not break into fractional vortices, we obtain the metallic superfluid state.

Consider now a $U(1) \times U(1)$ system with two similarly charged fields. Employing the mapping between oppositely and similarly charged $U(1) \times U(1)$ systems—amounting to the transformation $\psi_1 \leftrightarrow \psi_1^*$—we immediately adapt the aforementioned results to the present case. The composite superfluid state is brought now by proliferation of composite $(1, 1)$ vortices [without proliferation of $(1, 0)$ and $(0, 1)$ elementary topological defects]. The resulting metallic superfluid state is associated with order in the phase difference sector, that is, in the composite field $\psi_1^* \psi_2$. The supertransport now corresponds to the dissipationless *counterflow* of two components and is referred to as *counterflow superfluidity*, or *supercounterfluidity*. The $\gamma_1 = \gamma_2$ case is rather special [cf. Equation (6.36) and the discussion in the following]: There is no net mass transfer in the counterflow of matter fields.

6.10.2   Counterflow Superfluidity in *N*-Component Charged Systems

The phenomenon of counterflow superfluidity is not at all limited to two-component systems. Here, we discuss classification of counterflow modes that arise only when the number of components is larger than two. In a generic *N*-component charged system, where the fields have similar charge modulus, the counterflow superfluidity is induced by proliferation of $(1, 1, 1, \ldots, 1)$ vortices [33], that is the vortices having unit winding number in each of the components. The special role of such vortices is ensured by the same physics of the magnetic screening as in the previously discussed two-component case. The counterflow superfluidity state resulting from the proliferation of these vortices is ordered in terms of composite fields $\psi_\alpha^* \psi_\beta$ ($\alpha \neq \beta$). An immediate observation then is that for $N > 2$, the total number of ordered composite fields is larger than the number $(N - 1)$ of independent flow modes. This is an excellent example of how the notion of "pairing" in context of the composite superfluid state is misleading, if taken literally. Hence, at $N > 2$, we are dealing with an "overcomplete" set of bilinear order parameters. In our case, the following $(N - 1)$ order parameters form a complete set: $\psi_1^* \psi_2, \psi_2^* \psi_3, \ldots \psi_{N-1}^* \psi_N$. Indeed, up to trivial prefactors, all other pairs reduce to products of the specified ones. For example, $\left(\psi_1^* \psi_2 \right) \left(\psi_2^* \psi_3 \right) = |\psi_2|^2 \psi_1^* \psi_3 \propto$ $\psi_1^* \psi_3$. Correspondingly, any winding number for any phase difference $\left(\theta_\alpha - \theta_\beta \right)$ can be expressed as a linear combination of windings of $(N - 1)$ independent phase differences $(\theta_1 - \theta_2), (\theta_2 - \theta_3), \ldots (\theta_{N-1} - \theta_N)$.[*] Likewise, any counterflow pattern in our system can be decomposed into a superposition of $(N - 1)$ pairwise counterflows associated with the earlier $(N - 1)$ composite order parameters. This is a manifestation of the fact, discussed in detail in Section 6.10.4, that, in general, coflow/counterflow superfluid orders are characterized by two linear spaces that are related to each other: the Q-dimensional linear space of proliferated topological defects and the $(N - Q)$-dimensional linear space of topological invariants.

6.10.3   Coflow/Counterflow Composite Superfluidity in Neutral Mixtures

Coflow/counterflow superfluid ordering can also be realized in interacting electrically neutral superfluid mixtures (6.2) as a result of entrainment (dissipationless drag) effect. A detailed discussion of the corresponding microscopic physics goes beyond the scope of the present chapter. We do mention, however, that the required microscopic conditions are not at all exotic. A rich family of counterflow orders is natural for lattice bosons—experimentally realized with ultracold atoms in optical lattices; see Chapter 15. In those systems, the supertransport in single- and few-component modes can be suppressed by the quantum phenomenon of Mott localization discussed in Chapter 9 (see also References [5,6]). For example, in a two-component lattice system with an appropriately strong on-site repulsion between the particles, the supertransport of particle number can be completely suppressed if the average site occupation numbers, $\nu_{1,2}$, satisfy the condition $(\nu_1 + \nu_2) \approx$ integer. The supercounterfluid state emerges either at low but finite temperature [if $(\nu_1 + \nu_2)$ is close—but not exactly equal—to an integer], or in the ground state (if $\nu_1 + \nu_2 =$ integer).

Macroscopically, the dissipationless drag ensures long-range attractive interaction between elementary topological defects. Therefore, when the drag exceeds a

---

[*] It should be emphasized that it is important to keep track of the 2π periodicity of the original phases when working with linear combinations of phases in mixtures.

certain threshold, a neutral system enters the state where only composite vortices proliferate. As an example, consider the two-component model (6.2), rewriting it in terms of the phase sum and phase-difference gradients (and setting $\rho_1 = \rho_2 = \varrho$ for simplicity):

$$F = \frac{\varrho - 2\rho_d}{4} [\nabla(\theta_1 - \theta_2)]^2 + \frac{\varrho}{4} [\nabla(\theta_1 + \theta_2)]^2. \tag{6.106}$$

The drag coefficient ρ_d can be either positive or negative. In the absence of vector potential, all topological defects have logarithmically diverging energy. As is seen from (6.106), changing ρ_d can render certain composite vortices—either $(1,1)$ or $(1,-1)$—much lower in energy than elementary vortices $(1,0)$ and $(0,1)$. Thus, composite defects will proliferate at lower temperatures, resulting in the coflow or counterflow composite superfluid state. Also, a rotating system can feature a transition to a vortex liquid state of composite vortices. This can take place below T_c for the onset of disordering state of elementary vortices [in the case of attractive interaction between $(1,0)$ and $(0,1)$ vortices]. Correspondingly, at a sufficiently strong drag, the following states are possible: (a) a mixture of two superfluids, (b) normal state, (c) a coflow composite superfluid [proliferated $(1,-1)$ composite topological defects], and (d) supercounterflow [proliferated $(1,1)$ composite topological defects].

The aforementioned microscopic Mott physics for lattice bosonic systems resulting in irreducible multicomponent orders with an arbitrarily large number of components can be easily realized experimentally. For example, to get the "triple" order $\langle \psi_1^* \psi_2 \psi_3 \rangle \neq 0$, it is sufficient to tune the site occupation numbers to satisfy the conditions $\nu_2 \approx \nu_3$ and $\nu_1 \approx (\nu_2 + \nu_3)/2 \approx 1$ and to make sure that strong on-site interactions between the particles favor equally only two Fock states on a lattice site: $|1,0,0\rangle$ and $|0,1,1\rangle$. The superflow pattern associated with this triple order involves the coflow of components 2 and 3, along with the counterflow (with respect to 2 and 3) of component 1. The corresponding picture of vortex proliferation will be discussed in the next subsection, in the context of the general classification scheme.

6.10.4   General Classification of N-Component Coflow/Counterflow Composite Superfluids

Consider an N-component system, in which the coflow/counterflow superfluid order is enforced by proliferation of a set of $Q < N$ independent vortices with windings

$$\left(M_1^{(j)}, M_2^{(j)}, \ldots, M_N^{(j)} \right), \qquad j = 1, 2, \ldots, Q. \tag{6.107}$$

The set of winding numbers forms an integer vector $\mathbf{M}$ and allows us to apply the linear algebra language to vortices by considering a vector space $\mathbf{M}$ over the field of integers. By independent, we understand a vortex that cannot be reduced to the linear superposition of other vortices in the set. For $Q > 1$, the set (6.107) is obviously not unique. By constructing linear combinations (with integer coefficients) of the vectors **M**, we create new sets of vortices behind the same multicomponent order. It turns out that at $N > 2$ (recall an example discussed in the previous section), the multicomponent order itself is also associated with a linear space. It is convenient to represent the latter as an $(N - Q)$-dimensional space of integer-valued topological invariants (winding numbers):

$$I = S_1 I_1 + S_2 I_2 + \cdots S_N I_N, \qquad I_j = (1/2\pi) \oint d\mathbf{l} \cdot \nabla \theta_j, \qquad (6.108)$$

where S_j are integers, the same as in Equation (6.103).

It is instructive to explicitly construct—in terms of the set (6.107)—the basis in the space of invariants (6.108), thereby relating the two linear spaces. To this end, observe that the necessary and sufficient condition for the invariant I to survive the proliferation of vortices (6.107) is

$$S_1 M_1^{(j)} + S_2 M_2^{(j)} + \cdots S_N M_N^{(j)} = 0 \qquad (\forall j = 1, 2, \ldots Q). \qquad (6.109)$$

Since, by definition, all vectors in (6.107) are linear independent, Q equations (6.109) define a $(N - Q)$-dimensional space of solutions. For integer vectors $\mathbf{M}^{(j)}$, the basis

$$\left\{ \left[S_1^{(\nu)}, S_2^{(\nu)}, \ldots, S_N^{(\nu)} \right] \right\}_{\nu = 1, 2, \ldots, (N-Q)} \qquad (6.110)$$

in the space of solutions can be selected in such a form that all S's are integers. Also, we assume that each basis vector $\left[S_1^{(\nu)}, S_2^{(\nu)}, \ldots, S_N^{(\nu)} \right]$ is primitive in the sense that the largest common divisor of its components is unity. Then the basis in the space of topological invariants is given by

$$\mathcal{I}^{(\nu)} = S_1^{(\nu)} I_1 + S_2^{(\nu)} I_2 + \cdots S_N^{(\nu)} I_N \qquad [\nu = 1, 2, \ldots, (N - Q)]. \qquad (6.111)$$

The set of equations (6.109)—"symmetric" with respect to S's and M's—allows us to proceed from it in both ways: from the space of vortices to the space of topological invariants and the other way around. For example, from (6.109), it follows that the irreducible "triple" order $\langle \psi_1^* \psi_2 \psi_3 \rangle \neq 0$ is associated with the 2D space of proliferated vortices; the basis of the latter is formed, for example, by the vortices $(0, 1, -1)$ and $(1, 1, 0)$. The necessity of having more than one type of proliferated vortices is clear from the fact that we have to suppress not only the single-component flows but also the "paired" flows, so that only the "triple" superflow survives.

Mathematically, the topological invariant $\mathcal{I}^{(\nu)}$ can take on any integer value, provided integer coefficients $S_j^{(\nu)}$ $(j = 1, 2, \ldots Q)$ are free of any common divisor larger than unity.[*]

---

[*] Physically, this reflects the fact that $\mathcal{I}^{(\nu)}$ represents winding numbers of the composite order parameter phase.

Problem 6.3 *Prove the aforementioned statement. Hint. By mathematical induction, the problem reduces to the rather simple case of $Q = 2$.*

In addition to the conserved windings $\mathcal{I}^{(v)}$, there are also constraints on various linear combinations of I_j's. For example, in a two-component supercounterfluid, the quantity $[(I_1 + I_2) \bmod 2]$ is conserved by proliferated $(1,1)$ vortices. Those "extra" conserved quantities, however, are not independent from the $\mathcal{I}^{(v)}$'s, as is seen by contradiction. The existence of $\tilde{I}$ independent of $\mathcal{I}^{(v)}$'s implies that we can change $\tilde{I}$ value while preserving the values of all $\mathcal{I}^{(v)}$'s. Both physically and mathematically, we can relate such a change to a certain vortex crossing a toroidal system. Moreover, this vortex must be linearly independent of the set (6.107), which is incompatible with conservation of all $\mathcal{I}^{(v)}$'s.

6.10.5 Multicomponent j-Current Models. Universality Class of the Phase Transition to Composite Superfluid State

Let us now discuss the phase transitions leading to the paired and metallic superfluid states. In Chapter 3, we introduced effective statistical models where the statistics of complex-number fields is mapped on the statistics of integer j-currents. Here, we utilize these models to describe the nature of transitions into the composite superfluid state.

The effective interactions between the fields leading to new superfluid phases can be modeled directly at the level of j-currents. For example, pairing phenomena can be induced by adding an attractive interaction $U_{12} < 0$ to the two-component version of the Villain model (3.23):

$$E = \sum_b E_b, \qquad E_b = \sum_{\sigma \leq \sigma'} U_{\sigma\sigma'} j_b^{(\sigma)} j_b^{(\sigma')}, \tag{6.112}$$

where $\sigma, \sigma' = 1, 2$ are the component indices. System stability requires that $|U_{12}| < 2\sqrt{U_{11} U_{22}}$ (to exclude current configurations with infinitely negative energy).

To illustrate the principle, consider the simplest example where the number of independent parameters is reduced by requiring interexchange symmetry, $1 \leftrightarrow 2$, implying $U_{11} = U_{22} = U$ (below $V \equiv -U_{12}$):

$$E_b = U\left[\left(j_b^{(1)}\right)^2 + \left(j_b^{(2)}\right)^2\right] - V j_b^{(1)} j_b^{(2)}. \tag{6.113}$$

At small enough U and $V \ll U$, the system is an interacting U(1)×U(1) superfluid mixture with weak intercomponent current–current interaction. In terms of j-current variables (dual to fields), the state is characterized by large j-loops and nonzero-mean-square-average winding numbers, in both components independently. When current–current interaction becomes large enough, the two-component j-current model undergoes the current-pairing transition—a dual representation of the transition associated with the proliferation of $(1,-1)$ vortices in the original U(1)×U(1) model.* To analyze this transition, it is convenient to rewrite the bond energy identically as

$$E_b = U\left[j_b^{(1)} - j_b^{(2)}\right]^2 + (2U - V)j_b^{(1)}j_b^{(2)}, \tag{6.114}$$

and take the limit of large U and V while keeping the difference $(2U - V)$ finite. The first term in (6.114) ensures that the values of two currents are the same up to irrelevant small-scale fluctuations; that is, large loops are only possible for the variable $J_b = (1/2)\left[j_b^{(1)} + j_b^{(2)}\right]$ that describes codirected currents. The statistics of J_b-currents is governed by the Villain model with $U_{\text{eff}} = 2U - V$. As $2U - V$ is decreased, one should observe a phase transition between the normal and paired superfluid states.

Duality between phase and current variables allows for the immediate characterization of the paired phase as a state with (long-range or algebraic) order in the "pair" phase field $(\theta_1 + \theta_2)$ while having no individual ordering in θ_1 and θ_2. Indeed, if we would also have the order in the single-component correlation function $\langle \psi^*(\mathbf{i})_\sigma \psi(\mathbf{k})_\sigma \rangle$, that would imply that typical configurations contain macroscopic loops of individual $j^{(\sigma)}$ currents.

As far as the universality class of the pairing transition is concerned, symmetry-wise we are talking of the U(1)×U(1) → U(1) phase transition, destroying order in the phase-difference field $(\theta_1 - \theta_2)$ on the background of robust order in the field $(\theta_1 + \theta_2)$. The picture of the composite-vortex proliferation suggests that the transition, if continuous, is in the U(1)-universality class. In the dual picture, the same conclusion can be reached by noting—see Figure 6.14—that critical large-scale fluctuations are in the form of oriented loops, similar to the XY model. Indeed, the "molecule-breaking" fluctuation shown in the left panel of Figure 6.14 can be converted into a closed-loop configuration in one of the components, shown in the right panel of Figure 6.14, by revealing one of the molecular loops in the background paired superfluid state (see middle panel) and "reassigning" lines belonging to the background molecular loops.

Further analogy can be drawn with the vortex-loop-proliferation picture for the transition from the U(1) supercounterfluid (or paired superfluid) to the fully symmetric normal state. In that case, we go from a state full of composite loops to a state where all kinds of vortex loops have proliferated. Across the transition, the composite vortex loops must split. An elementary splitting fluctuation process is shown in Figure 6.15. One can add/subtract the background composite vortices present at finite density. The elementary splitting process can be viewed as nucleation of a topological defect in the form of a vortex loop in the phase-difference sector. When such loops proliferate, the system becomes normal.

A superfluid mixture with an intercomponent drag has a symmetry associated with complex-conjugating one of the fields and simultaneously changing

---

* Recall that, in contrast to the superfluid "molecular" phase, we are talking of *statistical* pairing rooted in topological properties when, at large scales, current fluctuations of the two species are locked together.

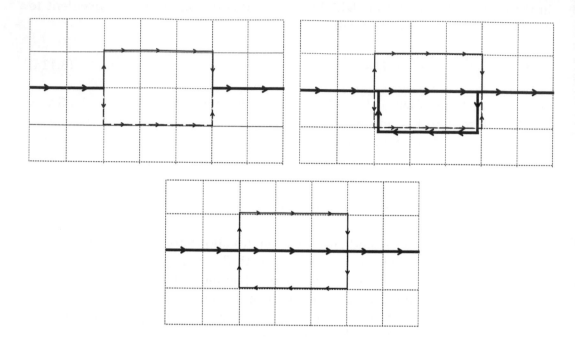

Figure 6.14   Large-scale j-current fluctuations leading to the transition to U(1)×U(1) double superfluid phase from the U(1) paired superfluid state. The current splitting configuration can be mapped onto an oriented loop in a single-component system by invoking the background loops present in the paired state. Solid and dotted lines represent two different components, while the thick bold line represents a molecular current consisting of one solid and one dotted line.

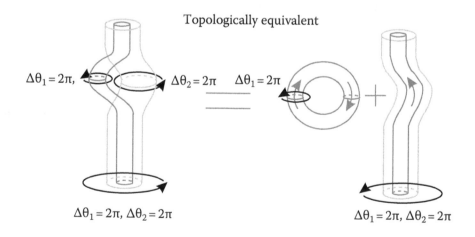

Figure 6.15   Fluctuating structure of the composite $(1, 1)$ vortex. On approach to the normal state from a state with order only in the phase-difference sector, the splitting fluctuations of the composite vortices can be mapped onto the process of generation of elementary vortex loops.

the sign of the drag coefficient. In the model (6.113), this symmetry manifests itself as follows: The model remains invariant under the transformation $V \to -V$, $j_b^{(2)} \to -j_b^{(2)}$. This symmetry immediately allows one to predict that, in the limit of large U and $-V$ and small $(2U + V)$, the system is in the counterflow superfluid state characterized by proliferated loops for the variable $J_b = \left(j_h^{(1)} - j_h^{(2)} \right)/2$, equivalent to the order in the phase-difference field $(\theta_1 - \theta_2)$ and no order in individual phase fields.

The following remark is worth making here. Further in the book, we will introduce a description of superfluid bosons in terms of Feynman path integrals and see that this description is analogous to j-current models. Within the path integral language, changing the sign of the j-current describing lattice bosons has a distinct physical meaning of replacing bosonic particles with bosonic holes, the latter being a special type of quasiparticles arising in Mott insulators. Correspondingly, in the composite states in such physical systems, the counterflow can be viewed as a coflow of particles of one component and holes of the other. By contrast, the supercounterfluid (metallic superfluid) state of a charged mixture arises purely at the level of classical fields: The notion of "holes" is irrelevant in this case.

The phase diagram of the model (6.113) is illustrated schematically in Figure 6.16. Since, in the dual language, U and V are proportional to temperature and inversely proportional to bare stiffness parameters, it is convenient to plot this diagram using $|V|/U$ and δ variables, where δ is a common scaling factor for U and V. This way, the vertical axis is representing the relative strength of interactions, while the horizontal axis is similar to temperature. In agreement with the picture developed earlier, the "pairing" transition takes place as δ, or temperature,

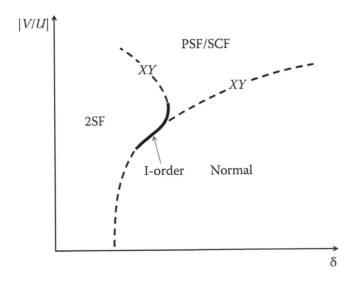

Figure 6.16   Schematic phase diagram of the two-component Villain j-current model. Here, 2SF represents the U(1)×U(1) superfluid state; PSF/SCF represents (depending on the sign of intercomponent coupling) the paired superfluid/supercounterfluid state.

is increased starting from the U(1)×U(1) state. In the vicinity of the point where all three lines meet, the 2SF boundary is the first-order line. This can be understood by introducing the "pair" field Φ and constructing the mean-field solution of the effective GL action for order parameters ψ_1, ψ_2, and Φ [34]:

$$F = r\left[|\psi_1|^2 + |\psi_2|^2\right] + r_m|\Phi|^2 + g\left[|\psi_1|^4 + |\psi_2|^4\right]$$
$$+ g_m|\Phi|^4 - g_p\left(\Phi^*\psi_1\psi_2 + \text{c.c.}\right). \tag{6.115}$$

Problem 6.4 *Minimize the GL free energy (6.115), setting, without loss of generality (explain why), $g = g_m = g_p = 1$, with respect to all fields to determine the lowest energy phase in the (r, r_m) plane. Determine parameters of the critical points where the first-order line starts and ends.*

Finally, we comment on the phase diagram of a general 3D charged mixture without applied magnetic field. We include mixed gradient terms representing current–current interactions (below $\gamma_{1,2} = 1$ and $\mathbf{v}_{1,2} = \nabla\theta_{1,2} - q\mathbf{A}$):

$$F = \frac{1}{2}\left\{\rho_1\mathbf{v}_1^2 + \rho_2\mathbf{v}_2^2 - \rho_d\left(\mathbf{v}_1 - \mathbf{v}_2\right)^2 + (\text{curl}\mathbf{A})^2\right\}. \tag{6.116}$$

One can integrate out $\mathbf{A}$ in the partition function (this form is quadratic in the vector potential) and formulate the remaining statistics in the j-current model representation. As demonstrated in Chapter 5, upon integration of $\mathbf{A}$, one arrives at the j-current model with long-range Coulomb current–current interaction (two component in the present case).

Numerical studies of the phase diagram of (6.116) with $\rho_d = 0$ indicate that (a) the critical value of charge q is required to stabilize the metallic superfluid state and (b) a finite segment of the transition line to the metallic superfluid state is of the first order [32]. In the different limit of small q but relatively large positive ρ_d (which favors $\mathbf{v}_1 \approx \mathbf{v}_2$), there is a phase transition associated with the proliferation of $(1, -1)$ vortices [35]. This results in a paired superconducting state characterized by the electric charge $2q$. In electronic systems, q is twice the electron charge; thus, the resulting composite state is a charge-4e superconductor.

References

1. A. F. Andreev and E. Bashkin, Three-velocity hydrodynamics of superfluid solutions, *Sov. Phys. JETP* **42**, 164 (1975).

2. E. Babaev, A. Sudbo, and N. W. Ashcroft, A superconductor to superfluid phase transition in liquid metallic hydrogen, *Nature* **431**, 666 (2004).

3. D. R. Tilley, The Ginzburg–Landau equations for pure two band superconductors, *Proc. Phys. Soc.* **84**, 573 (1964).

4. D. V. Fil and S. I. Shevchenko, Nondissipative drag of superflow in a two-component Bose gas, *Phys. Rev. A* **72**, 013616 (2005).

5. A. B. Kuklov and B. V. Svistunov, Counterflow superfluidity of two-species ultracold atoms in a commensurate optical lattice, *Phys. Rev. Lett.* **90**, 100401 (2003).

6. A. Kuklov, N. Prokof'ev, and B. Svistunov, Superfluid–superfluid phase transitions in a two-component Bose–Einstein condensate, *Phys. Rev. Lett.* **92**, 030403 (2004).

7. V. A. Moskalenko, Superconductivity in metals with overlapping energy bands, *Fiz. Met. Metalloved.* **8**, 503 (1959).

8. H. Suhl, B. T. Matthias, and L. R. Walker, Bardeen–Cooper–Schrieffer theory of superconductivity in the case of overlapping bands, *Phys. Rev. Lett.* **3**, 552 (1959).

9. E. Babaev, Vortices with fractional flux in two-gap superconductors and in extended Faddeev model, *Phys. Rev. Lett.* **89**, 067001 (2002).

10. E. Babaev, Phase diagram of planar $U(1) \times U(1)$ superconductors: Condensation of vortices with fractional flux and a superfluid state, cond-mat/0201547 (2002).

11. E. Babaev and N. W. Ashcroft, Violation of the London law and Onsager–Feynman quantization in multicomponent superconductors, *Nat. Phys.* **3**, 530 (2007).

12. E. Babaev, J. Jaykka, and J. M. Speight, Magnetic field delocalization and flux inversion in fractional vortices in two-component superconductors, *Phys. Rev. Lett.* **103**, 237002 (2009).

13. E. Babaev, L. D. Faddeev, and A. Niemi, Hidden symmetry and knot solitons in a charged two-condensate Bose system, *Phys. Rev. B* **65**, 100512 (2002).

14. E. Babaev, Non-Meissner electrodynamics and knotted solitons in two-component superconductors, *Phys. Rev. B* **79**, 104506 (2009).

15. J. Garaud, K. A. H. Sellin, J. Jaykka, and E. Babaev, Skyrmions induced by dissipationless drag in $U(1) \times U(1)$ superconductors, *Phys. Rev. B* **89**, 104508 (2014).

16. J. Garaud, J. Carlstrom, and E. Babaev, Topological solitons in three-band superconductors with broken time reversal symmetry, *Phys. Rev. Lett.* **107**, 197001 (2011).

17. L. Faddeev, *Quantisation of Solitons*, preprint IAS Print-75-QS70, (1975); and in *Einstein and Several Contemporary Tendencies in the Field Theory of Elementary Particles in Relativity*, Quanta and Cosmology, Vol. 1, M. Pantaleo, and F. De Finis (eds.), Johnson Reprint (1979).

18. L. Faddeev and A. J. Niemi, Stable knot-like structures in classical field theory, *Nature* **387**, 58 (1997).

19. E. Babaev, Neither a type-1 nor a type-2 superconductivity in a two-gap system, cond-mat/0302218 (2003).

20. E. Babaev and J. M. Speight, Semi-Meissner state and neither type-I nor type-II superconductivity in multicomponent systems, *Phys. Rev. B* **72**, 180502 (2005).

21. V. V. Moshchalkov, M. Menghini, T. Nishio, Q. H. Chen, A. V. Silhanek, V. H. Dao, L. F. Chibotaru, N. D. Zhigadlo, and J. Karpinski, Type-1.5 superconductors, *Phys. Rev. Lett.* **102**, 117001 (2009).

22. S. J. Ray, A. S. Gibbs, S. J. Bending, P. J. Curran, E. Babaev, C. Baines, A. P. Mackenzie, and S. L. Lee, Muon-spin rotation measurements of the vortex state in Sr2RuO4: Type-1.5 superconductivity, vortex clustering and a crossover from a triangular to a square vortex lattice, *Phys. Rev. B* **89**, 094504 (2014).

23. J. Carlstrom, J. Garaud, and E. Babaev, Semi-Meissner state and non-pairwise intervortex interactions in type-1.5 superconductors, *Phys. Rev. B* **84**, 134515 (2011).

24. J. Carlstrom, J. Garaud, and E. Babaev, Length scales, collective modes, and type-1.5 regimes in three-band superconductors, *Phys. Rev. B* **84**, 134518 (2011).

25. V. L. Ginzburg and L. D. Landau, To the theory of superconductivity, *Zh. Eksp. Teor. Fiz.* **20**, 1064 (1950).

26. M. Silaev and E. Babaev, Microscopic derivation of two-component Ginzburg–Landau model and conditions of its applicability in two-band systems, *Phys. Rev. B* **85**, 134514 (2012).

27. A. Gurevich, Limits of the upper critical field in dirty two-gap superconductors, *Physica C* **456**, 160 (2007).

28. J. Carlstrom, E. Babaev, and J. M. Speight, Type-1.5 superconductivity in multiband systems: the effects of interband couplings, *Phys. Rev. B* **83**, 174509 (2011).

29. M. Silaev and E. Babaev, Microscopic theory of type-1.5 superconductivity in multiband systems, *Phys. Rev. B* **84**, 094515 (2011).

30. E. Babaev, J. Carlstrom, and J. M. Speight, Type-1.5 superconducting state from an intrinsic proximity effect in two-band superconductors, *Phys. Rev. Lett.* **105**, 067003 (2010).

31. J. Garaud, J. Carlstrom, E. Babaev, and J. M. Speight, Chiral CP^2 skyrmions in three-band superconductors *Phys. Rev B* **87**, 014507 (2012).

32. A. Kuklov, N. Prokof'ev, B. Svistunov, and M. Troyer, Deconfined criticality, runaway flow in the two-component scalar electrodynamics and weak first-order superfluid-solid transitions, *Ann. Phys.* **321**, 1602 (2006).

33. J. Smiseth, E. Smorgrav, E. Babaev, and A. Sudbo, Field- and temperature induced topological phase transitions in the three-dimensional N-component London superconductor, *Phys. Rev. B* **71**, 214509 (2005).

34. A. Kuklov, N. Prokof'ev, and B. Svistunov, Commensurate two-component bosons in an optical lattice: Ground state phase diagram, *Phys. Rev. Lett.* **92**, 050402 (2004).

35. E. V. Herland, E. Babaev, and A. Sudbo, Phase transitions in a three dimensional U(1) × U(1) lattice London superconductor: metallic superfluid and charge-4e superconducting states, *Phys. Rev. B* **82**, 134511 (2010).

Part III

Quantum-Mechanical Aspects: Macrodynamics

CHAPTER 7

Quantum-Field Perspective

Up to this point in our discussion of the superfluidity phenomenon, we have deliberately not mentioned quantum mechanics: this is in contrast with the conventional approach that usually starts with the notion of the condensate wavefunction. In previous chapters, we have shown that classical systems, whether they are dynamical or purely statistical, can be in the superfluid state. Thus, an important connection to follow in the quantum case is the development of the classical-field component that then undergoes the superfluid transition. This can be achieved naturally within the coherent states approach advocated by Langer [1]. As we will see shortly, this connection is rather straightforward and, surprisingly, brings about the counterintuitive (from the classical-particle perspective) question: "Do we ever have an insulating quantum solid state in bosonic systems?" In Chapter 9, we answer this question positively.

7.1 Coherent States, Operator of Phase

This section is devoted to coherent states that are instrumental in establishing the connection between the quantum many-body states and emergent classical-field behavior.

7.1.1 Coherent States for a Single Mode

In the second quantization language, one introduces bosonic creation and annihilation operators, which change by one the number of particles in a given single-particle quantum state (or mode) $\varphi_i(\mathbf{r})$ and obey the following rules:

$$b_i|n_i\rangle = \sqrt{n_i}|n_i - 1\rangle, \qquad b_i^\dagger|n_i\rangle = \sqrt{n_i + 1}|n_i + 1\rangle, \tag{7.1}$$

$$\left[b_i, b_j\right] = \left[b_i^\dagger, b_j^\dagger\right] = 0, \qquad \left[b_i, b_j^\dagger\right] = \delta_{ij}, \tag{7.2}$$

where $|n_i\rangle$ stands for the state of n_i particles occupying the same mode $\varphi_i(\mathbf{r})$. Coherent states, $|\alpha_i\rangle$, are introduced as appropriate superpositions of states with different particle numbers such that they are eigenstates of the annihilation operator

$$b_i|\alpha_i\rangle = \alpha_i|\alpha_i\rangle. \tag{7.3}$$

By Hermitian conjugation $\langle \alpha_i | b_i^\dagger = \alpha_i^* \langle \alpha_i |$. Using the properties of the annihilation operator, one readily obtains

$$|\alpha_i\rangle = e^{-|\alpha_i|^2/2} \sum_{n_i=0}^{\infty} \frac{\alpha_i^{n_i}}{\sqrt{n_i!}} |n_i\rangle. \tag{7.4}$$

Problem 7.1 *Derive Equation (7.4) from the properties of the annihilation operator.*

The set of coherent states is complete, the completeness relation being

$$\pi^{-1} \int d\alpha_i |\alpha_i\rangle\langle \alpha_i| = 1, \tag{7.5}$$

which is verified by direct integration over the phase and modulus of α_i. We will see shortly that this property is extremely helpful in practice because one can use it to "decouple" any sequence of bosonic operators and convert the calculation of matrix elements to integration over complex numbers. For example,

$$\langle \Psi_1 | b_i^\dagger b_i | \Psi_2 \rangle = \frac{1}{\pi^2} \int d\alpha_i d\beta_i \langle \Psi_1 | \alpha_i \rangle \langle \alpha_i | b_i^\dagger b_i | \beta_i \rangle \langle \beta_i \| \Psi_2 \rangle$$

$$= \frac{1}{\pi^2} \int d\alpha_i d\beta_i \alpha_i^* \beta_i \langle \Psi_1 | \alpha_i \rangle \langle \alpha_i | \beta_i \rangle \langle \beta_i | \Psi_2 \rangle. \tag{7.6}$$

Since coherent states are not orthogonal,* one also needs an overlap integral

$$\langle \alpha_i | \beta_i \rangle = e^{-[|\alpha_i|^2 + |\beta_i|^2 - 2\alpha_i^* \beta_i]/2}, \tag{7.7}$$

to start using the formula. Note also that $|\langle \alpha_i | \beta_i \rangle|^2 = e^{-|\alpha_i - \beta_i|^2}$; that is, the modulus squared of the overlap integral vanishes as a Gaussian when parameters of two states are separated.

Problem 7.2 *Prove the completeness relation.*

Though one cannot introduce eigenstates of the creation operator (for the obvious reason that the basis state with the lowest value of n_i contributing to such a state is no longer present in the superposition after the action of $b_i^\dagger$), the following asymptotic relation is true:

$$b_i^\dagger |\alpha_i\rangle \approx \alpha_i^* |\alpha_i\rangle, \quad (|\alpha_i| \gg 1): \tag{7.8}$$

that is, for coherent states with large $|\alpha_i|^2$, the creation operator is a near-diagonal matrix. Correspondingly, in the same limit,

$$n_i |\alpha_i\rangle = b_i^\dagger b_i |\alpha_i\rangle \approx |\alpha_i|^2 |\alpha_i\rangle. \tag{7.9}$$

---

* A fundamental reason behind the nonorthogonality is that the set of coherent states is *overcomplete*. The overcompleteness of the set also implies that a representation of an arbitrary state as a superposition of coherent states is not unique.

Both properties follow from the structure of the modulus squared of the expansion coefficients, $e^{-|\alpha_i|^2}|\alpha_i|^{2n_i}/n_i!$, which coincides with the Poisson distribution. For large $|\alpha_i|$, it takes the form the Gaussian distribution centered at $n_i \approx |\alpha_i|^2$ with dispersion $|\alpha_i| \ll |\alpha_i|^2$. The significance of relations (7.8) and (7.9) is that in the limit of large occupation numbers one may neglect, to the first approximation, the quantum nature of bosonic creation and annihilation operators and replace them with complex numbers.

Problem 7.3 *Prove the validity of Equation (7.8). Show that $b^\dagger|\alpha\rangle = \alpha^*|\alpha\rangle + |x\rangle$ with $\langle x|x\rangle = 1$.*

Problem 7.4 *Let the Hamiltonian of noninteracting bosons occupying an energy level ϵ be*

$$H = \epsilon b_\epsilon^\dagger b_\epsilon.$$

Show that the time evolution of the coherent state $|\alpha_\epsilon\rangle$ is given by

$$e^{-iHt}|\alpha_\epsilon\rangle = |e^{-i\epsilon t}\alpha_\epsilon\rangle:$$

that is, the state remains coherent. Through exact analogy between the noninteracting bosons occupying the same energy level and the single-particle harmonic oscillator problem, one can introduce coherent harmonic oscillator states and solve for their evolution.

7.1.2 Coherent States for Fields

So far, we have been discussing the case when particles were occupying a single quantum state/mode $\varphi_i(\mathbf{r})$. It is easy to generalize all notions to the case of the quantum field by considering an orthonormal basis of single-particle states $\{\varphi_i(\mathbf{r})\}$, such as plane waves, and defining the field annihilation operator as

$$\hat{\psi}(\mathbf{r}) = \sum_i b_i \varphi_i(\mathbf{r}). \tag{7.10}$$

Its commutation relations follow directly from Equation (7.2) and are essentially the same—the only nonzero relation is now $\left[\hat{\psi}(\mathbf{r}), \hat{\psi}^\dagger(\mathbf{r}')\right] = \delta(\mathbf{r} - \mathbf{r}')$.

Now, let us construct a state based on a direct product of coherent states for each mode,

$$|\{\alpha_i\}\rangle = \prod_i |\alpha_i\rangle, \tag{7.11}$$

and act on it with $\hat{\psi}(\mathbf{r})$. We get

$$\hat{\psi}(\mathbf{r})|\{\alpha_i\}\rangle = \psi(\mathbf{r})|\{\alpha_i\}\rangle, \tag{7.12}$$

where

$$\psi(\mathbf{r}) = \sum_i \alpha_i \varphi_i(\mathbf{r}); \tag{7.13}$$

that is, (7.11) is an eigenstate of the field annihilation operator. At this point, we realize that $\{\alpha_i\}$ are the Fourier series coefficients of $\psi(\mathbf{r})$ in a particular basis

$$\alpha_i = \langle \varphi_i | \psi \rangle, \tag{7.14}$$

and thus $\psi(\mathbf{r})$ itself can be used to fully characterize the state (7.11). This parameterization is more convenient because it eliminates the need to mention any particular single-particle basis

$$\hat{\psi}(\mathbf{r}) | \psi \rangle = \psi(\mathbf{r}) | \psi \rangle. \tag{7.15}$$

Introducing invariant infinitesimal volume in the space of configurations of $\psi(\mathbf{r})$,

$$\mathcal{D}\psi = \prod_i \pi^{-1} d\alpha_i, \tag{7.16}$$

we write the completeness relation as

$$\int \mathcal{D}\psi | \psi \rangle \langle \psi | = 1. \tag{7.17}$$

Problem 7.5 *Show that $\mathcal{D}\psi$ (7.16) is invariant with respect to the choice of the complete set of single-particle states.*

An important remark is in order here. From Equation (7.15), it follows that an arbitrary coherent state [e.g., one with all $|\alpha_k|^2$ coefficients being small in the plane-wave basis $\varphi_k(\mathbf{r})$] can be *identically* viewed as a coherent state where all particles occupy *the same mode* $\psi(\mathbf{r})$, since one is free to consider $\psi(\mathbf{r})$ as one of the basis functions in some orthogonal set. Correspondingly, the single-particle density matrix of a coherent state factorizes as $\rho(\mathbf{r}, \mathbf{r}') = \psi^*(\mathbf{r})\psi(\mathbf{r}')$, with its largest and only nonzero eigenvalue being unity. Despite these facts, in general, it would be fundamentally misleading to call an arbitrary $|\psi\rangle$ state a "condensate" and link it to superfluidity (a mistake explicitly made in Reference [1]) without analysis of the phase field $\arg[\psi(\mathbf{r})]$. Being in the classical-field limit does not yet imply long-range order since the off-diagonal correlation function, $\rho(\mathbf{R}) = \int d\mathbf{r}\psi^*(\mathbf{r})\psi(\mathbf{r} + \mathbf{R})$, may still decay exponentially with the distance $\mathbf{R}$. The apparent paradox with the Onsager–Penrose definition of the condensate through the eigenvector corresponding to the largest eigenvalue of the single-particle density matrix [13] is resolved by considering time-averaged quantity

$$\langle \rho(\mathbf{r}, \mathbf{r}') \rangle = t^{-1} \int_0^t d\tau \rho(\mathbf{r}, \mathbf{r}', \tau), \tag{7.18}$$

which in the $t \to \infty$ limit also demonstrates vanishing off-diagonal correlations for some states.

Overlap integrals for coherent states of quantum fields can also be expressed in terms of $\psi(\mathbf{r})$. By the standard identity of the complete orthonormal set,

$$\sum_i \left[|\alpha_i|^2 + |\beta_i|^2 - 2\alpha_i^*\beta_i \right] \equiv \int d\mathbf{r} \left[|\psi_a(\mathbf{r})|^2 + |\psi_b(\mathbf{r})|^2 - 2\psi_a^*(\mathbf{r})\psi_b(\mathbf{r}) \right], \tag{7.19}$$

leading to [see (7.7)]

$$\langle\psi_a|\psi_b\rangle = \prod_i\langle\alpha_i|\beta_i\rangle = \exp\left\{-\frac{1}{2}\int d\mathbf{r}\left[|\psi_a(\mathbf{r})|^2 + |\psi_b(\mathbf{r})|^2 - 2\psi_a^*(\mathbf{r})\psi_b(\mathbf{r})\right]\right\}. \quad (7.20)$$

We conclude this section by establishing a useful relation. Consider the formula

$$\psi_a^*(\mathbf{r})\langle\psi_a|\psi_b\rangle = \left[\psi_b'(\mathbf{r}) + \frac{\delta}{\delta\psi_b'(\mathbf{r})}\right]\langle\psi_a|\psi_b\rangle \qquad (7.21)$$

obtained from (7.20) by straightforward variational differentiation with standard notations for real and imaginary parts of a complex number, $\psi = \psi' + i\psi''$. With Equation (7.21), one can calculate matrix elements of the creation operator acting on the coherent state as follows (using completeness relation):

$$\langle\Psi|\hat{\psi}^\dagger(\mathbf{r})|\psi_b\rangle = \int\mathcal{D}\psi\langle\Psi|\psi\rangle\langle\psi|\hat{\psi}^\dagger(\mathbf{r})|\psi_b\rangle = \int\mathcal{D}\psi\langle\Psi|\psi\rangle\psi^*(\mathbf{r})\langle\psi|\psi_b\rangle$$

$$= \int\mathcal{D}\psi\langle\Psi|\psi\rangle\left[\psi_b'(\mathbf{r}) + \frac{\delta}{\delta\psi_b'(\mathbf{r})}\right]\langle\psi|\psi_b\rangle = \left[\psi_b'(\mathbf{r}) + \frac{\delta}{\delta\psi_b'(\mathbf{r})}\right]\langle\Psi|\psi_b\rangle,$$

$$(7.22)$$

where $|\Psi\rangle$ is an arbitrary state.

7.1.3 Operator of Phase

For the generic quantum state, the notion of the operator of phase is hardly useful at all. The situation changes radically when the bosonic field is adequately described by a set of coherent states that under coarse graining exhibit topological order in the phase field. Quantization of the phase field then leads to one of the most fundamental relations in the theory of superfluidity introduced by P. Anderson, that is, an uncertainty relation between the phase and particle number.

In the subspace of coherent states with topological order, the operator of phase is defined by

$$\hat{\Phi}(\mathbf{r})|\psi\rangle = \Phi(\mathbf{r})|\psi\rangle: \qquad (7.23)$$

that is, it has a common set of eigenstates with the annihilation operator $\hat{\psi}(\mathbf{r})$. The notion of order and the *coarse-grained* phase field $\Phi(\mathbf{r})$ for coherent states is introduced in exactly the same way as it was done for classical field by applying the coarse-graining procedure Q to $\psi(\mathbf{r})$. The coarse-graining procedure is crucial in view of the nonuniqueness of the expansion over coherent states, meaning that in general the definition (7.23) is intrinsically ambiguous. The ambiguity is removed by the *orthogonality* of coherent states corresponding to different configurations of the coarse-grained phase field $\Phi(\mathbf{r})$. The orthogonality will be shown below.

To establish the commutation relation between the phase and particle density, we express the matrix element of the density operator as

$$\langle\psi_a|\hat{n}(\mathbf{r})|\psi_b\rangle = \psi_a^*(\mathbf{r})\psi_b(\mathbf{r})\langle\psi_a|\psi_b\rangle \equiv -i\frac{\delta}{\delta\Phi_b(\mathbf{r})}\langle\psi_a|\psi_b\rangle \equiv i\frac{\delta}{\delta\Phi_a(\mathbf{r})}\langle\psi_a|\psi_b\rangle. \quad (7.24)$$

With this formula at hand, we rewrite

$$\langle\psi_a|\left[\hat{\Phi}(\mathbf{r}_1),\hat{n}(\mathbf{r}_2)\right]|\psi_b\rangle = \langle\psi_a|\hat{\Phi}(\mathbf{r}_1)\int\mathcal{D}\psi|\psi\rangle\langle\psi|\hat{n}(\mathbf{r}_2)|\psi_b\rangle - \Phi_b(\mathbf{r}_1)\langle\psi_a|\hat{n}(\mathbf{r}_2)|\psi_b\rangle$$

as a functional derivative [compare with the derivation of formula (7.22)]

$$\langle\psi_a|\left[\hat{\Phi}(\mathbf{r}_1),\hat{n}(\mathbf{r}_2)\right]|\psi_b\rangle = -i\frac{\delta}{\delta\Phi_b(\mathbf{r}_2)}\Phi_b(\mathbf{r}_1)\langle\psi_a|\psi_b\rangle + i\Phi_b(\mathbf{r}_1)\frac{\delta}{\delta\Phi_b(\mathbf{r}_2)}\langle\psi_a|\psi_b\rangle$$

$$= -i\langle\psi_a|\psi_b\rangle\frac{\delta\Phi_b(\mathbf{r}_1)}{\delta\Phi_b(\mathbf{r}_2)} = -i\delta(\mathbf{r}_1-\mathbf{r}_2)\langle\psi_a|\psi_b\rangle. \quad (7.25)$$

Thus, the commutation relation between the phase and density fields is given by

$$\left[\hat{\Phi}(\mathbf{r}_1),\hat{n}(\mathbf{r}_2)\right] = -i\delta(\mathbf{r}_1-\mathbf{r}_2), \quad (7.26)$$

which, upon integration over the coordinate $\mathbf{r}_2$ over a volume around point $\mathbf{r}_1$ in which the phase field may be considered constant, leads to the answer

$$\left[\hat{\Phi}(\mathbf{r}_1),\hat{N}\right] = -i. \quad (7.27)$$

Problem 7.6 *Establish Equation (7.24) by functional differentiation of the overlap integral (7.20).*

The significance of Equation (7.27) is that it establishes that particle number and phase are canonically conjugated in the coarse-gained description, similarly to the particle coordinate and momentum on the lattice. This, in particular, implies that Φ and N cannot share the same eigenstate; that is, states with well-defined phase have uncertain particle number and vice versa.

Let us show now that properties of the phase operator are such that for TLRO states,

$$\hat{I} = \oint d\mathbf{r}\cdot\nabla\hat{\Phi}(\mathbf{r}) \quad (7.28)$$

(periodic boundary conditions are implied) is a conserved quantity because it commutes with the Hamiltonian. Clearly, $\hat{I}$, similarly to $\hat{\Phi}(\mathbf{r})$, commutes with the annihilation operator $\hat{\psi}(\mathbf{r})$. On the other hand,

$$\langle\psi_a|\left[\hat{\psi}^\dagger(\mathbf{r}_1),\nabla\hat{\Phi}(\mathbf{r}_2)\right]|\psi_b\rangle = -\langle\psi_a|\psi_b\rangle\frac{\delta}{\delta\psi_b'(\mathbf{r}_1)}\nabla\Phi_b(\mathbf{r}_2). \quad (7.29)$$

The derivation of this formula is based on Equations (7.21) and (7.22), and it goes along identical lines leading to Equation (7.25). By integrating both sides over the variable $(\mathbf{r}_2)$, we obtain

$$\langle\psi_a|\left[\hat{\psi}^\dagger(\mathbf{r}_1),\hat{I}\right]|\psi_b\rangle = -\langle\psi_a|\psi_b\rangle \frac{\delta I_b}{\delta\psi_b'(\mathbf{r}_1)} \equiv 0, \qquad I_b = \oint d\mathbf{r}\cdot\nabla\Phi_b(\mathbf{r}). \qquad (7.30)$$

This means that $\hat{I}$ commutes with any operator involving a finite number of $\hat{\psi}$ and $\hat{\psi}^\dagger$ fields, including the Hamiltonian; that is, $\hat{I}$ is a conserved quantity within the subspace spanned by topologically ordered coherent states.

The eigenstates of $\hat{I}$ are linear superpositions of coherent states that are characterized by the same value of the topological invariant under coarse graining. One can prove that in $d \geq 2$, the states $|I\rangle$ and $|I'\rangle$ $(I' \neq I)$ are orthogonal in the thermodynamic limit. Indeed, the overlap,

$$|\langle\psi_a|\psi_b\rangle|^2 = e^{-\int d\mathbf{r}|\psi_a(\mathbf{r})-\psi_b(\mathbf{r})|^2}, \qquad (7.31)$$

remains finite only if the difference $|\psi_a(\mathbf{r}) - \psi_b(\mathbf{r})|^2$ is of the order of V^{-1} almost everywhere, or large differences (say, in the form of vortex loops) are localized in space. In both cases, the value of the topological invariant remains the same under coarse graining. Thus, the exponential in (7.31) must be macroscopically large.

An immediate implication of this orthogonality is the fact that the operator of phase is Hermitian in the subspace of topologically ordered states. By the same token, local operators and perturbations cannot induce transitions between states with a different topological invariant. The only "loop hole" in our arguments is the possibility of invoking states with ever-increasing sizes of topological defects in the form of vortex lines. When the defect and system sizes become comparable, the coarse-graining procedure can no longer properly separate states with different values of I. This mechanism ultimately leads to thermally activated (at high enough temperature) or quantum tunneling (at low temperature) transitions between persistent current states in finite-size systems.

Now let us turn to a rather special case of 1D superfluidity. Clearly, the orthogonality of the topologically ordered states with different values of topological invariant I is a *necessary* condition for I to be a good quantum number, and thus for superfluidity. In one dimension, Equation (7.31) shows that the two topologically ordered coherent states are orthogonal with the macroscopic accuracy only if their ψ's differ substantially at a macroscopically large length. We thus conclude that superfluidity in 1D is possible only if the state of the system is a superposition of special topologically ordered coherent states for which the gradient of phase is macroscopically slow. This immediately rules out 1D superfluidity at a finite temperature, because thermal fluctuations can lead to finite phase gradients, since the latter cost only finite energy.

One can notice that the aforementioned proof of asymptotic orthogonality of the topologically ordered coherent states applies also to two states with the same I, provided the two have macroscopically different fields of the coarse-grained phase. And this is crucially important, because, as we already pointed out, it is only due to this orthogonality that the operator of phase (7.23) is not ill defined.

As mentioned previously, the macroscopic properties of the superfluid state are identical between classical and quantum systems provided that we identify the central parameter in the classical theory, γ, with the quantum expression $\hbar/m$. With thus introduced operator of phase, we readily establish this correspondence. Since the operator of phase is a direct analog of the classical coarse-grained phase, all the relations for classical superfluids apply to quantum superfluids as well, upon replacing the classical coarse-grained phase with the operator of phase (and considering averages, if necessary) and noticing that in all the quantum relations involving velocities/current densities, the ratio $\hbar/m$ plays exactly the same role as the parameter γ in the theory of classical matter fields. We encourage the reader to go through all the theory of Chapters 1 and 2 and rederive it in terms of quantum operator of phase. For most derivations, generalization to the quantum case is straightforward. More delicate are the relations obtained on the basis of Poisson brackets. As a remedy, one can rely on the general correspondence between Poisson brackets of classical Hamiltonian mechanics and commutators of corresponding quantum operators. This way, the relation (7.27), explicitly derived earlier, would emerge as a quantum counterpart of the corresponding Poisson bracket (2.104). However, direct quantum mechanical derivations for commutators are instructive and, in fact, are very close to those for classical-field Poisson brackets, provided one uses the formalism of coherent states and tricks of functional derivatives. The way we obtained Equation (7.27) yields a typical example of such a derivation. With coherent states and functional derivatives, the mathematical expressions one deals with when doing classical-field Poisson brackets and matrix elements of quantum-field commutators become essentially similar.

Along with the operator of phase for the field, it is often useful to introduce the operator of phase for a single mode i with large occupation number $n_i \gg 1$ (at $n_i \sim 1$, the operator of phase is ill defined). In a direct analogy with (7.23), we require that, for the operator of phase $\hat{\varphi}_i$ and a coherent state, the following is true:

$$\hat{\varphi}_i |\alpha_i\rangle = \varphi_i |\alpha_i\rangle, \qquad \varphi_i = \arg \alpha_i. \qquad (7.32)$$

Similar to the case of the coarse-grained field of phase, the definition (7.32) is not ambiguous only if coherent states corresponding to substantially different values of φ_i are (almost) orthogonal. This is the reason for the requirement $n_i \gg 1$ (equivalent to $|\alpha_i|^2 \gg 1$), under which the overlap integral (7.7) for two coherent states with different phases is exponentially small in the parameter $|\alpha_i|^2 \gg 1$. This parameter also guarantees that

$$\langle \alpha_i | \hat{\varphi}_i \approx \langle \alpha_i | \varphi_i, \qquad (7.33)$$

as well as the fact that $\hat{\varphi}_i$ is a Hermitian operator. In a direct analogy with (7.27), one shows that (here and below, we suppress the mode subscript)

$$[\hat{\varphi}, \hat{n}] = -i. \tag{7.34}$$

For future purposes, it is useful to express creation and annihilation operators of the given mode in terms of operators $\hat{\varphi}$ and $\hat{n}$:

$$b = e^{i\hat{\varphi}} \sqrt{\hat{n}}, \qquad\qquad b^{\dagger} = \sqrt{\hat{n}} e^{-i\hat{\varphi}}. \tag{7.35}$$

With Equation (7.34) implying $\left[e^{i\hat{\varphi}}, \hat{n}\right] = e^{i\hat{\varphi}}$, we readily check that this parameterization indeed reproduces bosonic commutation relations. By the very nature of the approximation allowing one to introduce the operator of phase, the parameterization implicitly means that the operators $\hat{n}$ can be represented as $\hat{n} = \bar{n} + \hat{\delta n}$, where $\bar{n}$ is a certain integer and the variance of the operator $\hat{\delta n}$ is small compared to $\bar{n}$ on the subspace of states relevant to the problem in question. This, in particular, means that one can expand the square root in powers of $\hat{\delta n}/\bar{n}$ up to the leading nontrivial term:

$$\sqrt{\hat{n}} = \sqrt{\bar{n} + \hat{\delta n}} \approx \sqrt{\bar{n}}\left(1 + \hat{\delta n}/2\bar{n}\right). \tag{7.36}$$

7.2   Harmonic Hydrodynamic Hamiltonian

The low-lying excitations of a superfluid are phonons. General quantum-mechanical theory of phonons in a superfluid can be constructed on the basis of harmonic hydrodynamic Hamiltonian, H_{hh}. The form of H_{hh} readily follows from the following considerations: the low-lying modes in a superfluid are those associated with slow spatial variations of the superfluid phase and density, the two being *not independent*, because corresponding operators are canonically conjugated [see Equation (7.26)]. Since we are talking of perturbations of the ground-state that can be arbitrarily weak—the small parameter controlling the weakness is the characteristic wavevector of the perturbation—an effective Hamiltonian describing those perturbations should have the form of Taylor expansion of the ground-state energy in terms of the field of density deviation, $\hat{\delta n}(\mathbf{r})$, and the field of the gradient of phase, $\nabla\hat{\Phi}(\mathbf{r})$. The zero-order term of the expansion is a constant; the first-order term in $\nabla\hat{\Phi}$ is absent by symmetry. The first-order term in $\hat{\delta n}$ is proportional to the shift of the total number of particles. It commutes with the Hamiltonian and thus is dynamically irrelevant. Keeping only the leading dynamically relevant terms, we arrive at H_{hh} of the form ($\hbar = 1$):

$$H_{hh} = \int d\mathbf{r}\left\{\frac{n_s}{2m}[\nabla\hat{\Phi}(\mathbf{r})]^2 + \frac{1}{2\varkappa}[\hat{\delta n}(\mathbf{r})]^2\right\}. \tag{7.37}$$

For H_{hh} to be consistent with the ground-state thermodynamics, the expansion coefficients n_s and $\varkappa$ must be nothing but the ground-state superfluid density and compressibility. At this point, it is useful to recall standard thermodynamic relations for compressibility at $T = 0$ in terms of the ground-state energy, E; pressure, p; number density, n; and chemical potential, μ:

$$\varkappa^{-1} = \frac{1}{V}\left(\frac{\partial^2 E}{\partial n^2}\right)_V = \frac{\partial \mu}{\partial n}, \qquad \varkappa = \frac{\partial n}{\partial \mu} = \frac{\partial^2 p}{\partial \mu^2} \qquad (T = 0). \tag{7.38}$$

In the Fourier representation (from now on, we set $V = 1$)

$$\hat{\Phi}_{\mathbf{k}} = \int d\mathbf{r}\, e^{-i\mathbf{k}\cdot\mathbf{r}} \hat{\Phi}(\mathbf{r}), \qquad \hat{\delta n}_{\mathbf{k}} = \int d\mathbf{r}\, e^{-i\mathbf{k}\cdot\mathbf{r}} \hat{\delta n}(\mathbf{r}), \tag{7.39}$$

the Hamiltonian (7.37) decouples into the sum of independent harmonic Hamiltonians, $H_{\pm\mathbf{k}}^{(hh)}$ describing modes of a given (modulo sign, as explained in the following) wavevector:

$$H_{\pm\mathbf{k}}^{(hh)} = \frac{n_s k^2}{2m}\left(\hat{\Phi}_{\mathbf{k}}^2 + \hat{\Phi}_{-\mathbf{k}}^2\right) + \frac{1}{2\varkappa}\left(\hat{\delta n}_{\mathbf{k}}^2 + \hat{\delta n}_{-\mathbf{k}}^2\right). \tag{7.40}$$

The independence of modes with different wavevectors is seen from the commutation relation

$$\left[\hat{\Phi}_{\mathbf{k}_1}, \hat{\delta n}_{-\mathbf{k}_2}\right] = -i\delta_{\mathbf{k}_1,\mathbf{k}_2} \tag{7.41}$$

following from (7.26). For a given mode $\mathbf{k}$, operators $\hat{\Phi}_{\mathbf{k}}$ and $\hat{\delta n}_{-\mathbf{k}}$ play, respectively, the roles of coordinate and momentum (or momentum and coordinate, depending on the taste!) because they obey the canonical commutation relation $\left[\hat{\Phi}_{\mathbf{k}}, \hat{\delta n}_{-\mathbf{k}}\right] = -i$. The wavevectors $\mathbf{k}$ and $-\mathbf{k}$ are mixed in Equation (7.40) because the corresponding Fourier components are not independent. Indeed, the fact that operators $\hat{\Phi}(\mathbf{r})$ and $\hat{\delta n}(\mathbf{r})$ are Hermitian implies

$$\hat{\Phi}_{-\mathbf{k}} = \hat{\Phi}_{\mathbf{k}}^\dagger, \qquad \hat{\delta n}_{-\mathbf{k}} = \hat{\delta n}_{\mathbf{k}}^\dagger. \tag{7.42}$$

The parameterization in terms of creation and annihilation operators (of acoustic phonons with a given wavevector, as we will see shortly),

$$\hat{\Phi}_{\mathbf{k}} = \sqrt{\frac{1}{2\varkappa\omega_k}}\left(\hat{b}_{\mathbf{k}} + \hat{b}_{-\mathbf{k}}^\dagger\right), \qquad \hat{\delta n}_{\mathbf{k}} = i\sqrt{\frac{\omega_k \varkappa}{2}}\left(\hat{b}_{\mathbf{k}} - \hat{b}_{-\mathbf{k}}^\dagger\right), \tag{7.43}$$

respects (7.41) and (7.42) and leads to the canonical form of the harmonic Hamiltonian:

$$H_{\pm\mathbf{k}}^{(hh)} = \omega_k\left(\hat{b}_{\mathbf{k}}^\dagger \hat{b}_{\mathbf{k}} + \hat{b}_{-\mathbf{k}}^\dagger \hat{b}_{-\mathbf{k}} + 1\right), \tag{7.44}$$

with linear dispersion law:

$$\omega_k = ck, \qquad c = \sqrt{n_s/m\varkappa}. \tag{7.45}$$

We see that we are dealing with acoustic phonons, c being the sound velocity.

Inverting (7.39) and using (7.43), we express the operators $\hat{\Phi}(\mathbf{r})$ and $\hat{\delta n}(\mathbf{r})$ in terms of phonon creation/annihilation operators:

$$\hat{\Phi}(\mathbf{r}) = \sum_{\mathbf{k}} \sqrt{\frac{1}{2\omega_k \varkappa}} \left(\hat{b}_{\mathbf{k}} + \hat{b}_{-\mathbf{k}}^{\dagger} \right) e^{i\mathbf{k}\cdot\mathbf{r}}, \qquad \hat{\delta n}(\mathbf{r}) = i \sum_{\mathbf{k}} \sqrt{\frac{\omega_k \varkappa}{2}} \left(\hat{b}_{\mathbf{k}} - \hat{b}_{-\mathbf{k}}^{\dagger} \right) e^{i\mathbf{k}\cdot\mathbf{r}}.$$

$$\tag{7.46}$$

It is instructive to compare the quantum approach of this section to the classical analysis (in Chapter 1) of the normal modes on top of the ground-state solution of the Gross–Pitaevskii equation (GPE). Both treatments lead to acoustic phonons in the $k \to 0$ limit. The small parameters the treatments are based upon are, however, fundamentally different. In a specific model of a classical matter field, one can rely on the smallness of the amplitude of perturbation, thus obtaining the normal modes of a given model for an *arbitrarily large k*. Such an approach is simply impossible for a generic strongly interacting quantum superfluid, because the quantization of elementary excitations means that there is a lower bound for an amplitude of an elementary excitation of a given wavevector. In a general case, the harmonic approximation based on Taylor-expanding the energy operator fails at finite k not to mention that the elementary excitations may become either substantially damped due to decay processes or even ill defined (overdamped). The characteristic example is the phonon spectrum of helium-4 featuring an end point, k_{end}, beyond which there are no phonons. That is why in this section, we have been relying on the smallness of the wavevector k, which, in accordance with Equation (7.46), guarantees the smallness of the phase gradient and density deviations (both scaling as $\propto \sqrt{k}$).

On the other hand, by confining analysis to $k \to 0$, one gains the possibility of treating an arbitrary superfluid state rather than a specific one. It is thus useful to outline a direct classical-field counterpart to the quantum treatment. One starts with the same Taylor expansion of energy in terms of the gradient of the phase and density deviation [Equation (7.37)]. The crucial step is in identifying canonical variables to promote this Taylor expansion to the Hamiltonian functional. This is done by observing that there is a direct Poisson-bracket equivalent of the commutator (7.26), meaning that the phase and density fluctuation form a canonical pair. The rest of the analysis coincides with the treatment of this section, up to replacing creation and annihilation operators of phonons with their classical counterparts—complex-valued canonical variables.

7.3 Superfluid Thermodynamics at Low Temperature

In a translation-invariant system, one can easily argue that the superfluid density equals the total density in the ground state. Indeed, one way to construct a superfluid ground state moving with velocity $\mathbf{v}$ is to go to the moving reference frame, in which case the total momentum of the state is $Vnm\mathbf{v}$. From uniqueness of the ground state, this is the only available option. From the general relations (2.19), (2.26), we then conclude that

$$n_s(T = 0) = (Vnm v)/(Vmv) = n. \tag{7.47}$$

Whenever translation invariance is broken, either explicitly (by external potentials or disorder) or spontaneously (e.g., in crystals), this relation is no longer valid, and, generally, $n_s(T = 0) < n$.

At low finite temperature, the lowest free-energy state is that of a dilute gas of phonons described by the Bose-distribution function $N_k = \left(e^{\omega_k/T} - 1\right)^{-1}$ (since the number of phonons is not conserved, their chemical potential is zero). Calculating the system energy (for definiteness, we consider a 3D case; we keep $\hbar = 1$, but restore V)

$$E(T) - E_0 = V \int \frac{d^3k}{(2\pi)^3} \omega_k N_k \tag{7.48}$$

with the energy dispersion (7.45) leads to

$$E(T) - E_0 = V \frac{\pi^2 T^4}{30c^3}. \tag{7.49}$$

The specific heat is given by

$$C(T) = V \frac{2\pi^2 T^3}{15c^3}. \tag{7.50}$$

Note that quantum mechanics is crucial here to suppress occupation numbers at energies $\omega_k \gg T$. The corresponding classical expression would suffer from the famous ultraviolet divergence because in this case $N_k^{(\text{class})} = T/\omega_k$.

To derive the temperature dependence of the superfluid density in the translation-invariant system (a more general expression will be derived later in Chapter 12), we start with Equation (2.19), which relates current density to the system momentum in the laboratory reference frame. On the other hand, by going to the reference frame moving with the superfluid velocity, we can relate the system momentum $\mathbf{P} = mV(n_s - n)\mathbf{v}_s \equiv -mVn_n\mathbf{v}_s$ to the momentum of the phonon gas that is moving in the opposite direction. By Galilean invariance [see Equation (1.84)] in the reference frame of the static superfluid component, the dispersion relation of elementary excitations transforms as $\omega_\mathbf{k} \to \omega_k + \mathbf{k} \cdot \mathbf{v}_s$, and the total momentum is given by

$$\mathbf{P} = V \int \frac{d^3k}{(2\pi)^3} \mathbf{k} \, N_{\omega_k + \mathbf{k} \cdot \mathbf{v}_s}. \tag{7.51}$$

The leading nonvanishing term here is proportional to $\mathbf{v}_s$:

$$\mathbf{P} = V \int \frac{d^3k}{(2\pi)^3} \mathbf{k}(\mathbf{k} \cdot \mathbf{v}_s) \frac{dN_{\omega_k}}{d\omega_k} \equiv -mV\mathbf{v}_s \int \frac{d^3k}{(2\pi)^3} \frac{k^2}{3m} \left(-\frac{dN_{\omega_k}}{d\omega_k}\right). \tag{7.52}$$

By relating the two expressions for normal component, we arrive at the Landau formula:

$$n_n = \int \frac{d^3k}{(2\pi)^3} \frac{k^2}{3m} \left(-\frac{dN_{\omega_k}}{d\omega_k} \right).$$ (7.53)

Substituting the phonon dispersion relation in this expression, we find

$$n_s = n - \frac{2\pi^2 T^4}{45mc^5}.$$ (7.54)

7.4   Density Matrix

A direct quantum analog of the off-diagonal correlator $\langle \psi^*(\mathbf{r}')\psi(\mathbf{r}) \rangle$ is the single-particle density matrix:

$$\rho(\mathbf{r}',\mathbf{r}) = \left\langle \hat{\psi}^\dagger(\mathbf{r}')\hat{\psi}(\mathbf{r}) \right\rangle.$$ (7.55)

In translation-invariant system, the density matrix depends only on the difference between the two points, $\rho(\mathbf{r}',\mathbf{r}) \equiv \rho(\mathbf{r}'-\mathbf{r})$, and its Fourier transform is nothing but the average occupation number in momentum representation:

$$\rho_{\mathbf{k}} = \left\langle \hat{\psi}_{\mathbf{k}}^\dagger \hat{\psi}_{\mathbf{k}} \right\rangle = \langle \hat{n}_{\mathbf{k}} \rangle = n_{\mathbf{k}}.$$ (7.56)

Using the basis of exact eigenstates of the system, we can also write the density matrix as

$$\rho(\mathbf{r}',\mathbf{r}) = \sum_\alpha e^{-E_\alpha/T} \int d\mathbf{r}_2 \ldots d\mathbf{r}_N \, \Psi_\alpha^*(\mathbf{r}',\mathbf{r}_2,\ldots,\mathbf{r}_N)\, \Psi_\alpha(\mathbf{r},\mathbf{r}_2,\ldots,\mathbf{r}_N).$$ (7.57)

One immediately sees the crucial connection between the density matrix and superfluidity. The emergence of the classical-field component and topological order (in the single-particle channel) requires that occupation numbers for $k \to 0$ become large. This implies that bosonic particles are at least delocalized with respect to each other; otherwise, $n_{\mathbf{k}}$ would remain essentially finite. Now, one way to look at Equation (7.57) is to say that ρ quantifies what happens with the typical system state when one selects one particle/coordinate and takes it to arbitrary large distances. Density matrix delocalization then translates to the possibility of having diverging $n_{\mathbf{k}\to0}$. In the following section, we will present arguments that an exponential decay of $\rho(\mathbf{r})$ necessarily implies that the ground state is normal. Later, in Chapter 8, we will prove a stronger statement that divergence of $\int d\mathbf{r}\rho(\mathbf{r}) = n_{k=0}$ is required for superfluidity.

As discussed in Chapter 2, the asymptotic decay of the off-diagonal correlation function is governed by fluctuations of phase [see Equation (2.62)]. Formula (2.64) is equally valid for the quantum case with the notable difference that the phase correlator (2.61) now takes the form

$$G(\mathbf{r}) = V^{-1} \sum_{k<k_*} G_{\mathbf{k}} e^{i\mathbf{k}\cdot\mathbf{r}}, \qquad G_{\mathbf{k}} = \frac{N_k + 1/2}{\omega_k \varkappa},$$ (7.58)

which is easy to obtain by using quantized expressions for the phase fields (7.43), (7.46). For large occupation numbers $N_k \approx T/\omega_k \gg 1$, we recover the classical result (2.61). The quantum expression for the asymptotic decay of the density matrix takes the form

$$\rho(r) = n_{k_*} \exp\left[-\int\limits_{k<k_*} \frac{d^d k}{(2\pi)^d} \left(1 - e^{i\mathbf{k}\cdot\mathbf{r}}\right) \frac{N_k + 1/2}{\omega_k \varkappa}\right], \qquad r \to \infty. \qquad (7.59)$$

It is valid at distances much larger than the healing length; at low temperature, it extends into the region $r \ll c/T$ where the classical expression (2.62) is no longer appropriate. In 3D, the difference between the $T = 0$ and finite-T expressions at $r \to \infty$ accounts for the condensate density dependence on temperature (Kerr's law):

$$n_0(T) = n_0 \exp\left[-\int \frac{d^3 k}{(2\pi)^3} \frac{N_k}{\omega_k \varkappa}\right] = n_0 \exp\left[-\frac{T^2}{12 c^3 \varkappa}\right]. \qquad (7.60)$$

7.5 Functional Integral Representation for Bosonic Fields

In this section, we will develop a formalism of functional integral that will allow us to get an idea of when and how quantum bosonic fields can mimic the behavior of classical matter fields. In Chapters 11 and 12, we will use this formalism to develop a systematic microscopic treatment of the weakly interacting Bose gas.

7.5.1 Evolution Operator

Consider a typical Hamiltonian describing identical bosons of mass m with the pairwise interaction potential $U(\mathbf{r} - \mathbf{r}')$. In the second quantization formalism, it is described by

$$H[\hat{\psi}] = \int d\mathbf{r}\, \hat{\psi}^\dagger(\mathbf{r}) \left[-\frac{\nabla^2}{2m} - \mu(\mathbf{r})\right] \hat{\psi}(\mathbf{r}) + \frac{1}{2} \int d\mathbf{r}\, d\mathbf{r}'\, \hat{\psi}^\dagger(\mathbf{r}) \hat{\psi}^\dagger(\mathbf{r}') U(\mathbf{r} - \mathbf{r}') \hat{\psi}(\mathbf{r}') \hat{\psi}(\mathbf{r}). \qquad (7.61)$$

The most basic problem one has to solve in quantum mechanics is the time evolution of some initial state $|\Psi(0)\rangle$:

$$|\Psi(t)\rangle = e^{-iHt} |\Psi(0)\rangle. \qquad (7.62)$$

To arrive at the functional integral representation, we use the following standard trick: the evolution operator is written as a product of $P \to \infty$ identical factors,

$$e^{-iHt} = e^{-i\epsilon H} e^{-i\epsilon H} \cdots e^{-i\epsilon H} = \prod_{j=1}^{P} e^{-i\epsilon H},$$

with $\epsilon = t/P \to 0$, and the completeness relation (7.17) is used to insert its left-hand side between all factors. By implementing these steps, we obtain

$$|\Psi(t)\rangle = \left(\prod_{j=0}^{P} \int \mathcal{D}\psi_j\right) |\psi_P\rangle\langle\psi_P|e^{-i\epsilon H}|\psi_{P-1}\rangle\langle\psi_{P-1}|e^{-i\epsilon H}|\psi_{P-2}\rangle\langle\psi_{P-2}|$$

$$\dots|\psi_1\rangle\langle\psi_1|e^{-i\epsilon H}|\psi_0\rangle\langle\psi_0|\Psi(0)\rangle. \tag{7.63}$$

Next, we make use of $\epsilon \to 0$ and compute matrix elements of exponentials by expanding the latter to the first nontrivial term:

$$\langle\psi_j|1 - i\epsilon H[\hat{\psi}]|\psi_{j-1}\rangle = \left(1 - i\epsilon E\left[\psi_j^*, \psi_{j-1}\right]\right)\langle\psi_j|\psi_{j-1}\rangle \approx e^{-i\epsilon E\left[\psi_j^*, \psi_{j-1}\right]}\langle\psi_j|\psi_{j-1}\rangle. \tag{7.64}$$

The energy functional is obtained from the Hamiltonian by replacing operators with the complex-number field according to

$$E\left[\psi_j^*, \psi_{j-1}\right] = H\left[\hat{\psi}^\dagger(\mathbf{r}) \to \psi_j^*(\mathbf{r}), \hat{\psi}(\mathbf{r}) \to \psi_{j-1}(\mathbf{r})\right]. \tag{7.65}$$

This result follows directly from properties of coherent states [see Equation (7.15)] and its Hermitian conjugate version. Since $E\left[\psi_j^*, \psi_{j-1}\right]$ and $H\left[\hat{\psi}^\dagger, \hat{\psi}\right]$ [see Equation (7.61)] have an identical structure upon replacing operators with fields, we will proceed with the notation $H\left[\psi_j^*, \psi_{j-1}\right] \equiv E\left[\psi_j^*, \psi_{j-1}\right]$ for the energy functional.

Combining everything and taking into account that, in accordance with (7.19),

$$\prod_{j=1}^{P}\langle\psi_j|\psi_{j-1}\rangle = \exp\left\{-\frac{1}{2}\sum_{j=1}^{P}\int d\mathbf{r}\left[\psi_j^*\left(\psi_j - \psi_{j-1}\right) - \left(\psi_j^* - \psi_{j-1}^*\right)\psi_{j-1}\right]\right\}, \tag{7.66}$$

we arrive at the expression for the evolution operator

$$e^{-iHt} = \int \mathcal{D}\psi e^{iS} |\psi(t)\rangle\langle\psi(0)|, \tag{7.67}$$

where

$$\mathcal{D}\psi = \prod_{j=0}^{P}\mathcal{D}\psi_j, \qquad \psi(0) \equiv \psi_0, \qquad \psi(t) \equiv \psi_P, \tag{7.68}$$

$$S = \sum_{j=1}^{P}\left\{\frac{i}{2}\int d\mathbf{r}\left[\psi_j^*\left(\psi_j - \psi_{j-1}\right) - \left(\psi_j^* - \psi_{j-1}^*\right)\psi_{j-1}\right] - \epsilon H\left[\psi_j^*, \psi_{j-1}\right]\right\}. \tag{7.69}$$

Given that ε is arbitrary small, it is tempting to introduce a smooth time-dependent field

$$\psi(\mathbf{r}, \tau) \equiv \psi_j(\mathbf{r}), \qquad j = \text{int}(t/P), \tag{7.70}$$

replace finite differences with time derivatives,

$$\psi_j - \psi_{j-1} \rightarrow \varepsilon\dot\psi, \tag{7.71}$$

convert summation to integration (with $H\left[\psi_j^*, \psi_{j-1}\right] \rightarrow H[\psi] \equiv H\left[\psi^*, \psi\right]$), and formulate the classical-field action in the compact form

$$S[\psi] = -\int_0^t dt \left\{ \int d\mathbf{r}\, \mathrm{Im}\,\psi^*\dot\psi + H[\psi] \right\}. \tag{7.72}$$

Unfortunately, the field $\psi(\mathbf{r}, t)$ defined by Equation (7.70) is neither smooth nor even continuous, and Equations (7.70) through (7.72), if taken literally, are meaningless. Nevertheless, Equations (7.70) through (7.72) can be used as a convenient convention for encoding the actual discrete structure of the action (7.69), provided the following decoding rules are obeyed: First, the term $\mathrm{Im}\,\psi^*\dot\psi$ should be identically rewritten in the symmetric form $\mathrm{Im}\,\psi^*\dot\psi \rightarrow (i/2)\left(\dot\psi^*\psi - \psi^*\dot\psi\right)$. Then, the terms with time derivatives should be replaced with finite differences, in accordance with the structure of Equation (7.66). Finally, care should be taken of $H[\psi^*, \psi]$, replacing $\psi^*(\mathbf{r}, t) \rightarrow \psi_j^*(\mathbf{r})$, $\psi(\mathbf{r}, t) \rightarrow \psi_{j-1}(\mathbf{r})$. Note that the latter rules apply also to ψ^* and ψ factors in the terms with time derivatives, *before* replacing the derivatives with finite differences.

7.5.2 Partition Function

In quantum statistics, the quantity of prime interest is the partition function $Z = \mathrm{Tr}\,e^{-\beta H'}$, which can be viewed as the trace of the evolution operator in imaginary time, $t \rightarrow -i\beta$, with $H \rightarrow H'$, where H' contains the standard set of extra terms with additive constants of motion. The construction of the functional-integral representation of the imaginary-time evolution operator follows essentially the same steps as in real time, leading to the imaginary-time action, S_E, called *Euclidean action*. The new element that appears in connection with the partition function is the necessity to deal with the trace. Doing traces with coherent states is based on a simple fact that for any operator A,

$$\mathrm{Tr}\,A = \int \mathcal{D}\psi \langle\psi|A|\psi\rangle. \tag{7.73}$$

The proof is as follows: Insert the left-hand side of the completeness relation (7.17) under the sign of trace. Change the orders of tracing and integration. Take into account that for any vector $|b\rangle$, $\mathrm{Tr}\,A|b\rangle\langle b| = \langle b|A|b\rangle$.

Essentially repeating the derivation of the real-time formalism, we get

$$Z = \int \mathcal{D}\psi \prod_{j=1}^P \langle\psi_j|(1 - \varepsilon H')|\psi_{j-1}\rangle = \int \mathcal{D}\psi \prod_{j=1}^P e^{-\varepsilon H'[\psi_j^*, \psi_{j-1}]}\langle\psi_j|\psi_{j-1}\rangle, \tag{7.74}$$

with $\psi_0 \equiv \psi_P$ and $\mathcal{D}\psi \equiv \prod_{j=1}^{P} \mathcal{D}\psi_j$. And then, using (7.66), we arrive at

$$Z = \int \mathcal{D}\psi e^{S_E}, \tag{7.75}$$

$$S_E = -\sum_{j=1}^{P} \left\{ \int d\mathbf{r}\, \psi_j^*(\psi_j - \psi_{j-1}) + \varepsilon H'\left[\psi_j^*, \psi_{j-1}\right] \right\}. \tag{7.76}$$

Here, we used the freedom of modifying the form of the first term by making use of periodic boundary conditions in β and summation over all j's:

$$\sum_{j=1}^{P} \left(\psi_j^* - \psi_{j-1}^*\right)\psi_{j-1} \equiv -\sum_{j=1}^{P} \psi_j^*\left(\psi_j - \psi_{j-1}\right). \tag{7.77}$$

As in the real-time case, we can pretend that we are dealing with a smooth (β-periodic) time-dependent field,

$$\psi(\mathbf{r}, \tau) \equiv \psi_j(\mathbf{r}), \qquad j = \text{int}(\tau/P) + 1, \tag{7.78}$$

and formally replace the finite difference with the corresponding derivative,

$$\psi_j - \psi_{j-1} \to \varepsilon \dot{\psi}. \tag{7.79}$$

By replacing summation with integration, we arrive at the classical-field Euclidean action:

$$S_E[\psi] = -\int_0^\beta d\tau \left\{ \int d\mathbf{r}\, \psi^* \dot{\psi} + H'[\psi] \right\}. \tag{7.80}$$

Similar to its real-time counterpart, this action is just a convention allowing us to present (7.76) in a more compact form. The rules of restoring (7.76) from (7.80) are the same.

We employ this representation to derive the diagrammatic technique for bosonic systems in Chapter 11. What is also worth mentioning here is that thermo-dynamic correlation functions (more details are provided in Chapter 11) require computing the trace of the time-ordered operator sequence

$$\mathcal{K} = Z^{-1} \text{Tr}\left[e^{-\tau_1 H'}\hat{O}_1 e^{-(\tau_2 - \tau_1)H'} \dots e^{-(\tau_{m-1} - \tau_m)H'}\hat{O}_m e^{-(\beta - \tau_m)H'} \right], \tag{7.81}$$

where $\hat{O}_i$ can be either $\hat{\psi}(\mathbf{r}_i)$ or $\hat{\psi}^\dagger(\mathbf{r}_i)$, and $\tau_1 < \tau_2 \cdots < \tau_m < \beta$. By repeating identi-cally all steps in the derivation of the functional integral for Z one readily arrives at the formula

$$\mathcal{K} = Z^{-1} \int \mathcal{D}\psi e^{S_E} O_1 O_2 \dots O_m, \tag{7.82}$$

where O_i are now complex-number fields $\psi(\mathbf{r}_i, \tau_i)$ or $\psi^*(\mathbf{r}_i, \tau_i)$.

7.6   Classical-Field Limit: Gross–Pitaevskii Equation

We have mentioned briefly that coherent states with large occupation numbers allow one to replace operators with c-numbers and ignore the quantum nature of bosonic fields in the leading approximation. In functional integral language, the evolution of such a state can be associated with a *smooth extremal trajectory*, $\psi_0(\mathbf{r}, t)$, of the action (7.69). Fluctuations about $\psi_0(\mathbf{r}, t)$ correspond to subleading quantum corrections. Note that while the action (7.69) is essentially discrete in terms of the time-dependence, its extremal trajectory is smooth. Because of its smoothness, the extremal trajectory can be obtained also from the action (7.72). And this is, perhaps, the most important aspect behind introducing action (7.72). In the classical-field limit, this action becomes the action of the corresponding classical field.

The requirement $\delta S / \delta \psi(\mathbf{r}, \tau) = 0$ for the extremal trajectory $\psi_0(\mathbf{r}, t)$ leads to the GPE:

$$i \frac{\partial \psi_0(\mathbf{r}, t)}{\partial t} = \left[-\frac{\nabla^2}{2m} - \mu(\mathbf{r}) + \int d\mathbf{r}' |\psi_0(\mathbf{r}', t)|^2 U(\mathbf{r} - \mathbf{r}') \right] \psi_0(\mathbf{r}, t). \tag{7.83}$$

Alternative derivation can be done by considering the coherent state $|\Phi_t\rangle = |\psi(\mathbf{r}, t)\rangle$ such that occupation numbers in the plane-wave harmonics are large for momenta $k < k_c$ and zero otherwise, and looking at its evolution according to the Schrödinger equation. After an infinitesimally small time dt, the state will evolve to

$$|\Phi_{t+dt}\rangle = \left(1 - i\hat{H} dt \right) |\Phi_t\rangle. \tag{7.84}$$

We now act with the field operator on the evolving state

$$\hat{\psi}(\mathbf{r}) |\Phi_{t+dt}\rangle = \psi(\mathbf{r}, t)(1 - i\hat{H} dt) |\Phi_t\rangle - i[\hat{\psi}(\mathbf{r}), \hat{H}] dt |\Phi_t\rangle, \tag{7.85}$$

and, using an explicit form of the commutator [see the note below Equation (7.10)],

$$\left[\hat{\psi}(\mathbf{r}), H \right] = \left[-\frac{\nabla^2}{2m} - \mu(\mathbf{r}) + \int d\mathbf{r}' \hat{\psi}^\dagger(\mathbf{r}') U(\mathbf{r} - \mathbf{r}') \hat{\psi}(\mathbf{r}') \right] \hat{\psi}(\mathbf{r}), \tag{7.86}$$

we observe that the only nontrivial term to evaluate is $\hat{\psi}^\dagger(\mathbf{r}') |\Phi_t\rangle$. The low-momentum ($k < k_c$) part of the field operator is acting on harmonics with large occupation numbers and leads to $\hat{\psi}^\dagger_{k<k_c}(\mathbf{r}') |\Phi_t\rangle \approx \psi^*(\mathbf{r}, t) |\Phi_t\rangle$. There will also be an admixture of states with large momenta and small occupation numbers. Let us ignore these states for the moment and discuss next their role in regularizing ultraviolet properties of the GPE, when it is a valid description of the long-wave physics. Then, we find (within the accuracy of order dt^2, the time argument of the state in the last term of (7.85) can be changed from t to $t + dt$)

$$\hat{\psi}(\mathbf{r}) |\Phi_{t+dt}\rangle \approx \left\{ \psi(\mathbf{r}, t) - i dt \left[-\frac{\nabla^2}{2m} - \mu(\mathbf{r}) + \int d\mathbf{r}' |\psi(\mathbf{r}', t)|^2 U(\mathbf{r} - \mathbf{r}') \right] \psi(\mathbf{r}, t) \right\} |\Phi_{t+dt}\rangle, \tag{7.87}$$

meaning that the evolution of the coherent state follows that of the classical field $\psi(\mathbf{r}, t)$ obeying Equation (7.83), since the expression in curly brackets is nothing but $\psi(\mathbf{r}, t + dt)$.

In equilibrium statistics, the classical-field limit corresponds to extremum of the τ-dependence of the field $\psi(\mathbf{r}, \tau)$, meaning that the leading contribution to Z comes from *time-independent* configurations of the field: $\psi(\mathbf{r}, \tau) \equiv \psi(\mathbf{r})$, the τ-dependent configurations being responsible for subleading quantum corrections. In the classical-field limit, $S_E \to -H'[\psi]/T$, and Equation (7.75) takes the form of Gibbs distribution with the classical-field Hamiltonian $H'[\psi]$ associated with the classical-field action (7.72) and the GPE.

It appears then that it is very natural for quantum bosonic fields to behave as classical fields when dealing with modes having large occupation numbers (for brevity, we will call them "classical" modes). We already explained in Chapter 1 that at low enough temperature, the classical component will always undergo a transition to the superfluid state. Thus, an important question to ask is: When does the classical-field description work, and what is the role of modes with occupation numbers of order unity or less? We will call them "quantum" modes to stress that their occupation numbers are sensitive to adding a "quantum of unity"; often, these modes are best represented by the classical-particle picture, but one should not forget that it takes quantum mechanics to relate classical fields to classical particles.

In Chapters 11 and 12, we provide a detailed discussion and derivation of conditions under which classical-field description is an accurate *quantitative* framework for dealing with bosonic systems. Here, we simply state the final results. In a weakly interacting gas of bosons, GPE can be used at sufficiently low temperature, as well as for essentially non-equilibrium states, provided modes with large occupation numbers play the dominant role; there is no requirement whatsoever of being below the critical temperature or having a condensate. Since GPE describes the evolution of *all* excited modes with large occupation numbers, it can be used for studies of such problems as vortex dynamics and superfluid turbulence, condensate formation, and long-wave dynamics of weakly interacting 2D and 1D systems at finite (but low enough) temperature, including the normal state.

Regarding the interaction potential, it is required that either the interaction potential is weak (i.e., the scattering amplitude is well described by the Born approximation) or the gas is dilute, in $d = 2, 3$ (i.e., $na_s^d \ll 1$ where a_s is the s-wave scattering length and n is the gas density). These conditions guarantee that interaction effects do not significantly alter the GPE dynamics or induce localization into the solid phase.[*] This does not mean that interactions can be neglected altogether; they do play a crucial role in GPE in determining the spectrum of low-energy modes, in stabilizing the superfluid state against weak disorder, etc. However, GPE fails when na^d is approaching unity—for example, in helium.

---

[*] From the classical-field perspective, localization phenomena revealing the classical-particle aspects of quantum fields is the most nontrivial quantum effect, associated, in particular, with the previously discussed discrete character of the fields $\psi(\mathbf{r}, t)$ and $\psi(\mathbf{r}, \tau)$ in the functional-integral representation.

The role of quantum modes in the physics of GPE (even though they are not part of it!) is twofold. First, they renormalize effective parameters governing GPE. For nonperturbative interaction, the low-energy scattering amplitude requires an exact solution of the quantum two-body problem. In a dilute gas, this accounts for replacing the Born amplitude, $g = \int d\mathbf{r} U(\mathbf{r})$, with the s-wave scattering length; for example, in three dimensions, $g = 4\pi a_s/m$. Indeed, the GPE description in this case is valid only at length scales much larger than the potential radius r_0 because modes with large occupation numbers correspond to low momenta $k \ll k_T = \sqrt{mT} \ll 1/r_0$ and thus quantum modes are crucial for the proper derivation of the effective low-energy coupling. A systematic derivation of this result is provided in Chapter 12. These considerations lead to the *effective* GPE for a dilute gas

$$i\frac{\partial \psi(\mathbf{r},t)}{\partial t} = \left[-\frac{\nabla^2}{2m} - \mu(\mathbf{r}) + g|\psi(\mathbf{r},t)|^2 \right] \psi(\mathbf{r},t), \qquad (7.88)$$

which was discussed extensively in Chapter 1. Second, quantum modes are essential for regularizing GPE properties at short scales and, in particular, establishing the finite-temperature equilibrium. Without regularization, any initial state $\psi(\mathbf{r}, t = 0)$ will ultimately evolve to the state containing $T = 0$ condensate. Indeed, the finite-temperature classical distribution function $n_k^{(\text{class})} = T/\epsilon(k)$ implies infinite kinetic energy due to *ultraviolet catastrophe*, and thus, equilibrium is only possible at $T \to 0$. As we will see in Chapter 13, the evolution toward the ordered state from initially disordered (turbulent) states is quite nontrivial.

To achieve equilibrium at finite T, the GPE dynamics must be supplemented with the quantum kinetic equation based on the Golden Rule and the matching condition for the distribution function set at $n_k \gg 1$, where the two alternative descriptions are both quantitatively accurate.[*]

7.7 Superfluidity and Compressibility: The XY Model Paradox

The fundamental commutation relation between the phase and particle number [see (7.26) or (7.27)] forces us to conclude that superfluids have no energy gap with respect to adding/removing particles to/from the system and thus change their particle number in response to the chemical potential. The canonical commutation relation leads to the uncertainty principle $\Delta\Phi\Delta N \geq 1/2$, which implies that the state with well-defined phase has to be viewed as a broad superposition of particle-number states, for example, a wave packet of the form

$$\Psi_\Phi \propto \sum_N e^{i\Phi N} w(N - N_0)\Psi_N, \qquad (7.89)$$

where $w(x)$ is a smooth envelope function with $d\ln(w)/dx \ll 1$. To ensure that the wave packet remains stable in time, the energies of the number states Ψ_N and

---

[*] For more details on matching quantum- and classical-field kinetic descriptions, see Chapter 13.

$\Psi_{N\pm1}$ should be the same with macroscopic accuracy. In general, this does not yet imply finite compressibility as long-range forces (e.g., in a Coulomb system) may render energy density to scale with the system size. In compressible superfluids, particle-number fluctuations are of the order of $\Delta N \propto \sqrt{N}$, which guarantees stability of the phase wave packets over macroscopic time scales. This would be the typical case for systems with short-range interactions.

At first glance, this conclusion contradicts with the properties of the XY model that is in the superfluid phase at low temperature. As discussed in Equation (3.2), the XY model can be obtained from the classical complex field ($|\psi|^4$) model on a lattice by taking the limit $U \to \infty$, $\mu/U = 1$. In this limit, the modulus of the field is fixed at $|\psi| = 1$, and the classical-field compressibility $\varkappa = 1/U$ is zero. By interpreting $|\psi|^2$ as the density of the underlying particle system, we seem to arrive at the paradoxical counterexample of an incompressible superfluid.

The loophole of the paradox is in the order of taking the limits. Note that an integer particle number is an ill-defined concept within the classical-field approach, and thus, it is extremely important to trace how the limits of large occupation numbers and strong interactions should be properly taken. We start with the quantum-mechanical Hamiltonian behind the classical complex-field model, the so-called Bose–Hubbard Hamiltonian:

$$H = -t \sum_{<ij>} b_j^\dagger b_i + \frac{U}{2} \sum_i n_i^2 - \mu \sum_i n_i = -t \sum_{<ij>} b_j^\dagger b_i + \frac{U}{2} \sum_i \left(n_i - \frac{\mu}{U}\right)^2 + const.$$

(7.90)

Here, t is the matrix element describing hopping transitions of bosons between the nearest-neighbor sites, and $n_i = b_i^\dagger b_i$ is the on-site occupation number.

The classical-field limit can be approached by considering the limit of "strong" interactions $U/t \gg 1$ and large chemical potentials $\mu/U \gg 1$ to ensure that the average occupation number, or filling factor, is large $\langle n_i \rangle = \nu \approx \mu/U \gg 1$ and fluctuations in the occupation numbers are small in relative units (i.e., $|\delta n_i|/\nu \ll 1$)—note that they can still be much larger than unity. Observing that we are in the domain of applicability of parameterization (7.35) through (7.36), we use it to approximate the original Hamiltonian with the so-called "quantum rotor" model:

$$H = -J \sum_{<ij>} \cos\left(\hat{\phi}_i - \hat{\phi}_j\right) + \frac{U}{2} \sum_i \left(\hat{\delta n}_i - \sigma\right)^2.$$

(7.91)

Here $J = t\nu$, $\hat{\delta n}_i = \hat{n}_i - \bar{n}$, while $\bar{n}$ and σ are the integer and fractional parts of μ/U, respectively. In the representation of eigenstates of operators $\hat{\phi}_j$, the quantum rotor model reads

$$H = -J \sum_{<ij>} \cos\left(\phi_i - \phi_j\right) - \frac{U}{2} \sum_i \left(\frac{d}{d\phi_i} + i\sigma\right)^2.$$

(7.92)

When the second term can be neglected, the quantum rotor model becomes the classical XY model.

We immediately see that, in the limit of $U \to \infty$, one cannot ignore the particle-number quantization. For integer ν, or $\sigma = 0$, and $U \gg zJ$ (where z is the lattice coordination number), the ground state is an incompressible Mott insulator; that is, it is fundamentally different from the XY model. If ν is half-integer ($\sigma = 1/2$) and U goes to infinity, we are dealing with the strongly correlated compressible superfluid liquid of hard-core bosons in the half-filled band (the role of large occupation numbers is reduced to the renormalization of the effective hopping parameter $t \to J/2$). The appropriate superfluid XY model limit is reached only for $U < zJ$ when the second term in the Hamiltonian (7.91) does allow large particle-number fluctuations in absolute units but keeps them much smaller than ν; that is, when the following inequalities are satisfied:

$$zt/\nu \ll U \ll zt\nu, \qquad \text{or} \qquad \mu \gg zt, \ U \ll \sqrt{zt\mu}. \tag{7.93}$$

This, in particular, means that the compressibility of XY model remains finite, $\varkappa_{XY} \approx 1/U \gg 1/J$, if the limit is taken in compliance with quantum mechanics.

Formally, at the purely classical level, we notice that one can take the classical complex field model, parameterize it with $\psi_i = |\psi| \chi_i e^{i\Phi_i}$, and consider the following limit at fixed $J = t|\psi|^2$ and μ: as t goes to zero and $|\psi|^2$ goes to infinity, the interaction vanishes as $U = \mu/|\psi|^2$. One can easily check that the answer can be expressed as

$$H_\psi = -J \sum_{\langle ij \rangle} \chi_i \chi_j \cos(\Phi_i - \Phi_j) + \frac{\mu|\psi|^2}{2} \sum_i \left(\chi_i^2 - 1\right)^2 \longrightarrow -J \sum_{\langle ij \rangle} \cos\left(\Phi_i - \Phi_j\right). \tag{7.94}$$

This limit can only be interpreted as an infinitely compressible (!) classical XY model because typical fluctuations of the field modulus are diverging as $\Delta|\psi|^2 \sim \sqrt{zJ/U}$.

Problem 7.7 *Perform the limit $t \to 0$, $\mu = JU/t \to \infty$ for fixed U and J for the classical field model on a lattice.*

7.8 Connectivity of the Ground-State Wavefunction

An important connection between the delocalization of the density matrix and superfluidity was explored beautifully by Kohn [2] and Leggett [3]. To simplify the conceptual understanding of the argument, consider the case of two particles confined to the 1D ring of radius R and strongly attracting each other ("hydrogen atom"-type system). In the ground state, the wavefunction can be written as $\Phi_0(r_1 - r_2)$, where Φ_0 describes a tight bound state of the pair; that is, it falls off exponentially fast as the particle coordinates are separated (for bosons, it is also symmetric, sign positive, and $2\pi R$-periodic). Now, the question is this: What is the wavefunction of the first rotating state Φ_1? The most straightforward answer would be that $\Phi_1 = e^{ipr_M/\hbar}\Phi_0(r_1 - r_2)$, where $r_M = (r_1 + r_2)/2$ is the center of mass coordinate and $p = \hbar/R$ is the molecule momentum. Essentially, we attempt here not to pay any special attention to the internal structure of the compact molecule and treat it as a point particle described by only one coordinate r_M.

The fundamental problem with this form of Φ_1 is that it is not single valued! Indeed, if we fix r_2 and move r_1 around, we will accumulate a phase factor $(2\pi R)/2R = \pi$ when we complete the circle. Somewhat counterintuitively, it is the structure of the wavefunction of the relative motion that has to be modified to remedy the problem; that is, the correct answer is

$$\Phi_1 = e^{ipr_M/\hbar} \tilde{\Phi}_0(r_1 - r_2). \tag{7.95}$$

For slow rotation, $\tilde{\Phi}_0$ and Φ_0 are nearly identical at short distances for energetic reasons. However, at a large distance $r = |r_1 - r_2| \sim \pi R$, where $\Phi_0(r)$ is exponentially small, one can modify it at will by paying a negligibly small energy penalty. Thus, to ensure that the wavefunction is single valued, $\tilde{\Phi}_0$ is modified to have a zero and a sign change at $r = \pi R$. This consideration can be immediately generalized to the bound state of N-particles, when the first rotating state is given by

$$\Phi_1 = \tilde{\Phi}_0(r_1, r_2, \ldots, r_N) \exp\left[i \frac{\sum_{j=1}^{N} r_j}{NR}\right] \tag{7.96}$$

with $\tilde{\Phi}_0$ having hyperplanes of zeros in places where one of the coordinates is as far from the center of mass coordinate as possible. Note that the velocity circulation in such a state is (h/mN) and much smaller than (h/m) for large N; that is, in the limit of $N \to \infty$, the system can be set in rotation with arbitrary small velocity via the center of mass coordinate. This rather obvious outcome nevertheless requires us to consider making zeros in the multidimensional wavefunction; that is, the center of mass rotation is not possible within the subspace of functions that involve only adding phase factors to the ground state $\Phi_0(r_1, r_2, \ldots, r_N)$.

Another simple, but equally important, observation concerns the structure of the wavefunction of a particle sitting in a tight localized state $\Phi_0(r - r_0)$ created by the external potential well $U(r - r_0)$. Consider how the state is changed when we impose the twisted boundary condition (equivalent to a slow rotation; see Chapter 1). The lowest energy state will accommodate the twisting phase φ by creating a zero in the wavefunction at the point where $\Phi_0(r - r_0)$ reaches its smallest value (exponential in R). This means that energy has vanishingly small sensitivity to the twisting boundary condition and the particle moves along with the rotating frame. To complete the picture, we note that in a macroscopic system, the center of mass coordinate $\mathbf{r}_M$ is generically localized by the wall potential ($M = \sum_{i=1}^{N} m \to \infty$ and $\sum_{i=1}^{N} \langle V_{\text{wall}}(\mathbf{r}_M) \rangle \to \infty$).

Let us introduce the notion of the "disconnected" state [2,3] characterized by the wavefunction that gets vanishingly small when one, two, or any finite number of coordinates is moved around a macroscopic circle. More specifically, for a typical configuration of fixed $\{\mathbf{r}_2, \ldots, \mathbf{r}_N\}$ in the disconnected state, the wavefunction of the remaining coordinate $\Phi_0(\mathbf{r}_1)$ is localized. Furthermore, the same has to be true for any finite number of coordinates, $\Phi_0(\mathbf{r}_1, \ldots, \mathbf{r}_\nu)$, though typically localizing a large cluster of particles is getting progressively easier. There are, of course, notable exceptions to this rule: for example, when particles pair into molecules, $\Phi_0(\mathbf{r}_1)$ may be localized while $\Phi_0(\mathbf{r}_1,\mathbf{r}_2)$ is not, leading to the paired superfluid state (electronic superconductors belong to this category). For brevity, we assume below that delocalization, if any, happens already at the single-particle level and thus multiparticle states are delocalized automatically. All considerations can be easily generalized to the case when delocalization happens first at the ν-particle level ($\nu > 1$). If $\Phi_0(\mathbf{r}_1)$ is delocalized (assume that its normalization volume diverges with the system size), the state is called "connected."

The energy of the disconnected state will not respond to the twisting phase. As discussed in Chapter 1, this means vanishing superfluid density, since at $T = 0$, we have $n_s^{(\varphi)} = \left(mL^{2-d}/\hbar^2\right)\partial^2 E/\partial\varphi^2$. Here, we assume periodic boundary conditions; otherwise, one may assume that the system is confined to a narrow cylindrical annulus with the twisting phase φ applied in the azimuthal direction. The mechanism of accommodating φ without energy penalty is exactly as described earlier: zeros are created in places where $\Phi_0(\mathbf{r}_1,\mathbf{r}_2,\ldots,\mathbf{r}_N)$ is negligible. Finally, for the disconnected state, the single-particle density matrix

$$\rho(\mathbf{r}',\mathbf{r}) = \int d\mathbf{r}_2\ldots \int d\mathbf{r}_N \Psi_0^*(\mathbf{r}',\mathbf{r}_2,\ldots,\mathbf{r}_N)\Psi_0(\mathbf{r},\mathbf{r}_2,\ldots,\mathbf{r}_N) \tag{7.97}$$

is supposed to decay fast with $|\mathbf{r}'-\mathbf{r}|$ because it can be viewed as averaging of the product of localized states $\left\langle \Phi_0^*(\mathbf{r}_1 = \mathbf{r}')\Phi_0(\mathbf{r}_1 = \mathbf{r})\right\rangle$ over typical many-body configurations.

Contrarily, creating zeros in the connected state cannot be done without energy penalty since one has to force $\Phi_0(\mathbf{r}_1)$ to go to zero at the cross-sectional area S. By denoting $n_{\min} = \min\left\{|\Phi_0(\mathbf{r}_1)|^2\right\}$, one can roughly estimate the energy cost per unit area by optimizing the sum of potential and kinetic terms $l\left[n_{\min}nU + \left(\hbar^2 n_{\min}/ml^2\right)\right]$ with respect to the size of the modified region l, which happens to be (not surprisingly!) of the order of the healing length $l \sim \hbar/\sqrt{nUm}$. This leads to an estimate of the energy penalty as $\delta E \sim S\hbar^2 n_{\min}/ml$. A large energy cost for adding zeros to the ground-state wavefunction precludes setting the system in rotation with arbitrarily small velocity circulation. Only when the rotating wavefunction is single valued with respect to *each* coordinate, $\Phi_1 = e^{i(\sum_{i=1}^{N} r_i)/R}\Phi_0$ (i.e., the velocity circulation is quantized), can zeros be avoided. The energy of the lowest rotating state $\left(S\hbar^2\pi n_s\right)/(mR)$ is indeed much smaller than δE unless $n_{\min}$ is vanishing with the system size. This consideration can be used to introduce a criterion for avoiding hyperplanes of zeros in the lowest rotating state in terms of the minimum value of the density matrix, $n_{\min} \propto \rho_{\min}^2$. We find that the energy cost balance is in the favor of superfluid behavior when

$$\rho_{\min} > R^{-1/2}. \tag{7.98}$$

Both $d = 2$ and $d = 3$ systems pass this criterion for scaling of the density matrix with the system size in the superfluid state. Note also that according to this criterion, the normalization integral of the density matrix is divergent in the superfluid state.

The close relationship between the wavefunction connectivity and superfluid order in the density matrix provides one more way to prove that superfluid ground states have no gap for adding/removing particles from the system. By definition, $|\Psi(\mathbf{r}_1, \mathbf{r}_2, \ldots, \mathbf{r}_N)|^2$ is the probability density to observe particles at the specified positions. Let us take a typical configuration and fix all coordinates except one, $\mathbf{r}_1$. For connected states, $f(\mathbf{r}_1) = \Psi(\mathbf{r}_1)/\Psi\left(\mathbf{r}_1^{(in)}\right)$ does not decay exponentially when $\mathbf{r}_1$ is taken arbitrarily far from the initial position $\mathbf{r}_1^{(in)}$; moreover, since in super-fluid states the density matrix is nonintegrable, configurations with arbitrary large $\left|\mathbf{r}_1 - \mathbf{r}_1^{(in)}\right|$ are dominating in the normalization integral. At this point, we note that states obtained by shifting $\mathbf{r}_1$ to infinity contain two independent (even in the presence of Coulomb interactions) remote regions doped with an extra hole and a particle, respectively. Thus, adding/removing an extra particle to/from the system cannot be distinguished (in the thermodynamic limit) from typical fluctuations already present in the ground state.

7.9   Popov Hydrodynamic Action

Now, we derive Popov hydrodynamic action that describes equilibrium and hydro-dynamic properties of a superfluid in the long-wave low-temperature limit. At the harmonic level, Popov action is equivalent to the harmonic Hamiltonian (7.37). Here, we are interested in going beyond the harmonic approximation. Especially important is that Popov action allows one to systematically treat vortices and the vortex–phonon interaction in the universal hydrodynamic limit.

Our derivation is based on the imaginary-time action for the partition func-tion. The final result is readily translated into the real-time action by using the general fact that any real-time action, $S(t)$, is related to its imaginary-time (Euclidean) counterpart, $S_E(\tau)$, by the formula[*] $S(t) = -iS_E(\tau \to it)$.

Starting from the microscopic action (7.80) [in the following, we explain why the pseudo-continuous representation is sufficient], we proceed in Popov's spirit, but with the significant difference that allows us to avoid certain technical prob-lems of Popov's original treatment.[†] Rather than employing Popov's additive

---

[*] Strictly speaking, the Euclidean action is based on the thermodynamic Hamiltonian H' rather than the bare Hamiltonian H, and the direct correspondence between $S(t)$ and $S_E(\tau)$ is for H' rather than H. However, the dynamic implications of this circumstance are trivial because the difference between H and H' is only in terms of the additive constants of motion.

[†] Popov's derivation was based on an additive decomposition, $\psi = \psi_0 + \psi_1$, of the field ψ into the low-, ψ_0, and high-momentum, ψ_1, parts. To proceed, he made restrictive assumptions (inadequate for a general case) concerning the structure of the slow field. Namely, $\psi_0 = \sqrt{n}e^{i\Phi}$, where Φ is the field of the superfluid phase and n is the *total* density, and the fluctuations of n are strongly suppressed: $n = n_0 + \delta n$, with $|\delta n| \ll n_0$ where n_0 was a constant referred to as "condensate."

decomposition, we *factor* the hydrodynamic modes by representing the field as [cf. classical-field Equation (2.69)]

$$\psi = \tilde{\psi} e^{i\Phi}, \tag{7.99}$$

where $\Phi \equiv \Phi(\mathbf{r}, \tau)$ is the "slow" field of the superfluid phase. Potential double counting of hydrodynamic modes is removed by the requirement that the phase of the "fast" field $\tilde{\psi}$ does not contain slowly varying in spacetime terms. This constraint not only enforces the off-diagonal long-range order in the field $\tilde{\psi}$, but also excludes all configurations of $\tilde{\psi}$ that have nonzero winding numbers. With the parameterization (7.99), we have

$$S[\psi] = S\left[\tilde{\psi}\right] - \int_0^\beta d\tau \int d^d r \left\{ \left[i\dot{\Phi} + \frac{(\nabla\Phi)^2}{2m} \right] |\tilde{\psi}|^2 + \frac{i\nabla\Phi}{2m} \cdot \left[\left(\nabla\tilde{\psi}^* \right) \tilde{\psi} - \tilde{\psi}^* \nabla\tilde{\psi} \right] \right\}. \tag{7.100}$$

Now, we comment on why working with the pseudo-continuous Equation (7.80)—rather than with the exact discrete Equation (7.76)—does not lead to ambiguity. As opposed to the field ψ, the coarse-grained phase field Φ is continuous in τ (due to the previously discussed orthogonality of topologically ordered coherent states). Even though this field is not smooth, the procedure does not lead to any trouble since the algebraic behavior of small finite differences is the same as that of derivatives. As far as the field $\tilde{\psi}$ in Equation (7.100) is concerned, it is discontinuous, and all the previously discussed rules of translation into discrete action (7.76) apply to the terms involving $\tilde{\psi}$.

The crucial observation made by Popov was that introducing the new parameterization is associated with the transformation

$$\begin{cases} \mu \to \mu - i\dot{\Phi} - (\nabla\Phi)^2/2m, \\ \mathbf{v}_n \to \mathbf{v}_n - \nabla\Phi/m. \end{cases} \tag{7.101}$$

Combined with the condition of small gradients and time derivatives of Φ, this transformation yields a simple thermodynamic way of finding the result of integration over the fast field $\tilde{\psi}$. First, we split the system into macroscopically large spacetime cells $\Delta\tau\Delta V$, while assuming that the size of the cell is much smaller than the typical scale of spacetime variation of the field Φ. Second, we utilize the fact that integration over $\tilde{\psi}$ within one cell—by construction, different cells are independent with respect to this integration, being linked only by the field Φ—corresponds to the ensemble that is canonical with respect to the phase-winding number and grand canonical with respect to the total number of particles and momentum within each cell [its parameters are given by Equation (7.101)]. [Within one cell, $\dot{\Phi}$ and $\nabla\Phi$ are considered to be constant.] For the partition function, we thus have (ν labels the spacetime cells)

$$Z = \int \mathcal{D}\psi e^{S_E[\psi]} = \int \mathcal{D}\Phi \prod_v Z_v, \qquad (7.102)$$

$$Z_v = \int_{\text{cell } v} \mathcal{D}\tilde{\psi} e^{S_E[\psi]} = \exp\left[\Delta\tau \Delta V p_0\left(\mu - i\dot{\Phi} - \frac{(\nabla\Phi)^2}{2m}, \mathbf{v}_n = -\frac{\nabla\Phi}{m}, \mathbf{v}_s = 0\right)\right], \quad (7.103)$$

where $p_0(\mu, \mathbf{v}_n, \mathbf{v}_s)$ is the pressure at $T = 0$. With Equation (2.39), we can somewhat simplify the expression:

$$p_0\left(\mu - i\dot{\Phi} - \frac{(\nabla\Phi)^2}{2m}, \mathbf{v}_n = -\frac{\nabla\Phi}{m}, \mathbf{v}_s = 0\right) = p_0\left(\mu - i\dot{\Phi}, \mathbf{v}_n = 0, \mathbf{v}_s = \frac{\nabla\Phi}{m}\right). \qquad (7.104)$$

Summation over the cells in the exponential can now be replaced with integration, yielding [for brevity, from now on, we shall omit the argument $\mathbf{v}_n = 0$]

$$Z = \int \mathcal{D}\Phi e^{S_{\text{hd}}[\Phi]}, \qquad S_{\text{hd}}[\Phi] = \int_0^\beta d\tau \int d^d r \, p_0\left(\mu - i\dot{\Phi}, \mathbf{v}_s = \frac{\nabla\Phi}{m}\right). \qquad (7.105)$$

What we arrived at is nothing but the Euclidean form of Popov hydrodynamic action (2.115). This brings us to the (rather intuitive) conclusion that the quantum hydrodynamic behavior of superfluids is described—within functional integral formalism—by the classical hydrodynamic action.

In almost all practical cases of application of the Euclidean form of Popov hydrodynamic action, the spacetime gradients of the field Φ are small, allowing one to expand p_0 at the point $(\mu, \mathbf{v}_s = 0)$, thus getting the Euclidean counterparts of Equations (2.119) through (2.122):

$$S_{\text{hd}}[\Phi] = S_{\text{hd}}^{(\text{top})}[\Phi] + S_{\text{hd}}^{(\text{harm})}[\Phi] + S_{\text{hd}}^{(\text{n.h.})}[\Phi], \qquad (7.106)$$

$$S_{\text{hd}}^{(\text{top})}[\Phi] = -in \int d^d r \int_0^\beta \dot{\Phi} d\tau, \qquad (7.107)$$

$$S_{\text{hd}}^{(\text{harm})}[\Phi] = -\int_0^\beta d\tau \int d^d r \left[\frac{\varkappa}{2}\dot{\Phi}^2 + \frac{n_s}{2m}(\nabla\Phi)^2\right] \qquad (7.108)$$

$$S_{\text{hd}}^{(\text{n.h.})}[\Phi] = \int_0^\beta d\tau \int d^d r \left[\frac{i\varkappa_s}{2m}\dot{\Phi}(\nabla\Phi)^2 - \frac{i\varkappa^3\lambda}{6}\dot{\Phi}^3 + \cdots\right]. \qquad (7.109)$$

Action (7.108) is equivalent to the harmonic hydrodynamic Hamiltonian (7.37).

Based on Equation (7.105), we can also derive the canonical form of the hydrodynamic functional integral involving the coarse-grained field of density, n along with the field Φ. To this end, consider the general mesoscopic thermodynamic relation

$$e^{-\Omega/T} = \int e^{-(F-\mu N)/T} dN. \qquad (7.110)$$

In the limit of $V \to \infty$ and $T \to 0$, Equation (7.110) can also be written in the form of the functional integral over the field $n(\tau, \mathbf{r})$ of the coarse-grained density:

$$e^{\int d\tau \, d\mathbf{r} \, p_0(\mu)} = \int e^{-\int d\tau \, d\mathbf{r} \, [\mathcal{E} - \mu n]} \mathcal{D}n. \tag{7.111}$$

With the replacement $\mu \to \mu - i\dot{\Phi}$, $\mathbf{v}_s = \nabla\Phi/m$, the relation (7.111), used in the functional-integral representation (7.105), yields the canonical representation:

$$Z = \int \mathcal{D}n \, \mathcal{D}\Phi \, e^{S_{\text{hd}}[n,\Phi]}, \qquad S_{\text{hd}}[n,\Phi] = -\int d\tau \, d\mathbf{r} \left\{ in\dot{\Phi} + \mathcal{E}[n,\Phi] - \mu n \right\}. \tag{7.112}$$

Note that μ is a constant parameter having no spacetime dependence. We thus see that (7.112) is the Euclidean counterpart of the classical canonical hydrodynamic action (2.114).

The Euclidean counterpart of the expanded (in density deviations η and spacetime phase gradients) canonical action (2.123) is

$$S_{\text{hd}}[\eta,\Phi] = -\int d\tau \, d\mathbf{r} \left[i(n+\eta)\dot{\Phi} + \frac{n_s}{2m}(\nabla\Phi)^2 + \frac{1}{2\varkappa}\eta^2 + \frac{\varkappa_s/\varkappa}{2m}\eta(\nabla\Phi)^2 + \frac{\lambda}{6}\eta^3 \right]. \tag{7.113}$$

7.10 Long-Range Asymptotics of Off-Diagonal Correlators

With Popov hydrodynamic action, we can readily generalize the theory of long-range behavior of off-diagonal correlators in such a way that quantum and classical fluctuations of the phase are treated on equal footing. Physically, it is clear that quantum fluctuations contribute nontrivially to the long-range off-diagonal correlations only at a sufficiently low temperature, when the typical wavelength of thermally excited phonons is much larger than the zero-point healing length. Otherwise, the long-range asymptotic *evolution* of off-diagonal correlators is governed exclusively by the classical-field fluctuations of the phase, and the theory of Section 2.1.5 can be used without modifications to describe them.

The theory starts with Equation (7.99), the quantum-field analog of Equation (2.69), and it almost literally repeats all the steps of classical-field derivation of Section 2.1.5, up to the replacements, $\psi(\mathbf{r}) \to \psi(\mathbf{r},\tau)$, $\varphi(\mathbf{r}) \to \Phi(\mathbf{r},\tau)$. As a result, for the m-particle correlator; see also Equation (7.82)

$$K = \left\langle \psi_1 \psi_2^* \psi_3 \psi_4^* \dots \psi_{2m-1} \psi_{2m}^* \right\rangle, \tag{7.114}$$

we get (the vector $\vec{X}$ stands for all the spacetime coordinates)

$$K\left(\vec{X}\right) = K\left(\vec{X}'\right) e^{\Lambda(\vec{X}') - \Lambda(\vec{X})}, \tag{7.115}$$

$$\Lambda\left(\vec{X}\right) - \Lambda\left(\vec{X}'\right) = \sum_{s<j} (-1)^{s+j} \Xi\left(\mathbf{r}_{sj}, \tau_{sj}; \mathbf{r}'_{sj}, \tau'_{sj}\right), \tag{7.116}$$

where $\mathbf{r}_{sj} = \mathbf{r}_s - \mathbf{r}_j$, $\tau_{sj} = \tau_s - \tau_j$ (the same for primed variables), and

$$\Xi(\mathbf{r},\tau;\mathbf{r}',\tau') = \langle \Phi(\mathbf{r},\tau)\Phi(0,0) - \Phi(\mathbf{r}',\tau')\Phi(0,0) \rangle$$

$$= T \int \sum_{\xi} \left[e^{i(\mathbf{k}\cdot\mathbf{r}-\xi\tau)} - e^{i(\mathbf{k}\cdot\mathbf{r}'-\xi\tau')} \right] \langle |\Phi_{\mathbf{k},\xi}|^2 \rangle d^d k/(2\pi)^d. \qquad (7.117)$$

Here, $\Phi_{\mathbf{k},\xi}$ is the Fourier component of the field $\Phi(\mathbf{r},\tau)$,

$$\Phi(\mathbf{r},\tau) = \sqrt{T/V} \sum_{\mathbf{k},\xi} e^{i(\mathbf{k}\cdot\mathbf{r}-\xi\tau)} \Phi_{\mathbf{k},\xi}, \qquad (7.118)$$

with the Matsubara frequency ξ running over all integer multiples of $2\pi T$.

Hence, the long-range spacetime evolution of the correlator (7.114) is governed by the function $\Xi(\mathbf{r},\tau;\mathbf{r}',\tau')$, which is defined through the average $\langle |\Phi_{\mathbf{k},\xi}|^2 \rangle$. The latter is readily found from the harmonic part of the Euclidean form of Popov hydrodynamic action (7.108):

$$S_{\mathrm{hd}}^{(\mathrm{harm})} = -\frac{1}{2} \sum_{\xi,\mathbf{k}} \left[(n_\mathrm{s}/m)k^2 + \varkappa\xi^2 \right] |\Phi_{\mathbf{k},\xi}|^2, \qquad (7.119)$$

leading to

$$\langle |\Phi_{\mathbf{k},\xi}|^2 \rangle = \left[(n_\mathrm{s}/m)k^2 + \varkappa\xi^2 \right]^{-1}. \qquad (7.120)$$

Replacing the variable τ with an equivalent "distance" $c\tau$, with c the sound velocity (7.45), we map the theory onto a $(d+1)$-dimensional classical counterpart of Chapter 2, with a finite size $L_\beta = c\beta$ in one of the directions; the corresponding classical $(d+1)$-dimensional parameter $T/\gamma n_s$ being replaced with $\sqrt{m/\varkappa n_s}$. Introducing the $(d+1)$-dimensional distance vector $\vec{x} = (\mathbf{r}, c\tau)$, we then adapt the results of Section 2.1.5.

The $|\vec{x}'|, |\vec{x}| \ll c\beta$ limit corresponds to $L_\beta = \infty$, and here, using Equation (2.77), we have the following:

$$\Xi(|\vec{x}|,|\vec{x}'|) = \sqrt{\frac{m}{\varkappa n_s}} \times \begin{cases} (1/2\pi)\ln(|\vec{x}'|/|\vec{x}|), & d=1, \\ (1/4\pi)\left(|\vec{x}|^{-1} - |\vec{x}'|^{-1}\right), & d=2, \\ (1/4\pi^2)\left(|\vec{x}|^{-2} - |\vec{x}'|^{-2}\right), & d=3 \end{cases} \quad (|\vec{x}'|,|\vec{x}| \ll c\beta). \quad (7.121)$$

The $|\vec{x}'|, |\vec{x}| \gg c\beta$ limit corresponds to a purely classical d-dimensional regime described directly by Equation (2.77). Formally, this follows from Equation (7.117) by the observation that only the zeroth frequency survives, the contributions of nonzero frequencies being exponentially small in $|\vec{x}|/c\beta$.

In the aforementioned treatment, all the spacetime distances were supposed to be appropriately large (for the long-range limit to take place). Alternatively, if only some of the coordinates in the m-particle correlation function are in the asymptotic regime, only these coordinates should be kept in (7.115).

7.11   Cooper Pair Problem

We have seen that the emergence of the classical-field component in bosonic systems requires large occupation numbers for long-wave harmonics of the quantum field. This requirement seems impossible to satisfy in fermionic systems due to the Pauli principle preventing multiple occupation of the same quantum state. This is indeed true at the single-particle level. The way out is to consider fermionic complexes. After all the hydrogen atom is a bosonic particle consisting of two fermions (ignoring that the proton itself is a composite particle consisting of quarks), helium-4 atom consists of six fermions, etc. The composite particle approach is most trivial when the binding energy is so large that for all practical purposes (Bell's FAPP principle), we can ignore the internal structure of the composite particle and see it as relevant only for determining an effective particle mass and interparticle forces. For the rest, we treat composites as point objects.

In essence, binding of fermions into pairs or other even-number clusters (for simplicity, we will talk only about pairs below) is all one needs to proceed with the theory of superfluidity/superconductivity in the fermionic system because the order-parameter field is developed for the pair field, which has bosonic properties. It also means that the coarse-grained description of the electronic superconductor at large distances is exactly the same as for charged (with the particle charge $q = 2e$) bosons. The crucial question then is this: Under what conditions do fermions form pairs? While the case of strong attractive interaction leading to deep bound states (as in the ^{4}He atom) seems trivial, for a long time, the answer was not at all obvious for electrons in metals because the strongest force in condensed matter is the Coulomb repulsion between charges of the same sign. Moreover, as we know from quantum mechanics of one particle in the 3D potential well of radius R, a critical strength of the attractive potential is required to create a bound state. The solution of a simple Cooper problem [4] changes this naïve thinking entirely.

Consider two electrons, with opposite spins, residing on top of the occupied Fermi surface characterized by the Fermi momentum k_F, Fermi energy $\epsilon_F = k_F^2/2m$, Fermi velocity $v_F = k_F/m$, and Fermi surface density of states per spin component $\rho_F = mk_F/2\pi^2\hbar^3$, where m is the electron mass. Let these two electrons interact with each other by a short-range attractive potential $U(\mathbf{r}_1 - \mathbf{r}_2) = u\delta(\mathbf{r}_1 - \mathbf{r}_2)$ that is "switched on" only when the total kinetic energy of the pair, $k_1^2/2m + k_2^2/2m$, is below some threshold value $2\epsilon_F + 2\omega_0$ with $\omega_0 \ll \epsilon_F$; in other words, it is operating only when the pair is within the narrow energy band of width $2\omega_0$ above the Fermi surface. The passive role of other electrons in the Cooper problem is solely to prevent the first two having momenta below k_F because the corresponding states are already occupied. To avoid an apparent contradiction between the Pauli principle derived for identical particles and the simplifying assumption that only the first two particles are interacting, we modify the dispersion relation to

$$\epsilon_k = \begin{cases} k^2/2m & \text{for } k > k_F \\ \infty & \text{for } k \leq k_F \end{cases}. \tag{7.122}$$

The Cooper problem then reduces to the solution of the Schrödinger equation for the two-body Hamiltonian:

$$\langle \mathbf{k}_4, \mathbf{k}_3 | \hat{H}_C | \mathbf{k}_1, \mathbf{k}_2 \rangle = \left(\epsilon_{k_1} + \epsilon_{k_2} \right) \delta_{\mathbf{k}_3,\mathbf{k}_1} \delta_{\mathbf{k}_4,\mathbf{k}_2} + (u/V)\delta_{\mathbf{k}_3+\mathbf{k}_4,\mathbf{k}_1+\mathbf{k}_2}. \tag{7.123}$$

Since the total momentum $\mathbf{k}_1 + \mathbf{k}_2 = \mathbf{P}$ is conserved, the problem is effectively a single-particle one for the relative momentum variable $\mathbf{q} = (\mathbf{k}_1 - \mathbf{k}_2)/2$. For simplicity, we consider here the case of zero total momentum $P = 0$ when

$$\langle \mathbf{q}' | \hat{H}_C(P=0) | \mathbf{q} \rangle = 2\epsilon_q \delta_{\mathbf{q},\mathbf{q}'} + u/V. \tag{7.124}$$

The spherically symmetric solution $\phi(q)$ for the relative motion wavefunction obeys an equation

$$\left(2\epsilon_q - E \right) \phi(q) = -(u/V) \sum_{\mathbf{q}'} \phi(q') \tag{7.125}$$

or

$$\phi(q) = -\frac{uA}{2\epsilon_q - E}, \quad \text{with} \quad A = \sum_q \frac{\phi(q)}{V} = -A \sum_q \frac{u}{V(2\epsilon_q - E)}. \tag{7.126}$$

The eigenvalue equation (replacing sum with the momentum integral and mentioning explicitly all restrictions on the dispersion law and interaction)

$$-\frac{1}{u} = \frac{1}{V} \sum_q \frac{1}{2\epsilon_q - E} = \int\limits_{\epsilon_q > \epsilon_F}^{\epsilon_q < \epsilon_F + \omega_o} \frac{d\mathbf{q}}{(2\pi\hbar)^3} \frac{1}{q^2/m - E}, \tag{7.127}$$

predicts a bound state with the total energy below the Fermi surface $E = 2\epsilon_F - 2\Delta < 2\epsilon_F$ for arbitrary small value of the coupling constant $u < 0$! Indeed, near the Fermi surface, the integral can be expressed as $(\rho_F/2) \int_0^{\omega_0} d\xi/(\xi+\Delta)$, where $\xi = \epsilon_q - \epsilon_F \approx (q - k_F)v_F$ is a new integration variable, revealing a logarithmic divergence for $\Delta = 0$. The resulting binding energy Δ is exponentially small for small $|u|$

$$-\frac{2}{u\rho_F} = \int\limits_0^{\omega_0} d\xi/(\xi+\Delta) = \ln\left(\frac{\omega_0 + \Delta}{\Delta}\right) \longrightarrow \Delta \approx \omega_0 e^{-2/|u|\rho_F}. \tag{7.128}$$

One cannot miss an analogy between this solution and the problem of bound states in the 2D potential well. In both cases, the log divergence of the integral has its origin in the constant density of states for low-energy single-particle states. Putting it differently, the "life" on the Fermi surface is 2D no matter that the actual system is 3D.

Finite values of $\mathbf{P}$ can be done similarly, provided one takes proper care of the restriction on the momentum space $q^2 \pm \mathbf{P} \cdot \mathbf{q} > k_F^2 - P^2/4$. The binding energy $\Delta(P)$ decreases with P and vanishes at relatively small pair momentum $P \sim \Delta(0)/v_F$, indicating critical velocities much smaller than v_F.

Problem 7.8 *In the limit of small g and $\omega_0 \ll \epsilon_F$, calculate the critical velocity of the Cooper pair above which the bound state solution disappears.*

Going from the Cooper-pair problem to a realistic many-body Hamiltonian with weak short-range attraction between fermions does not change the final conclusion that the system develops long-range coherence and superfluidity in the pair channel at sufficiently low temperature. This is the famous result derived by Bardeen, Cooper, and Shrieffer (BCS) [5]. Indeed, for weak interactions, the log-divergent integral over virtual pair states away from the Fermi surface is not altered. In Chapter 12, we provide more details on the structure of the microscopic BCS theory using the diagrammatic approach. It should be noted though that the size of the bound state predicted by the Cooper problem $\xi_C \sim v_F/\Delta$ for small g is much larger than interparticle distance, meaning that in the system of identical fermions, one should rather speak of pair correlation functions than of well-defined, localized-in-space, pairs of fermions (see Section 7.13).

7.12   Where Does the Coulomb Repulsion Go?

By discovering that pairing occurs for arbitrarily weak attraction still does not address the issue of the repulsive Coulomb forces. The answer is both simple and subtle. The proper discussion is outside the scope of our book; thus, we will only briefly mention the key points. To begin with, attraction at the level of bare Hamiltonian is not at all required (!) for pairing because effective interactions between quasiparticles at the Fermi surface are screened by many-body effects and generically acquire complicated dependence on the scattering angle. Consider the scattering vertex for two quasiparticles with $\mathbf{P} = 0$ and $|\mathbf{k}_1| = |\mathbf{k}_2| = k_F$ as a function of angle θ between $\mathbf{k}_1$ and $\mathbf{k}_3$. This can be written as Fourier series $V(\theta) = \sum_l u_l P_l(\cos(\theta))$ where $P_l(x)$ is the Legendre polynomial of order l. By angular momentum conservation, it is easy to see that one has to deal with the effective Cooper problem characterized by coupling u_l in each orbital momentum channel l independently. If any of the couplings u_l is negative, there will be a guaranteed pairing in this channel [6,7] (provided other attractive channels are weaker). Thus, strong overall repulsion may prevent s-pairing, but there remain plenty of possibilities for pairing with nonzero orbital momentum of the pair [7]. The corresponding transition temperature may be extremely low, though, because $u_l \rho_F$ are typically getting very small for large l. It is actually possible to confirm the previous picture quantitatively with controlled accuracy for the dilute Fermi gas with short-range repulsive potential [8,9] and show explicitly that this system has a pairing instability in the p-channel.

Pairing with higher orbital momentum does not explain most electronic superconductors though. The crucial observation by Fröhlich [10] was that electron–phonon coupling leads to the effective attraction between electrons if they are close enough to the Fermi surface. More precisely, electrons must have excitation energies below the typical phonon energy ω_0 (this is the origin of the "mysterious" energy cutoff used in the Cooper problem). Essentially, any effective interaction between electrons mediated by virtual exchange of other excitations is attractive for electrons at the Fermi surface. To simplify the picture of real materials, assume that we have strong repulsive interaction $g_r > 0$ acting between electrons at all energies comparable to ϵ_F to be combined with a weaker attractive coupling $g_a < 0$, which acts only at quasiparticle energies below ω_0.

At this point, we employ an extremely useful point of view on Equations (7.127) and (7.128) provided by the renormalization group approach. Let us introduce dimensionless coupling $g = u\rho_F/2$ and define *renormalized* $g(\omega)$ as

$$1/g(\omega) = 1/g + \int_{\omega}^{\omega_0} d\xi/\xi, \tag{7.129}$$

with explicit solution in the form

$$g(\omega) = \frac{g}{1 + g\ln[\omega_0/\omega]}. \tag{7.130}$$

For negative g, we see that $g(\omega)$ increases in modulus as the cutoff ω gets smaller until it becomes of order unity, $|g|\ln[\omega_0/\omega] \approx 1$; that is, the dimensionless coupling constant renormalizes to strong coupling at low energies. The condition for strong coupling coincides with Equation (7.128). One may also rewrite Equation (7.128) as (for $\omega \gg \Delta$)

$$-\left[1/g + \int_{\omega}^{\omega_0} d\xi/(\xi + \Delta)\right] \approx -1/g(\omega) = \int_{0}^{\omega} d\xi/(\xi + \Delta) \tag{7.131}$$

and formulate the problem in a closed form in terms of the renormalized coupling and reduced energy cutoff.

Applying now Equation (7.129) to the situation when the coupling constant depends on energy, we observe that repulsive interaction is renormalized to a smaller value [11]

$$\tilde{g}_r = \frac{g_r}{1 + g_r \ln(\epsilon_F/\omega_0)} < g_r, \tag{7.132}$$

when we arrive at the energy scale ω_0. The dimensionless coupling g_r is of order unity in metals, while $\omega_0 \ll \epsilon_F$ by nearly two orders of magnitude. Thus, the suppression of Coulomb repulsion in the Cooper channel is substantial. At lower energies, we combine $\tilde{g}_r$ with attractive g_a to obtain an effective coupling:

$$\lambda = \tilde{g}_r + g_a. \tag{7.133}$$

Negative λ implies superconducting instability at low temperatures. Since Coulomb forces play a central role in the formulation of the phonon subsystem in metals as well, the resulting electron–phonon coupling is also relatively strong, and it is not at all unusual to have $|g_a| > \tilde{g}_r$.

Curiously enough, on one hand, it is important that nuclei in solids are mobile and provide a finite frequency range for effective attractive interactions between electrons (phonon frequencies scale with the atomic mass as $M^{-1/2}$) ultimately leading to finite $T_c \propto \omega_0 e^{-1/\lambda}$; on the other hand, it is equally important to have nuclei heavy enough to suppress Coulomb repulsion $g_r \rightarrow \tilde{g}_r$ below the g_a threshold.

7.13   BCS–BEC Crossover

One may wonder whether pairing of fermions in the BCS regime of weak interactions when the pair size v_F/Δ is much larger than the interparticle distance is a novel phenomenon compared to the Bose–Einstein condensation (BEC) of compact bi-fermion molecules or if the difference is merely quantitative. The most transparent answer to this question was given by Leggett [12] who considered a "zero-range" attractive potential; that is, a Fermi gas with the interparticle distance vastly larger than the potential well range but finite (or even infinite) $k_F a$ where a is the s-wave scattering length (see Figure 7.1). The scattering length can be made arbitrarily large because the potential well is designed to be at the edge of forming a bound state. By tuning the scattering length through the resonance, one can go from the BCS regime of weakly interacting fermions at small negative $k_F a$ to the BEC regime of weakly interacting compact molecules at small positive $k_F a$ with the unitary point $k_F a = \infty$ in between.

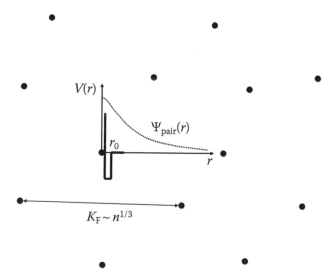

Figure 7.1   Fermi gas with the pairwise potential $V(r)$ featuring a potential well of radius r_0 deep enough to be at the edge of having a bound state for two particles in vacuum. Zero-range limit is obtained by taking the gas density to zero and the s-wave scattering length a to infinity, so that $k_F a$ remains a finite number. At length scales much shorter than the interparticle distance, the pair wavefunction is diverging as $1/r$.

The most remarkable property of the zero-range limit is its universality, since interactions are characterized only by the $k_F a$ parameter no matter what the microscopic origin of the large scattering length is—be it a potential well of an arbitrary shape, as in Figure 7.1, or a Feshbach resonance between the open and closed channels [13]. Introduced theoretically as a convenient and conceptually transparent model system, resonant fermions in the zero-range limit provide an accurate description of a two-species ultracold fermionic gas under the conditions of Feshbach resonance and are also directly relevant to the equation of state of neutron matter (at nonrelativistic densities) [14,15]. Moreover, at the unitary point defined by $1/k_F a = 0$, one obtains an amazing strongly correlated experimental system with $1/k_F$ and ϵ_F being the only remaining units of length and energy. Interested readers may find further details in the review article [13].

To discuss superfluid properties, one may start from the variational ground-state wavefunction in the standard BCS form (we present it here just to help the reader see more clearly the structure of the argument made; additional details can be found in Section 12.8):

$$\Psi_G = \prod_{\mathbf{k}} \left(u_{\mathbf{k}} + v_{\mathbf{k}} f_{\mathbf{k}\uparrow}^\dagger f_{\mathbf{k}\downarrow}^\dagger \right)|0\rangle, \tag{7.134}$$

where $f_{\mathbf{k}\sigma}^\dagger$ is creating a fermion with momentum $\mathbf{k}$ and spin σ and $|0\rangle$ is the vacuum state without particles. The amplitudes $u_{\mathbf{k}}$ and $v_{\mathbf{k}}$ satisfy the relation $u_{\mathbf{k}}^2 + v_{\mathbf{k}}^2 = 1$ and can be obtained by minimizing the total energy. The result is

$$u_k^2 = \frac{E_k + \xi_k}{2E_k}, \quad E_k = \sqrt{\Delta^2 + \xi_k^2}, \tag{7.135}$$

with $\xi_k = \epsilon_k - \epsilon_F$ and Δ obtained from the self-consistent gap equation:

$$\frac{m}{2\pi\hbar^2 a} = \int \frac{d\mathbf{k}}{(2\pi\hbar)^3} \frac{E_k - \epsilon_k}{E_k \epsilon_k}. \tag{7.136}$$

An explicit expression for Ψ_G in the second-quantization language is convenient for performing calculations, but it is not considered suitable for visualizing what is happening in real space. In any case, the setup was developed for the weakly interacting state with large Cooper pairs. It was noticed by A. Leggett that the most probable component of the wavefunction containing exactly N particles has a very transparent structure in real space:

$$\Psi_{G,N}(\mathbf{r}_1,\ldots,\mathbf{r}_N) = [\phi(\mathbf{r}_1,\mathbf{r}_2)\phi(\mathbf{r}_3,\mathbf{r}_4)\cdots\phi(\mathbf{r}_{N-1},\mathbf{r}_N)]_{AS}. \tag{7.137}$$

To construct it, one takes the product of $N/2$ localized singlet pair orbitals, $\phi(\mathbf{r}_{i-1},\mathbf{r}_i)$, and applies the antisymmetrization procedure (denoted by the subscript AS). The remarkable result is that the spatial dependence of the orbital at short distances, $\phi(\mathbf{x}) = (2\pi\hbar)^{-3}\int d\mathbf{k}(v_k/u_k)e^{i\mathbf{k}\cdot\mathbf{x}}$, automatically reproduces the correct shape of the molecular bound state $\phi_{\mathrm{mol}}(\mathbf{x}) = e^{-x/a}/\sqrt{2\pi a x}$ for small positive $k_F a$. Since in this limit different molecular orbitals essentially do not overlap, the ground state smoothly approaches that of the molecular BEC. This consideration is a strong microscopic argument in favor of a single unified picture of superfluidity in bosonic and fermionic systems.

References

1. J. S. Langer, Coherent states in the theory of superfluidity, *Phys. Rev.* **167**, 183 (1968); Coherent states in the theory of superfluidity. II. Fluctuations and irreversible processes, *Phys. Rev.* **184**, 219 (1969).

2. W. Kohn, Theory of the insulating state, *Phys. Rev.* **133**, A171 (1964).

3. A. J. Leggett, Can a solid be "superfluid"?, *Phys. Rev. Lett.* **25**, 1543 (1970).

4. L. N. Cooper, Bound electron pairs in a degenerate Fermi gas, *Phys. Rev.* **104**, 1189 (1956).

5. J. Bardeen, L. N. Cooper, and J. Schrieffer, Theory of superconductivity, *Phys. Rev.* **108**, 1175 (1957).

6. L. D. Landau, On the theory of the Fermi liquid, *Sov. Phys. JETP* **8**, 70 (1959).

7. L. Pitaevskii, On the superfluidity of liquid ^{3}He, *Sov. Phys. JETP* **10**, 1267 (1960).

8. M. Yu. Kagan and A. V. Chubukov, Possibility of a superfluid transition in a slightly nonideal Fermi gas with repulsion, *JETP Lett.* **47**, 614 (1988).

9. M. A. Baranov, D. V. Efremov, M. Yu. Kagan, H. W. Capel, and M. S. Marenko, Strong coupling corrections in a superfluid Fermi gas with repulsion, *JETP Lett.* **59**, 290 (1994).

10. H. Fröhlich, Theory of the superconducting state. I. The ground state at the absolute zero of temperature, *Phys. Rev.* **79**, 845 (1950).

11. V. V. Tolmachev and S. V. Tiablikov, A new method in the theory of superconductivity, *Sov. Phys. JETP* **7**, 46 (1958).

12. A. J. Leggett, Cooper pairing in spin-polarized Fermi systems, *J. Phys. Colloq.* **41**, C7–19 (1980).

13. S. Giorgini, L. P. Pitaevskii, and S. Stringari, Theory of ultracold atomic Fermi gases, *Rev. Mod. Phys.* **80**, 1215 (2008).

14. G. A. Baker, Jr, Neutron matter model, *Phys. Rev. C* **60**, 054311 (1999).

15. H. Heiselberg, Fermi systems with long scattering lengths, *Phys. Rev. A* **63**, 043606 (2001).

CHAPTER 8

Path Integral Representation

Quantum mechanics is often contrasted with its classical counterpart. This distinction is nearly "washed out" when quantum statistics for bosonic systems are formulated in terms of Feynman's path integrals [1]. These provide a unique possibility of projecting subtle quantum mechanical properties into the language of classical statistics of closed trajectories in $(d+1)$-dimensional space. We note, however, that the corresponding projection and an intuition associated with it require that the statistics of path integrals be sign-positive (which is not the case for the generic fermionic system). Indeed, for the positive definite statistics, one can appeal to properties of "typical" configurations contributing to the answer and then estimate probabilities of atypical configurations to occur. If configurations contribute to the statistics with the sign or, in a more general case, with phase factors, then the notion of typical and relevant configurations is misleading—configurations with large modulus and opposite sign cancel each other to a large degree in the final answer, and only their combined effect has a physical meaning.

We choose to introduce path integrals for lattice models first because they emerge naturally from the interaction representation of the statistical operator expanded into a series over hopping terms. It is also much easier to prove several important results about superfluid states starting from the lattice model: Continuous space results are easily recovered by considering the limit of the lattice constant taken to zero. Finally, lattice path integrals provide a suitable formulation for exact numeric algorithms to simulate them. Continuous path integrals will be discussed next with special emphasis on similarities and crucial differences between the lattice and continuous models.

8.1 Lattice Path Integrals

To be specific, we consider an extended Bose–Hubbard model

$$H = -t \sum_{\langle ij \rangle} b_j^\dagger b_i + \frac{1}{2} \sum_{i,j} U_{ij}\, n_i n_j - \sum_i \mu_i n_i, \tag{8.1}$$

which is a generalization of the Bose–Hubbard Hamiltonian (7.90) to include arbitrary pairwise potential U_{ij} and, in general, an arbitrary inhomogeneous external field μ_i. In what follows, we denote the hopping terms as K, split the Hamiltonian into the kinetic and potential energy parts, $H = K + V$, and work with the Fock basis set of site occupation numbers $|\alpha\rangle = |\{n_i\}\rangle$. In this basis, V is diagonal,

$V_{\alpha\gamma} = V_\alpha \delta_{\alpha\gamma}$, and K is off-diagonal, $K_{\alpha\alpha} = 0$. The statistical operator can then be written identically as

$$
e^{-\beta H} = e^{-\beta V} T_\tau \exp\left\{ -\int_0^\beta d\tau K(\tau) \right\}, \tag{8.2}
$$

where $K(\tau) = e^{\tau V} K e^{-\tau V}$ and T_τ is a shorthand notation for the imaginary-time ordering of operators in the exponent:

$$
e^{-\beta H} = e^{-\beta V}\left(1 - \int_0^\beta d\tau\, K(\tau) + \int_0^\beta \int_0^{\tau_2} d\tau_1 d\tau_2\, K(\tau_2) K(\tau_1) + \cdots \right.
$$
$$
\left. + (-1)^m \int_0^\beta \cdots \int_0^{\tau_2} d\tau_1 \ldots d\tau_m\, K(\tau_m)\ldots K(\tau_1)\ldots \right). \tag{8.3}
$$

This expression is all we need to proceed.

The partition function is obtained by taking the trace of Equation (8.3) over the Fock basis. Each m-th order term in the expansion takes the form of a product of matrix elements after one inserts sums over the complete basis set, $\sum_{\alpha_k} |\alpha_k\rangle\langle\alpha_k|$, or identity operators, between all K operators (to simplify notations we write $\langle\alpha_i|V|\alpha_i\rangle = V_i$, $\langle\alpha_i|K|\alpha_{i+1}\rangle = K_{i,i+1}$, and $d^m\tau \equiv d\tau_1 d\tau_2 \ldots d\tau_m$)

$$
(-1)^m d^m\tau\, e^{-(\beta-\tau_m)V_m} K_{m,m-1}\, e^{-(\tau_m-\tau_{m-1})V_{m-1}} \ldots e^{-(\tau_2-\tau_1)V_1} K_{1,0}\, e^{-\tau_1 V_0}, \tag{8.4}
$$

with the periodic boundary condition in imaginary time $\alpha_m \equiv \alpha_0$ to reflect the trace condition. Consider a particular sequence of basis states $\alpha_m, \alpha_{m-1}, \ldots, \alpha_1, \alpha_0$. Since K operators change the state of the system by shifting one boson to the nearest-neighbor site only, a couple of occupation numbers are different between the two consecutive basis states. It is convenient then to characterize the sequence of states in this expression by specifying an "evolution" or "imaginary-time trajectory" of occupation numbers $\{n_i(\tau)\}$. For example, the trajectory shown in Figure 8.1 describes one of the fourth order terms and contributes $t^4 d^4\tau 1 \cdot 2 \cdot \sqrt{2} \cdot \sqrt{2} \exp\left\{ -\int_0^\beta dt V(t) \right\}$ to the answer, where $V(t)$ is the potential energy of the state $|\{n_i(t)\}\rangle$. We also explicitly mention values of the hopping matrix elements $\langle n_i - 1, n_j + 1| - t b_j^\dagger b_i |n_i, n_j\rangle = -t\sqrt{n_i(n_j+1)}$. In other words, the trace of Equation (8.3) can be written as a sum/integral over all possible paths $\{n_i(\tau)\}$ such that $n_i(\beta) = n_i(0)$

$$
Z = \sum_{\{n_i(\tau)\}} W[\{n_i(\tau)\}], \tag{8.5}
$$

$$
W[\{n_i(\tau)\}] = (-1)^m d^m\tau \prod_{i=0}^{m-1} K_{i+1,i} \exp\left\{ -\int_0^\infty dt V(t) \right\}, \tag{8.6}
$$

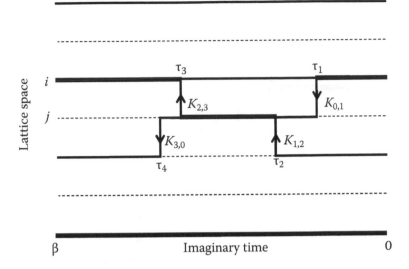

Figure 8.1   Occupation number trajectory for the partition function with four kinks. The line width is proportional to the occupation number.

with strict rules relating trajectory shape to its contribution to the partition function, or weight $W\left[\{n_i(\tau)\}\right]$, which depends on the initial state, expansion order, time ordering of hopping events (kinks), and the corresponding product of matrix elements. The path integral language at this point is nothing but a convenient way of visualizing each term in the perturbative expansion (8.3).

The trajectory weight is always sign-positive if t is positive, that is, if matrix elements of K are negative or the lattice is bipartite (in this case only even values of m contribute to the answer). The other important observation is that due to the particle number conservation, the many body trajectory can be always decomposed into the set of closed (in the time direction) single-particle trajectories, or worldlines.* World lines can "wind" around the β-circle several times before closing on themselves because particles are identical. World lines with nonzero β-windings are said to form exchange cycles and will be shown shortly to be responsible for the nonclassical response of the system to rotation.

Let us now look at Green's function of the system

$$G\left(\tau_M-\tau_I, \mathbf{j}, \mathbf{i}\right)=\left\langle T_\tau b_{\mathbf{j}}^{\dagger}\left(\tau_M\right) b_{\mathbf{i}}\left(\tau_I\right)\right\rangle=Z^{-1}\operatorname{Tr}\left[e^{-\beta H}T_\tau b_{\mathbf{j}}^{\dagger}\left(\tau_M\right) b_{\mathbf{i}}\left(\tau_I\right)\right]. \qquad (8.7)$$

If we employ the same approach and expand the statistical operator in powers of K, we will obtain, term after term, equations similar to (8.4) with the only difference that there will be additional operators $b_{\mathbf{j}}^{\dagger}$ at point τ_M and $b_{\mathbf{i}}$ at point τ_I. Turning back to the graphical representation of series in terms of occupation number paths, we observe that on the time interval (τ_I, τ_M) there is one more particle present in the system than on the (τ_M, τ_I) interval (on the β-circle). It means, in

---

* On a lattice, this decomposition is not unique; when on some site $n_i(\tau) > 1$, one is free to decide which world line will leave this site first.

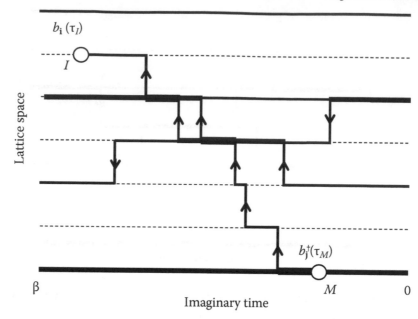

Figure 8.2   A typical trajectory for the Green's function. The end points of the correlation function are labeled as $\mathcal{I}$ and $\mathcal{M}$. The line width is proportional to the occupation number.

particular, that when the $\{n_i(\tau)\}$ evolution is decomposed in terms of single particle world lines, there will always be one *open* world line originating at point $\tau_{\mathcal{M}}$ and terminating at point $\tau_{\mathcal{I}}$ (see Figure 8.2). For brevity, these special points will be labeled as $\mathcal{I}$ and $\mathcal{M}$ throughout the text. According to the standard rules for bosonic creation and annihilation operators, there are additional factors in the trajectory weight $\sqrt{n_i(\tau_{\mathcal{I}} - 0)}$ and $\sqrt{n_j(\tau_{\mathcal{M}} - 0) + 1}$ associated with the $\mathcal{I}$ and $\mathcal{M}$ points, respectively.

In what follows, we will use shorthand notations: Z-path for trajectories contributing to the partition function and G-path for trajectories contributing to Green's function. The corresponding Z- and G-configuration spaces are continuously connected since to go from one to another all we have to do is to "draw" a short piece of a new world line or "erase" a small part of the existing trajectory. In Section 8.7, we will show how this feature is utilized in the construction of the worm algorithm for efficient Monte Carlo (MC) simulations of bosonic path integrals.

8.2   Lattice Gauge Field and the Winding Number Formula

Lattice gauge field is introduced to the model by adding phase factors to hopping amplitudes

$$t b_j^\dagger b_i + \text{H.c.} \longrightarrow \left(t e^{i a_{ij}} \right) b_j^\dagger b_i + \text{H.c.} \tag{8.8}$$

The physics of the model cannot be changed by adding phase factors to bosonic operators, $b_i \to e^{i\theta_i} b_i$, where θ_i is an arbitrary field specified on lattice sites, since this can be ascribed to the redefinition of the Fock basis set. Thus, gauge fields related by the transformation

$$a_{ij} \to a_{ij} + \theta_i - \theta_j \tag{8.9}$$

should be regarded as physically identical. A convenient way of parameterizing lattice bonds is to use i, α instead of $\langle ij \rangle$, where α enumerates bonds attached to the site i with the shorthand notation $\mathbf{g}_\alpha = \mathbf{j} - \mathbf{i}$ for the vector connecting the nearest-neighbor sites along the bond α.

To connect with the standard way of introducing gauge field in continuous space $\nabla \to \nabla - i\mathbf{a}$, one has to consider small phase fields that are smooth functions of $\mathbf{i}$, that is, $|a_{i,\alpha}| \ll 1$ and $|a_{i+g,\alpha} - a_{i,\alpha}| \ll |a_{i,\alpha}|$. It is a simple exercise then to show that in the long-wave limit the lattice gauge model is approaching the continuous one. In this limit, the continuous version of Equation (8.9) takes a familiar form $a_\alpha \to a_\alpha + \mathbf{g}_\alpha \cdot \nabla \theta$.

Problem 8.1 *Consider a simple cubic lattice with a smooth gauge phase field $\varphi_{i\alpha} \equiv l\mathbf{a}(\mathbf{r})$, where l is the lattice constant, $\mathbf{r} = l\mathbf{i}$, and $\mathbf{a}$ is a smooth vector field. By considering the limit $l \to 0$ and $t \to \infty$ in such a way that $tl^2 = 1/2m = \text{const}$, show that the continuous space result for the kinetic energy density is recovered, $K \to -\frac{1}{2m} b_r^\dagger (\nabla + i\mathbf{a})^2 b_r - 6t b_r^\dagger b_r$. [The last term defines a "counting zero" for the chemical potential and has no physical significance.]*

Of particular interest to us is the system's response to a weak homogeneous gauge field $a_{i,\alpha} = \varphi_\alpha \cdot \mathbf{g}_\alpha / L_\alpha$, where L_α is the system size in the α-direction. We immediately recognize that using Equation (8.9) with $\theta_i = \varphi_\alpha \cdot i_\alpha / L_\alpha$, this field can be gauged out completely in a system with closed (hard wall) boundary conditions. If we try to do the same in a system with periodic boundary conditions in α-direction, we will not be able to remove the gauge field on bonds crossing the system boundary since for these bonds the θ field is already fixed by the requirement that a is zero everywhere else. Thus, on all bonds crossing the system boundary, we will have $a_\alpha = \theta_{i_\alpha = L_\alpha} = \varphi_\alpha$. As discussed in Chapter 1, it is customary to refer to this specific choice of the gauge as "twisted boundary condition" with the twist phase φ_α.

Let us see now how a system's response to the homogeneous gauge field (equivalent to setting the system in motion) is transcribed into the geometrical properties of particle world lines. To this end (for Z-paths), we introduce the notion of the winding number in the space direction $\hat{\alpha}$. Suppose we draw a plane, or cross-section, perpendicular to the $\hat{\alpha}$-axis and count how many times in a given trajectory particles have crossed this plane from left to right, $N_\to^\alpha$. Similarly, we count all crossings from right to left, $N_\leftarrow^\alpha$. The difference $W_\alpha = N_\to^\alpha - N_\leftarrow^\alpha$ is the total flux of particles through the cross-section, which is called a winding number.[*] Similarly, one can introduce winding numbers in other directions to define an integer

---

[*] This winding number should not be confused with the winding number of the phase [cf. Equations (3.14) and (3.15)].

vector **W**. The continuity of world lines (or, equivalently, particle number conservation) guarantees that W_α does not depend on the position of the plane along the $\hat{\alpha}$-axis, that is, W_α is the *global topological* property of the trajectory that cannot be changed by small local modifications of the shape. We immediately recognize that the phase factor associated with the Z-path is given by

$$e^{i\sum_\alpha \varphi_\alpha W_\alpha}. \tag{8.10}$$

Indeed, W_α is counting the imbalance $N_\rightarrow^\alpha - N_\leftarrow^\alpha$ between the number of hopping transitions on bonds crossing the system boundary and each contributing a factor $e^{\pm i\varphi_\alpha}$. [According to Equation (8.4), the trajectory weight is a product of hopping matrix elements.] We readily see here the one-to-one correspondence with the phase factor obtained for the high-temperature expansion configuration of the classical XY model, Equation (3.14).

One can arrive at the same final result by noticing that the trajectory phase is given by the sum of phases of all hopping transitions in a given many-body trajectory since hopping matrix elements multiply in the weight [see Equation (8.4)]. Thus, the phase factor is

$$\exp\left\{ i \sum_\alpha \sum_{g_{tr}} \varphi_\alpha g_\alpha / L_\alpha \right\}, \tag{8.11}$$

where the sum runs over all displacements in a given trajectory (or its decomposition into world lines). The continuous space version is a straightforward generalization of the sum to the integral, $\exp\left\{ i \sum_\alpha \int_{tr} \varphi_\alpha dr_\alpha / L_\alpha \right\}$. Both Equation (8.11) and its continuous version remain valid as written when the constants $\varphi_\alpha / L_\alpha$ are replaced with an arbitrary field $a_{i,\alpha}$. We will use this fact later in the derivation of the nonclassical moment of inertia formula. We see that for $\varphi_\alpha = \varphi = \text{const}$, the trajectory phase is proportional to the total "displacement" of the trajectory in the direction of $\hat{\alpha}$. In a system with periodic boundary conditions, this displacement has to be a multiple of L_α in the α-direction leading to the winding number expression (8.10).

We are ready to derive an important relation between the statistics of winding numbers and the superfluid stiffness. We split the sum in the partition function (8.5) into contributions from trajectories with the same winding number

$$Z = \sum_\mathbf{W} e^{i\sum_\alpha \varphi_\alpha W_\alpha} Z_\mathbf{W}, \tag{8.12}$$

and use it to calculate $\partial^2 F / \partial\varphi_\alpha \partial\varphi_\gamma|_{\varphi=0}$. At this point, the derivation is identical to what was done in Section 3.2 for the XY model resulting in

$$\left. \frac{\partial^2(-T\ln Z)}{\partial\varphi_\alpha \partial\varphi_\gamma} \right|_{\varphi=0} = -T \left. \frac{\partial}{\partial\varphi_\alpha} \frac{\sum_\mathbf{W}(iW_\gamma)e^{i\sum_\delta \varphi_\delta W_\delta} Z_\mathbf{W}}{Z} \right|_{\varphi=0}$$

$$= T\left[\langle W_\alpha W_\gamma \rangle - \langle W_\alpha \rangle \langle W_\gamma \rangle \right]. \tag{8.13}$$

We thus obtain an exact relation between the average winding number squared and the superfluid stiffness tensor:

$$(\Lambda_s)_{\alpha\gamma}^{(\varphi)} = \frac{TL_\alpha L_\gamma}{V}\langle W_\alpha W_\gamma\rangle. \tag{8.14}$$

In the d-dimensional system with the isotropic superfluid stiffness and cubic shape

$$\Lambda_s^{(\varphi)} = \frac{T}{dL^{d-2}}\langle \mathbf{W}^2\rangle. \tag{8.15}$$

This formula was first introduced by Pollock and Ceperley [2] (in three dimensions, there is no difference between the $\Lambda_s^{(\varphi)}$ and the superfluid stiffness).

8.3 Continuous-Space Path Integrals

Consider now an interacting Hamiltonian for N *distinguishable* particles in continuous space

$$H = K + V = -\frac{1}{2m}\sum_{i=1}^{N}\nabla_i^2 + \frac{1}{2}\sum_{i\neq j}^{N}U\left(\mathbf{r}_i - \mathbf{r}_j\right) + \sum_{i=1}^{N}[v\left(\mathbf{r}_i\right) - \mu]. \tag{8.16}$$

We start with the standard expression for the partition function and choose to work in the real-space representation. To simplify notations, we will use the dN-dimensional vector $\mathbf{R} = (\mathbf{r}_1, \mathbf{r}_2, \ldots, \mathbf{r}_N)$ to specify coordinates of all particles. Then,

$$Z = \int d\mathbf{R}_0 \langle \mathbf{R}_0 | e^{-\beta H} | \mathbf{R}_0\rangle \equiv \int d\mathbf{R}_0\, \rho(\mathbf{R}_0, \mathbf{R}_0, \beta), \tag{8.17}$$

which can be interpreted as a trace of the evolution operator in imaginary time β. Next, we split the exponential into a product of M identical terms $e^{-\beta H} = e^{-\epsilon H}e^{-\epsilon H}\ldots e^{-\epsilon H} = \left[e^{-\epsilon H}\right]^M$, with $\epsilon = \beta/M$, and introduce real-space matrix elements of the evolution operator in a shorter time τ to write the trace as a product

$$Z = \prod_{k=0}^{M-1}\left(\int d\mathbf{R}_i\right)\rho(\mathbf{R}_0, \mathbf{R}_1, \epsilon)\,\rho(\mathbf{R}_1, \mathbf{R}_2, \epsilon)\ldots\rho(\mathbf{R}_{M-1}, \mathbf{R}_M, \epsilon), \tag{8.18}$$

where $\mathbf{R}_M \equiv \mathbf{R}_0$ to satisfy the trace condition. This procedure is known as imaginary time "slicing." In the absence of interactions, these matrix elements describe an evolution of a free particle between two points in the dN-dimensional space

$$\rho^{(0)}(\mathbf{R}_i, \mathbf{R}_{i+1}, \epsilon) = \langle \mathbf{R}_i | e^{-\epsilon K} | \mathbf{R}_{i+1}\rangle = \left(\frac{m}{2\pi\epsilon}\right)^{dN/2}\exp\left\{-\frac{m(\mathbf{R}_{i+1} - \mathbf{R}_i)^2}{2\epsilon}\right\}. \tag{8.19}$$

When interactions are included in the picture, an exact expression for finite ϵ is no longer possible in the general case. However, the freedom of choosing ϵ can be used to make errors in the calculation of the evolution operator matrix elements arbitrarily small. We will not elaborate here on advanced schemes that make the corresponding errors vanish as high powers of ϵ (an extensive discussion of this issue can be found in Reference [3]; see also Reference [4]). The procedure discussed in the following introduces errors that vanish as ϵ^3 and thus becomes exact in the $\epsilon \to 0$ limit. All we must do is use an approximate factorization of exponentials, $e^{-\epsilon H} \approx e^{-\epsilon V/2} e^{-\epsilon K} e^{-\epsilon V/2}$ leading to

$$\rho(\mathbf{R}_i, \mathbf{R}_{i+1}, \epsilon) \approx \rho^{(0)}(\mathbf{R}_i, \mathbf{R}_{i+1}, \epsilon) e^{-\epsilon[V(\mathbf{R}_i)+V(\mathbf{R}_{i+1})]/2}. \tag{8.20}$$

Substituting this form back to Equation (8.18), we obtain

$$Z = \left(\frac{m}{2\pi\epsilon}\right)^{dNM/2} \prod_{k=0}^{M-1} \left(\int d\mathbf{R}_i\right) \exp\left\{-\sum_{i=0}^{M-1} \left[\frac{m(\mathbf{R}_{i+1} - \mathbf{R}_i)^2}{2\epsilon} + \epsilon V(\mathbf{R}_{i+1})\right]\right\}. \tag{8.21}$$

The collection of points $\mathbf{R}_i$ forms a many-body trajectory in discrete imaginary time $\tau_i = \epsilon i$ (see also Section 8.7). Correspondingly, the product of integrals (with normalization factors) over particle positions on all time slices represents the path integral with the shorthand notation $\oint \mathcal{D}\mathbf{R}(\tau)$. In the limit of $\epsilon \to 0$, it is convenient to use the notion of "velocity" along the trajectory $\dot{\mathbf{R}}(\tau) = \lim_{\epsilon \to 0}[\mathbf{R}(\tau) - \mathbf{R}(\tau - \epsilon)]/\epsilon$ and replace the sum over time slices with the integral to express Equation (8.21) as

$$Z = \oint \mathcal{D}\mathbf{R}(\tau) \exp\left\{-\int_0^\beta \left[\frac{m\dot{\mathbf{R}}^2}{2} + V(\mathbf{R})\right] d\tau\right\}. \tag{8.22}$$

This completes the derivation of the continuous path integral for distinguishable particles. Note, however, that typical trajectories contributing to (8.22), while being continuous, have little to do with smooth functions of time because, in the $\epsilon \to 0$ limit, one finds that $\langle[\mathbf{R}(\tau) - \mathbf{R}(\tau - \epsilon)]^2\rangle \sim \epsilon$, that is, the trajectory is diffusive at short time scales [cf. the discussion of singular behavior of $\psi(\tau)$ in functional integrals discussed following Equation (7.72)].

For indistinguishable particles, only minor modifications are required. In the bosonic case, Equation (8.22) remains valid as written although the configuration space of allowed trajectories $\mathbf{R}(\tau)$ is now enlarged to include exchange cycles. The periodic boundary condition $\mathbf{R}(0) = \mathbf{R}(\beta)$ for distinguishable particles requires $\mathbf{r}_i(0) = \mathbf{r}_i(\beta)$ and the many-body path is a collection of N individual single-particle trajectories/world lines. For identical particles, the periodic boundary condition in time is satisfied when the set of coordinates at time $\tau = \beta$ equals to an arbitrary permutation, $\{\mathbf{r}_i(\beta)\} = \mathcal{P}\{\mathbf{r}_i(0)\}$, of the corresponding $\tau = 0$ set. It means, in particular, that when the many-body path is decomposed into world lines (see Figure 8.3), some of them may wind on the β-cylinder several times before closing on themselves, forming exchange cycles. The only difference between fermionic

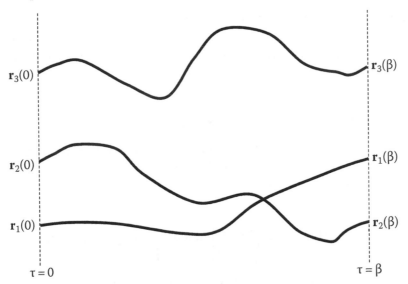

Figure 8.3 Continuous-space world lines with exchange cycle formed by two particles. The $\tau = \beta$ set of coordinates is obtained by permuting coordinates $\mathbf{r}_1(0)$ and $\mathbf{r}_2(0)$ places. In the fermionic system, the corresponding configuration will contribute to the statistics with a negative sign.

and bosonic systems is in the sign rule: for fermions the trajectory contribution to the path integral involves an additional factor $(-1)^p$, where p is the parity of permutation $\mathcal{P}$.

One can easily notice that geometrical properties of lattice and continuous path integrals are very close. The only distinction in dimensions $d > 1$ is that continuous many-body paths have zero measure to intersect in the spacetime volume and thus their decomposition into single-particle world lines is unique. In any case, one may always think of continuous paths as the limiting case of the dilute lattice model (lattice spacing much smaller than the inter-particle separation and the de Broglie wavelength). This, in particular, allows us to skip the derivation of how a gauge field, introduced as $\nabla \rightarrow \nabla - i\mathbf{a}(\mathbf{r})$ in Equation (8.16), contributes to the trajectory phase factor [see discussion after Equation (8.11)]. In the most general form, it reads

$$\exp\left\{i \int_{tr} \mathbf{a} \cdot d\mathbf{R}\right\} \equiv \exp\left\{i \int_0^\beta \mathbf{a} \cdot \dot{\mathbf{R}}\, d\tau\right\}. \tag{8.23}$$

Since the meaning of winding numbers (and their properties) remains the same, Equations (8.14) and (8.15) for the superfluid stiffness hold true. Here, we would like to stress that the superfluid density is determined by macroscopic exchange cycles that span the entire system from one end to another with nonzero $\mathbf{W}$. Inter-particle exchange cycles, which happen only locally with significant probability, do *not* contribute to n_s since they cannot change the winding number.

Another point worth mentioning is that positive definite statistics of path integral trajectories for a rather general class of Hamiltonians (standard kinetic energy term plus arbitrary interparticle interactions) immediately translate into sign-positive density matrix.

8.4   Nonclassical Moment of Inertia Formula

Suppose now that the gauge field corresponds to that originating from the rotation about the $\hat{z}$-axis with angular velocity Ω. In this case, we have $\mathbf{a} = m\,\Omega \times \mathbf{r}$ giving rise to the trajectory phase factor

$$\exp\left\{ im \int_{tr} \Omega \times \mathbf{R} \cdot d\mathbf{R} \right\} = \exp\left\{ im\,\Omega \int_{tr} \mathbf{R} \times d\mathbf{R} \right\} = \exp\{i2mA_z\Omega\}, \qquad (8.24)$$

where A_z is the algebraic area produced by the closed trajectory when it is projected on the plane perpendicular to the rotation axis.

The rest is identical to what we did in the derivation of the winding number formula. One starts by writing

$$Z = \int dA_z e^{i2mA_z\Omega} Z_{A_z}, \qquad (8.25)$$

and proceeding with the evaluation of $\partial^2 F/\partial\Omega^2|_{\Omega=0} = I_z^{(s)} = I_z^{cl} - I_z$, where, by definition, $I_z^{(cl)} = \int d\mathbf{r}n(r)\,mz^2$ is the classical moment of inertia and $I_z^{(s)}$ is the nonclassical component. The result is [5]

$$I_z^{(s)} = \left.\frac{\partial^2(-T\ln Z)}{\partial\Omega^2}\right|_{\Omega=0} = 4m^2T\langle A_z^2\rangle, \qquad (8.26)$$

where we have used the fact that $\langle A_z\rangle = 0$ due to symmetry between the clockwise- and counterclockwise-spiraling world lines. In general, similarly to the superfluid stiffness, $I^{(s)}$ is a tensor quantity.

This formula is interesting in several ways. First, we observe that any quantum system, even in the normal state, has a nonclassical moment of inertia that does not scale to zero as an exponential function of the system size. Indeed, assuming that exchange cycles are absent and each particle can produce either a clockwise- or counterclockwise-spiraling world line with the projected area $\sim \lambda_T^2$ (here $\sim \lambda_T^2$ is the typical particle displacement in time β; e.g., the de Broglie wavelength in the gaseous state), we estimate $I_z^{(s)}$ to be $\sim m^2TN\lambda_T^4 \sim L^d$. Local exchange cycles do not change this scaling law but may significantly increase the prefactor. On one hand, in small clusters, the corresponding correction may have observable effects even when the same system in the thermodynamic limit is normal (see, e.g., Reference [6] for the experimental technique of measuring the moment of inertia of small droplets using spectroscopy of rotational spectrum [6] and the recent work on hydrogen clusters [7]). On the other hand, in the limit of large system sizes, we find a negligible correction to the classical moment of inertia that scales as $I_z^{(s)}/I^{(cl)} \sim 1/L^2$ and vanishes in the thermodynamic limit. We note also that approximate models of local motion that do not account for the finite area of projected particle world lines, such as two-level systems modeled by two sites with only one transition amplitude between them, do not contribute to $I^{(s)}$ at all. Thus, macroscopic exchange cycles are required to reduce the moment of inertia by a finite fraction.

In superfluid systems, $I^{(s)}/I^{(cl)}$ is finite in the thermodynamic limit. The relation between the superfluid density and moment of inertia within the hydrodynamic approach for a homogeneous (at least at mesoscopic scales) media is straightforward: $I^{(s)}/I^{(cl)} = n_s/n$. This relation can be obtained from the area formula (8.26) for a simple annulus geometry when the notion of winding numbers (introduced for systems with periodic boundary conditions) is well defined—in this particular case, in the azimuthal direction.

Problem 8.2 *Consider a cylinder of inner radius R, outer radius $R + d$, and length L_z rotating about the $\hat{z}$-axis. Assume that $d \ll R$ and consider d/R as a small parameter. Derive the $I_z^{(s)}/I_z^{(cl)} = n_s/n$ formula using winding numbers in the azimuthal direction to calculate $\langle A_z^2 \rangle$.*

8.5   Green's Function, Density Matrix, and Superfluidity

In Chapter 7, we briefly discussed an important connection between the superfluidity, the connectivity of the ground-state wavefunction, and the decay of the single-particle density matrix (assuming that pairing of particles is not involved). We will now extend this discussion to finite temperatures using path integrals for bosonic systems that present a clear topological connection between superfluidity and Green's function *regardless* of the system behavior at short scales. Unless we mention it specifically otherwise, all considerations apply equally to lattice and continuous systems.

In Figure 8.4, we show a typical path contributing to Green's function in the superfluid state. The crucial observation is that, in the thermodynamic limit, the

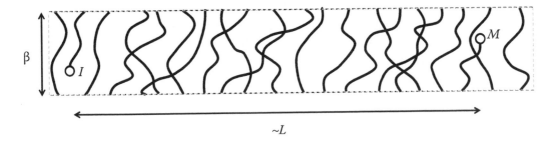

Figure 8.4   Green's function path in the superfluid state with the end points at a macroscopic distance.

statistics are dominated by configurations with $\mathcal{I}$ and $\mathcal{M}$ being separated by a macroscopic distance while β is finite (no matter how small). This result follows from the intimate connection between Green's function decay in space and statistics of winding numbers. Consider some Z-path with zero-winding number and perform a small modification transforming it to the G-configuration (by "drawing" or "erasing" a short piece of a world line). Since there is always a single-particle world line starting at $\mathcal{M}$ and terminating at $\mathcal{I}$, this will produce a spatial dependence of the winding number in the $\mathcal{I}-\mathcal{M}$ direction; $|W|$ is increased by one when the cross-section used to calculate the flux of trajectories cuts this world line. As $\mathcal{I}$ and $\mathcal{M}$ separate in space according to the statistics dictated by $G(\mathbf{r},\tau)$, the region with $W \neq 0$ keeps growing. To make a nonzero winding number in the Z-path, the ends of the disconnected world line must go around the system with periodic boundary conditions and reconnect, that is, they inevitably have to be found at a distance comparable to the system size. This observation agrees with the previous conclusion [see Equation (7.98)] that $\rho(r) = G(r,0)$ cannot decay faster $r^{-1/2}$ and its normalization integral diverges in the superfluid state. This, in particular, implies that at large distances, we always have either a perfect order or at least an algebraic order (slow power-law decay) in the density matrix. Both cases are found in nature; in 3D, superfluidity comes along with the spontaneous symmetry-breaking and finite condensate fraction; in 2D, superfluidity is accompanied by the slow power-law decay of $\rho(\mathbf{r})$.

An immediate consequence of the physical picture llustrated in Figure 8.4 is that, in momentum space, $G(\mathbf{p}=0,\tau)$ is τ-independent, disregarding small finite-size effects. Due to periodic boundary conditions in time, any value of $\tau_{\mathcal{M}}$ is equally probable unless it is correlated with $\tau_{\mathcal{I}}$. Since statistically $\mathcal{I}$ is found infinitely far from $\mathcal{M}$ (in the thermodynamic limit), such correlations are indeed absent. In Figure 8.5, we show the result of the path integral Monte Carlo (PIMC) simulation done for the ^{4}He superfluid at low temperature $T = 1$ K. The residual tiny τ dependence is a finite-size effect in the system with about $N = 1000$ atoms. This agrees with another rigorous statement about superfluid ground states—they are gapless with regard to adding/removing particles.

The other quantity of interest is the condensate density $n_0 = G(\mathbf{p} = 0, \tau = 0)$. On one hand, in the superfluid state, the number of particles in the condensate is at least as large as $\sim N^{1-1/2d}$. On the other hand, we can reverse the main argument and present yet another proof that systems with finite condensate fraction are always superfluid (again, disregarding finite-size effects and the pathological case of the ideal Bose gas). Finite value of n_0 implies that the statistics of spatial positions of $\mathcal{I}$ relative to $\mathcal{M}$ are those of an unbiased random walk at large distances. According to this random walk analogy, $\mathcal{I}$ and $\mathcal{M}$ will come close to each other in $\propto L^d$ "steps" (displacements in time β) by sweeping the system volume across $\propto L^d/L^2$ times and thus changing the winding number value by $\propto \sqrt{L^{d-2}}$ by the central limit theorem. To see this argument more clearly, tile the infinite space with identical system-volume cells and estimate how far $\mathcal{I}$ can go away from $\mathcal{M}$ before hitting $\mathcal{M}$ (or its identical image). The typical number of steps in the $Z \to G \to G \cdots \to G \to Z$ cycle is L^d and the displacement is

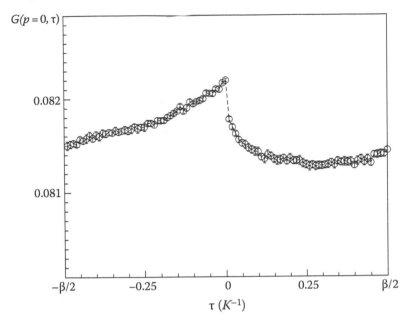

Figure 8.5 Green's function of the ⁴He superfluid at $T = 1\ K$ and $\mu = -7.35\ K$ in a finite system with about $N = 1000$ atoms normalized to $n = 0.02184$ nm^{-3}. Notice the small vertical scale. (Adapted from Reference [11].)

$WL \propto \sqrt{L^d}$. Since winding numbers introduce a systematic tilt of the world line trajectory in imaginary time, $\dot{R} \propto WL/L^d$, they increase the system's action by $\delta S_W \propto \dot{R}^2 L^d \propto W^2 L^2/L^d$. For $W \propto \sqrt{L^{d-2}}$, we find that $\delta S_W \sim 1$, meaning that typical winding number fluctuations are realized in a single random walk cycle $Z \to G \to G \cdots \to G \to Z$. This picture leads to $\langle W^2 \rangle \propto L^{d-2}$ and finite superfluid density.

Reversing the argument, finite superfluid density in 3D requires that typical winding numbers are $\sim \sqrt{L}$. This outcome is only possible if statistics of relative distances between $\mathcal{I}$ and $\mathcal{M}$ are flat at large scales, that is, finite superfluid fraction implies finite condensate fraction in 3D. This result can be obtained by mapping the power-law decay of the density matrix $\rho(r) \propto 1/r^b$ to the random walk motion in the presence of an effective L-periodic attractive potential $V_{\mathcal{I}\mathcal{M}}(r)/T \propto b\ln(2r/L)$ and observing that typical displacements of the trajectory in the $Z \to G \to G \cdots \to G \to Z$ cycle are $WL \propto \sqrt{L^{d-b}}$.

Problem 8.3 *Estimate typical particle displacement before it hits the origin of one of the system volume cells during a random walk in the presence of the L-periodic attractive potential $V(r)/T = b\ln(r/L)$. This problem is based on the calculation of the "trapping time," or inverse trapping rate, for diffusive particles. Hint: Trapping time is defined by the equation $dn(L)/dt = -n(L)/\tau_t$, where $n_L(t)$ is the concentration of particles away from traps. Within the quasistatic approach, each trap creates a current flux $j = 4\pi r^2 D[dn(r)/dr + n(r)d(V/T)/dr]$, which is obtained by solving*

$4\pi r^2 D[dn(r)/dr + n(r)d(V/T)/dr] = $ const *with boundary conditions* $n(r = L) = n(L)$ *and* $n(r = a) = 0.$ *Then,* $1/\tau_t = j/n(L).$

An exponential decay of $\rho(\mathbf{r})$ is then an unmistakable signature of the normal state. Here, two distinct scenarios must be considered, disregarding cases when transitions other than the normal-superfluid one take place at lower temperatures. One possibility is that the state is normal due to finite T and will eventually become superfluid at $T < T_c$. In this case, the decay of $\rho(r)$ is temperature dependent (if density is kept constant) because the cost of developing G-configurations with large exchange cycles at high temperature is dominated by the kinetic energy term in the action $\propto Tmn^{-2/d}$ per exchange. It is somewhat counterintuitive that even in the ideal normal Bose gas at temperatures $T \gg T_c$, the decay of the density matrix is exponential, $\rho(r) \propto r^{(1-d)/2}e^{-\sqrt{2m|\mu|}r}$, and *not* Gaussian as for the ideal gas of distinguishable particles, that is, effects of quantum statistics are still dominating at large distances.

Problem 8.4 *Consider an ideal Bose gas at $T \gg T_c$, where T_c is the condensation temperature and then show that the decay of the density matrix is exponential (up to power-law corrections) in space at large distances. At what distance does this law set in?*

The other possibility is that the cost of exchange cycles becomes temperature independent because the optimum of kinetic and potential energy contributions

$$\left[U\tau + mn^{-2/d}/\tau\right]_{\min\tau} \to \tau_{\text{opt}} \sim \sqrt{mn^{-2/d}/U} \qquad (8.27)$$

can be achieved at $\tau_{\text{opt}} \ll \beta$. Of course, this estimate is too rough and is based on the analysis of dimensions only. It does not take into account that the entire shape of the exchange trajectory, that is, in the solid phase, is optimized to minimize the trajectory weight (the optimal trajectory is often called an instanton). However, more refined considerations do not change the essence of the final conclusion that instanton solutions are confined in time. For temperatures such that $\tau_{\text{opt}} \ll \beta$, the exponential decay of the density matrix will become T-independent and characteristic of the ground state.

In the path integral language, it is also nearly obvious that the continuous superfluid-normal transition at finite temperature is always classical. This observation comes from the fact that we are dealing with closed-loop configurations in $(d + 1)$-dimensional box with $L \to \infty$ and $\beta = $ const. In the thermodynamic limit, and after coarse-graining that eliminates small exchange cycles, we arrive at the d-dimensional picture of oriented closed loops formed by large exchange cycles, which is identical to that of the XY model in the high-temperature expansion representation. In the next chapter, we will take this analogy further to discuss superfluid-Mott insulating quantum phase transitions in lattice models.

8.6 General Aspects of Path Integral Monte Carlo

Path integral formulation of quantum statistics has been proved to be an invaluable tool for *ab initio* numerical studies of quantum many-body systems, both on and off the lattice. It was pioneered by Ceperley and Pollock for the super-fluid transition of liquid ^{4}He [3,8]. At least for Bose systems, PIMC is the only presently known universal method capable of producing high accuracy results for various physical properties at finite temperature and in all dimensions, including the superfluid and condensate densities. [For Fermi systems and frustrated lattice models, its convergence can be severely affected by the sign problem to be discussed at the end of this section.] Before discussing specifics of PIMC, let us first briefly review key principles of the MC approach to the evaluation of complex multidimensional sums and integrals. Our description in the following is far from comprehensive and leaves out many details important for practical implementation of algorithms. Our modest goal is to explain the global framework of the MC approach, outline what it can achieve, and mention its most severe limitations.

Suppose one has to evaluate the ratio of two multidimensional sums/integrals of the form

$$\langle A \rangle = Z^{-1} \sum_\nu A(\nu) W(\nu), \qquad Z = \sum_\nu W(\nu), \tag{8.28}$$

where index ν represents the configuration space of a given system and functions A and W are arbitrary. There are no restrictions on the number of discrete and continuous variables necessary to fully parameterize the configuration space; moreover, different configurations may require different variables to specify it. For example, in the grand canonical ensemble, the fluctuating number of particles along with their states is part of the index ν. Similarly, in Equation (8.5), the configuration index ν includes the initial set of occupation numbers, the expansion order, the topology, and the times of hopping events. It is understood that, in general, the complexity of the calculation is such that neither an analytic approach nor a brute force numerical summation/integration is possible. The idea of the MC approach is to interpret Equation (8.28) as averaging of quantity A over the configuration space. For the moment, assume that W is positive definite; then the combination $P(\nu) = W(\nu)/Z$ may be interpreted as the configuration probability because it is positive and normalized to unity. Now, if ν is generated with probability $P(\nu)$, the full sum in Equation (8.28) can be replaced with the stochastic one,

$$\langle A \rangle \approx \sum_\nu^{MC} A(\nu)/N_{MC}, \tag{8.29}$$

where $N_{MC} = \sum_\nu^{MC} 1$ is the number of configurations included in the average. Formally, the procedure is exact in the limit of infinite simulation time, although the convergence is relatively slow and proportional to $1/\sqrt{N_{MC}}$, by the central limit theorem.

A generic algorithm for generating configurations with probabilities $P(\nu)$ was proposed by Metropolis et al. [9] when the time was ripe for the rise of computational physics. The key observation was that subsequent configurations to be included in the MC sum (8.29) may differ only slightly and "evolve" one from another according to the Markov-chain random process that consists of configuration updates. To guarantee that the simulation converges to the exact answer, the set of updates must be ergodic; given some configuration ν_0, any other configuration ν can be obtained in a finite number of updates with nonzero probability. The actual implementation of the algorithm is very simple (see the illustration in Figure 8.6):

1. Start from an arbitrary configuration ν.

2. Propose another configuration ν' that derives from ν by suggesting to change, add, or remove some of the variables in ν. The procedure of getting ν' from ν is called an *update*.

3. Accept the proposed change with probability $R_{\nu \to \nu'}$. At this junction, one must ensure that for any ν accepted to the sum, ν' is accepted $P(\nu')/P(\nu)$ times, on average. If this condition is satisfied, then the distribution of accepted configurations is given by $P(\nu)$, as required. For small changes, the ratio of probabilities can be made sufficiently large without problem in most cases. If the update is accepted, replace ν with ν' (otherwise keep ν unchanged).

4. Include $\mathcal{A}(\nu)$ into the sum, and proceed to step 2.

Technical details are hidden primarily in the structure of updates and acceptance probabilities. At this point, the algorithm is flexible, and one may design various updating schemes for ergodic sampling of a particular configuration space. This freedom may be used to maximize efficiency or minimize programming work. As a rule of thumb, the scheme must aim at generating ν' with probabilities leading to large $R_{\nu \to \nu'}$ without substantial CPU cost.

In the context of statistical physics, the MC method is capable of evaluating virtually any equilibrium property of the system such as energy, density, pressure, specific heat, superfluid density, structure factor, and Green's function. One important limitation, though, is that correlation functions are typically known in

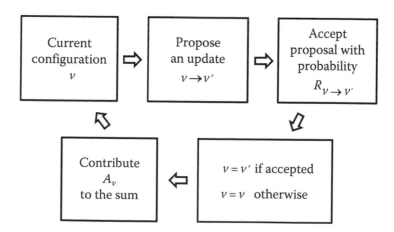

Figure 8.6   Metropolis–Rosenbluth–Teller algorithm consisting of a Markov-type chain of updates transforming system configurations $\nu \to \nu'$.

the imaginary-time domain, and their numeric analytic continuation to the real-time axis is an ill-defined mathematical problem. The other most severe limitation is the so-called sign problem. [Difficulties in simulating response functions on the real-time axis are related to it.] So far, we were assuming that $W(v)$ is positive. In this case, each term in the sum is "meaningful" and configurations with the largest weight dominate in the average. If $W(v)$ is alternating in sign, we rewrite Equation (8.28) identically as

$$\langle A \rangle = \frac{\langle AS \rangle_+}{\langle S \rangle_+}, \tag{8.30}$$

where $S = S(v)$ is the sign of $W(v)$ and the averages on the right-hand side are defined as

$$\langle AS \rangle_+ = Z_+^{-1} \sum_v A(v)S(v)|W(v)|, \quad Z_+ = \sum_v |W(v)|, \tag{8.31}$$

and similarly for $\langle S \rangle_+$. Both averages are formally identical to the previous case and are subject to the MC simulation, which is generating configurations with probabilities determined by the weight modulus. Thus,

$$\langle A \rangle = \frac{\sum_v^{MC} A(v)S(v)}{\sum_v^{MC} S(v)}. \tag{8.32}$$

The trouble shows up in cases when $\langle S \rangle_+ \to 0$ in the large system size limit (typically $\langle S \rangle_+$ approaches zero *exponentially* fast!), that is, when large contributions tend to nearly cancel each other in the sum. Finite $\langle A \rangle$ is obtained from the ratio of two quantities both converging to zero. Since both the numerator and denominator have finite statistical error bars, we essentially look at the ratio of error bars unless the simulation is run for an astronomically long time. There is no generic solution to the sign problem. It prevents MC methods from studies of such interesting systems as interacting fermions, frustrated bosonic and magnetic systems, and real-time dynamics. Sign problem also depends on the formulation, and it may disappear under mapping between alternative configuration spaces, for example, real space *vs.* momentum space basis, expansion in kinetic or potential energy, etc. To illustrate the point, consider an arbitrary quantum mechanical system and write the partition in the eigenfunction representation, $H\Psi_\alpha = E_\alpha \Psi_\alpha$:

$$Z = \sum_\alpha e^{-E_\alpha/T}.$$

This expression contains only positive definite terms. But, in general, there is no way of knowing eigenfunctions and eigenvalues for the complex many-body system in the first place(!), and this formal setup is thus not suitable for the simulation. Any solution of the sign problem for specific models is a breakthrough if it allows us to calculate answers in time that scales as some power of the system size. Some methods (e.g., the fixed node approximation for fermions [10]) "trade" sign problem for a systematic (variational) error.

Returning to the path integral representation for Bose systems, we notice that this formulation is sign-positive for continuous-space Hamiltonian and lattice models with positive product of all hopping matrix elements $(-K)_{0,1} \ldots (-K)_{n-1,n}$. The last condition is automatically satisfied if tunneling amplitudes between lattice sites are all negative, or the number of positive amplitudes in the product is always even, as happens for models with the n.n. hopping on bipartite lattices. Equations (8.5), (8.6), and (8.21) specify the corresponding configuration spaces and the weights associated with them. In essence, there is no fundamental difference between the lattice and continuous-space formulations (the latter can always be considered as the limiting case of the former, while the former can often be recovered as the low-energy low-temperature limit of the latter in a deep periodic potential). For this reason, we will concentrate on the continuous-space formulation disregarding technical differences between the two cases, which can be found in the literature [11–14].

The configuration space of PIMC consists of all possible many-body trajectories that satisfy periodic boundary conditions for identical particles in imaginary time. These include both Z- and G-paths to ensure efficient sampling of off-diagonal correlations and allow simulations in the grand canonical ensemble. The updating scheme is discussed in the next section. Surprisingly enough, the simplest properties one can evaluate in PIMC, apart from the particle number that is already part of the Z-path definition, are the superfluid density n_s and Green's function $G(\mathbf{r}, \tau)$, or $G(\mathbf{p}, \tau)$ (and thus the condensate density and the density matrix). If configurations are accepted with probability $P(\nu)$, then the next crucial thing required to set the simulation running is an expression $A(\nu)$ for the quantity of interest. For Green's function, it is just

$$G(\nu) = e^{i\mathbf{p}(\mathbf{r}_M - \mathbf{r}_I)} \delta\left(\tau - (\tau_M - \tau_I)\right).$$ (8.33)

For winding numbers, one needs to know the total displacement of the Z-path

$$W_\alpha(\nu) = \sum_{i=0}^{M-1} (R_{i+1,\alpha} - R_{i,\alpha})/L_\alpha.$$ (8.34)

Within the simplest approximate scheme (8.21), the potential and kinetic energy expressions are

$$V(\nu) = M^{-1} \sum_{i=0}^{M-1} V(\mathbf{R}_i),$$ (8.35)

$$K(\nu) = \frac{dNM}{2\beta} - \frac{m}{2\beta\epsilon} \sum_{i=0}^{M-1} (\mathbf{R}_{i+1} - \mathbf{R}_i)^2.$$ (8.36)

They can be derived from $\langle E \rangle = -\partial \ln Z / \partial \beta = -M^{-1} \partial \ln Z / \partial \epsilon$, which is the most common approach for defining a physical observable on an arbitrary configuration space (see also Problem 8.5). Interested reader may want to look at References [3,11] for analogous expressions for the structure factor, pressure, etc., as well as for expressions that have better convergence properties and more accurate approximations of the short-time many-body evolution operator. For ^{4}He, essentially all important equilibrium properties of a system of several thousand particles down to very low temperature can be addressed by PIMC with accuracy of the order of 1% (this accuracy is limited by our approximate knowledge of renormalized interparticle interactions, not by the simulation itself). For bosonic atoms in optical lattice in the strongly correlated regime, it is possible to simulate an experimental system "as is," that is, for the same particle number, lattice, trap, and interaction parameters.

Problem 8.5 *Derive Equation (8.36) from* $\langle K \rangle = T \partial \ln Z / \partial \ln m$. *This approach can be used to derive expressions for various contributions to energy when more elaborate schemes are employed to describe system evolution in short imaginary time. What is the kinetic (hopping) energy expression in the lattice path integral formulation?*

8.7   Worm Algorithm

In this section, we shall briefly discuss an updating scheme that performs efficient sampling of the Z- and G-paths within the PIMC approach. We do not write explicit expressions for proposal and acceptance probabilities; these rather technical details can be found in References [11–14] and are not essential for understanding the scheme. All one has to know at this point about the MC protocol is that configurations are contributing to the stochastic sum with probabilities proportional to $P(\nu)$.

The updating scheme itself can be described as a "draw-and-erase" procedure. It cannot be implemented efficiently within the Z-path space because, for closed trajectories, one must insert and remove world lines forming winding exchange cycles as a whole in a single update. This leads to macroscopically small acceptance ratios and, thus, severe inefficiency. Within the worm algorithm, we interpret the end points of the disconnected path as "pencil" and "eraser" tips, which can be used to "draw" and "erase" pieces of the world line. The following set of updates forms an ergodic set (when we mention updates performed with one of the ends, similar updates for another end follow immediately from the time-reversal symmetry).

Open/close: This pair of updates takes us back and forth between the Z- and G-spaces by selecting an existing world line and erasing a small part of it (in *open*) or drawing a small piece of world line between the end points to close the loop. These updates are illustrated in Figure 8.7.

Insert/remove: This pair of updates also switches back and forth between the Z- and G-paths by drawing a small piece of a new world line (*insert*) or erasing a small piece of world line between the end points (*remove*). This pair of updates is illustrated in Figure 8.8.

Draw/erase: Once in the G-configuration space, the algorithm is proposing to move $\mathcal{I}$ and $\mathcal{M}$ points around in spacetime. This motion is accompanied by drawing and erasing additional segments of the world line connecting the end points (see Figure 8.9).

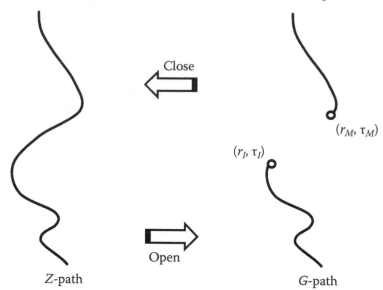

Figure 8.7   A complementary pair of *open* and *close* updates.

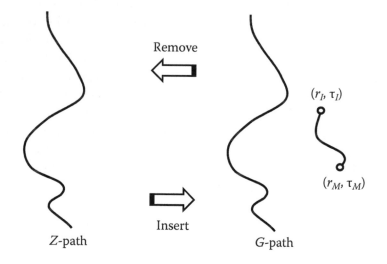

Figure 8.8   A complementary pair of *insert* and *remove* updates.

Reconnection: This is the only update that is different from continuous transformations of lines. Without changing the time position of the end point, we place it on the neighboring world line and reconnect the two involved world lines locally in such a way that the rest of the path remains intact. We illustrate the *reconnection* update in Figure 8.10.

Ergodicity follows from a simple observation that a finite number of steps is required to erase any initial trajectory and to draw any other one, line after line. Exchange cycles of arbitrary length are introduced either by drawing a world line with multiple windings around the β-cylinder or by a chain of reconnection

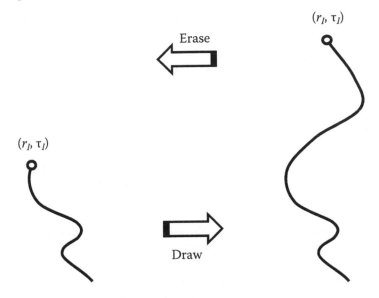

Figure 8.9   *Draw* and *erase* updates performed in the G-path space.

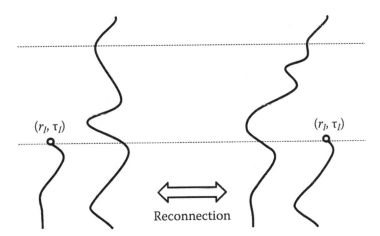

Figure 8.10   *Reconnection* update. The trajectory is modified between the two dashed lines only.

updates. The efficiency of the worm algorithm scheme is ultimately linked to the fact that it works directly with the correlation function of the order-parameter field.

References

1. R. Feynman, Space-time approach to non-relativistic quantum mechanics, *Rev. Mod. Phys.* **20**, 367 (1948).

2. E. L. Pollock and D. M. Ceperley, Path-integral computation of superfluid densities, *Phys. Rev. B* **36**, 8343 (1987).

3. D. M. Ceperley, Path integrals in the theory of condensed helium, *Rev. Mod. Phys.* **67**, 279 (1995).

4. S. A. Chin, Symplectic integrators from composite operator factorizations, *Phys. Lett. A* **226**, 344 (1997).

5. P. Sindzingre, M. L. Klein, and D. M. Ceperley, Path-integral Monte Carlo study of low-temperature ^{4}He clusters, *Phys. Rev. Lett.* **63**, 1601 (1989).

6. S. Grebenev, J. P. Toennies, A. F. Vilesov, Superfluidity within a small helium-4 cluster: The microscopic andronikashvili experiment, *Science* **279**, 2083 (1998).

7. F. Mezzacapo and M. Boninsegni, Superuidity and quantum melting of p-H2 clusters, *Phys. Rev. Lett.* **97**, 045301 (2006).

8. D. M. Ceperley and E. L. Pollock, Path-integral computation of the low-temperature properties of liquid ^{4}He, *Phys. Rev. Lett.* **56**, 351 (1986).

9. N. Metropolis, A. W. Rosenbluth, M. N. Rosenbluth, A. H. Teller, and E. Teller, Equations of state calculations by fast computing machines, *J. Chem. Phys.* **21**, 1087 (1953).

10. D. M. Ceperley and B. J. Alder, Quantum Monte Carlo, *Science* **231**, 555 (1986).

11. M. Boninsegni, N. V. Prokofev, and B. V. Svistunov, Worm algorithm and diagrammatic Monte Carlo: A new approach to continuous-space path integral Monte Carlo simulations, *Phys. Rev. E* **74**, 036701 (2006).

12. N. V. Prokof'ev, B. V. Svistunov, and I. S. Tupitsyn, Worm algorithm in the quantum Monte Carlo simulations, *Phys. Lett. A* **238**, 253 (1998).

13. N. V. Prokof'ev, B. V. Svistunov, and I. S. Tupitsyn, Exact, complete and universal continuous-time world line Monte Carlo approach to the statistics of discrete quantum systems, *Sov. Phys. JETP* **87**, 310 (1998).

14. N. Prokof'ev and B. Svistunov, Worm algorithm for problems of quantum and classical statistics. In: *Understanding Quantum Phase Transitions* (ed. L. D. Carr), Taylor & Francis, Boca Raton, FL (2010).

CHAPTER 9

Supersolids and Insulators

In this chapter, we will discuss the conditions under which ground states of quantum systems fail to develop the classical-field component and enter an insulating state. There are several reasons for this to happen. If interparticle interactions are strong and particles are heavy, bosons localize into the crystalline structure. One may argue that a solid with small zero-point fluctuations is essentially a classical state. However, in this case, we employ the classical-*particle* picture, which, somewhat ironically, cannot be connected to the classical-*field* picture without invoking quantum mechanics in a non-perturbative way. Bosons may be localized into an insulator also by a deep external lattice potential if their number is commensurate with the lattice period (in a commensurate state, the number of particles per unit lattice cell—the so-called filling factor ν—is integer). For integer ν, an insulating state that does not break any of the lattice symmetries is called a Mott insulator (MI); otherwise, the most common insulating structure is a lattice solid. Finally, an insulating state may result from placing bosons in a strong disordered potential. The generic name for such a state is the Bose glass (BG). The crucial difference between the BG and MI is that the former is compressible, while the latter has a gap for creating particle or hole excitations. Under special circumstances (e.g., in the limit of extremely large [but integer] site occupation numbers and disorder in the interaction potential or hopping amplitudes [the so-called off-diagonal disorder]), the system Hamiltonian becomes particle–hole symmetric. This symmetry leads to a new disordered insulating state, the Mott glass (MG), which has no gap for exciting particles and holes and yet has zero compressibility. Finally, superfluidity and crystalline order may coexist in one and the same material, leading to the supersolid state.

9.1 Supersolid State from the Classical-Field Perspective

Strangely enough, the question about existence of the insulating ground state, which is so easy to answer positively from the classical-*particle* perspective, is much harder to deal with starting from the classical-field point of view. For classical particles at $T = 0$, all we must do is minimize the total potential energy with respect to particle positions. This naturally leads to the regular crystalline arrangement of particles for realistic interatomic potentials.

The order in particle positions is quantitatively characterized by the structure factor, or the density-density correlation function (the volume is set equal to unity),

$$D(\mathbf{R}) = \int \langle n(\mathbf{r})n(\mathbf{r}+\mathbf{R})\rangle d\mathbf{r} - \langle n\rangle^2 \equiv \int \langle \delta n(\mathbf{r})\delta n(\mathbf{r}+\mathbf{R})\rangle \, d\mathbf{r}, \qquad (9.1)$$

where $n(\mathbf{r}) = \sum_{i=1}^{N} \delta(\mathbf{r}-\mathbf{r}_i)$ is the density at point $\mathbf{r}$. The average density squared is subtracted to underline that in liquid phases with uncorrelated density fluctuations, $\delta n(\mathbf{r}) = n(\mathbf{r})-\langle n\rangle$, at large distances one expects $D_L(R \to \infty) \to 0$. In a 3D solid (and also in a 2D solid at $T = 0$), the structure factor features undamped oscillations at arbitrary large distances.* At low but finite temperature, the description of solids must include crystal vibrations (phonons), vacancy and interstitial excitations (their number densities are different because vacancies are typically less expensive to produce than interstitial atoms), and dislocation loops. However, all excitations, except for acoustic phonons, have exponentially small concentration at low T, due to finite energies required for their formation.

In the ground state of a classical solid, particles sit at local minima of potential energy field, U, created by other particles and cannot conduct matter current. Quantum mechanics forces one to delocalize coordinates at least over distances of the order of $x \sim \sqrt{\hbar}/(U''_{xx}m)^{1/4}$, to keep potential and kinetic energies comparable while satisfying the uncertainty relation $p \sim \hbar/x$. Next, one notices that the exchange-symmetric wavefunction of the solid ground state implies that any given single-particle coordinate is "delocalized" between all the sites of the crystal lattice, and the question arises whether this "delocalization" automatically implies superfluid order. If based on the classical-field paradigm only (cf. Section 1.2.7), the answer to this question is both straightforward and striking: All ground states, including solids, are superfluid at sufficiently low temperature.† In other words, the only alternative to the liquid (homogeneous or modulated by an external potential) is the supersolid state.

Most generally, supersolid is defined as a state that simultaneously breaks translation and $U(1)$ symmetries (in $d + 1 > 2$) and has long-range order both in the structure factor (9.1) and in the density matrix (2.40), (7.55). Algebraic (topological) orders in 2D at finite temperature may be included here too, though they do not break the system symmetry. To exclude trivial cases—such as a two-component system with one component being responsible for the solid structure and the other for the superfluid response, similar to superconducting metals—the definition of supersolid also requires that one and the same species of atoms are involved in developing both orders. The excluded trivial cases are no different from the superfluid states modulated by the external periodic potential.

Supersolids are generically incommensurate states [1]; that is, the number of particles per unit cell of the crystalline structure can be an integer only by a zero

---

* In 2D solids at $T \neq 0$, the structure factor $D(R)$ decays algebraically, because of thermally excited long-wave phonon modes. In 1D, solids exist only at $T = 0$. In a quantum 1D solid ground state, the structure factor $D(R)$ decays algebraically, because of zero-point quantum fluctuations of long-wave phonons.

† The only (and rather specific) exception is the case of the 1D disordered classical field, where exponentially rare, but exponentially weak, links can lead to the breakdown of macroscopic homogeneity of the system and, as a result, to the superfluid stiffness vanishing in the macroscopic limit.

probability chance! This is because part of the particle density is described by the classical field component for which the notion of particle-number quantization is irrelevant. Equivalently, one may argue that supersolids, just like any other superfluid state, have finite *isochoric compressibility*—the particle-density response to a small variation in the chemical potential under the condition of constant volume and fixed geometry of crystal lattice.* (If the volume is kept constant, the geometric structure of crystal lattice is protected by the topology of crystalline planes pinned to the sample walls.) Thus, only by zero-chance accident can the average density of a supersolid be commensurate with the lattice. This, in turn, means that there are vacancies and/or interstitial atoms delocalized in the supersolid ground state. Typically, vacancies have lower energies; for brevity, we will ignore interstitial atoms in the discussion that follows, though in systems featuring particle–hole symmetric points both vacancies and interstitials must be treated on equal footing. These delocalized vacancies are often referred to as "zero-point vacancies" to underline that they are not produced by deliberately taking atoms away from the lattice. Instead, the lattice period and the unit cell volume, Ω_0, are automatically adjusted to have a non-integer $\langle n \rangle \Omega_0$ in the fully equilibrated ground state.

Within the classical-field paradigm, the supersolid state is not exotic, being nothing else but a liquid with a density wave; see Section 1.2.7. Within GPE, one must choose sufficiently strong interaction parameter $V\left(nr_0^3\right) \geq \left(\gamma/r_0^2\right)$, where r_0 is the potential range and n is the average field density, and look for the density-wave instability. This scenario was first elaborated by Gross in his pioneering paper [2]. For the soft-sphere repulsive potential with amplitude V_0 and interaction radius a, the corresponding numerical study was performed in Reference [3]. For large enough $\left(nr_0^3\right)$, the density profile in the ground state features very narrow peaks arranged into a perfect crystalline structure. Nevertheless, the system has finite n_s determined by the field density in between the peaks. Within the classical-field description, the notion of integer particle number is lost, so it makes no sense to speak of a commensurate crystal without vacancies. Since $\langle n \rangle \Omega_0$ is a real (not integer!) number, the state is filled with vacancies. Full quantum-mechanical solution of the soft-sphere (and its modifications) systems based on path integral Monte Carlo (PIMC) simulations [4] also finds a robust supersolid phase when the number of particles per unit cell is large, >5, but the limit can be pushed down to $2 \div 3$ particles [5].

9.2 Supersolid State from the Quantum-Particle Perspective

We thus find that the classical-particle and classical-field predictions are simply incompatible with each other and only an appropriate quantum-mechanical treatment can deal with the insulating ground state. Before we proceed, let us discuss two microscopic scenarios for the supersolid ground state based on the quantum-particle approach.

---

* See the discussion of properties of connected states at the end of Section 7.8.

Andreev and Lifshitz start with the calculation of the vacancy-formation energy [6]. In a nearly classical solid, removing an atom from the lattice site will cost some finite amount of energy, E_c, relative to the chemical potential of the system. At zero pressure, E_c is dominated by the change in the potential energy of attraction between the atoms, which is lost when the atom is removed. In a quantum solid, E_c also includes the kinetic energy of the removed particle. Finally, one must account for the reduced kinetic energy of the remaining atoms since they have a larger volume to move around. Within the tight-binding model, which assumes that the matrix element $(-t)$ for moving an atomic cloud to the vacant site is relatively small (much smaller than the Debye energy ω_D), the kinetic energy change can be described by the Hamiltonian

$$H_K = -\sum_{ij} t_{ij}\, v_j^\dagger v_i, \tag{9.2}$$

where $v_j^\dagger$ is the vacancy creation operator on site j. In most cases, the sum is dominated by the n.n. terms since the hopping matrix elements t_{ij} decay exponentially with the distance between the sites. In the Fourier representation, we have (assuming, for simplicity, one vacancy position per lattice site)

$$H_K = \sum_{\mathbf{p}} \epsilon_{\mathbf{p}}\, v_{\mathbf{p}}^\dagger v_{\mathbf{p}}, \qquad \epsilon_{\mathbf{p}} = -\sum_{\mathbf{g}} t_{\mathbf{g}}\, e^{i\mathbf{p}\cdot\mathbf{g}}, \tag{9.3}$$

with the vector $\mathbf{g}$ enumerating distances between the sites. Thus, vacancies in the plane-wave state with $\mathbf{p} = 0$ have reduced energies

$$E_v = E_c + \epsilon_0. \tag{9.4}$$

If atoms are heavy and interactions between them are strong, then the kinetic energy contribution accounting for vacancy delocalization is exponentially small. This leaves us with positive E_v. However, in a quantum solid composed of light atoms interacting by weak forces, one may encounter a situation when $\sum_{\mathbf{g}} t_{\mathbf{g}} > E_c$, and the vacancy-formation energy goes negative, $E_v < 0$. At this point, the commensurate crystalline structure is proved to be unstable, and one must consider an alternative ground state. Some possibilities include (1) another crystalline structure (in which case, the same vacancy energy analysis must be repeated for the new structure), (2) a superfluid liquid state (a very likely scenario since liquids typically have smaller density at the same pressure), and (3) the supersolid phase viewed as a lattice gas of vacancies. For the last picture to be valid, vacancies must form a repulsive Bose gas. If this is the case, then an estimate for the equilibrium concentration of vacancies follows from the minimization of the energy density functional $E = E_v n_v + U_{vv} n_v^2$, where U_{vv} is determining the strength of the repulsive vacancy–vacancy interaction. The result is $n_v = |E_v|/2U_{vv}$. Provided that $n_v \ll \langle n \rangle$, one can go one step further and estimate the supersolid–normal-solid transition temperature from the ideal Bose gas relation $T_c \approx 3.31 n_v^{2/3}/m_v$, where $m_v = (1/3)d^2\epsilon(\mathbf{p})/d\mathbf{p}^2|_{p=0}$ is the vacancy effective mass (assumed spherically symmetric for simplicity). The $T = 0$ condensate fraction is then approximately n_v.

The main difficulty with the Andreev and Lifshitz scenario is that it is based on competing kinetic and potential energies in the quantum solid and, thus, there are no small parameters in the problem to make reliable analytic calculations and precise predictions for the existence of the supersolid phase in a given micro-scopic Hamiltonian. Next, there is no reason for the vacancy concentration n_v to be small. Finally, it is not at all typical to have vacancies that effectively repel each other. Though two vacancies cannot occupy the same site, they experience attrac-tive interactions at larger distances similar to the interatom forces. In the purely classical rigid lattice, E_c accounts for the potential energy between the missing atom and the rest of the system. For two vacancies at a distance $\mathbf{g}$, this part of the potential energy increase is $2E_c - [-V(\mathbf{g})]$—to avoid double counting of the interaction energy between the two missing atoms; that is, it is attractive under standard circumstances. An attractive Bose gas is unstable against collapse. For the present case, it would mean either that the vacancy-doped crystal will purge itself of vacancies (while preserving the total number of particles at the expense of forming dislocation loops) or that the genuine ground state is a liquid. In the latter case, doping a metastable crystal with vacancies will promote a collapse into the liquid state. Energy changes originating from the lattice deformation make things more complex, but the point is made that it should not be taken for granted that vacancy–vacancy interaction is repulsive.

Despite these inherent difficulties, it is possible to name a couple of realistic continuous-space Hamiltonians that are guaranteed to have a supersolid phase. The first example is based on our previous discussion of how supersolids can be viewed as superfluids with a density wave. For the classical-field picture to provide an accurate description of a quantum model, we must ensure that there are many particles per primitive cell of the crystal structure. Thus, if we take a Hamiltonian with weak soft-sphere potential and increase system density to ensure that the dimensionless coupling parameter $g = V(nr_0^3)(mr_0^2/\hbar^2)$ is large enough, we will enter the supersolid phase. It is worth noting, however, that, if we keep increasing n, the crystal will ultimately become insulating at certain $g \gg 1$, due to Mott physics (see Section 9.4) brought on by exponential suppression of particle tunneling between well-separated condensate droplets (which the crystal structure reduces to at $g \to \infty$).

The other example is an immediate consequence of the physics behind the theorem that, in a 2D system with dipole interactions $V(\mathbf{r} - \mathbf{r}') = D/|\mathbf{r} - \mathbf{r}'|^3$, first-order transitions between phases with different densities $n_1 \neq n_2$—such as generic liquid-solid transitions—are forbidden (see, e.g., Reference [7]). This is because an elementary calculation of the surface energy between the two phases involves a negative term $-D(n_1 - n_2)^2 \ln(R/r_0)$ that diverges with the sample size R (here, r_0 is a microscopic scale of order of the interatomic distance). It is thus guaranteed that, on approach to the first-order line (say, from the superfluid liquid phase), one can construct a state that is lower in energy than liquid—by embedding sufficiently large solid domains into it. At $T = 0$, large solid domains will form a secondary lattice superstructure leading to a global supersolid phase (in fact, a whole set of supersolid phases [8]).

In both examples, however, the validity of the argument is based on a large number of particles per unit cell. At the time of writing, we are not aware of any realistic Hamiltonian in continuous space with the supersolid state having one particle per cell. On the other hand, it is relatively easy to write a trial wavefunction featuring such a supersolid state. As was noted by Reatto [9] and Chester [10], an analogy between the trial Jastrow wavefunction

$$\Psi_G^{(J)} \propto \exp\left\{-\sum_{i<j=1}^{N} u\left(\mathbf{r}_i - \mathbf{r}_j\right)\right\}, \tag{9.5}$$

and the partition function of a classical system of interacting particles with pairwise potential $\mathcal{V}(r) = 2Tu(r)$ at finite temperature suggests that, among (9.5), there are states that feature spontaneous crystalline order, as classical systems are known to crystallize at sufficiently low temperature. States (9.5) are also superfluid since they can be shown to be connected [9] and incommensurate, that is, with gapless zero-point vacancies. The last property immediately follows from mapping to a classical solid state at finite temperature, which always has finite concentration of vacancies.

One might think that the Jastrow function holds the key to microscopic understanding of supersolidity. However, there are at least two fundamental problems associated with Equation (9.5). First, Equation (9.5) describes the ground state of the Hamiltonian with rather pathological pair- and three-body interactions between particles and zero compressibility [11]. Second, though it provides very accurate description of local properties of solid ^{4}He, including reliable estimates of large energy gaps for vacancies, we know for sure that vacancies are gapless in any Jastrow state [9]. Hence, the Jastrow function cannot describe an insulating solid ground state.

As the "supersolid" name suggests, on a macroscopic scale, the state combines features specific to both solids and superfluids. On one hand, the solid component is responsible for broken translation symmetry, finite shear modulus, standard phonon spectrum, ability to hold pressure gradients, etc. On the other hand, the superfluid component is responsible for the nonclassical moment of inertia and finite superfluid density, additional phonon mode, etc. It would be a mistake to suggest that, if two vessels, connected by a narrow channel, are filled with a supersolid, then the pressures in the two vessels will equilibrate. Formally, this situation is no different (apart from time scales involved) from that of a much more familiar classical solid at finite temperature in the absence of plastic flow. The pressure difference between the two vessels will change only in the amount proportional to the concentration of vacancies (superfluid component). The resulting quasi-equilibrium state will correspond to *a constant chemical potential within the superfluid component*, not constant pressure. This is because there is no local mechanism within the solid subsystem to change the topology of connections between the lattice points and relax the stress. The ultimate equilibrium state, of course, is achieved only after reconfiguration of atomic planes on a macroscopic scale, a process that is kinetically frozen in the absence of topological crystalline defects such as dislocations.

Finally, we would like to comment on the difference between the continuous-space and lattice supersolids. In a typical lattice system, the solid-structure period derives from the lattice period and thus is not a continuous variable. Under these conditions, supersolid phases can be introduced in a rather trivial way by doping insulating solids with holes or particles, provided the latter do not undergo phase separation. While continuous solids can lower energy by eliminating vacancies and interstitials using complete atomic layers and adjusting the solid period accordingly, such a mechanism is forbidden when the solid structure is pinned to the lattice. The net result is that vacancies and interstitials can be introduced by hand and their presence is not necessarily of the "zero-point" origin. The other crucial difference between continuous and pinned lattice solids is in the absence of long-range elastic forces between the defects, including vacancies and interstitials. Nearly all lattice supersolids studied in the literature belong to the category of doped insulators, which is fundamentally different from the zero-point vacancy picture developed for continuous-space systems, such as ^{4}He.

9.3   Existence of Insulators

As discussed previously, insulators do not exist within the classical-field picture. One may also add that the classical-particle picture of the insulating state is incomplete because it neglects small tunneling-exchange processes that, in principle, may lead to particle delocalization. There are several ways of proving that insulating quantum states do exist. We present below two arguments approaching the problem from different angles. We also discuss a compelling numeric proof of the existence of bosonic insulating ground states.

The first argument is based on the theorem that superfluid states have no energy gap to add/remove particles to/from the system one by one. This, in particular, implies that vacancies are gapless excitations. Looking back at the Andreev and Lifshitz calculation of the vacancy-formation energy, we notice that the kinetic energy of vacancy delocalization can be made exponentially small for large interatomic potentials and heavy particle masses. The kinetic energy of the zero-point motion around a given lattice site $\sim \hbar\sqrt{\mathcal{V}''(a)/m}$, where $\mathcal{V}''(a)$ is the second derivative of the interatomic potential at the lattice distance a, can also be made much smaller than the classical-vacancy energy E_c for large potentials and heavy particles because $E_c \sim \mathcal{V}(a)$. In other words, one can always make quantum mechanical corrections to the vacancy-formation energy much smaller than $\mathcal{V}(a)$, leaving us with $E_v \approx E_c > 0$, which is only possible in the nonsuperfluid state.

Paradoxical speculations that particle exchange between the lattice sites *necessarily* leads to the supersolid state—even in the absence of vacancies—are countered as follows. Local exchange cycles do not contribute to the superfluid response at all: According to the winding number formula (8.15), only macroscopic exchange cycles spanning the entire system determine n_s. Furthermore, exchange cycles with nonzero winding number can be visualized (in the path-integral representation)

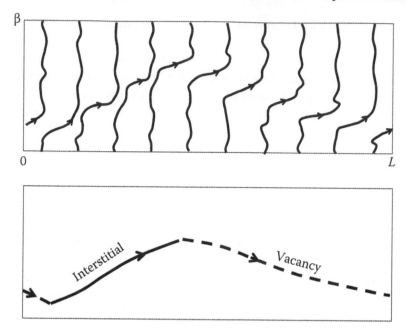

Figure 9.1   Visualizing a winding number trajectory as a propagating vacancy–interstitial pair.

at macroscopic distances as propagating vacancy–interstitial pairs; see Figure 9.1. For gapped vacancy and interstitial excitations, such a trajectory is exponentially (in the system size) improbable. This brings us back to the statement that, in a super-solid, either vacancies or interstitials or both are gapless and the probability for a continuous-space supersolid crystal to be commensurate is vanishing.

The second argument is based on explicit construction of a crystalline many-body wavefunction Ψ_G that can be shown to have exponentially decaying density matrix. Our function Ψ_G is not translation symmetric, but this is an irrelevant issue for a crystalline state: The state breaks translation symmetry spontaneously and thus can be pinned by vanishingly small (in the thermodynamic limit) external fields, which have the same periodicity as the solid lattice. The function Ψ_G then should be viewed as the wavefunction of pinned crystal (see also References [12,13]).

So, the claim is that

$$\Psi_G \propto \sum_{P} \prod_{j=1}^{N} \varphi_{Pj}\left(\mathbf{r}_j\right) \tag{9.6}$$

is a normal state for Gaussian site orbitals $\varphi_k(\mathbf{x}) = Ce^{-(\mathbf{x}-\mathbf{R}_k)^2/4\sigma^2}$, where the sum $\sum_P$ is over all permutations P of particle labels and $\{\mathbf{R}_j\}$ form a set of regular lattice points with the lattice spacing $a \gg \sigma$. We do *not* require that site orbitals be orthogonal. This makes the calculation of the density matrix more elaborate but is necessary to satisfy the requirement $\Psi_G > 0$.

Since overlap integrals between orbitals are exponentially small, $\langle \varphi_j | \varphi_i \rangle = s_{i,j} \propto e^{-R_{ij}^2/8\sigma^2}$, and have quadratic dependence on the distance between the lattice points R_{ij} in the exponent, one can easily identify the largest terms contributing to the density matrix $\rho(\mathbf{r}, \mathbf{r}')$ at large separation between the coordinates $\mathbf{r}$ and $\mathbf{r}'$; see Figure 9.2. First, we identify sites closest to the density matrix points (sites $\mathbf{R}_1$ and $\mathbf{R}_4$ in Figure 9.2). Without loss of generality, given the freedom of relabeling integration variables, any of the remaining terms in $\Psi_G(\mathbf{r}, \mathbf{r}_2, ... \mathbf{r}_N)$ can be written as $\Phi = \varphi_1(\mathbf{r}) \prod_{j=2}^{N} \varphi_j(\mathbf{r}_j)$. Next we select the term in $\Psi_G(\mathbf{r}', \mathbf{r}_2, ... \mathbf{r}_N)$ that has the largest overlap with Φ. The best we can do for the variable $\mathbf{r}_4$ is to associate it with the n.n. site of $\mathbf{r}_4$. Let it be site $\mathbf{R}_3$. This forces us to look for the best overlap for variable $\mathbf{r}_3$, $\mathbf{r}_2$, etc. The process will continue with the smallest possible shifts in the lattice site associations until we hit point $\mathbf{R}_1$. The rest of the variables should keep their associations intact since for them the overlap is nearly perfect. If we represent association shifts and the initial $\mathbf{R}_1 \to \mathbf{r}$ and $\mathbf{r}' \to \mathbf{R}_4$ associations by arrows, they then form a trajectory originating from $\mathbf{r}'$ and terminating at $\mathbf{r}$.

Every n.n. shift in associations results in a small factor s_a. Thus, the optimal contribution corresponds to the shortest possible trajectory between the end

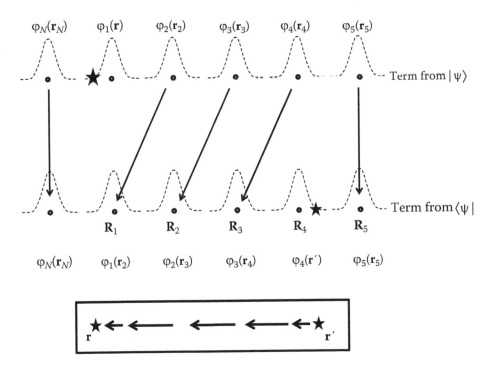

Figure 9.2   Graphical representation of the optimal contribution to the density matrix $\rho(\mathbf{r}, \mathbf{r}')$ at large separation between $\mathbf{r}$ and $\mathbf{r}'$ shown by stars. For any term in the sum for $|\Psi\rangle \equiv \Psi_G(\mathbf{r}, ...)$, we identify a term in the sum for $\langle \Psi | \equiv \Psi_G(\mathbf{r}', ...)$ that gives the largest contribution after integration over coordinates $\mathbf{r}_2$, $\mathbf{r}_3$, ... $\mathbf{r}_N$. Arrows show shifts in the permutation over lattice points between the two terms and associations for the density matrix points.

points $\mathbf{r}'$ and $\mathbf{r}$ and scales as $e^{-\zeta|\mathbf{r}'-\mathbf{r}|}$ where $\zeta = -\ln(s_a)$ is a large positive parameter. Further analysis is simplified by the observation that replacing two n.n. shifts with a perfect overlap and a double shift results in additional exponential suppression of the overlap due to nonlinear dependence of $-\ln(s_{i,j})$ on distance. Simple combinatoric considerations show that double shifts make an exponentially small correction to the ζ coefficient $\zeta \to \zeta - 1/e^{2\zeta+1})$ and cannot change the exponential decay of the density matrix predicted by the optimal configuration.

Problem 9.1 *The statistical weight of a chain consisting of $N - 2m$ links of unit length and m links of double length is given by $W_{N,m} = e^{-\zeta(N-2m)-4\zeta m} = e^{-\zeta N - 2\zeta m}$ with $\zeta \gg 1$. Show that the total weight of all possible configurations with $m = 0, 1, \ldots N/2$ is approximately given by $e^{-N(\zeta-1/e^{2\zeta+1})}$; that is, double links do not change an exponential dependence of W on N and can be neglected altogether unless $W < \exp\{-\zeta e^{2\zeta+1}\}$, which is "zero for all practical purposes" already for $\zeta > 5$!*

Accounting for the statistics of different single-shift trajectories between the end points does not change an exponential law either. Contribution of rare back-and-forth shape fluctuations can be calculated using combinatorics similar to the problem of replacing single links with double links (see text above and Problem 9.1). The outcome of this analysis is that shape fluctuations are indeed rare and account for a negligible ($\sim 2s_a/e$) correction to ζ. Note that lattice discreteness is essential here and an analogy with the elastic string for dealing with shape fluctuations would be misleading (Problem 9.2). This concludes our proof that the state (9.6) is not superfluid because its density matrix decays exponentially. We remind the reader, however, that the discussion of the system's response to the twisted boundary condition *cannot* be done without involving states with hyperplanes of zeros (see Section 7.8). In other words, it would be wrong to study it within the ansatz (9.6).

Problem 9.2 *Assume that large fluctuations of the trajectory in the transverse direction are possible and use an analogy with the elastic string model to calculate them. Since the exponential of the trajectory contribution to $\rho(\mathbf{r}, \mathbf{r}')$ is proportional to the trajectory length, introduce a dimensionless variable x measuring the distance along the optimal trajectory in lattice units; also introduce trajectory deviations (in lattice units) in the perpendicular directions $\mathbf{y}(x)$; then write the trajectory contribution as*

$$w[\mathbf{y}(x)] = \exp\left[-\zeta \int_O^N \sqrt{1+(d\mathbf{y}/dx)^2}\,dx\right] \to \exp\left[-\zeta N - \frac{\zeta}{2}\int_O^N (d\mathbf{y}/dx)^2\,dx\right], \quad (9.7)$$

where $\mathbf{y}(0) = \mathbf{y}(N) = 0$, further assuming that deviations are smooth. The mean-square fluctuations can be found from $\langle y^2 \rangle \equiv N^{-1}\int dx \langle y^2(x)\rangle_w$, where the last average over shape fluctuations is taken with the $w[\mathbf{y}(x)]$ distribution. Find $\langle y^2 \rangle$ using Fourier representation $\mathbf{y}_k = \sqrt{2/N}\int \sin(\pi kx/N)\mathbf{y}(x)dx$, where k is an integer, and explain why the result $\langle y^2 \rangle \gg 1$ for $N \gg \zeta$ is misleading for lattice trajectories with large ζ.

Finally, compelling evidence of the existence of bosonic insulating ground states can be readily obtained by first-principles numerical simulations. To avoid objections about system size and finite temperature effects, equilibration, etc., we choose a 1D system of bosons described by the Lieb–Liniger Hamiltonian

$$H^{(1D)} = \sum_{i=1}^{N} \frac{p^2}{2m} + U \sum_{i<j} \delta\left(r_i - r_j\right). \tag{9.8}$$

It is conventional to characterize interactions using a dimensionless parameter $\bar{\gamma} = 2/na^{(1D)}$ where $a^{(1D)} = 2/mU$ is the 1D scattering length and n is the system density. The regime of strong interaction is achieved for large $\bar{\gamma}$. For any finite $\bar{\gamma}$, the ground state of the system is superfluid and, in our example, we choose $\bar{\gamma} = 10$. The simulation of the density matrix is performed with PIMC, described in Chapter 8. In 1D, one can easily simulate systems at temperatures down and below the lowest excitation mode, as well as study progressively larger system sizes to eliminate finite-size effects.* It is also universally accepted that, in 1D systems, quantum fluctuations are more pronounced.

In Figure 9.3, we illustrate how superfluid properties build up in the Lieb–Liniger system as one goes to lower and lower temperature. Clearly, the slope of the exponential decay is strongly T-dependent at higher temperatures. This is a

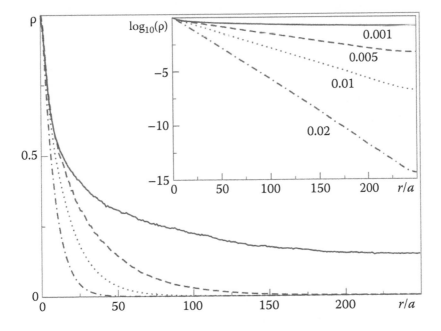

Figure 9.3   Density matrix of the Lieb–Liniger model at $\bar{\gamma} = 10$ and $L/a^{(1D)} = 500$ for various temperatures $T/T_0 = 0.02$, 0.01, 0.005, and 0.001 where $T_0 = 2\pi n^2/m$. The same data are shown in the inset using a logarithmic scale.

---

* The surface-to-volume ratio in higher-dimensional systems is much larger: in 3D, one can hardly reduce it to be below 0.1.

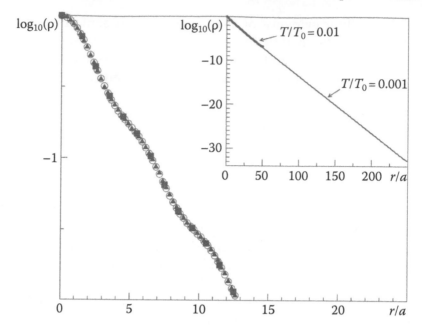

Figure 9.4 Density matrix of the Lieb–Liniger gas in external potential at $\bar{\gamma} = 10$ for various temperatures and system sizes: ($T/T_0 = 0.01$, $L/a^{(1D)} = 100$), triangles; ($T/T_0 = 0.001$, $L/a^{(1D)} = 100$), open circles; ($T/T_0 = 0.001$, $L/a^{(1D)} = 500$), squares. In the inset, we show that the exponential decay for $T/T_0 = 0.001$ continues all the way to $L/2 = 250$ without any sign of the slope change.

strong indication that the ground state may be superfluid, and, indeed, it is superfluid as evidenced by the algebraic decay of the density matrix at the lowest temperature (below the lowest sound-wave excitation).

When the Lieb–Liniger system is subject to the commensurate external potential $V_{ext}(r) = (2\pi n/m)\cos(2\pi nr)$, it loses its superfluid properties and becomes an MI (we discuss this state in more in Section 9.4 below). Though the strength of the potential is modest (the overlap between the Gaussian functions localized at the neighboring sites is about 0.14), the behavior of the density matrix is radically different from the liquid state. In Figure 9.4, we show three curves calculated at different temperatures and system sizes. Clearly, changing temperature by a factor of 10, or system size by a factor of 5, makes no difference whatsoever (periodic boundary conditions influence the data only in the immediate vicinity of the point $L/2$). Undoubtedly, the ground state of the system is insulating. There are numerous other examples of numeric simulations (say, of ^{4}He at elevated pressure) that demonstrate behavior similar to Figure 9.4. They also present solid theoretical evidence for the insulating crystalline ground state.

9.4 Mott Insulators

So far, we have discussed superfluid-to-normal phase transition taking place at a certain finite temperature. There is an important class of lattice systems that undergo a continuous superfluid-to-normal *quantum* phase transition—the transition that happens in the ground state as the Hamiltonian parameters are changed. In this section, we discuss the simplest example of this kind, the transition to the state of MI, which is found in the Bose–Hubbard model introduced in Chapter 7; see Equation (7.90). For weak on-site interactions $U \ll \nu zt$, where $\nu = \langle n \rangle$ is the average occupation number, or filling factor, the ground state is superfluid. If ν is of the order unity, we are dealing with weakly interacting lattice gas, in which case the temperature of superfluid-to-normal phase transition temperature can be estimated as the Bose–Einstein condensation/condensate (BEC) temperature of a noninteracting system. The corresponding relation is

$$\nu = \int_{BZ} \frac{d\mathbf{p}}{(2\pi)^3} N_{\epsilon_{\mathbf{p}} - \epsilon_0}. \tag{9.9}$$

Here, $\epsilon_{\mathbf{p}}$ is the tight-binding dispersion relation [see (9.3)], N_E is the Bose function, and the integral is over the first Brillouin zone. For large occupation numbers, as discussed in Chapter 4, an adequate description is provided by the classical-lattice-field theory.

Regardless of U, the ground state is also superfluid for any non-integer filling factor. And this is quite a specific property of the model (7.90), which has only on-site interactions. If repulsive interactions extend at least to the n.ns., checkerboard insulating solids take place at half-integer filling factors, provided U is large enough. In a general case of extended intersite interactions, an insulating solid state can take place at any *commensurate* filling (and appropriately large U).[*]

In the limit of $U/\nu zt \gg 1$ and as long as we are concerned with low-energy degrees of freedom, most of the particle density is locked to the lattice, so that the system can be viewed as a gas of hard-core particles (holes) with effective particle (hole) density $\nu_{\text{eff}}^{(p)} = \nu - [\nu]$ ($\nu_{\text{eff}}^{(h)} = [\nu] + 1 - \nu$) and effective hopping $t_{\text{eff}} \approx t([\nu] + 1)$, where $[\nu]$ is the integer part of the filling factor. This result immediately follows from considering the lowest possible potential energy configurations and neglecting all configurations with energies higher by $\approx U$. The best state for the integer occupation number would correspond to $n_i = [\nu]$; the remaining particles can be placed randomly with the only requirement that nowhere $n_i - [\nu] > 1$. In this subspace, the matrix element of moving a particle to the n.n. site is enhanced by the bosonic factors $\sqrt{[\nu] + 1}\sqrt{[\nu] + 1} = [\nu] + 1$. This picture, in particular, implies that when ν is approaching an integer value from above (below), the description of the system is rather simple: one deals with the dilute gas problem, which can be solved using well-developed theory based on controlled approximations; see Chapter 12.

For integer ν, the situation is radically different [14]. At $U/\nu zt \gg 1$, the lowest energy configuration in terms of on-site occupation numbers is unique and has exactly ν particles on each site with a large gap, $\approx U$, separating it from states containing sites with occupation numbers $\nu \pm 1$. This is a classical picture of the Mott ground state; see Figure 9.5. The superfluid–MI quantum phase transition driven by increasing/decreasing the chemical potential—thus decreasing the

---

[*] The filling is called commensurate if the filling factor is a rational number.

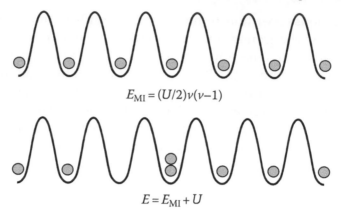

$$E_{MI} = (U/2)\nu(\nu-1)$$

$$E = E_{MI} + U$$

Figure 9.5   Classical picture of the Mott insulating ground state and the lowest particle–hole excitation for $\nu = 1$.

number of holes/particles—is trivial in nature. It is identical in its description to what happens in the weakly interacting Bose gas when the particle density goes to zero and we are left with a pure vacuum state. This picture holds not only in the limit of strong interactions but everywhere along the superfluid–MI phase boundary with the exception of just one point at the tip of the phase diagram discussed in Figure 9.6.

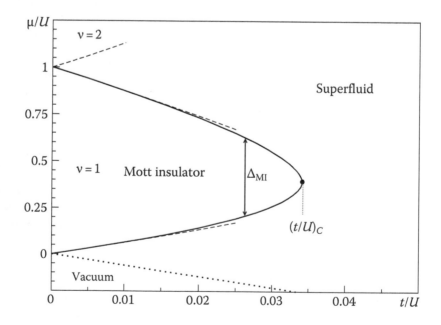

Figure 9.6   Superfluid–MI phase diagram of the 3D Bose–Hubbard model. The $\nu = 1$ phase boundaries were derived from MC simulations [17]. Dashed lines show results of second-order perturbation theory [Equation (9.11)]. The vacuum line at $-zt/U$ is exact.

It is possible to perform very accurate calculations of the superfluid–MI boundary in the chemical potential by doing high-order perturbation theory in hopping, starting from the classical Mott state [15,16]. In the zero's order, MI is doped with particles (holes) when the chemical potential reaches values μ_+ (μ_-)

$$\mu_+^{(0)} = \nu U, \qquad \mu_-^{(0)} = (\nu - 1)U. \qquad (9.10)$$

The first-order correction accounts for the energy gain due to delocalization of particles and holes within the tight-binding approximation discussed earlier. Since the effective hopping matrix element for doped particles (holes) is given by $(\nu+1)t$ (νt), for the simple cubic lattice with the coordination number $z = 6$, this results in shifts of the critical chemical potentials $\mu_+^{(1)} = \nu U - z(\nu+1)t$ and $\mu_-^{(1)} = (\nu-1)U + z\nu t$. Within the same tight-binding approximation, Equation (9.3), we find effective masses of particle and hole excitations as $m_p = 1/[2(\nu+1)ta^2]$ and $m_h = 1/[2\nu ta^2]$, respectively.

The second-order perturbation theory in t includes the change of the MI ground-state energy E_{MI}, dressing of quasiparticles by vacuum fluctuations, and second-order hopping transitions to next-n.n. sites

$$\mu_+^{(2)} = \nu U - z(\nu+1)t + \frac{(zt)^2}{U}\left[\frac{\nu(5\nu+4)}{2z} - \nu(\nu+1)\right],$$

$$\mu_-^{(2)} = (\nu-1)U + z\nu t - \frac{(zt)^2}{U}\left[\frac{(\nu+1)(5\nu+1)}{2z} - \nu(\nu+1)\right], \qquad (9.11)$$

$$m_p = \frac{1}{2(\nu+1)ta^2}\left[1 - \frac{2z\nu t}{U}\right],$$

$$m_h = \frac{1}{2\nu ta^2}\left[1 - \frac{2z(\nu+1)t}{U}\right]. \qquad (9.12)$$

As t/U is increased, the two boundaries approach each other and ultimately meet at the tip of the Mott insulating lobe. High-order perturbation theory extrapolated to infinite order and Monte Carlo (MC) simulations [17] provides a very accurate (up to four digits) description of the superfluid–MI transition; see Figure 9.6.

Problem 9.3 *Within the second-order perturbation theory in t for the insulating Mott state with the integer occupation number ν, calculate the change of the ground-state energy per particle. Now add one particle/hole and calculate (1) the energy change due to suppressed vacuum fluctuations on neighboring sites, (2) the effective next-n.n. hopping, and (3) the resulting dispersion relation $\epsilon(\mathbf{p})$; determine the band bottom $\epsilon(\mathbf{p} = 0)$ and the effective mass m_{eff}.*

Knowing the MI energy gap $\Delta_{MI} = \mu_+ - \mu_-$ and masses of particle and hole excitations, one has a complete characterization of low-temperature thermodynamics of the Mott state that is continuously connected to the classical lattice gas.

Indeed, the density of quasiparticle excitations is exponentially small at $T \ll \Delta$, and interactions between them can be neglected. Otherwise, the calculation is no different from dealing with the thermodynamics of semiconducting materials: one assumes first that grand-canonical ensembles of particles and holes are independent and then equates their densities to determine the chemical potential.

On approach to the tip of the phase diagram in Figure 9.6, the description of the superfluid–MI boundary in terms of a gas of quasiparticles is getting more difficult and works only on progressively larger length scales. Also, as the MI gap closes, the effective parameters of the theory—the quasiparticle mass and scattering length—are strongly renormalized in the vicinity of the tip. The general rule in such cases is to understand properties of the quantum critical point at the tip, since these control the scaling laws for other properties [18].

To discuss quantum criticality, one must supplement the notion of the spatial correlation length ξ with the corresponding correlation length in time direction, ξ_τ, or, equivalently, characteristic frequency/energy scale. The relation between the two is controlled by the so-called dynamic critical exponent, z, such that $\xi_\tau \propto \xi^z$.

On approach to the critical point characterized by the dimensionless parameter

$$\delta = t/U - (t/U)_c \tag{9.13}$$

(an analog of reduced temperature for classical phase transitions), both $\xi \propto |\delta|^{-\nu}$ and $\xi_\tau \propto |\delta|^{-z\nu}$ diverge, defining the correlation volume $\xi^d \xi_\tau \propto \xi^{d+z} \propto |\delta|^{-\nu(d+z)}$. We use conventional notation for critical exponents, defined in Chapter 3. We do not review here standard scaling and hyperscaling arguments that establish relations between various critical exponents, and simply note that, for the most part, they are straightforward generalizations from d- to the $(d+z)$-dimensional space. For example, the hyperscaling relation gives the singular part of the classical free-energy density as $f \propto \xi^{-(d+z)} \propto \delta^{\nu(d+z)}$. Scaling properties of the superfluid stiffness and compressibility immediately follow from the geometric nature of winding numbers in spacetime, meaning that their mean-square averages are scale-invariant properties at criticality. From $\Lambda_s = \langle \mathbf{W}^2 \rangle / d\beta L^{d-1}$ and $\varkappa = \langle (\delta N)^2 \rangle \beta / L^d$, we immediately conclude that (by replacing system size and inverse temperature with the corresponding correlation lengths)

$$\Lambda_s \propto \xi^{-(d+z-2)} \propto \delta^{\nu(d+z-2)}, \qquad \varkappa \propto \xi^{-(d-z)} \propto \delta^{\nu(d-z)}. \tag{9.14}$$

On the MI side of the transition, the most important quantities characterizing the state are the energy gap $\Delta_{\mathrm{MI}} \propto \xi^{-z}$, the scattering length for quasiparticle excitations, $a_s \propto \xi$, and the effective quasiparticle masses $m_{p,h} \propto 1/(\Delta_{\mathrm{MI}} a_s^2) \propto \xi^{z-2}$. We stress that these are parameters within the MI phase in the vicinity of the tip, and the dynamic critical exponent z is referring to the multicritical point only. The scaling laws are based on the analysis of dimensions and irrelevance of the microscopic scale.

Let us employ the path-integral representation to visualize and interpret the picture of increasing quantum fluctuations inside the Mott state on approach to the tip while keeping the particle number exactly at $\langle v \rangle = 1$. As is customary in the physics of condensed quantum matter, we "subtract" an ideal vacuum state from consideration, which, in our case, is the classical Mott state represented by straight particle trajectories on every site; see Figure 9.7. Evidently, particle–hole fluctuations in the MI state are represented by closed spacetime loops, which proliferate in density and size on approach to the superfluid state. In a dense "loop soup," the difference between particle and hole excitations is washed out and, at the tip of the Mott lobe, a new emergent particle–hole symmetry sets in at large scales; that is, the statistical weights of trajectory pieces propagating forward and backward in time become equivalent. Indeed, the tip of the lobe is a special multicritical point, where the lines describing asymptotically weakly interacting gases of holes and particles meet each other. We immediately recognize an exact correspondence between this picture and statistics of closed loops in the classical XY [or generic $U(1)$] model in $(d+1)$-dimensional spacetime volume, where proliferation of loops signals the onset of the superfluid behavior. Note that space and time directions here become equivalent, which is a specific property of integer filling. We thus establish the fact that the quantum phase transition from superfluid to MI in d-dimensional system with integer filling belongs to the $(d+1)$-dimensional $U(1)$ universality class [14].

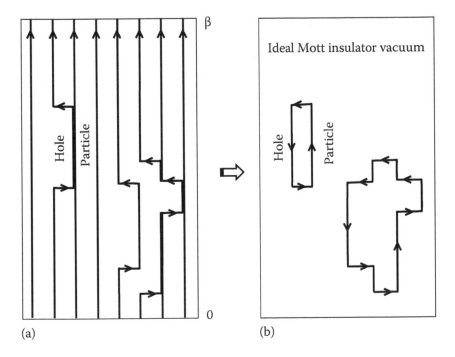

(a)   (b)

Figure 9.7   Particle–hole excitations in the Mott state (a). By subtracting trajectories for the ideal vacuum state, one obtains an identical picture (b).

The same conclusion about the nature of the criticality at the tip of the lobe can be derived by considering Popov's hydrodynamic action in the superfluid state [Equations (7.106) through (7.108)], which, for convenience, we reproduce here in the imaginary-time representation

$$S = \int d\mathbf{r} \int d\tau \left\{ i\langle n \rangle \dot{\Phi}(\mathbf{r}, \tau) + \frac{\varkappa}{2} [\dot{\Phi}(\mathbf{r}, \tau)]^2 + \frac{\hbar^2 \Lambda_s}{2} [\nabla \Phi(\mathbf{r}, \tau)]^2 \right\}. \tag{9.15}$$

As mentioned previously in Chapter 7 (see also the discussion of supercurrent states in 1D rings), the first term is topological in nature because it is based on the full-time derivative. For the phase variable subject to periodic boundary conditions, an integration over time may result only in a multiple of 2π or zero. Correspondingly, the contribution to the action of the vortex loop in the phase field is given by

$$i\gamma = i2\pi \int_A d\mathbf{r} \langle n \rangle, \tag{9.16}$$

where A is the projected (algebraic) volume of the spacetime loop. This formula plays a central role in the study of disordered lattice bosons below. In a clean system at integer filling factor, however, γ is always equal to a multiple of 2π and thus can be omitted in the exponent. More accurately, γ changes by exactly a multiple of 2π only if the boundary of the projected volume A is translated by exactly one lattice period. Nevertheless, this is sufficient for the conclusion of irrelevance of γ, since residual contributions are essentially microscopic and, as such, should be attributed to the surface terms associated with the boundary of A. We are left then with the standard classical complex-field action that belongs to the $(d+1)$-dimensional $U(1)$ universality class. It means that we are dealing with the $z = 1$ case with space and time directions appearing in the action on equal footing.

Hence, both path-integral representation and Popov's hydrodynamic action bring us to the central result [14] that, at the diagram tip, the physics in the long-wave limit is the Lorentz invariant, due to emergent spacetime and particle–hole symmetries. The key "relativistic" parameter is the sound velocity, c, which remains finite and nonsingular at the transition point because $c = \sqrt{\Lambda_s/\varkappa}$ and $\varkappa \propto \Lambda_s \propto 1/\xi^{(d-1)}$. The Lorentz-invariant spectrum of elementary excitations in the Mott phase can be expressed through the energy gap $\Delta_{\mathrm{MI}} \propto 1/\xi$, using standard formulae

$$mc^2 = \Delta_{\mathrm{MI}}/2, \qquad E_{p,h}(p) = \sqrt{m^2 c^4 + c^2 p^2}. \tag{9.17}$$

This is a remarkable example of the relativistic field theory realized in one of the simplest lattice models. What makes it particularly special is that it can be realized in tabletop experiments with cold atomic gases [19].

The spacetime symmetry at the tip of the MI lobe implies that the only relevant energy scale in the vicinity of the tip is given by the inverse correlation length. Therefore, the insulating gap (and thus the asymptotic shape of the lobe) is given by*

$$|\mu_\pm - \mu_c| = \frac{\Delta_{MI}}{2} \propto \xi^{-1} \propto \begin{cases} |\delta|^\nu & (d \geq 2), \\ e^{-\text{const}/\sqrt{|\delta|}} & (d = 1). \end{cases} \tag{9.18}$$

The same argument leads to the expression for the critical temperature of the superfluid transition on the approach to the lobe tip:

$$T_c(\mu, \delta) \propto \frac{1}{\xi(\mu, \delta)}. \tag{9.19}$$

An important circumstance here is the dependence of the correlation length on chemical potential. Close enough to the $\mu = \mu_c$ line (i.e., at an essentially constant integer filling), this dependence is negligible:

$$\frac{1}{\xi(\mu, \delta)} \sim \frac{1}{\xi(\mu_c, \delta)} \propto \begin{cases} \delta^\nu & (d \geq 2), \\ e^{-\text{const}/\sqrt{\delta}} & (d = 1) \end{cases} \qquad \left[|\mu - \mu_c| \lesssim \frac{1}{\xi(\mu_c, \delta)}\right]. \tag{9.20}$$

When the deviation from constant filling becomes substantial, the quantity $|\mu - \mu_c|$ starts to control the correlation length:

$$\xi(\mu, \delta) \sim \frac{1}{|\mu - \mu_c|} \qquad \left[|\mu - \mu_c| \gtrsim \frac{1}{\xi(\mu_c, \delta)}\right]. \tag{9.21}$$

For these large $|\mu - \mu_c|$, we have $T_c \sim |\mu - \mu_c|$.

As we cross the superfluid–MI boundary in the vicinity of the tip by varying μ, or doping the system with particles or holes, we are dealing with the dilute Bose gas of very light particles/holes and large effective coupling constant $g = 4\pi a_s/m \sim \xi^2$ (in $d = 3$). This description holds only for tiny doping levels $\delta n \ll \Delta_{MI}/g \propto 1/\xi^3$. To derive it, we require that the chemical potential of the quasiparticle gas is smaller than Δ_{MI} to ensure the validity of the dilute gas approximation. These results are readily verified in numerical simulations [20].

9.5   Superfluidity and Disorder

In previous chapters, we learned that superfluids are robust against momentum transfer to rough substrates, walls, and static impurities because their flows are topologically protected, provided they have the ability to develop the classical-field component. We will discuss below the conditions under which disorder can be ignored and what happens when it is strong enough to drive the system into the normal state.

---

* Note a very specific needle shape of the lobe in 1D, enforced by the exponential divergence of correlation length at the Berezinskii–Kosterlitz–Thouless (BKT) transition point.

9.5.1 Harris Criterion

We start our considerations from the case of weak disorder. There are unlimited possibilities to introduce various models of disorder, for example, in the form of randomly distributed point scatterers, a superposition of a macroscopic number of external potential waves with random amplitudes and phases, porous media, etc. Theoretically and conceptually, the simplest model of disorder is that of an external potential $V(\mathbf{r})$ described by random Gaussian statistics and finite correlation length in space. More precisely, ensemble-averaged properties of the disordered external potential are given by

$$\langle V(\mathbf{r}) \rangle = 0, \qquad \langle V(\mathbf{r}_1) V(\mathbf{r}_2) \rangle = v^2 c_{12} \equiv v^2 c_{12} (\mathbf{r}_1 - \mathbf{r}_2), \qquad (9.22)$$

where c_{12} is a dimensionless function with $c_{12}(0) = 1$ and finite decay (correlation) length l_c. Quantitatively, we define the correlation length from $\int d\mathbf{r} c_{12}(\mathbf{r}) = l_c^d$. All other multipoint correlations are reduced to pairwise averages using Wick's theorem; for example,

$$\langle V(\mathbf{r}_1) V(\mathbf{r}_2) V(\mathbf{r}_3) V(\mathbf{r}_4) \rangle = v^4 \left[c_{12} c_{34} + c_{13} c_{24} + c_{14} c_{23} \right]. \qquad (9.23)$$

The first requirement in Equation (9.22) is trivially satisfied for any model of disorder by shifting the chemical potential. Weak-disorder limit is defined as a random potential in which localized states with energy v are spread out over a distance much larger than l_c.

The first problem we address is whether the nature of the normal-to-superfluid transition—the $U(1)$-universality class—changes in the presence of weak disorder. An argument developed by Harris [21] is based on the central limit theorem (CLT). Since weak disorder can be relevant only on length scales much larger than l_c, the prime quantity to look at is the value of the potential integrated over a large volume $W \gg l_c^d$. An estimate for $V_W = W^{-1} \int_W d\mathbf{r} V(\mathbf{r})$ immediately follows from the CLT for any random potential

$$P(V_W) \propto e^{-V_W^2/2\delta^2}, \qquad \delta^2 = v^2 l_c^d / W. \qquad (9.24)$$

This result is exact for the Gaussian model (9.22) and (9.23).

Problem 9.4 *Prove Equation (9.24) for the Gaussian model of disorder. Recall that Wick's theorem leads to the following identity*

$$\left\langle e^{F[V(\mathbf{r})]} \right\rangle = e^{(1/2)\langle F[V(\mathbf{r})]F[V(\mathbf{r})] \rangle}, \qquad (9.25)$$

for any linear functional $F[V(\mathbf{r})]$.

Thus, in a large volume W, it is typical to find that the average of the external potential in this volume fluctuates by an amount $\sim v \sqrt{l_c^d / W}$. On the other hand, such fluctuations are indistinguishable from the chemical potential shift in this volume and zero, on average, random potential. At this point, we recall that detuning chemical potential from the critical point by $|\mu - \mu_c|$, keeping all other parameters fixed, introduces a finite correlation length $\xi \propto |\mu - \mu_c|^{-\nu}$. The Harris criterion then is to check whether effective chemical potential shifts induced locally by fluctuations of disorder potential inside the correlation volume ξ^d are smaller than $|\mu - \mu_c|$. If this is the case, then weak disorder is irrelevant for criticality. By substituting $W = \xi^d$ into δ [Equation (9.24)], we find that $\delta \propto \xi^{-d/2}$, which is asymptotically much smaller than $|\mu - \mu_c| \propto \xi^{-1/\nu}$ if the correlation length exponent satisfies the Harris criterion

$$\nu > 2/d. \tag{9.26}$$

A convenient way to express this relation for the classical phase transition is to use the specific heat exponent

$$\alpha < 0, \tag{9.27}$$

where we have used the hyperscaling relation $\alpha = 2 - d\nu$ in d-dimensions. The Harris criterion suggests (but cannot guarantee; see below) that the classical (finite-temperature) $U(1)$ universality class is not modified by weak uncorrelated disorder because in 2D, the divergence of the correlation length is exponential in $\sqrt{|\mu - \mu_c|}$ and faster than any power law; in $d = 3$ we have $\nu = 0.6717 > 2/3$; and in $d = 4$ (at the upper critical dimension) where $\nu = 1/2$, it is satisfied due to logarithmic corrections to power laws.

The Harris criterion equally applies to classical and quantum criticality because the relevant comparison is done at the level of control parameters in the Hamiltonian. However, one should not be misled by this statement because disorder is far more dangerous in changing the quantum phase transition than the classical one! Physically, this is rather obvious because static disorder is infinitely correlated in the time direction. Mathematically, for quantum phase transitions, the correlation length exponent in (9.26) is defined for the $(d + z)$-dimensional universality class, and this makes all the difference.

It should be emphasized that the Harris, argument does not provide a guaranteed criterion of *irrelevance* of disorder. Indeed, dealing with a specific mechanism of how disorder can become relevant, the argument provides only a sufficient—rather than necessary—condition for the relevance of disorder. In this context, most illustrative is the superfluid–MI quantum phase transition. As we will see later, the transition is completely destroyed by disorder in all spatial dimensions. Meanwhile, the Harris criterion reveals the relevance of disorder (in on-site potential) for the superfluid–MI transition only in $d \leq 2$ (see Section 9.6).

The Harris criterion is modified if disorder has long-range spatial correlations. Consider Gaussian disorder with the nonintegrable power-law decay of the correlation function $c_{12}(r) \propto r^{-d+\psi}$. The formula for the typical disorder fluctuation in a volume $W = \xi^d$ now takes the form $\delta^2 = \left(v^2/W\right) \int_W d\mathbf{r}\, g(\mathbf{r})$ leading to $\delta \sim v\xi^{-(d-\psi)/2}$ and a modified criterion

$$\nu > 2/(d - \psi). \tag{9.28}$$

Since in three dimensions $\nu = 0.6717$ is extremely close to 2/3, the Harris criterion (9.28) is easily violated by correlated disorder. The best studied case is superfluidity of ^{4}He in Aerogel [22,23], which is a highly porous glass consisting of silica strands of few nm in diameter and a typical distance between the strands of up to 200 nm. The porosity depends on the fabrication process. Long-range correlations are reflected in the fractal structure of strands and are well represented by the diffusion-limited cluster aggregation (DLCA) model. One indeed finds both experimentally and theoretically (within the DLCA model) [24] that in this case the critical exponents for the normal-to-superfluid transition change from the conventional $U(1)$ universality class, for example, $\nu \approx 0.73$.

9.5.2 Bose Glass

We now turn to the interplay between superfluidity and localization in the ground state, that is, to the nature of the quantum phase transition between disordered insulator (BG) and the superfluid liquid as the disorder strength or interactions between bosons are changed. To have a meaningful discussion of localization effects, the random potential distribution should not feature a substantial δ-function at the lowest possible value. Indeed, consider a counterexample of randomly distributed hard-sphere impurities with small excluded volume fraction. In this case, particles will remain delocalized at the lowest energies. However, the moment we allow a tiny, but nonzero, probability of having negative values of the potential, we create a small concentration of localized states. In the discussion to follow, we assume that the lowest single-particle energy states in the disorder field are localized. The uncorrelated Gaussian disorder model (9.22) and (9.23) satisfies this requirement.

An ideal Bose gas at $T = 0$ is immediately localized by disorder because all particles go to the lowest-energy bound state.* This state is pathological in the thermodynamic limit because it admits infinite local particle-number density. When interactions are included, the rise of the local mean-field chemical potential $\propto gn(r)$ forces bosons to occupy higher and higher bound energy states until all particles are accommodated. This leads to the picture of localized condensates, well separated from each other in space; see Figure 9.8.

Within the classical-field description, the state remains superfluid since the field amplitude is finite and positive everywhere, though the superfluid density is predicted to be exponentially small for large separation between the lakes, R_l. However, the classical-field approach is invalid on large scales. For isolated condensate lakes, the spectrum of energy levels describing different particle numbers is quantized. If the lake size, determined by the localization length at energy μ, is r_l and compressibility is $1/g$, then the energy gap between the successive levels (for the grand-canonical Hamiltonian) is $\delta E = E_{N_i+1} - E_{N_i} - \mu \sim g/r_l^d$. This must be compared with the tunneling amplitude between the lakes, which is exponential in $\Gamma \propto e^{-R_l/r_l}$. At $\Gamma \lesssim g/r_l^d$, the particle-number quantization/localization in lakes

---

* As discussed previously, ideal bosons do not support stable nondissipative current states in the presence of *any* disorder, that is, even in the absence of localization.

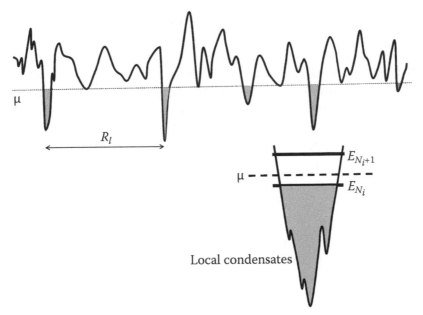

Figure 9.8   Local structure of the BG phase for the weakly interacting Bose gas in a disordered potential. The deepest bound states are occupied by bosons up to the global chemical potential.

is of crucial importance, so that the interlake coherence does not obey classical-field description. For $\Gamma \ll g/r_l^d$, particle exchange between lakes is blocked by interactions, in close analogy with the standard picture of Anderson localization for single-particle states. The result is an insulating BG phase with no phase coherence on large scales.

Statistically, the BG phase shares many similarities with fermionic systems because the low-lying excitations are "particles" and "holes" with the hard-core repulsion constraint. To construct an excitation with energy ϵ, one must find two lakes with the energy levels $E_{N_i} = \mu - \epsilon_h$ and $E_{N_j+1} = \mu + \epsilon_p$ and transfer a particle from lake "i" to lake "j." Since the bottom of the potential well is a continuous variable for disorder in the chemical potential, the probability density of having an energy level in a given lake in the vicinity of μ is a constant determined by the level spacing times $\sim r_l^d/g$. The density of states at the chemical potential involves also the concentration of lakes leading to

$$\rho_\mu \sim \frac{r_l^d}{g} \cdot \frac{1}{R_l^d} \sim \frac{r_l^d}{g R_l^d}. \tag{9.29}$$

The energy of the BG phase at finite temperature is then found from the standard fermionic expression

$$E(T) - E(0) = 2\rho_\mu \int_0^\infty \epsilon n_\epsilon \, d\epsilon = \frac{\pi^2}{6}\rho_\mu T^2, \qquad n_\epsilon = \left(e^{\epsilon/T} + 1\right)^{-1}. \tag{9.30}$$

The corresponding specific heat is linear in the $T \to 0$ limit [14], which is typical for structural glasses.

The region of validity of the T^2-law is limited by the lake size, which may be large for weak disorder. Fermi statistics provides an accurate description of particle-number fluctuation as long as temperature is much smaller than the energy-level spacing for adding/removing particles to/from the lake. For $T > g/r_l^d$, the appropriate picture is that of independent particle-number fluctuations controlled by the local compressibilities, $H = \sum_i C_i (\delta N_i)^2$ with $C_i \sim g/r_l^d$. In the limit of high temperatures, when $\langle (\delta N_i)^2 \rangle \gg 1$, it describes a collection of classical harmonic modes with the average energy $T/2$ per mode

$$E(T) - E(0) \sim \frac{T}{R_l^d} \qquad \left(T \gg g/r_l^d \right). \tag{9.31}$$

We note that Equations (9.30) and (9.31) utilize particle-number fluctuations in condensate lakes as well-defined localized degrees of freedom. In the superfluid state, density and phase fluctuations become inseparable and ultimately lead to the spectrum of collective sound excitations.

9.5.3 BG-to-Superfluid Transition in a Weakly Interacting System

As a generic second-order quantum phase transition, the transition from BG to superfluid features universal properties at criticality. However, if interaction is weak, critical universality takes place only in a very narrow region of phase diagram—in close vicinity of the critical line, the universal properties developing only at a very large distance compared to characteristic microscopic length scales. Meanwhile, the position of the critical line and mesoscopic properties are dictated by specific physics of weakly interacting systems—most notably, the fact that the latter can be viewed as linked condensate lakes. Here, we outline corresponding theory estimating critical parameters.

Typical parameters of condensate lakes can be calculated for the weakly interacting gas using Equation (9.24) by counting the number of particles that occupy potential wells of volume W [25,26]. If the well of size r_l is to support a bound state with binding energy $|E| \sim 1/mr_l^2$, it must be deeper than $1/mr_l^2$ (here, we focus on the most relevant $d = 3$ case). Exponential dependence of the distribution (9.24) on V_W and W ensures that these parameters should be set at the smallest allowed values mentioned earlier. By substituting in Equation (9.24) the lake volume with $r_l^3 \sim (m|E|)^{-3/2}$ and V_W^2 with $|E|^2$, we arrive at the probability density for finding a bound state with energy $-|E|$

$$P(E) \propto e^{-\sqrt{|E|/E_D}}, \qquad E_D \sim m^3 v^4 l_c^6. \tag{9.32}$$

The number of particles that can be accommodated on energy levels up to the chemical potential is then approximately given by the integral

$$\frac{N}{V} \sim \int_{-\infty}^{\mu} \left(\frac{\mu - E}{g}\right) P(E)\, dE = \frac{1}{g} \int_{0}^{\infty} z P(\mu - z)\, dz. \tag{9.33}$$

The analysis is simplified due to sharp exponential decay of $P(E)$ for $|E| \gg E_D$. For fixed particle density and $g \to 0$, the dominant contribution comes from $|\mu| \gg E_D$ leading to

$$|\mu| e^{-\sqrt{|\mu|/E_D}} \sim ng. \tag{9.34}$$

In this regime, the typical lake size $r_l \sim (m|\mu|)^{-1/2}$ is much smaller than the separation between lakes $R_l \sim r_l e^{\frac{1}{3}\sqrt{|\mu|/E_D}}$. The corresponding state is insulating because the tunneling rate is proportional to the exponential of $-R_l/r_l$ reflecting the decay of the bound-state wavefunction in space, while the energy-level spacing in the well is only a power-law function $\propto 1/r_l^3$. This prohibits delocalized particle-number fluctuations.

The transition to the superfluid state happens when condensate lakes start to overlap, $R_l \sim r_l$ or, equivalently [25,26],

$$|\mu| \sim ng \sim E_D \propto v^4. \tag{9.35}$$

One consequence of the $g \sim v^4$ law for the BG–superfluid phase boundary is the remarkable ability of bosonic systems to screen weak disorder potentials and to support superfluidity. In practice, even relatively strong disorder $v > 1/ml_c^2$ has only tiny, barely detectable, effect on the superfluid properties of systems with extremely small gas parameter $na_s^3 \sim 10^{-6}$ [27,28].

The physics of weak links between condensate lakes has to be treated with special care in 1D systems because exponentially rare exponentially weak links create effective "road blocks" that cannot be avoided. The proper discussion goes outside the scope of our book, and we refer interested reader to Reference [29] where it is explained that the superfluid-to-insulator transition may occur at a nonuniversal value of Luttinger parameter.

9.6   Theorem of Inclusions

It turns out that, under rather general assumptions about disorder properties, one can prove a simple theorem imposing severe constraints on the topology of the ground-state phase diagram of a disordered system. Our discussion in this section is not limited to superfluid systems. The theorem of inclusions [27,30] and its consequences apply to an arbitrary phase transition, in any dimension, and to any type/nature of disorder. In subsequent sections, we will use it to discuss phase diagrams of disordered lattice bosons.

At this point, it is important to introduce the notion of generic disorder as a random potential that has a nonzero probability density of realizing any value of the potential within the bounds; that is, $P(V(\mathbf{r})) \neq 0$ for any $V(\mathbf{r}) \in [-\Delta, \Delta]$. The theorem of inclusions then states the following:

In a system with generic disorder and in the vicinity of the transition line between phases A and B—and if the transition is not of the Griffiths type—one can always find arbitrary large inclusions of phase A (B) in phase B (A).

The proof of the theorem is based on the following observation about properties of generic disorder: Since the probability density of having any value of $V(\mathbf{r})$ is nonzero, the probability density of having an arbitrary realization of disorder field satisfying the same bound in any finite volume is also finite. This rather trivial observation implies that one model of disorder characterized by $P_1(V(\mathbf{r}))$ and a set of multipoint correlation functions can mimic another (any other) model of disorder characterized by $P_2(V(\mathbf{r}))$ and a different set of correlation functions in an arbitrary large, but finite, volume. The probability of observing events when a rare disorder realization for one model reproduces a typical realization for another model decreases exponentially with the volume of the corresponding region.

Consider now the phase transition line between A and B depicted in Figure 9.9. In a given model of disorder (e.g., with uniform probability density and intermediate-range spatial correlations), the transition from A to B happens at the point Δ_c as the disorder bound is increased. Even though $\Delta > \Delta_c$ in phase B, there exist finite regions in which disorder never exceeds a value $\Delta' < \Delta_c$; that is, it mimics the same model with a smaller disorder bound. Since the size of the rare region can be arbitrarily large and $\Delta_c - \Delta'$ is finite, we conclude that in this region, one will observe phase A locally.

The same argument cannot be used to construct inclusions of B in A, because, in phase A, the disorder bound is smaller than Δ_c. However, one can make use of the transition line dependence on other properties of disorder such as the shape of the distribution function (shown to the right in Figure 9.9), the correlation length, and the shape of c_{12}. Formally, there are infinitely many parameters controlling properties of the disorder field, which we collectively denote by $\vec{\zeta}$ and represent by the vertical axis in Figure 9.9. Any of these parameters can be used to shift the transition point between A and B so that $\Delta_c\left(\vec{\zeta}_2\right) < \Delta < \Delta_c\left(\vec{\zeta}_1\right)$. Now we proceed with constructing large domains of phase B inside A using statistically rare realizations of disorder that mimic properties of model $\left(\Delta', \vec{\zeta}_2\right)$ with $\Delta_c\left(\vec{\zeta}_2\right) < \Delta' < \Delta$ in this domain. This completes the proof of the theorem for the transition line with meaningful dependence of Δ_c on $\vec{\zeta}$; in what follows, we will call such transitions generic to distinguish them from the Griffiths type singularity, discussed next.

The only scenario that does not allow one to construct B-domains in A is realized when Δ_c does not depend on *any* properties of the disorder field except its bound; this would correspond to having a vertical transition line in Figure 9.9. [Having a minimum of Δ_c at $\vec{\zeta}_1$ in the infinite-dimensional continuous space of parameters is of zero measure.] Independence of Δ_c on $\vec{\zeta}$ means that the nature of the transition is solely determined by the possibility of creating a special regular

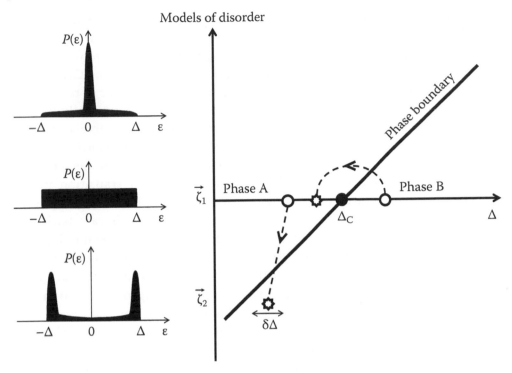

Figure 9.9   An illustration for the theorem of inclusions: Rare regions in a disordered system in the vicinity of a generic phase boundary. The particular position of the transition point between phases A and B depends on all the characteristics of disorder, such as the distribution function of random fields (shown to the left) and spatial two-point and multipoint correlation functions. Dashed lines with arrows originate from points describing disorder properties in the thermodynamic limit. The end points of the dashed lines correspond to disorder properties in an arbitrarily large, but finite, domain representing a rare statistical fluctuation in the same system as at the origin of the line.

field with amplitude Δ_c that would cause a phase transition in a regular system. This is a familiar example of the Griffiths type phase transition, which works as follows: Suppose that in a pure system a phase transition from A to C takes place when a regular external field V_{ext} exceeds some critical value Δ_c. The Griffiths transition from A to B in the disordered system then proceeds by means of creating large rare domains (with the domain size diverging and the probability vanishing at the transition point) of ideal phase C inside the disordered phase A. In other words, if not for these exponentially rare domains, the two phases look identical to each other locally.

   Despite its simplicity, the theorem of inclusions immediately imposes severe restrictions on the allowed types of phase transitions. For example, it prohibits transitions between the fully gapped (A) and superfluid (B) phases. If the hypothetical transition line is generic, then we immediately arrive at the self-contradicting conclusion that A must be gapless. Indeed, in the superfluid phase, there is no gap for adding/removing particles to/from the system. Since, in the vicinity of the transition line, one can always find arbitrary large super-fluid domains inside A, there is no gap with respect to adding/removing particles to/from the phase A either. If the hypothetical transition is of the Griffiths type, then we must conclude that B is not superfluid because it consists of rare (possibly superfluid) lakes without global phase coherence due to exponentially large sep-aration between them. Moreover, this consideration forces one to accept that all transition lines between fully gapped and gapless phases in disordered systems are of the Griffiths type and the corresponding gapless phases are insulating.

Our last observation concerns the superfluid–insulator transition in a sys-tem with short-range interactions when disorder is in the chemical potential. This transition cannot be of the Griffiths type because in this case both phases would be either insulating or superfluid. For the generic line shown in Figure 9.9, assuming that A is insulating, we see that the local chemical potential in the superfluid domain can be shifted continuously within a finite interval of width $\Delta\left(\vec{\zeta}_1\right) - \Delta_c\left(\vec{\zeta}_2\right)$. Since the energy gap for adding/removing particles to/from the superfluid domain vanishes as its size is increased, the density of states, ρ_μ, gener-ated by shifting these energy levels within a finite interval, is nonzero. This proves that an insulating phase is compressible, not just gapless.

Finite compressibility across the superfluid–BG boundary leads to the impor-tant relation for the dynamic critical exponent [14]. Starting from the exact compressibility formula in terms of particle-number fluctuations, we obtain the critical contribution by noting that for quantum critical modes $1/T \propto L^z$ and $\left\langle (\delta N)^2 \right\rangle \sim 1$:

$$\varkappa = \frac{L^d}{T}\left\langle (\delta N)^2 \right\rangle \longrightarrow \varkappa_{\text{crit}} \sim L^{d-z}. \tag{9.36}$$

By insisting that $\varkappa_{\text{crit}}$ stays finite in the thermodynamic limit, we get $z = d$. This is an exact relation. Indeed, on approach to the critical point at $\mu = \mu_c$, the compress-ibility can be formally decomposed into critical and regular (nonsingular) parts, $\varkappa(\mu) = \varkappa_{\text{crit}}(\mu - \mu_c) + \varkappa_{\text{reg}}(\mu)$ with $\varkappa_{\text{crit}}(\mu - \mu_c) \propto (\mu - \mu_c)^{\nu(d-z)}$. One may speculate that finite $\kappa(\mu_c)$ can be due to the regular part while the critical part vanishes at $\mu = \mu_c$. However, this possibility is ruled out by observing that finite $\varkappa$ in BG is due to localized single-particle and single-hole states while such states do not exist in the superfluid phase. Thus, the finite compressibility at $\mu = \mu_c$ is entirely due to critical modes.

9.7 Superfluidity in Disordered Lattice Models

An interplay between disorder and interactions is more subtle when interactions are strong because interactions themselves can localize particles into the crys-talline state or the MI. To be specific, in this section, we will focus on the dis-ordered Bose–Hubbard model (8.1) as the prototypical system. Consider the case of integer filling factor $\langle n \rangle = 1$ and spatially uncorrelated disorder in the chemical potential with μ_i uniformly distributed on the interval $(\mu-\Delta, \mu+\Delta)$. In the presence of weak disorder and extremely small interactions, the ground state is localized in the BG phase. As U is increased [see Equation (9.35)], the system undergoes the BG–superfluid transition and enters the superfluid phase, which persists to large values of U, at which point the MI physics comes into play. It is evident that, in the limit $U \gg t, \Delta$, the MI ground state with large gaps for particle and hole excitations is stable against disorder. This means that superfluidity will eventually disappear; that is, the insulating behavior is *re-entrant* in U [31].

Even more nontrivial is re-entrant behavior as a function of Δ starting from the MI state. For $\Delta > \Delta_{MI}/2$, the MI state becomes unstable against formation of domains containing a nearly pure system with local chemical potentials violating the MI bounds μ_+ and μ_- [14], large enough to dope them with just one particle. According to the theorem of inclusions, this Griffiths type transition is the only allowed possibility. The phase that emerges for $\Delta > \Delta_{MI}/2$ is a compressible insulator as follows from considerations identical to those presented in the next to last paragraph of the previous section; that is, it is a BG. As disorder is increased further, the domains doped with particles and holes become more and more numerous. At this juncture, we face two alternative scenarios:

1. Domains doped with particles and holes start to overlap* and the system enters the SF phase. This is an interesting example of how disorder promotes superfluid properties by destroying the competing insulating phase. At a higher disorder amplitude, it localizes again into the BG phase.

2. Localization takes place without entering the superfluid phase.

In the following, we show that, in 1D, the first scenario is unavoidable in the vicinity of the MI tip [32]. Interestingly enough, numerical studies establish that the first scenario also takes place in other physical dimensions [31]. For very large values of U, the second scenario takes place.

The picture that emerges from this discussion and the theorem of inclusions is that, for integer filling factor, the phase diagram in the $(\Delta/t, U/t)$ plane consists of the superfluid region surrounded from all sides by the BG phase. The MI–BG boundary at large U/t obeys the $\Delta = \Delta_{MI}/2$ law and is of the Griffiths type. This fixes the topology of the phase diagram. The actual shape of the superfluid–BG boundary and its re-entrant nature as a function of Δ are obtained from numerical simulations; see the 3D example in Figure 9.10.

9.7.1 Superfluid–BG Transition in 1D and 2D

Let us examine in detail the microscopic mechanism that leads to the BG phase separating superfluid and MI states and re-entrant behavior in 1D [32]. Our starting point is the Euclidean superfluid hydrodynamic action (9.15) in the presence of disorder, which we rewrite identically using the 2D "coordinate" $\mathbf{r} = (x, c\tau)$

---

* The largest domains may be viewed as superfluid lakes embedded into disordered MI phase.

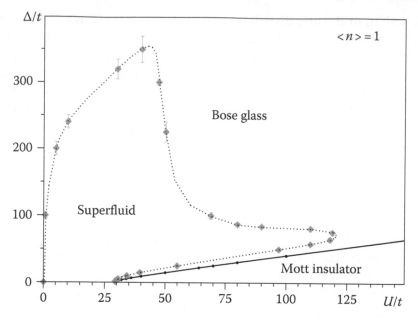

Figure 9.10   Phase diagram of the disordered 3D Bose–Hubbard model at unity filling. (Adapted from Reference [27].) The quantum phase transition in a pure system between superfluid and MI phases occurs for $\Delta = 0$. In the presence of disorder, a compressible, insulating BG surrounds the superfluid from all sides, as dictated by the theorem of inclusions [30] (dotted line is to guide the eye). The transition between MI and BG is of the Griffiths type [27]. As $U/t \to 0$, the superfluid-to-BG transition line has an infinite slope [25,26].

$$S = \int d\mathbf{r} \left\{ in(x) \frac{\partial \Phi(\mathbf{r})}{\partial y} + \frac{K}{2\pi} [\nabla \Phi(\mathbf{r})]^2 \right\}, \tag{9.37}$$

where $c = \sqrt{\hbar^2 \Lambda_s / \varkappa}$ is the velocity of sound, $K = \pi \sqrt{\Lambda_s \varkappa}$ is the Luttinger liquid parameter, and $n(x)$ is the average particle density at the point x. Due to periodic boundary conditions in y-direction, the first term, $\mathrm{Im}S$, is nonzero only if the phase field contains topological defects in the form of xy-vortices [in the space-time representation, they are also known as (phase slippage) instantons, and, in this section, we will use both names interchangeably]. Each time the line of integration over y crosses a vortex with charge q_i centered at point X_i, the value of the y-integral in $\mathrm{Im}S$ changes by $2\pi q_i$. Summing over all defects, we obtain

$$\gamma = \mathrm{Im}S = 2\pi \sum_i q_i \int_0^{X_i} dx\, n(x). \tag{9.38}$$

The lower limit of integration can be set at any space point with the same final result for any neutral, $\sum_i q_i = 0$, vortex configuration. The ith term in the sum should be regarded as the phase of the ith instanton. Since, in the partition function, one must sum over all possible instanton positions, the instanton phases may interfere constructively or destructively; in the latter case, the instanton contributions to the action may be neglected [33].

In a pure system at some generic (incommensurate) filling factor, we immediately see that the mechanism of destructive phase interference between instantons at different space locations makes instantons irrelevant and preserves the superfluid state. This is yet another way of expressing momentum conservation law in a pure homogeneous system. At an integer filling factor, we observe an opposite phenomenon when all phase factors in (9.38) are multiples of 2π. In this case, all instantons are in phase and we are dealing with the familiar classical xy-model that has the standard BKT transition between the superfluid and MI phases at $K = K_{MI} = 2$. Finally, at a fractional finning factor, $\langle n \rangle = s/q$, the phase factors are always multiples of 2π only for vortices with charge q (or multiple of q), which leads to the BKT transition at a critical value of the Luttinger parameter $K_q = 2/q^2$. These conclusions are in agreement with the original results by Haldane [34].

We now turn to the analysis of a system with integer filling factor in the presence of weak disorder. As before, we will assume that disorder is uncorrelated between lattice sites and its amplitude has a bound Δ. For weak disorder and in the long-wave limit, the linear static response of the local density to the local change in the chemical potential is controlled by compressibility, which remains finite in the superfluid phase. From Equation (9.38), we see that the phase factor for the vortex–antivortex pair is determined by the total particle-number fluctuation on the interval of length $R = X_1 - X_2$, which can be estimated as

$$\int_{X_1}^{X_2} dx\, n(x) \approx \kappa \int_{X_1}^{X_2} dx\, V(x) \propto \varkappa \Delta \sqrt{aR}, \tag{9.39}$$

where in the past relation we have used formula (9.24) adapted for the lattice model with $l_c = a$ and $v \sim \Delta$.

Recall that the BKT transition is described by the renormalization group (RG) equation based on vortex pairs of large sizes. On approach to the transition point, the correlation length diverges exponentially $\xi \propto \exp(-C/\sqrt{|p|})$ where $p = U/t - (U/t)_c$ is a dimensionless parameter controlling detuning from the superfluid–MI critical point. By a posteriori analysis, one can verify that any regular dependence of the critical point (and detuning) on disorder strength can be neglected in this relation. The value of the energy gap in the MI phase is inversely proportional to the correlation length because the dynamic critical exponent is unity, $\Delta_{MI} \propto 1/\xi$. From this consideration, we immediately see that disorder amplitudes much smaller than Δ_{MI} have no effect on the RG flow toward MI phase because corresponding phase terms are negligibly small: $2\pi\varkappa\Delta\sqrt{a\xi} \propto \sqrt{a/\xi} \ll 1$. On the other hand, for $\Delta > \Delta_{MI}/2$, the insulating state must be BG according to the rare region Griffiths scenario [14] and the theorem of inclusions [27]. We thus conclude that, for Δ in the range $\Delta_{MI} < \Delta \ll \sqrt{\Delta_{MI} U}$, the system is guaranteed to be in the insulating BG phase for $p > 0$.

For larger disorder amplitudes, the RG flow should be modified because of the destructive phase interference eliminating contributions from vortex pairs separated by spatial distances larger than

$$R_* \sim \left(4\pi^2\varkappa^2\Delta^2 a\right)^{-1} \propto a(U/\Delta)^2. \tag{9.40}$$

This means that vortex pairs become more and more elongated in the vertical (time) direction under the RG flow. The critical value of K_c for proliferation of vertical vortex pairs can be determined using standard energy vs. entropy considerations: The contribution of the vortex pair of size L to the action is $2K\ln L$, while its entropy is $3\ln L$ leading to the critical value $K_{BG} = 3/2$, in agreement with Reference [35]. The RG equations for vertical vortex pairs at length scales $R > \xi$ are derived from the effective action in close analogy with the standard KT treatment. The new feature is associated with the fact that vortex pairs couple only to currents perpendicular to the vector connecting two vortices in the pair, meaning that vertical pairs do not renormalize compressibility $\varkappa$ (linear response coefficient for $\partial\Phi/\partial y$). Thus, the quantities subject to the RG flow are the superfluid stiffness $\Lambda_s = K^2/\pi^2\varkappa$ and sound velocity $c = K/\pi\varkappa$. As before, the renormalization of the superfluid stiffness by vertical vortex pairs with sizes corresponding to the length scale $R/a = e^{-\lambda}$ is proportional to the probability of finding such pairs in an area R^2 multiplied by the square of the pair size. This leads to the standard equation

$$d\Lambda_s(\lambda)/d\lambda \propto g_v(\lambda) \tag{9.41}$$

with

$$dg_v(\lambda)/d\lambda = -(2K - 3)g_v(\lambda). \tag{9.42}$$

Since the second equation is invariant under multiplication of the $g_v(\lambda)$ by a constant factor, any coefficient in the first equation can be scaled out. This allows us to replace Λ_s with K in Equation (9.41) by noting that K remains finite at the transition point. This leads to the familiar pair of KT equations

$$dK/d\lambda = g_v, \qquad dg_v/d\lambda = -2(K - 3/2)g_v, \tag{9.43}$$

describing the flow of the system toward smaller values of K on length scales $\gg R_*$.

In the vicinity of the superfluid–MI transition in the pure system and on the MI side of the critical point, both ξ and R_* are much larger than all other microscopic scales. In this regime, the RG equations can be initialized on a large length scale l where the mesoscopic value of K is close to but above 2 and $g_v(l) \ll 1$ (see Figure 9.11). For $\xi(p) > R_*$, the flow is essentially indistinguishable from that in a pure system, though the actual insulating phase is BG. As R_* gets comparable to $\xi(p)$, the flow around R_* crosses over to the disordered scenario [Equation (9.43)]. The fate of $K(\infty)$ is determined by the fine competition at the crossover scale,

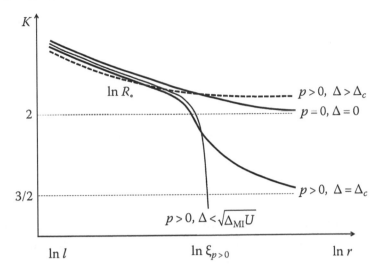

Figure 9.11   An illustration of the RG flow in the vicinity of the superfluid–MI critical point in the pure system. Here, positive $p = U/t - (U/t)_c$ corresponds to the MI phase in the absence of disorder.

with the BG phase taking place when, right after the crossover, K is smaller or slightly above the critical value of 3/2. As R_* gets smaller than $\xi(p)$, the system enters the superfluid phase. It is guaranteed that, for weak disorder such that $t \gg \Delta \gg \sqrt{\Delta_{MI}U}$, the system is superfluid, because, in this case, at the crossover scale, we have $K(R_*) > 2$, which is larger than 3/2 by a finite amount that cannot be renormalized down substantially by small values of g_v; see Figure 9.11.

We thus conclude that, as disorder amplitude is increased starting from the MI phase with small gap $\Delta_{MI} \ll t$, the system first enters the BG phase at $\Delta = \Delta_{MI}/2$ (the transition line is undetectable numerically and experimentally), then goes superfluid at $\Lambda \sim \sqrt{\Delta_{MI}U}$, and finally localizes again though the re-entrant superfluid–BG transition at large disorder strength $\Delta \gg t$. The overall shape of the phase diagram in 1D is qualitatively similar to that presented in Figure 9.10, though with less dramatic "sleeve" of the superfluid phase at large values of U/t [36].

The mechanism of disorder-induced superfluidity in the vicinity of the superfluid–MI point rests on the fact that $K_{MI} = 2$ is *larger* than $K_{BG} = 3/2$. For fractional filling factors $\langle n \rangle = s/q$, the superfluid–MI point corresponds to values of the Luttinger liquid parameter $K_q = 1/q^2$. Those are *smaller* than K_{BG}, meaning that weak disorder is relevant and localizes particles into BG in a finite region in parameter space corresponding to $K_{BG} > K(U/t) > K_q$. In this case, disorder works against superfluidity with extreme efficiency.

Nearly identical considerations can be also employed to establish relevance of disorder for the transition between the superfluid and insulating states in the vicinity of the Mott transition in 2D [37]. Again, we start with the analysis of the phase term, Equation (9.16), and, similarly to Equation (9.39), estimate its value for a vortex-loop configuration in the $(d+1)$-dimensional phase field

$$i2\pi \int_A d\mathbf{r}\, n(\mathbf{r}) \approx i2\pi\varkappa(\xi) \int_{A\sim\xi^d} d\mathbf{r}\, V(\mathbf{r}) \propto i2\pi\varkappa(\xi)\Delta\xi^{d/2}. \qquad (9.44)$$

In the last relation, we used the CLT to find the fluctuation of the chemical potential in the projected loop area $A \sim \xi^2$ and explicitly mentioned possible compressibility dependence on the length scale in the critical region. If we assume that compressibility goes to zero, according to the Josephson relation for the $(d+1)$-dimensional $U(1)$-universality class, $\varkappa(\xi) \propto 1/U\xi^{d-1}$, we find that, on approach to the superfluid–MI transition point, vortex loops of typical size ξ acquire a phase term that does not depend on the loop size in 2D. More precisely, it means that different length scales make equal contributions to the phase term and, thus, loops of size ξ acquire phases that scale logarithmically with their size:

$$\gamma \sim (\Delta/U)\ln(\xi/a) \qquad (d=2). \qquad (9.45)$$

The final conclusion then is that the RG flow toward MI state will eventually be stopped by randomness of vortex-loop phases. In higher dimensions, one must invoke additional microscopic mechanisms leading to finite compressibility of states with $\Delta > \Delta_{MI}$ in the thermodynamic limit.

The Harris criterion for the superfluid–MI transition point with disorder in the chemical potential must be based on Equation (9.21)—thus reflecting the fact that μ is a variable tangential to the phase boundary at the diagram tip. As a result, we readily see that the Harris argument guarantees relevance of disorder only in $d = 1, 2$. It does not, however, properly cover higher dimensions, in which disorder is relevant as well (to the extent that a direct transition from MI to SF is completely destroyed by an arbitrarily weak disorder).

9.7.2 Mott Glass

In this subsection, we will examine the nature of the insulating state emerging in disordered systems with exact particle–hole symmetry. By definition, it must be a lattice model since particle–hole symmetry is absent in continuous condensed-matter Hamiltonians. A typical physical example is provided by the quantum rotor model [Equation (7.91)], with zero "chemical potential" σ:

$$H = -\sum_{\langle ij\rangle} J_{ij}\cos\left(\phi_i - \phi_j\right) + \sum_i \frac{U_i}{2}\delta n_i^2. \qquad (9.46)$$

Both the exchange coupling J and the on-site coupling U are subject to disorder fluctuations and may vary from bond/site to bond/site, $J_{ij} \in \left(J-\Delta_J, J+\Delta_J\right)$, $U_i \in (U-\Delta_U, U+\Delta_U)$. To simplify the discussion, we assume that disorder is present only in the exchange coupling, while all considerations and conclusions are equally valid for the generic model. For unrestricted density fluctuations,

$\delta n = 0, \pm 1, \pm 2, \ldots$, the quantum rotor model is invariant with respect to the particle–hole transformation, $\delta n \to -\delta n$. (Recall that the quantum rotor model describes the limit of large occupation numbers in the Bose–Hubbard Hamiltonian.)

In the pure system, we observe the standard superfluid–MI transition at some critical ratio $U/J = (U/J)_c$. When the theorem of inclusions is applied to the disordered model, we conclude that a direct transition between fully gapped MI and superfluid phases is prohibited and the two must be separated by some insulating gapless state. In what follows, we will call it the MG state because, as we will see shortly, it is incompressible and thus different from the compressible BG. The Griffiths line between MI and MG states is determined by the condition

$$(J + \Delta_c)/U = (J/U)_c \qquad \Rightarrow \qquad \Delta_c = U(J/U)_c - J, \tag{9.47}$$

which allows one to construct large domains with U/J values corresponding to the pure-system critical point. In the MG phase with $\Delta_J > \Delta_c$, one can always find arbitrary large superfluid lakes that can be doped with particles and holes. This ensures that MG has no gap in the energy spectrum. However, it remains incompressible because the energy required to add/remove a particle scales as l^{-d} and vanishes only when the lake size goes to infinity, which is an exponentially rare event. We thus expect that the density of low-energy states, $\rho(\epsilon)$, vanishes exponentially with $\ln \rho(\epsilon) \propto -\epsilon^{-\delta}$, where positive $\delta > 0$ is related to the size dependence of the probability distribution of large lakes, $\ln P \propto -l^{d\delta}$.

Even more remarkable is the prediction that the superfluid–MG transition in 1D is in the same universality class as the superfluid–MI transition in a pure system. Indeed, the phase factor in the superfluid hydrodynamic action plays no role in this case, because time-averaged on-site occupation numbers remain spatially independent and commensurate with the lattice due to exact local particle–hole symmetry. We are left then with the standard $U(1)$-symmetric action in one dimension, which satisfies the Harris criterion. Though phase factors remain irrelevant in higher dimensions, the universality class of the superfluid–MG transition apparently changes from that for $U(1)$ models in $(d + 1)$ dimensions, because the Harris criterion is violated. In particular, numerical studies present evidence that the dynamic critical exponent is intermediate between 1 and 2 [38].

9.8   Superfluidity of Crystalline Defects

So far, when discussing superfluidity in $(d < 3)$-dimensional systems, we did not pay any special attention to what happens in the remaining dimensions. For cold atomic systems with tight confinement in one or two spatial directions, helium films on various substrates, or superconducting wires, this is an accurate approximation because these systems are surrounded either by vacuum or an insulating material composed of a different species of atoms/molecules. There are, however, cases when superfluidity of $(d < 3)$-dimensional structures is inseparable from properties of the insulating bulk system. In other words, the same particles that form an insulating state in the bulk are responsible for superfluidity in the

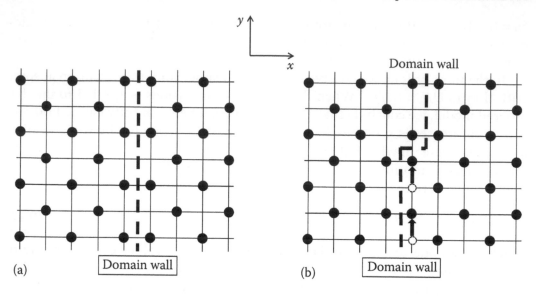

Figure 9.12   (a) Domain wall in the checkerboard solid. (b) When an atom is moved along the wall by one atomic distance, the wall position shifts in the perpendicular direction and the kink moves along the wall by two atomic distances.

lower-dimensional structure. A typical example is provided by the domain wall in the checkerboard solid [39,40] shown in Figure 9.12. The solid structure in the bulk is formed when the n.n. repulsive interaction between hard-core bosons on a square lattice (at half-integer filling factor), V, is strong enough; more precisely, $V > V_c = 2t$. Realistic systems are rarely perfect on the macroscopic scale. Domain walls form naturally in the out-of-equilibrium cooling process across the transition temperature, but they can also be part of the ground state when the boundary conditions are frustrated. The structure shown in Figure 9.12 will automatically appear in the checkerboard solid on a lattice with periodic boundary conditions and an odd number of sites in the x-direction.

A general question to ask about extended defects in quantum solids is this: Can their transport properties be fundamentally different from those of the matrix they are embedded in? It turns out that frequently, but not always, the answer is "yes." An insulating state is typically the result of competition between kinetic and increasing potential energies. One can then imagine cases when, in some finite vicinity of the liquid-to-solid transition in the bulk system, even small changes in the arrangement of atoms change the balance in favor of particle delocalization. The most obvious scenario would be that the potential barriers for moving atoms along a structural defect are substantially lower than in the bulk. For example, when an atom in Figure 9.12 is shifted to the n.n. position in the bulk, the potential energy goes up by $3V$. Shifting an atom by one atomic distance along the wall will increase potential energy only by V. Clearly, these considerations crucially depend on both the bulk and the defect structure as well as the type of the motion considered.

It is also easy to see from this example an intricate connection between the motion of atoms along the defect and the motion of the defect itself in the bulk. If we take the classical picture shown in Figure 9.12a and shift up by one atomic distance all atoms nearest and to the left of the dashed line, then the wall position will shift to the left in the transverse direction; see Figure 9.12b. As we will see shortly, this brings about interesting physics relating superfluidity to roughening and quantum plasticity.

Finally, it is worth mentioning that superfluidity along structural defects forming an interconnected 3D network is yet another way of obtaining the supersolid state. It does break both translation and $U(1)$ symmetries even though it cannot be regarded as being in the thermal equilibrium. Nevertheless, since extended defects such as domain walls, grain boundaries, and dislocations are topologically protected, the lifetime of their networks may easily exceed all relevant experimental time scales.

9.8.1 Domain Walls

In general, there is no guarantee that domain walls, or grain boundaries in solids, have superfluid properties. It is the right combination of light atomic mass, weak interatomic forces, and degree of structural frustration that allows atoms at the defect core to establish long-range coherence. For the domain wall in the checkerboard solid shown in Figure 9.12a, one can easily argue, using the particle-spin analogy, that, at least for $V - V_c \ll t$, the superfluidity of atoms along the domain wall is guaranteed. The mapping between hard-core bosons and spins is trivially obtained by identifying the two allowed occupation number (Fock) states on each site, $|n_i = 0\rangle$ and $|n_i = 1\rangle$, with $|\downarrow\rangle$ and $|\uparrow\rangle$ spin-1/2 states. In this language, the Hamiltonian can be identically rewritten as the anisotropic XXZ model

$$H_{XXZ} = \sum_{\langle ij \rangle} -2t\left(\hat{s}_i^x \hat{s}_j^x + \hat{s}_i^y \hat{s}_j^y\right) + V\hat{s}_i^z \hat{s}_j^z. \tag{9.48}$$

The checkerboard order at $V > 2t$ is recognized as the antiferromagnetic Néel state that replaces the superfluid or the XY-ferromagnetic order at $V < 2t$ through rotation of the vector order parameter, $\vec{S}$, on the O(3) sphere from equator to pole at the SU(2)-symmetric transition point.

The order parameter across the domain wall changes its direction from along $\hat{z}$ and $-\hat{z}$ and, thus, has to be transformed within the wall. In continuum approximation, the optimal energy configuration is determined by competition between the gradient $2t(\nabla \vec{S})^2$ and easy-axis anisotropy $(V - V_c)(S^z)^2$ terms leading to large wall sizes for $V - V_c \ll t$ and smooth rotation of $\vec{S}$. This picture predicts that, in the middle of the wall, $\vec{S}$ is pointing at the equator or, in the original particle language, there is superfluid xy-order within the wall. Thus, despite the fact that all bulk excitations are gapped, there is long-range phase coherence and gapless sound excitations along the wall.

As the interparticle interaction is increased beyond V_W ($V_W = 3.57(3)t$ in 2D and $V_W = 2.683(3)t$ in 3D [40]), the wall size shrinks and it becomes energetically favorable to transform the order parameter across the wall by suppressing its amplitude to zero in the middle. This signals the onset of the superfluid–insulator transition within the wall and disappearance of gapless sound excitations. In the limit of large V/t, we recover the classical picture of the Ising-type domain wall shown in Figure 9.12a. For 2D walls in 3D solids, we do not see any obvious reason for relating superfluidity to structural transformations of walls and grain boundaries at the superfluid–insulator transition point.

Similar considerations apply to any domain wall, twin, or generic grain boundary in solids with one notable difference—there is no guarantee that they will have superfluid properties even at the solid melting line. Most solids are formed in a strong first-order transition, and the resulting crystalline structures are typically very efficient in localizing atoms. For example, even in helium crystals at the melting curve, the smallest exchange cycles are suppressed relative to the liquid phase by several orders of magnitude (this is despite large zero-point motion of individual atoms). As a figure of merit, one may mention the energy scale for the tunneling matrix element of ^{3}He impurity in ^{4}He solid, $J_{34} \sim 10^{-4}$ K [41,42]. It is more than five orders of magnitude smaller than the Debye frequency characterizing large zero-point motion of atoms, and it leads to the effective mass of ^{3}He impurity, which is about four orders of magnitude heavier than helium atom in vacuum. In other words, structural order emerging from first-order transitions in most solids is so strong that grain boundaries and other extended defects are likely to end up in the insulating state. Helium is a unique element in this regard since it is the only one remaining in the liquid phase at ambient pressure. Though experimental evidence for superfluidity of grain boundaries in ^{4}He is rather incomplete [43,44], PIMC simulations do indicate that this possibility is real [45].

One-dimensional walls in 2D solids (at least in some cases) allow for an interesting connection between superfluidity and roughening. Though, naively, these two phenomena refer to completely different types of motion—along and perpendicular to the wall direction—they may have exactly the same microscopic origin, rendering one impossible without the other. An illustration in Figure 9.12b explains that the lowest energy transformation, which moves the wall in the x-direction by creating a kink–antikink pair, is obtained by shifting atoms along the wall; that is, it is inseparable from the mass transfer along the wall. This suggests the intriguing possibility that superfluidity comes along with the roughening transition. By rough, we understand a line with large (diverging in the thermodynamic limit) mean-square fluctuations of the line shape $(\Delta x)^2$. More quantitatively, if the spacetime wall coordinate* is described by $x(y, \tau)$, then

$$(\Delta x)^2 = \left\langle L^{-1} \int_0^{L_y} dy \, [x(y, \tau) - \bar{x}]^2 \right\rangle, \qquad \bar{x} = L^{-1} \int_0^{L_y} dy x(y, \tau), \qquad (9.49)$$

---

* Here, we deal with the $(d+1)$-dimensional classical representation of bosonic statistics.

where the average is taken over the statistical ensemble. The line is in the so-called "smooth" state when $(\Delta x)^2$ is system-size independent.

Roughening is crucial for understanding why the superfluid transition in the 1D wall happens at the Luttinger parameter $K = 1/2$ corresponding to half-integer filling factor [39]. Since the checkerboard order in the bulk doubles the size of the unit cell, one might infer, from the classical picture of a straight wall shown in Figure 9.12a, that the effective filling factor for the wall, n^{1D}, (number of particles per period along the wall) is integer. Notice, however, that bulk potential is shifted by half a period when the wall is displaced by one lattice spacing in x-direction, meaning that, for rough walls, the average potential still has translation symmetry of the original lattice and the effective filling factor remains half-integer. This significantly increases the parameter range in which superfluidity survives, from $K \geq 2$ to $K \geq 1/2$. Numerically, one indeed finds that superfluid and roughening transitions coincide.

In a broader context, the conspiracy between superfluidity and roughening yields an instructive example of emergent duality of orders. The two orders in question are the (topological) superfluid order and order in the field $x(y, \tau)$ (a smooth state of the domain wall implies long-range order in $x(y, \tau)$). This system also provides an interesting example of how fractionalized degrees of freedom occur in the nominally $(d > 1)$-dimensional system. In the smooth insulating phase, the lowest energy excitations are kinks and antikinks. An immediate observation is that shifting an atom by one lattice spacing shifts kinks by two lattice spacing (see Figure 9.12). Thus, if we consider a gauge field on y-bonds, or twisted boundary conditions in the y-direction with phase φ, then the phase accumulated by a kink making two complete loops in the y-direction will be only φ; that is, kinks are quasiparticles with the fractional "gauge charge" of 1/2, or "spinons." Fractionalized degrees of freedom can be directly detected by analyzing exponentially rare imaginary-time trajectories with nonzero winding numbers $W_y = \pm 1$ (also known as instantons). Recall that winding numbers are defined as the time-integrated flux of particle trajectories across any plane perpendicular to the y-direction; see Section 8.2. Let us define $W_y(\tau)$ as the flux of trajectories through any such plane during time τ; then $W_y = W_y(\tau = \beta)$. If kinks are spinons, then the instanton trajectory will split into two distinct fractional instantons connecting near-degenerate ground states (distinguished by the wall location). The resulting $W_y(\tau)$ trajectory has a well-defined plateau at $W_y(\tau) = 1/2$ and two transitions $0 \to \pm 1/2$ and $\pm 1/2 \to \pm 1$ [40].

9.8.2   Screw and Edge Dislocations in ^{4}He

Formally, there are infinitely many distinct dislocation lines in the solid structure. It was conjectured by Shevchenko [46] that edge dislocations in ^{4}He may possess superfluid cores, but, in the absence of first-principle simulations, it is impossible to guess which edge dislocations, if any, have this property. Somewhat surprisingly, MC simulations find most edge dislocations to be insulating and observe that the core is split into two partials. Split core configuration

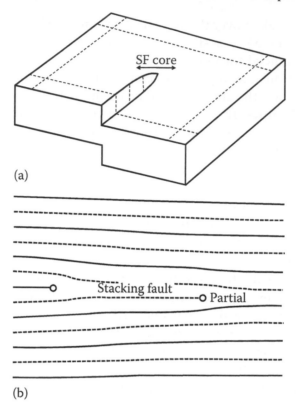

(a)

(b)

Figure 9.13   Illustrations of (a) screw and (b) split-core edge dislocations with the super-fluid cores/partials.

substantially reduces frustration in the atomic arrangement and locks atoms in the insulating state [47]. Instead, the most basic screw dislocation oriented along the c-axis of the hcp structure [see Figure 9.13a] was found to have a clear-cut superfluid core [48]. At the melting curve, the effective 1D superfluid density was found to be about $n_s^{1d} \approx 1\text{Å}^{-1}$, corresponding to a "superfluid tube" at the core of diameter ≈ 6 Å. The corresponding Luttinger liquid parameter, $K \approx 4.9(5)$, is large enough to guarantee that the superfluidity of screw dislocation is stable against weak disorder and commensurate potentials.

Since screw dislocations are very common in crystals and often facilitate crystal growth, it is not unreasonable to expect that, in real samples of solid ^{4}He, we do have a network of superfluid "pipes" penetrating the insulating bulk and capable of conducting dissipationless mass currents of ^{4}He atoms. The only serious objection would be that one needs more than one type of dislocations with superfluid cores to form a fully connected 3D network. Fortunately, edge dislocations with the Burgers vector along the hcp c-axis and the dislocation core along the x-axis in the basal plane also demonstrate superfluid properties [49], though with much smaller superfluid density. The core of this edge dislocation is split into two partials separated by an extremely large distance (possibly as large as $150 \div 200$ Å), and the superfluid signal is "marginal" in the sense that the corresponding Luttinger liquid parameter was found to be close to the critical value $K = 2$ (with large uncertainty) when a commensurate potential becomes a relevant perturbation. Large separation between two half-cores is explained by unmeasurably small surface energy of the stacking fault defect in the hcp structure leading to a very weak linear potential confining the partials; see Figure 9.13b. The unusual split-core structure may have interesting consequences for how edge dislocations move, anneal, and are pinned in quantum crystals, but these have not been fully explored yet.

Superfluidity in cores of edge dislocations leads to an interesting "quantum metallurgy" phenomenon (the term was coined by A. Dorsey). In the absence of vacancies (i.e., at temperatures well below the vacancy activation gap), non-superfluid edge dislocations cannot climb in the direction perpendicular to their axis and the Burgers vector. Indeed, climb would correspond to the growth of an atomic layer, which, in turn, requires mass transfer to the growth spot. In insulating solids, thermally activated vacancies provide the only mechanism of atomic mass transport. This, in particular, means that insulating crystals are *isochorically incompressible*, $\chi \equiv (dn/d\mu)_V \approx 0$; in other words, the density n of the crystal should demonstrate no response to infinitesimal, quasistatic changes in μ (note an analogy with incompressibility of MIs in externally imposed lattices).*

If cores of edge dislocations in the network are superfluid, then atoms can be efficiently transported throughout the sample and superfluid edge dislocations would be able to climb (or, *superclimb*). Correspondingly, the crystal density will react continuously to small changes of the chemical potential, and one would observe the effect of anomalous isochoric compressibility of a material, which is insulating in the absence of structural defects. This is precisely the phenomenon observed by Ray and Hallock in the experiment where ^{4}He atoms (superfluid liquid) were fed into the crystal through implanted Vycor rods [50,51]. In their setup, the chemical potential μ is the physical quantity relevant to the external perturbation applied to the crystal. Also, the increase in the crystal density was correlated with the ability of solid samples to conduct atoms between the Vycor rods, in agreement with the edge dislocation superclimb scenario.

9.8.3 Superglass States

As mentioned previously, extended structural defects are metastable excited states of the system protected by crystal topology. One may wonder to what degree this picture holds as the number of defects increases and we enter a state where one can barely recognize the original solid structure, not to mention a state created from the superfluid liquid by large instantaneous density change (volume squeezing)— a structural glass. It may happen that such states do not live long and, on relevant experimental time scales, they relax to a polycrystalline state with clearly identifiable mesoscopic solid structure. An alternative scenario would be that even glass

---

* As long as the creation of single vacancies and interstitials is forbidden by finite energy gaps, the only way the density of a crystal can react dynamically to a small change in the chemical potential, $\delta\mu$, is by creating/removing crystalline layers. This requires nucleation times exponentially large in $|\delta\mu|^{-1}$.

produced from the density-quenched liquid remains stable on the experimental time scale similar to numerous amorphous materials that are often among the strongest solid materials in nature!

The conventional wisdom is that, at low temperature, glasses are insulating states with regard to motion of their own atoms: If atoms could easily move around, then they would quickly find that the polycrystalline state is lower in energy. This reasoning is not precise in that it is assuming that motion leading to atomic delocalization is the same as the type of motion that leads to formation of polycrystalline seeds. In other words, jamming of structural relaxation does not exclude the possibility of the superfluid motion in the same sample. One can imagine a system where these two types of motion occur on vastly different time scales so that, under realistic experimental conditions, the glass remains structurally stable and superfluid. The resulting state breaks translation invariance (the time average density profile is inhomogeneous) and features the superfluid order; that is, it formally satisfies the definitions of the supersolid state. However, since glasses lack order in the density–density correlation function, a more appropriate name would be "superglass" (SG). At the phenomenological level, one may view the SG state as a special limit of Tisza's two-fluid picture with the normal component freezing into the glass.

The first observation of SG was reported in numerical PIMC simulations of solid ^{4}He samples prepared by fast temperature quench from the normal liquid state at relatively high density $n = 0.0359\,\text{Å}^{-3}$ [52] (for comparison, the density of the solid at the $T = 0$ melting line is $n_m = 0.0287\,\text{Å}^{-3}$). A similar outcome is achieved when samples are prepared by squeezing the low-temperature superfluid liquid into a smaller volume. When only local updates of atomic trajectories are employed in simulations, thus prepared samples of about 800 atoms remain in the SG phase with inhomogeneous density profile and no long-range order in the density–density correlation function. Though a direct connection with experiments is hard to make with regard to the sample preparation protocols, broad metastability limits for the SG state of solid ^{4}He are strongly suggested by simulations.

Further support of the concept comes from the possibility to "design" a quantum system that is guaranteed to have the SG phase [11]. The idea is based on two key observations: First, for any Jastrow wavefunction $\Psi_G^{(J)}$ [see Equation (9.5)], one can name a Hamiltonian, H_J, with contact two- and three-body interactions that has $\Psi_G^{(J)}$ as its ground state. One may directly check that $\Psi_G^{(J)}$ satisfies the Schrödinger equation for

$$H_J = \sum_{i=1}^{N} \frac{\mathbf{p}_i^2}{2m_i} + U(\mathbf{r}_1,\ldots,\mathbf{r}_N),$$

$$U = -\sum_{i \neq j} \frac{\hbar^2}{2m_i} \nabla^2 u\left(\mathbf{r}_i - \mathbf{r}_j\right) + \sum_{i \neq j, i \neq k} \frac{\hbar^2}{2m_i} \nabla u\left(\mathbf{r}_i - \mathbf{r}_j\right) \cdot \nabla u\left(\mathbf{r}_i - \mathbf{r}_k\right), \quad (9.50)$$

with zero eigenvalue [the spectrum of (9.50) is nonnegative]. Second, as mentioned earlier, the square of the Jastrow wavefunction can be straightforwardly interpreted as the Boltzmann–Gibbs measure for a classical system with pair potential $V = uT/2$. Moreover, this correspondence extends to all eigenstates that are common for the Hamiltonian (9.50) and the Fokker–Planck operator governing the evolution of the classical-probability distribution within the framework of the stochastic Langevin dynamics

$$m_i \frac{d^2 \mathbf{r}_i}{dt^2} = -\frac{1}{2} \sum_{i \neq j} \nabla_i u \left(\mathbf{r}_i - \mathbf{r}_j \right) + \eta_i(t), \tag{9.51}$$

where $\eta_i(t)$ is the thermal Gaussian white noise characterized by the correlation function $\langle \eta_i^\alpha(t) \eta_i^\beta(t') \rangle = m_i \delta_{ij} \delta_{\alpha\beta} \delta(t - t')$. We omit here the derivation, which can be found in standard statistical mechanics text; see Reference [11] for details.

This correspondence can be used to compute both static and dynamic correlation functions in the quantum system at zero temperature by using known results for its classical counterpart. The last observation directly links the well-studied jamming phenomenon in classical systems to the long-lived metastable amorphous density profile in the quantum counterpart. As far as the superfluidity is concerned, it is guaranteed by construction—Jastrow states have a finite condensate fraction. Admittedly, there are serious drawbacks in the theoretical construction (9.50) originating from pathological properties of the resulting Hamiltonian, which always has the same ground-state energy equal to zero regardless of the system density (infinite compressibility), quadratic dispersion relation for elementary excitations, and the fact that the SG state always forms on top of the supersolid ground state. Nevertheless, this construction (1) proves existence of SG as a matter of principle, and (2) can be made more physical by adding weak standard pairwise potentials to Equation (9.50) in a hope that the overall picture does not change.

9.8.4 Shevchenko State

In this subsection, we will discuss superfluid properties of the 3D network formed by interconnected 1D liquid channels. Let us start with a fictitious system of interconnected pipes running along bonds of the simple cubic lattice. Assume that pipes of diameter $d_0 = 1$ mm are filled with low-temperature ($T \ll T_\lambda$) ^{4}He and the bond length L is such that $L \gg d_0^2/a$ where a is the interatomic distance, say, $L = 10^6$ km. On one hand, it is readily seen (an explicit expression is presented below, see Equation (9.53)) that the thermodynamic transition temperature to the superfluid state in this setup with quasi-1D geometry of pipe dimensions between the nodes is strongly suppressed relative to the λ-point in bulk ^{4}He. On the other hand, it is clear that at, say, $T < T_\lambda/2$, the network will support frictionless low-velocity flows and persistent currents with decay times larger than the time of the Universe. Indeed, at this temperature, ^{4}He already has strong local order (definitely on the $d_0 = 1$ mm scale) in the phase field preventing large vortices from nucleating and proliferating in the space occupied by helium.

Both conclusions are correct and not in contradiction with each other, because we are dealing with an extreme situation when one of the kinetic time scales diverges so fast that, in an experimental system, the ergodicity hypothesis—on which the equilibrium thermodynamic ensemble calculations are based—is violated. Consider kinetic energy of the current state corresponding the phase winding 2π around one elementary plaquette isolated from the rest of the system. An elementary calculation based on the integration of the energy density $\epsilon = (n_s/2m)\left(\pi^2/4L^2\right)$ gives

$$E_{\mathrm{pl}} = \frac{n_s a \pi^3}{8m} \frac{d_0^2}{La} \approx T_\lambda \frac{d_0^2}{La}, \tag{9.52}$$

a value that is much smaller than T_λ for $L \gg d_0^2/a$ (for the parameters mentioned previously, this energy is in the micro-Kelvin range). In the network, currents are not confined to a single plaquette. The minimal phase defect is a small "vortex ring" type configuration depicted in Figure 9.14. Though the notion of a vortex is meaningless at length scales $\sim L$, and the best description is provided by phase gradients along bonds, the topology of phase windings is still easiest to picture using vortex lines. In any case, Equation (9.52) sets the scale for the energy cost of having circulating currents in the network. As long as $E_{\mathrm{pl}} \ll T$, the thermodynamic equilibrium state remains normal because, on all scales, the closed-contour integrals $\oint (\nabla \varphi) \cdot d\mathbf{r}$ are frequently taking nonzero values.

The transition temperature to the superfluid state can be estimated from the condition $E_{\mathrm{pl}} \sim T$ when plaquette currents become too expensive thermodynamically. This condition leads to

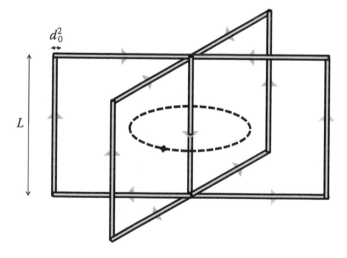

Figure 9.14   The minimal phase defect in the network is obtained by imagining a vortex ring winding around the pipe and creating $\pm 2\pi$ phase windings in four elementary plaquettes.

$$T_c \sim T_\lambda \frac{d_0^2}{La} \propto \frac{1}{L} \qquad \left(L \gg d_0^2/a \right). \qquad (9.53)$$

The same estimate for T_c is obtained from the condition that 1D phase fluctuations on the length scale L are reduced to a value of order unity: By introducing an effective 1D superfluid stiffness $\Lambda^{(1D)} \sim n_s d_0^2/m$, we find that $\left\langle (\varphi(0) - \varphi(L))^2 \right\rangle \sim TL/\Lambda_s \sim O(1)$. Once phase fluctuations are frozen out along the pipes, the 3D order in the network sets in. By counting the superfluid liquid in the pipes, one can see that $n_s(T = 0) \propto 1/L^2 \propto T_c^2$.

Academically speaking, the universality class of the transition is not altered by the quasi-1D microscopic geometry; in particular, the superfluid fraction $n_s(T)/n$ curve starts with an infinite derivative at T_c. However, under specified conditions, the amplitudes of universal features (and the size of the temperature interval where the universality develops) are extremely small. Furthermore, the thermodynamic transition itself is of purely academic interest because it is impossible to observe experimentally. At temperatures below T_λ, the notion of thermal equilibrium quickly becomes irrelevant because any given distribution of persistent currents is kinetically frozen and cannot change in response to temperature variations or slow rotation. In other words, in terms of observed behavior, the state is no different than the genuine superfluid, though typically one finds large circulating currents on most plaquettes! This is the essence of the state introduced by Shevchenko [46] to describe superfluid properties of the dislocation network, in the temperature interval $T_c \ll T \ll T_\lambda$, under the hypothesis that edge dislocations in ^{4}He have superfluid cores.

The only real difference between our fictitious "pipelines" example and dislocation network is that, for dislocations, the actual pipe diameter is of the order of the interatomic distance, $d_0 \sim a$. This immediately implies two things: First, the condition $L \gg d_0^2/a$ is reduced to $L \gg a$ and thus is easily satisfied in experimental samples except, maybe, the glassy ones. Correspondingly, we typically expect $T_c \ll T_\lambda$ and a broad temperature interval where the Shevchenko state may be realized. Second, narrow superfluid channels at low temperature should be described by the Luttinger liquid theory with relatively small (order unity) values of K; that is, their superfluid properties are rather fragile, and the stability of circulating currents is no longer protected by the robust superfluid order on large length scales across the channel. Long relaxation times, $\tau_{LL}(T)$, for equilibration/redistribution of plaquette currents leading to the Shevchenko state are calculated in the next subsection. Here, we simply mention that one finds a power-law dependence

$$\tau_{LL}(T) \propto \left(\frac{T_*}{T} \right)^{2K-1}, \qquad (9.54)$$

where $T_* \sim mc^2$ is the characteristic energy scale based on the 1D sound velocity c, below which the Luttinger liquid behavior sets in. For dislocations in solid helium, we have $T_* \sim T_\lambda$. At temperatures $T \sim T_\lambda$, one finds relatively short relaxation times, ensuring normal state behavior at experimental time scales τ_{exper}. However, at low temperatures, say, for $K \approx 5$ found for screw dislocations, it is possible to observe a dynamic crossover to the superfluid state when $\tau_{LL}(T) \gg \tau_{\text{exper}}$. The crossover might be rather sharp and easily accessible experimentally for large values of K; note that typical values of the Luttinger liquid parameter for ultracold atomic systems (in the absence of the optical lattice) are very large, of the order of one hundred, or larger! [For the fictitious "helium pipelines" system, the corresponding Luttinger parameter is $K \sim (d_0/a)^2$, and the dynamic crossover occurs right below T_λ.]

9.8.5   Stability of Supercurrent States in 1D

We are interested here in a generic (nonintegrable) 1D system with periodic boundary conditions (i.e., of the ring shape with circumference L) in the absence of translation invariance. Since the total momentum of the system is not conserved, one naturally expects that current states will either decay (at finite temperature) or form superpositions composed of states with different velocity circulation. In the following, we employ the superfluid hydrodynamic action and an equivalent effective Hamiltonian formulation to calculate matrix elements connecting quaside-generate current states and the corresponding relaxation rates in such rings.

9.8.5.1   Avoided Crossing between Current States at $T = 0$

The only low-lying *normal modes* of our mesoscopic superfluid 1D ring are phonons. Along with the occupation numbers of phonons, the low-energy states of the system are characterized by topological quantum number (the winding number of the superfluid phase) $I = (2\pi)^{-1} \int_0^L dx [\partial \varphi(x, \tau)/\partial x]$. At $I \neq 0$, we are dealing with a (mesoscopic) supercurrent state. Below, we will be interested only in the supercurrent excitations, assuming that all phonon occupation numbers are either identically zero or fixed.* In this case, there is one-to-one correspondence between the topological number I and the expectation value of the supercurrent, J, so we can use the latter to label the states.

Consider two states, $|J\rangle$ and $|-J\rangle$, which carry opposite currents and are equivalent otherwise. In the simplest case, these would be the states with opposite Is. Alternatively, the states may have phase windings I and $-(I + 1)$, for example, in rings subject to the gauge flux π. Upon elimination of the gauge phase by the gauge transformation, such states are characterized by half-integer windings $\pm(I + 1/2)$. It is worth mentioning that the latter situation occurs automatically in (1) a standard fermionic system with an even particle number and (2) lattice spin models with antiferromagnetic interactions and lattice bosons with positive hopping amplitude at odd number of sites.

---

* Note a fundamental difference between the supercurrent excitations and phonons. While the latter are normal modes (quasiparticles), the former are characterized by a global quantum number and, in this sense, are *quasiground states*.

Changing winding number I requires a process that results in

$$\int_0^L \frac{\partial \varphi(x,\tau_2)}{\partial x} dx - \int_0^L \frac{\partial \varphi(x,\tau_1)}{\partial x} dx = 2\pi m, \tag{9.55}$$

where $m = I' - I$ is some integer. Hence, the process involves a nonzero closed-contour integral in the $\mathbf{r} = (x,\tau)$-plane: $\oint \nabla \varphi \cdot d\mathbf{r} = 2\pi m$. In other words, it is described by the vortex-type configuration in the $(1+1)$-dimensional phase field with the vortex charge m. We refer to this process as a *(phase slippage) instanton*.[*]

In a translation-invariant system, m-instantons can happen at any spacetime point, but, since their imaginary parts are uniformly distributed, they cancel each other exactly. Nonzero effects are expected only when one of the following conditions is satisfied:

1. There is an impurity site on the ring.

2. The system is disordered.

3. The system is commensurate *and* the product $m\nu$ is an integer number.

We will see shortly that matrix elements connecting current states in the superfluid system, $\Delta_{I',I} = \Delta_{I'-I}$, are macroscopically smaller than the supercurrent-energy scale set by $2\pi c/L$, and thus only energy levels that are in a macroscopically sharp resonance with each other are significantly affected by instantons. In Figure 9.15, we present an illustration of the low-energy spectrum: a state $|I, \{n_q\}\rangle$ with the phase winding I and phonon occupation numbers $\{n_q\}$ of momentum modes $q = 2\pi k c/L$, where k is a nonzero integer, has an energy $E = E_I + \sum_q c|q|n_q$. Note that the first term can be written identically as $E_I = 2\pi^2 \Lambda_s I^2/L = 2\pi K c I^2/L$; see definitions below Equation (9.37). We start by analyzing the case of non-integer IK, when current states are not in resonance with phonon excitations, and one may proceed with instanton transitions in the absence of other excitations.

A vortex of charge $m = I' - I$ in the phase field reduces the configuration weight by a factor $e^{-m^2 K \ln(L/\Lambda)} \propto (\Lambda/L)^{m^2 K}$, where $\Lambda \sim T_*/c$ is a microscopic length scale defining the onset of the Luttinger liquid behavior. Integrating over the instanton position in real space, we find that an integral $\int_0^L e^{i2\pi mn(x)} dx$ is of the order unity in case (a), is proportional to $\sqrt{L}$ in case (b), and $\propto L$ in case (c). An integral over the instanton position in time is trivially proportional to the time itself because, in the single-instanton approximation, the transition amplitude between states $|I\rangle$ and $|I'\rangle$ is given by $-it\Delta_{I'-I}$. This is nothing but the standard protocol of

---

[*] In the Luttinger liquid theory—and especially when the quantitative analysis is based on the nomenclature of noninteracting fermions—the instanton is also known as the backscattering process. It should be emphasized, however, that the term "backscattering" should not be taken literally as an elementary (i.e., perturbative) single-particle-scattering process. In a 1D fermionic system with a finite strength of interparticle interaction, the backscattering is a collective phenomenon, fully equivalent to the phase slippage instanton in a generic 1D superfluid.

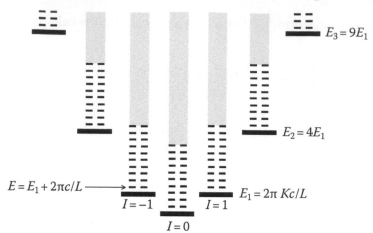

Figure 9.15   Schematic illustration of the low-energy spectrum in 1D system. Bold lines represent quasiground states with nonzero-integer phase windings I. Phonon excitations are shown by dashed lines. Highly excited phonon states have exponentially large multiplicity; with spectrum nonlinearities and phonon–phonon interactions taken into account, the degenerate levels quickly spread out into a quasicontinuum of states.

defining nondiagonal matrix elements of effective Hamiltonians, $\langle a | e^{-iHt} | b \neq a \rangle \to$ $\langle a | e^{-iH_{\text{eff}}t} | b \neq a \rangle \approx -it \langle a | H_{\text{eff}} | b \neq a \rangle$, from the system evolution on time scales much shorter than $1/(H_{\text{eff}})_{ab}$. Combining all pieces together, we find

$$\Delta_{I'-I} \sim V_\Lambda \left(\frac{\Lambda}{L} \right)^{m^2 K} \times \begin{cases} 1 & \text{single impurity} \\ \sqrt{L/\Lambda} & \text{disorder} \\ L/\Lambda & \text{integer } m\nu \end{cases} , \tag{9.56}$$

where the amplitude V_Λ characterizes scattering properties of the potential responsible for violation of translation invariance at the length scale Λ. With the critical values of $m^2 K$ for all the three cases (recall that they are 1, 3/2, 2, respectively), we see that the condition $\Delta_{I'-I} \ll (E_{I=1}, c/L)$ is always satisfied.

At the critical point, one must account for renormalization of the Luttinger liquid parameter with the scale. This leads to the following integral for the vortex contribution to the action

$$K_c \ln(L/\Lambda) + \int\limits^{\ln(L/\Lambda)} [K(\lambda) - K_c] d\lambda. \tag{9.57}$$

At this point, we simply substitute here the solution of the KT RG flow derived in Chapter 4 for $w = K - K_c = 1/\lambda$. Thus, the integral in expression (9.57) equals to $\ln[\ln(L/\Lambda)]$. Interestingly, for all the three cases mentioned previously, we find the same expression for the tunneling amplitude at the superfluid–insulator transition

$$\Delta_{I'-I} \sim V_\Lambda \frac{\Lambda}{L \ln^{m^2}(L/\Lambda)} \qquad (K = K_c). \tag{9.58}$$

We are in a position now to formulate an effective Hamiltonian for the low-energy subspace of states with phase windings (for definiteness, we will consider in the following the case representative of a single impurity located at point x_0)

$$H_I = \sum_I \frac{2\pi cK}{L}(I - \varphi_0/2\pi)^2 c_I^\dagger c_I + \sum_{I',I}\left[\Delta_{I'-I}(x_0)c_{I'}^\dagger c_I + \text{H.c.}\right]. \tag{9.59}$$

In the first term, we introduced explicitly the twist phase (of the twisting boundary condition or gauge vector potential), φ_0. The phase φ_0 is defined on the interval $(-\pi, \pi)$ since multiples of 2π can be absorbed into the shift of the integer index I. Operators $c_I^\dagger/c_I$ create/delete a state with quantum number I. Formally, the Hamiltonian (9.59) is that of a particle hopping among the sites of a 1D lattice subject to parabolic confinement. For one thing, $\Delta_{0,1} \neq 0$ is responsible for avoiding the energy-level crossing at $\varphi_0 = \pi$. In the vicinity of the crossing point, the lowest energy levels are well described by the two-level-system solution

$$\epsilon_+ = \frac{\pi Kc}{2L} \pm \sqrt{|\Delta_{0,1}|^2 + \delta^2}, \qquad \delta = \frac{Kc(\varphi_0 - \pi)}{L}. \tag{9.60}$$

This result can be used to calculate the persistent *ground-state* current arising in response to the twisting phase, $j = -d\epsilon_-/d\varphi_0$, which, in the absence of phase slippage instantons, is a saw-function $j = Kc(I - \varphi_0/2\pi)/L$ (the value of integer I in this expression corresponds to the lowest-energy state) with abrupt sign changes at $\varphi_0 = \pm\pi$. With $\Delta_{0,1} \neq 0$, one finds a smooth periodic curve. The transition matrix element also defines the experimental time scale, $\tau_{\exp}$, separating adiabatic *vs.* nonadiabatic processes imposing the phase twist: for $\tau_{\exp}\Delta_{0,1}^{-1} \gg 1$, the induced current will follow the ground-state curve, while in the opposite limit, the current will keep increasing linearly with φ_0.

9.8.5.2   Relaxation Rates at Finite Temperature

At finite temperature, we are dealing with excited states of the system containing numerous phonon excitations. As pictured in Figure 9.15, apart from the lowest modes, the spectrum of many-body levels is so dense that it is appropriate to replace it with a continuum of states. Under these conditions, one can always find states $|I, \{n_q\}\rangle$ and $|I', \{n_q'\}\rangle$ that are in near resonance with each other; that is, the energy difference between them is much smaller than the mixing matrix element. Since energy conservation requires that $E_I - E_I'$ is transferred to the phonon subsystem, we need an explicit expression for coupling between supercurrent and phonon excitations. To find the corresponding effective Hamiltonian, we notice that the instanton phase in

$$\Delta_{I'-I}(x_0)c_{I'}^\dagger c_I + \text{H.c.} = |\Delta_{I'-I}|e^{i\gamma(x_0)}c_{I'}^\dagger c_I + \text{H.c.} \tag{9.61}$$

accounts only for the static part of the particle density. More generally, the instanton phase involves also density fluctuations; that is, in Equation (9.61), we must make a substitution

$$|\Delta_{I'-I}|e^{i\gamma(x_0)} \longrightarrow V_\Lambda e^{i\gamma(x_0)}e^{i\tilde{\gamma}_\Lambda(x_0)}, \tag{9.62}$$

where (recall that $m = I' - I$)

$$\tilde{\gamma}_\Lambda(x_0) = 2\pi m \int^{x_0} \delta n(x)\,dx = -m\sqrt{K} \sum_{q\neq0}^{|q|<2\pi/\Lambda} \sqrt{\frac{2\pi}{L|q|}}\,\mathrm{sgn}(q)\left(\hat{b}_q - \hat{b}_{-q}^\dagger\right)e^{-iqx_0} \tag{9.63}$$

is an operator acting on phonon variables with momenta $q < 2\pi/\Lambda$. The second quantization form is obtained by direct substitution of Equation (7.43). Alternatively, the effective Hamiltonian can be derived in a mathematically rigorous way by mapping the fermionic backscattering Hamiltonian to the equivalent sine–Gordon model; see References [34,53,54]. Corresponding derivation goes beyond the scope of our book.

The ground-state result [Equation (9.56)] is recovered by averaging the $e^{i\tilde{\gamma}_\Lambda(x_0)}$ operator over the ground state,

$$\left\langle e^{i\tilde{\gamma}_\Lambda(x_0)}\right\rangle_{T=0} = \exp\left\{-m^2 K \int_{2\pi/L}^{2\pi/\Lambda} \frac{dq}{q}\right\} = \left(\frac{\Lambda}{L}\right)^{m^2 K}, \tag{9.64}$$

leading to $\Delta_{I'-I}(x_0) = V_\Lambda e^{i\gamma(x_0)}\left\langle e^{i\tilde{\gamma}_\Lambda(x_0)}\right\rangle_{T=0}$ in agreement with Equation (9.56). This allows us to reformulate the effective Hamiltonian for low temperatures $T \ll 2\pi c/\Lambda$ in a cutoff-independent form through the ground-state matrix element as

$$H_I = \sum_I \frac{2\pi c K}{L}(I - \varphi/2\pi)^2 c_I^\dagger c_I + \sum_{I',I}\left[\Delta_{I'-I}(x_0)\frac{e^{i\tilde{\gamma}(x_0)}}{\left\langle e^{i\tilde{\gamma}(x_0)}\right\rangle_{T=0}}c_{I'}^\dagger c_I + \mathrm{H.c.}\right]. \tag{9.65}$$

Indeed, in this expression, contributions from phonon modes with energies $\gg T$ cancel out between the numerator and denominator. This Hamiltonian can be used to calculate corrections to the low-energy spectrum, for example, avoided crossings for current states, which include mixing of excited phonon modes. For integer KI, such mixing is, in fact, extremely important; we refer the reader to Reference [33] for further details.

Problem 9.5 *Perform the average in Equation (9.64). Note that different modes can be averaged independently and for any operator $\hat{o}$, which is linear in bosonic fields, the following identity holds $\left\langle e^{\hat{o}}\right\rangle = e^{\left\langle \hat{o}^2\right\rangle/2}$ (see also previous problem).*

Consider now an incoherent finite-temperature transition between current states I and $I-1$ with the persistent current energy $E_I - E_{I-1} = \xi$, dissipated into the phonon subsystem. The corresponding transition rate can be calculated by the Fermi Golden Rule:

$$\tau^{-1} = |V_\Lambda|^2 \int_{-\infty}^{\infty} dt \, e^{i\xi t} \left\langle e^{-i\tilde{\gamma}^\dagger(x_0,t)} e^{i\tilde{\gamma}(x_0,0)} \right\rangle, \tag{9.66}$$

where $\tilde{\gamma}(x_0,t)$ is the operator $\tilde{\gamma}(x_0)$ (9.63) in the interaction representation. Averaging exponentials containing linear functions of bosonic fields is standard (see Problems 9.4 and 9.5) and leads directly to

$$\tau^{-1} = |V_\Lambda|^2 \int_{-\infty}^{\infty} dt \, e^{i\xi t - F(t)}, \tag{9.67}$$

where

$$F(t) = 2K \int_0^{2\pi c/\Lambda} \{ \coth(x/2T)[1 - \cos(xt)] + i \sin(xt) \} \frac{dx}{x}, \tag{9.68}$$

or, equivalently,

$$F(t) = 2K \int_0^{2\pi c/\Lambda} \frac{\cosh(x/2T) - \cosh[x(it - 1/2T)]}{\sinh(x/2T)} \frac{dx}{x}. \tag{9.69}$$

Shifting the time integration contour into the complex plane, $t \to t - i/2T$, we arrive at

$$\tau^{-1} = |V_\Lambda|^2 e^{\xi/2T} \int_{-\infty}^{\infty} dt \, e^{i\xi t - \tilde{F}(t)}, \tag{9.70}$$

with

$$\tilde{F} = 2K \int_0^{2\pi c/\Lambda} \frac{\cosh(x/2T) - \cos(xt)}{\sinh(x/2T)} \frac{dx}{x} = 2K \ln \left[\frac{\gamma_E c}{\Lambda T} \cosh(\pi T t) \right], \tag{9.71}$$

where $\gamma_E = 0.577215$ is the Euler constant. Introducing

$$\Delta_1(T) = V_\Lambda \left(\frac{T}{T_*} \right)^K \qquad (T_* = \gamma_E c/\Lambda) \tag{9.72}$$

and evaluating the remaining integral over time, we finally obtain

$$\tau^{-1} = \frac{|\Delta_1(T)|^2 \Omega}{\xi^2 + \Omega^2} e^{\xi/2T} \frac{|\Gamma[1 + K + i\xi/2T]|^2}{\Gamma[1 + 2K]} \qquad (\Omega = 2\pi KT). \tag{9.73}$$

Though our derivation is valid for an arbitrary value of the ξ/T ratio, in the case under consideration, we have $|\xi| = 2\pi c K/L \ll T$ and thus may safely put $\xi = 0$ everywhere:

$$\tau^{-1} = \frac{|\Delta_1(T)|^2}{T} \frac{\Gamma^2[1+K]}{2\pi\Gamma[1+2K]} \propto T^{2K-1}. \tag{9.74}$$

This concludes our derivation of the current relaxation time mentioned previously in Equation (9.54).

References

1. N. Prokof'ev and B. Svistunov, Supersolid state of matter, *Phys. Rev. Lett.* **94**, 155302 (2005).

2. E. P. Gross, Unified theory of interacting bosons, *Phys. Rev.* **106**, 161 (1957).

3. C. Josserand, Y. Pomeau, and S. Rica, Coexistence of ordinary elasticity and Super-fluidity in a model of a defect-free supersolid, *Phys. Rev. Lett.* **98**, 195301 (2007).

4. M. Boninsegni, F. Cinti, P. Jain, M. Boninsegni, A. Micheli, P. Zoller, and G. Pupillo, Supersolid droplet crystal in a dipole-blockaded gas, *Phys. Rev. Lett.* **105**, 135301 (2010).

5. F. Cinti, T. Macri, W. Lechner, G. Pupillo, and T. Pohl, Defect-induced super-solidity with soft-core bosons, *Nat. Commun.* **5**, 3235 (2014).

6. A. F. Andreev and I. M. Lifshitz, Quantum theory of crystal defects, *Sov. Phys. JETP* **29**, 1107 (1969).

7. B. Spivak and S. A. Kivelson, Phases intermediate between a two-dimensional electron liquid and Wigner crystal, *Phys. Rev. B* **70**, 155114 (2004).

8. B. Spivak, Talk given at the workshop *Quantum Coherent Properties of Spins II*, PITP, UBC, Vancouver, British Columbia, Canada, December 4–9 (2009).

9. L. Reatto, Bose–Einstein condensation for a class of wave functions, *Phys. Rev.* **183**, 334 (1969).

10. G.V. Chester, Speculations on Bose–Einstein condensation and quantum crystals, *Phys. Rev. A* **2**, 256 (1970).

11. G. Biroli, C. Chamon, and F. Zamponi, Theory of the superglass phase, *Phys. Rev. B* **78**, 224306 (2008).

12. O. Penrose, On the quantum mechanics of helium II, *Phil. Mag.* **42**, 1373 (1951).

13. O. Penrose and L. Onsager, Bose–Einstein condensation and liquid helium, *Phys. Rev.* **104**, 576 (1956).

14. M. P. A. Fisher, P. B. Weichman, G. Grinstein, and D. S. Fisher, Boson localization and the superfluid-insulator transition, *Phys. Rev. B* **40**, 546 (1989).

15. J. K. Freericks and H. Monien, Strong-coupling expansions for the pure and disordered Bose–Hubbard model, *Phys. Rev. B* **53**, 2691 (1996).

16. N. Teichmann, D. Hinrichs, M. Holthaus, and A. Eckardt, Bose–Hubbard phase diagram with arbitrary integer filling, *Phys. Rev. B* **79**, 100503 (2009).

17. B. Capogrosso-Sansone, N. V. Prokof'ev, and B. V. Svistunov, Phase diagram and thermodynamics of the three-dimensional Bose–Hubbard model, *Phys. Rev. B* **75**, 134302 (2007).

18. S. Sachdev, *Quantum Phase Transitions*, Cambridge University Press, Cambridge, U.K. (1999).

19. M. Greiner, O. Mandel, T. Esslinger, T. W. Haensch, and I. Bloch, Quantum phase transition from a superfluid to a Mott insulator in a gas of ultracold atoms, *Nature* **415**, 39 (2002).

20. S. Trotzky, L. Pollet, F. Gerbier, U. Schnorrberger, I. Bloch, N. V. Prokov'ev, B. Svistunov, and M. Troyer, Suppression of the critical temperature for superfluidity near the Mott transition, *Nat. Phys.* **6**, 998 (2010).

21. A. B. Harris, Effect of random defects on the critical behaviour of Ising models, *J. Phys. C* **7**, 1671 (1974).

22. M. H. W. Chan, K. I. Blum, S. Q. Murphy, G. K. S. Wong, and J. D. Reppy, Disorder and the superfluid transition in liquid ^{4}He, *Phys. Rev. Lett.* **61**, 1950 (1988).

23. J. Yoon, D. Sergatskov, J. Ma, N. Mulders, and M. H. W. Chan, Superfluid transition of ^{4}He in ultralight aerogel, *Phys. Rev. Lett.* **80**, 1461 (1998).

24. M. Nikolaou, M. Wallin, and H. Weber, Critical scaling properties at the superfluid transition of ^{4}He in aerogel, *Phys. Rev. Lett.* **97**, 225702 (2006).

25. P. Lugan, D. Clement, P. Bouyer, A. Aspect, M. Lewenstein, and L. Sanchez-Palencia, Ultracold Bose gases in 1D disorder: From Lifshits glass to Bose–Einstein condensate, *Phys. Rev. Lett.* **98**, 170403 (2007).

26. G. M. Falco, T. Nattermann, V. L. Pokrovsky, Weakly interacting Bose gas in a random environment, *Phys. Rev. B* **80**, 104515 (2009).

27. V. Gurarie, L. Pollet, N. V. Prokof'ev, B. V. Svistunov, and M. Troyer, Phase diagram of the disordered Bose–Hubbard model, *Phys. Rev. B* **80**, 214519 (2009).

28. S. Pilati, S. Giorgini, and N. Prokof'ev, Superfluid transition in a Bose gas with correlated disorder, *Phys. Rev. Lett.* **102**, 150402 (2009).

29. L. Pollet, N. V. Prokof'ev, and B. V. Svistunov, Asymptotically exact scenario of strong-disorder criticality in one-dimensional superfluids, *Phys. Rev. B* **89**, 054204 (2014).

30. L. Pollet, N. V. Prokof'ev, B. V. Svistunov, and M. Troyer, Absence of a direct super-fluid to Mott insulator transition in disordered Bose systems, *Phys. Rev. Lett.* **103**, 140402 (2009).

31. W. Krauth, N. Trivedi, and D. Ceperley, Superfluid-insulator transition in disordered Boson systems, *Phys. Rev. Lett.* **67**, 2307 (1991).

32. B. V. Svistunov, Superfluid Bose-glass transition in weakly disordered commensurate one-dimensional system, *Phys. Rev. B* **54**, 16131 (1996).

33. V. A. Kashurnikov, A. I. Podlivaev, N. V. Prokofev, and B. V. Svistunov, Supercurrent states in one-dimensional finite-size rings, *Phys. Rev. B* **53**, 13091 (1996).

34. F. D. M. Haldane, Effective harmonic-fluid approach to low-energy properties of one-dimensional quantum fluids, *Phys. Rev. Lett.* **47**, 1840 (1981).

35. T. Giamarchi and H. J. Schulz, Anderson localization and interactions in one-dimensional metals, *Phys. Rev. B* **37**, 325 (1988).

36. N. V. Prokof'ev and B. V. Svistunov, Comment on "one-dimensional disordered bosonic Hubbard model: A density-matrix renormalization group study," *Phys. Rev. B* **80**, 4355 (1998).

37. S. G. Söyler, M. Kiselev, N. V. Prokof'ev, and B. V. Svistunov, Phase diagram of commensurate two-dimensional disordered Bose–Hubbard model, *Phys. Rev. Lett.* **107**, 185301 (2011).

38. N. V. Prokof'ev and B. V. Svistunov, Superfluid-insulator transition in commensurate disordered bosonic systems: Large-scale worm algorithm simulations, *Phys. Rev. Lett.* **92**, 015703 (2004).

39. E. Burovski, E. Kozik, A. Kuklov, N. Prokof'ev, and B. Svistunov, Superfluid interfaces in quantum solids, *Phys. Rev. Lett.* **94**, 165301 (2005).

40. Ş. G. Söyler, B. Capogrosso-Sansone, N. V. Prokof'ev, and B. V. Svistunov, Superfluid-insulator and roughening transitions in domain walls, *Phys. Rev. A* **76**, 043628 (2007).

41. R. A. Guyer, R. C. Richardson, and L. I. Zane, Excitations in quantum crystals (a survey of NMR experiments in solid helium), *Rev. Mod. Phys.* **43**, 532 (1971).

42. M. G. Richards, J. H. Smith, P. S. Tofts, and W. J. Mullin, Frequency dependence of T_1 for ^{3}He in Solid ^{4}He, *Phys. Rev. Lett.* **34**, 1545 (1975).

43. S. Sasaki, R. Ishiguro, F. Caupin, H.J. Maris, and S. Balibar, Superfluidity of grain boundaries and supersolid behavior, *Science* **313**, 1098 (2006).

44. S. Balibar and F. Caupin, Supersolidity and disorder, *J. Phys. Cond. Mat.* **20**, 173201 (2008).

45. L. Pollet, M. Boninsegni, A. B. Kuklov, N. V. Prokof'ev, B. V. Svistunov, and M. Troyer, Superfluidity of grain boundaries in solid ^{4}He, *Phys. Rev. Lett.* **98**, 135301 (2007).

46. S. I. Shevchenko, One-dimensional superfluidity in Bose crystals, *Fiz. Nizk. Temp.* **13**, 115 (1987).

47. L. Pollet, M. Boninsegni, A. B. Kuklov, N. V. Prokof'ev, B. V. Svistunov, and M. Troyer, Local stress and superfluid properties of solid ^{4}He, *Phys. Rev. Lett.* **101**, 097202 (2008).

48. M. Boninsegni, A. B. Kuklov, L. Pollet, N. V. Prokof'ev, B. V. Svistunov, and M. Troyer, Luttinger liquid in the core of a screw dislocation in helium-4, *Phys. Rev. Lett.* **99**, 035301 (2007).

49. Ş. G. Söyler, A. B. Kuklov, L. Pollet, N. V. Prokof'ev, and B. V. Svistunov, Underlying mechanism for the giant isochoric compressibility of solid ^{4}He: Superclimb of dislocations, *Phys. Rev. Lett.* **103**, 175301 (2009).

50. M. W. Ray and R. B. Hallock, Observation of unusual mass transport in solid hcp ^{4}He, *Phys. Rev. Lett.* **100**, 235301 (2008).

51. M. W. Ray and R. B. Hallock, Observation of mass transport through solid ^{4}He, *Phys. Rev. B* **79**, 224302 (2009).

52. M. Boninsegni, N. Prokof'ev, and B. Svistunov, Superglass phase of ^{4}He, *Phys. Rev. Lett.* **96**, 105301 (2006).

53. S. Coleman, Quantum sine-Gordon equation as the massive Thirring model, *Phys. Rev. D* **11**, 2088 (1975).

54. S. T. Chui and P. A. Lee, Equivalence of a one-dimensional fermion model and the two-dimensional Coulomb gas, *Phys. Rev. Lett.* **35**, 315 (1975).

CHAPTER 10

Dynamics of Vortices
and Phonons: Turbulence

This chapter, based substantially on Reference [1], is written in coauthorship with Evgeny Kozik.

A plethora of striking phenomena exhibited by superfluids is based on the quantization of velocity circulation. It is quite remarkable that quantized circulation is perfectly consistent with the hydrodynamics of ideal classical incompressible fluid (ICIF). Indeed, vortex filaments with arbitrary velocity circulation around them introduced in an ideal fluid are preserved under evolution governed by the Euler equation. In this respect, very instructive is the example of Kelvin waves (KWs)—waves of precessing distortions on a single vortex line—which were predicted by Thompson (Lord Kelvin) in the context of classical hydrodynamics well before the discovery of superfluidity. However, such vortex lines in realistic classical fluids are rather artificial; only in a superfluid at $T \to 0$ does any form of rotational motion necessarily involve quantized vortex lines.

From the fundamental point of view, of special interest is the low-temperature limit, when the normal component—an extremely dilute gas of phonons—can be either completely neglected or taken into account perturbatively. In this case, the quantized vortex lines play the leading role in the superfluid hydrodynamics, which results in a substantial simplification of governing equations, but, interestingly, not the dynamics itself. On the contrary, as we will see, even the zero-point relaxation dynamics of the vortex tangle are very involved.

At $T = 0$, vortices and phonons are the only long-wave hydrodynamic modes. To the first approximation, the two are decoupled from each other because they coexist at drastically different dynamic scales: the velocity of a vortex loop is proportional to the inverse curvature/torsion radius and thus is negligible compared to the sound velocity for large loops (much larger than the healing length). This means that to a very good approximation—controlled by the parameter

$$\beta = \frac{\gamma}{Rc} \ll 1, \tag{10.1}$$

where c is the speed of sound and R is the vortex-line curvature/torsion radius—the motion of the vortex-line filament corresponds to that in an ICIF.

10.1 Basic Relations of Vortex Dynamics

10.1.1 Kelvin–Helmholtz Theorem and Biot–Savart Equation

Let $T = 0$. Then, superfluid hydrodynamics is identical to that of ICIF with the constraint on the allowed velocity circulation—the only possible rotational motion is in the form of vortex filaments with a fixed value of velocity circulation $\kappa = 2\pi\gamma$ around them. Our goal is to obtain a closed dynamical equation for vortex filaments in this regime. The most intuitive way is to resort to the Kelvin–Helmholtz theorem of fluid mechanics,[*] as we do here, which immediately leads to the conclusion that vortices move with the local fluid velocity. More formally (and generally), one can derive the vortex dynamics directly from Popov hydrodynamic action. We will consider this approach later and, in particular, employ it to introduce Hamiltonian formalism.

If vortex lines are the only degrees of freedom excited in the fluid (i.e., the fluid density is constant except near the vortex core), and if the typical curvature radius and interline separation are much larger than the vortex core size, a_0, then the instant velocity field at distances much larger than a_0 from the vortex lines is defined by the form of the vortex-line configuration. Indeed, away from the vortex core, the density is practically constant and the hydrodynamic continuity equation reduces to

$$\nabla \cdot \mathbf{v} = 0. \tag{10.2}$$

Then, taking into account that $\nabla \times \mathbf{v} = 0$ everywhere except for the vortex lines and that for any (positively oriented) contour Γ surrounding one vortex line

$$\oint_\Gamma \mathbf{v} \cdot d\mathbf{l} = 2\pi\gamma, \tag{10.3}$$

we observe that the *instant* velocity $\mathbf{v}(\mathbf{r}, t)$ satisfies the same spatial equations as the *static* velocity field, which minimizes energy of the corresponding "pinned" vortex configuration considered in Section 1.2.4. Hence, we can employ the analysis of Section 1.2.4 to relate $\mathbf{v}(\mathbf{r}, t)$ to the instant configuration of the vortex lines. The result is the Biot–Savart formula (1.122)

$$\mathbf{v}(\mathbf{r}, t) = \frac{\gamma}{2} \int \frac{d\mathbf{s} \times (\mathbf{r} - \mathbf{s})}{|\mathbf{r} - \mathbf{s}|^3}, \tag{10.4}$$

with the radius vector $\mathbf{s} \equiv \mathbf{s}(t)$ running along all the vortex filaments at a given time t.

According to Kelvin–Helmholtz theorem, each element of the vortex line moves with a velocity corresponding to the velocity field averaged over a small contour (of the size, say, of order a_0) surrounding this element. This fact is very important, since the velocity field (10.4) is singular at all points on the vortex line, and the Kelvin–Helmholtz theorem yields a simple regularization prescription: Take a small circular contour centered at the vortex-line element in the plane perpendicular to the vortex line and average the velocity field over this contour to eliminate the local rotational component.

---

[*] The Kelvin–Helmholtz theorem states that in an ideal fluid, the velocity circulation around a closed contour moving with the fluid remains constant.

As is clear from (10.4), the singularity of the velocity field, at some point $s = s_0$ in the vortex line, comes from the integration over the near vicinity of the point s_0. To isolate the singularity, we expand the function $s(\xi)$ around the point s_0:

$$s = s_0 + s'_\xi \xi + s''_{\xi\xi} \xi^2/2 + \cdots ,$$

(10.5)

where ξ is the (algebraic) arc length measured from the point s_0. Then, substituting expansion (10.5) into (10.4), we get

$$\mathbf{v}(\mathbf{r} \to s_0) = \frac{\gamma}{4} s'_\xi \times s''_{\xi\xi} \int_{-\xi_*}^{\xi_*} \frac{d\xi}{|\xi|} + \text{regular part},$$

(10.6)

where ξ_* is some upper cutoff parameter on the order of the curvature radius at the point s_0. The integral in the right-hand side is divergent at $\xi \to 0$. To regularize it, we note that (1) by its very origin, the expression (10.4) is meaningful only at $|\xi| > a_0$ and (2) the part of the vortex line with $|\xi| \ll a_0$ makes a small contribution to the net velocity on the contour of radius $\sim a_0$. Hence, with logarithmic accuracy, we can adopt the regularization $|\xi| > a_0$, that is,

$$\int_{-\xi_*}^{\xi_*} \frac{d\xi}{|\xi|} \to 2\ln(\xi_*/a_0).$$

(10.7)

Moreover, by fine-tuning the value of a_0 in accordance with a model-specific behavior at distances $\sim a_0$, the logarithmic accuracy of the regularization (10.7) can be improved to the accuracy $\sim a_0/\xi_*$. We will discuss this option in detail in Section 10.1.3 and utilize it in Section 10.2.3.

We thus arrive at the Biot–Savart equation for the vortex-line motion

$$\dot{s} = \frac{\gamma}{2} \int \frac{ds_0 \times (s - s_0)}{|s - s_0|^3},$$

(10.8)

with the integral regularized in accordance with (10.7).

Equation (10.8) has two important constants of motion:

$$\mathcal{E} = \int \frac{ds \cdot ds_0}{|s - s_0|},$$

(10.9)

$$\vec{\mathcal{P}} = \int s \times ds.$$

(10.10)

Up to dimensional factors, they represent the liquid energy and momentum, respectively; see Section 1.2.4.

Problem 10.1 *Show that (10.9) and (10.10) are conserved by (10.8).*

10.1.2   Local Induction Approximation

In the absence of significant enhancement of nonlocal interactions by polarization of the vortex tangle, the regular part in (10.6) is smaller than the first term
containing a large logarithm. In such cases, it is often* safe to neglect the second
term in (10.6) and proceed within the local induction approximation (LIA):

$$\dot{s} = \beta s'_\xi \times s''_{\xi\xi},$$

$$(10.11)$$

where

$$\beta = \frac{\gamma}{2}\ln(R/a_0),$$

$$(10.12)$$

with the typical curvature radius R treated as a constant. Recalling that the parameter ξ in (10.6) is the arc length, one must ensure that the equation of motion
(10.11) preserves natural parameterization in the course of evolution. This is
established by directly checking that (10.11) implies

$$\frac{d}{dt}\sqrt{ds \cdot ds} = 0.$$

$$(10.13)$$

Problem 10.2 *Prove (10.13).*

From Equation (10.13), it trivially follows—by integrating over ξ—that the
total line length is conserved. The total line length in LIA plays the same role
as the energy (10.9) in the genuine Biot–Savart equation. One may also see that
Equation (10.11) conserves $\vec{P}$ (10.10).

Problem 10.3 *Show that with the same accuracy as that of Equation (10.11), $\mathcal{E}$ (10.9)
is proportional to β times line length conserved by Equation (10.13).*

Problem 10.4 *Prove that Equation (10.11) conserves $\vec{P}$ (10.10) exactly.*

The standard constants of motion—the energy (line length), momentum, and
angular momentum—are not the only quantities conserved by Equation (10.11).
For example, the integral of the square of the curvature radius is also conserved:

$$\frac{d}{dt}\int \left(s''_{\xi\xi}\right)^2 d\xi = 0.$$

$$(10.14)$$

Problem 10.5 *Prove Equation (10.14).*

Problem 10.6 *Using LIA, demonstrate that at $T = 0$, a vortex ring of radius R prop-
agates (in an infinite container) translationally in the direction perpendicular to its
plane with the constant velocity $v_R = (\gamma/2)\ln(R/a_0)/R$.*

---

* But not always (!); see the theory of the pure KW Kelvin-wave cascade in the following.

It turns out that the constant of motion (10.14) is just one of an *infinite* set of constants of motion implied by the *integrability* of LIA. Betchov [2] revealed certain interesting properties of LIA, for example, Equation (10.14), by rewriting Equation (10.11) in terms of intrinsic variables of the vortex line: curvature, ζ, and torsion, τ. Hasimoto [3] further advanced these ideas by discovering that for the complex variable $\psi(\xi, t)$, such that

$$\zeta = |\psi|, \qquad \tau = \frac{\partial \Phi}{\partial \xi} \tag{10.15}$$

(Φ is the phase of ψ), the LIA equation (10.11) is equivalent to the nonlinear Schrödinger equation (time is measured in units β^{-1})

$$i \frac{\partial \psi}{\partial t} = -\frac{\partial^2 \psi}{\partial \xi^2} - \frac{1}{2} |\psi|^2 \psi. \tag{10.16}$$

The 1D nonlinear Schrödinger equation is known to be an integrable system featuring an infinite number of additive constants of motion, the explicit form of which is given by ($n = 1, 2, 3, \cdots$)

$$I_n = \int_{-\infty}^{\infty} \varphi_n(\xi) d\xi, \qquad \varphi_1 = \frac{1}{4} |\psi|^2, \qquad \varphi_{n+1} = \psi \frac{d}{d\xi} (\varphi_n/\psi) + \sum_{n_1 + n_2 = n} \varphi_{n_1} \varphi_{n_2}. \tag{10.17}$$

The integrability of LIA renders any nontrivial relaxation kinetics impossible, unless the vortex-line reconnections are involved to break the conservation of I_n's. We will see in the following that this process plays a key role in the theory of superfluid turbulence (ST).

10.1.3 Lagrangian and Hamiltonian Formalisms for Vortex Lines

Within the incompressible-liquid approximation, the vortex-line dynamics must obey the least-action principle with respect to the action (2.127). That is, the Biot–Savart equation of motion (10.4) should be an extremal trajectory of the action (2.127). To make sure that this is the case and, most importantly, to proceed to the Hamiltonian formalism, we must explicitly express (2.127) in terms of the vortex-line shape. In view of the condition $\Delta \Phi_0 = 0$, the answer for the second term corresponds to Equation (1.119). We thus need to consider the first term of (2.127), which we represent as (the time integration is omitted for briefness)

$$\int d^3 r \dot{\Phi}_0 = \int du \, \dot{\mathbf{s}} \cdot \frac{\delta}{\delta \mathbf{s}} \int d^3 r \Phi_0. \tag{10.18}$$

Here the vector function $\mathbf{s} \equiv \mathbf{s}(u, t)$, parameterized by u, specifies the vortex-line configuration at time t. To calculate the variational derivative in the right-hand side of (10.18), we consider the variation of the phase field due to the distortion of the vortex line by $\delta s(u)$ and then observe that[*]

$$\delta \int d^3 r \Phi_0 = \int d^3 r \delta \Phi_0 = \frac{1}{3} \int d^3 r \nabla \cdot (r \delta \Phi_0). \tag{10.19}$$

The variation $\delta \Phi_0(\mathbf{r})$ can be viewed as being produced by two vortex lines with opposite circulation quanta separated by $\delta s(u)$. Because of the quantization of circulation, the field $\delta \Phi_0(\mathbf{r})$ has to experience a jump of 2π. Without loss of generality, we locate the jump at the infinitesimal surface δS that has the two vortex lines at its edge. Away from the surface δS, the field $\delta \Phi_0(\mathbf{r})$ is well defined, allowing us to apply the Gauss theorem to represent the integral (10.19) as a surface integral: $\int d^3 r \nabla \cdot (r \delta \Phi_0) = 2\pi \int_{\delta S} \mathbf{r} \cdot d\vec{S}$. The infinitesimal surface element can be represented as $d\vec{S} = \delta s \times ds$, reducing the surface integral to a line integral: $\int_{\delta S} \mathbf{r} \cdot d\vec{S} = \int \delta s \cdot (ds \times s) = \int \delta s \cdot (s'_u \times s) \, du$. Here we also used an infinitesimal smallness of δs to replace $\mathbf{r}$ with $\mathbf{s}$. Having explicitly related $\delta \int d^3 r \Phi_0$ with δs, we thus obtain an expression for the variational derivative in (10.18),

$$\frac{\delta}{\delta s} \int d^3 r \Phi_0 = \frac{2\pi}{3} s'_u \times s, \tag{10.20}$$

bringing us to

$$\int d^3 r \dot{\Phi}_0 = \frac{2\pi}{3} \int \dot{s} \cdot (ds \times s). \tag{10.21}$$

and resulting in the following expression for the vortex-line Lagrangian ($n_s = n$ because of Galilean invariance):

$$L_{\text{vor}} = \frac{2\pi n}{3} \int (\dot{s} \times s) \cdot ds - \frac{\pi \gamma n}{2} \int \frac{ds \cdot ds_0}{|s - s_0|}. \tag{10.22}$$

The corresponding least-action trajectory satisfies the equation

$$s'_u \times \dot{s} = \frac{\gamma}{2} \int \frac{ds_0 [(s - s_0) \cdot s'_u]}{|s - s_0|^3} - \frac{\gamma}{2} \int \frac{(s - s_0)(ds_0 \cdot s'_u)}{|s - s_0|^3}. \tag{10.23}$$

It is easy to check that if $s(u, t)$ satisfies the Biot–Savart equation of motion (10.4), then it also satisfies (10.23). However, a solution of (10.23) is not necessarily a solution of (10.4). The reason behind this fact is very simple. Geometrically, two evolving curves, $s(u, t)$ and $s_1(u, t)$, are indistinguishable, if $\dot{s}_1 = \dot{s} + s'_u f(u, t)$, where $f(u, t)$ is an arbitrary function of u and t. Indeed, the shift of all points of the line along the line does not change the shape and position of the line. It is this parameterization freedom that is being respected by Equation (10.23)—the least-action principle is invariant with regard to the choice of generalized coordinates, but not respected by Equation (10.4), implying that the points of the vortex line are the "material" objects moving with the velocity of surrounding fluid.

---

[*] $\int d^3 r \mathbf{r} \cdot \nabla \delta \Phi_0 = (1/2) \int d^3 r (\nabla r^2) \cdot \nabla \delta \Phi_0 = -(1/2) \int d^3 r \, r^2 \delta (\Delta \Phi_0) = 0.$

Problem 10.7 *Derive Equation (10.23) from the Lagrangian (10.22).*

Consider a more special case, very important for applications, when the shape of a vortex line allows parameterization

$$\mathbf{s}(z,t) = \hat{\mathbf{z}}z + \boldsymbol{\rho}(z,t), \tag{10.24}$$

where z is the Cartesian coordinate along a certain fixed z-axis, $\hat{\mathbf{z}}$ is the unit vector along this axis, and $\boldsymbol{\rho}(z,t) = (x(z,t), y(z,t))$. A necessary and sufficient condition for this parameterization is the existence of Cartesian basis in which coordinates x and y are single-valued functions of coordinate z. In this parameterization, the shape of the line is uniquely defined by a 1D two-component vector field $\boldsymbol{\rho}(z)$, and the time evolution is thus exhaustively described by the function $\boldsymbol{\rho}(z,t)$. We will see that it is especially convenient to use the complex field $w(z) = x(z) + iy(z)$ instead of $\boldsymbol{\rho}(z)$. The equation of motion for $w(z,t)$ could be obtained by substituting (10.24) into* (10.23). A more fundamental way is to employ the Lagrangian (10.22) and derive the corresponding Lagrangian for w. The observation that $(2/3) \int dt \int (\dot{\mathbf{s}} \times \mathbf{s}) \cdot d\mathbf{s} = \int dt \int dz \dot{\boldsymbol{\rho}} \cdot (\hat{\mathbf{z}} \times \boldsymbol{\rho}) = i \int dt \int dz w^* \dot{w}$, together with a straightforward change of variables in the energy integral, yields

$$\frac{1}{\pi n} L_{\text{vor}}[w] = i \int dz w^* \dot{w} - H, \tag{10.25}$$

$$H = \frac{\gamma}{2} \int \frac{[1 + \operatorname{Re} w'^*(z_1) w'(z_2)] dz_1 dz_2}{\sqrt{(z_1 - z_2)^2 + |w(z_1) - w(z_2)|^2}}. \tag{10.26}$$

According to the standard rules of transformation from the Lagrangian to Hamiltonian picture, Equation (10.25) implies that $w(z)$ and $w^*(z)$ are the canonical variables with respect to the Hamiltonian (10.26), with the equation of motion

$$i\dot{w} = \frac{\delta H}{\delta w^*}. \tag{10.27}$$

The Hamiltonian (10.26) is singular at $z_1 \to z_2$ and thus needs to be regularized. Let us introduce r_* such that

$$a_0 \ll r_* \ll \lambda, \tag{10.28}$$

where λ is the typical length scale of the vortex-line structure, and write

$$H = H_{\text{loc}} + H_{\text{n.l.}} + \mathcal{O}(r_*/\lambda), \tag{10.29}$$

$$H_{\text{n.l.}} = (\gamma/2) \int_{|z_1 - z_2| > r_*} \frac{[1 + \operatorname{Re} w'^*(z_1) w'(z_2)] dz_1 dz_2}{\sqrt{(z_1 - z_2)^2 + |w(z_1) - w(z_2)|^2}}, \tag{10.30}$$

$$H_{\text{loc}} = 2\beta \int dz \sqrt{1 + |w'(z)|^2}, \tag{10.31}$$

$$\beta = \gamma \ln(r_*/a_*). \tag{10.32}$$

---

* Note that $\mathbf{s}(z,t)$ does not obey the Biot–Savart Equation (10.4) because the parameterization (10.24) is different from the "material" parameterization implied by (10.4).

Here the value of $a_* \sim a_0$ is fine-tuned to eliminate a factor of order unity in the logarithm. The absence of such a factor in Equation (10.32) should not be confused with a lack of control on first sublogarithmic corrections. In view of the first inequality in (10.28), the Hamiltonian $H_{n.l.}$ is of purely hydrodynamic nature, while the Hamiltonian H_{loc} takes care of *both* hydrodynamic and microscopic (system-specific) features at distances smaller than r_*. In view of the condition $\lambda \gg a_0$, implied by Equation (10.28), the microscopic specifics of the system are completely absorbed by the proper choice of the value of a_*. Indeed, whatever the physics at distances of the order of the vortex-core radius is, the leading contribution from these distances to the energy of a smooth vortex line should be directly proportional to the line length; this is precisely what is expressed by Equation (10.31).

The freedom of choosing a particular value of r_* within the range (10.28) can be used to introduce the LIA by requiring that

$$\ln(r_*/a_*) \gg \ln(\lambda/r_*). \tag{10.33}$$

In this case, the Hamiltonian (10.29), with

$$\beta = \gamma \ln(\lambda/a_*) = \gamma \Lambda, \tag{10.34}$$

captures the leading—as long as the interline interactions are not relevant—logarithmic contribution, compared to which the nonlocal Hamiltonian (10.30) can be omitted with logarithmic accuracy guaranteed by the parameter $\Lambda^{-1} \ll 1$.

A less obvious technical trick is to "formally" set

$$r_* = a_* \tag{10.35}$$

to *nullify* the Hamiltonian H_{loc}. Doing so might seem to violate the range of applicability of the essentially hydrodynamic Hamiltonian $H_{n.l.}$. Nevertheless, it is readily seen by inspection that, under the condition $\lambda \gg a_0$, the resulting theory is equivalent to the theory (10.29) through (10.32) up to negligibly small corrections of the order of a_0/λ. The resulting Hamiltonian

$$H_{psd} = (\gamma/2) \int\limits_{|z_1 - z_2| > a_*} \frac{[1 + \mathrm{Re}\, w'^*(z_1)\, w'(z_2)]\, dz_1\, dz_2}{\sqrt{(z_1 - z_2)^2 + |w(z_1) - w(z_2)|^2}}, \tag{10.36}$$

in direct analogy with the pseudo-potential method in the scattering theory, has the status of the *pseudo*-Hamiltonian in the following sense: While being a rather inadequate model at the scales $\sim a_0$, it accurately accounts for the long-wave motion of the vortex lines (including all subleading corrections to LIA coming from distances $\sim a_0$) by an appropriate choice of a_*. In Section 10.2.1, we will describe an explicit procedure of extracting the value of a_* from a model-specific dispersion relation of KWs.

10.1.4 Phenomenology of Interaction with Normal Component: Superfluid Magnus Force

In the phenomenological approach, one follows the dynamics of an element of the vortex core described by the radius vector $\mathbf{s}(\xi, t)$ (ξ is the arc length) by considering all the forces exerted on it by the surrounding fluid in the hydrodynamic limit.

The crucial role in the motion of a quantized vortex is played by the *superfluid Magnus force*. The force arises due to the classical Magnus effect, whereby the circulation of fluid around an object superimposed on its translational motion creates a difference of absolute fluid velocities along the opposite sides of the object, resulting in excess pressure on the side with lower velocity. Correspondingly, the circulation of superfluid results in the superfluid Magnus force acting on the vortex core element $\mathbf{s}$ per unit line length given by the standard formula for the cylindrical geometry:

$$\mathbf{f}_M = \kappa \rho_s \mathbf{s}' \times [\dot{\mathbf{s}} - \mathbf{v}_s(\mathbf{s})], \tag{10.37}$$

where $\mathbf{v}_s$ is the superfluid velocity field of the surrounding fluid given by Equation (10.4) plus an externally imposed superflow if any and κ is the velocity circulation ($\kappa = 2\pi\gamma$ for superfluids).

At $T = 0$, the Magnus force is the only force acting on the vortex. Note that $\mathbf{f}_M$ does not depend on the vortex core radius. As a result of taking the hydrodynamic limit $a_0 \to 0$, the inertia of the vortex core vanishes, leading to the equation of motion in the form $\mathbf{f}_M = 0$. The condition $\mathbf{f}_M = 0$ is equivalent to the Kelvin–Helmholtz theorem $\dot{\mathbf{s}} = \mathbf{v}_s(\mathbf{s})$ (see Section 10.1.1) and thereby leads to the Biot–Savart equation (10.8). This is not surprising since, at $T = 0$, this approach is equivalent to the direct solution of the hydrodynamic equations of motion in view of the aforementioned equivalence of superfluid hydrodynamics to that of ICIF, as soon as quantized vortices are introduced in ICIF as an initial condition. The advantage of the phenomenological approach is that it allows us to straightforwardly incorporate the effects of interaction with the normal component into the equations of motion.

At finite temperatures, when a vortex line moves relative to the normal component, the scattering of elementary excitations, composing the normal component, on the vortex results in an exchange of momentum between the vortex and the normal component, which gives rise to the *mutual friction*. It is important to emphasize that this scattering is *not* due to the interaction of the quasiparticles with the vortex core itself but rather is due to their deflection by the singular velocity field of the vortex. As a result, the corresponding scattering cross section does not vanish in the limit of $a_0 \to 0$ and is captured by hydrodynamics. The chirality of the velocity field around the vortex allows the mutual friction force to also have a component perpendicular to the direction of relative motion. Thus, one can write down the total drag force per unit length acting on a vortex-line element in the form

$$\mathbf{f}_D = D\mathbf{s}' \times [\mathbf{s}' \times [\mathbf{v}_n(\mathbf{s}) - \dot{\mathbf{s}}]] + D'\mathbf{s}' \times [\mathbf{v}_n(\mathbf{s}) - \dot{\mathbf{s}}], \tag{10.38}$$

where D and D' are empirical coefficients. The first term in the right-hand side of Equation (10.38) results in dissipation of energy, whereas the second term effectively leads to a "drift" of the vortex along the normal-fluid velocity.

Finally, the equations of motion for the vortex-line element $\mathbf{s}$ follow from Newton's law $\mathbf{f}_M + \mathbf{f}_D = 0$:

$$\dot{\mathbf{s}} = \mathbf{v}_s(\mathbf{s}) + \alpha \mathbf{s}' \times [\mathbf{v}_n(\mathbf{s}) - \mathbf{v}_s(\mathbf{s})] + \alpha'[\mathbf{s}' \times [\mathbf{v}_n(\mathbf{s}) - \mathbf{v}_s(\mathbf{s})]] \times \mathbf{s}'. \qquad (10.39)$$

Here α and α' are the phenomenological dimensionless mutual friction coefficients.

A remark is in order here. In the hydrodynamic limit for the normal component, which requires that the relevant length scales are much larger than the excitation mean-free path λ_{mf}, the normal component is governed by the Navier–Stokes equation. In this case, the problem of slow motion of a cylindrical body through the fluid is singular, known as the Stokes paradox. The drag force exerted on the cylinder contains a denominator $\ln(\lambda_O/R)$, which diverges logarithmically in the limit of small velocities. Here, R is the cylinder radius and $\lambda_O = \lambda_O(u) = \nu/u$ is the so-called Oseen length (ν being the kinematic viscosity and u the velocity of the cylinder relative to the fluid) at which the nonlinear term in the Navier–Stokes equation becomes larger than the viscous one. Thus, the mutual friction coefficients α, α', generally speaking, depend on vortex velocities via such a logarithmic factor with ν the kinematic viscosity of the normal component, $u = |\mathbf{v}_n(\mathbf{s}) - \dot{\mathbf{s}}|$, and R replaced by the mean-free path of the excitations λ_{mf} as the lower cutoff scale of the viscous hydrodynamics, so that $\alpha, \alpha' \to 0$ in the limit $|\mathbf{v}_n(\mathbf{s}) - \dot{\mathbf{s}}| \to 0$. However, the hydrodynamic limit for the normal component is a rather strong requirement (in contrast to that for the superfluid component) and is easily violated under realistic experimental conditions. In particular, in ST, where one is dealing with a hierarchy of length scales in the vortex-tangle structure, certain scales can be in the hydrodynamic regime of the normal component, while for the smaller ones, the excitations are essentially ballistic. In addition, λ_{mf} quickly increases at low temperatures and, for example, for $T < 0.5\,\mathrm{K}$ in ^{4}He, becomes on the order of the system size. More specifically, the hydrodynamic regime for the normal component is broken, and thus, the (logarithmic) dependence of α, α' on the vortex-line velocity is absent for vortex structures with the wavelength $\lambda < \lambda_{mf}$ or as soon as $\lambda_O < \lambda_{mf}$. In the opposite limit of $\lambda \gg \lambda_{mf}$, $\lambda_O \gg \lambda_{mf}$, this dependence is usually neglected due to the slowness of the logarithm.

The coefficient α describes the dissipation (and pumping in case of induced counterflow) of energy of quantized vortices, that is, the decrease (increase) of the vortex-line length. In ^{4}He, $\alpha < 1$ at all temperatures but very close to T_c. For instance, at $T = 2.1\,K$, one already finds $\alpha \approx 0.5$, and thus dissipation can typically be considered as a perturbation. The qualitative role of α' is rather minor: it describes renormalization of the vortex velocity due to its drift with the normal component (and thus $\alpha' \leq 1$ in principle). In addition, in ^{4}He, $\alpha' \ll \alpha$ and thus its effects are usually neglected altogether. In the superfluid ^{3}He-B, the situation is somewhat different, as α does not drop to $\alpha = 0.5$ until about $T \sim 0.5\,T_c$, and thus the overdamped regime of vortex dynamics is easily attainable. The coefficient

α'—although $\alpha' < \alpha$ for most temperatures—also remains of the order of α in ^{3}He-B except very close to T_c and thus must be taken into account in quantitatively accurate calculations.

We discuss a microscopic calculation of the universal contribution to α due to phonons at $T \ll T_c$ in Section 10.3.

Problem 10.8 *At finite temperatures, the lifetime of a quantized vortex ring is finite as a result of its shrinking due to mutual friction. Show that the distance traveled by a vortex ring of radius R in an otherwise stationary fluid is given by $R(1 - \alpha')/\alpha$.*

10.2   Kelvin Waves

If a straight vortex line is distorted, its further evolution by the Biot–Savart equation (10.8) implies precession and propagation of the distortions, or KWs. They play a crucial part in the theory of ST in the $T \to 0$ limit. In this section, we will develop the theory of KWs (or kelvons), including the theory of kelvon scattering and KW cascade.

10.2.1   Bilinearized Hamiltonian: Kelvons

Suppose the amplitude of the straight vortex-line distortion is small enough, so that the following condition is satisfied:

$$\alpha(z_1, z_2) = \frac{|w(z_1) - w(z_2)|}{|z_1 - z_2|} \ll 1. \tag{10.40}$$

By expanding (10.36) up to the leading order in the dimensionless function $\alpha(z_1, z_2)$, we obtain

$$H_0 = (\gamma/4) \int\limits_{|z_1-z_2|>a_*} \frac{dz_1 dz_2}{|z_1 - z_2|} \left[2\mathrm{Re}\, w'^*(z_1)\, w'(z_2) - \alpha^2(z_1, z_2) \right]. \tag{10.41}$$

The Hamiltonian H_0 describes the linear properties of KWs. It is diagonalized by the Fourier transformation $w(z) = L^{-1/2} \sum_k w_k e^{ikz}$ (L is the system size; periodic boundary conditions for w are assumed):

$$H_0 = \sum_k \omega_k w_k^* w_k, \tag{10.42}$$

yielding Kelvin's dispersion law

$$\omega_k = \frac{\gamma}{2} \left[\ln\frac{1}{ka_*} + C_0 + \mathcal{O}\big((ka_*)^2\big) \right] k^2, \tag{10.43}$$

with

$$C_0 = 2 \int\limits_1^\infty \frac{dx}{x} \left[\cos x + \frac{\cos x - 1}{x^2} \right] + 2 \int\limits_0^1 \frac{dx}{x} \left[\cos x - \frac{1}{2} + \frac{\cos x - 1}{x^2} \right] \approx -2.077. \tag{10.44}$$

Generally speaking, the last term $\mathcal{O}\big((ka_*)^2\big)$ should be omitted, since it goes beyond the range of applicability of the universal pseudo-Hamiltonian (10.36). In different microscopic models, this term depends on the physics of the vortex core. With logarithmic accuracy, the constant C_0 can be omitted as well at the expense of replacing a_* with an order-of-magnitude estimate for the vortex core radius.

The practical utility of Equation (10.43) with the sublogarithmic correction C_0 (10.44) for a given microscopic model—*different* from the pseudo-Hamiltonian (10.36)—is as follows. Equations (10.43) and (10.44) can be used to *calibrate* the value of a_* and thus fix the pseudo-Hamiltonian, by solving for the KW spectrum independently and casting the answer in the form (10.43) and (10.44). Likewise, for a realistic, strongly correlated system like ^{4}He, the appropriate value of a_* in the pseudo-Hamiltonian (10.36) can be calibrated by an experimentally measured kelvon dispersion law.

Although the problem of the KW cascade generated by decaying ST is purely classical, it is technically convenient to approach it quantum mechanically*— by introducing kelvons, or quanta of KWs. In accordance with the canonical quantization procedure, we understand w_k as the annihilation operator of the kelvon with momentum k and correspondingly treat $w(z)$ as a quantum field. The Hamiltonian functional (10.36) is *proportional* to the energy—with the coefficient $\pi\gamma\rho$, where ρ is the mass density—but not *equal* to it. If we were to use the genuine quantum mechanical description instead of an effective one (the latter is absolutely acceptable at large occupation numbers; i.e., when the problem is classical, and $\hbar$ drops out from the final answers), then we would be dealing with annihilation operators, $\hat{a}_k = \hat{w}_k/\sqrt{\hbar/\pi\gamma\rho}$, and the Hamiltonian, $\hat{H} = (\pi\gamma\rho)H$. By choosing the units $\hbar = \pi\gamma = \rho = 1$, we simply ignore these coefficients to simplify the description until the final answers are obtained.

In the quantum approach, the notion of the number of kelvons naturally arises. This number is *conserved* in view of the global $U(1)$ symmetry of the Hamiltonian, $w \to e^{i\varphi}w$ reflecting the rotational symmetry of the original problem. This symmetry is also responsible for the conservation of the angular momentum component along the vortex-line direction. Thus, the number of kelvons is related to the total angular momentum component: one kelvon carries a single (negative) quantum, $-\hbar$, of angular momentum around the vortex line relative to the macroscopic angular momentum of the straight vortex line.

Problem 10.9 *Prove the previous statement.*

Problem 10.10 *Demonstrate that* $(1 - \alpha')/\alpha$ *gives the characteristic number of oscillations of a KW before its magnitude decays substantially. Show that the power dissipated per unit line length from a KW with the wavenumber k and amplitude b_k is given (neglecting α') by* $\Pi(k) \sim \alpha\gamma\rho\omega_k^2 b_k^2$.

---

* In quantum mechanics, perturbative kinetics are readily captured by the Fermi Golden Rule and, if higher-order corrections are necessary, by diagrammatic field-theoretical techniques.

The geometrical nature of the field $w(z)$ leads to special symmetries of the Hamiltonian (10.36) that play an important role in vortex dynamics. Along with a natural set of Noether's constants of motion, which—apart from their rather specific expressions in terms of $w(z)$—are nothing but components of the total linear and angular momenta of the fluid, the geometry brings about fundamental constraints on the kinetics of KWs. More specifically, it constrains the elementary kelvon-scattering processes (Section 10.2.3) to a degree sufficient to fix the structure of the kinetic equation (Section 10.2.4) and even to guarantee the locality of kelvon scattering in the wavenumber space (Section 10.2.5).

10.2.2   Emission of Kelvin Waves by Relaxing Vortex Angle

With a slight modification of the time dependence of the phase, the standard self-similar solution of the linear Schrödinger equation—its Green's function—applies also to the nonlinear equation (10.16). Indeed, the absolute value of the solution remains homogeneous in ξ, in which case the nonlinearity is immediately absorbed into the time-dependent phase term. The result is

$$\psi(\xi,t) = \frac{A}{\sqrt{t}}\exp\left[i\left(\frac{\xi^2}{4t} + \frac{A^2}{2}\ln|t|\right)\right]. \tag{10.45}$$

In accordance with (10.15), this solution implies

$$\zeta = A/\sqrt{t}, \tag{10.46}$$

$$\tau = \xi/2t. \tag{10.47}$$

The physical meaning of the solution (10.46) and (10.47), first revealed by Buttke [4], is the relaxation of the vortex angle, the value of which is controlled by the parameter A. This solution gives an accurate description (within LIA) of relaxation of two vortex lines after their reconnection, as long as the curvature in the relaxation region remains much larger than the curvatures of the two lines away from the region, so that the distant parts of the two lines can be treated approximately as straight lines. By dimensional argument, the self-similarity regime should be achieved very rapidly upon the reconnection, with the velocity of propagation of the fastest KWs (with wavevectors $\sim a_0$).

With the solution (10.46) and (10.47), one can explicitly see that reconnections lift the integrability constraints. At $\xi \to \infty$, we have $\varphi_n \sim \xi^n$, meaning that a single reconnection renders all the integrals I_n (10.17) divergent.

Let us look at the asymptotic form of the solution (10.46) and (10.47) in Cartesian coordinates, taking the direction of the z-axis along one of the two lines, $z \to +\infty$ corresponding to the asymptotic limit:

$$x(z,t) + iy(z,t) = \left(4At^{3/2}/z^2\right)e^{iz^2/4t}, \qquad z \gg \sqrt{t}. \tag{10.48}$$

Equation (10.48) reveals a helical KW structure moving away from the reconnection region. It is important that while arbitrarily small wavelengths are present in the solution (10.48), the integral for the total line lengths comes from the largest length scale $z \sim \sqrt{t}$. This provides support for two crucial points in the scenario of self-reconnections-driven KW cascade: (1) Reconnections push KWs to smaller length scales; (2) apart from higher-order corrections, the line length associated with the curvature radius R cannot originate from line distortions with the curvature radius much smaller than R (locality of the cascade in the wavenumber space).

When the angle between the two reconnecting lines is larger than a certain critical value (which proves to be rather close to π; i.e., the lines must be almost antiparallel), helical structures on the two sides of the relaxing angle also intersect and reconnect, ultimately producing a cascade of vortex loops of ever-increasing size.

10.2.3 Three-Kelvon Elastic Scattering

As long as the parameter $\alpha(z_1, z_2)$ (10.40) is much smaller than unity, the kinetics of KWs are perturbative. They reduce to elementary processes of kelvon scattering, the overall picture being accurately described by the corresponding kinetic equation. The 1D character of the problem and conservation of the total number of kelvons imply that the leading kinetic channel is a three-kelvon elastic scattering. Indeed, the dominant process, on one hand, involves the minimal possible number of kelvons (the scattering matrix element is getting progressively smaller due to higher powers of the small parameter $\alpha(z_1, z_2)$ for multikelvon processes) and, on the other hand, must result in a nontrivial change of kelvon momenta. Conservation of the total number of quasiparticles implies that the allowed scattering processes are two-body, three-body, four-body, and so on, elastic scattering events. If the momenta of two 1D quasiparticles before the scattering event were (k_1, k_2), then, by conservation of momentum and energy, the only possible outcomes of the scattering event are $\left(k_1' = k_1,\ k_2' = k_2\right)$ and $\left(k_1' = k_2, k_2' = k_1\right)$. Both are equivalent to the initial state. That is why the three-kelvon elastic scattering turns out to be the leading scattering channel.

The three-kelvon scattering is described by the effective interaction Hamiltonian

$$H_{\text{int}} = \sum_{k_1,\ldots,k_6} \delta(\Delta k) V_{1,2,3}^{4,5,6} a_6^\dagger a_5^\dagger a_4^\dagger a_3 a_2 a_1, \qquad (10.49)$$

where the three-point vertex V is symmetrized with respect to the corresponding momenta permutations (in sub- and superscripts, we replace the six momenta with their integer labels), $\delta(k)$ is equivalent to $\delta_{k,0}$, and $\Delta k = k_1 + k_2 + k_3 - k_4 - k_5 - k_6$. The central goal of this section is the derivation of $V_{1,2,3}^{4,5,6}$. We start with the pseudo-Hamiltonian (10.36). The condition (10.40) allows us to expand (10.36) in powers of $\alpha(z_1, z_2) \ll 1$: $H = E_0 + H_0 + H_1 + H_2 + \cdots$, where H_0 is given by Equation (10.41) and describes noninteracting kelvons (the constant term E_0 can be ignored in this discussion), while the higher-order terms are responsible for interactions between them. The terms that will prove relevant are

$$H_1 = \frac{\gamma}{16} \int\limits_{|z_1 - z_2| > a_*} \frac{dz_1 dz_2}{|z_1 - z_2|} \left[3\alpha^4 - 4\alpha^2 \text{Re}\, w'^*(z_1)\, w'(z_2) \right], \tag{10.50}$$

and

$$H_2 = \frac{\gamma}{32} \int\limits_{|z_1 - z_2| > a_*} \frac{dz_1 dz_2}{|z_1 - z_2|} \left[6\alpha^4 \text{Re}\, w'^*(z_1)\, w'(z_2) - 5\alpha^6 \right]. \tag{10.51}$$

The effective three-kelvon scattering vertex, $V_{1,2,3}^{4,5,6}$, has two distinctive contributions; see Figure 10.1. The first one consists of terms generated by the two-kelvon vertex A corresponding to the Hamiltonian H_1 in the second order of the perturbation theory. All these terms are similar to each other; we explicitly specify just one of them: $A_{1,2}^{4,7} G(w_7, k_7) A_{7,3}^{5,6}$, with $G(\omega, k) = 1/(\omega - \omega_k)$ being the free-kelvon propagator, $w_7 = w_1 + w_2 - w_4$, and $k_7 = k_1 + k_2 - k_4$. The second contribution is the bare three-kelvon vertex B corresponding to the Hamiltonian H_2.

Explicit expressions for vertices A and B directly follow from (10.50) and (10.51) after the Fourier transform $w(z) = L^{-1/2} \sum_k w_k e^{ikz}$:

$$A = (6D - E)/8\pi, \tag{10.52}$$

$$D_{1,2}^{3,4} = \int\limits_{a_*}^{L} (dx/x^5) \left(1 - \left[\begin{smallmatrix}1\end{smallmatrix}\right] - \left[\begin{smallmatrix}2\end{smallmatrix}\right] - \left[\begin{smallmatrix}3\end{smallmatrix}\right] - \left[\begin{smallmatrix}4\end{smallmatrix}\right] + \left[\begin{smallmatrix}3\\2\end{smallmatrix}\right] + \left[\begin{smallmatrix}43\end{smallmatrix}\right] + \left[\begin{smallmatrix}4\\2\end{smallmatrix}\right] \right),$$

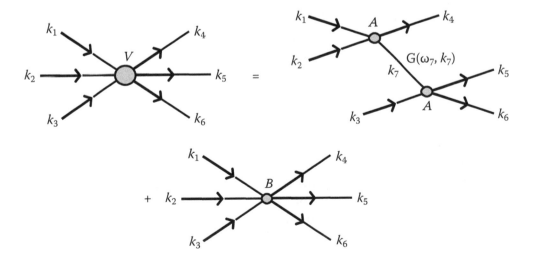

Figure 10.1   The effective vertex of the three-kelvon scattering. The vertices A and B come from the terms H_1 (10.50) and H_2 (10.51) and are given by (10.52) and (10.53), respectively. $G(\omega, k)$ is the free-kelvon propagator.

$$E^{3,4}_{1,2} = \int_{a_*}^{L} (dx/x^3)\left\{k_4 k_1 \left(\left[{}^4\right] + \left[{}_1\right] - \left[{}^{43}\right] - \left[{}^4_2\right]\right) + k_3 k_1 \left(\left[{}^3\right] + \left[{}_1\right] - \left[{}^{43}\right] - \left[{}^3_2\right]\right)\right.$$

$$\left. + k_3 k_2 \left(\left[{}^3\right] + \left[{}_2\right] - \left[{}^{43}\right] - \left[{}^3_1\right]\right) + k_4 k_2 \left(\left[{}^4\right] + \left[{}_2\right] - \left[{}^{43}\right] - \left[{}^3_2\right]\right)\right\},$$

$$B = (3P - 5Q)/4\pi, \tag{10.53}$$

$$P^{4,5,6}_{1,2,3} = \int_{a_*}^{L} (dx/x^5) k_6 k_2 \left\{\left[{}_2\right] - \left[{}^5_2\right] - \left[{}_{23}\right] + \left[{}^5_{23}\right] - \left[{}^4_2\right] + \left[{}^{45}_2\right] + \left[{}^4_{23}\right] - \left[{}^6_1\right] + \left[{}^6\right] - \left[{}^{56}\right]\right.$$

$$\left. - \left[{}^6_3\right] + \left[{}^{56}_3\right] - \left[{}^{46}\right] + \left[{}^{456}\right] + \left[{}^{46}_3\right] - \left[{}_{12}\right]\right\},$$

$$Q^{4,5,6}_{1,2,3} = \int_{a_*}^{L} (dx/x^7)\left\{1 - \left[{}^4\right] - \left[{}_1\right] + \left[{}^4_1\right] - \left[{}^6\right] + \left[{}^{46}\right] + \left[{}^6_1\right] - \left[{}^{46}_1\right] - \left[{}^5\right] + \left[{}^{45}\right] + \left[{}^5_1\right] - \left[{}^{45}_1\right]\right.$$

$$+ \left[{}^{65}\right] - \left[{}^{456}\right] - \left[{}^{56}_1\right] + \left[{}_{23}\right] - \left[{}_3\right] + \left[{}^4_3\right] + \left[{}_{13}\right] - \left[{}^4_{13}\right] + \left[{}^6_3\right] - \left[{}^{46}_3\right] - \left[{}^6_{13}\right]$$

$$\left. + \left[{}^5_2\right] + \left[{}^5_3\right] - \left[{}^{45}_3\right] - \left[{}^5_{13}\right] + \left[{}^6_2\right] - \left[{}^{65}_3\right] + \left[{}_{12}\right] + \left[{}^4_2\right] - \left[{}_2\right]\right\}.$$

Here $[\vdots]$'s encode cosine functions with various arguments, $\left[{}_1\right] = \cos k_1 x$, $\left[{}^4_1\right] = \cos(k_4 - k_1)x$, $\left[{}^{45}_1\right] = \cos(k_4 + k_5 - k_1)x$, $\left[{}^{45}_{12}\right] = \cos(k_4 + k_5 - k_1 - k_2)x$, etc.

The final expression for the effective vertex of the three-kelvon scattering turns out to be quite involved. However, for the purposes of qualitative analysis, we will need only the following three general properties, which can be established without explicitly performing all the integrations: (1) the fact that the ultimate result is free from logarithmic ultraviolet (UV) divergences (characteristic of the Biot–Savart dynamic equation), (2) uniformity of V with respect to simultaneously rescaling all the momenta, and (3) momentum-space locality; that is, an appropriately rapid decay of V when some of the six momenta tend to zero and the others kept fixed.

Property (1) follows from the integrability of LIA. Indeed, the UV physics is spatially local and thus is captured by LIA; the integrability of LIA implies the absence of real kelvon scattering events in all possible channels.* Given (1), property (2) is readily seen by inspection; see the next subsection. Being associated with geometric symmetries of the problem, property (3) is rather subtle—even in terms of its exact formulation (this will be addressed later, in the context of the theory of the KW cascade and geometric symmetries of the vortex-line dynamics).

An important question is that of the accuracy (systematic error) of the previously formulated kinetic description. To address this issue, we process the integrals (10.52) and (10.53) as follows: We introduce a characteristic wavelength $\lambda \equiv \lambda(k_1, k_2, k_3, k_4, k_5, k_6)$, the particular functional form being not important. Then, from each of the integrals (10.52) and (10.53), we subtract the corresponding LIA

---

* Mathematically, this means an exact cancelation of same-order contributions to each effective scattering amplitude.

contributions, which are easily obtained by the same procedure that led to Equations (10.52) and (10.53) from the local Hamiltonian (10.31) defined by

$$\beta_\lambda = \gamma \Lambda_\lambda, \qquad \Lambda_\lambda = \ln(\lambda/a_*). \tag{10.54}$$

As a result of subtraction, up to system-specific terms $\sim (a_*/\lambda)^2$ (to be neglected in the following), we obtain a_*-independent convergent integrals. Thereby, we arrive at a convenient decomposition: $A = \Lambda_\lambda A^{(0)} + A^{(1)}$, $B = \Lambda_\lambda B^{(0)} + B^{(1)}$, where $A^{(0)}$ and $B^{(0)}$ correspond to LIA. On the technical side, the decomposition solves the problem of handling the logarithmic divergences, whereas physically it clarifies the form of the dependence on the vortex-core details, the latter being entirely captured by LIA.

By expanding the propagator G in powers of inverse Λ_λ,

$$G = \Lambda_\lambda^{-1}\left[G^{(0)} + \Lambda_\lambda^{-1} G^{(1)} + \Lambda_\lambda^{-2} G^{(2)} + \mathcal{O}\left(\Lambda_\lambda^{-3}\right)\right], \tag{10.55}$$

for the vertex $V = AGA + B$ (see Figure 10.1), we have

$$V = \Lambda_\lambda V^{(0)} + V^{(1)} + V^{(2)}/\Lambda_\lambda + \mathcal{O}\left(\Lambda_\lambda^{-2}\right), \tag{10.56}$$

with

$$V^{(0)} = A^{(0)} G^{(0)} A^{(0)} + B^{(0)},$$
$$V^{(1)} = 2A^{(0)} G^{(0)} A^{(1)} + A^{(0)} G^{(1)} A^{(0)} + B^{(1)},$$
$$V^{(2)} = A^{(0)} G^{(2)} A^{(0)} + 2A^{(0)} G^{(1)} A^{(1)} + A^{(1)} G^{(0)} A^{(1)}.$$

Note that, by construction, all the quantities $A^{(i)}$, $B^{(i)}$, $G^{(i)}$, and $V^{(i)}$ are a_*-independent, since the dependence on a_*—up to the neglected terms $\sim (a_*/\lambda)^2$—comes exclusively through Λ_λ. The term $\Lambda_\lambda V^{(0)}$ is precisely the effective vertex that follows from the LIA Hamiltonian (10.31); it necessarily obeys $V^{(0)} \equiv 0$ by integrability of LIA. The leading contribution to V is the a_*-independent $V^{(1)}$. Thus, the short-range physics enters the answer only as a correction, small in the parameter $1/\Lambda_\lambda$.

Now we are in a position to specify the systematic error due to uncertainties in microscopic details. In the case when it is possible to calibrate the pseudo-Hamiltonian, that is, to find (analytically or experimentally; see Section 10.1.3) the accurate value of a_*, the systematic error is of the order of $(a_*/\lambda)^2$. Otherwise, we are forced to set $a_* \sim a_0$, meaning that the parameter $\Lambda_\lambda \gg 1$ is known with uncertainty $\delta\Lambda_\lambda \sim 1$, which translates into the systematic uncertainty of V, given by $V^{(2)}\delta\Lambda_\lambda/\Lambda_\lambda^2 \sim 1/\Lambda_\lambda^2$. We thus arrive at the important conclusion that even when the microscopic details, such as the vortex-core shape, are completely unknown, by naively setting $\Lambda_\lambda = \ln(\lambda/a_0)$, we introduce only a small relative error $1/\Lambda_\lambda^2$.

10.2.4   Kinetic Equation, Kelvin-Wave Cascade

When written in terms of averaged over the statistical ensemble kelvon occupation numbers $n_k = \langle a_k^\dagger a_k \rangle$, the kinetic equation is given by

$$\dot{n}_1 = \frac{1}{(3-1)!3!} \sum_{k_2,\ldots,k_6} \left(W_{4,5,6}^{1,2,3} - W_{1,2,3}^{4,5,6} \right).$$

(10.57)

Here, $W_{1,2,3}^{4,5,6}$ is the probability per unit time of the elementary three-kelvon scattering event $(k_1, k_2, k_3) \to (k_4, k_5, k_6)$, and the combinatorial factor compensates for multiple counting of the same scattering event. The probabilities $W_{1,2,3}^{4,5,6}$ can be straightforwardly obtained by the Fermi Golden Rule applied to the Hamiltonian (10.49):

$$W_{1,2,3}^{4,5,6} = 2\pi \left| (3!)^2 V_{1,2,3}^{4,5,6} \right|^2 f_{1,2,3}^{4,5,6} \delta(\Delta\omega)\delta(\Delta k),$$

$$f_{1,2,3}^{4,5,6} = n_1 n_2 n_3 (n_4 + 1)(n_5 + 1)(n_6 + 1),$$

(10.58)

$$\Delta\omega = \omega_1 + \omega_2 + \omega_3 - \omega_4 - \omega_5 - \omega_6.$$

Here the combinatorial factor $(3!)^2$ accounts for the addition of equivalent amplitudes. We are particularly interested in the classical-field limit of Equation (10.58), corresponding to $n_k \gg 1$, when one can retain only the leading terms in n_k:

$$\dot{n}_1 = 216\pi \sum_{k_2,\ldots,k_6} \left| V_{1,2,3}^{4,5,6} \right|^2 \delta(\Delta\omega)\delta(\Delta k) \left(\tilde{f}_{4,5,6}^{1,2,3} - \tilde{f}_{1,2,3}^{4,5,6} \right),$$

(10.59)

$$\tilde{f}_{1,2,3}^{4,5,6} = n_1 n_2 n_3 (n_4 n_5 + n_4 n_6 + n_5 n_6).$$

Problem 10.11 *Obtain Equation (10.59) from the Fermi Golden Rule.*

The kinetic equation (10.59) supports an energy cascade, provided two conditions are met: (1) the kinetic time is getting progressively smaller (vanishes) in the limit of large wavenumbers and (2) the collision term is local in the wavenumber space; that is, the relevant scattering events are only those where all the kelvon momenta are of the same order of magnitude. In the following, we make sure that both conditions are satisfied: the condition (1) can be checked by a dimensional estimate, provided (2) is true. The condition (2) will be proven in Section 10.2.5. Under these conditions, one can establish the cascade spectrum by a simple dimensional analysis of the kinetic equation.

The locality of the collision term implies that the integral in (10.59) builds up around $(k_2,\ldots,k_6) \sim k_1$, which dramatically simplifies the scaling structure of the kinetic equation

$$\dot{n}_k \propto k^5 \cdot |V|^2 \cdot \omega_k^{-1} \cdot k^{-1} \cdot n_k^5,$$

(10.60)

where the factors estimate the corresponding terms/sums in the same order as they appear in (10.59). At $k_1 \sim \cdots \sim k_6 \sim k$, we have $|V| \propto k^6$, as it is seen by nondimensionalizing the integrals in Equations (10.52) and (10.53). Hence,

$$\dot{n}_k \propto \omega_k^{-1} n_k^5 k^{16}.$$

(10.61)

The energy flux (per unit vortex-line length), θ_k, at the momentum scale k is defined as

$$\theta_k = L^{-1} \sum_{k'<k} \omega_{k'} \dot{n}_{k'}, \tag{10.62}$$

implying the estimate $\theta_k \sim k \dot{n}_k \omega_k$. Combined with (10.61), this yields $\theta_k \sim n_k^5 k^{17}$. The cascade requirement that θ_k be k-independent, that is, $\theta_k \equiv \theta$, leads to the spectrum

$$n_k \propto \left\langle \hat{w}_k^\dagger \hat{w}_k \right\rangle = Q k^{-17/5}. \tag{10.63}$$

The value of the spectrum amplitude Q in (10.63) controls the energy flux that the cascade transports. The relation between θ and Q is

$$\theta = C_\theta \gamma^3 \rho Q^5. \tag{10.64}$$

The dimensionless coefficient C_θ in this formula can, in principle, be obtained by a numerical calculation.

Restoring all the dimensional coefficients, we get

$$\left\langle \hat{w}_k^\dagger \hat{w}_k \right\rangle = \frac{\hbar n_k}{\pi \gamma \rho} = Q k^{-17/5}. \tag{10.65}$$

It is useful to express the spectrum in terms of the typical geometrical amplitude, b_k, of the KW turbulence at the wavevector $\sim k$. By definition of the field $\hat{w}(z)$, we have $b_k^2 \sim L^{-1} \sum_{q \sim k} \langle \hat{w}_k^\dagger \hat{w}_k \rangle \sim k \langle \hat{w}_k^\dagger \hat{w}_k \rangle$. Hence, using Equation (10.64), we obtain

$$b_k \sim \left(\theta / \gamma^3 \rho \right)^{1/10} k^{-6/5}. \tag{10.66}$$

One can also convert (10.65) into the curvature spectrum. For the curvature $\zeta(\xi) = \partial^2 s / \partial \xi^2$, the spectrum is defined as Fourier decomposition of the integral $I_c = \int |\zeta(\xi)|^2 d\xi$. The smallness of α, Equation (10.40), allows one to write $I_c \approx \int dz \langle \hat{w}''^\dagger(z) \hat{w}''(z) \rangle = \sum_k k^4 n_k \propto \sum_k k^{3/5}$, arriving thus at the curvature exponent 3/5.

10.2.5 Geometric Symmetries: Locality of Kelvin-Wave Cascade

So far, we have been heavily relying on the assumption of locality of the kinetic processes in the wavenumber space. Now we shall prove this assumption for the KW-cascade spectrum (10.65) by invoking additional symmetries associated with the geometric character of the complex-valued field $w(z) = x(z) + iy(z)$. We have already employed the rotation symmetry about the z-axis and translation symmetry along rather standard lines in the theory of fields. More specific symmetries, based essentially on the geometric nature of $w(z)$, are the translations in the direction of a unit vector $\hat{n} = (n_x, n_y, 0)$ perpendicular to the z-axis and the rotations about $\hat{n}$. We will refer to them as *shift* and *tilt* symmetries, respectively.

Shifting the line by distance l results in the transformation

$$w(z) \rightarrow w(z) + nl, \tag{10.67}$$

where $n = n_x + in_y$. The invariance of the Hamiltonian with respect to this transformation implies

$$n^* \int dz \, \dot{w} + \text{c.c.} = 0, \tag{10.68}$$

which, by linear independence of n and n^*, means the conservation of

$$P_\perp = \int dz \, w. \tag{10.69}$$

Up to a dimensional factor, $P_\perp$ is the xy-component of the fluid momentum. In kelvon terminology, $P_\perp$ is, up to normalization, the amplitude of kelvon condensate. Its conservation implies that kelvon condensate does not interact with the rest of the system, and vice versa: The dynamics of kelvons are insensitive to the condensate amplitude. This fact—which is by no means surprising since the kelvon condensate is a mere shift of the whole vortex line—proves to be crucially important for understanding the structure of the kelvon kinetic processes, as discussed in the following.

An infinitesimal tilt by an angle $\delta\phi$ (rotation about the $\hat{n}$-axis) results in the transformation

$$w(z) \rightarrow w(z) - w'(z)\delta z + \delta w, \tag{10.70}$$

with $\delta w \equiv \delta x + i\delta y$ and δz obeying $\delta \mathbf{r} = [\hat{n} \times \mathbf{r}]\delta\phi$, where $\delta \mathbf{r} = (\delta x, \delta y, \delta z)$. By substituting

$$\delta z = \frac{i}{2}(w^* n - wn^*)\delta\phi, \qquad \delta w = -izn\,\delta\phi, \tag{10.71}$$

into (10.70), we write the tilt transformation as

$$w(z) \rightarrow w(z) - i\left[w'(w^* n - wn^*)/2 + zn\right]\delta\phi. \tag{10.72}$$

The invariance of the Hamiltonian with respect to the transformation (10.72) implies

$$\int dz \, \dot{w}^* \left[w'(w^* n - wn^*) + 2zn\right] + \text{c.c.} = 0, \tag{10.73}$$

and, by linear independence of n and n^*, leads to the constant of motion

$$L_\perp = \int dz \left[2wz - |w|^2 w'\right], \tag{10.74}$$

which, up to a dimensional factor, is the kelvon contribution to the xy-component of the angular momentum of the fluid.

Now let us turn to the question of kinetics locality in the wavenumber space. We observe that the leading contribution to the collision term for interactions between dramatically different wavenumber scales is *spatially local*. Indeed, the separation of spatial and temporal scales between the short- and long-wave dynamics* (substantial change of the line configuration at wavenumber λ can happen only over a time scale inversely proportional to the frequency of modes at this wavenumber) means that fast short-wave harmonics propagate, to a first approximation, in a static background geometry determined by slow long-wave modes. For fields of purely geometrical nature considered here, the tilt and shift symmetries rule out coupling to the amplitude of the long-wave field and its first derivative, the angle (i.e., the short-wave harmonics can only couple to *spatially local, instantaneous*, higher-order spatial derivatives of the smooth curve, such as *curvature*). This enforces the key constraint on the asymptotic form of the integrand in the collision term (10.59) when one k_i, say k_2, is much smaller than k_1: the wavenumber $k_2 \ll k_1$ can only enter the collision term through the curvature (or higher derivatives) of the long-wave component. For the spectrum (10.65), the curvature is actually dominated by *large* wavenumbers [see the text Equation (10.66)], while the contribution from long-wave harmonics is negligible. This immediately implies that the collision term builds up at $k_2, \ldots, k_6 \sim k_1$ and is therefore local.

10.3   Vortex–Phonon Interaction

In this section, we will develop a systematic approach to the problem of interaction between phonons and vortices in the hydrodynamic regime; that is, when any physical length scale is much larger than the vortex core size a_0, which allows us, as before, to describe vortices as geometric lines. Another benefit of the hydrodynamic regime is that it comes with the "nonrelativistic" small parameter β, Equation (10.1). It is this parameter that allows us to derive an accurate Hamiltonian for vortex–phonon interactions. To employ a physically transparent description in terms of normal modes, we confine ourselves to the case of weak nonlinearity. For kelvons, this implies the existence of small parameter α [Equation (10.40)], when the amplitudes b_k of KWs at the typical wavelength $\lambda \sim k^{-1}$ are much smaller than λ, $\alpha_k = b_k k \ll 1$. For phonons, this requires that $\eta \ll n$, where η is the number density fluctuation in a sound wave and n is the average number density.

We start with introducing qualitative ideas behind the derivation. The condition (10.1), meaning that the typical vortex-line velocities are much less than the speed of sound, along with $\eta \ll n$ ensures that the vortex–phonon coupling results only in small corrections to the dynamics of noninteracting vortex and phonon subsystems. Therefore, a perturbative approach is applicable, provided the interaction energy is written in terms of canonical variables. Normally, canonical variables are obtained from the solution of the interaction-free dynamics. However, when separating vortex degrees of freedom from phonons, one naturally

---

* Not to be confused with kinetics, which can be *arbitrarily* slower than dynamics. The perturbative kinetics with time-independent coupling between the modes are essentially Markovian.

neglects the compressibility of the fluid, which we did in this chapter, because finite compressibility leads only to higher-order "relativistic" corrections to the vortex dynamics described by the Biot–Savart equation (10.8) or the Hamiltonian (10.36) written in terms of the geometrical configuration of the vortex lines. In the absence of vortices, the phonon modes come from the bilinear Hamiltonian for density fluctuations $\eta(\mathbf{r}, t)$ and the phase field $\varphi(\mathbf{r}, t)$. As it turns out, when finite compressibility of a superfluid is taken into account, the positions of vortices and the fields $\eta(\mathbf{r}, t), \varphi(\mathbf{r}, t)$ are no longer forming a set of canonical variables because of the mixing terms in the Lagrangian. This renders the problem quite peculiar: By introducing the interaction, one must simultaneously reconsider the canonical variables.

It is easy to see that the standard vortex parameterization introduced in Section 10.1.1 fails to capture the physics of vortices when phonons are present. In Equation (10.4), the geometrical position of the vortex line unambiguously determines the instantaneous velocity field configuration at arbitrary distances, which is inconsistent with the finite velocity of propagation of excitations in a compressible medium. From this simple physical argument, we can already guess the proper form of canonical variables: the long-distance part of the velocity field produced by vortex loops in Equation (10.4) should belong to phonons.

The small parameters allow us to obtain an asymptotic expansion of canonical variables using a systematic iterative procedure. Physically, the procedure respects the retardation effects in the adjustment of the superfluid velocity field to the evolving vortex configuration. It qualitatively changes the structure of the Hamiltonian with respect to the terms responsible for radiation of sound and the relativistic corrections to vortex dynamics.

10.3.1 Hamiltonian of Kelvon–Phonon Interaction

The starting point in the derivation is the Popov hydrodynamic action introduced in Chapter 2 (see also Chapter 7 for quantum derivation), with the parameterization (2.124) and (2.125) for the field of the phase. In view of Galilean symmetry, we have $n_s \equiv n$ and $\varkappa_s \equiv \varkappa$. In accordance with the results of Section 2.2.4, the Lagrangian of the problem reads

$$L = \int d^3 r \left[-n\dot{\Phi}_0 - \eta\dot{\varphi} - \eta\dot{\Phi}_0 \right] - H, \tag{10.75}$$

where $H = H_{\text{vor}} + H_{\text{ph}} + H'_{\text{int}}$,

$$H_{\text{vor}} = \frac{n}{2} \int d^3 r \, |\nabla \Phi_0|^2, \tag{10.76}$$

$$H_{\text{ph}} = \int d^3 r \left[\frac{n}{2} |\nabla \varphi|^2 + \frac{1}{2\varkappa} \eta^2 \right], \tag{10.77}$$

$$H'_{\text{int}} = \frac{1}{2} \int d^3 r \left[\eta |\nabla \Phi_0|^2 + 2\eta \nabla \varphi \cdot \nabla \Phi_0 \right]. \tag{10.78}$$

Here we set $\gamma = 1$. This allows us (see the discussion in Section 2.2.4) to make no distinction between the matter density n and mass density ρ, the latter being a natural hydrodynamic quantity.

The coupling between the vortex variable Φ_0 and the density waves $\{\eta, \varphi\}$ comes from H'_{int}, as well as from the time derivative term $\int d^3 r \eta \dot{\Phi}_0$, both being the first-order corrections to the noninteracting parts. Following the standard perturbative approach, we first neglect this coupling altogether to find the noninteracting normal modes. The vortex part of the Lagrangian, $L_{vor} = -n \int d^3 r \dot{\Phi}_0 - H_{vor}$, has been considered already in Section 10.1.3, with the result that the functional $H_{vor} \equiv H_{vor}[w, w^*]$ plays the role of the Hamiltonian for the canonical complex-valued field $w(z) = x(z) + iy(z)$. Here we need only the bilinear term of the Hamiltonian H_{vor} expanded with respect to $\alpha_k \ll 1$, since kelvon interactions are irrelevant for purposes of this section. For the sake of uniformity, we change the notation for kelvons, writing the noninteracting kelvon Hamiltonian (10.42) and dispersion relation (10.43) as*

$$H_{vor} \approx \sum_k \varepsilon_k a_k^\dagger a_k, \qquad \varepsilon_k = [\ln(1/ka_0) + C_0] k^2 / 2. \qquad (10.79)$$

The noninteracting sound waves are described by the Lagrangian

$$L_{ph} = -\int d^3 r \eta \dot{\varphi} - H_{ph}[\eta, \varphi]. \qquad (10.80)$$

To explicitly comply with the rotation symmetry of the vortex part, we work in the cylindrical coordinate system. We assume that the system is contained in a cylinder of radius R with the symmetry axis along the z-direction, the direction of the vortex line, and that the system is periodic along z with the period L. In cylindrical geometry, the phonon fields $\eta(r, \theta, z)$, $\varphi(r, \theta, z)$ are parameterized by phonon creation and annihilation operators c_s, $c_s^\dagger$ as

$$\eta = \sum_s \sqrt{\omega_s \varkappa / 2} \left[\chi_s c_s + \chi_s^* c_s^\dagger \right],$$

$$\varphi = -i \sum_s \sqrt{1/2\omega_s \varkappa} \left[\chi_s c_s - \chi_s^* c_s^\dagger \right], \qquad (10.81)$$

$$\chi_s = \chi_s(r, \theta, z) = \mathcal{R}_{mq_r}(r) Y_m(\theta) \mathcal{Z}_{q_z}(z),$$

where $\mathcal{R}_{mq_r}(r) = (\pi q_r / R)^{1/2} J_m(q_r r)$, $Y_m(\theta) = (2\pi)^{-1/2} \exp(im\theta)$, $\mathcal{Z}_{q_z}(z) = L^{-1/2} \exp(iq_z z)$, s stands for $\{q_r, m, q_z\}$, and $J_m(x)$ are the Bessel functions of the first kind. The phonon Hamiltonian then reads

$$H_{ph} = \sum_s \omega_s c_s^\dagger c_s, \qquad \omega_s = cq, \qquad (10.82)$$

with $q = \sqrt{q_r^2 + q_z^2}$ and c the sound velocity.

---

* As discussed previously, perturbative kinetic problems for classical fields are readily solved by the quantum-mechanical Golden Rule formalism. That is why the creation and annihilation operators are being introduced here.

We are ready to address the coupling between phonons and vortices. In terms of the obtained variables, the Lagrangian (10.75) takes on the form

$$L = \sum_k i\dot{a}_k a_k^\dagger + \sum_s i\dot{c}_s c_s^\dagger - T - H, \tag{10.83}$$

where $T = \int d^3r\eta\Phi_0 \equiv T\{a_k, \dot{a}_k, a_k^\dagger, \dot{a}_k^\dagger, c_s, \dot{c}_s, c_s^\dagger, \dot{c}_s^\dagger\}$ and $H = H_{vor} + H_{ph} + H'_{int}$. The coupling term T plays a special role in the Lagrangian (10.83). This term is linear in time derivatives of the variables and thus cannot contribute to the energy in accordance with Lagrangian formalism. However, because of the time derivatives in T, the equations of motion in terms of $\{a_k, a_k^\dagger\}, \{c_s, c_s^\dagger\}$ do not agree with the Hamiltonian formalism. This implies that the chosen variables become noncanonical in the presence of the interaction, and therefore H *in terms of these variables* cannot be identified with the Hamiltonian.

One thus must seek a variable transformation $\{a_k, a_k^\dagger\}, \{c_s, c_s^\dagger\} \rightarrow \{\tilde{a}_k, \tilde{a}_k^\dagger\}, \{\tilde{c}_s, \tilde{c}_s^\dagger\}$ that restores the canonical form of the Lagrangian, $L = \sum_k i\dot{\tilde{a}}_k \tilde{a}_k^\dagger + \sum_s i\dot{\tilde{c}}_s \tilde{c}_s^\dagger - H\{\tilde{a}_k, \tilde{a}_k^\dagger, \tilde{c}_s, \tilde{c}_s^\dagger\}$. The canonical variables are obtained by the following iterative procedure: The T-term is expanded with respect to $\alpha_k \ll 1$, $\beta \ll 1$ and $\eta \ll n$ yielding $T = T^{(1)} + T^{(2)} + \cdots$. Then the variables are modified as $a_k \rightarrow a_k + a_k^{(1)}$, $c_s \rightarrow c_s + c_s^{(1)}$, where $a_k^{(1)}(\{a_k, a_k^\dagger, c_s, c_s^\dagger\})$ and $c_s^{(1)}(\{a_k, a_k^\dagger, c_s, c_s^\dagger\})$ are chosen to eliminate $T^{(1)}$ in (10.83). As a result, $T \rightarrow 0 + T'^{(2)} + \cdots$, where the prime indicates that the structure of the remaining terms has changed under the transformation. At the next step, $T'^{(2)}$ is eliminated by $a_k \rightarrow a_k + a_k^{(2)}$, $c_s \rightarrow c_s + c_s^{(2)}$, and so on. By construction, the canonical variables are given by $\tilde{a}_k = a_k + a_k^{(1)} + a_k^{(2)} + \cdots$, $\tilde{c}_s = c_s + c_s^{(1)} + c_s^{(2)} + \cdots$, and likewise for their conjugates. In what follows, we will retain the first few terms and ignore the rest as higher-order corrections.

The explicit form of T is obtained by reproducing the steps used in the derivation of Equation (10.21), starting with the analog of (10.18),

$$\int d^3r\eta\Phi_0 = \int d\zeta \dot{\mathbf{s}} \cdot \frac{\delta}{\delta\mathbf{s}} \int d^3r\eta\Phi_0. \tag{10.84}$$

The analog of (10.19) is

$$\delta \int d^3r\eta\Phi_0 = \int d^3r\eta\delta\Phi_0 = \int d^3r\nabla \cdot (\nabla Q\delta\Phi_0), \tag{10.85}$$

where we introduce a scalar field $Q(\mathbf{r})$ such that $\Delta Q(\mathbf{r}) = \eta(\mathbf{r})$. Observing that the only difference between the rightmost integrals in (10.19) and (10.85) is in replacing $\mathbf{r}$ with $3\nabla Q(\mathbf{r})$, we skip the rest of the steps and write the final answer [similarly, we must replace vector $\mathbf{s}$ with $3\nabla Q(\mathbf{r} = \mathbf{s})$]. We thus get

$$\int d^3r\eta\Phi_0 = 2\pi \int \dot{\mathbf{s}} \cdot (d\mathbf{s} \times \nabla Q). \tag{10.86}$$

In the canonical representation $s(z) = \hat{z}z + \rho(z)$ (see Section 10.1.3), we then have

$$T = 2\pi \int dz \left[\left(\hat{z} + \frac{\partial \rho}{\partial z} \right) \times \nabla Q(\rho(z), z) \right] \cdot \dot{\rho}(z). \tag{10.87}$$

After switching to the cylindrical coordinates, $\rho = (\rho_0 \cos \phi, \rho_0 \sin \phi, 0)$, in accordance with small parameter $q_r|\rho| \sim \beta k|\rho| \ll 1$, we expand the radial functions, retaining only the leading term for each particular angular momentum m. Noticing that

$$\left[\dot{\phi} \rho_0 \frac{\partial}{\partial r} - \frac{\dot{\rho}_0}{\rho_0} \frac{\partial}{\partial \theta} \right] \mathcal{R}_{mq_r}(r) Y_m(\theta) \Bigg|_{r=\rho_0, \theta=\phi} \propto \begin{cases} d(w^m)/dt, & m \geq 0 \\ d(w^{*|m|})/dt, & m < 0 \end{cases}, \tag{10.88}$$

we obtain (in the Heisenberg picture for operators)

$$T \approx \sum_{s, k_1 \ldots k_m} \left[-iA_{s,k_1 \ldots k_m} c_s \frac{d}{dt} \left(a_{k_1} \ldots a_{k_m} \right) - iB_{s,k_1 \ldots k_m} c_s \frac{d}{dt} \left(a_{k_1}^\dagger \ldots a_{k_m}^\dagger \right) \right] + \text{H.c.}, \tag{10.89}$$

where the sum is over all s with $m \neq 0$ and

$$A_{s,k_1 \ldots k_m} = -\Theta(m) A_s \delta_{k_1 + \cdots + k_m, -q_z},$$

$$B_{s,k_1 \ldots k_m} = (-1)^{|m|} \Theta(-m-1) A_s \delta_{k_1 + \cdots + k_m, q_z}, \tag{10.90}$$

$$A_s = \frac{\sqrt{q/c}}{2^{|m|/2+1}|m|!} \frac{(2\pi)^{\frac{2-|m|}{2}} n^{\frac{1-|m|}{2}} q_r^{|m|+\frac{1}{2}} q^{-2}}{L^{(|m|-1)/2} R^{1/2}},$$

where $\Theta(m) = \begin{cases} 1, & m \geq 0 \\ 0, & m < 0 \end{cases}$. Thus,

$$a_k = \tilde{a}_k \tag{10.91}$$

(terms that do not contain phonon operators and thus result only in relativistic corrections to the kelvon spectrum and kelvon–kelvon interactions were omitted), and

$$c_s = \tilde{c}_s + (1 - \delta_{m,0}) \sum_{k_1 \ldots k_m} \left[A_{s,k_1 \ldots k_m} \tilde{a}_{k_1}^\dagger \ldots \tilde{a}_{k_m}^\dagger + B_{s,k_1 \ldots k_m} \tilde{a}_{k_1} \ldots \tilde{a}_{k_m} \right], \quad s = \{q_r, m, q_z\}. \tag{10.92}$$

Finally, the vortex–phonon Hamiltonian is given by the energy (10.76) through (10.78) in terms of canonical variables $\{\tilde{a}_k, \tilde{a}_k^\dagger\}, \{\tilde{c}_s, \tilde{c}_s^\dagger\}$.

Let us first obtain the interaction terms responsible for sound radiation. This phenomenon is quite atypical because the "coupling energy" between the "bare" vortex and phonon variables $\{a_k, a_k^\dagger\}$ and $\{c_s, c_s^\dagger\}$ plays no role in the interaction Hamiltonian! The radiative vertices arise solely from the transformation of the phonon-precession term (10.82) when under the transformation to canonical variables, (10.92), yielding

$$H_{\text{int}}^{(\text{rad})} = \sum_{s, \{k_i\}} (1 - \delta_{m,0}) \left[w_s A_{s,k_1 \ldots k_m} \tilde{a}_{k_1}^\dagger \ldots \tilde{a}_{k_m}^\dagger \tilde{c}_s + w_s B_{s,k_1 \ldots k_m} \tilde{a}_{k_1} \ldots \tilde{a}_{k_m} \tilde{c}_s^\dagger \right] + \text{H.c.} \tag{10.93}$$

Correspondingly, the energy term $\propto \int d^3 r \eta |\nabla \Phi_0|^2$ in (10.78), which has the same operator structure as (10.93), may be neglected because it is smaller in parameter $\beta \ll 1$. The term $\propto \int d^3 r \eta \nabla \varphi \cdot \nabla \Phi_0$ in Equation (10.78) contributes to the amplitudes of elastic and inelastic scattering of phonons. In addition, it leads to a macroscopically small splitting of the phonon spectrum due to the superimposed fluid circulation. Up to small "relativistic" corrections, the variable transformation does not change the spectrum of normal modes: the zero-order Hamiltonians are still given by (10.79) and (10.82), in which $\{a_k, a_k^\dagger\}, \{c_s, c_s^\dagger\}$ are simply replaced with $\{\tilde{a}_k, \tilde{a}_k^\dagger\}, \{\tilde{c}_s, \tilde{c}_s^\dagger\}$.

Note that the interaction (10.93) explicitly conserves the angular momentum: A real process of the emission of a phonon with the angular momentum $(-m)$ requires an annihilation of m kelvons since a kelvon carries one quantum of (negative) angular momentum (see Section 10.2.1).

Vortex–phonon Hamiltonian in the plane-wave basis for phonon fields: To describe the processes of elastic and inelastic phonon scattering on vortices, we find it more convenient to work in the plane-wave basis for the phonon fields:

$$\eta(\mathbf{r}) = \sum_{\mathbf{q}} \sqrt{\frac{\omega_q \pi}{V}} \left[e^{i\mathbf{q}\mathbf{r}} c_{\mathbf{q}} + e^{-i\mathbf{q}\mathbf{r}} c_{\mathbf{q}}^\dagger \right],$$

$$\varphi(\mathbf{r}) = -\sum_{\mathbf{q}} \frac{i}{2\sqrt{\pi V \omega_q}} \left[e^{i\mathbf{q}\mathbf{r}} c_{\mathbf{q}} - e^{-i\mathbf{q}\mathbf{r}} c_{\mathbf{q}}^\dagger \right], \tag{10.94}$$

where $\omega_q = cq$ and $V = L^3$ is the system volume. The harmonic phonon Hamiltonian is given by Equation (10.82) with s now labeling different wavenumbers $\mathbf{q}$.

A calculation analogous to the one that led to Equations (10.91) and (10.92) yields the following relation for canonical kelvon

$$a_k = \tilde{a}_k + \gamma_{kk_1 \mathbf{q}} \tilde{a}_{k_1} \left[\tilde{c}_{\mathbf{q}} + \tilde{c}_{-\mathbf{q}}^\dagger \right] + \text{smaller terms},$$

$$\gamma_{kk_1 \mathbf{q}} = -\sqrt{\frac{q}{8\rho c V}} \frac{q_x^2 + q_y^2}{q^2} \delta_{k_1 + q_z, k}, \tag{10.95}$$

(as mentioned previously, we omit terms that do not contain phonon operators and thus result only in "relativistic" corrections to the kelvon spectrum and kelvon–kelvon interactions) and phonon variables

$$c_{\mathbf{q}} = \tilde{c}_{\mathbf{q}} + \sigma_{\mathbf{q}k}^* \tilde{a}_k + \sigma_{\mathbf{q}k} \tilde{a}_{-k}^\dagger + i \mu_{\mathbf{q}k_1 k_2}^* \tilde{a}_{k_1} \tilde{a}_{k_2} + i \mu_{\mathbf{q}k_1 k_2} \tilde{a}_{-k_1}^\dagger \tilde{a}_{-k_2}^\dagger + \text{smaller terms}$$

$$\sigma_{\mathbf{q}k} = i \sqrt{\frac{2\pi q}{c}} \frac{q_x + i q_y}{2Lq^2} \delta_{k, q_z},$$

$$\mu_{\mathbf{q}k_1 k_2} = -i \sqrt{\frac{q}{32\rho c V}} \frac{(q_x + i q_y)^2}{q^2} \delta_{k_1 + k_2, q_z}. \tag{10.96}$$

Here and below, we assume summation over repeating indices in the products of terms.

Similarly to the cylindrical case, up to relativistic corrections, the variable transformations (10.95)–(10.96) applied to Equations (10.79) and (10.82) do not change the spectrum of normal modes and generate nontrivial contribution to the interaction Hamiltonian:

$$H_{\text{int}}^{(1)} = \left[S_{\mathbf{q}k}^* \tilde{c}_{\mathbf{q}}^\dagger \tilde{a}_k + S_{\mathbf{q}k} \tilde{c}_{\mathbf{q}}^\dagger \tilde{a}_{-k}^\dagger + i M_{\mathbf{q}k_1 k_2}^* \tilde{c}_{\mathbf{q}}^\dagger \tilde{a}_{k_1} \tilde{a}_{k_2} + i M_{\mathbf{q}k_1 k_2} \tilde{c}_{\mathbf{q}}^\dagger \tilde{a}_{-k_1}^\dagger \tilde{a}_{-k_2}^\dagger \right.$$
$$\left. + G_{kk_1 \mathbf{q}} \left(\tilde{a}_k^\dagger \tilde{a}_{k_1} \tilde{c}_{\mathbf{q}} + \tilde{a}_k^\dagger \tilde{a}_{k_1} \tilde{c}_{-\mathbf{q}}^\dagger \right) \right] + \text{H.c.} + \text{smaller terms,} \tag{10.97}$$

where

$$S_{\mathbf{q}k} = \omega_{\mathbf{q}} \, \sigma_{\mathbf{q}k},$$
$$M_{\mathbf{q}k_1 k_2} = \omega_{\mathbf{q}} \, \mu_{\mathbf{q}k_1 k_2}, \tag{10.98}$$
$$G_{kk_1 \mathbf{q}} = \varepsilon_k \, \gamma_{kk_1 \mathbf{q}}.$$

Another relevant contribution to the interaction Hamiltonian comes from the vortex–phonon coupling energy [Equation (10.78)]

$$H_{\text{int}}^{(2)} = \int d\mathbf{r} \, \eta \nabla \varphi \cdot \mathbf{v}_{\text{v}}, \tag{10.99}$$

where $\mathbf{v}_{\text{v}} = \mathbf{v}_{\text{v}}(\mathbf{r})$ is the velocity field produced by the vortex line. Written in canonical variables, this term takes the form (the terms $\propto \tilde{a}\tilde{c}\tilde{c}, \tilde{a}\tilde{c}^\dagger \tilde{c}^\dagger$ are omitted)

$$H_{\text{int}}^{(2)} = \left[R_{\mathbf{q}_1 \mathbf{q}_2} \tilde{c}_{\mathbf{q}_1} \tilde{c}_{\mathbf{q}_2}^\dagger + R_{\mathbf{q}_1 - \mathbf{q}_2}^* \tilde{c}_{\mathbf{q}_1} \tilde{c}_{\mathbf{q}_2} + T_{k\mathbf{q}_1 \mathbf{q}_2} \tilde{a}_k^\dagger \tilde{c}_{\mathbf{q}_1} \tilde{c}_{\mathbf{q}_2}^\dagger \right]$$
$$+ \text{H.c.} + \text{smaller terms,} \tag{10.100}$$

where

$$R_{\mathbf{q}_1 \mathbf{q}_2} = \frac{i\pi}{L^2} \sqrt{\frac{\omega_{\mathbf{q}_1}}{\omega_{\mathbf{q}_2}}} \frac{(\mathbf{q}_1 \times \mathbf{q}_2 \cdot \hat{\mathbf{z}})}{|\mathbf{q}_1 - \mathbf{q}_2|^2} \delta_{q_1^z, q_2^z},$$

$$T_{k\mathbf{q}_1 \mathbf{q}_2} = \frac{1}{2V} \sqrt{\frac{\pi L}{\rho}} \mathbf{Q}_{\mathbf{q}_1, \mathbf{q}_2} \cdot \left[\hat{\mathbf{z}} \left[(\mathbf{q}_2 - \mathbf{q}_1) \cdot (\hat{\mathbf{x}} + i\hat{\mathbf{y}}) \right] + k (\hat{\mathbf{x}} + i\hat{\mathbf{y}}) \right] \delta_{k+q_2^z, q_1^z},$$

$$\mathbf{Q}_{\mathbf{q}_1, \mathbf{q}_2} = \frac{\left(\sqrt{\frac{\omega_{\mathbf{q}_2}}{\omega_{\mathbf{q}_1}}} \mathbf{q}_1 + \sqrt{\frac{\omega_{\mathbf{q}_1}}{\omega_{\mathbf{q}_2}}} \mathbf{q}_2 \right) \times (\mathbf{q}_2 - \mathbf{q}_1)}{|\mathbf{q}_2 - \mathbf{q}_1|^2}. \tag{10.101}$$

The complete interaction Hamiltonian up to the most relevant order in $\beta \ll 1$, $\alpha_k \ll 1$, and $|\eta|/n \ll 1$ is given by $H_{\text{int}} = H_{\text{int}}^{(1)} + H_{\text{int}}^{(2)}$. Note that, in this form, the Hamiltonian explicitly conserves the momentum along the vortex line (along the z-axis), but not the transverse momentum. The change of transverse momentum during the scattering is transferred to the vortex line as a whole (an analog of the Mössbauer effect), which results in a macroscopically small displacement of the vortex line from its initial position. The corresponding Goldstone mode can be straightforwardly taken into account if necessary.

10.3.2   Sound Emission by Kelvin Waves

Since $\varepsilon_k \sim (a_0 k)\omega_k$, the total momentum transferred to phonons in a radiation event should be small in order to satisfy the energy conservation law. Thus, radiating phonons in a process annihilating one kelvon on an infinite line is kinematically prohibited. The leading radiation process then is the emission of the $m = -2$ (quadrupole) phonon mode, while the events involving more than two kelvons are suppressed by $\alpha_k \ll 1$. First-order processes leading to the two-phonon emission come from the $\propto \int d^3 r \eta \nabla \varphi \cdot \nabla \Phi_0$ term in (10.78). Their amplitude is suppressed by the relativistic parameter $\beta \ll 1$.

The qualitative observation that the sound emission is dominated by the quadrupole radiation is already sufficient to obtain the formula for the power Π_k radiated by KWs at the wavenumber scale k per unit vortex-line length. By general hydrodynamic considerations, Π_k must be proportional to the square of the third-order time derivative of the quadrupole moment, $\propto \varepsilon_k^6 b_k^2 b_{-k}^2$, and inversely proportional to the fifth power of the sound velocity c. The rest of the dimensional coefficients are restored unambiguously, since the only remaining time and length scales are set by the wavevector k and the velocity circulation quantum κ; the dimension of mass is coming from the fluid mass density ρ. This yields

$$\Pi_k \sim \frac{\kappa^2 \rho}{c^5 k} \varepsilon_k^6 b_k^2 b_{-k}^2 \sim \frac{\varepsilon_k^6 k}{c^5 \rho} n_k n_{-k}. \tag{10.102}$$

Expressing this result in terms of kelvon occupation numbers n_k is a matter of convenience. We do it to facilitate comparison with the exact result obtained in the following using the Fermi Golden Rule even though the process of sound radiation behind Equation (10.102) is purely classical. Also, the fact that the velocity circulation is quantized ($\kappa = 2\pi\gamma$) has no qualitative effect on the physics of radiation: The estimate (10.102) works for a vortex line with an arbitrary circulation κ in a generic ICIF.

Based on the Hamiltonian (10.93), we can obtain an accurate formula for the kelvon energy decay rate due to the sound radiation fixing, in particular, the order-of-unity dimensionless coefficients lacking in (10.102). The rate is given by $\dot{n}_k = -\sum_{s,k_1} W_{s,k,k_1}$, where W_{s,k,k_1} is the probability of the event $|0_s, n_k, n_{k_1}\rangle \to |1_s, n_k - 1, n_{k_1} - 1\rangle$ per unit time. Applying the Fermi Golden Rule to W_{s,k,k_1} with the interaction (10.93) and replacing sums with integrals, we obtain

$$\dot{n}_k = -\frac{(\kappa/2\pi c)^5}{15\pi\rho} [\ln(1/a_* k) + C_0]^5 k^{10} n_k^2. \tag{10.103}$$

The total power radiated by kelvons per unit vortex-line length is then obtained from $\Pi_k = -\sum_{k' \sim k} \varepsilon_{k'} \dot{n}_{k'}/L$.

A remark is in order here: In ST, kelvons do not live on infinite straight vortex lines, as we assumed throughout this section, but are propagating along vortex structures with a typical curvature radii R_0 much larger than the kelvon wavelength, $R_0 \gg \lambda$. This, strictly speaking, breaks momentum conservation and allows the lowest-order processes converting *one* kelvon into a phonon. However, the probabilities of such processes are suppressed *exponentially*—the kelvon-curvature coupling contains an exponentially small factor $\sim \exp(-R_0 k)$, which arises from the convolution of the smooth vortex-line profile with the oscillating kelvon mode.

10.3.3   Phonon Scattering by Vortices

Elastic phonon scattering: We now consider the scattering geometry shown in Figure 10.2. Generally, the vortex line has a distribution of kelvon occupation numbers $\{n_k\}$. The elastic scattering differential cross section can be found as

$$\frac{d\sigma}{d\phi} = (V/c)\frac{d}{d\phi}\sum_{\mathbf{q}_2} W_{\mathbf{q}_1 \to \mathbf{q}_2}, \tag{10.104}$$

where $W_{\mathbf{q}_1 \to \mathbf{q}_2}$ is the probability per unit time of a scattering event in which a phonon changes its wavenumber state from $\mathbf{q}_1$ to $\mathbf{q}_2$ such that

$$\mathbf{q}_1 = q_1(\sin\theta, 0, \cos\theta),$$
$$\mathbf{q}_2 = q_2(\sin\theta\cos\phi, \sin\theta\sin\phi, \cos\theta).$$

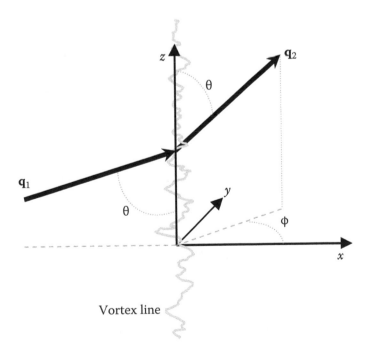

Figure 10.2   Geometry of the elastic phonon scattering.

The probability $W_{\mathbf{q}_1 \to \mathbf{q}_2}$ is given by the Golden Rule, $W_{\mathbf{q}_1 \to \mathbf{q}_2} = 2\pi\delta\left(\omega_{\mathbf{q}_1} - \omega_{\mathbf{q}_1}\right)$ $\left|V_{\mathbf{q}_1 \to \mathbf{q}_2}\right|^2$.

In the Born approximation, the transition amplitude $V_{\mathbf{q}_1 \to \mathbf{q}_2}$ is composed of the elementary processes shown diagrammatically in Figure 10.3. Note that the term $V_{\mathbf{q}_1 \to \mathbf{q}_2}^{(a)}$ alone is insufficient to describe the full scattering process. The remaining terms are necessary to capture the virtual deformation of the vortex line (produced by a kelvon with $k = q_{1z}$). It is easily seen that physically $V_{\mathbf{q}_1}^{(b-e)}$ accounts for the oscillatory motion of the vortex line induced during the scattering: If the vortex were pinned by an external potential, these terms would have to vanish. The transition amplitudes in Figure 10.3 are given by

$$V_{\mathbf{q}_1 \to \mathbf{q}_2}^{(a)} = R_{\mathbf{q}_1 \mathbf{q}_2} + R_{\mathbf{q}_2 \mathbf{q}_1}^* = \frac{i\pi\gamma}{L^2} \frac{\sqrt{q_1 q_2}\,(q_1 + q_2)\sin^2\theta \sin\phi}{q_1^2 + q_2^2 - 2q_1 q_2\left(\sin^2\theta \cos\phi + \cos^2\theta\right)}\delta_{q_1^z, q_2^z},$$

$$(10.105)$$

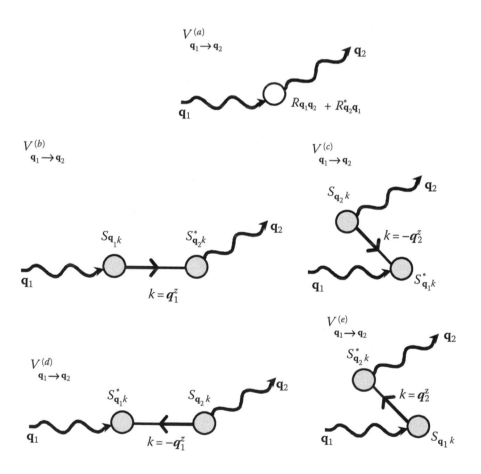

Figure 10.3 Diagrammatic representation of various contributions to the transition amplitude of elastic phonon scattering on a vortex line. The wavy and straight lines represent phonons and kelvons, respectively.

$$V_{\mathbf{q}_1 \to \mathbf{q}_2}^{(b)} = \frac{S_{\mathbf{q}_1 k} S_{\mathbf{q}_2 k}^*}{\omega_q - \varepsilon_k} (n_k + 1), \tag{10.106}$$

$$V_{\mathbf{q}_1 \to \mathbf{q}_2}^{(c)} = \frac{S_{\mathbf{q}_2 k} S_{\mathbf{q}_1 k}^*}{-\omega_q - \varepsilon_k} (n_{-k} + 1), \tag{10.107}$$

$$V_{\mathbf{q}_1 \to \mathbf{q}_2}^{(d)} = \frac{S_{\mathbf{q}_2 k} S_{\mathbf{q}_1 k}^*}{\omega_q + \varepsilon_k} n_{-k}, \tag{10.108}$$

$$V_{\mathbf{q}_1 \to \mathbf{q}_2}^{(e)} = \frac{S_{\mathbf{q}_1 k} S_{\mathbf{q}_2 k}^*}{-\omega_q + \varepsilon_k} n_k. \tag{10.109}$$

Note that at small scattering angles the amplitude $V_{\mathbf{q}_1 \to \mathbf{q}_2}^{(a)}$ diverges and the perturbation theory is not applicable. However, the small-angle scattering does not play a role in the dissipative component of the mutual friction force. Since $\varepsilon_k/\omega_q \sim a_0 q_{1z} \lesssim \beta \ll 1$ for $|k| = |q_{1z}| = |q_{2z}|$, we can neglect the kelvon energy in the denominators of Equations (10.105) through (10.109). Thus, the terms proportional to $\{n_k\}$ cancel out in the total transition amplitude,

$$V_{\mathbf{q}_1 \to \mathbf{q}_2} = R_{\mathbf{q}_1 \mathbf{q}_2} + \left[S_{\mathbf{q}_1 k} S_{\mathbf{q}_2}^* - \text{c.c.} \right] / \omega_q, \tag{10.110}$$

meaning that, alternatively, the amplitude can be obtained by retaining only the terms of type $V_{\mathbf{q}_1 \to \mathbf{q}_2}^{(a-c)}$ but evaluating them in kelvon vacuum. Substituting the vertex expressions into Equation (10.110) and doing the algebra yields

$$V_{\mathbf{q}_1 \to \mathbf{q}_2} = \frac{i\pi\gamma}{L^2} \frac{\sin^2\theta \sin\phi \cos\phi + \cos^2\theta \sin\phi}{1 - \cos\phi} \delta_{q_{1z}, q_{2z}}. \tag{10.111}$$

Finally, replacing the sum in Equation (10.104) with the integral,

$$\frac{d\sigma}{d\phi} = (V/c) \int \frac{q_2 dq_2 L^2}{2\pi} \delta\left(\omega_{\mathbf{q}_1} - \omega_{\mathbf{q}_1}\right) \left| V_{\mathbf{q}_1 \to \mathbf{q}_2} \right|^2, \tag{10.112}$$

we arrive at

$$\frac{d\sigma}{d\phi} = Lq \frac{\pi\gamma^2}{2c^2} \frac{\left[\sin^2\theta \sin\phi \cos\phi + \cos^2\theta \sin\phi \right]^2}{(1 - \cos\phi)^2}. \tag{10.113}$$

Due to the divergence at $\phi \to 0$, this formula is only applicable in cases where the small-angle scattering is not dominating the answer (i.e., the singularity is integrable), for example, in the calculation of the dissipative mutual friction force, discussed in the following, which immediately follows from Equation (10.113). In contrast, the calculation of the nondissipative component is essentially nonperturbative.

Inelastic scattering: Employing the same formalism, one can describe the inelastic phonon scattering: that is, scattering accompanied by generation of KWs (we will not present here the complete calculation). In the diagrammatic language, the transition amplitude $V_{\mathbf{q}_1 \to \mathbf{q}_2, k}$, where k is the kelvon wavenumber, is shown in Figure 10.4. Note that diagrams with reversed virtual kelvon propagators are omitted because they are accounted for by evaluating the diagrams remaining over the kelvon vacuum.

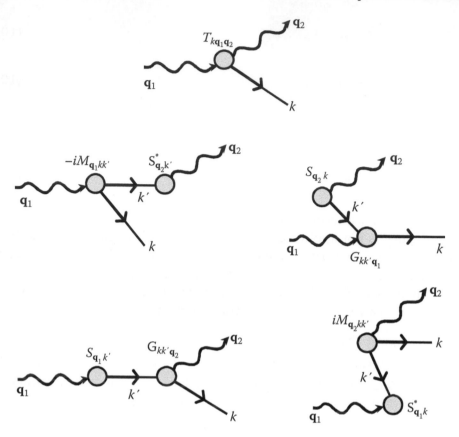

Figure 10.4   Inelastic phonon scattering by vortex lines. The sum over all the diagrams determines the total transition amplitude. The wavy and straight lines represent phonons and kelvons, respectively. The virtual kelvons are represented by vacuum propagators.

10.3.4   Microscopic Derivation of the Dissipative Mutual Friction Force

The formalism developed previously allows the derivation of the universal phonon contribution to the dissipative component of the mutual friction force directly from the hydrodynamic Hamiltonian. At $T \ll T_c$, the phonon contribution dominates in the mutual friction coefficient α. At temperatures $T \sim T_c$, the mutual friction is due mainly to excitations with wavelengths of the order of the healing length. These excitations are gapped both in ^{4}He (rotons) and ^{3}He (fermionic quasiparticles) and lead to an exponential dependence of α on temperature. We will not consider here the nonuniversal $T \sim T_c$ regime.

The momentum transferred by phonons to a vortex-line element (oriented along $\hat{z}$) per unit length per unit time according to the scattering cross section (10.113) is given by

$$\mathbf{f} = \frac{c}{V} \sum_{\mathbf{q}} n_{\mathbf{q}} \left[[\hat{z} \times \mathbf{q}] \times \hat{z} \right] \frac{1}{L} \int \frac{d\sigma}{d\phi} (1 - \cos \phi) d\phi, \tag{10.114}$$

where n_q is the phonon distribution function. Note that, at small scattering angles, the integrand is a regular function. However, it would not be the case if, despite the symmetry of the cross section (10.113) with respect to the sign of ϕ, we were to formally write down the momentum transferred in the perpendicular to $\mathbf{q}$ direction. This implies that the force component perpendicular to $\mathbf{f}$ and $\hat{z}$, the so-called *Iordanskii force*, requires more careful analysis of the small-angle scattering, which is beyond the scope of our book.

When the normal component drifts with the velocity $\mathbf{v_n}$, the phonon distribution is given by $n_q \approx n_q^{(0)} - \left(\partial n_q^{(0)} / \partial \omega_q \right)(\mathbf{q} \cdot \mathbf{v_{nL}})$, where $n_q^{(0)} = \left[\exp\left(\omega_q / T \right) - 1 \right]^{-1}$ and $\mathbf{v_{nL}} = \mathbf{v_n} - \mathbf{v_L}$ are the normal fluid velocities relative to the velocity of the vortex-line element $\mathbf{v_L}$. Replacing the sum over $\mathbf{q}$ with an integral and doing the integrals yields

$$\mathbf{f} = C \frac{\gamma^2 T^5}{c^7} [[\hat{z} \times \mathbf{v_{nL}}] \times \hat{z}], \tag{10.115}$$

$$C = \frac{19\pi}{210} \int_0^\infty \frac{e^x x^5}{(e^x - 1)^2} dx \approx 35.368. \tag{10.116}$$

This derivation assumes that phonons can be considered to be propagating ballistically as is typically the case in experiments at low temperatures so that the Stokes-like logarithmic corrections (see Section 10.1.4) to the friction force due to the normal-fluid viscosity are absent.

Following Section 10.1.4, the force $\mathbf{f}$ results in a motion additional to $\mathbf{v_L}$ of the vortex-line element with velocity $\mathbf{v_{Lf}} = \hat{z} \times \mathbf{f}/2\pi\gamma\rho_s$, where with our accuracy $\rho_s \approx \rho$. By definition of the mutual friction coefficient α, Equation (10.39), $\mathbf{v_{Lf}} = \alpha\hat{z} \times \mathbf{v_{nL}}$, which yields (for practical convenience, we restore k_B and substitute $\gamma \to \hbar/m_0$, where m_0 is the particle mass)

$$\alpha = \frac{C}{2\pi} \frac{(k_B T)^5}{\hbar^3 m_0 \rho c^7}. \tag{10.117}$$

10.4 Superfluid Turbulence

ST, also known as quantum/quantized turbulence, is a tangle of quantized vortex lines in a superfluid. ST can be created in a number of ways: (1) in the counterflow of normal and superfluid components; (2) by vibrating objects; (3) as a result of macroscopic motion of a superfluid (referred to as *quasiclassical* turbulence), in which case ST can mimic, at large enough length scales, classical-fluid turbulence; (4) by the technique of ion injection; (5) in the process of (strongly) nonequilibrium Bose–Einstein condensation, in which case it is a manifestation of the generic Kibble–Zurek effect, etc.

From the mid-1950s until the mid-1990s, ST was intensively studied in the context of counterflow of normal and superfluid components with prominent contributions by Vinen (an equation for qualitative description of growth/decay kinetics), extensive experimental studies, and pioneering microscopic simulations of vortex-line dynamics by Schwartz. The counterflow setup naturally implies finite density of the normal component and thus a relatively simple—by comparison with the $T = 0$ case—relaxation mechanism: The normal component exerts an appreciable (the mutual friction coefficient $\alpha \sim 1$) drag force on a vortex filament leading to the decay of the total vortex-line length. We shall consider the counterflow turbulence in more detail in Section 10.4.1.

In contrast to the counterflow setup, for the nonequilibrium process of BEC in a weakly interacting gas, the $\alpha \ll 1$ regime is quite characteristic in view of the following circumstance: The condensation kinetics is a classical-field phenomenon and its essential physics is quantitatively captured by the time-dependent GPE. Hence, the most natural setup to study it is to start with the initial condition featuring large occupation numbers for bosonic modes when the entire evolution is accurately described by GPE from the very outset. In the idealized situation of the classical field, the final temperature asymptotically approaches zero in view of the UV catastrophe. The ST generated in this case is nonstructured; this will be discussed in Section 10.4.2.

Another situation where the $\alpha \ll 1$ regime occurs naturally is the quasiclassical turbulence considered in more detail in Section 10.4.3. Unlike counterflow turbulence, it is due to a macroscopic turbulent motion and is generated by purely classical fluid dynamics, that is, by mechanical agitation such as stirring, at arbitrary temperatures without the help of mutual friction.

At $\alpha \ll 1$ ($\alpha \propto T^5$ at $T \to 0$), the dynamics of vortex lines in a significant range of length scales becomes approximately conservative. In this case, relaxation of turbulent motion usually involves certain types of *cascades*. In both classical and superfluid turbulence, the key role is played by the cascade of energy in the wavenumber space toward high wavenumbers. For this relaxation regime to occur, the following conditions must be satisfied:

1. A substantial separation between the energy-carrying scale λ_{en} and the scale $\lambda_{cutoff} \ll \lambda_{en}$, where the dissipation of energy to other degrees of freedom takes place; this defines the cascade *inertial range* $\lambda_{en} > \lambda > \lambda_{cutoff}$.

2. The kinetics are local in the wavenumber space; that is, the energy exchange takes place mainly between the adjacent length scales.

3. The "collisional" kinetic time $\tau_{coll}(\lambda)$, defined as the time between elementary events of energy exchange at a given scale λ, gets progressively shorter down the scales.

These requirements determine the main qualitative features of the cascade: The decay is governed by the slowest kinetics at λ_{en}, where the energy flux ε in the wavenumber space is formed, while the faster kinetic processes at shorter scales adjust instantly to this flux value and support the transfer of energy toward λ_{cutoff}, where it is dissipated into heat. Thus, the cascade is a (quasi-)steady-state regime in which the energy flux ε is constant through the length scales and the variation of ε in time is described by the slowest time scale $\sim \tau_{coll}(\lambda_{en})$.

With vortex lines, one can think of numerous different cascades. Perhaps the most obvious one is the Richardson–Kolmogorov (RK) cascade of eddies, which can be realized in a superfluid (even at $T = 0$) due to its ability to emulate the classical-fluid turbulent motion by the motion of a polarized vortex-line tangle (see Section 10.4.3). Clearly, this type of cascade is fundamentally impossible in the nonstructured tangle. It was believed for a while that nonstructured ST decays at $T = 0$ via the Feynman's cascade of vortex rings. Feynman had conjectured that reconnections of vortex lines produce vortex rings, with subsequent decay of each ring into a pair of smaller rings, and so forth. It can be shown, however, that this conjecture is inconsistent with simultaneous conservation of energy and momentum for the daughter rings as soon as coupling between separate rings becomes negligible (see Section 10.4.2), which necessarily happens below the length scale of interline separation.

Another option is a cascade of KWs supported by their nonlinear interactions. While such a cascade is indeed possible (see Section 10.2.4), it is subject to a specific constraint on the maximal energy flux. The origin of this constraint is the approximate integrability of the vortex dynamics controlled by the large parameter $\Lambda = \ln(\lambda/a_*)$, where λ is the characteristic kelvon wavelength. For $\Lambda \gg 1$, the leading term in the KW dynamics is given by the LIA (see Section 10.1.2), which does not support the KW cascade. This circumstance predetermines the existence of intermediate cascades that are required to transfer energy to wavenumbers high enough for the pure KW cascade to take over. These intermediate cascades must involve reconnections to lift the constraint imposed by the integrability of LIA.

In the nonstructured tangle, there is only one type of reconnection-induced cascade. It is supported by two subsequent events taking place at a given wavelength λ: (1) emission of a vortex ring with the radius $\sim \lambda$ by local self-intersection of a kinky vortex line and (2) reabsorption of the ring by the tangle. Both events are accompanied by energy transfer to KWs at shorter wavelengths (still on the order of λ), because KWs ought to have their wavelength smaller than the loop radius. For the self-reconnection cascade to occur, the vortex line must be kinky (loosely speaking, fractal) at all relevant length scales. Due to the integrability of LIA, the fractal structure is provided by the cascade itself: In the absence of self-crossings at a given length scale, the amplitude of KWs keeps growing due to self-crossings at larger length scales. It is worth emphasizing that in contrast to Feynman's scenario, where the vortex rings are the primary energy carriers, in the local self-crossing scenario, the rings play only a supportive role in a sequence of the vortex tangle rearrangements. Clearly, the crucial role is played by KWs with large amplitudes (they carry the energy) and reconnections (which directly promote the KW cascade).

In the polarized tangle emulating the RK classical-fluid cascade, there are two more varieties of reconnection-supported KW cascades. One is when reconnections are between bundles of quasiparallel vortex lines, and the other one is due to reconnections between the two neighboring lines in a bundle. In Section 10.4.4, we will present arguments that, in the theoretical limit of $\Lambda \to \infty$, all the three reconnection-driven cascades are necessary to cross over from RK to the pure KW regime, the crossover being a series of three distinct cascades: bundle driven, neighbor reconnection driven, and local self-crossing driven. The total extent of the crossover regime in the wavenumber space is predicted to be $\sim \Lambda/10$ decades. For a realistic value $\Lambda \lesssim 15$ (for ^{4}He), the crossover takes about one decade in length scales; one can hardly expect to see a sharp distinction between the three different regimes in this case.

At $T = 0$, the reconnection-supported cascade(s) ultimately crosses over to the pure KW cascade and the latter is cut off by phonon emission. The elementary process of phonon emission converts two kelvons with almost opposite momenta into one bulk phonon.

In what follows, we shall confine ourselves explicitly to the fundamental regime of homogeneous isotropic turbulence occupying a large volume, which allows us to ignore boundary effects and tangle diffusion.

10.4.1   Counterflow Superfluid Turbulence: Vinen's Equation

A vortex tangle is generated by thermal counterflow (see Section 2.2.2) if the counterflow velocity v_{ns} exceeds a certain critical value v_{c1}. To a reasonable approximation, such tangles are homogeneous and isotropic as indicated by substantial experimental evidence. The key variable characterizing the tangle is the vortex-line density L, which is the average length of all vortex lines in a unit volume. One can obtain a phenomenological equation describing the evolution of L using simple qualitative considerations based on the general equations of vortex-line motion.

As an illustration of the turbulence generation mechanism, consider a vortex ring of radius R placed in a counterflow of velocity v_{ns} with its plane perpendicular to the flow. At $T = 0$, the vortex ring would move through the fluid with the constant velocity $(\gamma/2)\ln(R/a_*)/R$ (see Problem 10.6). At finite temperature, an element of the ring acquires a velocity component perpendicular to the ring's direction of propagation, which results in either shrinking or expanding of the ring. From Equation (10.39), it follows that (see Problem 10.8)

$$\dot{R} = \alpha[v_{ns} - (\gamma/2)\ln(R/a_*)/R]. \tag{10.118}$$

Thus, at $T > 0$, the vortex ring is always unstable and either collapses or grows up to the size of the container depending on its initial radius R_0. Rings with $v_{ns} > (\gamma/2)\ln(R_0/a_*)/R_0$ expand; otherwise, they collapse.

An estimate for the critical velocity v_{c1} readily follows from Equation (10.118). The topological nature of vorticity ensures that there is typically a number (no matter how small) of vortices remaining in the macroscopic container after the system is prepared and allowed to relax; the so-called *remanent vorticity*. The ends of these vortices are pinned by the walls, thus preventing remanent vorticity from further relaxation. The typical curvature radius of these vortices is given by the size of the container channel b used in the counterflow experiments (short KWs on vortex lines ultimately relax and decay into phonons). Hence, counterflow velocities $v_{\mathrm{ns}} > v_{\mathrm{c1}} \sim (\gamma/2)\ln(b/u_*)/b$ lead to the growth of the long-wavelength structures until the neighboring lines cross each other and reconnect, leading to the generation of vortex rings of the radius $\sim b$, which also start to grow and reconnect forming a random (nonstructured) tangle.

There is an additional important condition for generation of turbulence by counterflow, which is standard for any oscillator. Our estimate of v_{c1} assumes that a structure on a vortex line does not relax substantially, due to the mutual friction during its natural (rotation) period, which allows it to be amplified by the counterflow each time the plane of the loop aligns perpendicular to the flow. At sufficiently large mutual friction coefficients, the net effect of dissipation can dominate the pumping irrespective of the value of v_{ns}. Indeed, if a wave on a vortex line does not survive a single oscillation, its growth due to pumping becomes impossible. Thus, the transition to turbulence is only possible if $(1 - \alpha')/\alpha > 1$ (see Problem 10.10). In practice, this condition is always satisfied in ^{4}He except very close ($\sim 1\%$ distance) to T_c. In ^{3}He-B, in contrast, temperatures $T/T_c \lesssim 0.6$ are necessary for generation of turbulence.

Upon viewing the tangle as a set of randomly oriented vortex loops of the typical radius R given by the scale of interline separation, $\sim 1/L^{1/2}$ (also called the *main scale*), Equation (10.118) describes two major mechanisms responsible for evolution of the vortex line density in the tangle. The first term in the right-hand side corresponds to pumping of L by the counterflow and gives a contribution to $\dot{L}$ proportional to $\alpha v_{\mathrm{ns}} L^{3/2}$, while the second term is responsible for decay of L due to mutual friction and the corresponding contribution to $\dot{L}$ is proportional to $-\alpha\beta_L L^2$, where $\beta_L = (\gamma/2)\ln(1/a_*L^{1/2})$.

In addition to the mutual friction–driven mechanisms of decay, there is an inertial one, which persists in the $T \to 0$ limit and is due to reconnections between vortex lines. As discussed in Section 10.2.2, a reconnection at the length scale R and the subsequent relaxation of the vortex angle lead to the transfer of energy—which is, to a good approximation, proportional to the line length (see Problem 10.3) in nonstructured tangles—to shorter length scales. Since the rate of reconnections in the characteristic volume $\sim R^3$ is given by the natural frequency β_L/R^2 of the structure, the line-length flux toward short scales per unit volume is $\sim \beta_L/R^4$. The line-length flux is ultimately dissipated at shorter scales, where the mutual friction is progressively more efficient (see Problem 10.10). Thus, reconnections lead to a contribution to $\dot{L}$ proportional to $-\beta_L L^2$.

With these estimates at hand, we formulate the phenomenological equation for the evolution of the vortex-line density (originally suggested by Vinen):

$$\dot{L} = \chi_1 \alpha v_{\mathrm{ns}} L^{3/2} - \chi_2 \alpha \beta_L L^2 - \chi_3 \beta_L L^2, \tag{10.119}$$

where χ_1, χ_2, χ_3 are constants of order unity, which cannot be determined by qualitative analysis. Setting $\dot{L} = 0$ gives the steady-state vortex-line density for a given counterflow velocity v_{ns},

$$L = \left[\frac{\chi_1 \alpha}{(\chi_2 \alpha + \chi_3)\beta_L} \right]^2 v_{\mathrm{ns}}^2. \tag{10.120}$$

Qualitative considerations that led us to Vinen's equation (10.119) can be used in a more general case of arbitrary nonstructured vortex tangle even in the absence of a counterflow, $v_{ns} = 0$, or at $\alpha \to 0$ (regardless of the method used to generate the initial line density L). In the latter case, only the last term in the right-hand side survives, but the derivation of Vinen's equation becomes more subtle because of the fractal vortex-line structure, which notably affects the total line length L. In the following section, we will consider the characteristics of a nonstructured vortex tangle and the underlying relaxation mechanisms at $T \to 0$ in more detail.

At counterflow velocities exceeding the second critical value v_{c2}, the normal component of the fluid becomes turbulent as well, leading to a substantial increase in L.

10.4.2 Decay of Nonstructured Tangle, $T = 0$ Case, and Self-Reconnection-Driven Kelvin-Wave Cascade

As we discussed previously, turbulence in the superfluid component is only possible when $(1 - \alpha')/\alpha > 1$. Depending on the generation method, the resulting vortex tangle can be either (1) structured in the sense that one can locally identify a preferred direction that a considerable number of neighboring vortex lines align with on average or (2) nonstructured. In the latter case, considered here, the orientation of a vortex line is approximately independent from that of its neighbor. Whether the tangle is structured or nonstructured depends on where most of the energy of the turbulent motion is stored: The case where the coupling between neighboring vortex lines dominates corresponds to structured (quasiclassical) tangles, which display a macroscopic (i.e., happening at length scales much larger than the interline separation) motion of the superfluid. This case will be discussed in the following section. In nonstructured tangles, the coupling is negligible, and the energy is mostly stored in the total length of vortex lines (correspondingly, we will be referring to the energy as the line length throughout this section). The motion of neighboring lines is then independent from each other (i.e., self-induced) and, to a good approximation, is governed by the LIA (see Section 10.1.2). The competition between the two regimes will also be considered in more detail in Section 10.4.4.

As a result of LIA motion, the neighboring vortex lines can cross and undergo a reconnection. This leads to an energy flux toward the shorter length scales; see Section 10.2.2. At $\alpha \sim 1$, this flux is dissipated by mutual friction at length scales somewhat smaller but similar to R. This is seen from the fact that the power dissipated per unit line length from structures of size R, given by $\Pi(R) \sim \alpha \gamma^3 \rho_s \Lambda^2 / R^2$ (see Problem 10.10), is of the same order as the energy flux (per unit line length) produced by reconnections, $\sim \gamma^3 \rho_s \Lambda^2 / R^2$. In this situation, the vortex lines remain smooth at length scale shorter than R. However, as the temperature is decreased, the dissipation due to mutual friction becomes appreciable only at shorter and shorter distances, opening an increasing range of length scales over which the vortex dynamics are approximately conservative. This leads to the emergence of a hierarchy of short-wavelength structures on vortex lines. We find it instructive to consider the most general (from the point of view of the vortex-tangle structure) case of zero temperatures first and discuss mutual friction effects at $T > 0$ later.

As discussed previously, Feynman's cascade, based on reconnections leading to smaller and smaller rings, is inconsistent with simultaneous conservation of energy and momentum for individual vortex rings. Only when coupling between vortex lines is appreciable relative to the LIA motion [Equation (10.11)] are Feynman's reconnections allowed to take place because the energy (10.9) is stored in both the vortex-line length and the interline interactions. However, once the line coupling becomes negligible (e.g., at the scale of interline separation), making the energy and the momentum of the rings additive, the Feynman scenario stops working. Thus, in disordered tangles, this scenario has no inertial range. In quasiclassical tangles, Feynman's mechanism is an important constituent of the complex nonlinear dynamics characteristic of the RK regime.

We start with introducing the crucial notion of a smoothed line length, $\mathcal{L}(\lambda_1)$, which is the length of a (fractal) line upon smoothing out of all structures with length scales smaller than λ_1. The mathematical expression relating line lengths with two well-separated cutoffs reads

$$\ln \mathcal{L}(\lambda_1) \sim \ln \mathcal{L}(\lambda_0) + \int_{\lambda_1}^{\lambda_0} (b_\lambda/\lambda)^2 \, d\lambda/\lambda, \tag{10.121}$$

where b_λ is the amplitude of KWs with the wavelength λ. In particular, if λ_0 is the largest wavelength in the problem, then $\mathcal{L}(\lambda_0)$ is the length of the most smooth vortex line. With $\mathcal{L}(\lambda)$, we can estimate the number of local self-crossings at the scale λ per unit time as

$$N_\lambda \sim \frac{\mathcal{L}(\lambda)}{\lambda} \Omega_{b_\lambda}(\lambda) \omega_\lambda, \tag{10.122}$$

where $\mathcal{L}(\lambda)/\lambda$ gives the number of statistically independent line segments for such crossings to occur, $\Omega_{b_\lambda}(\lambda)$ is the probability of the KW amplitude within a line element of length $\sim \lambda$, given the spectrum b_λ, to fluctuate to a value $\sim \lambda$ necessary to produce a self-crossing, and ω_λ is the kelvon-precession frequency.

To estimate $\Omega_{b_\lambda}(\lambda)$, we can consider the limit of $b_\lambda \ll \lambda$, in which kelvons are independent harmonic modes and, correspondingly, the statistics of fluctuations of the amplitude are Gaussian. This readily yields

$$\Omega_{b_\lambda}(\lambda) \sim (b_\lambda/\lambda) e^{-(\lambda/b_\lambda)^2}. \tag{10.123}$$

We leave out of this formula a dimensionless factor of order unity in front of λ, which defines the actual critical amplitude required for self-crossing in a typical geometry.

For any cascade, the fundamental notion is the flux of conserved quantity. In our case, it is the flux of the vortex-line length. In contrast to a standard cascade, in which the integral for conserved quantity comes from a single length scale, we are dealing in our case with fractal structures when the line length is spread over all the scales of distance in the inertial range. Moreover, different scales are not *entirely* independent in the sense that, for fractals, the short-range structures with their energy are absorbed into the long-wave structures. More precisely, the line length flux from long waves to shorter ones is carried out by short-wave structures developing on top of the long-wave modes. The crucial observation is that, with respect to longer wavelengths, the short wavelengths play a passive role in the energy balance because their contribution to the line length is simply proportional to the length of a smoothed line. Correspondingly, one can speak of the *smoothed* line-length flux, $Q(\lambda_1)$, where λ_1 is the smoothing parameter in Equation (10.121). By definition of the cascade, the quantity Q is one and the same at any wavelength scale λ, as long as $\lambda \gg \lambda_1$.

According to the self-reconnections scenario,

$$Q(\lambda_1) \sim N_\lambda \mathcal{R}_\lambda(\lambda_1), \tag{10.124}$$

where $\mathcal{R}_\lambda(\lambda_1)$ is the length of a smoothed circle of radius $\sim\lambda$. In direct analogy with (10.121), we have

$$\ln \mathcal{R}_\lambda(\lambda_1) \sim \ln \lambda + \int\limits_{\lambda_1}^{\lambda} (b_{\lambda'}/\lambda')^2 \, d\lambda'/\lambda'. \tag{10.125}$$

From (10.125) and (10.121), there follows a useful relation

$$\mathcal{L}(\lambda)\mathcal{R}_\lambda(\lambda_1) \sim \lambda \mathcal{L}(\lambda_1). \tag{10.126}$$

With N_λ from Equation (10.122), $\Omega(\lambda)$ from Equation (10.123), an estimate $\omega_\lambda \sim \beta/\lambda^2$, and the relation (10.126), the expression for the line flux (10.124) takes the form

$$Q(\lambda_1)/\mathcal{L}(\lambda_1) \sim \left(\beta b_\lambda/\lambda^3\right) e^{-(\lambda/b_\lambda)^2}. \tag{10.127}$$

The left-hand side of this relation is a function of λ_1, while the right-hand side is a function of λ, meaning that both sides are actually constants. For the right-hand side this implies

$$(b_\lambda/\lambda)^2 \sim \frac{\left(b_{\lambda_0}/\lambda_0\right)^2}{1 + \left(b_{\lambda_0}/\lambda_0\right)^2 \ln(\lambda_0/\lambda)}. \tag{10.128}$$

[Since up to logarithmic corrections we have $b_\lambda \sim \lambda$, we do not distinguish between $\ln b_\lambda$ and $\ln \lambda$.] Expression (10.128) yields the KW cascade spectrum in terms of the characteristic amplitude b_λ. As far as the function $\mathcal{L}(\lambda)$ is concerned, from (10.128) and (10.121), we find

$$\mathcal{L}(\lambda) = \mathcal{L}(\lambda_0) \left[1 + \left(\frac{b_{\lambda_0}}{\lambda_0}\right)^2 \ln \frac{\lambda_0}{\lambda} \right]^\nu, \tag{10.129}$$

where ν is a constant of order unity the particular value of which cannot be established by our order-of-magnitude analysis.

It is clear from Equation (10.66) that no matter how large is the energy flux (per unit vortex-line length) $\theta \propto Q$ transported by the self-crossings regime is, at sufficiently high wavenumbers, the pure KW cascade will be capable of supporting it. As soon as the collision kinetics become fast enough to reduce the amplitudes of KWs, the reconnections are inhibited, in view of Equation (10.123). Thus, the self-crossings-driven regime will inevitably be replaced by the pure KW cascade at some scale λ_*. Note, however, that due to a small difference between the KW spectra in the two regimes, Equations (10.66) and (10.128), the crossover between them is likely to be extended in the wavenumber space. A rough estimate of λ_* can be obtained by setting $b_k \sim k^{-1} \sim \lambda_*$ in Equation (10.66). The result clearly depends on the cascade energy flux θ, which is specific to the physics at the energy-containing scale. In nonstructured tangles, the energy flux is formed by reconnections at the scale of interline separation R when $\theta_{ns} \sim \gamma^3 \rho \Lambda^2 / R^2$, with $\Lambda = \ln(R/a_*)$, leading to

$$\lambda_* \sim R/\Lambda \qquad \text{(nonstructured tangles).} \tag{10.130}$$

According to this rough estimate, the inertial range for the regime driven by local self-crossings is only about $\Lambda/10$ decades, which for realistic values of Λ could turn out to be an insignificant range without a distinct spectral signature.

10.4.3 Classical-Fluid Turbulent Regime: Vorticity Equation

Even at zero temperature, the superfluid dynamics support a turbulent regime that is indistinguishable from its classical-fluid counterpart. That may seem surprising since the only degrees of freedom in a superfluid at $T = 0$ are quantized vortex lines, which are very different from the classical eddies responsible for turbulence in ICIF. Nonetheless, collective dynamics of quantized vortex lines can effectively mimic classical vorticity at large scales. Recall that, in Chapter 1, we derived the *coarse-grained* velocity profile of a rapidly rotated superfluid identical to that of the solid-body rotation and accomplished by formation of a dense array of vortex lines aligned along the rotation axis. In terms of the Biot–Savart equation (10.8), these dynamics arise from strong coupling between the vortex lines in a dense array. It makes a vortex bundle behave as a single classical object. Therefore, by standard classical turbulence generation methods (for example, stirring), one can produce vorticity fields indistinguishable from those of a normal fluid (under coarse-graining up to length scales larger than the typical interline separation R); the underlying vortex tangle is organized in polarized "bundles" of vortex lines.

The hallmark of the ICIF hydrodynamics is the RK cascade. Although understood intuitively on the basis of the vortex-array argument, the ability of quantized vortex lines to mimic the classical cascade is not immediately obvious. In what follows, we will rigorously derive this behavior from the Biot–Savart equation (10.8).

To this end, we rewrite Equation (10.8) in terms of vorticity $\mathbf{w} = \nabla \times \mathbf{v}$ in the momentum space:

$$\mathbf{w_k} = \int \mathbf{w(r)} e^{-i\mathbf{k \cdot r}} d^3 r = \kappa \int e^{-i\mathbf{k \cdot s}} d\mathbf{s}. \tag{10.131}$$

The result is *identical* to the vorticity equation for an ICIF,

$$\frac{\partial \mathbf{w_k}}{\partial t} = \mathbf{k} \times \int \frac{d^3 q}{(2\pi)^3} q^{-2} \left[\mathbf{w_{k-q}} \times \left[\mathbf{w_q} \times \mathbf{q} \right] \right]. \tag{10.132}$$

Remarkably, the circulation quantum κ drops out from this expression—a clear manifestation that the superfluid hydrodynamics is indistinguishable from the classical Euler equation with respect to which the quantization of circulation is nothing but an imposed initial condition, preserved during the evolution in view of the Kelvin–Helmholtz theorem.

Given Equation (10.132), we now formulate conditions under which the vortex-tangle dynamics are equivalent to the ICIF turbulence: (1) the energy must be concentrated at a sufficiently small wavenumber scale $k_{en} \ll R^{-1}$, and (2) the decay scenario must be local in momentum space to ensure that quantized nature of vorticity is irrelevant for the long-wavelength behavior. These conditions are not restrictive because (1) is automatically satisfied if turbulence is generated by classical means due to large (compared to κ) values of velocity circulation, and (2) is required for the existence of the RK cascade in classical fluids as well.

10.4.4 Crossover from Richardson–Kolmogorov to Kelvin-Wave Cascade

At $T = 0$, answering the question of how the vortex tangle looks at the level of individual lines when one zooms in down to scales of order R is not easy. The only solid fact is that energy dissipation by sound radiation at $\lambda_{ph} \ll R$ is supported by the energy flux transported by the pure KW cascade. At intermediate scales, one finds a sequence of cascades governed by different mechanisms that connect the classical cascade of eddies to the KW cascade on individual vortex lines. The corresponding theory crucially relies on the large parameter $\Lambda = \ln R/a_*$ and can be found in the review article [1]. In what follows, we briefly describe the relevant energy transfer mechanisms, present final results (without derivation) for KW spectra, and estimate crossover scales between different cascades. We refer the reader to [1] for technical details of the derivations.

It would be natural to assume that the classical cascade supporting RK energy flux ε extends down to the scale of interline separation R, at which point the discrete nature of vorticity becomes essential and the classical-fluid picture is replaced with kinetics of KWs. If the RK cascade were to reach the scale R, the typical vortex-line curvature at this scale would be of the order of R ensuring that vortex reconnections can occur in the tangle geometry. Estimating the energy flux ε_{rec} processed by the reconnections at this scale, we observe that $\varepsilon_{rec}/\varepsilon \sim \Lambda^2 \gg 1$; that is, the resulting flux is larger than the one supplied from the larger scales. Such a situation is, of course, forbidden by the energy conservation. We are thus forced to conclude that some transformation of the classical cascade must take place at length scales larger than R controlled by $\Lambda \gg 1$.

Since the dynamics of each individual vortex line are controlled by the circulation quantum κ, the independence of Equation (10.132) on κ leads to an important observation that any large-scale (classical) motion necessarily implies strong coupling of lines in the bundle and that the crossover to the quantized regime happens when self-induced motion starts to dominate over the interline coupling. To deduce the corresponding crossover scale r_0, let us formally decompose the integral (10.4) into the self-induced part, $\mathbf{v}^{SI}(\mathbf{s})$, for which the integration is restricted to the vortex line containing the element $\mathbf{s}$, and the remaining contribution induced by all the other lines, $\mathbf{v}^I(\mathbf{s})$,

$$\mathbf{v}(\mathbf{s}) = \mathbf{v}^{SI}(\mathbf{s}) + \mathbf{v}^I(\mathbf{s}). \tag{10.133}$$

By definition, at length scales $r \gg r_0$, the turbulent motion mimics that of ICIF in the form of dense coherently moving arrays of vortex lines bent with the curvature radius $\sim r$. On one hand, the velocity field of this configuration must obey the Kolmogorov law

$$v_r \sim (\varepsilon r)^{1/3}, \qquad r \gg r_0, \tag{10.134}$$

where ε is the energy flux per unit mass of the fluid formed by the energy-carrying eddies and transferred by the cascade (the subscript r stands for a typical variation of a field over a distance $\sim r$). On the other hand, the value of v_r is fixed by the velocity circulation around a contour of radius r, namely, $v_r r \sim \kappa n_r r^2$, where n_r is the areal density of vortex lines responsible for vorticity at the scale r. [Scale invariance requires that, on top of vorticity at the scale r, there exists a fine structure of vortex bundles of shorter distances, so that, mathematically, $n_r r^2$ is the difference between the large number of vortex lines crossing the area of the contour r in opposite directions.] This leads to the density-flux relation

$$n_r \sim \left[\frac{\varepsilon}{\kappa^3 r^2}\right]^{1/3}, \qquad r \gg r_0. \tag{10.135}$$

The underlying dynamics of a vortex line in the bundle is governed by v_r^I and v_r^{SI}. While, by definition, $v_r^I \sim v_r$, the self-induced part is determined by the curvature radius r according to the LIA, Equation (10.11),

$$v_r^{SI} \sim \Lambda_r \frac{\kappa}{r}, \tag{10.136}$$

where $\Lambda_r = \ln(r/a_*)$. Here and throughout this section, we work with logarithmic accuracy and, correspondingly, we are allowed to replace Λ_r with $\Lambda = \ln(R/a_0*)$. At length scales where $v_r^I \gg v_r^{SI}$, the vortex lines in the bundle move coherently with the same velocity $\sim v_r^I$. However, at the scale $r_0 \sim \left(\Lambda^3 \kappa^3/\varepsilon\right)^{1/4}$, the self-induced motion of the vortex line becomes comparable to the collective motion, $v_r^{SI} \sim v_r^I$. Below this scale, individual vortex lines start to behave independently of each other and thus r_0 defines the lower cutoff for the inertial region of the Kolmogorov spectrum (10.134).

Since r_0 is the size of the smallest classical eddies, the areal density of vortex lines at this scale is given by the typical interline separation, $n_{r_0} \sim 1/R^2$. In other words, vortex bundles at the scale r_0 consist of nearly parallel vortex lines separated by R. With Equation (10.135), we arrive at relations

$$r_0 \sim \Lambda^{1/2} R, \tag{10.137}$$

$$R \sim \left(\Lambda \kappa^3 / \varepsilon \right)^{1/4}. \tag{10.138}$$

The crossover can also be understood in slightly more visual terms. Let us introduce an effective number of vortex lines N_r in a bundle of size r. This number is obtained as an algebraic sum of the number of lines going through the bundle cross section in opposite directions and is related to previously introduced density by $N_r = n_r r^2 \sim \varepsilon^{1/3} r^{4/3} / \kappa$. In view of Equations (10.134) and (10.136), the number of lines in a bundle relative to Λ determines whether it behaves as a classical eddy or a set of independent vortex lines: For $N_r \gg \Lambda$, the coupling between the lines dominates in dynamics. Thus, in the theoretical limit of $\Lambda \gg 1$, the bundles still contain a large number of vortices at the scale where the classical regime breaks down.

Let us now discuss the cascade mechanism that supersedes the RK cascade of eddies at length scales immediately adjacent to r_0.

Reconnections of bundles: At the crossover scale r_0, the relative orientation of two lines (with their short-wavelength structures smoothed out) becomes uncorrelated if they are a distance $\gtrsim r_0$ apart. At the same time, individual lines start moving according to their own geometric shape, as prescribed by Equation (10.11). Therefore, reconnections, at least between lines belonging to different bundles, are inevitable.

As before, the key quantity in our considerations is the energy transferred to a lower scale after one reconnection of vortex lines at the scale $k^{-1} \ll r_0$. Crossing of entire bundles results in reconnections between all their vortex lines and generation of KWs with a smaller but adjacent wavelength λ. Moreover, coherent motion of bundles translates into coherent generation of waves on vortex lines within the same bundle. Thus, at the scale $k^{-1} \lesssim r_0$, adjacent vortex lines remain nearly parallel; that is, vortex lines at the scale k^{-1} also form bundles. Similarly, these bundles reconnect and transport energy to a lower scale, where further bundle reconnections occur, and so on down to the scale comparable to the interline separation where this self-similar regime is cut off.

The spectrum of KWs b_k in this regime can be obtained from the condition $\tilde{\varepsilon}_k \equiv \varepsilon$, where $\tilde{\varepsilon}_k$ is the energy flux per unit mass transported by reconnections at the scale k^{-1}. The final result for the spectrum of KWs in the bundle-crossing regime has the form

$$b_k \sim r_0^{-1} k^{-2}, \tag{10.139}$$

and corresponds to a small typical crossing angle $kb_k \sim 1/kr_0 \ll 1$. At the wavelength $\sim \lambda_b = \Lambda^{1/4}R$, the wave amplitudes become of the order of the interline separation $b_k \sim R$, and the notion of bundles loses meaning—at this scale, the cascade of bundles is cut off and replaced with reconnections of individual lines.

Reconnections of adjacent lines: In view of Equation (10.139), the vortex lines are only slightly bent at the crossover scale λ_b because $b_{\lambda_b}/\lambda_b \sim \Lambda^{-1/4} \ll 1$. This poses an interesting question of what is driving the cascade at $k^{-1} < \lambda_b$, where the mechanism of self-reconnections is strongly suppressed, while the kinetics of kelvon scattering are too slow to carry the flux ε. To answer this, we observe that each n.n. small-angle reconnection performs a sort of *parallel processing* of the energy distribution for *each* of the wavelength scales $\in [\lambda_c, \lambda_b]$. By equating the energy flux transferred by small-angle n.n. reconnections at a given wavenumber k to ε, one arrives at the spectrum

$$b_k \sim R(\lambda_b k)^{-1/2} \sim k^{-1}\sqrt{\lambda_c k}, \tag{10.140}$$

where $\lambda_c \sim R/\Lambda^{1/4}$. The rise of the relative amplitude $b_k k \propto k^{1/2}$ implied by Equation (10.140) describes the process of building up the fractal structure on the vortex line that culminates at λ_c. Note that in this regime the energy is contained at the low end of the inertial range. We refer to this phenomenon as *deposition* of energy. Formally, it looks different from the standard cascade setup, but its presence has no effect on the rest of the inertial range and thus on the overall cascade efficiency.

Transition to the KW cascade: Once the crossover from the classical tangle decay to lines self-reconnections at $\sim \lambda_c$ is complete, the cascade picture at shorter scales is qualitatively similar to that of the nonstructured tangle described in Section 10.4.2. The self-reconnection regime continues in the range $\lambda_* \ll \lambda \ll \lambda_c$ with the spectrum $b_k \sim k^{-1}$ until it is replaced with the cascade driven by nonlinear kelvon kinetics. The spectrum of KW amplitudes in the nonlinear cascade is given by Equation (10.66). The value of λ_* is then determined from the energy-flux matching condition $\varepsilon = \theta/\rho R^2$ with $\theta \sim \rho/\lambda_*^2$ [see also Equation (10.138)]

$$\lambda_* = R/\Lambda^{1/2} \qquad \text{(structured tangles).} \tag{10.141}$$

Finally, at $T = 0$, KWs decay by emitting phonons. This dissipation mechanism is negligibly weak all the way down to wavelengths of the order of λ_{ph}, where the rate of energy radiation by the vortex lines becomes comparable to ε. The scale $\lambda_{ph} \ll \lambda_*$ estimated in the next section defines the lower cutoff for the KW cascade.

10.4.5   Dissipative Cutoff: Vortex-Line Length and Spectrum of Kelvin Waves

The vortex-line length density L is accessible in experiments (e.g., [6–8]). At first sight, the information contained in L apart from the overall "strength" of the turbulent motion is very limited—even the connection between L and the total energy density requires additional knowledge about the tangle polarization. If the vortex lines were smooth, L would be trivially related to the interline separation by $L = R^{-2}$. However, the presence of a fine wave structure on the lines may result in the line density exceeding R^{-2} by a large factor (or even diverging in the limit of fractal lines without a cutoff). This increase is related to the spectrum of KWs by Equation (10.121). Most importantly, by varying the temperature, one can control the degree of "smoothness" of vortex lines, thereby probing the distribution of KWs on them. In particular, the $L(T)$ dependence directly reflects transitions between different decay mechanisms across the inertial range and provides information about the structure of the tangle at $T = 0$.

The temperature dependence of L comes from the cutoff scale $\lambda_{\text{cuttoff}}(\alpha)$ with T-dependent mutual friction coefficient $\alpha(T) \propto T^5$ $(T \to 0)$. At the cutoff scale, the energy dissipated into the environment becomes comparable to the cascade energy. This prevents the buildup of the KW spectrum at shorter wavelengths and is equivalent to imposing a high-momentum cutoff in Equation (10.121):

$$\ln[L(\alpha)/L_0] = \int_{k_0}^{k_{\text{cutoff}}(\alpha)} (b_k k)^2 \, dk/k. \qquad (10.142)$$

Here, $k_{\text{cutoff}} \sim 1/\lambda_{\text{cutoff}}$, k_0 is the smallest wavenumber of the KW cascade (not to be confused with the smallest wavenumber of the RK cascade) at which the concept of well-defined KWs remains meaningful, and L_0 is the "background" line density corresponding to k_0-smoothed lines.

At $T = 0$, the mutual friction is absent and the cascade is cut off by the radiation of sound (with the possible exception of ^{3}He-B, where the bound states inside the vortex cores can play a role)* at $k_{\text{cutoff}} = k_{\text{ph}}$. In both nonstructured and classical tangles, the sound is radiated at very large wavenumbers where turbulence is decaying, according to the purely nonlinear KW cascade with the spectrum given by Equation (10.66). By increasing temperature, one decreases $k_{\text{cutoff}}(T) < k_{\text{ph}}$ in Equation (10.142), and thus one can study properties of the KW cascade spectrum, including qualitative changes as different cascade regimes are cut off; see Figure 10.5. The existence of a well-defined dissipative cutoff is based on the fact that the cascade is supported by rare kinetic events in the sense that the collision time $\tau_{\text{coll}} \equiv \tau_{\text{coll}}(\varepsilon, k)$ is much longer than the KW oscillation period, $\tau_{\text{per}} \equiv \tau_{\text{per}}(k)$. The same is true for the dissipative time at low temperature, $\tau_{\text{dis}} \equiv \tau_{\text{dis}}(\alpha, k) \sim \tau_{\text{per}}/\alpha$. The cascade is cut off when the energy dissipation rate at a given wavenumber scale becomes comparable to the energy being transferred by the cascade to higher wavenumbers. This defines $k_{\text{cutoff}}(\varepsilon, \alpha)$.

At finite T, the dissipative dynamics of a vortex-line element are described by Equation (10.39) [we will neglect α' in the following, based on the reasoning discussed in Section 10.1.4]. At $\alpha \sim 1$, the superfluid and normal components are strongly coupled, and the cascade is cut off before it enters the quantized regime; that is, $\lambda_{\text{cuttoff}} \gtrsim r_0$. At small mutual friction, the characteristic number of KW

---

* In neutral superfluids.

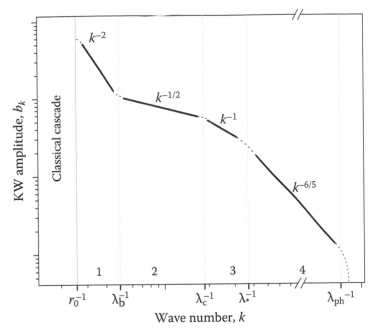

Figure 10.5   Spectrum of KWs on vortex lines in the limit of $\Lambda \gg 1$. The inertial range consists of a chain of cascades driven by different mechanisms: (1) reconnections of vortex-line bundles, (2) reconnections between n.n. vortex lines in a bundle, (3) self-reconnections on single vortex lines, and (4) nonlinear dynamics of single vortex lines without reconnections. Regimes (3) and (4) are familiar in the context of nonstructured-vortex-tangle decay.

oscillations before the wave is substantially damped (Problem 10.10) becomes large $\sim 1/\alpha \gg 1$. In the range $\lambda \ll r_0$, the normal component flow is already laminar and thus the field $\mathbf{v_n}$ has no structure at such small scales. Moreover, as long as $\alpha \ll 1$, the disturbance of the normal component caused by the vortex-line motion at short scales can be neglected. Therefore, $\mathbf{v}_n(\mathbf{s})$ can be treated as a constant in Equation (10.39), which makes it irrelevant for the question of KW dissipation. In this case, Equations (10.39) and (10.11) give the rate at which the amplitude b_k decays due to the mutual friction,

$$\dot{b}_k \sim -\alpha \omega_k b_k. \tag{10.143}$$

Since the energy per unit line length associated with the wave is $E_k \sim \kappa \rho \omega_k b_k^2$, the power dissipated (per unit line length) at the scale $\sim k^{-1}$ is given by (Problem 10.10)

$$\Pi(k) \sim \alpha \kappa \rho \omega_k^2 b_k^2. \tag{10.144}$$

Using this expression in combination with established cascade spectra, one can determine the $k_{\text{cutoff}}(\varepsilon, \alpha)$ function [and thus $L(\varepsilon, \alpha)$] for various regimes.

Since most relations in this and the previous section are written as order-of-magnitude estimates, a quantitative comparison with experiments necessarily involves numerous coefficients of the order of unity. To reduce the number of fitting parameters for meaningful comparison, one demands that the spectra and the dissipative cutoff are continuous functions at the crossover scale. This leaves seven fitting parameters over the entire inertial range (see [1] for further details).

Let us discuss the qualitative form of $L(\alpha)$ in connection with the results of measurements of quasiclassical tangles in ^{4}He, depicted in Figure 10.6, where $\alpha_b, \alpha_c, \alpha_*$ are mutual friction coefficients at the crossover scales between the cascades $\lambda_b, \lambda_c,$ and λ_*, respectively. Regardless of microscopic details, the experimental data reveal highly nontrivial $\alpha \ll 1$ physics: The saturation toward the $\alpha \to 0$ regime occupies five (!) decades in α—the sharpest change taking place when α is as low as $\sim 10^{-3}$. That the data are consistent with the decay scenario of Section 10.4.4 can hardly be attributed to a large number of fitting parameters because the curve features a prominent *multiscale* behavior. Mathematically, all fitting parameters reduce to order-of-unity dimensionless numbers. The hierarchy of scales is due solely to different powers of the large parameter Λ; these powers are predicted by the theory and not controlled by the fitting procedure.

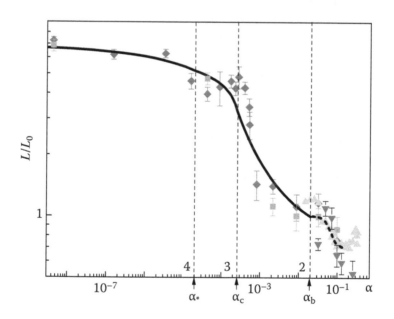

Figure 10.6   The vortex-line density as a function of the mutual friction coefficient α. (Adapted from Reference [5].) The low-temperature experimental data (squares and diamonds) are from Walmsley et al. [6]. The high-temperature measurements of Stalp et al. [7] and Chagovets et al. [8] are represented by triangles and inverted triangles, respectively. The form of $\alpha(T)$, necessary to convert $L(T)$ into $L(\alpha)$, is taken according to Samuels and Donnelly [9] at $T \gtrsim 0.5$K (roton scattering) and according to Equation (10.117) at $T \lesssim 0.5$K (phonon scattering). The solid line is a theoretical fit.

As α decreases from $\alpha \sim 1$ to values significantly smaller than unity, the line density L increases only by a factor close to unity (~ 1.5 in the experiment), which reflects the formation of the regime (1) driven by reconnections of the vortex bundles. During the crossover from (1) to (2), the increase in L is minimal, leading to a *shoulder* on the $L(\alpha)$ curve. Only well inside region (2), the increase in L becomes pronounced, and, at the crossover to (3), the $L(\alpha)$ function achieves its maximal slope, determined by the development of fractal structures on the vortex line necessary to support the cascade of self-reconnections. As the cutoff moves along the interval (3) toward higher wavenumbers, the slope of $L(\alpha)$ becomes less steep, due to the decrease of the characteristic amplitude of KW turbulence. When the cutoff finally enters the regime (4), the $L(\alpha)$ curve gradually levels off.

At $T = 0$ the only dissipative mechanism in vortex tangles is the emission of sound (with the possible exception of ^{3}He-B, where the bound states inside the vortex cores can play some role), and thus $\lambda_{\text{cutoff}} = \lambda_{\text{ph}}$. In both nonstructured and classical tangles, the sound is radiated at very large wavenumbers where turbulence is decaying according to the purely nonlinear Kelvin-wave cascade with the spectrum given by Equation (10.66). In nonstructured tangles, the energy flux is given by expression $\theta_{\text{ns}} \sim \kappa^3 \rho \Lambda^2 / R^2$, which (using Equation (10.102) for the power of sound radiation) leads to

$$\lambda_{\text{ph}} \sim \Lambda^{24/31} \left[\kappa / c\,R\right]^{25/31} R \qquad \text{(nonstructured tangles).} \qquad (10.145)$$

For classical-type tangles $\theta_{\text{c}} \sim \varepsilon \rho R^2$, or, with Equation (10.138) taken into account, $\theta_{\text{c}} \sim \kappa^3 \rho \Lambda / R^2 \sim \theta_{\text{ns}} / \Lambda$. Thus, for the cascade cutoff we obtain

$$\lambda_{\text{ph}} \sim \Lambda^{27/31} \left[\kappa / c\,R\right]^{25/31} R \qquad \text{(polarized tangles).} \qquad (10.146)$$

References

1. E. V. Kozik and B. V. Svistunov, Theory of decay of superfluid turbulence in the low-temperature limit, *J. Low. Temp. Phys.* **156**, 215 (2009).

2. R. Betchov, On the curvature and torsion of an isolated vortex filament, *J. Fluid Mech.* **22**, 471 (1965).

3. H. Hasimoto, A soliton on a vortex filament, *J. Fluid Mech.* **51**, 477 (1972).

4. T. F. Buttke, A numerical study of superfluid surbulence in the self-induction approximation, *J. Comput. Phys.* **76**, 301 (1988).

5. E. Kozik and B. Svistunov, Scanning superfluid-turbulence cascade by its low-temperature cutoff, *Phys. Rev. Lett.* **100**, 195302 (2008).

6. P. M. Walmsley, A. I. Golov, H. E. Hall, A. A. Levchenko, and W. F. Vinen, Dissipation of quantum turbulence in the zero temperature limit, *Phys. Rev. Lett.* **99**, 265302 (2007).

7. S. R. Stalp, J. J. Niemela, W. F. Vinen, and R. J. Donnelly, Dissipation of grid turbulence in helium II, *Phys. Fluids* **14**, 1377 (2002).

8. T. V. Chagovets, A. V. Gordeev, and L. Skrbek, Effective kinematic viscosity of turbulent He II, *Phys. Rev. E* **76**, 027301 (2007).

9. D. C. Samuels and R. J. Donnelly, Dynamics of the interactions of rotons with quantized vortices in helium II, *Phys. Rev. Lett.* **65**, 187 (1990).

Part IV

Weakly Interacting Gases

Weakly interacting Bose gas (WIBG) has been playing a very special role in the theory of superfluidity as a system allowing first-principle analytic description. Bogoliubov's seminal work of 1947, addressing the ground-state properties of WIBG at the mean-field level, and Beliaev's pioneering work of 1958, introducing systematic the diagrammatic approach, rendered WIBG a textbook superfluid. The experimental breakthrough by Cornell, Wieman, and Ketterle, who attained (in 1995) experimentally BEC in WIBG of alkaline atoms, gave rise to an explosive growth of the physics of ultracold atomic systems.

   The present text is likely to be the first book on the principles of superfluidity that does not begin with the theory of WIBG. Having said that, we make a caveat that the previous statement is not entirely correct. The classical matter field introduced axiomatically in Part I corresponds to the matter-field regime of WIBG. The trick of introducing the classical field axiomatically—rather than tracing its origin microscopically—allowed us to circumvent a significant amount of technical aspects, including knowledge of the basics of quantum mechanics (which is not directly relevant to the phenomenon of superfluidity). Indeed, it will not be an exaggeration to say that all the "quantumness" of WIBG that is relevant to superfluidity reduces to its having a component—associated with single-particle modes with large occupation numbers—behaving as a classical complex-valued field of Part I. Nevertheless, microscopic derivation of this fact as well as systematic treatment of essentially quantum-statistical aspects of the system require an elaborate theory.

Apart from the purely technical challenges of *ab initio* description of weakly interacting bosonic systems—especially in addressing more subtle issues such as control of systematic errors, subleading corrections to the mean-field results, adequate treatment of low-dimensional systems in the absence of genuine long-range order, or overcoming problems of infrared divergence—there is yet another aspect that justifies deferring the analysis of WIBG until the moment when a general theory of superfluidity is fully developed: namely, a pronounced specificity of the system. Certain properties of WIBG—such as closeness of the values of the superfluid and condensate densities away from the fluctuation region; closeness of the condensate and total densities in the low-temperature limit; large, as compared to the interparticle distance, healing length; dramatic dependence of the linear part of the quasiparticle spectrum on temperature (at a fixed density); superfluid phase transition being mostly driven by statistics of an ideal system; and extremely specific nonequilibrium kinetics of Bose–Einstein condensation—might prove rather misleading if WIBG were literally taken as a prototypical superfluid. Rather, these properties, distinguishing WIBG from strongly interacting superfluids, are fundamentally interesting on their own and are very important in the context of ultracold atomic systems.

The Hamiltonian of WIBG is a generic Hamiltonian for single-component non-relativistic bosons of mass m, interacting via pair potential, $U(\mathbf{r}_1 - \mathbf{r}_2)$. In terms of the field-operator $\hat{\psi} \equiv \hat{\psi}(\mathbf{r})$ (we use "hats" to avoid confusion with classical fields), the Hamiltonian reads ($\hbar = 1$)

$$H = H_0 + H_{\text{int}} + H_1, \tag{P.1}$$

where

$$H_0 = -\frac{1}{2m} \int \hat{\psi}^\dagger \Delta \hat{\psi} \, d\mathbf{r}, \tag{P.2}$$

$$H_{\text{int}} = \frac{1}{2} \int \hat{\psi}(\mathbf{r}_2)^\dagger \, \hat{\psi}(\mathbf{r}_1)^\dagger \, U(\mathbf{r}_1 - \mathbf{r}_2) \, \hat{\psi}(\mathbf{r}_1) \hat{\psi}(\mathbf{r}_2) \, d\mathbf{r}_1 d\mathbf{r}_2, \tag{P.3}$$

$$H_1 = \int \hat{\psi}^\dagger [V(\mathbf{r}) - \mu] \hat{\psi} \, d\mathbf{r}. \tag{P.4}$$

For generality, and especially keeping in mind realistic systems of ultracold atoms that are normally trapped, we introduce the term H_1 with the external potential $V(\mathbf{r})$; this term also contains the chemical potential μ, since we will be working with the grand canonical ensemble. By definition, the pair interaction described by the term (P.3) is supposed to be weak. It is important, however, that, in 2D and 3D, the weakness of interaction does not necessarily require weakness of the pair potential. A potential is weak—and is referred to as *Born potential*, when its characteristic amplitude U_{char} satisfies the condition $U_{\text{char}} \ll 1/mR_0^2$, where R_0 is the typical size of the potential. In 3D and 2D, in addition to the obvious Born case, the interaction is *effectively* weak in the dilute limit, when both the size of the potential, R_0, and the s-scattering length, a, are much smaller than the interparticle distance. The s-scattering length of a 3D (2D) potential is equal to the radius of the hard-sphere (hard-disk) potential, yielding the same s-scattering amplitude (with the negative sign, if the amplitude is negative). Normally, $|a| \sim R_0$, but there are special cases when $|a| \ll R_0$, or $|a| \gg R_0$. In the dilute regime, the leading interaction-induced terms in all the answers for WIBG are universal in the sense that they can be expressed in such a form that the only parameter characterizing the strength of interaction is a, the quantitative accuracy of the theory controlled by the parameter $an^{1/d} \ll 1$. With diagrammatic technique, accurate derivation of this fact requires summation of an infinite (ladder) diagrammatic series, which leads to certain subtleties (avoiding double counting in higher-order diagrams, freedom of trading higher-order corrections between effective interaction and other entities, etc.).

In terms of the final results, we confine ourselves with the most typical case of the dilute gas, when the size of the potential R_0 is much smaller than the distance between the particles:

$$R_0 \ll n^{-1/d}. \tag{P.5}$$

Theoretically, the case of WIBG at $R_0 \gtrsim n^{-1/d}$ is quite simple, because the weakness of interaction here literally implies a Born-type potential. However, it is only with the condition (P.5) that the final answers become insensitive to the details of the interaction potential. The second condition, which is assumed in our final results, is

$$\lambda_T \gg R_0, \tag{P.6}$$

where λ_T is the de Broglie wavelength. Given the inequality (P.5), the condition (P.6) is satisfied *automatically* as long as one is interested in essentially quantum properties of the system, implying $\lambda_T^d n \gtrsim 1$. An extension of the theory to the Boltzmann high-temperature regime $\lambda_T \lesssim R_0$, where answers become sensitive to the particular form of the interaction potential, is of no interest in the context of the theory of superfluidity, since the system here is deeply normal.

In Chapter 12, we develop an accurate microscopic theory of equilibrium properties of WIBG in one, two, and three dimensions. The accuracy of the theory of Chapter 12 comes at a price, however. The treatment relies on the formalism of diagrammatic technique (the basics of which are reviewed in Chapter 11). For those readers who are interested in the final results only, we summarize the main results in the beginning of the chapter. At the end of Chapter 12, we present BCS theory for weakly interacting Fermi gas.

In Chapter 13, we consider strongly nonequilibrium kinetics of BEC formation in WIBG. The essential part of the BEC-formation process deals with the modes with large occupation numbers and thus is described by the classical-field component of the quantum field (with the effective interaction already discussed in Chapter 12). For that reason, Chapter 13 is technically simpler than Chapter 12.

CHAPTER 11

Green's Functions and Feynman's Diagrams

In this chapter, we will briefly review the key notions and crucial steps in the derivation of Feynman's diagrammatic technique. Our description is compact—a substantial portion of theory is presented as problems—but self-contained. It will be used next to address properties of weakly interacting gases.

11.1 Matsubara Representation

In equilibrium field-theoretical formalism, the central role is played by field operators in Matsubara representation:

$$\hat{\Psi}_\alpha(\tau,\mathbf{r}) = e^{\tau H}\hat{\psi}_\alpha(\mathbf{r})e^{-\tau H}, \qquad \hat{\bar{\Psi}}_\alpha(\tau,\mathbf{r}) = e^{\tau H}\hat{\psi}_\alpha^\dagger(\mathbf{r})e^{-\tau H}, \tag{11.1}$$

where $\hat{\psi}_\alpha$ is the annihilation field operator, α is the spin subscript (if required), H is the grand canonical Hamiltonian, and $\tau \in [-\beta,\beta]$ is the imaginary-time variable. Note that $\hat{\bar{\Psi}}_\alpha$ is not a Hermitian conjugate of $\hat{\Psi}_\alpha$. The quantities of interest are correlators of Matsubara operators, or Matsubara Green's functions, obtained by averaging products of $\hat{\bar{\Psi}}_\alpha$ and $\hat{\Psi}_\alpha$ over the Gibbs distribution. Matsubara Green's functions give access to any physical property of interest in the equilibrium system, including the thermodynamic functions. A perturbative expansion for Green's function of an interacting system reduces to a series of integrals formed by single-particle Green's functions of the noninteracting system and interaction potentials.

11.1.1 Single-Particle Green's Functions

The single-particle Matsubara Green's function is defined as

$$G_{\alpha\beta}(\tau_1,\mathbf{r}_1;\tau_2,\mathbf{r}_2) = -\langle T_\tau \hat{\Psi}_\alpha(\tau_1,\mathbf{r}_1)\hat{\bar{\Psi}}_\beta(\tau_2,\mathbf{r}_2)\rangle, \tag{11.2}$$

where $\langle\ldots\rangle$ stands for averaging over the grand canonical Gibbs distribution:

$$\langle(\ldots)\rangle \equiv Z^{-1}\,\mathrm{Tr}\,e^{-\beta H}(\ldots), \qquad Z = \mathrm{Tr}\,e^{-\beta H}, \tag{11.3}$$

with the grand canonical partition function Z. The τ-*ordering* symbol T_τ requires that all imaginary-time-dependent operators be rearranged in the order of increasing imaginary times, from right to left. In the case of fermions, there is also a sign rule, $(-1)^P$, stating that the global sign changes depending on the parity P of the permutation that leads to the ordered sequence of field operators. The idea behind the time ordering in Equation (11.2) is purely technical: It turns out that

387

by combining $\langle \hat{\Psi}_\alpha(\tau_1, \mathbf{r}_1) \hat{\Psi}_\beta(\tau_2, \mathbf{r}_2) \rangle$ with $\langle \hat{\Psi}_\beta(\tau_2, \mathbf{r}_2) \hat{\Psi}_\alpha(\tau_1, \mathbf{r}_1) \rangle$ into a single mathematical function, one gets an object for which there exists a powerful computational tool.

As can be explicitly checked, $G_{\alpha\beta}(\tau_1, \mathbf{r}_1; \tau_2, \mathbf{r}_2) \equiv G_{\alpha\beta}(\tau, \mathbf{r}_1, \mathbf{r}_2)$, where $\tau = \tau_1 - \tau_2$, and [in what follows, the upper (lower) sign is for fermions (bosons)]

$$G_{\alpha\beta}(\tau, \mathbf{r}_1, \mathbf{r}_2) = \begin{cases} \pm Z^{-1} \, \mathrm{Tr} \, e^{-(\tau+\beta)H} \, \hat{\psi}_\beta^\dagger(\mathbf{r}_2) e^{\tau H} \, \hat{\psi}_\alpha(\mathbf{r}_1), & \tau < 0, \\ -Z^{-1} \, \mathrm{Tr} \, e^{-\tau H} \, \hat{\psi}_\beta^\dagger(\mathbf{r}_2) e^{(\tau-\beta)H} \, \hat{\psi}_\alpha(\mathbf{r}_1), & \tau > 0. \end{cases} \tag{11.4}$$

From this expression, it is readily seen that Green's function is antiperiodic (periodic) on the imaginary-time interval:

$$G(-\tau) = \mp G(\beta - \tau), \qquad \tau > 0. \tag{11.5}$$

Problem 11.1 *Derive Equations (11.4) and (11.5).*

Green's function directly relates to the system density:

$$n(\mathbf{r}) = \pm \sum_\alpha G_{\alpha\alpha}(\tau = -0, \mathbf{r}, \mathbf{r}). \tag{11.6}$$

This expression leads to the standard routine of extracting other thermodynamic functions from $n(\mu, T)$ (for a uniform system) within the diagrammatic technique. Pressure is obtained from the general thermodynamic relation

$$p(\mu, T) = \int_{-\infty}^{\mu} n(\mu', T) \, d\mu', \tag{11.7}$$

while the rest of thermodynamic properties can be derived from the grand canonical potential $\Omega(V, \mu, T) = -V p(\mu, T)$.

In view of β-antiperiodicity (periodicity) of Green's function, it can be expanded into Fourier series ($m = 0, \pm 1, \pm 2, \dots$):

$$G(\tau) = T \sum_{m=-\infty}^{\infty} e^{-i\xi_m \tau} G(\xi_m), \qquad G(\xi_m) = \int_0^\beta e^{i\xi_m \tau} G(\tau) \, d\tau, \tag{11.8}$$

with half-integer (integer) harmonic frequencies

$$\xi_m = \begin{cases} 2m\pi T, & \text{bosons}, \\ (2m+1)\pi T, & \text{fermions}. \end{cases} \tag{11.9}$$

In a spatially uniform system, the coordinate dependence of G reduces to $\mathbf{r} = \mathbf{r}_1 - \mathbf{r}_2$. In this case, Fourier integral representation,

$$G(\mathbf{r}) = \int e^{i\mathbf{k}\cdot\mathbf{r}} G(\mathbf{k}) \frac{d^d k}{(2\pi)^d}, \qquad G(\mathbf{k}) = \int e^{-i\mathbf{k}\cdot\mathbf{r}} G(\mathbf{r}) \, d^d r, \tag{11.10}$$

proves convenient. In particular, Green's function of a uniform ideal normal (i.e., without symmetry breaking) gas takes the form

$$G_{\alpha\beta}^{(0)}(\xi_m, \mathbf{k}) = \frac{\delta_{\alpha\beta}}{i\xi_m - \epsilon(k) + \mu}, \tag{11.11}$$

with the energy dispersion relation $\epsilon(k) = k^2/2m_*$, where m_* is the particle mass.

Problem 11.2 *Derive Equation (11.11).*

11.1.2 Many-Particle Green's Functions: The Generating Functional

The single-particle Green's function (11.2), also referred to as *two-point* correlator, can be generalized to an s-point correlator:

$$K_s(X_1, X_2, \ldots, X_s) = \langle T_\tau \check{\Psi}_1(X_1) \check{\Psi}_2(X_2) \cdots \check{\Psi}_s(X_s) \rangle, \tag{11.12}$$

where $X_j \equiv (\tau_j, \mathbf{r}_j, \alpha_j)$, $j = 1, 2, \ldots, s$, and $\check{\Psi}$ is a generic notation for both $\hat{\Psi}$ and $\hat{\bar{\Psi}}$. As previously discussed, the sign rule $(-1)^P$ is imposed for fermions. As a powerful technical tool, we construct a *generating functional* $Q[\lambda, \lambda^*]$, based on an auxiliary complex-valued field $\lambda(X)$ such that all s-point correlators for bosons (11.12) are expressed as variational derivatives of $Q[\lambda, \lambda^*]$ at $\lambda(X) \equiv 0$:

$$K_s(X_1, X_2, \ldots, X_s) = \frac{\delta^s Q[\lambda, \lambda^*]}{\delta \check{\lambda}_1(X_1) \delta \check{\lambda}_2(X_2) \ldots \delta \check{\lambda}_s(X_s)}\bigg|_{\lambda \equiv 0}, \tag{11.13}$$

where

$$\check{\lambda}_j(X_j) = \begin{cases} \lambda^*(X_j), & \text{if } \check{\Psi}_j(X_j) = \hat{\Psi}(X_j), \\ \lambda(X_j), & \text{if } \check{\Psi}_j(X_j) = \hat{\bar{\Psi}}(X_j). \end{cases} \tag{11.14}$$

We will look at the bosonic case first. [When dealing with fermions, one must introduce a generating functional in terms of the Grassmann field $\lambda(X)$ in combination with the $(-1)^P$ sign rule for imaginary-time ordering.]

An explicit expression for $Q[\lambda, \lambda^*]$ is based on the notion of τ-exponential, which we introduce now. Let $\tilde{H}_\tau$ be a certain Hamiltonian-type operator that, in contrast to the genuine Hamiltonian, depends on the variable τ as a parameter. Given $\tilde{H}_\tau$, the corresponding τ-exponential is defined by the formula ($\tau_b \geq \tau_a$ is implied)

$$T_\tau e^{-\int_{\tau_a}^{\tau_b} \tilde{H}_\tau d\tau} = 1 + \sum_{s=1}^{\infty} \frac{(-1)^s}{s!} \int_{\tau_a}^{\tau_b} d\tau_1 \int_{\tau_a}^{\tau_b} d\tau_2 \cdots \int_{\tau_a}^{\tau_b} d\tau_s \, T_\tau \tilde{H}_{\tau_1} \tilde{H}_{\tau_2} \cdots \tilde{H}_{\tau_s}. \tag{11.15}$$

Two other definitions are equivalent to (11.15):

$$T_\tau e^{-\int_{\tau_a}^{\tau_b} \tilde{H}_\tau d\tau} = 1 + \sum_{s=1}^{\infty} (-1)^s \int_{\tau_a}^{\tau_b} d\tau_1 \int_{\tau_a}^{\tau_1} d\tau_2 \int_{\tau_a}^{\tau_2} d\tau_3 \cdots \int_{\tau_a}^{\tau_{(s-1)}} d\tau_s \, \tilde{H}_{\tau_1} \cdots \tilde{H}_{\tau_s}, \qquad (11.16)$$

and

$$T_\tau e^{-\int_{\tau_a}^{\tau_b} \tilde{H}_\tau d\tau} = \lim_{n \to \infty} (1 - \varepsilon \tilde{H}_{\tau_n}) \cdots (1 - \varepsilon \tilde{H}_{\tau_2})(1 - \varepsilon \tilde{H}_{\tau_1}), \qquad (11.17)$$

where

$$\varepsilon = (\tau_b - \tau_a)/n, \quad \tau_j = \varepsilon j, \quad j = 1, 2, 3, \ldots, n. \qquad (11.18)$$

Problem 11.3 *Prove the equivalence of all the three definitions of the τ-exponential, (11.15), (11.16), and (11.17). Hint: A good starting point is to observe that, with τ-ordering in place, the noncommutativity of operators is no longer an issue.*

Problem 11.4 *Prove that for $\tau_c \geq \tau_b \geq \tau_a$,*

$$\left[T_\tau e^{-\int_{\tau_b}^{\tau_c} \tilde{H}_\tau d\tau} \right] \left[T_\tau e^{-\int_{\tau_a}^{\tau_b} \tilde{H}_\tau d\tau} \right] = T_\tau e^{-\int_{\tau_a}^{\tau_c} \tilde{H}_\tau d\tau}. \qquad (11.19)$$

Problem 11.5 *Suppose that $\tilde{H}_\tau$ is τ-independent: $\tilde{H}_\tau \equiv \tilde{H}$. Show that in this case*

$$T_\tau e^{-\int_{\tau_a}^{\tau_b} \tilde{H}_\tau d\tau} = e^{(\tau_a - \tau_b)\tilde{H}}. \qquad (11.20)$$

The expression for the generating functional reads (from now on we omit spin subscripts for clarity)

$$Q[\lambda, \lambda^*] = \frac{\text{Tr}\left[T_\tau e^{-\int_0^\beta \tilde{H}_\tau d\tau} \right]}{\text{Tr}\, e^{-\beta H}}, \qquad \tilde{H}_\tau = H - \int [\lambda^*(\tau, \mathbf{r}) \hat{\psi}(\mathbf{r}) + \text{H.c.}] d^d r. \qquad (11.21)$$

Problem 11.6 *Prove that $Q[\lambda, \lambda^*]$, defined by Equation (11.21), features the required property (11.13) and (11.14). Hint: The representation (11.17) of the τ-exponential is the most relevant for the proof.*

Expression (11.21) plays a central role in constructing the diagrammatic technique within the functional-integral formalism used to represent traces in the numerator and denominator of (11.21).

11.2 Diagrammatic Technique for Normal Bosonic Systems

In this section, we develop Matsubara diagrammatic technique for normal (i.e., without condensate) bosonic systems. In essence, the technique is a perturbative expansion in powers of weak interaction potential on top of equilibrium statistics of ideal, normal Bose gas. We also introduce Dyson summation as a powerful nonperturbative tool. The extension of the technique to the case of a superfluid system, as well as the case of a dilute gas with strong interaction potential—being accomplished by Dyson-type summation—will be developed in the next chapter.

11.2.1 Functional Integral Representation

The traces in the numerator and denominator of Equation (11.21) can be expressed as functional integrals over $(d + 1)$-dimensional complex-valued classical field, thereby mapping the original d-dimensional quantum-field problem onto a certain $(d + 1)$-dimensional *pseudo*-classical counterpart. The prefix "pseudo" is to warn that the action of the resulting classical problem turns out to be complex valued, so that corresponding expressions should not be literally associated with usual classical statistics. The mapping itself is identical to what we did using coherent states in Section 7.5 to arrive at Equations (7.72) and (7.80). This leads to the following expression for the generating functional:

$$Q[\lambda, \lambda^*] = \frac{\int e^{\tilde{S}[\psi]} \mathcal{D}\psi}{\int e^{S[\psi]} \mathcal{D}\psi}, \tag{11.22}$$

with

$$S[\psi] = \int_0^\beta d\tau \left\{ \int \psi \frac{\partial \psi^*}{\partial \tau} d^d r - H[\psi^*, \psi] \right\}, \tag{11.23}$$

$$\tilde{S}[\psi] = S[\psi] + \int_0^\beta d\tau \int d^d r \, (\lambda^* \psi + \lambda \psi^*). \tag{11.24}$$

Problem 11.7 *Show that Equations (11.22) through (11.24) imply*

$$\langle T_\tau \hat{\Psi}(X_1) \cdots \hat{\Psi}(X_n) \hat{\Psi}^\dagger(X_{n+1}) \cdots \hat{\Psi}^\dagger(X_{n+m}) \rangle = \langle \psi(X_1) \cdots \psi(X_n) \psi^*(X_{n+1}) \cdots \psi^*(X_{n+m}) \rangle,$$

$$\tag{11.25}$$

where, by definition,

$$\langle \psi(X_1) \cdots \psi(X_n) \psi^*(X_{n+1}) \cdots \psi^*(X_{n+m}) \rangle$$

$$= \int \psi(X_1) \cdots \psi(X_n) \psi^*(X_{n+1}) \cdots \psi^*(X_{n+m}) e^{S[\psi]} \mathcal{D}\psi / \int e^{S[\psi]} \mathcal{D}\psi. \tag{11.26}$$

Problem 11.8 *On the basis of Equations (11.25) and (11.26), derive the relation*

$$\langle \psi^*(\mathbf{r}_1, \tau + 0) \psi(\mathbf{r}_2, \tau) - \psi^*(\mathbf{r}_1, \tau) \psi(\mathbf{r}_2, \tau + 0) \rangle = -\delta(\mathbf{r}_2 - \mathbf{r}_1), \tag{11.27}$$

explicitly revealing that ψ is a discontinuous function of τ.

11.2.2 Green's Functions for an Ideal Normal Bose Gas: Wick's Theorem

In the case of an ideal (i.e., noninteracting) normal (i.e., without condensate) Bose gas—no matter whether uniform or with an external potential—the generating functional has a very simple structure. Specifically,

$$Q^{(0)}[\lambda, \lambda^*] = e^{-\int \lambda^*(X_1) G^{(0)}(X_1, X_2) \lambda(X_2) \, dX_1 dX_2}, \tag{11.28}$$

where the superscripts remind that we are dealing with a noninteracting system and $G^{(0)}$ is—and cannot be anything else, by the definition of the generating functional—the single-particle Green's function: $G^{(0)}(X_1, X_2) = -\langle \psi(X_1)\psi^*(X_2)\rangle$. [For definiteness, here and in what follows, we write correlators in terms of classical fields rather than τ-ordered Matsubara operators, utilizing the fact that the two representations are identical in view of Equation (11.25).]

The derivation of Equation (11.28) will be given in the following. Meanwhile, let us formulate the most important property of (11.28), known as Wick's theorem. The theorem states that the correlator (11.26) with $m = n$ is equal to the sum of all possible products of n two-point correlators $\langle \psi(X_a)\psi^*(X_b)\rangle$ constructed by "pairing" each of the fields $\psi(X_a)$, $a = 1, 2, \ldots, n$ with one of the fields $\psi^*(X_{n+b})$, $b = 1, 2, \ldots, n$. All correlators with $m \neq n$ are identically zero by U(1) symmetry of the problem.

Problem 11.9 *Show that Equation (11.28) leads to Wick's theorem.*

The derivation of Equation (11.28) is based on the Gaussian (bilinear in the fields) form of the action

$$S[\psi] = \int dX \, \psi^* \hat{A} \psi, \tag{11.29}$$

where $\hat{A}$ is a certain operator in the space of (β-periodic in τ) functions $\psi(X)$. Technically, the problem simplifies dramatically for Hermitian operator $\hat{A}$. Formally, this is not the case in view of the *anti*-Hermitian first term in the action (7.76). However, we can use the trick of analytical continuation by writing

$$\hat{A} = i\alpha \frac{\partial}{\partial \tau} - \hat{H}^{(0)}, \qquad \alpha = i, \tag{11.30}$$

and noticing that, for real α, the operator is Hermitian. Hence, we can solve the problem for real α and then set $\alpha = i$ in the final answer.

The crucial property of bilinear action (11.29) with Hermitian $\hat{A}$ is the existence of a special field $\phi_0(X)$ such that for *any* field $\psi(X)$, the functional $\tilde{S}[\psi + \phi_0]$ decouples into a sum of two independent terms, of which one is $S[\psi]$ and the other is a certain functional of ϕ_0:

$$\tilde{S}[\psi + \phi_0] = S[\psi] + S_*[\phi_0]. \tag{11.31}$$

It is straightforward to see that the desired field $\phi_0(X)$ is the solution of the equation

$$\hat{A}\phi_0 = -\lambda, \tag{11.32}$$

and, correspondingly,

$$S_*[\phi_0] = -\int dX \, \lambda^* \hat{A}^{-1} \lambda. \tag{11.33}$$

Problem 11.10 *Derive Equations (11.31) through (11.33) using standard tools of the theory of inner-product vector spaces.*

The property (11.31) allows one to use the trick of shifting the field ψ in the numerator of (11.22) by fixed (for a given λ) function ϕ_0:

$$\psi \to \psi + \phi_0, \qquad \mathcal{D}(\psi + \phi_0) = \mathcal{D}\psi. \tag{11.34}$$

The trick works as follows:

$$Q^{(0)}[\lambda, \lambda^*] = \frac{\int e^{\tilde{S}[\psi + \phi_0]}\,\mathcal{D}\psi}{\int e^{S[\psi]}\,\mathcal{D}\psi} = e^{S_*[\phi_0]}\frac{\int e^{S[\psi]}\,\mathcal{D}\psi}{\int e^{S[\psi]}\,\mathcal{D}\psi} = e^{S_*[\phi_0]}. \tag{11.35}$$

With Equation (11.33) taken into account, Equation (11.35) reproduces the structure of Equation (11.28). This completes the proof, since, as mentioned earlier, the kernel $G^{(0)}(X_1, X_2)$ in the integral in (11.28) has to be the single-particle Green's function, by definition of the generating functional, Equation (11.13). The fact that Green's function is the kernel of the resolvent of $\hat{A}$ (with $\alpha = i$ for the actual Green's function),

$$\left(\frac{\partial}{\partial \tau_1} + \hat{H}^{(0)}_{\mathbf{r}_1}\right)G^{(0)}(\tau_1 - \tau_2, \mathbf{r}_1, \mathbf{r}_2) = -\delta(\mathbf{r}_1 - \mathbf{r}_2)\delta(\tau_1 - \tau_2), \qquad G^{(0)}(\tau) = G^{(0)}(\tau + \beta), \tag{11.36}$$

is an instructive *by-product* of the proof. Equation (11.36) explains the origin of the term "Green's function" used in the context of two-point correlators rather than resolvents of linear operators.

Problem 11.11 *For a uniform ideal normal Bose gas, derive the result (11.11) from (11.36).*

A comment is in order here: In our derivation of Equation (11.28), we did not explicitly use the requirement that the gas be normal, so it might seem that the requirement is not relevant. The situation is more subtle than that. For the non-interacting system, the grand canonical ensemble, crucial for introducing the formalism of functional integration, becomes pathological below the Bose–Einstein condensation temperature. The physical reason for that is infinite compressibility of the noninteracting condensate leading to divergence of the total particle number at fixed chemical potential and temperature below the critical one.

11.2.3 Perturbative Expansion

The functional integral representation of Green's functions, Equations (11.25) and (11.26), together with Wick's theorem, forms the basis for constructing perturbative expansion. One starts with decomposing the Hamiltonian of the system into two parts:

$$H = H_0 + H_{\text{pert}}, \tag{11.37}$$

where the nonperturbed part H_0 is bilinear in the field operators and the perturbation H_{pert} is arbitrary (it may contain some bilinear terms). Correspondingly, the action splits into two parts:

$$S = S_0 + S_{pert} = S_0 - \int_0^\beta d\tau\, H_{pert}[\psi^*(\tau+0), \psi(\tau)]. \tag{11.38}$$

The next step is to expand the exponential $\exp\{S_{pert}\}$ [in both the numerator and denominator of Equation (11.26)] in Taylor series. If H_{pert} is a sum of different functionals, then $\exp\{S_{pert}\}$ factors into a product of elementary exponentials corresponding to different H_{pert} terms (we remind in passing that we are dealing with complex functions, not operators, so that all terms commute), and each elementary exponential is expanded individually. In practice, various terms in H_{pert} are polynomials of $\psi(\tau)$ and $\psi^*(\tau+0)$, taken at different spatial points and convoluted with interaction potential kernels. In what follows, we confine ourselves to these types of perturbations. Moreover, in this section, we explicitly discuss only the case of pairwise interaction (most relevant for our purposes):

$$H_{pert} = \frac{1}{2} \int \mathcal{U}(\mathbf{r}_1 - \mathbf{r}_2)|\psi(\mathbf{r}_1)|^2|\psi(\mathbf{r}_2)|^2 d\mathbf{r}_1 d\mathbf{r}_2. \tag{11.39}$$

For briefness, we omit here the variable τ and use the convention $|\psi(\tau)|^2 \equiv \psi^*(\tau+0)\psi(\tau)$. One may "visualize" this term graphically with a *diagrammatic element* (see Figure 11.1) that plays the role of a building block for Feynman diagrams.

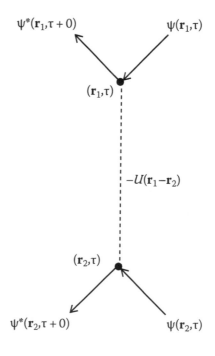

Figure 11.1   Diagrammatic element originating from pair interaction.

The interaction potential (with the minus sign, because H_{pert} enters S_{pert} with the negative sign) is represented by a dashed line connecting points 1 and 2 with spacetime coordinates $(\mathbf{r}_1, \tau)$ and $(\mathbf{r}_2, \tau)$, respectively. With each of the two ends of the dashed line, we associate one incoming and one outgoing arrow, representing (respectively) the fields ψ and ψ^* at the corresponding spacetime point. The factor of $1/2$ is *not* part of the element description. The integrations over coordinates and times are assumed to be done *after* the functional integration over the fields (i.e., with respect to the functional integration, the coordinates and times of the element are fixed).

After expanding the exponential $\exp\{S_{\text{pert}}\}$ in the numerator and denominator of (11.26), the functional integration over field ψ is done by Wick's theorem. Indeed, each term in the expansion has the form of a polynomial function, $P(\psi, \psi^*)$, integrated over $\mathcal{D}\psi$ with the weighing function $e^{S_0[\psi]}$. We note that

$$\int P\, e^{S_0[\psi]}\, \mathcal{D}\psi = \mathcal{Z}_0\, \langle P \rangle_0, \qquad \mathcal{Z}_0 = \int e^{S_0[\psi]}\, \mathcal{D}\psi, \tag{11.40}$$

where $\langle \ldots \rangle_0$ stands for averaging with action S_0. The normalization factor $\mathcal{Z}_0$ (the partition function of the noninteracting system) drops out from the final answer because it cancels between the numerator and denominator in (11.26). The average $\langle P \rangle_0$ is evaluated by Wick's theorem, with the result being conveniently represented *graphically* by a set of Feynman diagrams.

First, we introduce Feynman diagrams for the denominator of (11.26), which is the partition function of the system (if we keep $\mathcal{Z}_0$). In this case, the order-s polynomial P corresponds to the s-th term of the Taylor expansion of $\exp\{S_{\text{pert}}\}$. It is a product of s identical diagrammatic elements (with different spacetime coordinates) and the global factor of $1/(2^s s!)$. Clearly, the 0-order term is just unity. The $s = 1$ term has one element. Its evaluation requires finding the average of the four-field product. By Wick's theorem, the result is the sum of two products of two Green's functions:

$$\langle \psi^*(\mathbf{r}_1, \tau + 0)\psi(\mathbf{r}_1, \tau)\psi^*(\mathbf{r}_2, \tau + 0)\psi(\mathbf{r}_2, \tau) \rangle_0 = (-G_{11}^{(0)})(-G_{22}^{(0)}) + (-G_{12}^{(0)})(-G_{21}^{(0)}). \tag{11.41}$$

[Here, we use a shorthand notation $G_{ij} \equiv G(X_i, X_j)$.] The result must be multiplied by $-\mathcal{U}(\mathbf{r}_1 - \mathbf{r}_2)$ and integrated over $\mathbf{r}_1$, $\mathbf{r}_2$, and τ. The entire procedure is conveniently represented graphically (see Figure 11.2). One starts with the diagrammatic element (Figure 11.1) and observes that each term in the sum of products generated by Wick's theorem can be associated with a particular way of pairing incoming and outgoing ends of the arrows into directed lines. The directed line itself is associated with Green's function (with minus sign) and is referred to as a *propagator*. In other words, a propagator going from vertex j to vertex i is ascribed a factor $-G_{ij}^{(0)}$.

Clearly, the aforementioned rule of constructing diagrams out of elements by pairing incoming and outgoing arrows into directed lines applies to any order of expansion. As an example, in Figure 11.3, we present some second-order

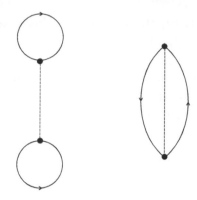

Figure 11.2   First-order diagrams for the partition function.

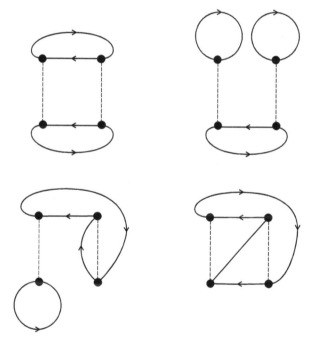

Figure 11.3   Connected second-order diagrams for the partition function.

terms. The diagrams in this figure are the so-called *connected* diagrams, meaning that each element in the graph is connected to the rest of the elements by a
chain of either dashed or solid lines. There are also *disconnected* diagrams in the
partition function expansion. For example, disconnected second-order diagrams
correspond to the products of two first-order diagrams shown in Figure 11.2.
The notion of connected diagrams is crucial for the diagrammatic formalism
because disconnected diagrams factor into products of sums of connected ones.
We will see in the following that this factoring, in both numerator and denominator of (11.26), ultimately results in an elegant cancelation—and thus complete
elimination from the technique—of disconnected diagrams.

Another important observation is combinatorial symmetry of diagrams. In Figure 11.3, we deliberately omitted labels on the vertices and showed only topologically different diagrams. Generally speaking, a topologically identical—and thus having the same value upon integration—diagram can be obtained in more than one way, for example, by exchanging the interaction element labels and changing the assignment of incoming (or outgoing) ends of elements to form propagators, as long as the transformation corresponds to another way of implementing Wick's pairing. It is natural then to introduce a rule that one must count only diagrams with different topologies and mention separately the combinatorial symmetry factor ascribed to each diagram, along with the aforementioned factor $1/(2^s s!)$.

Problem 11.12 *Determine combinatorial factors for all the diagrams in Figures 11.2 and 11.3.*

Turning to the expansion of the numerator in (11.26), we see that apart from the pair-interaction elements (shown in Figure 11.1), we now also have a product of $2n$ fields, each taken at a different spacetime point. By the analogy with Figure 11.1, we introduce $2n$ *external* vertices with either incoming or outgoing arrows, representing ψ and ψ^*, respectively. The diagrammatic rules are then essentially the same as for the denominator. Wick's theorem translates into converting—in all possible ways—pairs of arrows into propagators. As an illustration, in Figure 11.4, we present connected zeroth- and first-order diagrams for the case of the two-point correlator.

Now we are in a position to prove a crucial statement concerning disconnected diagrams in the numerator of Equation (11.26). As in the case of the denominator, we adopt a rule that we classify diagrams by their topology and account for their combinatorial factors. As opposed to the denominator, each order-s diagram for the numerator has a specific combinatorial factor of

$$C_s^{s_0} = \frac{s!}{s_0!\,(s-s_0)!},\tag{11.42}$$

where $s_0 = 0, 1, 2, \ldots, s$ is the total number of interaction lines in the diagram connected to at least one external vertex. Correspondingly, $s_1 = s - s_0$ is the number

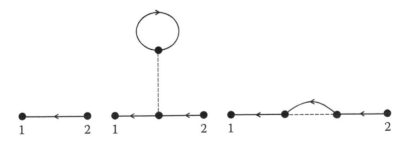

Figure 11.4 Connected zeroth- and first-order diagrams for the numerator of Equation (11.26) for $\langle \psi(X_1)\psi^*(X_2)\rangle \equiv -G_{12}$.

of interaction lines that remain disconnected from external vertices. The factor
(11.42) is the number of ways of selecting s_0 elements out of the set of s elements.
With the equality

$$\frac{1}{2^s s!} C_s^{s_0} = \frac{1}{2^{s_0} s_0!} \frac{1}{2^{s_1} s_1!} \tag{11.43}$$

taken into account, we observe that the sum of *all* the diagrams—compared to the
sum of diagrams with $s_1 = 0$—contains an extra global factor, which is nothing but
the diagrammatic expansion for the partition function.

Hence, the diagrammatic expansion for the $2n$-point correlator (11.26) reduces
to the sum of $s_1 = 0$ diagrams for the numerator, that is, the sum of diagrams that
do not have parts disconnected from the external vertices.

Most importantly, along with eliminating the denominator and dramatically
reducing the number of diagrams within a given order of expansion, the aforementioned result leads to the radical simplification of the structure of combinatorial prefactors. Since each interaction line is now connected to some piece of the
diagram, a permutation of labels of interaction elements, as well as an exchange in
the assignment of incoming (or outgoing) arrows for each element to form propagators, yields a new pattern of Wick's pairing. Thus, for each diagram of order s,
irrespective of its topology, the combinatorial factor is always $(2^s s!)$ and cancels
exactly with the global factor $1/(2^s s!)$. We arrive at the remarkable property that
all topologically distinct diagrams for correlators have no extra factors associated
with them. In its turn, this property opens up the possibility of performing partial
summations of infinite geometrical series: in particular, the Dyson summation,
which plays a crucial role in the theory of weakly interacting Bose gas.

11.2.4   Diagrammatic Rules/Frequency and Momentum Representations

Let us summarize the diagrammatic rules for correlators (11.25) and (11.26)
derived in the previous section. In the following, we assume $n = m$, since otherwise the correlator is zero.

The diagrams consist of propagators (denoted with solid lines) and interaction
lines (denoted with dashed lines). Each of the two ends of an interaction line is also
the incoming and the outgoing end of either the same or two different propagators.
For a $2n$-point correlator, there are $2n$ external spacetime vertices; n of them stand
for the variables $(X_1, X_2, \ldots, X_n)$ and are associated with some incoming propagators; the other n external vertices represent the variables $(X_{n+1}, X_{n+2}, \ldots, X_{2n})$ and
are associated with some outgoing propagators. Each of the two ends of the propagator is connected to either the external vertex or the end of the interaction line.
The diagrammatic series includes all the diagrams—distinguished only by their
topology—such that every interaction line is connected to at least one external
vertex by a chain of other lines. The value of each diagram is determined by the
following rule: The value of the diagram is the integral over variables involved in
the characterization of each of the interaction elements (see Figure 11.1), while
the integrand is based on the product of functions associated with all lines.

Each interaction line contributes a factor $-\mathcal{U}(\mathbf{r}_{2i-1}-\mathbf{r}_{2i})$ where $i = n+1,\ldots,s$; propagators going from X_b (tail end) to X_a (arrow end) contribute a factor $-G^{(0)}(X_a, X_b)$ where $(a,b) = 1,\ldots,2s+2n$.

Frequency representation: Due to translation invariance in imaginary time, one may find it more convenient to work in the frequency representation for propagators (and correlators), Equation (11.8). Clearly, changing the representation does not affect the topological structure of the diagram. Now each propagator is ascribed a frequency $\xi_m = 2m\pi T$ ($m = 0,\pm1,\pm2,\ldots$) and, correspondingly, contributes a factor $-G^{(0)}(\xi_m)$. The integration over time variables of interaction lines translates into the frequency conservation law, stating that, for each interaction element, the sum of frequencies for two incoming propagators equals the sum of frequencies for two outgoing propagators. Each summation over frequencies that remain undetermined by external vertices and the conservation law comes with a factor of T.

Momentum representation: In a uniform system, translation invariance implies $G(X_1,X_2) \equiv G(X_1 - X_2)$, rendering momentum representation (11.11) attractive for practical calculations. The rules in momentum representation are similar to those in frequency representation. The main difference is that momenta are now ascribed not only to propagators but also to interaction lines. An interaction line with the momentum $\mathbf{q}$ brings a factor $-\mathcal{U}_{\mathbf{q}}$, a Fourier harmonic of the potential

$$\mathcal{U}_{\mathbf{q}} \equiv \mathcal{U}_{-\mathbf{q}} = \int e^{-i\mathbf{q}\cdot\mathbf{r}}\mathcal{U}(\mathbf{r})d^d r, \qquad \mathcal{U}(\mathbf{r}) \equiv \mathcal{U}(-\mathbf{r}). \qquad (11.44)$$

The momentum conservation law applies now to all three-line vertices: With each interaction line (ascribed some direction), the sum of incoming momenta should equal the sum of outgoing momenta. The direction of the interaction line is chosen arbitrarily: In view of $\mathcal{U}_{\mathbf{q}} \equiv \mathcal{U}_{-\mathbf{q}}$, this does not affect the result. Each summation over momenta, which remain undetermined by external vertices and the conservation law, comes with a standard factor of $(2\pi)^{-d}$.

11.2.5 Dyson Summation and Skeleton Diagrams

Dyson summation (along with its generalizations, such as ladder summation and screening) is a powerful tool that allows one to extend the domain of applicability of the diagrammatic technique beyond straightforward perturbative expansion, by summing up *infinite* series of diagrams of a certain type.

Consider diagrammatic expansion for the single-particle Green's function and group diagrams as shown in the first line of Figure 11.5. The circle stands for the sum of all *irreducible* contributions. By irreducible contribution, we mean a part of a diagram that cannot be cut into two disconnected pieces by cutting a single propagator. The sum of all irreducible parts (times minus unity) is referred to as *self-energy*, $\Sigma(X_1,X_2)$. In Figure 11.6, we show the first- and second-order diagrams contributing to the self-energy. With this grouping of the diagrams, we see that the sum of all graphs following the first self-energy term is nothing but the Green's function G, so that the series can be equivalently represented by the second line

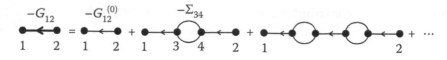

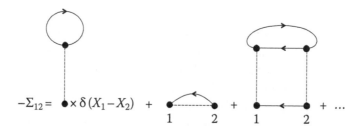

Figure 11.5 Dyson summation. The bold line denotes Green's function (times minus unity); the circle stands for the sum of all irreducible parts, referred to as self-energy (times minus unity).

$$-\Sigma_{12} = \quad \bullet \times \delta(X_1 - X_2) \quad + \quad \underset{1 \qquad 2}{\bullet \text{-----} \bullet} \quad + \quad \underset{1 \qquad 2}{\bullet \longleftarrow \bullet} \quad + \cdots$$

Figure 11.6 Diagrammatic expansion for the self-energy.

in Figure 11.5. (Note that this is a direct consequence of having no order-specific factors in the diagram contribution—only then can we decompose a graph into repeated pieces that have meaning of their own.) Analytically, the Dyson equation reads

$$G(X_1, X_2) = G^{(0)}(X_1, X_2) + \int G^{(0)}(X_1, X_3)\Sigma(X_3, X_4)G(X_4, X_2)\,dX_3\,dX_4. \qquad (11.45)$$

The Dyson trick leading to Equation (11.45) is similar to the one used in mathematics to sum geometric series, even beyond the convergence radius of the latter. Moreover, in a uniform system, where $G(X_1, X_2) \equiv G(X_1 - X_2)$ and $\Sigma(X_1, X_2) \equiv \Sigma(X_1 - X_2)$, the Dyson series literally reduces to the geometric series in the frequency-momentum representation. Correspondingly, the Dyson equation becomes algebraic:

$$G(\xi, \mathbf{k}) = \frac{1}{[G^{(0)}(\xi, \mathbf{k})]^{-1} - \Sigma(\xi, \mathbf{k})}. \qquad (11.46)$$

With the Dyson trick, we have eliminated reducible diagrams, at the expense of having to restore Green's function from the self-energy. With essentially the same or similar trick, one can—and in certain cases *must*—go further, and perform a summation of a certain infinite subclass of diagrams within a given diagram. We consider here a particular example leading to the so-called skeleton diagrams.

Reading the first equality in Figure 11.5 from right to left and thinking of the spacetime points X_1 and X_2 as if they are some *internal* vertices of a higher-order diagram to which our series are attached, we conclude that a whole infinite series of diagrams can be replaced with just one diagram in which the spacetime points X_1 and X_2 are connected by a *bold* rather than a thin line. Clearly, this procedure of introducing "bold" lines by eliminating series having the structure of the Dyson equation applies to any propagator in any self-energy diagram. The diagrams built of bold lines are called skeleton diagrams. The number of skeleton diagrams of a given order is smaller than the number of perturbative diagrams, since a skeleton diagram cannot contain parts attached to the rest of the graph by only two propagators—all such contributions are already accounted for by bold lines. Alternatively, instead of using bold lines representing the exact Green's function, one can introduce partially "dressed" lines corresponding to keeping only a finite number of terms in the expansion for self-energy.

11.2.6   Fermionic Systems

Feynman diagram rules for fermions are exactly the same as for bosons with only one additional ingredient: namely, the sign rule stating that the diagram sign alternates between the diagrams with odd and even numbers of closed fermionic loops. This rule comes from the anticommutation relation for fermionic field operators.

The derivation itself is essentially identical to that done for bosons within the functional integral and generating functional formalisms, with one notable difference: Coherent states for fermions can be introduced only if the vector space is defined over the field of the so-called Grassmann variables. Discussion of their properties, though simple, goes beyond the scope of our book. We only mention that formal procedures for manipulating Grassmann variables—that include formal rules for "integration" and "differentiation"—are such that nearly all steps in the derivation can be literally repeated one to one, except the additional sign rule, which now has its origin in the anticommutation relation for multiplying Grassmann variables, $\xi_1 \xi_2 = -\xi_2 \xi_1$.

CHAPTER 12

Thermodynamics of Weakly Interacting Bose Gas, BCS Theory

This chapter deals for the most part with a systematic microscopic description of equilibrium weakly interacting Bose gas (WIBG) in one, two, and three dimensions. Our analysis is based on Beliaev's diagrammatic technique [1] for Matsubara Green's functions, regularized with small but finite Bogoliubov's symmetry-breaking terms. This approach was discussed in detail in Reference [2]; the analytic part of this chapter is based on that paper. At the end of the chapter, we will present the mean-field theory for weakly interacting Fermi gas, or the Bardeen–Cooper–Schrieffer (BCS) solution.

In the so-called fluctuation region, the perturbative diagrammatic approach does not provide an adequate description for the strongly nonlinear fluctuations of the long-wave classical-field component, and we have to augment the analytic perturbative treatment of the short-wave part of the system with the numeric treatment of the long-wave classical-field part. The latter amounts to employing universal—to all weakly interacting U(1) systems—scaling functions, naturally describing both the fluctuation and critical behavior of the system. The fluctuation region of WIBG is of a distinct fundamental interest on its own as an example of a system where the order parameter describing the superfluid phase transition is of simple microscopic origin.* Our treatment of the fluctuation region is based on the analysis of dimensions and first-principle data of References [3–7].

Our main results are summarized in Section 12.1. The results, in particular, reflect a specificity of the system, which is worth discussing prior to embarking on developing the formalism. This indicates why a systematic treatment of WIBG—a theory that allows one to estimate the systematic errors—turns out to be rather technical and also sheds light on the necessity and physical origin of the tricks that will be used to regularize the otherwise singular perturbation theory.

Despite its success in treating WIBG, Beliaev's technique, in its original form, is known to have serious infrared problems [8–10]. Moreover, being explicitly based on the condensate part of the field operator, the formalism cannot be directly applied to low-dimensional systems where the genuine condensate is absent. A number of recipes exist on how to deal with this inherent infrared problem of Beliaev's technique, all of them taking advantage of the bimodality of correlation properties of superfluid WIBG. The off-diagonal correlation functions, like the single-particle density matrix (see Figure 12.1), demonstrate an initial drop and then remain almost constant until very large distances are reached; all

---

* In strongly correlated quantum systems, introducing the classical-field order parameter requires a substantial coarse graining.

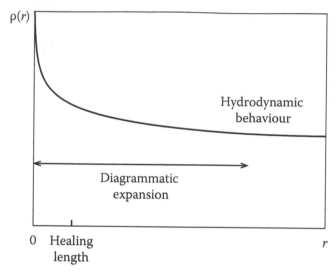

Figure 12.1   Sketch of the density matrix dependence on distance in the WIBG at temperatures below the fluctuation region. After the initial drop, the density matrix levels off at the healing length and undergoes slow decay up to a distance where the superfluid hydrodynamic approach becomes applicable. The U(1) symmetry-breaking terms make the density matrix converge to a finite value (in any dimension) at the hydrodynamic length scales.

the infrared features of the theory, including further decay of the off-diagonal correlators in low-dimensional systems, develop at distances much larger than the healing length. This picture implies that all the infrared problems of WIBG are of essentially universal—to all superfluids—hydrodynamic character and thus can be treated accordingly. A physically transparent, and convenient-to-implement, idea—stemming from Bogoliubov's quasiaverages (see Chapter 1)—is to cut off the infrared divergencies by introducing controllably small terms that explicitly break the U(1) symmetry:

$$H \to H - \int \left(\eta^* \hat{\psi} + \eta \hat{\psi}^\dagger \right) d^d r. \tag{12.1}$$

The magnitude of the parameter η is such that the effect of the symmetry-breaking terms on the local thermodynamic properties of the system is negligibly small while all infrared singularities in Beliaev's technique are eliminated. More specifically, we assume that the scale of η is defined by the highest-order terms of diagrammatic expansion being taken into account—the higher the order, the smaller the magnitude of η—in such a way that the systematic error associated with distorting the Hamiltonian by the term (12.1) is commensurate with the error due to neglecting higher-order terms of the expansion. Clearly, the solution to the infrared problem comes at the expense of suppressing long-range fluctuations of the phase of the order parameter (with opening a gap in the spectrum at small momenta) and thus distorting the long-range behavior of correlation functions. Nevertheless, this distortion is essentially irrelevant since it takes place at large distances where the behavior (generic to all superfluids) of correlators is governed by hydrodynamic fluctuations of the phase of the order parameter, with a simple universal theoretical description. Moreover, in all final answers, one can explicitly take the limit of vanishing symmetry-breaking terms.

Yet another complicating circumstance is the pathology of the grand canonical ensemble for an ideal Bose gas below the critical point—its physical origin being an infinite compressibility of the Bose–Einstein condensate in the absence of interaction. In its simplest form, the diagrammatic technique is a perturbation theory on top of a noninteracting system, naturally formulated in the grand canonical ensemble. Taking, for definiteness, a uniform macroscopic ideal system, we observe that, at the critical point, the chemical potential—negative definite in the normal phase—becomes exactly equal to zero. Going below the critical point would require having a positive μ, but because there is no potential-energy penalty that limits condensate density, that would immediately lead to a divergence of the number of condensate particles. Hence, with an ideal gas, we do not have a straightforward perturbative access to positive values of μ. Meanwhile, the chemical potential of WIBG below the critical point (and even slightly above it!) is positive, meaning that, to appropriately treat the system, one must rely on nonperturbative tools of the diagrammatic technique. Specifically, one must formally treat as a perturbation—on top of the normal ideal system—not only the interaction but also the chemical potential [as well as the symmetry-breaking terms (12.1)] and then utilize the Dyson summation of infinite series to arrive at an appropriate skeleton diagrammatic technique.

12.1   Main Results

Here we present the most important results, which will be derived systematically later, assuming that conditions (P.5) and (P.6) are met. These relations work for both continuous-space and lattice systems; in the latter case, the mass m is understood as the effective mass of low-energy motion. The equilibrium state of the system is conveniently parameterized by two independent variables, the temperature T and the effective chemical potential $\tilde{\mu}$ (which is always negative: $\tilde{\mu} \equiv -|\tilde{\mu}|$), related to the genuine chemical potential by

$$\mu = 2nU - |\tilde{\mu}|, \tag{12.2}$$

where n is the number density and U is the (effective) coupling constant we will discuss below.

12.1.1   Coupling Constants

The interaction is characterized by an (effective) coupling constant U. In one dimension, U is the zero-momentum Fourier component of the bare potential $\mathcal{U}(\mathbf{r})$. In two and three dimensions, U is naturally expressed in terms of the s-wave scattering length a; the latter being defined as the radius of hard-disk/hard-sphere potential, $\mathcal{U}_{\text{hard}}(r) = 0$ $[\mathcal{U}_{\text{hard}}(r) = \infty]$ at $r \geq a$ $[r < a]$, with low-energy scattering properties identical to those of the potential $\mathcal{U}(\mathbf{r})$. A representational freedom allows one to use slightly different expressions for U, without sacrificing the order of accuracy. Our choices for three and two dimensions are

$$U = \frac{4\pi a}{m} \quad (d = 3), \qquad U = \frac{4\pi/m}{\ln(2/a^2 m\epsilon_0) - 2\gamma} \quad (d = 2), \tag{12.3}$$

where $\gamma = 0.5772\ldots$ is Euler's constant. A particular expression for U comes from fixing the specific form of an auxiliary function $\Pi(k)$.* The latter enters thermodynamic relations in such a way that the values of thermodynamic functions remain insensitive—up to irrelevant higher-order corrections—to slightly changing U by adopting a slightly different Π. Equations (12.3) correspond to

$$\Pi(k) = -\frac{1}{2\epsilon(k)} \quad (d = 3), \qquad \Pi(k) = -\frac{1}{2}\frac{1}{\epsilon(k) + \epsilon_0} \quad (d = 2), \tag{12.4}$$

where $\epsilon(k)$ is the particle dispersion relation. In one dimension, $\Pi \equiv 0$. Different choices of the (Π, U) pair are readily connected with each other by Equation (12.61).

In the 2D case, the coupling constant U and, correspondingly, function $\Pi(k)$ slowly evolve with changing effective chemical potential and temperature:

$$\epsilon_0 \equiv \epsilon_0(\tilde{\mu}, T) = \begin{cases} (e/2)|\tilde{\mu}| & (T < T_c), \\ T & (T > T_c). \end{cases} \tag{12.5}$$

It is well within representational freedom to multiply the right-hand side of (12.5) by a constant of order unity, provided the same expression for ϵ_0 is used in both (12.3) and (12.4). The order-unity factor $e/2$ is introduced to cast the zero-temperature equations of state in the form (12.122) adopted in the literature.

Our results remain valid for the *quasi*-2D case (as well as for the quasi-1D case). An account of the effective interaction for quasi-2D/quasi-1D cases is given in Section 12.2.6. The expressions for the effective two-body elastic interactions in terms of s-scattering length in 2D and 3D (and the expression for the effective interaction in the quasi-1D case discussed in Section 12.2.6) remain valid for dilute metastable gases. Therefore, our thermodynamic results apply to these systems as well. One should remember, however, that finite lifetime is a natural source of extra systematic errors for any equilibrium theory.

12.1.2   Thermodynamics in the Superfluid Region

Thermodynamic quantities are obtained in parametric form, as functions of the pair of parameters $\tilde{\mu}$ and T. The crucial role is played by the relation for the number density:

$$n(\tilde{\mu}, T) = |\tilde{\mu}|/U + n', \qquad n' = \sum_{\mathbf{k}} \left[\frac{\epsilon(k) - E(k)}{2E(k)} - |\tilde{\mu}|\Pi(k) + \frac{\epsilon(k)}{E(k)} N_E \right], \tag{12.6}$$

---

* Equation (12.53) directly relates U to the bare potential $\mathcal{U}$ and Π; one can resort to this relation for evaluating U in cases when finding the scattering length a is not easier than solving (12.53).

where $\epsilon(k)$ is the single-particle dispersion law, and

$$E(k) = \sqrt{\epsilon(k)[\epsilon(k) + 2|\tilde{\mu}|]}, \qquad N_E = \left(e^{E/T} - 1\right)^{-1} \tag{12.7}$$

are (respectively) the Bogoliubov-type quasiparticle dispersion relation and the Bose–Einstein occupation number. In Equation (12.6) and later throughout the text, where we sum over momenta, we set the system volume equal to unity to omit a trivial normalization factor. The pressure is given by

$$p = \frac{\tilde{\mu}^2}{2U} + 2n'|\tilde{\mu}| + (n')^2 U - \frac{1}{2}\sum_k \left\{ E(k) - \epsilon(k) - |\tilde{\mu}| - \tilde{\mu}^2\Pi(k) + 2T\ln\left[1 - e^{-E(k)/T}\right]\right\}. \tag{12.8}$$

The expression for entropy,

$$s = \sum_k \left[\frac{E(k)}{T}N_E - \ln\left(1 - e^{-E(k)/T}\right)\right], \tag{12.9}$$

corresponds to noninteracting bosonic quasiparticles with (temperature-dependent) dispersion $E(k)$. The formula for the energy density then follows by the general relation $\varepsilon = \mu n + Ts - p$:

$$\varepsilon = \frac{\tilde{\mu}^2}{2U} - n'\tilde{\mu} + n'^2 U + \frac{1}{2}\sum_k \left[E(k)(2N_E + 1) - \epsilon(k) + \tilde{\mu} - \tilde{\mu}^2\Pi(k)\right]. \tag{12.10}$$

At zero temperature, all the integrals can be evaluated analytically, as discussed in Section 12.3.3.

In continuous space, where $\epsilon(k) = k^2/2m$, the formula for the superfluid density can be cast into Landau form

$$n_s = n + \frac{1}{d}\sum_k \frac{k^2}{m}\frac{\partial N_E}{\partial E}. \tag{12.11}$$

Due to the condition (P.5), the answer for a lattice differs *only* by a global factor equal to the ratio between the bare and effective masses.

12.1.3 Single-Particle Density Matrix, Quasicondensate

In the superfluid region, the single-particle density matrix can be parameterized as

$$\rho(\mathbf{r}) = \tilde{\rho}(\mathbf{r})e^{-\Lambda(\mathbf{r})}, \qquad \tilde{\rho}(\mathbf{r}) = n - \sum_k \left(1 - e^{i\mathbf{k}\cdot\mathbf{r}}\right)\frac{\epsilon}{E^2}\left[\frac{E - \epsilon - |\tilde{\mu}|}{2} + (E + |\tilde{\mu}|)N_E\right],$$

$$\Lambda(\mathbf{r}) = \frac{|\tilde{\mu}|}{n}\sum_k \left(1 - e^{i\mathbf{k}\cdot\mathbf{r}}\right)\frac{E - \epsilon}{2E^2}[1 + 2N_E]. \tag{12.12}$$

By definition, $\rho(\infty)$ is the condensate density. The function $\tilde{\rho}(\infty)$ is finite in any dimension. In 1D and at finite temperature in 2D, the function $\Lambda(\mathbf{r})$ diverges at $r \to \infty$, consistently with the general fact of absence of condensate in those systems. Nevertheless, at distances at which the function $\tilde{\rho}(\mathbf{r})$ saturates to its asymptotic value $\tilde{\rho}(\infty)$, the function $\Lambda(\mathbf{r})$ is still much smaller than unity. This characteristic feature of the weakly interacting system allows one to speak of the quasiconden- sate of the density $n_{qc} = \tilde{\rho}(\infty)$. Up to the distances at which Λ becomes ~ 1, the correlation properties of condensed and quasicondensed systems are essentially the same.

Physically, it is very important that

$$n_{qc} = |\tilde{\mu}|/U + \text{higher-order corrections}, \tag{12.13}$$

since it reveals the classical-field nature of the spectrum $E(k)$, Equation (12.7). Equation (12.13) shows that this spectrum corresponds to the normal modes of the Gross–Pitaevskii complex-valued classical field of the density n_{qc} (see Chapter 1).

Details of the general approach to long-range off-diagonal many-particle correlation functions are described in Section 12.5.

12.1.4   Thermodynamics in the Normal Region

The density $n \equiv n(\tilde{\mu}, T)$ in the normal region is given by

$$n = \sum_{\mathbf{k}} \left[e^{\tilde{\epsilon}(k)/T} - 1 \right]^{-1}, \qquad \tilde{\epsilon}(k) = \epsilon(k) + |\tilde{\mu}|. \tag{12.14}$$

The pressure is obtained by

$$p = n^2 U - T \sum_{\mathbf{k}} \ln\left[1 - e^{-\tilde{\epsilon}(k)/T} \right]. \tag{12.15}$$

For the entropy and energy densities, the relations are

$$s = \sum_{\mathbf{k}} \left[\frac{\tilde{\epsilon}(k)}{T} N_{\tilde{\epsilon}} - \ln\left(1 - e^{-\tilde{\epsilon}(k)/T} \right) \right], \qquad \epsilon = U n^2 + \sum_{\mathbf{k}} \epsilon(k) N_{\tilde{\epsilon}}. \tag{12.16}$$

12.1.5   Accuracy Control and Fluctuation Region

A necessary condition for the previously outlined relations to apply is the small- ness of the parameter

$$\gamma_0 = \begin{cases} \sqrt{na^3} & (d = 3), \\ mU & (d = 2), \\ \sqrt{mU/n} & (d = 1). \end{cases} \tag{12.17}$$

At $T \leq nU$, the parameter γ_0 directly controls the systematic error of the theory, and the condition $\gamma_0 \ll 1$ is sufficient for the theory to be accurate. At higher temperatures, the condition (12.17) is only *necessary*, since there exists the fluctuation region where the fluctuations of the order parameter are essentially nonlinear and are not captured by the perturbation theory. The closeness to the fluctuation region is described by the dimensionless parameter

$$x = \frac{\mu - \mu_c^{(d)}(T)}{(m^d T^2 U^2)^{\frac{1}{4-d}}}. \qquad (12.18)$$

In 2D and 3D systems, $\mu_c^{(d)}(T)$ is the critical value of the chemical potential for a given temperature (explicit expressions can be found in Section 12.7); in 1D systems, where there is no finite-temperature phase transition (see Section 12.4.4 for more detail), $\mu_c^{(1)} \equiv 0$. The theory applies as long as $|x| \gg 1$, getting progressively less accurate with decreasing $|x|$. At $|x| \lesssim 1$, the theory fails to properly describe condensate and superfluid densities, the values of which are defined by the fluctuating classical-field order parameter. The description of other thermodynamic quantities is better since the fluctuation contributions to them are of a higher order than the leading (and sometimes even the subleading) terms.

Parametric estimates for systematic errors away from and within the fluctuation region can be obtained on the basis of dimensional analysis of diagrams. Explicit estimates for major thermodynamic quantities are given in Section 12.4 (in the superfluid region) and Section 12.6.3 (in the normal region).

In the fluctuation region and its vicinity, the theory can be fixed by incorporating accurate description of fluctuation contributions with dimensionless scaling functions—universal to all weakly interacting U(1) models and available from numeric simulations. This theory is developed in Section 12.7.

12.2   Beliaev's Diagrammatic Technique

In this section, we shall upgrade the Matsubara diagrammatic technique for the normal Bose gas to Beliaev's technique for a superfluid system.

12.2.1   Diagrammatic Expansion

We take the Hamiltonian of an ideal gas, H_0, Equation (P.2), to be our nonperturbed Hamiltonian. All other terms of the Hamiltonian—the pair interaction H_{int}, Equation (P.3); the external and chemical potential term H_1, Equation (P.4); and the symmetry-breaking terms (12.1)—are treated as perturbations. The prime diagrammatic expansion is thus based on the propagator

$$G^{(0)}(\tau_1, \tau_2, \mathbf{r}_1, \mathbf{r}_2) = \sum_{\xi, \mathbf{k}} G^{(0)}(\xi, \mathbf{k}) e^{i\mathbf{k}\cdot(\mathbf{r}_1 - \mathbf{r}_2) - i\xi(\tau_1 - \tau_2)}, \qquad (12.19)$$

$$G^{(0)}(\xi, \mathbf{k}) = [i\xi - \epsilon(k)]^{-1}. \qquad (12.20)$$

We typically assume a parabolic single-particle dispersion relation $\epsilon(k) = k^2/2m$, but our final answers (excluding formulae for superfluid properties that explicitly invoke Galilean invariance) are valid for arbitrary $\epsilon(k)$, with a parabolic dependence on momentum in the long-wave limit, for example, for the tight-binding spectrum. We use the Matsubara imaginary-frequency representation: $\xi \equiv \xi_s = 2s\pi T$ $(s = 0, \pm 1, \pm 2, \ldots)$. For graphical representation of diagrammatic expansions and relations, we introduce a set of objects in Figure 12.2 that depict the single-particle propagator, the condensate, and various terms in the Hamiltonian.

The nonperturbative response to H_{int} and H_1 is accounted for by the Dyson summation. First, we consider diagrams that feature only one incoming or outgoing line—we call them tails. Given our starting point with no condensate in the nonperturbed system, such particle-number-changing diagrams exist only due to the symmetry-breaking field η. The frequency of the line connecting η to the rest of the diagram is zero, by frequency conservation. The Dyson summation of all tails attached to a given point replaces them with a single line, which we denote as $\psi_{in}(\mathbf{r})$ and $\psi_{out}(\mathbf{r})$ if the tails are incoming and outgoing, respectively (see Figure 12.3). Later, we write expressions in the frequency representation; if the frequency argument is not mentioned, it is implied that its value is zero. We also adopt a convention of integration over repeated

$$
\underset{-G^{(0)}(X-X')}{X \longleftarrow X'} \qquad \underset{\mu - V(\mathbf{r})}{\overset{\vdots}{\underset{\times \mathbf{r}}{}}} \qquad \underset{\eta^*(\mathbf{r})}{\mathbf{r}} \qquad \underset{\eta(\mathbf{r})}{\overset{\mathbf{r}}{}} \qquad \psi_{in}(\mathbf{r}) \qquad \psi_{out}(\mathbf{r}) \qquad \underset{-\mathcal{U}(\mathbf{r}-\mathbf{r}')}{\overset{\vdots \mathbf{r}'}{\underset{\vdots \mathbf{r}}{}}}
$$

Figure 12.2 Graphical objects representing the single-particle propagator $G^{(0)}(X - X')$, the external field $V(\mathbf{r}) - \mu$, the symmetry-breaking fields $\eta^*(\mathbf{r})$ and $\eta(\mathbf{r})$, the condensate lines $\psi_{in}(\mathbf{r})$ and $\psi_{out}(\mathbf{r})$, and the interaction $\mathcal{U}(\mathbf{r} - \mathbf{r}')$.

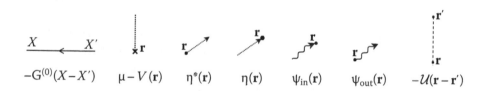

Figure 12.3 Diagrammatic expansion for (12.21) and (12.22).

coordinate/momentum/frequency arguments. The Dyson equation for $\psi_{in}(\mathbf{r})$ then reads as follows (see also Figure 12.3):

$$\psi_{in}(\mathbf{r}) = -G^{(0)}(\mathbf{r}-\mathbf{r}')\eta(\mathbf{r}') + G^{(0)}(\mathbf{r}-\mathbf{r}')[V(\mathbf{r}')-\mu]\psi_{in}(\mathbf{r}') + G^{(0)}(\mathbf{r}-\mathbf{r}')\Theta_{in}(\mathbf{r}').$$
(12.21)

Here Θ_{in} is the sum of all other diagrammatic elements attached to the first line that are not accounted for by the first two terms, that is, excluding diagrams with the field η and diagrams connected to the first solid line by the $[V(\mathbf{r})-\mu]$ vertex. The subscript "in" reminds that Θ_{in} has an extra incoming particle line. Similarly,

$$\psi_{out}(\mathbf{r}) = -\eta^*(\mathbf{r}')G^{(0)}(\mathbf{r}'-\mathbf{r}) + \psi_{out}(\mathbf{r}')[V(\mathbf{r}')-\mu]G^{(0)}(\mathbf{r}'-\mathbf{r}) + \Theta_{out}(\mathbf{r}')G^{(0)}(\mathbf{r}'-\mathbf{r}).$$
(12.22)

The fact that $G^{(0)}(\mathbf{r})$ is a real even function of its argument implies that

$$\psi_{out} = \psi_{in}^*, \qquad \Theta_{out} = \Theta_{in}^*.$$
(12.23)

The diagrammatic expansion for the tail is identical to that for the condensate wavefunction defined as an anomalous average

$$\psi_0(\mathbf{r}) = \langle\psi(\mathbf{r})\rangle \equiv \psi_{in}(\mathbf{r}).$$
(12.24)

Correspondingly, the condensate density is defined as $n_0(\mathbf{r}) = |\psi_0(\mathbf{r})|^2$. From now on, we will write ψ_0 for ψ_{in} and ψ_0^* for ψ_{out} and use the notions of condensate lines and tails on equal footing. Clearly, to speak of the condensate wavefunction in the limit of $\eta \to 0$ is meaningful only in those cases when the long-range fluctuations of the phase do not destroy it.

12.2.2 Normal and Anomalous Propagators

We now introduce exact, or "bold," particle propagators (denoted with full lines) and work with skeleton diagrams constructed out of these propagators, thereby performing Dyson's partial summation. We must emphasize a few differences between our approach and the standard Beliaev technique. Our bare (thin-line) propagators have zero chemical potential, and this, according to (12.21), immediately results in a constraint relating the condensate density to the chemical potential and Θ_{in} [see (12.36)]. Also, the chemical and external potentials have to be explicitly introduced into the standard Beliaev–Dyson equations for the normal and anomalous Green's functions. Within our approach, one arrives at these equations purely diagrammatically. To this end—proceeding in the frequency representation for the sake of definiteness—we introduce the normal Green's function, $G(\xi, \mathbf{r}_1, \mathbf{r}_2)$, defined as the sum—with a global minus sign, which is a mere convention—of all diagrams that have an incoming $G^{(0)}$-line with frequency ξ to point $\mathbf{r}_1$ and an outgoing $G^{(0)}$-line (with the same frequency) from point $\mathbf{r}_2$.

The anomalous Green's function, $F_{in}(\xi, \mathbf{r}_1, \mathbf{r}_2)$, by definition, has an incoming $G^{(0)}$-line with frequency ξ to point $\mathbf{r}_1$ and another incoming $G^{(0)}$-line with frequency $-\xi$, by conservation of frequency, to point $\mathbf{r}_2$. The anomalous Green's function $F_{out}(\xi, \mathbf{r}_1, \mathbf{r}_2)$ is a counterpart of the function $F_{in}(\xi, \mathbf{r}_1, \mathbf{r}_2)$: instead of two incoming, it has two outgoing $G^{(0)}$-lines—one with frequency ξ from point $\mathbf{r}_1$ and another with frequency $-\xi$ from point $\mathbf{r}_2$. The symmetry with respect to exchanging the end points of the anomalous Green's functions immediately implies the following relations:

$$F_{in}(\xi, \mathbf{r}_1, \mathbf{r}_2) = F_{in}(-\xi, \mathbf{r}_2, \mathbf{r}_1), \qquad F_{out}(\xi, \mathbf{r}_1, \mathbf{r}_2) = F_{out}(-\xi, \mathbf{r}_2, \mathbf{r}_1). \quad (12.25)$$

Since complex conjugation is equivalent to changing the sign of the Matsubara frequency and direction of propagation, we also have

$$[G(\xi, \mathbf{r}_1, \mathbf{r}_2)]^* = G(-\xi, \mathbf{r}_2, \mathbf{r}_1), \quad (12.26)$$

$$F_{in}^*(\xi, \mathbf{r}_1, \mathbf{r}_2) = F_{out}(\xi, \mathbf{r}_1, \mathbf{r}_2) = F_{out}(-\xi, \mathbf{r}_2, \mathbf{r}_1). \quad (12.27)$$

The physical meaning of G, F_{in}, and F_{out} follows from the structure of the two-point correlation functions in the imaginary-time-coordinate representation:

$$\langle \psi(\tau_1, \mathbf{r}_1) \psi^*(\tau_2, \mathbf{r}_2) \rangle = -G(\tau_1 - \tau_2, \mathbf{r}_1, \mathbf{r}_2) + \psi_0(\mathbf{r}_1) \psi_0^*(\mathbf{r}_2), \quad (12.28)$$

$$\langle \psi(\tau_1, \mathbf{r}_1) \psi(\tau_2, \mathbf{r}_2) \rangle = -F_{out}(\tau_1 - \tau_2, \mathbf{r}_1, \mathbf{r}_2) + \psi_0(\mathbf{r}_1) \psi_0(\mathbf{r}_2). \quad (12.29)$$

These relations can be readily checked by expanding the averages into diagrammatic series. The special case of (12.28), corresponding to $\tau_1 \to \tau_2 - 0$ and $\mathbf{r}_1 = \mathbf{r}_2$, relates the local density to the normal Green's function and the condensate density:

$$n(\mathbf{r}) = -G(\tau = -0, \mathbf{r}, \mathbf{r}) + n_0(\mathbf{r}). \quad (12.30)$$

The Beliaev–Dyson equations then read

$$G(\xi, \mathbf{r}_1, \mathbf{r}_2) = G^{(0)}(\xi, \mathbf{r}_1, \mathbf{r}_2) + G(\xi, \mathbf{r}_1, \mathbf{r}')[V(\mathbf{r}') - \mu]G^{(0)}(\xi, \mathbf{r}', \mathbf{r}_2)$$
$$+ G(\xi, \mathbf{r}_1, \mathbf{r}')\Sigma_{11}(\xi, \mathbf{r}', \mathbf{r}'')G^{(0)}(\xi, \mathbf{r}'', \mathbf{r}_2)$$
$$+ F_{in}(\xi, \mathbf{r}_1, \mathbf{r}')\Sigma_{20}(\xi, \mathbf{r}', \mathbf{r}'')G^{(0)}(\xi, \mathbf{r}'', \mathbf{r}_2), \quad (12.31)$$

$$F_{in}(\xi, \mathbf{r}_1, \mathbf{r}_2) = F_{in}(\xi, \mathbf{r}_1, \mathbf{r}')[V(\mathbf{r}') - \mu]G^{(0)}(-\xi, \mathbf{r}_2, \mathbf{r}')$$
$$+ F_{in}(-\xi, \mathbf{r}_1, \mathbf{r}')\Sigma_{11}(\xi, \mathbf{r}', \mathbf{r}'')G^{(0)}(-\xi, \mathbf{r}_2, \mathbf{r}'')$$
$$+ G(\xi, \mathbf{r}_1, \mathbf{r}')\Sigma_{02}(\xi, \mathbf{r}', \mathbf{r}'')G^{(0)}(-\xi, \mathbf{r}_2, \mathbf{r}''), \quad (12.32)$$

with the standard definition of self-energies Σ's as sums of diagrams that cannot be cut through a single G or F line. With the diagrammatic notation explained in Figure 12.4, Equations (12.31) and (12.32) are shown graphically in Figure 12.5 for a homogeneous system in momentum representation. The complexity of the theoretical solution is in the evaluation of the Θ and Σ functions.

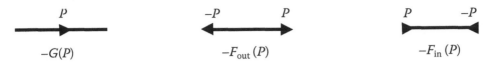

Figure 12.4 Symbols used for normal, G, and anomalous, F, propagators in a homogeneous system.

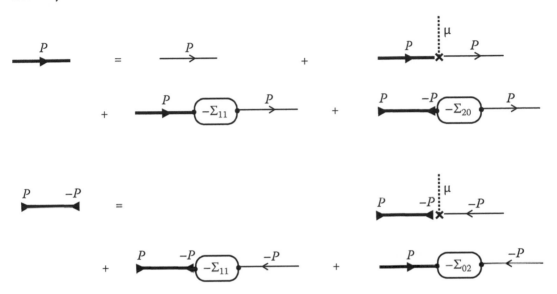

Figure 12.5 Beliaev–Dyson equations for a homogeneous system in the frequency-momentum representation: $P = (\xi, \mathbf{k})$.

The normal and anomalous self-energies have the same symmetry properties as corresponding normal and anomalous propagators:

$$[\Sigma_{11}(\xi, \mathbf{r}_1, \mathbf{r}_2)]^* = \Sigma_{11}(-\xi, \mathbf{r}_2, \mathbf{r}_1), \tag{12.33}$$

$$\Sigma_{20}^*(\xi, \mathbf{r}_1, \mathbf{r}_2) = \Sigma_{02}(\xi, \mathbf{r}_1, \mathbf{r}_2) = \Sigma_{02}(-\xi, \mathbf{r}_2, \mathbf{r}_1). \tag{12.34}$$

12.2.3 Chemical Potential and the Hugenholtz–Pines Relation

Since the bare Green's function at zero frequency is identical to the inverse Laplacian operator, one can cast (12.21) in the differential form

$$-\frac{\Delta}{2m}\psi_0(\mathbf{r}) + [V(\mathbf{r}) - \mu]\psi_0(\mathbf{r}) + \Theta_{in}(\mathbf{r}) = \eta(\mathbf{r}). \tag{12.35}$$

This equation reduces to the Gross–Pitaevskii equation at low enough temperature when the leading term in Θ_{in} is $\propto |\psi_0(\mathbf{r})|^2 \psi_0(\mathbf{r})$. In the homogeneous case at $\eta \to 0$, when ψ_0 and Θ_{in} are coordinate independent, Equation (12.35) simplifies to

$$\mu = \Theta_{in}/\psi_0 \equiv \Theta_{out}/\psi_0^* \quad (\eta \to 0). \tag{12.36}$$

There is also an exact relation between Σ_{11}, Σ_{20}, Θ_{in}, and ψ_0. Here (and only here!) we assume that all diagrams for Σ's are in terms of $G^{(0)}$ and condensate lines. Let $D_{in}^{(l)}$ be the sum of diagrams contributing to Θ_{in} with l incoming and $l-1$ outgoing condensate lines. Then (for $\xi = 0$) we have

$$\Sigma_{11}(\mathbf{r},\mathbf{r}')\psi_0(\mathbf{r}') = \sum_{l=1}^{\infty} l D_{in}^{(l)}. \tag{12.37}$$

Indeed, each diagram, with $l-1$ incoming condensate lines, contributing to $\Sigma_{11}(\mathbf{r},\mathbf{r}')$, produces—upon integration over $\mathbf{r}'$ with the weight $\psi_0(\mathbf{r}')$—a diagram contributing to $D_{in}^{(l)}$, and there are l such diagrams contributing to Σ_{11}. An identical argument leads to

$$\Sigma_{20}(\mathbf{r},\mathbf{r}')\psi_0^*(\mathbf{r}') = \sum_{l=2}^{\infty}(l-1)D_{in}^{(l)}. \tag{12.38}$$

By subtracting (12.38) from (12.37), we obtain

$$\Sigma_{11}(\mathbf{r},\mathbf{r}')\psi_0(\mathbf{r}') - \Sigma_{20}(\mathbf{r},\mathbf{r}')\psi_0^*(\mathbf{r}') = \Theta_{in}(\mathbf{r}). \tag{12.39}$$

In the homogeneous case and assuming η to be real, we have $\psi_0 = \psi_0^* = \text{const}$. Then, in the $\eta \to 0$ limit, Equations (12.39) and (12.36) can be combined to yield the Hugenholtz–Pines relation:

$$\mu = \Sigma_{11}(\xi = 0, \mathbf{k} = 0) - \Sigma_{20}(\xi = 0, \mathbf{k} = 0) \quad (\eta = \eta^* \to 0). \tag{12.40}$$

12.2.4 Beliaev–Dyson Equations in the Presence of Homogeneous Superflow

In order to discuss the superfluid properties of the homogeneous system, we add a phase factor to the symmetry-breaking field:

$$\eta(\mathbf{r}) = \eta_0 e^{i\mathbf{k}_0 \cdot \mathbf{r}} \quad (\text{Im}\eta_0 = 0, \ \text{Re}\eta_0 > 0), \tag{12.41}$$

which readily translates into the phase of the condensate wavefunction

$$\psi_0(\mathbf{r}) = \sqrt{n_0}e^{i\mathbf{k}_0 \cdot \mathbf{r}}. \tag{12.42}$$

The only difference from the previous discussion is that now we must associate a finite momentum $\pm\mathbf{k}_0$ carried by the condensate lines and modify the momentum conservation laws accordingly. Then (12.36) and (12.40) become (omitting the argument $\xi = 0$ for briefness)

$$\mu = \Theta/\sqrt{n_0} + k_0^2/2m - \eta_0/\sqrt{n_0}, \tag{12.43}$$

$$\Sigma_{11}(\mathbf{k}_0) - \Sigma_{20}(0) = \frac{\Theta}{\sqrt{n_0}} = \mu - \frac{k_0^2}{2m} + \frac{\eta_0}{\sqrt{n_0}}, \tag{12.44}$$

where $\Theta = \Theta_{in}e^{-i\mathbf{k}_0 \cdot \mathbf{r}} = \Theta_{out}e^{i\mathbf{k}_0 \cdot \mathbf{r}}$.

It is convenient (for the transparency of the expressions that follow) to combine frequency and momentum into a single "$(d+1)$-momentum" variable $P = (\xi, \mathbf{k})$ and to introduce an auxiliary momentum $P' = (-\xi, 2\mathbf{k}_0 - \mathbf{k})$. The symmetry between the two ends of the anomalous Green's functions and, equivalently, between the two ends of the anomalous self-energies is then expressed by (accounting for the momentum carried by the condensate lines)

$$F_{in/out}(P) = F_{in/out}(P'), \qquad \Sigma_{20/02}(P) = \Sigma_{20/02}(P'). \qquad (12.45)$$

[In a more comprehensive notation scheme, one must mention momenta of both incoming lines in $F_{in/out}(P, P')$.]

Complex conjugation of propagators and condensate lines changes the signs of their $(d+1)$-momenta. This property can be used to prove the symmetry relation for the Green's function

$$G^*(P)|_{\mathbf{k}_0} = G(-P)|_{-\mathbf{k}_0}. \qquad (12.46)$$

Similar symmetry relations take place for the anomalous Green's functions and all three self-energies.

In the momentum representation, inverting the direction of $\mathbf{k}_0$ does not change the analytical expression for propagators. Hence, by inverting the direction of all the lines, including the condensate ones, it follows that

$$F_{in}(P) = F_{out}(P) \equiv F(P), \qquad \Sigma_{20}(P) = \Sigma_{02}(P) \equiv \tilde{\Sigma}(P). \qquad (12.47)$$

(For the normal Green's function, $G(P)$, and the self-energy, $\Sigma_{11}(P)$, inverting momentum directions results in the same series and, in this sense, is trivial.)

We are now ready to formulate the pair of Beliaev–Dyson equations in the momentum representation. Taking symmetry properties into consideration and the shorthand notation $\Sigma \equiv \Sigma_{11}$, we obtain

$$G(P) = G^{(0)}(P) + G(P)[\Sigma(P) - \mu]G^{(0)}(P) + F(P)\tilde{\Sigma}(P)G^{(0)}(P), \qquad (12.48)$$
$$F(P) = F(P)[\Sigma(P') - \mu]G^{(0)}(P') + G(P)\tilde{\Sigma}(P)G^{(0)}(P'). \qquad (12.49)$$

The solution in terms of self-energies reads

$$G(P) = \frac{i\xi + \epsilon(|2\mathbf{k}_0 - \mathbf{k}|) + \Sigma(P') - \mu}{D(P)}, \qquad F(P) = -\frac{\tilde{\Sigma}(P)}{D(P)}, \qquad (12.50)$$

where

$$D = \tilde{\Sigma}^2(P) - [\epsilon(|2\mathbf{k}_0 - \mathbf{k}|) + \Sigma(P') - \mu + i\xi][\epsilon(k) + \Sigma(P) - \mu - i\xi]. \qquad (12.51)$$

With these relations at hand, one can calculate the current density induced by the phase gradient in the condensate wavefunction; see Section 12.3.4.

12.2.5 Low-Density Limit in 2D and 3D: Pseudopotentials

In two and three dimensions, the expansion in terms of the bare interaction potential can be (and, in most realistic cases, is) nonperturbative. The system is regarded as weakly interacting only because of the low density of particles. In one dimension—in contrast to 2D and 3D—the physics is perturbative in the high-density limit (at a fixed interaction potential). Theoretically, dealing with the strong bare potential implies summation of an infinite sequence of ladder diagrams and produces an effective interaction in the form of the four-point vertex, Γ; see Figure 12.6. The analytical relation behind Figure 12.6 reads

$$\Gamma(P_1, P_2, Q) = \mathcal{U}(\mathbf{q}) - \sum_K \mathcal{U}(\mathbf{q} - \mathbf{k}) G(P_1 + K) G(P_2 - K) \Gamma(P_1, P_2, K)$$

$$\equiv \mathcal{U}(\mathbf{q}) - \sum_K \Gamma(P_1 + K, P_2 - K, Q - K) G(P_1 + K) G(P_2 - K) \mathcal{U}(\mathbf{k}).$$

$$(12.52)$$

When the bare interaction lines are replaced with Γ's, the rest of the series becomes perturbative (excluding the critical region). On the technical side, working with Γ's is convenient as long as one does not intend to systematically take into account higher-order corrections, thus utilizing only the (simple and transparent) leading-order expression for $\Gamma(0,0,0)$. For the higher-order corrections, Equation (12.52) involves three different length scales—the size of the potential, R_0, the healing length of the Bose condensed system, and the de Broglie wavelength—while only the nonperturbative physics at the scale R_0 requires summation of the ladder diagrams. Hence, it makes sense to construct an object with a much simpler structure than Γ that captures the nonperturbative physics [and thus coincides with the leading-order expression for $\Gamma(0,0,0)$] and then to systematically investigate the difference between this object and Γ. To achieve this goal, let us define the *pseudopotential* $U(\mathbf{q})$ by the equation (see Figure 12.7)

$$U(\mathbf{q}) = \mathcal{U}(\mathbf{q}) + \sum_{\mathbf{k}} \mathcal{U}(\mathbf{q} - \mathbf{k})\Pi(\mathbf{k})U(\mathbf{k}) \equiv \mathcal{U}(\mathbf{q})$$

$$+ \sum_{\mathbf{k}} U(\mathbf{q} - \mathbf{k})\Pi(\mathbf{k})\mathcal{U}(\mathbf{k}), \qquad\qquad (12.53)$$

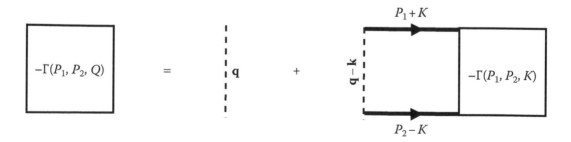

Figure 12.6 Ladder diagrams leading to Equation (12.52).

Figure 12.7 The diagrammatic expression for the pseudopotential (left-hand side) involves the bare potential (first term on the right-hand side) and a modified two-particle propagator $\Pi(\mathbf{k})$.

where $\Pi(\mathbf{k})$ is such that when k is much larger than the inverse healing length or thermal momentum, the value of $\Pi(\mathbf{k})$ approaches that of $-G(P_1 + K)G(P_2 - K)$ summed over the frequency of the $(d + 1)$-vector K. That is,

$$\Pi(\mathbf{k}) \to -\frac{1}{2\epsilon(k)} \quad \text{at} \quad k \to \infty. \tag{12.54}$$

In this case, $\Gamma(P_1, P_2, Q) \approx U(\mathbf{q})$, and one can expand the difference in a perturbative series. As a result, we arrive at the diagrammatic technique where instead of thin dashed lines standing for the bare interaction potential, we have bold dashed lines representing the pseudopotential U, with an additional requirement that whenever two (normal) Green's function lines are sandwiched between two pseudopotential lines, the former are supposed to be summed over the frequency difference (which is always possible in view of the frequency independence of the pseudopotential), and then Π has to be subtracted from the result of summation. Specifically, if P_1 and P_2 are the two external $(d + 1)$-momenta of the previously mentioned diagrammatic element, then the following replacement is supposed to take place for internal propagator lines:

$$\sum_{\xi^{(K)}} G(P_1 + K)G(P_2 - K) \to \sum_{\xi^{(K)}} G(P_1 + K)G(P_2 - K) + \Pi(\mathbf{k}), \tag{12.55}$$

where $\xi^{(K)}$ is the frequency of the $(d + 1)$-momentum K.

As long as we are not interested in the nonuniversal UV corrections, we can replace $U(\mathbf{q})$ with $U(0)$, the systematic error introduced by the replacement being controlled by the following dimensional estimate:

$$U(\mathbf{q}) = U(0)\left[1 + \mathcal{O}\left(q^2 R_0^2\right)\right]. \tag{12.56}$$

One may wonder how this estimate is reconciled with the momentum dependence of Γ, which is of the order qR_0 in 3D and of the order $1/\ln(1/qR_0)$ in 2D.

The solution is provided by (12.55), explaining that this dependence, which is both universal and perturbative, is taken into account by the second-order ladder-type diagram in U.

In 3D, a natural choice for $\Pi(\mathbf{k})$ is

$$\Pi(\mathbf{k}) = -\frac{1}{2\epsilon(k)} \qquad (d = 3), \tag{12.57}$$

leading to the simple expression

$$U(0) = \frac{4\pi a}{m} \qquad (d = 3), \tag{12.58}$$

with a the s-wave scattering length. In 2D, one cannot use (12.57) because of the infrared divergence of the integral in (12.53). A reasonable choice here is

$$\Pi(\mathbf{k}) = -\frac{1}{2}\frac{1}{\epsilon(k) + \epsilon_0} \qquad (d = 2), \tag{12.59}$$

where ϵ_0 is an arbitrary low-energy cutoff. The particular value of ϵ_0 is not important since final answers are not sensitive to it. Within the systematic error bars of the description discussed later, any choice of ϵ_0 such that $nU \leq \epsilon_0 \leq n/m$ is equally reasonable in terms of accuracy, provided the temperature is not much larger than n/m; moreover, replacing ϵ_0 with ck is also acceptable. (An optimal choice of ϵ_0 in the regime $T \gg n/m$ will be discussed in Section 12.6.2.) There is also no need to introduce Π in 1D. We will assume that formally $\Pi(\mathbf{k}) \equiv 0$ in 1D, in which case our final answers can be used as written in all spatial dimensions.

Given that (12.57) and (12.59) are not unique [there is a free parameter in (12.59)], it is instructive to explicitly relate two pseudopotentials, U_1 and U_2—corresponding to Π_1 and Π_2, respectively—to *each other*. We notice that (12.53) implies

$$U_2(\mathbf{q}) = U_1(\mathbf{q}) + \sum_{\mathbf{k}} U_1(\mathbf{q} - \mathbf{k})[\Pi_2(\mathbf{k}) - \Pi_1(\mathbf{k})]U_2(\mathbf{k}). \tag{12.60}$$

In view of (12.56), this simplifies [up to $\mathcal{O}(q^2 R_0^2)$ terms that we neglect in what follows] to

$$U_2 = U_1 + U_1 C_{12} U_2, \qquad C_{12} = \sum_{\mathbf{k}}[\Pi_2(\mathbf{k}) - \Pi_1(\mathbf{k})]. \tag{12.61}$$

If Π_1 and Π_2 are defined by (12.59) with different cutoffs $\epsilon_0^{(1)}$ and $\epsilon_0^{(2)}$, we have

$$C_{12} = \frac{m}{4\pi}\ln\frac{\epsilon_0^{(2)}}{\epsilon_0^{(1)}} \qquad (d = 2). \tag{12.62}$$

The right choice of the functions Π_1 and Π_2 implies that U_1 and U_2 differ only by subleading terms. Thus, we can expand the right-hand side of (12.61) in powers of $|U_1 C_{12}| \ll 1$:

$$U_2 = U_1 + U_1^2 C_{12} + \cdots. \tag{12.63}$$

Formally, the expansion in terms of the pseudopotential is perturbative only as long as we exclude the high-momentum contributions to the $(M > 3)$-body diagrams generating, upon complete summation, M-body scattering amplitudes. In a dilute gas, the corresponding diagrams are small (containing extra powers of the gas parameter na^3) and are neglected in this manuscript.

Expression (12.58) has the meaning of mapping a dilute 3D system with an arbitrary short-range interaction potential onto a system with the hard-sphere potential, $\mathcal{U}_{\text{hard}}(r) = 0$ [$\mathcal{U}_{\text{hard}}(r) = \infty$] at $r \geq a$ [$r < a$], so that the pseudopotentials of the two systems coincide. The same approach is possible (and popular) in 2D, the parameter a being called the 2D scattering length, since the mapping applies to scattering properties as well. For a given potential $\mathcal{U}$, the value of a can be obtained either from the asymptotic behavior of the pseudopotential U at appropriately small ϵ_0 or from the asymptotic behavior of the scattering amplitude $f(\mathbf{k}, \mathbf{k}')$ at $k, k' \to 0$. Both asymptotic behaviors are closely related to each other since the large-k limit of the kernel $\Pi(\mathbf{k})$ in (12.53) coincides with the large-k limit of the scattering kernel

$$\Pi_{\text{sc}}(k, k') = \frac{m}{k'^2 - k^2 + i0} \tag{12.64}$$

in the integral equation for $f(\mathbf{k}, \mathbf{k}')$:

$$f(\mathbf{k}, \mathbf{k}') = \mathcal{U}(\mathbf{k} - \mathbf{k}') + \sum_q \mathcal{U}(\mathbf{k} - \mathbf{q}) \Pi_{\text{sc}}(q, k') f(\mathbf{q}, \mathbf{k}')$$

$$\equiv \mathcal{U}(\mathbf{k} - \mathbf{k}') + \sum_q f(\mathbf{k}, \mathbf{q}) \Pi_{\text{sc}}(q, k') \mathcal{U}(\mathbf{q} - \mathbf{k}'). \tag{12.65}$$

Moreover, a direct relationship between U and $f(\mathbf{k}, \mathbf{k}')$ is readily obtained by noticing that (12.65) and (12.53) imply [cf. (12.60)]

$$f(\mathbf{k}, \mathbf{k}') = U(\mathbf{k} - \mathbf{k}') + \sum_q U(\mathbf{k} - \mathbf{q})[\Pi_{\text{sc}}(q, k') - \Pi(q)] f(\mathbf{q}, \mathbf{k}'), \tag{12.66}$$

which dramatically simplifies upon replacing $U(\mathbf{q})$ with $U(0)$, in accordance with (12.56). In this case, $f(\mathbf{k}, \mathbf{k}')$ is $\mathbf{k}$-independent, and for $f \equiv f(k')$, we get

$$f(k') = U + U f(k') \sum_q [\Pi_{\text{sc}}(q, k') - \Pi(\mathbf{q})]. \tag{12.67}$$

Substituting (12.59) and (12.64) into (12.67) and performing the integral, we find

$$f(k') = \frac{4\pi/m}{\ln(2m\epsilon_0/k'^2) + 4\pi/mU + i\pi} \qquad (d = 2). \qquad (12.68)$$

Comparing this to the known hard-disk result

$$f(k') = \frac{4\pi/m}{\ln(4/a^2 k'^2) - 2\gamma + i\pi} \qquad (d = 2), \qquad (12.69)$$

where $\gamma = 0.5772\ldots$ is Euler's constant, we conclude that

$$U = \frac{4\pi/m}{\ln(2/a^2 m\epsilon_0) - 2\gamma} \qquad (d = 2). \qquad (12.70)$$

In 3D, Equation (12.67)—with $\Pi(\mathbf{k})$ given by (12.57)—yields

$$f(k') = \frac{U}{1 - imk'U/4\pi} \qquad (d = 3), \qquad (12.71)$$

thus leading to (12.58) by comparison with known expression for the hard-sphere scattering amplitude.

12.2.6 Effective Interaction in Quasi-2D/Quasi-1D Cases

It is important to note that while our derivations of the effective interaction U in 2D were dealing with a pure 2D case, Equations (12.59) through (12.70) are also valid for the quasi-2D situation, when the reduced effective dimensionality (motion in the xy plane) is achieved by restricting the 3D motion by an appropriate potential $V \equiv V(z)$. The reason why Equations (12.59) through (12.70) remain valid is essentially the same as the reason the details of the short-range interaction potential drop out from the final answers, leaving the scattering length as the only relevant parameter. In a dilute 2D (and 3D) system, the length scales much shorter than the interparticle separation can be effectively excluded from the problem by introducing an appropriate boundary condition on the many-body wavefunction in terms of its behavior in the limit $|\mathbf{r}_i - \mathbf{r}_j| \ll n^{-1/d}$ (with $\mathbf{r}_i$ and $\mathbf{r}_j$ being coordinates of any two particles) corresponding to two-body scattering with an appropriate s-scattering length. The diagrammatic technique allows one to readily prove this statement, along with establishing important quantitative relations in the limiting case when the localization length of the quasi-2D confinement, l_*, is much larger than the size R_0 of the 3D interaction potential.

In the case of a quasi-2D system, the 3D Green's function of a noninteracting gas with zero chemical potential reads

$$G^{(0)}(\tau_1, \tau_2, \mathbf{r}_1, \mathbf{r}_2) = T \sum_{\xi, \mathbf{k}, \nu} G^{(0)}(\xi, \mathbf{k}, \nu)\varphi_\nu(z_1)\varphi_\nu^*(z_2) e^{i\mathbf{k}\cdot(\vec{\rho}_1 - \vec{\rho}_2) - i\xi(\tau_1 - \tau_2)}, \qquad (12.72)$$

$$G^{(0)}(\xi, \mathbf{k}, \nu) = [i\xi - \epsilon(k) - \tilde{\epsilon}_\nu]^{-1}. \qquad (12.73)$$

Here $\vec{\rho}_{1,2}$ is the xy projection of the 3D vector $\mathbf{r}_{1,2}$ and $\mathbf{k}$ is the corresponding 2D Fourier wavevector; φ_ν is the νth eigenfunction of the 1D motion in the potential $V(z)$, and $\tilde{\epsilon}_\nu$ is the corresponding energy eigenvalue. Without loss of generality, we assume that $\nu = 0, 1, 2, \dots$ and the ground-state energy is $\tilde{\epsilon}_0 = 0$. Equations (12.72) and (12.73) suggest working in the ξ-representation and employing the graphical elements shown in Figure 12.8. Ultimately, the dotted lines can be formally "detached" from Green's function and get associated with the interaction lines that connect to Green's function ends. Then the integration over the spatial variables can be performed, which will render the resulting technique very similar to the purely 2D case in momentum representation, with the only difference being that propagators will have an additional (discrete) variable, ν, and the interaction lines will depend on all four ν's. The next observation is that in the dilute regime and at not very high temperatures, the lines with $\nu \neq 0$ are important only in the ladder diagrams, yielding small perturbative contributions otherwise.[*] Moreover, even in ladder diagrams, the $\nu = 0$ lines play a special role, since it is only at $\tilde{\epsilon}_\nu = 0$ the two-propagator ladder elements yield the logarithmic behavior responsible for the infrared renormalization of interaction. This structure of the ladder diagrams (illustrated in Figure 12.9) brings us to the conclusion that with logarithmic accuracy (and up to higher-order perturbative corrections), the ladders *factorize* into products of purely 2D loops and 3D "braids"; see Figure 12.9. The braids are the ladders on their own. Denoting the sum of all the braid diagrams (for definiteness, with zero external momenta and frequency) with $\tilde{U}$ and replacing the $\nu = 0$ ladder

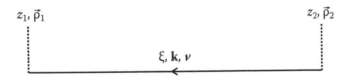

Figure 12.8   The free-particle propagator in the quasi-2D case. The two dotted lines at the ends 1 and 2 represent the factors $\varphi_\nu(z_1)e^{i\mathbf{k}\cdot\vec{\rho}_1}$ and $\varphi_\nu^*(z_2)e^{-i\mathbf{k}\cdot\vec{\rho}_2}$, respectively.

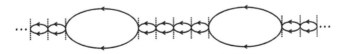

Figure 12.9   Schematic illustration of the structure of ladder diagrams in the quasi-2D case. The long loops denote the pairs of $\nu = 0$ propagators, and the short loops stand for the pairs of propagators in which at least one line has $\nu \neq 0$. For briefness, the interaction lines are not shown.

---

[*] This property can be taken as the precise meaning of the term "quasi-2D system."

loops with the 2D function Π (in a direct analogy with purely 2D case), we arrive at the following geometric series for the sum of the quasi-2D ladder:

$$U_{2D} = \tilde{U} + \tilde{U}Q\tilde{U} + \tilde{U}Q\tilde{U}Q\tilde{U} + \cdots, \tag{12.74}$$

$$Q \approx \int\limits_{k \leq k_*} \Pi(k)d^2k \approx -\frac{m}{4\pi}\ln\frac{k_*^2}{m\epsilon_0},$$

$$k_* \ll \min\left\{l_*^{-1}, R_0^{-1}\right\}, \tag{12.75}$$

which sums into

$$U_{2D} = \frac{4\pi/m}{4\pi/m\tilde{U} + \ln\left(k_*^2/m\epsilon_0\right)}. \tag{12.76}$$

At $l_* \sim R_0$, little can be said about the value of $\tilde{U}$ apart from the estimate (for nonperturbative potentials)

$$m\tilde{U} \sim 1 \qquad (l_* \sim R_0), \tag{12.77}$$

following by continuity from the purely 2D limit of $l_* \ll R_0$, in which case Equation (12.76) with $k_* \sim 1/R_0$ is supposed to reproduce Equation (12.70) with $a \sim R_0$. At $l_* \gg R_0$, the value of $\tilde{U}$ becomes insensitive to the details of the interaction potential and can be expressed in terms of its 3D scattering length a_{3D}. Indeed, in this case, the relevant momenta in the braids are much larger than l_*^{-1} and $\tilde{\epsilon}_\nu$ are much larger than the transverse confinement energy making the original homogeneous 3D nomenclature more convenient for evaluating the braids; that is, up to the relative corrections $\sim a_{3D}/l_*$ (coming from momenta $\sim l_*^{-1}$), the loops in the braids can be replaced with the 3D function Π, Equation (12.57). As a result, we get

$$\tilde{U} \approx U_{3D} \int [\varphi_0(z)]^4 dz = \frac{U_{3D}}{l_*} = \frac{4\pi a_{3D}}{ml_*} \qquad (l_* \ll R_0), \tag{12.78}$$

where we introduce an exact definition of l_* as

$$l_*^{-1} = \int [\varphi_0(z)]^4 dz \qquad (d = 2). \tag{12.79}$$

The factor $[\varphi_0(z)]^4$ originates from the four dotted lines separating the braids from the 2D loops; see Figure 12.9. All four z coordinates are set equal to each other because the typical scale of variation of $\varphi_0(z)$ is l_*, while the characteristic length scale in the braid is R_0.

Introducing an order-unity constant $\tilde{c}$ accounting for logarithmic accuracy on the momentum cutoff scale, we obtain

$$U_{2D} = \frac{4\pi/m}{l_*/a_{3D} + \ln\left(\tilde{c}/ml_*^2\epsilon_0\right)} \qquad (l_* \ll R_0). \tag{12.80}$$

If necessary, the precise value of $\tilde{c}$ can be found, for example, from accurately solving the scattering problem. In a very important (for ultracold atoms) case of parabolic potential $V(z) = m\omega_0^2 z^2/2$ (where $l_* = \sqrt{2\pi/m\omega_0}$), we can use the known result for the scattering amplitude [11],

$$f(k) = \frac{4\pi/m}{l_*/a_{3D} + \ln(2/l_*^2 k^2) + i\pi},$$ (12.81)

and, comparing it to Equations (12.68) and (12.80), see that in this case, $\tilde{c} = 1$. By comparing (12.80) to (12.70), we also find an explicit expression for the 2D scattering length for our quasi-2D problem:

$$a_{2D} = (2/\tilde{c})^{1/2} l_* e^{-(l_*/2a_{3D}+\gamma)} \qquad (l_* \ll R_0).$$ (12.82)

Similar analysis can be performed in the quasi-1D case. Assuming that the 1D motion is along the z-axis, the 1D analogs of Equations (12.72) and (12.73) are obtained by swapping z and $\vec{\rho}$ coordinates. As in 2D, it proves convenient to introduce a precise definition of localization length in terms of the wavefunction of the transverse motion, the 1D analog of (12.79) having the form

$$l_*^{-2} = \int [\varphi_0(\vec{\rho})]^4 d^2\rho \qquad (d=1).$$ (12.83)

A qualitative difference with the quasi-2D case is that the $\nu=0$ loops of the ladder diagrams (see Figure 12.9) do not have significant UV contributions, so that these loops are either perturbative or the whole theory becomes strongly interacting (this goes beyond the scope of our discussion here). This means that in quasi-1D, only the braids are subject to ladder summation. Moreover, for the same reason, the summation of braids is practically meaningful only at $R_0 \ll l_*$, since at $R_0 \geq l_*$, the only way to get a weak effective interaction is to deal with a weak (Born) 3D potential, implying that the braids become perturbative and the effective vertex of 1D interaction reduces to

$$U_{1D} = \int \mathcal{U}(\mathbf{r})[\varphi_0(\vec{\rho})]^4 d^3r \qquad \text{(Born approximation)}.$$ (12.84)

At $R_0 \ll l_*$, the summation of braids is essentially the same as in 2D, since within the leading approximation, we are dealing with a 3D theory without external potential. As a result, we obtain

$$U_{1D} = \frac{4\pi a_{3D}}{ml_*^2}\left[1 + \mathcal{O}\left(\frac{a_{3D}}{l_*}\right)\right].$$ (12.85)

The estimate of the subleading term $\sim a_{3D}/l_*$ readily follows from observation that at the momenta $\sim l_*^{-1}$, the braid loops become sensitive to the presence of confining potential, which qualitatively is equivalent to replacing the function Π of Equation (12.57) with, say, that of Equation (12.59) with $\epsilon_0 \sim 1/ml_*^2$; relation (12.61) then yields the estimate. From this consideration, it is also clear that the correction term has the form of the universal factor a_{3D}/l_* times an order-unity constant that depends on the particular type of confining potential. In particular, in the case of axisymmetric parabolic potential, this constant is approximately equal to $-1.46\sqrt{\pi}$ [12].

It is worth noting that the condition $R_0 \ll l_*$ is sufficient for deriving Equations (12.80) and (12.85) even when $a_{3D} \geq l_*$. However, such a situation, implying $a_{3D} \gg R_0$, is only possible when the 3D interaction is fine-tuned to the resonance. In its turn, the resonant interaction brings about the problem of metastability, since resonant bosons always form three-body bound states (the Efimov effect).

12.3 Thermodynamic Functions in the Quasicondensate Region

12.3.1 Basic Relations and Notions

Explicitly calculating the lowest-order diagrams shown in Figure 12.10 and utilizing (12.30), one finds

$$\Sigma(P) = -2G(r = 0, \tau = -0)U + 2n_0 U = 2nU, \tag{12.86}$$

$$\tilde{\Sigma}(P) = n_0 U. \tag{12.87}$$

We see that within the first approximation, both Σ and $\tilde{\Sigma}$ turn out to be momentum and frequency independent. [It is easy to check that the next-order diagrams inevitably introduce momentum and frequency dependence to self-energies and drastically change the structure of the theory.] At this level of accuracy, the chemical potential equals to $\mu = 2nU - n_0 U$, according to the Hugenholtz–Pines relation. As mentioned earlier, it is extremely convenient to use an effective chemical potential

$$\tilde{\mu} = \mu - 2nU \tag{12.88}$$

as a thermodynamic variable to characterize properties of the WIBG. [Note that $\tilde{\mu}$ is negative.] Within the same accuracy, we can substitute $\tilde{\Sigma}$ with $-\tilde{\mu} \equiv |\tilde{\mu}|$ and simplify expressions for G and F to

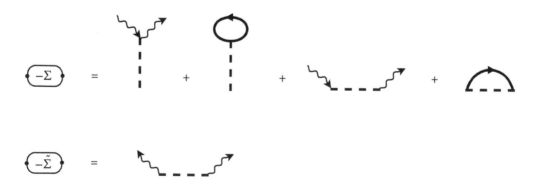

Figure 12.10   The lowest-order diagrams for the self-energy Σ in Equation (12.86) and the anomalous self-energy $\tilde{\Sigma}$ in Equation (12.87).

$$G(P) = -\frac{i\xi + \epsilon(k) + |\tilde{\mu}|}{\xi^2 + E^2(k)}, \qquad F(P) = \frac{|\tilde{\mu}|}{\xi^2 + E^2(k)}, \tag{12.89}$$

$$E^2(k) = \epsilon(k)[\epsilon(k) + 2|\tilde{\mu}|]. \tag{12.90}$$

A note is in order here: Later, we will calculate higher-order corrections to the chemical potential that are necessary for the construction of the accurate equation of state; see (12.93) and (12.95). However, it is still possible to use $\tilde{\mu}$ in the definitions of the propagators and the spectrum of elementary excitations while keeping the accuracy of the entire scheme intact; see Section 12.4, where we study systematic errors involved in approximations. Here we mention briefly that the anomalous average contribution to thermodynamic properties is always small, and thus, further corrections to F are negligible. The same is true for G at low temperature. At temperatures $T \gg n_0 U$, on the other hand, one can ignore tiny modifications in the spectrum because of an additional small parameter $n_0 U/T$.

For the total density, we have

$$n = n_0 - G(r = 0, \tau = -0) = n_0 + \sum_k \left[\frac{\epsilon(k) - \tilde{\mu}}{2E(k)}(1 + 2N_E) - \frac{1}{2} \right], \tag{12.91}$$

where

$$N_E = \left(e^{E/T} - 1\right)^{-1} \tag{12.92}$$

is the Bose–Einstein occupation number for the mode with the energy E. Formula (12.91) includes the leading and subleading terms. To get an expression for the chemical potential with the same degree of accuracy, we need to take into account higher-order diagrams and go beyond the leading-order expressions for Σ and $\tilde{\Sigma}$. This is because, in the absence of interaction, the chemical potential is identically zero, and thus, the leading term is proportional to U.

Three second-order diagrams contribute to Θ. The first one is the anomalous Green's function convoluted with the bare interaction potential (see Figure 12.11). The second one is the "sunrise" diagram with the proper correction (12.55) for the ladder structure, and the third one is similar to the sunrise diagram, but with propagators connecting different vertices (see the two diagrams in the second row in Figure 12.11). The latter two can be safely neglected away from the fluctuation regime because they involve an additional small parameter γ_0 or γ_T; see Section 12.4 for the analysis and definitions of the fluctuation region and parameters γ in different spatial dimensions. It is worth noting that consistently taking into account contributions of the two neglected diagrams in the condensed regime would require simultaneously going to higher orders in the expansions for self-energies (Figure 12.10). The latter, however, is impossible without sacrificing the attractive paradigm of independent quasiparticles with Bogoliubov dispersion.

Keeping the leading and the largest subleading diagrams for Θ, we get

$$\mu = -2G(\tau = -0, r = 0)U + n_0 \mathcal{U}(0) - \sum_q \mathcal{U}(\mathbf{q})F(\mathbf{q}, \tau = 0). \tag{12.93}$$

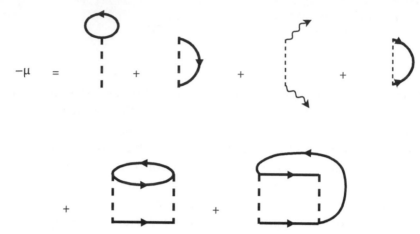

Figure 12.11   Diagrams contributing to the chemical potential μ of a condensed gas, up to the second order. The last two diagrams [which actually must be corrected according to (12.55)] are smaller than the previous ones and will be neglected; see the text for more detail.

We have no choice but to use the bare potential here because all ladder diagrams leading to the pseudopotential vertex are already absorbed in the anomalous Green's function, by construction of the latter. This feels unsatisfactory only at first glance since simple formal manipulations allow one to express (12.93) in terms of U alone. Let us introduce an auxiliary function $\Delta(\mathbf{q})$ defined by the following integral equation (shown graphically in Figure 12.12):

$$\Delta(\mathbf{q}) = F(\mathbf{q}, \tau = 0) + n_0 \Pi(\mathbf{q}) U(\mathbf{q}) - \sum_{\mathbf{k}} \Pi(\mathbf{q} - \mathbf{k}) U(\mathbf{q} - \mathbf{k}) \Delta(\mathbf{k}). \tag{12.94}$$

It can be used, in combination with the definition of the pseudopotential Equation (12.53), to transform the last two terms in (12.93):

$$n_0 \mathcal{U}(0) - \sum_{\mathbf{q}} \mathcal{U}(\mathbf{q}) F(\mathbf{q}, \tau = 0) = n_0 U(0) - \sum_{\mathbf{q}} U(\mathbf{q}) \Delta(\mathbf{q}) \approx (n_0 - \Delta_0) U, \tag{12.95}$$

where

$$\Delta_0 = \sum_{\mathbf{q}} \Delta(\mathbf{q}). \tag{12.96}$$

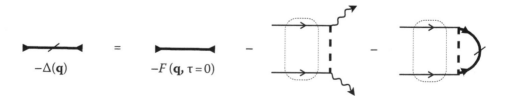

Figure 12.12   Definition of $\Delta(\mathbf{q})$.

The last approximate equality in (12.95) comes from $U(\mathbf{q}) \approx U(0)$ and the observation that $\Delta(\mathbf{q})$ vanishes at momenta much smaller than $1/R_0$.

The smallness of the parameter $\left| U \Pi k^d \right|$ at momenta $k \ll 1/R_0$ allows one to expand Δ [see (12.94)] in the series:

$$\Delta(\mathbf{q}) = [F(\mathbf{q}, \tau = 0) + n_0 \Pi(\mathbf{q}) U(\mathbf{q})] + \cdots. \tag{12.97}$$

For the effective chemical potential, we thus obtain

$$\tilde{\mu} = -(n_0 + \Delta_0) U, \tag{12.98}$$

where, within our order of accuracy, we can take

$$\Delta_0 = \sum_{\mathbf{q}} [F(\mathbf{q}, \tau = 0) + n_0 U \Pi(\mathbf{q})]. \tag{12.99}$$

We find it convenient to introduce a quantity with the dimension of density:

$$n_* = -\tilde{\mu}/U = n_0 + \Delta_0. \tag{12.100}$$

With this quantity, the form of certain thermodynamic relations simplifies. For example, using (12.100), we can replace n_0 with n_* in (12.91) and arrive at the result

$$n = n_* + n', \tag{12.101}$$

$$n' = \sum_{\mathbf{k}} \left[\frac{\epsilon(k) - E(k)}{2E(k)} + \tilde{\mu} \Pi(k) + \frac{\epsilon(k)}{E(k)} N_E \right]. \tag{12.102}$$

This completes the self-consistent theory because we obtain a closed set of relations that define $n = n(\mu, T)$ in the parametric form—given some $\tilde{\mu}$, or n_*, one calculates n from (12.101) and (12.102) and then determines μ from (12.88). The integral in (12.102) is convergent not only in 3D but also in 2D and 1D. Hence, at this point, the quantity η can be set equal to zero.

We emphasize the fact that all specific expressions derived in this section feature a two-parametric representational freedom, within the same order of accuracy. First, it is possible to use $\tilde{\mu}$ with or without higher-order corrections in the spectrum of elementary excitations and propagators. Second, there is a freedom in choosing the function $\Pi(k)$. For example, one could be tempted to absorb the first two terms in the integral in (12.102) into the definition of U, such that $n'(T = 0)$ becomes identically equal to zero, and $n_*(T = 0)$ equals the total number of particles. However, we do not see any merit in this protocol, because U becomes dependent on $\tilde{\mu}$ and cannot be considered as a fixed external parameter. Third, and finally, this and analogous "improvements" do not make the theory more accurate (or less accurate) since the difference is of the same order as omitted diagrams.

A remark is in order here concerning the 2D case, where the value of U cannot be defined irrespective of the system density. Even in this case, we can proceed with formally independent parameter ϵ_0 in (12.59) until we arrive at the final answers along with the estimates of neglected higher-order terms (the latter being essentially ϵ_0 dependent). After that, the value of ϵ_0 is selected in such a way that the order-of-magnitude values of neglected terms are minimal.

12.3.2 Pressure, Entropy, and Energy

To derive expressions for other thermodynamic properties, we start with pressure as a function of T and $\tilde{\mu}$. Using the general thermodynamic formula

$$n = \frac{\partial p(\mu, T)}{\partial \mu},\tag{12.103}$$

and adopting—throughout this subsection—the convention that T is treated as a fixed constant, so that partial derivatives with respect to either μ or $\tilde{\mu}$ can be replaced with ordinary ones, we write

$$p = p_c + \int_0^{\tilde{\mu}} n\frac{d\mu}{d\tilde{\mu}}d\tilde{\mu},\tag{12.104}$$

where

$$p_c = n_c^2 U - T\sum_k \ln\left[1 - e^{-\epsilon(k)/T}\right]\tag{12.105}$$

is the value of pressure at the mean-field critical density

$$n_c = \sum_k \left[e^{\epsilon(k)/T} - 1\right]^{-1}.\tag{12.106}$$

Equations (12.88) and (12.101) allow one to represent (12.104) as

$$p = p_c - n_c^2 U + \tilde{\mu}^2/2U - 2n'\tilde{\mu} + (n')^2 U + \int_0^{\tilde{\mu}} n'd\tilde{\mu}\tag{12.107}$$

(after a straightforward integration by parts). The integral in (12.107) is readily performed by noting that

$$\frac{\epsilon(k)}{E(k)} = -\frac{dE(k)}{d\tilde{\mu}},\qquad \frac{\epsilon(k)N_E}{E(k)} = -T\frac{d}{d\tilde{\mu}}\ln\left[1 - e^{-E(k)/T}\right],\tag{12.108}$$

and thus,

$$n' = -\frac{1}{2}\frac{d}{d\tilde{\mu}}\sum_{\mathbf{k}}\left[E(k) - \epsilon(k) + \tilde{\mu} - \tilde{\mu}^2\Pi(k)\right] - T\frac{d}{d\tilde{\mu}}\sum_{\mathbf{k}}\ln\left[1 - e^{-E(k)/T}\right], \qquad (12.109)$$

where $\epsilon(k)$ makes the first term convergent. The result of the integration is

$$\int_0^{\tilde{\mu}} n'd\tilde{\mu} = -\sum_{\mathbf{k}}\left\{\frac{E(k) - \epsilon(k) + \tilde{\mu} - \tilde{\mu}^2\Pi(k)}{2} + T\ln\frac{1 - e^{-E(k)/T}}{1 - e^{-\epsilon(k)/T}}\right\}, \qquad (12.110)$$

and we finally obtain Equation (12.8).

Whenever n, μ, and p are specified as functions of (T, x), where x is a quantity of arbitrary nature, the expressions for entropy per unit volume, s, and energy per unit volume, ε, are readily found from the following two generic thermodynamic relations:

$$s = \left(\frac{\partial p}{\partial T}\right)_x - n\left(\frac{\partial \mu}{\partial T}\right)_x, \qquad \varepsilon = \mu n + Ts - p. \qquad (12.111)$$

With $x \equiv \tilde{\mu}$, we thus obtain Equations (12.9) and (12.10).

12.3.3 Explicit Integrations: The $T=0$ Case

It is useful to note that the following two integrals (we use $\tilde{\mu} = -|\tilde{\mu}|$),

$$I_1^{(d)}(|\tilde{\mu}|) = \frac{1}{2}\sum_{\mathbf{k}}\left[E(k) - \epsilon(k) - |\tilde{\mu}| - \tilde{\mu}^2\Pi(k)\right], \qquad (12.112)$$

$$I_2^{(d)}(|\tilde{\mu}|) = \frac{1}{2}\sum_{\mathbf{k}}\left[\frac{\epsilon(k)}{E(k)} - 2|\tilde{\mu}|\Pi(k) - 1\right] = \frac{\partial I_1^{(d)}}{\partial |\tilde{\mu}|}, \qquad (12.113)$$

entering relations for thermodynamic quantities—see Sections 12.3.1 and 12.3.2—can be explicitly performed. [Here we assume that, in 3D and 2D, the kernel $\Pi(k)$ is fixed by expressions (12.57) and (12.59), respectively, and remind that, in 1D, it is zero.] Explicitly doing the integrals is especially useful at $T = 0$. In this case, all other integrals in the expressions for thermodynamic functions nullify, and the answers reduce to algebraic relations. In the grand canonical form, these relations read

$$|\tilde{\mu}| = \mu - 2UI_2^{(d)}(\mu), \qquad n = \frac{\mu}{U} - I_2^{(d)}(\mu), \qquad p = \frac{\mu^2}{2U} - I_1^{(d)}(\mu) \qquad (T = 0).$$

$$(12.114)$$

Here we take into account that, within our level of accuracy, we can replace $|\tilde{\mu}| \to \mu$ in the arguments of $I_1^{(d)}$ and $I_2^{(d)}$, since the integrals are responsible for subleading corrections, while the sub-subleading terms are beyond our control.

Performing the integrations, one finds the following:

$$I_1^{(d=1)}(|\tilde{\mu}|) = -\frac{2\sqrt{m}|\tilde{\mu}|^{3/2}}{3\pi}, \tag{12.115}$$

$$I_1^{(d=2)}(|\tilde{\mu}|) = \frac{m\tilde{\mu}^2}{8\pi}\left(\ln\frac{|\tilde{\mu}|}{2\epsilon_0} + \frac{1}{2}\right), \tag{12.116}$$

$$I_1^{(d=3)}(|\tilde{\mu}|) = \frac{8m^{3/2}|\tilde{\mu}|^{5/2}}{15\pi^2}, \tag{12.117}$$

and differentiating with respect to $|\tilde{\mu}|$, we find that

$$I_2^{(d=1)}(|\tilde{\mu}|) = -\frac{\sqrt{m|\tilde{\mu}|}}{\pi}, \tag{12.118}$$

$$I_2^{(d=2)}(|\tilde{\mu}|) = \frac{m|\tilde{\mu}|}{4\pi}\left(\ln\frac{|\tilde{\mu}|}{2\epsilon_0} + 1\right), \tag{12.119}$$

$$I_2^{(d=3)}(|\tilde{\mu}|) = \frac{4m^{3/2}|\tilde{\mu}|^{3/2}}{3\pi^2}. \tag{12.120}$$

In 2D, we have the freedom of fine-tuning ϵ_0 to simplify the form of the answer. A natural choice, especially convenient for the $T = 0$ limit, is to set $\epsilon_0 = (e/2)|\tilde{\mu}|$, in which case we have

$$I_2^{(d=2)} \equiv 0, \qquad I_1^{(d=2)} = -\frac{m\tilde{\mu}^2}{16\pi}. \tag{12.121}$$

At $T = 0$, this translates into compact grand canonical expressions:

$$|\tilde{\mu}| = \mu, \qquad n = \frac{\mu}{U}, \qquad p = \frac{\mu^2}{2U}\left(1 + \frac{mU}{8\pi}\right) \qquad (d = 2, \quad T = 0). \tag{12.122}$$

with

$$U = \frac{4\pi/m}{\ln(4/a^2 m\mu) - 2\gamma - 1} \qquad (d = 2, \quad T \le T_c). \tag{12.123}$$

It is seen that the simplicity of the form of (12.122) comes at the expense of more a sophisticated (fine-tuned) form of the effective interaction (12.123). If one opted, instead, to simplify the form for the effective interaction, getting rid of sublogarithmic terms, $U = (4\pi/m)/\ln\left(1/a^2 m\mu\right)$, then (12.122) would acquire the generic form (12.114). Needless to say, this does not change the *sum* of leading and subleading terms in the equations of state since, by construction, the result cannot depend on the specific choice of ϵ_0.

12.3.4   Superfluid Density

The standard way of calculating the superfluid density within the quasiparticle picture is based on Landau's formula for the normal component density, n_n, and the relation $n_s = n - n_n$. We employ a more general approach based on the current induced by the gradient of the condensate wavefunction. By definition, the superfluid density is the linear response coefficient relating the persistent current density to the gradient of the phase, φ, of the complex order-parameter field:

$$\mathbf{j} = \frac{n_s}{m} \nabla \varphi = n_s \frac{\mathbf{k}_0}{m}. \tag{12.124}$$

In the last equality, we assumed that $\varphi = \mathbf{k}_0 \cdot \mathbf{r}$. On the other hand, the average current can be calculated from the microscopic operator expression in terms of the condensate density and Green's function

$$\mathbf{j} = -\frac{i}{2m} \left\langle \left[\hat{\psi}^\dagger \nabla \hat{\psi} - \text{H.c.} \right] \right\rangle = \frac{\mathbf{k}_0}{m} n_0 + \sum_{\mathbf{k}} \frac{\mathbf{k}}{m} N(\mathbf{k}_0, \mathbf{k}), \tag{12.125}$$

where

$$N(\mathbf{k}, \mathbf{k}_0) = -G(\tau = -0, \mathbf{k})|_{\mathbf{k}_0}. \tag{12.126}$$

In the limit of $k_0 \to 0$, we obtain the required relation:

$$n_s = n_0 + \lim_{k_0 \to 0} \sum_{\mathbf{k}} \frac{\mathbf{k} \cdot \mathbf{k}_0}{k_0^2} N(\mathbf{k}_0, \mathbf{k}). \tag{12.127}$$

At first glance, Equation (12.127) does not resemble Landau's formula at all, since the first term is given by the condensate density, not n. However, the structure of the normal average contribution is such that it has a part that completes n_0 to the total density and a part that equals (with minus sign) the normal density component. To derive this result, we start with the expression for Green's function

$$G(\xi, \mathbf{k})|_{\mathbf{k}_0} = \frac{i\xi + \epsilon(\mathbf{k} - 2\mathbf{k}_0) - \tilde{\mu}}{\tilde{\mu}^2 + [i\xi + \epsilon(\mathbf{k} - 2\mathbf{k}_0) - \tilde{\mu}][i\xi - \epsilon(\mathbf{k}) + \tilde{\mu}]} \tag{12.128}$$

and solve for the roots of the denominator in (12.128),

$$R_{1,2} = \frac{\epsilon(k) - \epsilon(\mathbf{k} - 2\mathbf{k}_0)}{2}$$
$$\pm \sqrt{\epsilon(\mathbf{k})\epsilon(\mathbf{k} - 2\mathbf{k}_0) - [\epsilon(\mathbf{k}) + \epsilon(\mathbf{k} - 2\mathbf{k}_0)]\tilde{\mu} + \left[\frac{\epsilon(\mathbf{k}) - \epsilon(\mathbf{k} - 2\mathbf{k}_0)}{2}\right]^2},$$

to rewrite Green's function identically as

$$G = \frac{\alpha}{i\xi - R_1} + \frac{1 - \alpha}{i\xi - R_2}, \qquad \alpha = \frac{\epsilon(\mathbf{k} - 2\mathbf{k}_0) - \tilde{\mu} + R_1}{R_1 - R_2}.$$

Now we express frequency sums through the Bose functions to arrive at

$$N(\mathbf{k})|_{\mathbf{k}_0} = \alpha N(R_1) + (\alpha - 1)(1 + N(-R_2)). \tag{12.129}$$

The rest of the calculation is straightforward. In view of the limit to be taken in (12.127), it is sufficient to keep only the terms that are linear in $[\epsilon(\mathbf{k}) - \epsilon(\mathbf{k} - 2\mathbf{k}_0)] \approx 2\mathbf{k} \cdot \mathbf{k}_0/m$ in $R_{1,2}$, α, and $N(\mathbf{k})|_{\mathbf{k}_0}$. This leads to the result

$$n_s = n_0 + \frac{1}{d} \sum_{\mathbf{k}} \frac{k^2}{m} \frac{\tilde{\mu}^2}{E^2(k)} \left[\frac{1 + 2N_E}{2E(k)} - \frac{\partial N_E}{\partial E(k)} \right]. \tag{12.130}$$

Finally, using (12.91) to exclude n_0 and integrating by parts to simplify the expression, we arrive at Landau's formula, Equation (12.11).

Problem 12.1 *Complete all the algebra necessary to derive Equations (12.130) and (12.11).*

12.4   Expansion Parameters: Estimates for Higher-Order Terms

Let us analyze the structure of small parameters that control the applicability of the diagrammatic technique. As we will see *a posteriori*, it is sufficient to consider two characteristic limits: (1) the $T = 0$ case and (2) the finite-T contributions of diagrams with zero frequencies, that is, classical-field contributions. The analysis is based on dimensional estimates of the diagrams, the fact that an extra interaction vertex increases the total number of propagators by two and that the largest contributions come from small momenta (and frequencies), where the normal and anomalous propagators behave as

$$|G| \approx |F| \approx \frac{|\tilde{\mu}|}{\xi^2 + c^2 k^2}, \qquad c^2 = |\tilde{\mu}|/m, \tag{12.131}$$

and we do not need to distinguish between them. We will further assume that there is an infrared momentum cutoff $k_1 \ll k_0$ where

$$k_0 = \sqrt{mn_0 U} \tag{12.132}$$

is directly related to the inverse healing length. The symmetry-breaking field η also changes the behavior of propagators at $k \ll k_0$, but for our purposes, we do not need the explicit form of propagators at finite η because we will keep k_1 large enough to neglect effects originating from finite η.

At $T < T_c$, there are two kinds of generic vertices in higher-order diagrams, namely, (1) full vertices, where four propagators meet, and (2) vertices with one condensate line (vertices with two and three condensate lines are all accounted for by the lowest-order diagrams for the self-energies and cannot be part of the diagrammatic expansion). A naive definition of the diagram order as the total number of vertices proves inconvenient, since contributions of condensate vertices turn out to be larger than contributions of full vertices. Later, we will see that the difference is exactly compensated by the fact that condensate vertices (generically) come in pairs. This suggests defining the diagram order as the sum of the total number of full vertices and half the total number of condensate vertices (plus 1/2 for anomalous diagrams).

12.4.1   The $T = 0$ Case

At $T = 0$, the structure for the dimensionless factor associated with adding an extra full vertex to a diagram is

$$\gamma_0^{(\text{full})} \sim U \int \delta(\Delta k)\delta(\Delta \xi)\left[|G|d^d k d\xi\right]^2, \tag{12.133}$$

with $\delta(\Delta k)$ and $\delta(\Delta \xi)$ representing the δ-functions taking care of momentum and frequency conservation laws. The dimensional estimate, following from (12.133) and (12.131), reads

$$\gamma_0^{(\text{full})} \sim \frac{n^{1/2}(mU)^{3/2}}{k_1^{3-d}}. \tag{12.134}$$

The structure for the dimensionless factor associated with adding an extra pair of condensate vertices is

$$\gamma_0^{(\text{cnd})} \sim U^2 n \int [\delta(\Delta k)\delta(\Delta \xi)]^2 \left[|G|d^d k d\xi\right]^3, \tag{12.135}$$

and from the corresponding dimensional estimate, one can see that

$$\gamma_0^{(\text{cnd})} \sim \gamma_0^{(\text{full})} \frac{mnU}{k_1^2}. \tag{12.136}$$

Equations (12.134) and (12.136) clearly show the infrared problem of the theory in dimensions $d \leq 3$ (as usual, a zero power of the momentum cutoff should be understood as a logarithm) and emphasize the usefulness of the cutoff-enforcing field η. For the diagrammatic expansion to be consistent, we need k_1 to be large enough to guarantee $\gamma_0 \ll 1$. On the other hand, we do not want the field η to significantly affect the physics of the system and thus need $k_1 \ll k_0$. Hence, the smallest possible expansion parameter one can afford without distorting the physics corresponds to $k_1 \lesssim k_0$, in which case

$$\gamma_0^{(\text{full})} \sim \gamma_0^{(\text{cnd})} \sim \gamma_0 = \begin{cases} \sqrt{na^3} & (d = 3), \\ mU & (d = 2), \\ \sqrt{mU/n} & (d = 1). \end{cases} \tag{12.137}$$

Here we took into account that $|\tilde{\mu}(T = 0)| \sim nU$ and expressed U in terms of a in 3D. It is clear that γ_0, Equation (12.137), is the actual expansion parameter for all local thermodynamic quantities. These quantities, by referring to large length scales at $T = 0$, should mainly have contributions from wavevectors $\leq k_0$. Technically, it means that infrared divergencies of individual diagrams must cancel each other in the final answers and the resulting integrals become convergent at the wavevectors $k \sim k_0$, leading to a well-defined expansion in powers of γ_0 (12.137). [Since dimensional estimates do not distinguish between pure powers and powers with logarithmic prefactors, we extend the meaning of the term "in powers" to the powers with logarithmic prefactors.]

Note that all our relations between the thermodynamic quantities that do not vanish in the $T = 0$ limit (specifically, n, μ, p, ε, and n_0) are accurate up to the first-order corrections in γ_0, so that their systematic errors are of the order of γ_0^2 (with possible logarithmic prefactors). The relation for the entropy is accurate only up to the leading term, which actually has the same quasiparticle origin as the *subleading* terms in all nonvanishing at $T \to 0$ quantities. The same is true for the heat capacity that behaves similarly to entropy at $T \to 0$.

12.4.2 Finite-T Zero-Frequency Contributions

The dimensionless factor associated with adding an extra full vertex to a diagram consisting of zero-frequency propagators scales as

$$\gamma_T^{(\mathrm{full})} \sim (U/T) \int \delta(\Delta k) \left[|G| T d^d k \right]^2, \tag{12.138}$$

yielding the dimensional estimate

$$\gamma_T^{(\mathrm{full})} \sim \frac{m^2 T U}{k_1^{4-d}}. \tag{12.139}$$

For the pair of condensate vertices, we have

$$\gamma_T^{(\mathrm{cnd})} \sim (U/T)^2 n_0 \int [\delta(\Delta k)]^2 \left[|G| T d^d k \right]^3. \tag{12.140}$$

Similarly to the $T = 0$ case (assuming that $|\tilde{\mu}| \sim n_0 U$), we obtain

$$\gamma_T^{(\mathrm{cnd})} \sim \gamma_T^{(\mathrm{full})} \frac{m n_0 U}{k_1^2}. \tag{12.141}$$

In complete analogy with the $T = 0$ case, the largest possible k_1 (enforced by η) cannot exceed $\sqrt{m n_0 U}$. Assuming that $k_1 \lesssim \sqrt{m n_0 U}$ results in

$$\gamma_T^{(\mathrm{full})} \sim \gamma_T^{(\mathrm{cnd})} \sim \gamma_T \sim \gamma_0 \frac{T}{nU} \left(\frac{n}{n_0} \right)^{2-d/2}. \tag{12.142}$$

This expression shows that at $T \ll nU$, the parameter γ_0 dominates over γ_T, and thus, this temperature regime is equivalent to $T = 0$. In the crossover regime, $T \sim nU$, both γ_0 and γ_T are of the same order. In the regime $T \gg nU$, the classical-field parameter γ_T dominates and becomes of order unity when n_0 is sufficiently small. The latter situation corresponds to the fluctuation region, where the perturbative theory becomes inadequate, since it fails to properly describe the nonlinear long-wave fluctuations of the classical component of the quantum field. Before hitting the region $\gamma_T \sim 1$, one must switch to the description in terms of scaling functions, which are universal for all $U(1)$ weakly interacting theories; see Section 12.7.

At $T \gg nU$, all our relations for the thermodynamic quantities are accurate up to the first-order corrections in γ_T, so that their systematic errors are of the order of γ_T^2 (with possible logarithmic prefactors). Note that at $T \gg nU$, the leading terms for all the thermodynamic quantities, except for the chemical potential (and with a reservation for the density in 1D where the corresponding condition is $T \gg nU/\gamma_0$), are the same as for the ideal gas.

To summarize our analysis, we present an order-of-magnitude estimate of systematic uncertainties due to omitted higher-order terms for each thermodynamic quantity. It is convenient to start with μ and recall that the omitted term in (12.93) comes from the "sunrise" diagram for Θ. A dimensional analysis of this diagram gives

$$\Delta\mu/\tilde{\mu} \sim \max\left[\gamma_0^2, \gamma_T^2\right], \tag{12.143}$$

which transforms into a similar estimate for the total and superfluid density:

$$\Delta n/n_* \sim \max\left[\gamma_0^2, \gamma_T^2\right], \tag{12.144}$$

$$\Delta n_s \sim \Delta n. \tag{12.145}$$

This can be seen from the dimensional analysis of the second-order diagrams contributing to the self-energies. Relation (12.104) for the pressure implies

$$\Delta p \sim n\tilde{\mu}\max\left[\gamma_0^2, \gamma_T^2\right]. \tag{12.146}$$

In the case of entropy, it is convenient to start with the regime $T \geq nU$ when—using (12.111) to relate the uncertainty in s to Δp, Δn, and $\Delta\mu$—we find that

$$\Delta s \sim \frac{\tilde{\mu}}{T}n\gamma_T^2 \qquad (T \geq nU). \tag{12.147}$$

We see that at $T \sim nU$, in contrast to the behavior of other thermodynamic functions, the uncertainty in s scales only as the first power of γ_0 (simply because s itself becomes of order $\sim \gamma_0$). This scaling persists down to $T \to 0$:

$$\Delta s/s \sim \gamma_0 \qquad (T \leq nU), \tag{12.148}$$

because, at $T \ll nU$, the thermodynamics of the system corresponds to the generic low-temperature behavior of superfluids, where the leading temperature-dependent contributions are due to phonons. Hence, the first-order correction to the sound velocity, which is of the order of γ_0 as is seen from the expression for the energy, translates into the order-γ_0 correction to the entropy. At $T \ll nU$, when thermodynamics is exhausted by dilute noninteracting phonons, the correction to the sound velocity can be found directly from the zero-temperature compressibility, and the accuracy of the expression for s (and other thermodynamic quantities) can be improved. At $T \sim nU$, however, the order-γ_0 correction to the quasiparticle dispersion law goes beyond the Bogoliubov ansatz. Note also that improving the value of the sound velocity at $T \ll nU$, while rendering the answers for thermodynamic quantities more accurate, is inconsistent with retaining the Bogoliubov form of the spectrum for *all* the momenta.

Finally, Equation (12.10), in combination with the previous results, yields the estimate for the uncertainty of energy:

$$\Delta\varepsilon \sim \Delta p \sim n\tilde{\mu}\max\left[\gamma_0^2, \gamma_T^2\right]. \tag{12.149}$$

12.4.3 Fluctuational Contributions

In all spatial dimensions, there is an interval (in terms of temperature, if density is kept fixed, or in terms of chemical potential at fixed temperature, etc.) where γ_T becomes of order unity, and the systematic perturbative description breaks down. In 3D and 2D, this happens in the vicinity of the superfluid phase transition point. In 1D, there is no superfluidity in the strict sense of the word, and no phase transition occurs, but the picture remains very similar to that of the 2D and 3D case for a weakly interacting gas, as explained in Section 12.4.4. A detailed discussion of the fluctuation region, including accurate expressions for the critical points, will be presented later, in Section 12.7. Meanwhile, here we want to utilize the fact that results (12.143) through (12.149) allow one to make order-of-magnitude estimates of the nonperturbative fluctuation contributions to the thermodynamic functions. These estimates are quite important as they show the degree to which the perturbative results of the previous sections are inaccurate in the fluctuation region. As we will see, for some quantities, the fluctuation corrections are smaller than the leading ideal-gas contributions.

By continuity, the order-of-magnitude estimates for fluctuation contributions can be obtained from (12.143) to (12.149) by simply setting $\gamma_T \sim 1$, while for the quantities themselves, we can use their mean-field critical expressions. This way we arrive at the following results:

$$\frac{(\Delta\mu)_{\text{fluct}}}{\mu} \sim \frac{(\Delta n)_{\text{fluct}}}{n} \sim \begin{cases} \gamma_0^{2/3} & (d = 3), \\ \ln^{-1}(1/\gamma_0) & (d = 2), \\ 1 & (d = 1), \end{cases} \tag{12.150}$$

$$\frac{(\Delta p)_{\text{fluct}}}{p} \sim \frac{(\Delta\varepsilon)_{\text{fluct}}}{\varepsilon} \sim \frac{(\Delta s)_{\text{fluct}}}{s} \sim \begin{cases} \gamma_0^{4/3} & (d = 3), \\ \gamma_0 \ln(1/\gamma_0) & (d = 2), \\ \gamma_0^{1/2} & (d = 1). \end{cases} \tag{12.151}$$

For the superfluid and condensate densities, we get an obvious answer that in both cases, the fluctuation contributions are of the order of 100%.

12.4.4 Specifics of the 1D System

As long as a 1D system is weakly interacting, that is, $\gamma_T \ll 1$, the notion of the order-parameter field with well-defined amplitude and the two-component (normal + superfluid) description remain physically meaningful. Since superfluidity is a topological phenomenon, it can be destroyed only by topological defects— phase slips. At $\gamma_T \ll 1$, the phase slips are rare events that do not contribute significantly to the local thermodynamic quantities; in correlation functions, phase slips show up only at length scales much larger than $1/k_0$, where their effect can be described at the hydrodynamic level.

When the temperature reaches the characteristic scale of

$$T_{\text{fluct}}^{(1D)} \sim \gamma_0^{-1} n U \sim \gamma_0 \frac{n^2}{m}, \tag{12.152}$$

the parameter γ_T becomes of order unity. Apart from the lack of a genuine phase transition, the physics in the temperature range $T \sim T_{\text{fluct}}^{(1D)}$ is close to that of the fluctuation region in 2D and 3D. Only the long-wave classical-field subsystem of the original quantum field experiences strong nonlinear fluctuations. This leads to nonperturbative contributions to the system thermodynamics, which are universal for all weakly interacting 1D U(1) systems and can be described by universal scaling functions in direct analogy with the fluctuation regions in 2D and 3D. At $T \gg T_{\text{fluct}}^{(1D)}$, the low-momentum part of the classical-field component gets depleted, so that the nonlinearity of interactions becomes weak and accurately accounted for within the normal-gas mean-field picture.

12.5   Long-Range Off-Diagonal Correlations

Beliaev's diagrammatic technique allows one to calculate correlation functions up to distances much larger than the correlation radius $r_0 \sim 1/k_0$. However, addressing the asymptotic long-range behavior of off-diagonal correlation functions requires special techniques properly accounting for long-wave fluctuations of the Goldstone mode, that is, the phase of the superfluid order parameter. Since the diagrammatic and hydrodynamic descriptions overlap, one can utilize the general analysis of Section 7.10 to straightforwardly extend the diagrammatic calculation of off-diagonal correlators obtained up to large enough distances $r \gg r_0$ to arbitrarily large r's.

In our case, the zero-point compressibility κ in (7.120) is

$$\kappa = \left(\frac{d\mu}{dn}\right)_{T=0}^{-1} \approx U^{-1}. \tag{12.153}$$

The nonzero Matsubara frequencies are relevant only at $T \ll nU$. At $T \geq nU$, only the $\xi = 0$ term should be left in (7.117).

As a characteristic example, consider the single-particle density matrix $\rho(\mathbf{r}) = \langle \psi^*(\mathbf{r})\psi(0)\rangle$. Taking into account that $\rho(\mathbf{r}) \equiv n + G(r = 0, \tau = -0) - G(r, \tau = -0)$, from the diagrammatic expressions for G, we have

$$\rho(\mathbf{r}) = n - \sum_{\mathbf{k}} \left(1 - e^{i\mathbf{k}\cdot\mathbf{r}}\right)\left[\frac{\epsilon(k) - \tilde{\mu}}{2E(k)}(1 + 2N_E) - \frac{1}{2}\right] \quad \text{(at small and intermediate } r\text{'s).}$$

$$\tag{12.154}$$

We can use this expression in any dimension at distances significantly exceeding $1/k_0$. In 2D at finite temperature and in 1D at any temperature, expression

(12.154) ultimately becomes inaccurate, and we must rely on the previously described procedure of extrapolation. In the asymptotic regime, we have

$$\rho(r) = \rho(r')e^{-\Xi(r,r')}, \tag{12.155}$$

where, at $T \ll nU$,

$$\Xi(r,r') = \kappa^{-1} \sum_{\mathbf{k}} \left[e^{i\mathbf{k}\cdot\mathbf{r}'} - e^{i\mathbf{k}\cdot\mathbf{r}} \right] \frac{1 + 2N_{E=ck}}{2ck}, \tag{12.156}$$

with $c = \sqrt{n_s/m\kappa}$ the sound velocity at $T = 0$ (here $n_s = n$), whereas at $T \geq nU$,

$$\Xi(r,r') = \frac{mT}{n_s} \sum_{\mathbf{k}} \left[e^{i\mathbf{k}\cdot\mathbf{r}'} - e^{i\mathbf{k}\cdot\mathbf{r}} \right] \frac{1}{k^2}. \tag{12.157}$$

Within the relevant orders, Equation (12.154) corresponds to the expansion of the exponential in (12.155) in powers of Ξ, up to the term $\propto \Xi$ included. This immediately suggests that within the same accuracy, one can exponentiate Ξ-terms in (12.154) to extend its domain of applicability to much larger distances. The other advantage of proceeding this way is having a physical definition of the quasicondensate density as the amplitude of the order-parameter field in the long-wave limit. The equations that follow have the same accuracy as (12.154), although they do contain artificial higher-order terms, which arise from factorizing the correlation function. More specifically, we single out terms in (12.154) that on large distances reproduce Ξ and then exponentiate them:

$$\rho(\mathbf{r}) = [n - \rho_-(\mathbf{r}) - n\Lambda(\mathbf{r})] \longrightarrow \tilde{\rho}(\mathbf{r})e^{-\Lambda(\mathbf{r})}, \tag{12.158}$$

where $\tilde{\rho}(\mathbf{r}) = n - \rho_-(\mathbf{r})$, and

$$\rho_-(\mathbf{r}) = \sum_{\mathbf{k}} \left(1 - e^{i\mathbf{k}\cdot\mathbf{r}}\right) \frac{\epsilon}{E^2} \left[\frac{E - \epsilon + \tilde{\mu}}{2} + (E - \tilde{\mu})N_E \right], \tag{12.159}$$

$$\Lambda(\mathbf{r}) = -\frac{\tilde{\mu}}{n} \sum_{\mathbf{k}} \left(1 - e^{i\mathbf{k}\cdot\mathbf{r}}\right) \frac{E - \epsilon}{2E^2} [1 + 2N_E]. \tag{12.160}$$

We have added and subtracted $[1 + 2N_E]\left[\tilde{\mu}\epsilon/2E^2\right]$ to (12.154) to ensure that (1) both $\tilde{\rho}(\mathbf{r})$ and $\Lambda(\mathbf{r})$ are free from UV and infrared divergencies in all dimensions and (2) that only small momenta $k < k_0$ contribute to the phase correlator $\Lambda(\mathbf{r})$.

By comparing (12.160) and (12.156), we see that they coincide at large distances up to leading terms. It means that in 2D and in 1D at zero temperature, Equation (12.158) can be trusted up to exponentially large scales, while in 1D at finite temperature $T \geq |\tilde{\mu}|$, it works at least up to distances $\sim n_s/mT$.

We are now in a position to define the quasicondensate density as the limiting value of $\tilde{\rho}(\mathbf{r} \to \infty)$:

$$n_{qc} = n - \sum_{\mathbf{k}} \frac{\epsilon}{E^2} \left[\frac{E - \epsilon + \tilde{\mu}}{2} + (E - \tilde{\mu}) N_E \right]. \tag{12.161}$$

The physical meaning of this relation is the amplitude squared of the order-parameter field at large distances. There is a certain degree of freedom in attributing terms that do not result in the power-law or exponential decay of $\rho(\mathbf{r})$ to either $\rho_-(\mathbf{r})$ or $\Lambda(\mathbf{r})$. The idea behind our choice is threefold: (1) Equations (12.158) through (12.160) are valid as written in all spatial dimensions; (2) at large distances, $\Lambda(\mathbf{r})$ has the structure of hydrodynamic phase correlations; and (3) the final expressions have the simplest form possible within the same accuracy. It is also worth mentioning that n_{qc} is a quantity that controls all physical processes happening in the WIBG at short distances at low temperature, for example, m-body recombination rates. In all spatial dimensions, it plays the same role as the condensate density in the 3D system as long as one is interested in length scales not much larger than the healing length.

Problem 12.2 *Show that Equation (12.161) implies relation (12.13).*

Finally, the condensate density is defined using

$$n_0 = n_{qc} e^{-\Lambda(\infty)}. \tag{12.162}$$

It is not accidental that the ultimate long-wavelength property of the superfluid system is determined last. It was always an unpleasant feature of numerous mean-field treatments that the crucial parameter determining physics at short scales was linked to n_0.

12.6   Normal Region

We assume that the temperature is not very high, so that condition (P.6) is preserved and the quantity U remains the only parameter characterizing the interaction between the particles.

12.6.1   Thermodynamic Functions

In the normal region, where there are no anomalous correlators, we must deal only with the Dyson equation for Green's function G in terms of self-energy Σ_{11}. Within the leading-order approximation, $\Sigma_{11} = 2nU$ (in 2D, this formula implies an adequate choice of the parameter $\epsilon_0 \equiv \epsilon_0(T)$ for the effective interaction U, discussed in Section 12.6.2), and the expression for G yields result (12.14) for the number density. For the pressure, we find

$$p = \int_{-\infty}^{\tilde{\mu}} n \frac{d\mu}{d\tilde{\mu}} d\tilde{\mu} = \int_{-\infty}^{\tilde{\mu}} n \left(1 + 2U \frac{dn}{d\tilde{\mu}} \right) d\tilde{\mu} = \int_{-\infty}^{\tilde{\mu}} n \, d\tilde{\mu} + n^2 U. \tag{12.163}$$

Using

$$n = \sum_k N_{\tilde{\epsilon}} \equiv -T \frac{d}{d\tilde{\mu}} \sum_k \ln\left[1 - e^{-\tilde{\epsilon}(k)/T}\right], \tag{12.164}$$

we arrive at expression (12.15) for the pressure. Equations (12.14), (12.88), and (12.15) specify n, μ, and p as functions of $(T, \tilde{\mu})$. Utilizing (12.111), with $x \equiv \tilde{\mu}$, we obtain Equation (12.16) for the entropy and energy densities.

12.6.2   Effective Interaction in the Normal Regime in 2D

In the quasicondensate region, the parameter ϵ_0 defining the effective interaction U can vary over a wide range of values, thanks to the retained second-order correction that produces counterterms compensating for the arbitrary choice. Since in the normal region we confine ourselves to the first-order expression for the self-energy, $\Sigma = 2nU$, we need to choose the value of ϵ_0, which minimizes the omitted second-order contributions originating from the sunrise diagram; see Figure 12.13. Due to momentum independence of the pseudopotential line, the third (fourth) diagram is identical to the first (second) one; we thus consider only the first two diagrams and multiply the result by a factor of 2. The parameter ϵ_0 can be much smaller or much larger than T. In either case, the leading term comes from the logarithmic UV contribution from the product of two propagators, yielding the result

$$\Sigma^{(2)} = -2nU^2 \sum_{k > k_T} \left[\Pi(k) - \frac{m}{k^2}\right] + nU\mathcal{O}(mU), \tag{12.165}$$

where k_T is the thermal momentum. This leads to the expression

$$\Sigma = 2nU\left[1 + \frac{mU}{4\pi} \ln\frac{T}{\epsilon_0} + \mathcal{O}(mU)\right]. \tag{12.166}$$

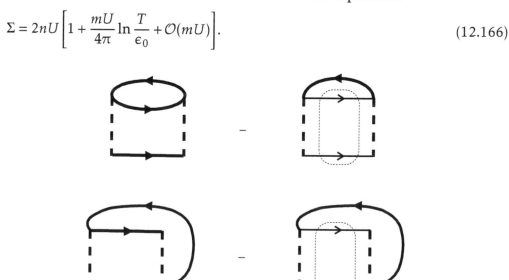

Figure 12.13   Second-order contributions to the self-energy in the normal regime, including the sunrise diagram.

Comparing the first two terms in the brackets with (12.62) and (12.63), we see that if we take the value of ϵ_0 substantially away from T, then the leading second-order correction to the expression $\Sigma = 2nU$ amounts to renormalizing the value of the pseudopotential U in such a way that it corresponds to $\epsilon_0 \sim T$. This brings us to the conclusion that $\epsilon_0 \sim T$ is the optimal choice for ϵ_0, in which case omitting the diagrams shown in Figure 12.13 is justified by the parameter $mU \ll 1$.

12.6.3 Expansion Parameters

At temperatures above the fluctuation region where the renormalized chemical potential remains small $|\tilde{\mu}| \ll T$, the expansion parameter is defined by the infrared behavior of the zero-frequency lines. The dimensionless factor, γ_T, associated with adding an extra full interaction vertex to a diagram is given by (12.138), but now with $G \sim T/\tilde{\epsilon}(k)$, resulting in the estimate

$$\gamma_T \sim \frac{TUm^{d/2}}{\tilde{\mu}^{2-d/2}} \sim \left(\frac{\tilde{\mu}_{\text{fluc}}}{\tilde{\mu}} \right)^{2-d/2}, \tag{12.167}$$

where $\tilde{\mu}_{\text{fluc}}$ is the size of the fluctuation region in terms of the chemical potential. At temperatures much higher than the degeneracy temperature $T_c \sim n^{2/d}/m$ where $\tilde{\mu} \approx -T \ln \left(n\lambda_T^d \right)$, all corrections are small in the parameter

$$\gamma_T \sim \frac{U}{|\tilde{\mu}| \lambda_T^d}. \tag{12.168}$$

12.7 Fluctuation Region

In the fluctuation region, the long-wave part of the classical-field component of the quantum field experiences strong (nonperturbative) fluctuations. It is exclusively due to these fluctuations that the diagrammatic technique for the quantum field loses an expansion parameter. The special role of the classical-field component is evident from the fact that on approach to the fluctuation region, the leading contribution to the expansion parameter γ_T is associated with zero-frequency propagators; see Sections 12.4.2 and 12.6.3. The diagrammatic technique built on zero-frequency propagators (with perturbative parts of self-energies included) and the interaction represented by pseudopotential U corresponds to the diagrammatic expansion of the Gibbs distribution

$$Z = \int e^{-H[\psi]/T} \mathcal{D}\psi \tag{12.169}$$

for the classical field (truncated in momentum space)

$$\psi(\mathbf{r}) = \sum_{k<k'} \psi_{\mathbf{k}} e^{i\mathbf{k}\cdot\mathbf{r}}, \tag{12.170}$$

with the Hamiltonian functional

$$H[\psi] = \int \left[\frac{1}{2m} |\nabla\psi|^2 + \frac{U}{2} |\psi|^4 - \mu' |\psi|^2 \right] d\mathbf{r}. \tag{12.171}$$

Here $k' \ll k_T = \sqrt{mT}$ is a truncation momentum (to be discussed later), and $\mu' \equiv \mu'(k')$ is the reduced chemical potential obtained from the bare chemical potential by subtracting relevant perturbative contributions—both classical and quantum—from all the harmonics with $k > k'$. The bottom line is that describing the quantum gas in the fluctuation region reduces to solving the classical-field problem (12.169) through (12.171), in its fluctuation region.

In 2D and 3D, numerical solutions to the problem (12.169) through (12.171)—in terms of scaling functions for thermodynamic quantities—are available in the literature and have already been used to accurately describe weakly interacting 2D and 3D quantum gases in the fluctuation region [3–7]. The same is possible in 1D, but to the best of our knowledge, it has so far not been done.

In the fluctuation region, an important quantity is the momentum $\tilde{k} \ll k_T$ separating weakly and strongly coupled classical modes. For a quantitatively accurate description of the fluctuating classical-field subsystem, the momentum k' separating classical modes of interest from the rest of the modes has to be much larger than $\tilde{k}$. However, as long as we are interested in the order-of-magnitude estimates, it is safe and convenient to set $k' \sim \tilde{k}$, so that all the modes we are dealing with in (12.169) through (12.171) are strongly coupled. To estimate $\tilde{k}$, we note that with $k' \sim \tilde{k}$, all three terms in the Hamiltonian (12.171) must be of the same order of magnitude, that is,

$$\tilde{k}^2/m \sim |\tilde{\mu}| \sim \tilde{n}U, \tag{12.172}$$

where

$$\tilde{n} \sim \sum_{k<\tilde{k}} |\psi_k|^2 = \sum_{k<\tilde{k}} n_k \tag{12.173}$$

is the long-wavelength contribution to the total density, and $\tilde{\mu} \equiv \mu'\left(k' \to \tilde{k}\right)$. For $\tilde{n}$ we have $\tilde{n} \sim \tilde{k}^d n_{\tilde{k}}$, and since $\tilde{k}$ is separating strongly coupled long-wave harmonics from slightly perturbed short-wave harmonics, the order-of-magnitude estimate for $n_{\tilde{k}}$ follows by continuity from the ideal system formula:

$$n_{\tilde{k}} \sim \frac{T}{\tilde{k}^2/2m - \tilde{\mu}} \sim \frac{T}{|\tilde{\mu}|}. \tag{12.174}$$

Substituting this back into (12.172) and (12.173) yields

$$\tilde{k} = \left(m^2 T U\right)^{\frac{1}{4-d}}, \qquad \tilde{n} \sim \left(m^d T^2 U^{d-2}\right)^{\frac{1}{4-d}}, \qquad |\tilde{\mu}| \sim \left(m^d T^2 U^2\right)^{\frac{1}{4-d}}. \tag{12.175}$$

The estimates (12.175) imply the following parameterization of the grand canonical equation of state in the fluctuation region and its vicinity:

$$n = n_c^{(d)}(T) + \left(m^d T^2 U^{d-2}\right)^{\frac{1}{4-d}} \lambda^{(d)}(x), \qquad x = \frac{\mu - \mu_c^{(d)}(T)}{\left(m^d T^2 U^2\right)^{\frac{1}{4-d}}}. \qquad (12.176)$$

Here $\lambda^{(d)}(x)$ is a dimensionless scaling function of the dimensionless scaling variable x. In 2D and 3D, the quantities $n_c^{(d)}(T)$ and $\mu_c^{(d)}(T)$ are the critical values of density and chemical potential at a given temperature. In 1D, where the phase transition is absent, one can set, without loss of generality, $\mu_c \equiv 0$, and correspondingly, $n_c(T) \equiv n(\mu = 0, T)$.

Similarly, superfluid and condensate densities—in the dimensions in which they are meaningful—are parameterized as

$$n_s = \left(m^d T^2 U^{d-2}\right)^{\frac{1}{4-d}} f_s^{(d)}(x) \qquad (d = 2, 3), \qquad (12.177)$$

and

$$n_0 = m^3 T^2 U f_0(x) \qquad (d = 3). \qquad (12.178)$$

The functions $\lambda(x)$, $f_s(x)$, and $f_0(x)$ are universal for all weakly interacting U(1)-symmetric systems in the given dimension (no matter quantum or classical, continuous space or lattice); the numerical data for them are available for $d = 2, 3$ (see Figures 12.14 through 12.16). By numerically solving the problem (12.169) through (12.171), it was established that, up to higher-order corrections in the parameter γ_0 [3–5],

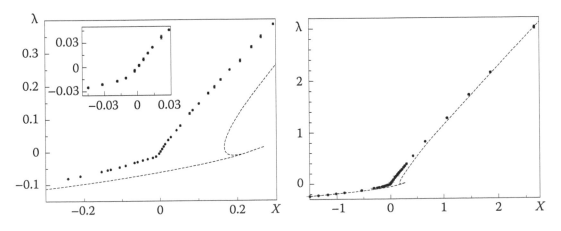

Figure 12.14   The function $\lambda(X)$ in 3D. Dashed lines correspond to the mean-field equations of state, Equations (12.6) and (12.14). (Adapted from Reference [7].)

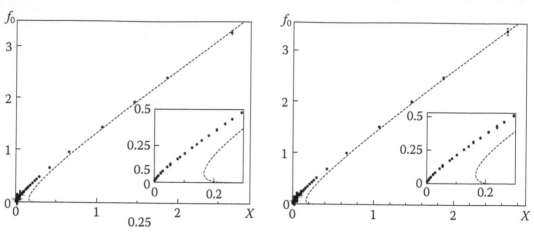

Figure 12.15 The superfluid and condensate densities as functions of X in 3D. Dashed lines correspond to the mean-field results, Equations (12.11) and (12.162). The initial parts of the plots (shown in the insets) correspond to the universal, up to numeric prefactors, U(1) critical behavior: $f_s(X) = A_s X^\nu$ and $f_0(X) = A_0 X^{\nu(1+\eta)}$, with $\nu = 0.6715$ and $\eta = 0.038$. The prefactors $A_s = 0.86(5)$ and $A_0 = 0.89(5)$ are universal for all 3D weakly interacting U(1) systems. (Adapted from Reference [7].)

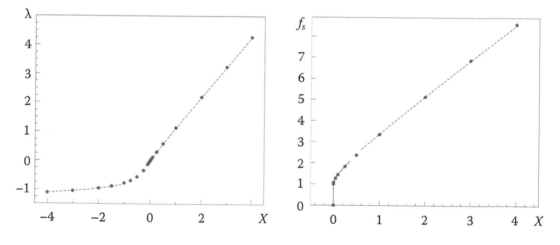

Figure 12.16 The functions $\lambda(X)$ and $f_s(X)$ in 2D. Dashed lines correspond to the mean-field results, Equations (12.6), (12.14), and (12.11). Solid line corresponds to the asymptotic ($X \to 0$) Kosterlitz–Thouless behavior $1/f_s + \ln f_s = \kappa' X$, with $\kappa' = 0.61(1)$ being universal for all 2D weakly interacting U(1) systems. (Adapted from Reference [6].)

$$n_c^{(3D)} = n_c^{(0)}(T) - Cm^3 T^2 U, \qquad C = 0.0142(4), \tag{12.179}$$

$$n_c^{(2D)} = \frac{mT}{2\pi} \ln\left(\frac{\xi}{mU}\right), \qquad \xi = 380 \pm 3, \tag{12.180}$$

with $n_c^{(0)}(T)$ the critical density for Bose–Einstein condensation in an ideal 3D gas. Analogous relations for $\mu_c^{(d)}(T)$ read [3–5]

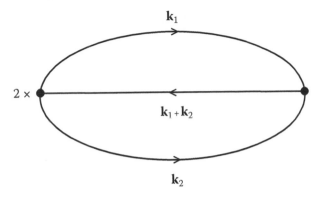

Figure 12.17 The sunrise diagram for the self-energy yielding an important subleading contribution to the chemical potential in the fluctuation region in the 3D case. The pseudopotential is represented by a dot rather than a line to emphasize its momentum independence, allowing one to combine two topologically different (but mathematically equal) diagrams into one diagram with a global factor of 2.

$$\mu_c^{(3D)} = 2U n_c^{(0)}(T) + \frac{m^3 T^2 U^2}{\pi^2} \ln\left(\frac{C_\mu}{\sqrt{m^3 T U^2}}\right), \qquad C_\mu = 0.4213(6), \qquad (12.181)$$

$$\mu_c^{(2D)} = \frac{mTU}{\pi} \ln\left(\frac{\xi_\mu}{mU}\right), \qquad \xi_\mu = 13.2 \pm 0.4. \qquad (12.182)$$

Problem 12.3 *Show that with logarithmic accuracy, the logarithmic term in the right-hand side of Equation (12.181) corresponds to the sunrise diagram, Figure 12.17.*

The constant C is universal for all 3D weakly interacting U(1) systems. The constants C_μ, ξ, and ξ_μ, entering logarithmic expressions and thus picking up some system-specific perturbative high-momentum contributions, are *semi* universal in the sense that knowing one of them for a certain weakly interacting system allows one to easily find its counterpart for another system.

Problem 12.4 *Establish the relations between C_μ, ξ, and ξ_μ of WIBG, on one hand, and their counterparts for the classical discrete complex-field model, on the other hand.*

The leading terms in relations (12.179) through (12.182), as well as order-of-magnitude estimates for the values of subleading terms, are readily obtained from the condition $\gamma_T \sim 1$; the accurate numerical treatment of the problem (12.169) through (12.171) is necessary to fix the values of the dimensionless constants.

Problem 12.5 *Show that up to the specific values of the dimensionless constants C, ξ, C_μ, and ξ_μ, the structure of Equations (12.179) through (12.182) follows from the condition $\gamma_T \sim 1$.*

By rewriting (12.180) and (12.179) in terms of the critical temperature as a function of density, we obtain

$$T_c^{(2D)} = \frac{2\pi n}{m \ln(\mathcal{E}/mU)},\tag{12.183}$$

$$T_c^{(3D)} = T_c^{(0)}\left(1 + C_0 a n^{1/3}\right), \qquad C_0 = 1.29 \pm 0.05,\tag{12.184}$$

where $T_c^{(0)}$ is the critical temperature of the ideal 3D gas. Now it is easy to estimate $\tilde{k}$. The relevant temperature for having strong fluctuations is given by

$$T_{\text{fluct}} \approx T_c \quad (d = 2, 3), \qquad T_{\text{fluct}} \sim nU/\gamma_0 \quad (d = 1).\tag{12.185}$$

Estimate (12.175) for $\tilde{k}$ at $T = T_{\text{fluct}}$ then yields

$$\tilde{k}/k_{T_{\text{fluct}}} \sim \begin{cases} \gamma_0^{2/3} & (d = 3), \\ \gamma_0^{1/2} & (d = 2), \\ \gamma_0^{1/2} & (d = 1). \end{cases}\tag{12.186}$$

We see that $\tilde{k} \ll k_{T_{\text{fluct}}}$ in all cases, which justifies the statement that *the physics of the fluctuation region is that of a (strongly interacting) classical field.*

Putting aside n_s and n_0, which are most sensitive to nonperturbative fluctuations of the classical field ψ, the next two quantities that are sensitive to fluctuations, especially in 1D and 2D, are the density and the chemical potential—see estimates (12.150). For other quantities—as is seen from (12.151)—neglecting fluctuation corrections will not result in a significant error. It is also important that up to a few dimensionless constants characterizing subleading contributions to critical values of p, ε, and s, the fluctuation corrections to these quantities are expressed in terms of the same function $\lambda^{(d)}(x)$ and its integral. Indeed, Equation (12.103) implies the following relation for p:

$$p = p_c^{(d)}(T) + n_c^{(d)}(T)\left[\mu - \mu_c^{(d)}(T)\right] + \left(m^{2d}T^4 U^d\right)^{\frac{1}{4-d}} \int_0^x \lambda^{(d)}(x')\,dx',\tag{12.187}$$

and then with (12.111), one obtains similar relations for s and ε.

12.8   Weakly Interacting Fermi Gas

Our description of Fermi gas will be very brief and at the mean-field level.* In particular, we will not discuss the UV regularization of strong potentials for the dilute gas since this is done along exactly the same lines as for bosons. Interested readers can look at the classic paper on this subject by Gor'kov and Melik-Barkhubarov [13]. Likewise, we will not venture into the discussion of pairing mechanisms

---

* That is, in the limit of appropriately small effective-coupling constant.

and emergence of attractive effective interactions in numerous condensed matter systems (this topic was touched upon in Chapter 5). One reason for doing this is exponential suppression of the critical temperature for weak interactions, meaning that the ratio between T_c and Fermi energy, ϵ_F, quickly becomes extremely small even for moderate-strength interactions. This leads to (a) an unmeasurably small fluctuation region (according to the Ginzburg-Levanuk criterion, this region is $\propto T_c^4/\epsilon_F^3$) and applicability of the mean-field treatment and (b) universal description in terms of T/T_c regardless of the underlying pairing mechanism. Finally, for simplicity, we will discuss below only the case of s-wave spin-singlet pairing as in the original BCS theory. Though generalizations to other pairing states are straightforward, proper discussion of such states quickly becomes rather technical and goes beyond the scope of our book.

Our starting point is the effective grand-canonical Hamiltonian

$$H_{\text{eff}} = \sum_{\mathbf{k},\sigma} \lambda_{k\sigma} \hat{a}_{\mathbf{k}\sigma}^\dagger \hat{a}_{\mathbf{k}\sigma} + \frac{g}{2} \sum_{\mathbf{k}_1\mathbf{k}_2\mathbf{q},\sigma\sigma'}^{FS} \hat{a}_{\mathbf{k}_1-\mathbf{q}\sigma'}^\dagger \hat{a}_{\mathbf{k}_2+\mathbf{q}\sigma}^\dagger \hat{a}_{\mathbf{k}_2\sigma} \hat{a}_{\mathbf{k}_1\sigma'}, \tag{12.188}$$

where $\sigma = \uparrow, \downarrow$ is the spin-1/2 subscript, $\lambda_k = \epsilon_k - \mu$ is the spherically symmetric dispersion relation characterized by the effective mass m_* at the Fermi surface $\lambda_{k_F} = 0$, and $g < 0$ is the momentum-independent, weak (the dimensionless parameter is defined below) attractive coupling constant between the fermions in close vicinity of the Fermi surface; that is, all momenta in the second term should satisfy the condition $|\lambda_k| < \omega_0$ with $\omega_0 \ll \mu$. This restriction is explicitly mentioned using the label "FS" on top of the momentum sum. The value of the high-energy cutoff ω_0 depends on the pairing mechanism; for electron–phonon coupling, ω_0 is approximately given by the Debye temperature of the crystal.

12.8.1 Symmetry-Breaking Terms and Anomalous Propagators

Explicit symmetry breaking in the Cooper channel is achieved by adding the following two terms to the system's Hamiltonian (the factor $-1/2$ is a matter of convenience):

$$-\frac{1}{2} \sum_{\alpha,\beta} \int d^d r_1 d^d r_2 \left[\eta_{\alpha\beta}(\mathbf{r}_1,\mathbf{r}_2) \hat{\psi}_\alpha^\dagger(\mathbf{r}_1) \hat{\psi}_\beta^\dagger(\mathbf{r}_2) + \eta_{\alpha\beta}^*(\mathbf{r}_1,\mathbf{r}_2) \hat{\psi}_\beta(\mathbf{r}_2) \hat{\psi}_\alpha(\mathbf{r}_1) \right], \tag{12.189}$$

where, without loss of generality (by fermionic symmetry),

$$\eta_{\beta\alpha}(\mathbf{r}_2,\mathbf{r}_1) = -\eta_{\alpha\beta}(\mathbf{r}_1,\mathbf{r}_2). \tag{12.190}$$

The details of the dependence of the function $\eta_{\beta\alpha}(\mathbf{r}_2,\mathbf{r}_1)$ on the distance $r = |\mathbf{r}_1 - \mathbf{r}_2|$ are not important. The function simply has to decay with r at a certain microscopic scale. What matters is the dependence on the direction $\hat{r} = (\mathbf{r}_1 - \mathbf{r}_2)/r$, which, on one hand, is supposed to respect the rotational symmetry of the problem (say, isotropy in a continuous space, or $\pi/2$-rotation symmetry on a square/cubic lattice), and, on the other hand, needs to conspire with the symmetry of the spin

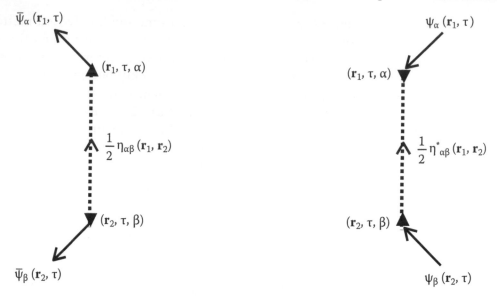

Figure 12.18   Diagrammatic elements associated with the symmetry-breaking terms
(12.189).

dependence to obey Equation (12.190). In our case of singlet pairing of spin-1/2
fermions described by spatially isotropic SU(2)-invariant Hamiltonian (12.188),
we take

$$\eta_{\sigma\sigma'}(\mathbf{r}_1,\mathbf{r}_2) = \mathcal{S}_{\sigma\sigma'}\eta(r), \tag{12.191}$$

$$\mathcal{S}_{\uparrow\uparrow} = \mathcal{S}_{\downarrow\downarrow} = 0, \qquad \mathcal{S}_{\uparrow\downarrow} = -\mathcal{S}_{\downarrow\uparrow} = 1. \tag{12.192}$$

The diagrammatic elements representing the symmetry-breaking terms
(12.189) are shown in Figure 12.18. The *direction* indicated by the central arrow on
a dotted line is not redundant: it is associated with the order of fermionic operators
and thus is relevant to the sign of the diagram. What requires a special discussion
is the rule for determining the global sign of the diagram. We will formulate the
rule at the very end, after introducing anomalous propagators and the skeleton
diagrammatics.[*]

Analogously to the bosonic case, the symmetry-breaking terms initialize
anomalous propagators and can be omitted in all the final expressions for thermodynamic quantities, as well as in all the short- and intermediate-range off-diagonal
correlators. In the skeleton diagrammatics, the crucial part is played by single-
particle anomalous propagators [$X \equiv (\mathbf{r},\tau,\alpha)$]

$$-F_{\text{in}}(X_1,X_2) = \langle T_\tau \hat{\bar{\Psi}}(X_1)\hat{\bar{\Psi}}(X_2)\rangle, \tag{12.193}$$

$$-F_{\text{out}}(X_1,X_2) = \langle T_\tau \hat{\Psi}(X_2)\hat{\Psi}(X_1)\rangle, \tag{12.194}$$

$$F_{\text{out}}(X_1,X_2) = [F_{\text{in}}(X_1,X_2)]^*. \tag{12.195}$$

---

[*] Where the sign rule has to deal with the direction of anomalous propagators.

Figure 12.19 Anomalous propagators.

Corresponding diagrammatic elements are shown in Figure 12.19. The order of operators in Equations (12.193) and (12.194) matters [cf. Equation (12.190)] because of the antisymmetric properties:

$$F_{\text{in}}(X_2, X_1) = -F_{\text{in}}(X_1, X_2), \qquad F_{\text{out}}(X_2, X_1) = -F_{\text{out}}(X_1, X_2), \qquad (12.196)$$

and that is why, generally speaking, we have to specify directions of the anomalous lines (with central arrows in corresponding elements, cf. Figure 12.18).

The fermionic normal and anomalous single-particle propagators have exactly the same meaning as the ones previously introduced for the Bose system except that (a) there is no need to separate out averages of field operators, as in Equations (12.28) and (12.29) and (b) propagators now carry an additional spin index. In our case of singlet pairing, the spin dependence of normal/anomalous single-particle propagator factors out in the form of a constant symmetric/antisymmetric tensor:

$$G_{\sigma\sigma'}(P) = \delta_{\sigma\sigma'} G(P), \qquad F_{\text{in}}^{(\sigma\sigma')}(P) = S_{\sigma\sigma'} F_{\text{in}}(P), \qquad F_{\text{out}}^{(\sigma\sigma')}(P) = S_{\sigma\sigma'} F_{\text{out}}(P).$$
$$(12.197)$$

Here we use a momentum-frequency representation, $P = (\xi, \mathbf{k})$, were ξ is the fermionic Matsubara frequency $\xi \equiv \xi_s = \pi T(2s + 1)$ [$s = 0, \pm 1, \pm 2, \dots$]. With the parameterization (12.197), the generic symmetry properties (12.195) and (12.196), and $[G_{\alpha\beta}(P)]^* = G_{\beta\alpha}(-P)$ become

$$F_{\text{out}}^*(P) = F_{\text{in}}(P) \equiv F(P), \qquad F(P) = F(-P), \qquad G^*(P) = G(-P). \qquad (12.198)$$

12.8.2 Gor'kov–Nambu Equations

Dyson summation leads to the system of two Gor'kov–Nambu equations shown diagrammatically in Figure 12.20. The pair of equations relates the normal and

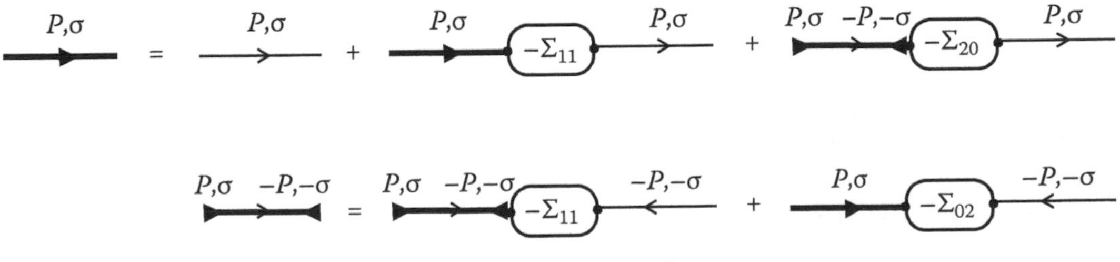

Figure 12.20 Gor'kov–Nambu equations for a homogeneous system.

anomalous Green's functions to the normal and anomalous self-energies and the single-particle propagator of the noninteracting system

$$G^{(0)}(P) = \frac{1}{i\xi - \lambda_k}. \tag{12.199}$$

The normal and anomalous self-energies share the symmetry properties with corresponding normal and anomalous propagators.

$$\Sigma_{11}^*(X_1, X_2) = \Sigma_{11}(X_2, X_1), \tag{12.200}$$

$$\Sigma_{20}^*(X_1, X_2) = \Sigma_{02}(X_1, X_2) = -\Sigma_{02}(X_2, X_1). \tag{12.201}$$

Along with generic equations (12.200) and (12.201), we also use (case-specific) analogs of (12.197), arriving thus at the parameterization

$$\Sigma_{11}^{\sigma\sigma'}(P) = \delta_{\sigma\sigma'}\Sigma(P), \qquad \Sigma_{20}^{(\sigma\sigma')}(P) = S_{\sigma\sigma'}\tilde{\Sigma}(P), \qquad \Sigma_{02}^{(\sigma\sigma')}(P) = S_{\sigma\sigma'}\tilde{\Sigma}^*(P), \tag{12.202}$$

$$\Sigma^*(P) = \Sigma(-P), \qquad \tilde{\Sigma}(P) = \tilde{\Sigma}(-P). \tag{12.203}$$

One immediately notices a close analogy with the Dyson–Beliaev equations (see Figure 12.20), except that the chemical potential term in the Fermi system is automatically included into the definition of the bare propagator (12.199). This can be done because a noninteracting Fermi system remains well defined for any finite value of μ. Moreover, the normal self-energy Σ can be safely omitted from the theory discussed here because its effects near the Fermi surface can be absorbed into a weakly renormalized expression for the dispersion relation and chemical potential, which, anyhow, are considered arbitrary in the effective Hamiltonian (12.188). This leaves us with

$$G(P) = G^{(0)}(P) - F(P)\tilde{\Sigma}(P)G^{(0)}(P), \qquad F(P) = G(P)\tilde{\Sigma}^*(P)G^{(0)}(-P). \tag{12.204}$$

(Taking into account relation $\sum_{\tilde{\sigma}} S_{\sigma\tilde{\sigma}}S_{\tilde{\sigma}\sigma'} = -\delta_{\sigma\sigma'}$, spin-index dependence drops out, leaving behind only the minus sign in the first equation.)

With the expression (12.199) for $G^{(0)}(P)$, the formal solution of (12.204) with respect to G and F is

$$G(P) = -\frac{i\xi + \lambda_k}{\xi^2 + \lambda_k^2 + |\tilde{\Sigma}(P)|^2}, \qquad F(P) = \frac{\tilde{\Sigma}^*(P)}{\xi^2 + \lambda_k^2 + |\tilde{\Sigma}(P)|^2}. \tag{12.205}$$

12.8.3 Skeleton Technique: Mean-Field Approximation

To proceed further, we need to use the skeleton diagrammatic technique to find $\tilde{\Sigma}(P)$ within the leading-order approximation. Graphically, the skeleton technique for superfluid fermions is significantly simpler than its bosonic counterpart: The obvious reason for that is the absence of nonzero single-operator quasi-averages (condensate lines). After boldification of propagators, the *only* place the dotted lines of symmetry-breaking terms enter the technique is the direct contribution

$$-\Sigma_{20}^{(\sigma,-\sigma)}(P) \quad = \quad \eta_{\sigma,-\sigma}(k) \quad + \quad$$

Figure 12.21   The leading contribution to Σ_{20}. Note the absence of the factor 1/2 in front of η. This is because both directions of the dotted lines (see Figure 12.18) are legitimate, contributing the same value. By contrast, only one direction of the anomalous propagator is allowed.

to anomalous self-energy; see Figure 12.21. A tricky aspect of the technique is the direction of the anomalous propagators. The boldification procedure does not fix the direction of the anomalous lines, rendering the notion of direction merely conventional. Meanwhile, changing the direction of a given anomalous line changes the value of the diagram by the factor (-1), suggesting that choosing the direction of anomalous propagators should be an essential part of the rule defining the overall sign on the diagram. An explicit analysis (requiring full-scale derivation of fermionic diagrammatics and thus going beyond the scope of this book) leads to the following very simple *sign-direction* prescription:

a. In any closed loop consisting of normal and anomalous propagators, all the anomalous propagators have to be of the same direction with respect to one of the two possible orientations of the loop. (It does not matter which of the two options is taken, because the number of anomalous propagators in any closed loop is even.)

b. In the "backbone" chain of propagators connecting the X_2-end (the second argument) to the X_1-end (the first argument) in all the diagrammatic expansions for normal and anomalous single-particle propagators/self-energies, the direction of all the anomalous propagators should coincide with the direction along the chain, that is, from the X_2-end to the X_1-end.

c. With the conventions (a) and (b) adopted, the sign rule remains the same as for the diagrammatics without anomalous lines. That is, each closed propagator loop contributes a factor of (-1).

The lowest-order expression for the anomalous self-energy, known as mean-filed approximation, is shown in Figure 12.21. At this point, we set the η-term in the right-hand side equal to zero. With our Hamiltonian (12.188), the frequency-momentum dependence of the mean-field anomalous self-energy is very simple:

$$\tilde{\Sigma}(\xi,k) = \begin{cases} \Delta, & \lambda_k < \omega_0, \\ 0, & \lambda_k \geq \omega_0, \end{cases} \tag{12.206}$$

where Δ is a constant related to $F(P)$ by

$$\Delta = -gT \sum_{s=-\infty}^{\infty} \sum_{\mathbf{q}}^{FS} F(\xi_s, \mathbf{q}). \tag{12.207}$$

Substituting this into (12.205), we find

$$G(P) = -\frac{i\xi + \lambda_k}{\xi^2 + E^2(k)}, \qquad F(P) = \frac{\Delta^*}{\xi^2 + E^2(k)} \qquad (\lambda_k < \omega_0), \qquad (12.208)$$

$$E(k) = \sqrt{\lambda_k^2 + |\Delta|^2}. \qquad (12.209)$$

The function $E(k)$ has the meaning of the spectrum of (fermionic) single-particle elementary excitations. This can be seen by finding the poles of G and F in the complex plane of the variable ξ. Note that the quasiparticle spectrum (12.209) features a gap $|\Delta|$ at the Fermi surface $\lambda_{k_F} = 0$.

12.8.4   Gap Equation and Critical Temperature

Substituting F from Equation (12.208), we obtain a closed equation for finding $|\Delta|$ as a function of temperature, the so-called gap equation:

$$\Delta(T) = -gT \sum_{s=-\infty}^{\infty} \sum_{\mathbf{q}}^{FS} \frac{\Delta(T)}{\xi_s^2 + \lambda_k^2 + |\Delta(T)|^2}, \qquad (12.210)$$

or

$$1 = -gT \sum_{s=-\infty}^{\infty} \int_{-\omega_0}^{\omega_0} \frac{d\lambda}{[\pi T(2s+1)]^2 + \lambda^2 + |\Delta(T)|^2}, \qquad (12.211)$$

where $\rho_F = m_* k_F / 2\pi^2$ is the Fermi surface density of states for one spin component (its energy dependence can be neglected provided $\omega_0 \ll \mu$). The sum over s can be taken analytically, leading to

$$1 = -g\rho_F \int_0^{\omega_0} \frac{\tanh\left[\sqrt{\lambda^2 + |\Delta|^2}/2T\right]}{\sqrt{\lambda^2 + |\Delta|^2}} d\lambda. \qquad (12.212)$$

Obviously, the nonzero solution exists only for negative values of the effective coupling constant g. The perturbative approach considered here is valid under the condition

$$|g|\rho_F \ll 1. \qquad (12.213)$$

At the phase transition point, the gap and anomalous averages become zero; that is, at $T = T_c$ we have

$$\int_0^{\omega_0} \frac{\tanh(\lambda/2T_c)}{\lambda} d\lambda = \frac{1}{|g|\rho_F}. \qquad (12.214)$$

With exponential accuracy in a large parameter $\omega_0/T_c \gg 1$, the integral can be taken analytically with the result $\ln(2\gamma\omega_0/\pi T_c)$, where $\gamma = 1.781\ldots$ is the Euler's constant. This leads to the BCS transition temperature

$$T_c = \frac{2\gamma}{\pi}\omega_0 e^{-1/|g|\rho_F}, \tag{12.215}$$

which is indeed exponentially small compared to ω_0, as assumed in the calculation. The gap equation (12.212) is particularly simple at zero temperature when it reads

$$\int_0^{\omega_0} \frac{d\lambda}{\sqrt{\lambda^2 + |\Delta|^2}} \approx \ln(2\omega_0/|\Delta|) = \frac{1}{|g|\rho_F}. \tag{12.216}$$

We thus have

$$|\Delta(T = 0)| = 2\omega_0 e^{-1/|g|\rho_F} \equiv \frac{\pi}{\gamma} T_c. \tag{12.217}$$

Note that the last expression does not involve the high-energy cutoff explicitly. This is a common feature of all weakly interacting fermionic superfluids—their low temperature properties are universal when expressed in terms of T_c. It is also worth noting that the exponential smallness of the gap compared to ω_0 implies that the conditions $G(\xi, k) \approx G^{(0)}(\xi, k)$ and $F(\xi, k) \approx 0$ are met already at very small momenta k, such that λ_k remains much smaller than ω_0.

This concludes our brief discussion of the BCS theory. Interested readers may look at the classic monograph [14] for additional details regarding thermodynamic properties of BCS systems, pairing mechanisms, role of weak disorder, etc.

References

1. S. T. Beliaev, Application of the methods of quantum field theory to a system of bosons, *Sov. Phys. JETP* **7**, 289 (1958); Energy spectrum of a non-ideal Bose gas, *Sov. Phys. JETP* **7**, 299 (1958).

2. B. Capogrosso-Sansone, S. Giorgini, S. Pilati, L. Pollet, N. Prokof'ev, B. Svistunov, and M. Troyer, The Beliaev technique for a weakly interacting Bose gas, *New J. Phys.* **12**, 043010 (2010).

3. V. A. Kashurnikov, N. V. Prokof'ev, and B. V. Svistunov, Critical temperature shift in weakly interacting Bose gas, *Phys. Rev. Lett.* **87**, 120402 (2001).

4. P. Arnold and G. Moore, BEC transition temperature of a dilute homogeneous imperfect Bose gas, *Phys. Rev. Lett.* **87**, 120401 (2001); P. Arnold, G. Moore, and B. Tomášik, T_c for homogeneous dilute Bose gases: A second-order result, *Phys. Rev. A* **65**, 013606 (2001).

5. N. Prokof'ev, O. Ruebenacker, and B. Svistunov, Critical point of a weakly interacting two-dimensional Bose gas, *Phys. Rev. Lett.* **87**, 270402 (2001).

6. N. V. Prokof'ev and B. V. Svistunov, Two-dimensional weakly interacting Bose gas in the fluctuation region, *Phys Rev. A* **66**, 043608 (2002).

7. N. Prokof'ev, O. Ruebenacker, and B. Svistunov, Weakly interacting Bose gas in the vicinity of the normal-fluid superfluid transition, *Phys. Rev. A* **69**, 053625 (2004).

8. A. A. Nepomnyashchii and Yu. A. Nepomnyashchii, Infrared divergence in field theory of a Bose system with a condensate, *Sov. Phys. JETP* **48**, 493 (1978).

9. V. N. Popov and A. V. Seredniakov, Low-frequency asymptotic form of the self-energy parts of a superfluid Bose system at $T = 0$, *Sov. Phys. JETP* **50**, 193 (1979).

10. Yu. A. Nepomnyashchii, Concerning the nature of the λ-transition order parameter, *Sov. Phys. JETP* **58**, 722 (1983).

11. D. S. Petrov, M. Holzmann, and G. V. Shlyapnikov, Bose–Einstein condensation in quasi-2D trapped gases, *Phys. Rev. Lett.* **84**, 2551 (2000).

12. M. Olshanii, Atomic scattering in the presence of an external confinement and a gas of impenetrable bosons, *Phys. Rev. Lett.* **81**, 938 (1998).

13. L. P. Gor'kov and T. K. Melik-Barkhudarov, Contribution to the theory of superfluidity in an imperfect Fermi gas, *Sov. Phys. JETP* **13**, 1018 (1961).

14. A. A. Abrikosov, L. P. Gor'kov, and I. E. Dzyaloshinskii, *Methods of Quantum Field Theory in Statistical Physics*, Dover Publications, New York (1975).

CHAPTER 13

Kinetics of Bose–Einstein Condensation

In this chapter,[*] we will analyze strongly nonequilibrium kinetics of Bose–Einstein condensate (BEC) formation in a weakly interacting Bose gas (WIBG). The process takes place as a result of self-evolution when occupation numbers in the initial state are large enough to ensure finite condensate density upon thermalization. In WIBG, the scenario of BEC formation involves several stages taking place at different wavenumber intervals. The key stages of the ordering process are of classical-field character and are described by the Gross–Pitaevskii equation (GPE), with turbulent initial condition.

As opposed to the case of slow external cooling, the Kibble effect of formation of topological defects—while being a signature feature of the ordering process—cannot be quantified using dimensional analysis alone. Typical separation between newly formed entangled vortex lines is dictated by the healing length of a small coherent fraction within the predominantly incoherent matter field. Estimating the coherent fraction is impossible without understanding a peculiar explosive initial kinetic stage of the process.

13.1   Weakly Turbulent State of a Degenerate Bose Gas

In this section, we will elaborate on the so-called weakly turbulent state of a degenerate Bose gas, which is the classical-field limit of the state with nearly independent single-particle modes. At large occupation numbers, each mode is equivalent to a harmonic of the corresponding classical field.

13.1.1   Weakly Turbulent State

Apart from general properties of coherent states established in Section 7.1, we will also need here statistical properties of states with large occupation numbers. Consider a *mixed* state of a bosonic mode and write its statistical operator in the form

$$W = \sum_{n} Q(n)|n\rangle\langle n|. \tag{13.1}$$

Let us consider that (1) typical n in the sum is much larger than unity and (2) Q is a smooth function of its argument.[†] The statistical operator in the form (13.1) is diagonal in the occupation number representation, which is characteristic of a bosonic mode weakly coupled to some environment (say, the rest of the modes), as

---

[*] Based on Reference [1].

[†] In practice, $Q(n)$ changes significantly on the scale of $\langle n \rangle$.

in the WIBG system. Under the condition $\langle n \rangle \gg 1$, this is true no matter whether one is dealing with equilibrium statistics or relaxational kinetics. It turns out that under the conditions (1) and (2), the statistical operator (13.1) is also *diagonal* in terms of coherent states

$$W \approx \frac{1}{\pi} \int d\psi \, Q(|\psi|^2) |\psi\rangle\langle\psi|, \tag{13.2}$$

where $\psi = \rho e^{i\varphi}$, $d\psi \equiv d\varphi \, \rho d\rho$. An easy way to prove (13.2) is to show that it implies (13.1). To this end, substitute explicit form (7.4) into the right-hand side of (13.2) and perform the integration over φ. The result is:

$$\frac{1}{\pi} \int d\psi \, Q(|\psi|^2) |\psi\rangle\langle\psi| = \int d(\rho^2) Q(\rho^2) \sum_n \frac{\rho^{2n} e^{-\rho^2}}{n!} |n\rangle\langle n|. \tag{13.3}$$

Changing the order of summation and integration, and noticing that the smoothness of the function $Q(\rho^2)$ in combination with having large typical n's in the sum allows one to replace

$$\rho^{2n} e^{-\rho^2} \rightarrow n! \, \delta(\rho^2 - n) \qquad (n \gg 1), \tag{13.4}$$

we arrive at Equation (13.1).

Suppose we have WIBG in a state (generally speaking, non-equilibrium) such that occupation numbers at momenta $k < k_*$ are much larger than unity, while occupation numbers at $k > k_*$ are zero. We also assume that there are no correlations between n_k values for different modes. Then, the statistical operator of the system is a direct product of statistical operators, W_k, for all modes

$$W = \prod_k W_k, \qquad W_k = \sum_{n_k} Q_k(n_k) |n_k\rangle\langle n_k|. \tag{13.5}$$

Functions $Q_k(n_k)$ are assumed to be smooth in the sense $d\ln Q_k/dn_k \ll 1$, but otherwise arbitrary.

The state (13.5) is a *mixture*, not a *pure* quantum state, because W is a sum of diagonal statistical operators. Without loss of generality, the mixed state (13.5) is equivalent to a statistical ensemble of pure states with well-defined momentum occupation numbers; $Q_k(n_k)$ is the probability of finding a particular occupation number n_k. We now interpret (13.5) as a statistical ensemble of *coherent states*. Indeed, by applying Equation (13.2) to each W_k, we get

$$W = \prod_k \int \frac{d\psi_k}{\pi} Q_k(|\psi_k|^2) |\psi_k\rangle\langle\psi_k|, \tag{13.6}$$

which can be written in a more compact form by using the coherent states of the field operator (see Section 7.1.2):

$$W = \int \mathcal{D}\psi \, Q^{(\text{wt})}(\psi) |\psi\rangle\langle\psi|, \tag{13.7}$$

where

$$\psi(\mathbf{r}) = \sum_{\mathbf{k}} \psi_{\mathbf{k}} \, e^{i\mathbf{k}\cdot\mathbf{r}}, \qquad\qquad\qquad\qquad (13.8)$$

$$\mathcal{D}\psi = \prod_{\mathbf{k}} \frac{d\psi_{\mathbf{k}}}{\pi} \equiv \prod_{\mathbf{k}} d(\rho_{\mathbf{k}}^2) \frac{d\varphi_{\mathbf{k}}}{2\pi} \qquad (\psi_{\mathbf{k}} \equiv \rho_{\mathbf{k}} e^{i\varphi_{\mathbf{k}}}), \qquad (13.9)$$

$$Q^{(\mathrm{wt})}(\psi) = \prod_{\mathbf{k}} Q_{\mathbf{k}}(\rho_{\mathbf{k}}^2). \qquad\qquad\qquad\qquad (13.10)$$

Since $\{Q_{\mathbf{k}}\}$ depend only on $\{\rho_{\mathbf{k}}\}$, all harmonics of the field ψ in the ensemble have independent and arbitrary phases. To emphasize this very important circumstance, we introduced the superscript "(wt)" for Q, which is the abbreviation for *weak turbulence*. The term comes from the theory of weakly nonlinear classical fields and stands for the regime (and kinetic approximations based on the same assumptions) when the phases of different Fourier harmonics of the field are approximately independent. In a weakly turbulent state of a macroscopic system, the correlation properties are very simple: they are described by the Wick's theorem.[*]

Problem 13.1 *Prove Wick's theorem for the ensemble of classical fields in the weakly turbulent state of a macroscopic system. Start with arguing that one can neglect (in the sums of products of Fourier components) terms containing more than two fields with the same wavevector.*

Time evolution of the degenerate coherent state with large occupation numbers in dynamically relevant harmonics is described by the GPE (see the derivation and discussion in Section 7.6). Recall that harmonics that are not immediately important for the system's evolution—normally, high-momentum harmonics weakly coupled to the rest of the system—can be formally eliminated from the Hilbert space for a long enough period of time. For GPE to be valid, in contrast to the previous section, there is no need to impose any restriction on the distribution function $Q(\psi)$. In particular, this function can be (or may become, as a result of evolution) radically different from $Q^{(\mathrm{wt})}(\psi)$ in Equation (13.10), reflecting, say, some short- or even long-range order in the real-space configuration of the field ψ. Thus, each coherent state in the ensemble (13.7) evolves following an independent classical-field trajectory $\psi(t)$ corresponding to the solution of the nonlinear Schrödinger equation (7.88).

13.2   Kinetic Equation in the Weak-Turbulence Regime

In the weak-turbulence regime, the (ensemble-averaged) evolution of the field ψ can be described within the *kinetic equation* (KE) that deals with the average occupation numbers $n_{\mathbf{k}} = |\psi_{\mathbf{k}}|^2$. [From now on, we will work with classical fields and

---

[*] As we remember from Chapter 11, Wick's theorem states that an ensemble average of a product of fields equals the sum of all possible products based on different pairwise averages.

understand $n_\mathbf{k}$'s as numbers, not operators.] The mathematical trick behind the kinetic description is replacing (in accordance with Wick's theorem) ensemble averages of products of $\psi_\mathbf{k}$'s and $\psi_\mathbf{k}^*$'s with sums of all possible products of $n_\mathbf{k}$'s. The applicability of Wick's theorem, in its turn, follows from the random-phase approximation that is justified by the fact that the frequencies of phase rotations, $\dot{\varphi}_\mathbf{k}$, are much higher than the typical inverse time associated with the nonlinear interaction. The criterion for the weak-turbulence regime to take place reads as follows:

$$k_*^2/m \gg U n_{k_*} k_*^3, \tag{13.11}$$

where k_* and n_{k_*} are the typical momentum and occupation number, respectively. Correspondingly, the parameter controlling the accuracy of the kinetic approach developed in the following is

$$\xi_{\text{wt}} = m U n_{k_*} k_* \ll 1. \tag{13.12}$$

To arrive at the KE, one starts with the nonlinear Schrödinger equation written in Fourier components (we use a shorthand notation in the following for the momentum subscripts: $\mathbf{k}_1 \to 1$, $\mathbf{k}_2 \to 2$, etc.)

$$i\dot{\psi}_1 = (k_1^2/2m)\psi_1 + U \sum_{\mathbf{k}_2 \mathbf{k}_3 \mathbf{k}_4} \psi_2^* \psi_3 \psi_4\, \delta_{\mathbf{k}_1; \mathbf{k}_3 + \mathbf{k}_4 - \mathbf{k}_2}, \tag{13.13}$$

and finds

$$\dot{n}_1 = \frac{\partial}{\partial t}\psi_1^* \psi_1 = 2U\,\mathrm{Im} \sum_{\mathbf{k}_2 \mathbf{k}_3 \mathbf{k}_4} \psi_1^* \psi_2^* \psi_3 \psi_4\, \delta_{\mathbf{k}_1; \mathbf{k}_3 + \mathbf{k}_4 - \mathbf{k}_2}. \tag{13.14}$$

Occupation number fluctuation from one configuration of a turbulent field to another are strong (of the order of 100%). That is why it is important to average Equation (13.14) over an ensemble of systems with statistically similar initial conditions. This leads to the time-dependent correlation function in the right-hand side of Equation (13.14)

$$Q_{1234} = \langle \psi_1^* \psi_2^* \psi_3 \psi_4 \rangle. \tag{13.15}$$

At this point, we cannot simply replace correlator (13.15) in the right-hand side of the averaged Equation (13.14) with its random-phase approximation, since the latter is real, and we will get identical zero. Hence, we must go to the higher-order (in the interaction constant U) corrections to the correlator Q to get the lowest-order nonvanishing contribution to Equation (13.14). To this end, we differentiate Q with respect to time and utilize Equation (13.13) to obtain

$$\dot{Q}_{1234} = i\,\Delta\varepsilon\, Q_{1234} + iU \sum_{\mathbf{k}_5 \mathbf{k}_6 \mathbf{k}_7} \Big[\langle \psi_5 \psi_6^* \psi_7^* \psi_2^* \psi_3 \psi_4 \rangle \delta_{\mathbf{k}_1; \mathbf{k}_6 + \mathbf{k}_7 - \mathbf{k}_5}$$
$$+ \langle \psi_5 \psi_6^* \psi_7^* \psi_1^* \psi_3 \psi_4 \rangle \delta_{\mathbf{k}_2; \mathbf{k}_6 + \mathbf{k}_7 - \mathbf{k}_5} - \langle \psi_5^* \psi_6 \psi_7 \psi_1^* \psi_2^* \psi_4 \rangle \delta_{\mathbf{k}_3; \mathbf{k}_6 + \mathbf{k}_7 - \mathbf{k}_5}$$
$$- \langle \psi_5^* \psi_6 \psi_7 \psi_1^* \psi_2^* \psi_3 \rangle \delta_{\mathbf{k}_4; \mathbf{k}_6 + \mathbf{k}_7 - \mathbf{k}_5} \Big], \tag{13.16}$$

where

$$\Delta\varepsilon \equiv \varepsilon_1 + \varepsilon_2 - \varepsilon_3 - \varepsilon_4, \qquad \varepsilon_k = k^2/2m. \tag{13.17}$$

Now we can apply Wick's theorem to correlators in the right-hand side of Equation (13.16). Since the four momenta, $\mathbf{k}_1, \mathbf{k}_2, \mathbf{k}_3$, and $\mathbf{k}_4$, are all different, each of the six-field averages produces only two nonzero terms (while in a general case, Wick's theorem would lead to six terms). For example,

$$\langle \psi_5 \psi_6^* \psi_7^* \psi_2^* \psi_3 \psi_4 \rangle = n_2 n_3 n_4 \, \delta_{\mathbf{k}_2 \mathbf{k}_5} (\delta_{\mathbf{k}_3 \mathbf{k}_6} \delta_{\mathbf{k}_4 \mathbf{k}_7} + \delta_{\mathbf{k}_3 \mathbf{k}_7} \delta_{\mathbf{k}_4 \mathbf{k}_6}). \tag{13.18}$$

We also assume that the momenta of the correlator Q_{1234} satisfy the condition $\mathbf{k}_1 + \mathbf{k}_2 = \mathbf{k}_3 + \mathbf{k}_4$, since only such terms contribute to Equation (13.14). As a result, we have

$$\dot{Q}_{1234} = i \, \Delta\varepsilon \, Q_{1234} + i R_{1234}, \tag{13.19}$$

where

$$R_{1234} = 2U[n_2 n_3 n_4 + n_1 n_3 n_4 - n_1 n_2 n_3 - n_1 n_2 n_4]. \tag{13.20}$$

The solution of the differential equation (13.19) can be written in the integral form

$$Q_{1234}(t) = i \lim_{\delta \to +0} \int_0^\infty e^{i(\Delta\varepsilon + i\delta)\tau} R_{1234}(t - \tau) \, d\tau. \tag{13.21}$$

The infinitesimal constant δ is introduced to regularize the integral. Formally, a general solution to Equation (13.19) requires adding to the right-hand side of (13.21) a term $\propto \exp(i \, \Delta\varepsilon \, t)$. However, any term that contains a phase that depends on the *absolute* time should also contain an arbitrary phase shift; under ensemble averaging over all possible phase shifts, such a term averages out.

Substituting (13.21) into (ensemble-averaged) equation (13.14), we see that the typical values of energies ε contributing to the sums over momenta are nothing but typical frequencies in the given range of wavenumbers: $\varepsilon \sim k_*^2/m$. The essence of the weak-turbulence regime, expressed by the criterion (13.11), is that this frequency corresponds exclusively to the harmonic phase rotation. The evolution of harmonic amplitudes is governed by terms that are higher order in the parameter (13.12), and thus it occurs at much longer time scales. Mathematically, this means that, in the integrals over momenta in the KE, we can make the replacement:

$$Q_{1234}(t) \to \pi R_{1234}(t) \delta(\Delta\varepsilon). \tag{13.22}$$

We thus arrive at the KE of the form (in the final expression, we replace summations over momenta with integrations, since the ensemble-averaged $n_\mathbf{k}$ is a smooth function of $\mathbf{k}$)

$$\dot{n}_1 = 4\pi U^2 \int \frac{d\mathbf{k}_2 d\mathbf{k}_3}{(2\pi)^6} \, \delta(\Delta\varepsilon) [(n_1 + n_2)n_3 n_4 - n_1 n_2(n_3 + n_4)], \tag{13.23}$$

where

$$\mathbf{k}_4 \equiv \mathbf{k}_1 + \mathbf{k}_2 - \mathbf{k}_3. \tag{13.24}$$

The dimensional analysis of kinetic equation (13.23) readily yields an estimate for characteristic kinetic time, τ_{kin}:

$$\tau_{\text{kin}}^{-1} \sim mU^2 k_*^4 n_{k_*}^2 \sim \xi_{\text{wt}}^2 (k_*^2/m). \tag{13.25}$$

We see that as long as the criterion (13.11) is satisfied, $\tau_{\text{kin}}^{-1} \ll k_*^2/m$ by a factor $\sim \xi_{\text{wt}}^2$. This quantifies the accuracy of the replacement (13.22).

It is instructive to compare the classical-field KE (13.23) with the full quantum KE

$$\dot{n}_1 = 4\pi U^2 \int \frac{d\mathbf{k}_2 d\mathbf{k}_3}{(2\pi)^6} \, \delta(\Delta\varepsilon) \left[(n_1 + 1)(n_2 + 1)n_3 n_4 - n_1 n_2 (n_3 + 1)(n_4 + 1) \right]$$

$$= 4\pi U^2 \int \frac{d\mathbf{k}_2 d\mathbf{k}_3}{(2\pi)^6} \, \delta(\Delta\varepsilon) \left[(n_1 + n_2 + 1)n_3 n_4 - n_1 n_2 (n_3 + n_4 + 1) \right], \tag{13.26}$$

where $n_{\mathbf{k}}$ is understood as the expectation value of occupation number of the single-particle mode with momentum $\mathbf{k}$.

Problem 13.2 *Derive Equation (13.26) from the Golden Rule for the rates of increasing/decreasing $n_{\mathbf{k}}$ by one in the two-particle scattering process.*

In quantum kinetics, the $(n_j + 1)$ factor in the collision term (originating from the number of particles in the final state) is interpreted as the sum of two processes: (1) *spontaneous* events that proceed at a rate insensitive to the occupation of state j and (2) *stimulated* events that have a rate proportional to n_j. An important qualitative reasoning behind this terminology is that from the classical-field perspective (it is the diametric opposite of the classical-particle perspective!), the stimulated processes are trivial; they are supposed to dominate when the behavior of the quantum field mimics that of its classical counterpart. The spontaneous processes are essentially quantum. Consistently with this general observation, the only difference between Equations (13.23) and (13.26) is that, in the classical-field description, there are no spontaneous processes, which is a legitimate approximation to (13.26) in the limit of large occupation numbers.

13.3 Self-Similar Analysis of Kinetic Equation

In this section, we will perform a generic analysis of the classical-field evolution in the weak-turbulence regime. We assume that the evolution is described by the KE of the form

$$\dot{n}_\varepsilon = \text{Coll}(\varepsilon, [n]), \tag{13.27}$$

where n_ε is the common occupation number for all modes with energy ε; that is, we assume equipartition over modes with the same energy, so that ε and time are the only parameters characterizing the distribution. This assumption is not as simple as it seems and, in principle, requires a justification/validation in each particular case.[*] Next, we assume that the collision term in the right-hand side of Equation (13.27) is scale invariant and *local* in the ε-space.[†] This allows us to introduce a dimensional estimate of the collision term,

$$\text{Coll}(\varepsilon, [n]) \sim \varepsilon^s n_\varepsilon^p, \tag{13.28}$$

where s and p are certain exponents. We also require that the KE conserves (with certain important reservations concerning the form of the solution; see in the following) the total number of particles

$$N \propto \int d\varepsilon \, \varepsilon^q n_\varepsilon \tag{13.29}$$

and total energy

$$E \propto \int d\varepsilon \, \varepsilon^{q+1} n_\varepsilon. \tag{13.30}$$

In what follows, we explore various kinetic regimes that depend on the exponents s, p, and q. Several physical reasons exist for being interested in the general mathematical analysis of the KE with arbitrary exponents in Equations (13.28) through (13.30). To begin with, the number of modes with the same energy depends on the dimension of space and the dispersion relation. In addition, one might consider interparticle potentials that feature a nontrivial energy dependence of the scattering amplitude in the $\varepsilon \to 0$ limit, the dominant contribution to the collision integral may come from scattering processes that involve more than two particles (e.g., when the two-body s-wave scattering amplitude is tuned to be zero), etc. Given the variety of available experimental systems, we choose to present a comprehensive analysis of possible kinetic scenarios well beyond the most obvious case of 3D WIBG in continuous space that corresponds to $q = 1/2$, $p = 3$, and $s = 2$.

Problem 13.3 *Show that $s = 2$ for the 3D WIBG in continuous space with two-body contact interaction potential.*

We will confine ourselves to the classical-field systems that feature simultaneously the BEC phenomenon and the ultraviolet catastrophe:

$$q > 0 \qquad \text{(BEC + UV catastrophe)}. \tag{13.31}$$

---

[*] For example, an isotropic solution of Equation (13.23) implied by the ansatz $n_\mathbf{k} \equiv n_\varepsilon$, while formally consistent with Equation (13.23), may not necessarily be kinetically stable.

[†] Note that the locality of the collision term (i.e., the situation when the main contribution to $\text{Coll}(\varepsilon, [n])$ comes from energies $\sim \varepsilon$) is the property that, speaking generally, depends on both the form of the collision term and the form of the solution in question.

Indeed, under the condition (13.31), the integral $\int d\varepsilon\, \varepsilon^{q-1}$ converges at the lower limit, implying BEC at low temperature, and diverges at the upper limit. The latter means that there is no equilibrium Gibbs distribution, $n_\varepsilon^{\text{(Gibbs)}} \to T/\varepsilon$ at $\varepsilon \to \infty$, with nonzero temperature and *finite* number of particles. The ultraviolet catastrophe implies that the system should asymptotically behave as a heat bath of zero temperature. In the $t \to \infty$ limit, all the particles are concentrated at $\varepsilon \to 0$, while all the energy goes to $\varepsilon \to \infty$ harmonics. Our goal then is to analyze kinetic scenarios leading to this asymptotic picture. We will see that in the space of parameters (s,p,q), there are three distinct regions characterized by qualitatively different evolutions of particle and energy distributions:

$$s < (p-1)(q+1) \qquad \text{(N cascades, E drifts)}, \qquad\qquad (13.32)$$

$$(p-1)(q+1) < s < (p-1)(q+2) \qquad \text{(both N and E drift)}, \qquad (13.33)$$

$$s > (p-1)(q+2) \qquad \text{(N drifts, E cascades)}. \qquad\qquad (13.34)$$

The precise meanings of the terms "drift" and "cascade" will become clear later. In the present context, "cascade" means an explosive evolution with a singularity developing at some finite time t_* and nonconservation of the cascading quantity taking place at $t > t_*$. In the regime (13.32), the time moment t_* corresponds to the onset of BEC (more accurately, the onset of quasicondensation, since formation of the genuine long-range order can occur only at macroscopically long times). Since the GPE behind KE (13.27) conserves N, the nonconservation of N at $t>t_*$ signals the inapplicability of Equation (13.27) to the description of evolution at $t > t_*$. At longer times, one must modify Equation (13.27) by explicitly introducing the condensate. In view of this fact, our analysis of regimes (13.32) and (13.34) will not go beyond $t \le t_*$. This analysis (but, generally speaking, not the result!) is "symmetric" with respect to N and E: If a certain qualitative behavior of N takes place in some region $\mathcal{A}$ of the parameter space (s,p,q), a similar behavior also takes place for E in the corresponding region $\mathcal{B}$. The rule establishing a direct relationship between $\mathcal{A}$ and $\mathcal{B}$ is as follows: If $\mathcal{A}$ is specified by some inequalities in terms of (s,p,q), then the inequalities specifying $\mathcal{B}$ are obtained by replacing $q \to q+1$ and simultaneously changing all inequality signs. In this sense, the regime (13.34) is a counterpart to the regime (13.32), while, in the regime (13.33), N and E behave qualitatively similarly. The symmetry between N and E applies also to the quantitative relations in terms of (s,p,q). If a self-similar solution in $\mathcal{A}$ is characterized by the exponents that can be generically expressed—by dimensional analysis—in terms of (s,p,q), then the exponents for the similar solution in $\mathcal{B}$ are obtained by the substitution $q \to q+1$.

Problem 13.4 *Explain the origin of the previously discussed symmetry between N and E.*

13.3.1 Power-Law Solution and Drift Scenario

The scale invariance of the collision term Equation (13.27) guarantees existence of the following power-law solution (provided it is consistent with the locality of the collision term):

$$n_\varepsilon(t) = \frac{C}{t^{1/(p-1)} \varepsilon^{s/(p-1)}}. \tag{13.35}$$

Problem 13.5 *Show that the exponents in (13.35) are fixed by the parametric estimate (13.28), while the value of C is sensitive to details of the collision term.*

The solution (13.35) is inconsistent with N and E being finite: With the power-law distribution, the integrals (13.29) and (13.30) inevitably diverge at least at one of the limits. This inconsistency, however, does not yet mean that the solution is absolutely irrelevant. In the $t \to \infty$ limit, the power-law distribution (13.35) can apply to an arbitrary large interval in the $(\ln \varepsilon)$-space, while the conservation of (13.29) and (13.30) is met by introducing appropriate (time-dependent) cutoffs. Specifically, if the condition (13.33) is satisfied, then the conservation of N implies the *lower* cutoff at the energy scale

$$\varepsilon_0^{(N)}(t) \propto t^{-\zeta}, \qquad \zeta = \frac{1}{s - (p-1)(q+1)} > 0, \tag{13.36}$$

while the conservation of E leads to the *upper* cutoff at the energy scale

$$\varepsilon_0^{(E)}(t) \propto t^{\lambda}, \qquad \lambda = \frac{1}{(p-1)(q+2) - s} > 0. \tag{13.37}$$

Problem 13.6 *Derive (13.36) and (13.37).*

With the cutoffs (13.36) and (13.37), the solution (13.35) makes perfect physical sense. It describes the range of energy scales between the particle-carrying scale $\varepsilon \sim \varepsilon_0^{(N)}(t)$ and the energy-carrying scale $\varepsilon \sim \varepsilon_0^{(E)}(t)$. The propagation (drift) of $\varepsilon_0^{(N)}(t)$ and $\varepsilon_0^{(E)}(t)$ toward $\varepsilon = 0$ and $\varepsilon = \infty$, respectively, completes the qualitative picture of the *drift* scenario. It takes an infinitely long time for both cutoffs to reach their final values implied by the UV catastrophe.

It is natural to expect that, in the vicinity of the cutoffs (13.36) and (13.37), the solution of the KE becomes (asymptotically) self-similar:

$$n_\varepsilon^{(N)}(t) \propto [\varepsilon_0^{(N)}(t)]^{-(q+1)} f^{(N)}\left(\varepsilon/\varepsilon_0^{(N)}(t)\right), \tag{13.38}$$

$$n_\varepsilon^{(E)}(t) \propto [\varepsilon_0^{(E)}(t)]^{-(q+2)} f^{(E)}\left(\varepsilon/\varepsilon_0^{(E)}(t)\right), \tag{13.39}$$

where $f^{(N)}$ and $f^{(E)}$ are dimensionless scaling functions. The time-dependent prefactors in (13.38) and (13.39) are fixed by conservation of N and E. The consistency of self-similar solutions (13.38) and (13.39) with the scale-invariant solution (13.35) at $\varepsilon_0^{(N)}(t) \ll \varepsilon \ll \varepsilon_0^{(E)}(t)$ implies

$$f^{(N)}(x) \propto x^{-s/(p-1)} \qquad \text{at} \qquad x \gg 1, \tag{13.40}$$

$$f^{(E)}(x) \propto x^{-s/(p-1)} \qquad \text{at} \qquad x \ll 1. \tag{13.41}$$

For the drift scenario to be valid, the condition (13.33) is absolutely crucial. As s approaches its lower bound $(p-1)(q+1)$, the exponent ζ diverges. By continuity, this suggests that at $s < (p-1)(q+1)$, there takes place an explosive propagation of the particle distribution front toward $\varepsilon = 0$. A similar argument applies to the case $s > (p-1)(q+2)$ where the divergence of λ at $s \to (p-1)(q+2)$ suggests an explosive propagation of the energy distribution front toward $\varepsilon = \infty$. In what follows, we will see that this is indeed the case.

Note that all relations of this section feature the previously mentioned symmetry between N and E.

13.3.2 Quasi-Steady-State Cascades

An alternative to the drift scenario is the *cascade* scenario, when the particle distribution evolution toward the $\varepsilon \to 0$ region is done by cascading along the energy scales. Such a scenario implies that the kinetic time as a function of energy is getting progressively shorter with decreasing ε, so that the bottleneck in the particle transport is associated with the kinetic time at the energy scale where the particles were originally distributed. In this subsection, we develop the theory of a quasi-steady-state Kolmogorov–Zakharov particle cascade. Its existence and form shed a considerable light on the BEC-formation kinetics. The theory of the energy cascade is similar, with all quantitative statements readily obtained by the symmetry between N and E mentioned earlier; see Problem 13.4.

Let us formally define the *particle flux* at the energy scale ε as $F_\varepsilon^{(N)}$:

$$F_\varepsilon^{(N)} = -\int_\varepsilon^\infty d\tilde{\varepsilon}\, \tilde{\varepsilon}^q \, \text{Coll}(\tilde{\varepsilon}, [n]). \tag{13.42}$$

Then, in accordance with the KE, the particle-conservation law can be written in the form of continuity equation

$$\dot{N}_\varepsilon = -F_\varepsilon^{(N)}, \tag{13.43}$$

where

$$N_\varepsilon = \int_\varepsilon^\infty d\tilde{\varepsilon}\, \tilde{\varepsilon}^q \, n_{\tilde{\varepsilon}} \tag{13.44}$$

is the number of particles with energies larger than ε. Clearly, the notions of the particle flux and cascade are meaningful only when *both* the collision term in the KE and the integral (13.42) are local, since only in this case can one really speak of transport (of the conserved quantity) in the space of energy scales. [Otherwise, Equations (13.42) and (13.43) are a mere tautology implied by the KE, irrespective of the conservation of N.] Especially important is the limiting case of (13.43),

$$\dot{N} = -F^{(N)}_{\varepsilon \to 0},$$

(13.45)

showing that the conservation or nonconservation of the total number of particles is related to the limiting behavior of flux at zero energy.

The analogs of Equations (13.42) through (13.45) for energy are

$$\dot{E}_\varepsilon = -F^{(E)}_\varepsilon, \qquad\qquad \dot{E} = -F^{(E)}_{\varepsilon \to \infty},$$

(13.46)

$$E_\varepsilon = \int_0^\varepsilon d\tilde{\varepsilon}\, \tilde{\varepsilon}^{q+1}\, n_{\tilde{\varepsilon}}, \qquad\qquad F^{(E)}_\varepsilon = -\int_0^\varepsilon d\tilde{\varepsilon}\, \tilde{\varepsilon}^{q+1}\, \mathrm{Coll}(\tilde{\varepsilon}, [n]).$$

(13.47)

The steady-state particle cascade is characterized by ε-independent flux $F^{(N)}$ in the inertial range of energy scales. Clearly, it requires permanent pumping of particles and thus goes beyond KE (13.27) in some region of energy space. Physically, the steady-state cascade is nearly identical to the quasi-steady-state cascade when the particle flux is essentially independent of ε and slowly depends on time. As we will discuss in the following, this regime can be realized without pumping in the case (13.32). The theory of the quasi-steady-state cascade is very close to that of the steady-state cascade, since in both cases, we are dealing with the function n_ε for which F_ε is ε-independent. To find n_ε, we notice that scale invariance of the collision term implies that this function is a power law: $n_\varepsilon \propto 1/\varepsilon^\beta$. However, after substituting this function into (13.42), we face a dilemma. We expect to get a nonzero constant flux F, while at the same time, our collision term should be identically zero, since we are looking for a stationary solution. This brings us to the conclusion that the integral (13.42) must be singular in the sense that the vanishing of the collision term is compensated for by the divergence of the integral. To make sure that this is indeed the case and, most importantly, to find an appropriate regularization, we will consider the collision term for a family of power-law functions $n_\varepsilon \propto 1/\varepsilon^{\beta+\delta}$ (where δ is assumed to be arbitrarily small). The analysis of dimensions yields

$$\mathrm{Coll}(\varepsilon, [n_\varepsilon = A/\varepsilon^{\beta+\delta}]) = C(\delta) A^p\, \varepsilon^{s-(\beta+\delta)p},$$

(13.48)

where A is some constant and $C(\delta = 0) = 0$. After substituting this into (13.42) and performing the integration, we arrive at the expression

$$F^{(N)}_\varepsilon = \frac{C(\delta)\, A^p\, \varepsilon^{q+s-(\beta+\delta)p+1}}{q+s-(\beta+\delta)p+1}.$$

(13.49)

We see that for $F_\varepsilon^{(N)}$ to be ε-independent in the limit of $\delta \to +0$ [the integral (13.42) is convergent only at $\delta > 0$], the exponent β must be

$$\beta^{(N)} = (q + s + 1)/p \qquad \text{(constant particle flux)}. \qquad (13.50)$$

At the same time, we find that the derivative $C'(\delta)$ should be finite and positive at $\delta = 0$, since it is related to the constant particle flux by

$$F^{(N)} = -(A^p/p)\lim_{\delta \to 0} C(\delta)/\delta = -(A^p/p)\,C'(0). \qquad (13.51)$$

Incidentally, Equation (13.51) establishes a procedure for calculating the particle-cascade flux.

In the cascade regime, most particles must be distributed at energies that are large compared to the inertial range. Hence, the consistency requirement for the cascade scenario requires that the integral $\int d\varepsilon\, \varepsilon^q/\varepsilon^{\beta^{(N)}}$ be convergent at $\varepsilon \to 0$ and, correspondingly, divergent at $\varepsilon \to \infty$. From (13.50), we see that this is the case if, and only if, the condition (13.32) is met.

Problem 13.7 *Estimate the characteristic kinetic time at the energy scale ε for the constant-flux stationary solution. Make sure that this time decreases with ε if, and only if, the condition (13.32) is met.*

For scale-invariant KE, it is expected that, in the region (13.32), there exists an asymptotic self-similar solution with the energy drift and a quasi-steady-state particle cascade. The qualitative form of the solution in this case is given by

$$n_\varepsilon(t) \propto [\varepsilon_0^{(E)}(t)]^{-(q+2)} f\left(\varepsilon/\varepsilon_0^{(E)}(t)\right), \qquad f(x) \propto x^{-\beta^{(N)}} \quad \text{at} \quad x \ll 1, \qquad (13.52)$$

with $\varepsilon_0^{(E)}(t)$ obeying the same Equation (13.37) as in the mutual drift scenario.

Problem 13.8 *Consider the quasi-steady-state energy cascade. In particular, find a counterpart of Equation (13.52) and show that it applies to the regime (13.34).*

13.3.3 Explosive Self-Similar Cascade Solution

The existence of the Kolmogorov–Zakharov cascade naturally leads to the explosive kinetic process it emerges from. Indeed, the particle cascade is based on vanishing relaxation time (divergent relaxation rate) in the $\varepsilon \to 0$ limit (from now on, we will consider the particle cascade; the energy-cascade considerations are similar). Hence, there can take place a fast kinetic process involving arbitrarily small energy scales in a finite time when, at $t = t_*$, the particle distribution front reaches $\varepsilon = 0$. This process paves the way—by populating all energy scales below some cutoff—for the consequent fast evolution toward the Kolmogorov–Zakharov cascade. By scale invariance of the KE, the explosive kinetics become self-similar in the region of small enough energies:

$$n_\varepsilon(t) \propto A(t)f(\varepsilon/\varepsilon_0(t)), \qquad (13.53)$$

whcre $f(x)$ is a dimensionless function of its argument, and $A(t)$ and $\varepsilon_0(t)$ are some (power-law) functions of time. Scale invariance dictates that $f(x)$ is a decaying power-law function at $x \to \infty$: $f(x) \propto 1/x^\alpha$. At $t = t_*$, we have $\varepsilon_0(t_*) = 0$, and, in view of locality of the collision term, we do not expect any singularity in the distribution $n_\varepsilon(t_*)$. This is possible only if

$$A(t) \propto 1/\varepsilon_0^\alpha(t). \tag{13.54}$$

When combined with the scaling properties of the collision term, the exponent α determines the evolution of ε_0 with time. To show this, we introduce a correction to the asymptotic expansion of $f(x)$:

$$f(x) \to \frac{c_0}{x^\alpha}\left(1 + \frac{c_1}{x^\sigma}\right), \qquad x \to \infty. \tag{13.55}$$

Here c_0 and c_1 are constants. The second term in (13.55) is responsible for time dependence of $n_\varepsilon(t)$ at $\varepsilon \gg \varepsilon_0$. As the time moment $t = t_*$ plays no special role at finite energies $\varepsilon \gg \varepsilon_0$, the solution $n_\varepsilon(t)$ should be regular in terms of $t - t_*$:

$$n_\varepsilon(t) = n_\varepsilon(t_*) + \dot{n}_\varepsilon(t_*)(t - t_*) + \cdots, \qquad \varepsilon \gg \varepsilon_0(t). \tag{13.56}$$

The requirement that (13.56) be consistent with (13.55) yields

$$\varepsilon_0(t) \propto (t_* - t)^{1/\sigma}. \tag{13.57}$$

Moreover, for the higher-order terms in Taylor expansion (13.56) to be proportional to integer powers of $(t_* - t)$, we must require

$$f(x) \to x^{-\alpha} \sum_{m=0} \frac{c_m}{x^{m\sigma}}, \qquad x \to \infty. \tag{13.58}$$

On the other hand, for consistency with KE (13.27)–(13.28), σ and α must be related as

$$\sigma = \alpha(p - 1) - s. \tag{13.59}$$

The value of exponent α depends on the structure of the collision term. Conservation laws for N and E—the only universal properties we have—translate only into integral constraints on $f(x)$ that do not fix the asymptotic behavior of $f(x)$ at $x \to \infty$ (see Problem 13.9). However, it is possible to establish important upper and lower bounds on α. The divergence of the particle-number integral (13.29) at the upper limit takes place only if

$$\alpha < q + 1. \tag{13.60}$$

Problem 13.9 *Show that the kinematics of the collision term in Equation (13.23) are such that the integral is local for any α satisfying (13.60).*

The natural requirement [see (13.55) and (13.57)] $\sigma > 0$ translates into

$$\alpha > \frac{s}{p-1}. \tag{13.61}$$

Note that Equations (13.60) and (13.61) imply the condition (13.32), as expected. Interestingly, the value of $\alpha = \beta^{(N)}$ is consistent with the inequalities (13.60) and (13.61), and this observation raises the question "Does this 'natural' situation indeed take place?"

Problem 13.10 *Demonstrate that if $\alpha = \beta^{(N)}$, then Equation (13.55) is inconsistent with locality of KE. Hint: With $\alpha = \beta^{(N)}$, the dominant term in the collision integral is identically zero.*

Having excluded $\alpha = \beta^{(N)}$, we can think of establishing an inequality between the two exponents. If $\alpha < \beta^{(N)}$, then one can consider the solution at $t = t_*$ as an *initial condition* and proceed with the evolution governed by the same kinetic equation.* Since in this situation $F^{(N)}_{\varepsilon \to 0} = 0$ [see (13.49)], we observe that we are still facing the problem of the transient regime that will result in $F^{(N)}_{\varepsilon \to 0} \neq 0$, necessary to arrive at the asymptotic self-similar solution with quasi-steady-state particle cascade [Equation (13.52)]. We are thus forced to invoke yet another explosive cascade, now with $\alpha > \beta^{(N)}$ and $F^{(N)}_{\varepsilon \to 0} = \infty$ at $t = t_*$. Taking the solution at $t = t_*$ as an initial condition to the KE (previous footnote applies), we naturally find an *antiwave* propagating from $\varepsilon = 0$ to higher ε's, leaving the desired quasi-steady-state cascade in its wake. From now on, we will be interested only in the case

$$\alpha > \beta^{(N)} = (q + s + 1)/p. \tag{13.62}$$

In addition to the upper bound (13.60), one can argue that

$$\alpha < \beta^{(E)} = (q + s + 2)/p, \tag{13.63}$$

where $\beta^{(E)}$ is the exponent of the steady-state energy cascade. Indeed, at $\alpha = \beta^{(E)}$, the particle flux nullifies (since the collision integral vanishes) and changes its sign for $\alpha > \beta^{(E)}$. It is hard to imagine a particle cascade with the *opposite* sign of the particle flux in its wake.

Problem 13.11 *Prove that conservation laws for N and E imply (respectively) that*

$$\int_0^\infty x^q \left[f(x) - \frac{c_0}{x^\alpha} \right] dx = 0, \qquad \int_0^\infty x^{q+1} \left[f(x) - \frac{c_0}{x^\alpha} \left(1 + \frac{c_1}{x^\sigma} \right) \right] dx = 0. \tag{13.64}$$

(In particular, make sure that the integrals converge at both limits.) Hints: Utilize the fact that, by locality of the collision term, the quantities N_ε and E_ε cannot be singular functions of time at $t = t_*$. Use Equation (13.58) and inequalities for α for proving the convergence of the integrals.

---

* To avoid confusion with the fact (discussed in the following) that Equation (13.27) becomes physically inadequate before t reaches t_*, we emphasize that here we are talking exclusively of the mathematical properties of Equation (13.27).

The self-similar regime (13.53) through (13.59) has a natural end point $t = t_*$, at which the distribution of particles at low enough energies has the power-law form $n_\varepsilon \propto 1/\varepsilon^\alpha$. This distribution is essentially nonequilibrium and starts to relax immediately after its formation. As can be readily obtained from dimensional analysis of the collision term [see also Equation (13.57)], the local kinetic time, $\tau_{kin}(\varepsilon)$, scales as $\sim \varepsilon^\sigma$, vanishing at $\varepsilon \to 0$. Therefore, at a later time, a wave in the distribution function will form propagating from $\varepsilon \to 0$, where the kinetic time is vanishingly small, toward higher energies where the distribution restructuring requires progressively larger kinetic times. In a general case, the character of distribution being formed in the wake of this wave cannot be established from the KE without taking into account qualitative features of the strongly turbulent regime that inevitably sets in at low enough energies. In fact, the strongly turbulent regime takes place *prior* to the time moment $t = t_*$ because the kinetic regime fails (in the low-energy part of the distribution) when the weak-turbulence condition $\xi_{wt}(\varepsilon_0) \sim [\varepsilon_0 \tau_{kin}(\varepsilon_0)]^{-1} \ll 1$ breaks at the $\varepsilon_0(t)$-front of the self-similar explosive cascade. The two typical options for the outcome of the strong-turbulence regime are the BEC formation and a collapse. For example, for the nonlinear Schrödinger equation with negative U, there is no stable homogeneous condensate—the system collapses to a point. At positive U, the homogeneous condensate is stable. Meanwhile, KE (13.23) is insensitive to the sign of U. The collapse regime at $U < 0$ is equivalent in effect to the nonconservation of particles—a drain in the low-energy region. That is why, at $U < 0$, it is natural for the back wave to lead to the formation of a quasi-steady-state particle cascade, while, in the case of BEC formation, it is natural to form a quasi-equilibrium Gibbs distribution. Here, we are interested in the latter case.

13.4 Strongly Nonequilibrium Bose–Einstein Condensation

Now we apply the analysis of the previous section to the BEC-formation kinetics in the 3D WIBG. We confine ourselves to the spatially homogeneous case with $q = 1/2$, $p = 3$, and $s = 2$ (see Problem 13.3). Hence, the condition (13.32) takes place, implying that the process starts with the explosive cascade (13.53) through (13.59):

$$n_\varepsilon(t) = A\,\varepsilon_0^{-\alpha}(t)\, f(\varepsilon/\varepsilon_0(t)) \qquad (t \le t_*), \qquad (13.65)$$

$$\varepsilon_0(t) = B\,(t_* - t)^{1/2(\alpha-1)}. \qquad (13.66)$$

Here, A and B are dimensional constants, related to each other by

$$ma^2 A^2 = C\hbar^3 B^{2(\alpha-1)}, \qquad (13.67)$$

where C is a numeric constant; we have restored Plank's constant and used the known expression for the pseudopotential of the dilute Bose gas in terms of the scattering length a: $U = 4\pi\hbar^2 a/m$. The dimensional part of relation (13.67) immediately follows from KE (13.23). The value of C has been found numerically: $C \approx 1.0$. From numeric simulations of KE (13.23), it was also determined that $\alpha \approx 1.234$ (Semikoz and Tkachev [2]; Lacaze et al. [3]).

Problem 13.12 *Relate A to the gas density and the characteristic energy scale, ε_{init}, of the initial weakly turbulent state of the field.*

At a certain time moment $t_0 < t_*$, the propagating front $\varepsilon_0(t)$ reaches energies at which the weak-turbulence parameter is of order unity: $\xi_{(wt)}(\varepsilon = \varepsilon_0(t_0)) \sim 1$. At $t \sim t_0$ and in the region $\varepsilon \sim \varepsilon_0(t_0)$, the KE description is replaced by the strong-turbulence regime. The dimensional analysis yields the following estimates for t_0 and $k_0 \sim \sqrt{2m\varepsilon_0(t_0)}$:

$$ t_* - t_0 \sim C_0 \left[\frac{\hbar^{2\alpha+1}}{ma^2A^2} \right]^{\frac{1}{2\alpha-1}} , \qquad k_0 \sim C_1 \left[\frac{m^\alpha aA}{\hbar^{2\alpha}} \right]^{\frac{1}{2\alpha-1}} , \tag{13.68} $$

where C_0 and C_1 are numeric constants. By direct simulation of the nonlinear Schrödinger equation, it has been found that [4]

$$ C_0 \sim 1, \qquad C_1 \sim 200. \tag{13.69} $$

The emergence of the strongly turbulent long-wave field at wavevectors $k \leq k_0$ marks the onset of a qualitatively new evolution process—the *coherent regime*. In this regime, the phases of complex amplitudes ψ_k become strongly correlated and the periods of their oscillations are comparable with the evolution times of the occupation numbers. Shortly after its emergence, the strongly turbulent field evolves into the state of *superfluid turbulence*, that is, the state that locally is identical to the condensate but on large scales is characterized as a vortex tangle with essentially nonequilibrium distribution of phonons. Such a state of a weakly interacting gas is also called *nonequilibrium quasicondensate*. Clearly, the typical time of the quasicondensate formation is given by the right-hand side of the first equation in (13.68), because no other temporal scales are associated with the regime of strong turbulence. It is also the shortest time of the entire evolution process. The typical separation between the just-formed vortex lines is of the order of the size of their cores and can be estimated as k_0^{-1}, as the only spatial scale associated with the strong turbulence.

Transformation of the strong turbulence into the quasicondensate state can be viewed as the limiting case of the generic Kibble picture regarding the formation of topological defects upon cooling a system below the point of the second-order phase transition. In contrast to the standard situation, where the cooling process is assumed to be slow compared to the short-range relaxation time, we do not have a separation of time scales here. The time of the quasicondensate formation is the shortest time of the entire evolution!

With respect to the weakly turbulent part of the field, the quasicondensate field plays the role of the genuine condensate. This allows us to continue describing the evolution of the former in terms of KE, which now should involve two entities: quasicondensate and noncondensate fractions. This description misses the transformation of quasicondensate into genuine condensate through relaxation of the vortex tangle, but this shortcoming is not restrictive since the corresponding physics is essentially identical to that described in Chapter 10.

It is convenient to isolate the quasicondensate from the noncondensate part of the distribution (for which we will be using the same symbol n_ε). The KE for n_ε then has the form

$$\dot{n}_\varepsilon = \text{Coll}([n], \varepsilon) - n_0(t)\,\text{Coll}_0([n], \varepsilon), \tag{13.70}$$

with n_0 the quasicondensate density. The conservation of the total number of particles requires that

$$\dot{n}_0(t) = -\int d\varepsilon\,\varepsilon^q\,\dot{n}_\varepsilon = n_0(t)\int d\varepsilon\,\varepsilon^q\,\text{Coll}_0([n], \varepsilon), \tag{13.71}$$

providing us with the equation describing the evolution of $N_0(t)$. As before, the evolution of n_ε can be understood from dimensional analysis relying on general properties of the collision terms—(effective) locality[*] and scale invariance. First, we notice that the estimate (13.28) for the first term in the right-hand side of (13.70) implies also that

$$\text{Coll}_0([n], \varepsilon) \sim \varepsilon^{s-q-1}\, n_\varepsilon^{p-1}, \tag{13.72}$$

since the latter term originates from the former one upon isolating the quasicondensate mode.

At $t = t_*$, we start with the power-law distribution $n_\varepsilon \propto 1/\varepsilon^\alpha$ that is (1) highly nonequilibrium and (2) features a short relaxation time at low energies. This suggests that the low-energy region quickly equilibrates and reaches Gibbs distribution with the time-dependent temperature $T(t)$; this temperature is infinite at $t = t_*$ and gets progressively smaller as the Gibbs distribution extends toward higher and higher energies. The self-similar solution of the KE corresponding to this scenario should have the following form:

$$n_\varepsilon(t) = A\,\tilde{\varepsilon}_0^{-\alpha}(t)\,\tilde{f}(\varepsilon/\tilde{\varepsilon}_0(t)), \qquad t > t_*, \tag{13.73}$$

$$\tilde{\varepsilon}_0(t) = \tilde{B}(t - t_*)^{1/\sigma}, \qquad \sigma = 2(\alpha - 1), \tag{13.74}$$

where $\tilde{B} \sim B$ and

$$\tilde{f}(x) \to f(x) \quad \text{at} \quad x \to \infty. \tag{13.75}$$

The fact that, in (13.73), we have the same amplitude A, exponent α, and the asymptotic behavior as in (13.65) naturally follows from the requirement that, at $\varepsilon \gg \tilde{\varepsilon}_0$, the distribution remains intact. As before, Equation (13.74) comes from the dimensional analysis of KE (13.70). The back wave at $\tilde{\varepsilon}_0(t)$ separates the power-law distribution of the explosive cascade from the Gibbs distribution at $\varepsilon \ll \tilde{\varepsilon}_0$:

$$\tilde{f}(x) \propto 1/x \quad \text{at} \quad x \to 0. \tag{13.76}$$

---

[*] Locality requires explicit separation of the field into quasicondensate and noncondensate fractions and two collision terms that treat N_0 as an "external" variable to the KE (13.70). Only in this sense, the collisions involving both condensate and noncondensate particles in $\text{Coll}_0([n], \varepsilon)$ can be treated as effectively local.

Problem 13.13 *On the basis of dimensional analysis of the collision term, estimate the kinetic time in the low-energy tail of the distribution (13.76). Use this estimate to prove the consistency of the quasi-equilibrium asymptotic behavior.*

From (13.73), (13.74), (13.76), and continuity of the distribution function at $\varepsilon \ll \tilde{\varepsilon}_0$, we deduce time dependence of the quasistatic Gibbs distribution temperature

$$T(t) \propto 1/\tilde{\varepsilon}_0^{\alpha-1}(t). \tag{13.77}$$

Time dependence of the condensate density follows from the particle-number conservation

$$N_0(t) = \int d\varepsilon\, \varepsilon^q \left[n_\varepsilon(t_*) - n_\varepsilon(t) \right] \propto \tilde{\varepsilon}_0^{q+1-\alpha}(t). \tag{13.78}$$

Taking into account (13.74), we obtain

$$N_0(t) \propto (t - t_*)^{\nu_1}, \qquad\qquad T(t) \propto 1/(t - t_*)^{\nu_2}, \tag{13.79}$$

where

$$\nu_1 = \frac{q+1-\alpha}{2(\alpha-1)} \approx 0.57, \qquad\qquad \nu_2 = 1/2. \tag{13.80}$$

In a general case of arbitrary s, p, and q, one has

$$\nu_1 = \frac{q+1-\alpha}{\alpha(p-1)-s}, \qquad\qquad \nu_2 = \frac{\alpha-1}{\alpha(p-1)-s}. \tag{13.81}$$

The self-similar solution (13.73) and (13.74) ceases to exist when the front $\tilde{\varepsilon}_0(t)$ reaches the characteristic scale $\varepsilon_{\text{init}}$ of the initial weakly turbulent state of the field. By that time, the numbers of quasicondensed and noncondensed particles are of the same order. Further evolution depends on whether the initial occupation numbers were much larger than unity (classical-field initial state), or not (quantum-field initial state). In the latter case, the noncondensed particles simply relax to the equilibrium state with the final temperature of the order of $\varepsilon_{\text{init}}$. In the former case, yet another self-similar solution develops that continues to add particles to the quasicondesate and is accompanied by a drift of the distribution to higher energies:

$$n_\varepsilon(t) = \varepsilon_{\text{drift}}^{-5/2}(t)\, q(\varepsilon/\varepsilon_{\text{drift}}(t)). \tag{13.82}$$

The exponent 5/2 in this solution is fixed by the conservation of energy. The asymptotic behavior of $q(x)$ at $x \to 0$ comes from the Gibbs distribution with time-dependent (decreasing) temperature: $q(x) \propto 1/x$. The argument for this behavior is the same as for the low-energy tail of the back wave (see Problem 13.13).

When the self-similar solution (13.82) extends to energies $\gg \varepsilon_{\text{init}}$, the total number of noncondensed particles becomes much smaller than N_0, meaning that, in Equation (13.70), the collision term $\text{Coll}([n], \varepsilon)$ can be omitted. With the same degree of accuracy, N_0 can be replaced in this equation with the total number of particles. Substituting (13.82) into our thus modified Equation (13.70), we find

$$\varepsilon_{\text{drift}}(t) \propto \sqrt{t}. \tag{13.83}$$

Combining then (13.83) with (13.82), we conclude that the total number of non-condensed particles decreases as $\propto t^{-1/2}$.

In a quantum gas, the self-similar solution (13.82) eventually becomes inapplicable. This happens when the occupation numbers at the back-wave front become comparable to unity. At this point, the quantum KE takes over, leading to the equilibration of the remaining noncondensate particles.

References

1. B. V. Svistunov, Highly non-equilibrium Bose condensation in a weakly interacting gas, *J. Moscow Phys. Soc.* **1**, 373 (1991); Yu. Kagan and B. V. Svistunov, Evolution of correlation properties and appearance of broken symmetry in the process of Bose–Einstein condensation, *Phys. Rev. Lett.* **79**, 3331 (1997); B. Svistunov, Kinetics of strongly non-equilibrium Bose–Einstein condensation. In: *Quantized Vortex Dynamics and Superfluid Turbulence* (eds. C. F. Barenghi, R. J. Donnelly, and W. F. Vinen), Lecture Notes in Physics, Vol. 571, Springer-Verlag, Berlin, Germany (2001).

2. D. V. Semikoz and I. I. Tkachev, Condensation of bosons in the kinetic regime, *Phys. Rev. D* **55**, 489 (1997).

3. R. Lacaze, P. Lallemand, Y. Pomeau, and S. Rica, Dynamical formation of a Bose–Einstein condensate, *Physica D* **152**, 779 (2001).

4. N. G. Berloff and B. V. Svistunov, Scenario of strongly Nonequilibrated Bose–Einstein condensation, *Phys. Rev. A* **66**, 013603 (2002).

Part V

Historical Overview: Nature and Laboratory

CHAPTER 14

Historical Overview

14.1 Liquid Helium, Superconductivity, Einstein's Classical Matter Field

The history of superphenomena begins a century ago when Kamerlingh Onnes produces the first liquid helium-4 (in 1908, at the University of Leiden) and uses it as a coolant to study (in 1911) the low-temperature resistance of mercury [1]. Until that time, the idea of working with solidified mercury had been to check the hypothesis that—thanks to the exceptional pureness of the system—the resistivity will tend to zero with vanishing temperature, as opposed to metals with impurities, which feature residual resistance at $T \to 0$. The sudden drop of resistance down to zero (within the experimental resolution) at 4.2 K comes as a complete surprise to Onnes. The term "superconductivity" (originally, "supraconductivity") will be coined by him 2 years later, in 1913, when persistent currents in toroidal samples are observed, proving the practical absence of resistance.

If the word "premature" can be used at all in connection with experimental discoveries, especially one so impressive and profound as the discovery of superconductivity, then we are taking that risk now to emphasize the negative impact that the observation of superconductivity had on revealing the physical nature of the effect, including the discovery of the most closely related, and much more simple microscopically, phenomenon of superfluidity.

The very same day Kamerlingh Onnes observes the onset of "nearly zero" resistivity of mercury at 4.2 K; he also notices that—with a further decrease of the temperature of the helium-4 cryostat—the helium suddenly stops boiling [2]. Surely enough, this striking phenomenon* had the potential to initiate a systematic study of the low-temperature properties of helium, which, in their turn, would naturally have led to the discovery of the phase transition at 2.177 K. It was the excitement of the mysterious superconductivity phenomenon that diverted Onnes and his successors from a deep study of helium-4.

An important historic deadline that the discovery of the λ-point in helium-4 failed to meet is Einstein's 1924–1925 paper (in two parts) [3] on what now is known as the theory of the (ideal) Bose gas and what Einstein at that time believes to apply to *any* gas of identical particles—atoms, molecules, and even electrons.[†] Einstein's work is based on the amazing theoretical observation by Bose [4] that Planck's law for black body radiation follows from Einstein's idea of *quanta* of the electromagnetic field, if the latter are treated as *indistinguishable particles*, distributed over phase-space cells of the size h^d (h is Planck's constant and d is the dimensionality of the corresponding configuration space). In the picture proposed

---

* Associated with the onset of the super heat conductivity in helium-4 below the λ-point.

† The difference between bosons and fermions was not revealed yet.

by Bose, the state of the phase-space cell is exhaustively described by the number of quanta in it (the "occupation number" in modern terminology); that is, one is not allowed to further label quanta within the cell, and that is the reason for the term "indistinguishability." Different cells are understood as statistically independent systems. Einstein conjectures that Bose statistics might apply equally to many-particle systems.

According to Einstein's truly revolutionary conjecture, the difference between the classical-field and classical-particle behavior comes exclusively from the values of the occupation numbers in the phase-space cells: Large, compared to unity, occupation numbers correspond to classical fields, while small occupation numbers correspond to classical particles. Einstein finds this idea to be perfectly consistent with the Maxwell–Boltzmann distribution of particles in the limit of small occupation numbers, and he reveals the conditions under which the occupation numbers become large. In addition to Bose's original picture, Einstein introduces the chemical potential term into the energy of a quantum, to account for the conservation of the total number of particles, while preserving the language of independent phase-space cells. In contrast to the thermal equilibrium state of the electromagnetic field, where the low-momentum phase-space cells always have large occupation numbers and thus demonstrate the classical-field behavior, the system of particles can have large occupation numbers only when the absolute value of its chemical potential is much smaller than temperature, while the $|\mu| \gg T$ case corresponds to the Maxwell–Boltzmann distribution in the entire phase space. Einstein observes that the condition $|\mu| \ll T$ inevitably takes place at low enough temperature; moreover (in the 3D case, to be precise), there exists a critical temperature, T_c, at which μ becomes equal to zero. At $T < T_c$, a finite fraction of the total number of particles has to occupy one and the same cell with zero momentum. Einstein calls the phenomenon "condensation" ("Bose–Einstein condensation [BEC]" in modern terminology).

One may speculate that if the λ-point in helium-4 had been well established by 1925, at least at the level of a singularity in specific heat,* it would have been readily associated with Einstein's condensation, especially given that the condensation temperature of 3.2 K for the ideal gas with the density and particle mass of helium-4 is reasonably close to $T_\lambda = 2.177$ K. It is astonishing that neither Einstein himself nor anyone else paid attention to the well-known fact that helium is the only substance that does not solidify down to Einstein's T_c, rendering it a natural suspect for the "condensation" phenomenon. While it is true that helium-4 is far from being an ideal gas and it would be naive to apply Einstein's theory directly, one would expect to see at least some traces of the condensation phenomenon in the thermodynamic functions of helium-4.

The late discovery of the λ-point was not the only unfortunate circumstance in the dramatic history of Einstein's paper. As we already mentioned, the

---

* Dana, an American postdoctoral fellow from Harvard, working in the Onnes laboratory in 1922, did observe a peculiar behavior of the specific heat [5]. However, the results of this work were published only after Onnes's death in 1926, and only for $T > 2.5$ K. Apparently, Dana was not sure whether or not the anomaly in the specific heat was an experimental artifact.

revolutionary aspect of Einstein's conjecture was not limited to the phenomenon of condensation *per se*. Starting from the purely statistical idea of Bose, Einstein proposed that it might be generalized to the classical-field *behavior* (including dynamics) of any many-particle system.[*] In Einstein's words[†]: "One can assign a scalar wave field to such a gas," and "It looks like there would be an undulatory field associated with each phenomenon of motion, just like the optical undulatory field is associated with the motion of light quanta" [3].

Einstein did not specify the equation for the scalar field that would have described the motion of the degenerate gas. Very soon, however, the equation is found by Schrödinger, in the context of the quantum mechanics of a single particle. But, in the light of Schrödinger's theory for many particles, Einstein's conjecture of the classical-field behavior of a gas of atoms appears as too simplistic. Indeed, Schrödinger states explicitly (in the paper summarizing his 1926 work on the new form of quantum theory [6]): "In the simple case of one material point moving in an external field of force the wave-phenomenon may be thought of as taking place in the ordinary three-dimensional space; in the case of a more general mechanical system it will primarily be located in the coordinate space (q-space, not pq-space) and will have to be projected somehow into original space." It then took decades to realize that, for bosonic weakly interacting or ideal particles under the condition of large occupation numbers, projecting from the multidimensional Hilbert space of quantum states onto the ordinary space indeed leads to the classical-field description—precisely the one envisioned by Einstein back in 1924–1925.

Subsequent years were marked by the tremendous success of quantum mechanics at the few-body level, and the community did not focus on investigating the nature of the crossover from single quanta to macroscopic classical-field phenomena. There was yet another unfortunate circumstance impeding the understanding of the phenomenon of BEC. In his 1927 thesis [7], Uhlenbeck concluded that Einstein's condensation could not take place in a finite-size system. Years later in 1938, London revisits the theory [8] (pointing to its relevance to the λ-transition in helium; see Section 14.3) and explains that Uhlenbeck's argument was nothing but an exaggerated concern about what today is known as finite-size rounding of phase transitions. Later in 1938, Uhlenbeck withdraws his objection. Ironically, the question of BEC in a finite-size system raised by Uhlenbeck was only one small step removed from a *gedanken* experiment that might readily have led to a deep understanding of the phenomenon by explicitly putting it in the Einstein classical-field context. This is the problem of BEC in an external potential, where the question of the condensate's shape immediately calls for the classical-field description. Furthermore, in a smooth external potential, the pathological nature (infinite compressibility) of a noninteracting gas is clearly seen, suggesting a further improvement of the model—the Schrödinger equation—with a nonlinear term accounting for interactions. In its turn, the nonlinear Schrödinger equation,

---

[*] To the best of our knowledge, this aspect of Einstein's paper was completely forgotten.

[†] Translated from German by M. Troyer and F. Werner.

if formulated at that time, could have led to the discovery of superfluidity on a purely theoretical basis.

Even after London's 1938 paper, the significance and correctness of Einstein's work was not fully understood. The theory of BEC in an ideal and, much later on, weakly interacting gas, has since become textbook material, but mostly in the context of quantum statistical mechanics. As we already mentioned, it took several decades to fully realize—with the development of an adequate quantum-field-theoretical formalism—Einstein's idea of classical-field description of quantum fields with large occupation numbers and to put the BEC phenomenon into this broader context. In 1963, a complete microscopic theory of the classical-field limit in quantum optics was developed by Glauber* [9]. Glauber's formalism of coherent states applies to any bosonic system. Its application to superfluidity was considered by Langer in 1968, who wrote [10]: "The coherent states are most useful for dealing with many-body systems which behave in some sense classically, that is, systems in which the boson modes are highly occupied. When this is true, the function $\psi(\mathbf{r})$ becomes a classical Schrödinger field which describes the complete many-body system in just the same way that the Maxwell field describes the classical limit of quantum electrodynamics." One might think that, at this point, the classical-field behavior of many-particle bosonic systems envisioned by Einstein was understood completely. However, quite unexpectedly for the modern reader, Langer continued [10]: "Our point is that, for many-particle Bose systems as opposed to many-photon systems, the validity of the classical description implies superfluidity." This prejudice, shared by many members of the community for quite a long time, is rooted in the dramatic and instructive history of understanding the nature of superfluidity.

14.2   Lambda Point, Abnormal Heat Transport, and Superfluidity

The phase transition in helium-4 at $T = 2.17\,\mathrm{K}$ is first officially[†] announced by Keesom and Wolfke in 1927 [11], on the basis of their observation of a sudden jump in the dielectric constant. The high- and low-temperature phases are called helium I and II, respectively. Keesom, Onnes's successor, continued to actively study the transition. In 1932, with Clusius, he observed the famous anomaly in the heat capacity that (following Ehrenfest's suggestion) gave the name "λ-point" to the point of the phase transition [12].

In 1935, Burton reports [13] the results of measurements of the viscosity of helium I and II carried out at the Cryogenic Laboratory in Toronto by Wilhelm, Misener, and Clark.[‡] (The fact that these results could be published by someone other than the authors reflects radical evolution of the accepted ethical behavior

---

[*] According to Glauber, the theory did not appear earlier "for the reasons which are partly mathematical and partly, perhaps, the accident of history."

[†] As we already mentioned, a number of features hinting at the transition were observed many years before that.

[‡] According to Griffin [14], the measurements were made by Misener.

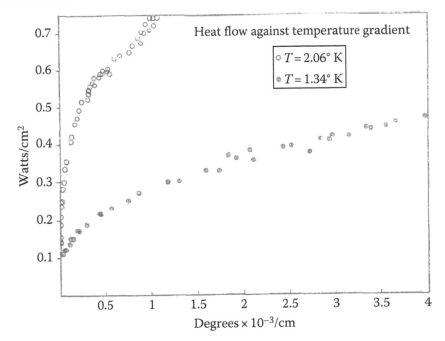

Figure 14.1   The heat transport plot by Allen et al. [17]. Within the experimental resolution, the heat flow remains finite when temperature gradient vanishes.

in science in the last century.) The viscosity is deduced from the decrement of the oscillation of a circular metal cylinder submerged into the liquid helium. "As the state of the liquid changed from II to I, there was a very distinct and abrupt change in the logarithmic decrement...." In the temperature interval from 2.3 K to 2.2 K, the viscosity coefficient drops by almost an order of magnitude (!) but remains finite. We now understand that, for this setup, the viscosity coefficient is nonzero due to the presence of the normal component.

Less than 3 years before the observation of mass supertransport in helium II, the system is found to feature anomalous heat transport [15–17], to the extent that helium II could be called "supra-heat-conducting" [16]. Following the initial qualitative observations by Keesom and Keesom [15] and Rollin* that heat conductivity increases abruptly and dramatically below the λ-point, the Keesoms quantify the effect, reporting the increase to be more than a factor of a million [16]. Subsequently, Allen, Peierls, and Zaki Uddin find the situation to be striking and having "no analogy in the behavior of other substances" [17]. By plotting the heat flow against the temperature gradient, they observe no asymptotic proportionality of the former to the latter (see Figure 14.1); that is, no finite coefficient of heat conductivity.

On January 8, 1938, *Nature* publishes two letters side by side: one by Kapitza and the other by Allen and Misener [18,19]. Both papers report striking results of

---

* B. V. Rollin, Thesis Oxford 1935. The result and its potential importance for obtaining information about helium II are mentioned in the footnote **) on page 269 in Physica, **3**, (1936).

similar but independent* studies of helium II flow through a narrow slit (Kapitza) and a narrow capillary (Allen and Misener). In both cases, the use of a narrow gap/capillary is motivated by the idea that the finite value of viscosity obtained in the Toronto experiment is due to the (well-known) dissipative effects of turbulence, so that the actual value could be even much lower. Kapitza begins with the conjecture that the abnormally high heat conductivity of helium II can be explained by convection, which, however, would require that the viscosity be much less than the one found in the Toronto experiment. Having observed that viscosity drops *at least* by a factor of 1500, Kapitza interprets it as a confirmation of the conjecture and goes further by putting forward a hypothesis that, if it were not for the turbulence in his experiment, the viscosity would be exactly zero. He concludes that the observed very low limit on the viscosity of helium II "is perhaps sufficient to suggest, by analogy with superconductors, that the helium below the λ-point enters a special state which might be called a 'superfluid' " [18].

In their letter, Allen and Misener explicitly cite and discuss the letter by Kapitza. While not questioning Kapitza's brilliant idea of superfluidity as a phenomenon intrinsically related to superconductivity, they make two further crucially important conclusions. First, they find that helium II is something more than a liquid with zero viscosity. "The observed type of flow, however, in which the velocity becomes almost independent of pressure, most certainly cannot be created as laminar or even as ordinary turbulent flow" [19]. We now know that this type of flow—the supercritical flow—is a hallmark of superfluids and superconductors. It reflects the fact that excitations responsible for large-scale turbulent motion are separated by a *macroscopic* gap from the low-lying normal modes.†
Finally, Allen and Misener present an estimate ruling out Kapitza's suggestion that the abnormal heat conductivity of helium II can be accounted for by undamped turbulent motion.

It thus becomes clear that in addition to vanishing viscosity—a striking phenomenon on its own—the hydrodynamic properties of helium II are dramatically different from what was known about ideal fluids before. The most direct experimental evidence of that difference comes quickly. On February 5, 1938, *Nature* publishes the letter by Allen and Jones, reporting an observation of the fountain effect [21] (see Figure 14.2), a phenomenon that would be impossible in simple classical fluids.

14.3   Theories Enter: London, Tisza, Landau, and Bogoliubov

On April 9, 1938, a letter by London appears in *Nature* [8] proposing that the λ-transition in helium is a phenomenon (modified by interactions) of a Bose–Einstein condensation: "Though actually the λ-point of helium resembles rather

---

* An instructive analysis of the two works, addressing the historical context, as well as the question of independence, can be found in publications by Griffin [14] and Balibar [20].

† Both the Landau and Bogoliubov theories of superfluidity, developed soon after the discovery, failed to understand/explain this phenomenon so central to superfluidity.

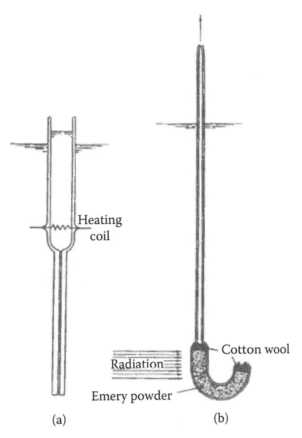

Figure 14.2   The original figures from the letter by Allen and Jones [21]. (a) is self-explanatory. In connection with (b), the following curious text was presented: "A more striking manifestation of the above effect was observed by one of us (J. F. A.) in collaboration with A. D. Misener. Observations were being made on the flow of liquid helium II through a tube packed with fine emery powder (b). The top of the tube was allowed to project several centimeters above the level of the liquid helium bath, and an electric pocket torch was flashed on the lower part of the tube containing the powder. A steady stream of liquid helium was observed to flow out of the top of the tube as long as the powder was irradiated."

a phase transition of second order, it seems difficult not to imagine a connection with the condensation phenomenon of the Bose–Einstein statistics." In the concluding part of his letter, London foresees the relevance of the BEC phenomenon to *all* anomalous properties of helium: "The conception here proposed might also throw a light on the peculiar transport phenomena observed with He II (enormous conductivity of heat, extremely small viscosity and also the strange fountain phenomenon recently discovered by Allen and Jones)." However, it is only the statistical aspect of BEC—most notably, the cusp in the specific heat—that London is able to directly relate to the experimental data on helium. The classical-field aspect emphasized by Einstein and, as we now know, playing the central part in the phenomenon of superfluidity is not even mentioned in London's letter. Ironically, London was among very few physicists at that time who were thinking of macroscopic quantum phenomena. His neglect of Einstein's classical-field idea, despite the urgent need for a completely new mechanism that could "throw a light on the peculiar transport phenomena," most likely means that in 1938, it was too late to think phenomenologically—given the breathtaking progress of quantum mechanics at the microscopic level—about the classical-field behavior of a multiparticle system. And it was too early—for objective and subjective reasons—to derive this behavior from first principles.

The next crucial step in theoretical understanding of helium II, and the very first success in explicitly addressing the unusual dynamics, was the two-fluid concept introduced by Tisza in a letter to *Nature* published on May 21, 1938 [22]. Tisza builds his picture on London's idea that BEC is relevant to the phenomenon of superfluidity. Like London, he avoids any explicit discussion of Einstein's idea of the classical-field aspect of BEC. Nonetheless, he conjectures that the condensate behaves as a single object in the sense that condensed atoms "do not take part in the dissipation of momentum" and, correspondingly, "the viscosity of the system is entirely due to the atoms in the excited states." Next, Tisza writes that "the fraction of substance consisting of atoms in the lowest energy state will perform—like a 'superfluid' liquid of viscosity $\mu \sim 0$—some sort of turbulent motion...." We see that Tisza uses the term "condensate" in a much broader sense than just particles with zero momentum (!) and in formal contradiction with himself: Turbulent motion can be associated only with a multitude of energy states. If we replace, throughout Tisza's letter, the word "condensate" with "vacuum" and substitute "elementary excitations" (or "normal modes") for "atoms in the excited states" and also make an assumption that the turbulent motion of the condensate requires a *macroscopic* amount of energy to excite, we will arrive at the Landau theory of superfluidity. The only important exception for this Tisza-to-Landau mapping should be done for "some sort of turbulent motion," which Tisza believed to be characteristic of a "condensate"; Landau's firm viewpoint was that the motion of the superfluid component was necessarily potential. This explains why Tisza's theory, despite being microscopically incorrect and based on false premises—one crucial flaw is the direct identification of the superfluid component with the condensate—yields a *qualitatively* correct phenomenological interpretation of the two-fluid hydrodynamics, explaining, in particular, the abnormal heat conductivity and the fountain effect.

In 1941, Landau publishes his famous theory of superfluidity. Before discussing the historical role and modern scientific status of the Landau theory, it is instructive to analyze Landau's criticism of Tisza's ideas presented in the beginning of the paper [23]: "Tisza's well-known attempt to consider helium II as a degenerate Bose gas cannot be accepted as satisfactory—even putting aside the fact that liquid helium is not an ideal gas, nothing could prevent the atoms in the normal state from colliding with the excited atoms; i.e., when moving through the liquid they would experience friction and there would be no superfluidity at all." Nowadays, this passage sounds naive since we know that collisions of condensate atoms with noncondensate atoms (or any elementary excitations for that matter) cannot *slow down* the flow of the condensate; they can only *deplete* the condensate. If we were to try to defend Landau by saying that the depletion was what he actually meant by "friction," we would still face a serious problem with the passage since it would read as an attack against the very idea of BEC in an interacting gas. Indeed, if collisions with noncondensate atoms (or any other excitations) deplete the condensate, then what prevents the condensate from being completely depleted in equilibrium? If, on the other hand, one admits that there are processes that balance the depletion processes,* then Landau's argument becomes inconclusive, since the very same processes can potentially protect the moving condensate from depletion. It is thus obvious that Landau's *general* criticism of Tisza's idea of relevance of condensate for the two-fluid picture lacked physical depth.[†] The reason Landau made the previous argument against Tisza's idea might simply be the fact that Landau's theory did not *require* resorting to BEC. What Landau (as well as others at that time, including Tisza) missed was that the gap between potential flow and vortex excitations—the central assumption in Landau's theory—might be (and now we know that it is) intrinsically connected with the presence of BEC (more generally, topological long-range order). The story is truly remarkable and ironic since by denying the role of the condensate, Landau was denying the importance of what is now called the superfluid *Ginzburg–Landau order parameter*.[‡] It is, however, less surprising in the historical context: as we've emphasized more than once, by that time, Einstein's idea of condensate behaving as a classical field, in which case its association with the order parameter can hardly be missed,[§] had been completely forgotten.

The circumstance clearly recognized by Landau was the necessity to have a gap for the vortex modes. In his paper [23], he first attempts to demonstrate that "between the lowest energy levels of vortex and potential motion there must be a certain energy interval." This inconsistent[¶] attempt, however, is not directly relevant to the main part of the paper based on the well-motivated *assumption* concerning the energy interval in question: "One may question which of these levels lie lower: apparently both cases are logically possible. The supposition that the normal level of potential motions lies lower than the beginning of the

---

* It is worth a reminder here that these processes are nothing else than the processes of *stimulated* scattering of noncondensate particles and that stimulated scattering, as opposed to the spontaneous one, is of purely *classical-field* nature!

[†] This should be distinguished, however, from pointing out other wrong statements made by Tisza (see below in this section).

[‡] Landau published his theory of the second-order phase transitions in 1937 [24], a year before the discovery of superfluidity. The cornerstone of the theory is the notion of order parameter, the emergent entity characterizing the degree of spontaneous symmetry breaking.

[§] And yet another peculiar historical twist: While failing to identify BEC with the order parameter of superfluidity in 1941, later (in 1950), together with Ginzburg, Landau introduced the notion—closest to BEC—of the complex order parameter for a superconductor [25].

[¶] It was established later by Feynman in Reference [26] that the existence of the gap requires particle indistinguishability since in a system of distinguishable particles, the gap is absent. More importantly, Landau missed the fact that the gap should be *infinite* for vorticity to be absent at *finite* temperatures!

spectrum of vortex motions leads to the phenomenon of superfluidity. Hence we must suppose that this very case exists in liquid helium." This quotation is central for understanding the scope of Landau theory and especially its qualitative difference from the London–Tisza approach. Landau does not claim tracing the origin of superfluidity from the first principles—*the vortex gap is postulated, not derived*. Thus, Landau theory is a consistent *effective* theory in which the vortex gap is the only essential postulate implied by the experimental evidence: In the absence of the gap, turbulence would destroy the superflow. This distinguishes Landau's approach from the phenomenological theory of a Galilean-invariant superfluid.

Landau's theory consists of two distinct parts. The first (semi-microscopic) part is the low-temperature theory, which, in essence, is quantized harmonic hydrodynamics—the quantum theory of phonons in an ideal fluid—enhanced by postulating the existence of yet another branch of elementary excitations, the so-called rotons. Later, guided by experimental data for thermodynamic properties, Landau realizes that rotons do not represent a separate branch of excitations, but are part of the single phonon branch with a pronounced minimum at high momenta. The second (macroscopic) part is the two-fluid hydrodynamics derived rigorously from the axioms of existence of the superfluid vacuum and Galilean invariance. By considering general hydrodynamical relations for velocities u_1 and u_2 of the first and second sound, respectively, in the $T \rightarrow 0$ limit where, in accordance with the first part of his theory, the normal component is nothing but a dilute gas of phonons, Landau obtains the famous limiting expressions: $u_1 = c$ and $u_2 = c/\sqrt{3}$, where c is the velocity of phonon excitation at zero temperature. Subsequently, together with Khalatnikov, Landau develops the kinetic theory of phonons and rotons. Excellent agreement with experiments leaves no doubt that the theory is correct.

By not explaining the microscopic origin of the vortex gap postulate, Landau theory remains fundamentally incomplete. In particular, it does not address the nature of the turbulent motion in a superfluid.* Furthermore, by failing to reveal a direct connection between the superfluid velocity potential and the gradient of the phase of the complex time-dependent (!) order parameter, Landau theory contains no hint at the effect discovered later by Josephson. Last but not least, Landau theory does not give a correct picture of the superfluid phase transition.

In 1947, Bogoliubov publishes his seminal paper on the microscopic theory of weakly interacting Bose gases [27]. Bogoliubov's work is remarkable in many respects. On the technical side, it introduces a new theoretical method, the famous Bogoliubov transformation. On the physical side, it contains a number of results and concepts that significantly advance the understanding of the superfluid phenomenon. It presents the first example of a system where many elements of Landau theory can be derived from microscopics. Bogoliubov also explains that Landau's criticism of Tisza's picture, based on the condensate depletion argument,

---

* As opposed to Tisza, who conjectured that some turbulent motion is inherent to the condensate, Landau assumes in Reference [23] that the motion of the superfluid liquid is always potential.

is not valid since Landau's "vacuum" can well be the BEC state; if the emergent quasiparticle dispersion relation is consistent with Landau criterion, then the system has a stable thermodynamic equilibrium between the moving condensate and the stationary gas of elementary excitations. The assumption that the gap between the low-lying elementary excitations and the vortex modes is *implicit* in Bogoliubov's theory. It simply reveals no excitations other than the Bogoliubov quasiparticles. Given the controllable accuracy of the treatment in the limit of weak interaction, this can be considered as a proof that there are no other low-lying modes.* Next, Bogoliubov's work reanimated and substantiated—in a less general form, though—Einstein's forgotten idea of the system of quantum particles behaving as a scalar classical field. The only (unfortunately) explicit classical field in Bogoliubov's work is the one describing the condensate. It is very characteristic that the phase of this field evolves in time. This, in principle, could have led to formulating a more general question of the classical-field behavior of a quantum field, keeping in mind Einstein's conjecture of a direct analogy between the degenerate Bose system and Maxwell's equations for the electromagnetic field. What's more, Bogoliubov makes an intriguing footnote (in his notation, f stands for the quasiparticle momentum): "We may remark that if we form the corresponding frequency $\omega = \hbar E(f)$ and put: $\hbar \to 0$, $f/\hbar = k = $ const we obtain the classical formula for the dispersion of the frequency deduced by Vlasov, cf. Journal of Physics, **9**, 25 (1945)." Bogoliubov did realize that what he has derived using quantum-field methods was nothing but the dispersion of normal modes for a nonlinear classical field! Objectively, he was very close to arriving at the Gross–Pitaevskii equation for a weakly perturbed condensate. Finally, Bogoliubov's work was an important step toward revealing in the theory of superfluidity the role of spontaneous symmetry breaking. In contrast to the microscopic Hamiltonian of the system, Bogoliubov's effective Hamiltonian does not possess U(1) symmetry; this feature is directly associated with condensate operators being treated as complex numbers. An important caveat is in order here: In his 1947 paper, Bogoliubov does not address explicitly the central role of symmetry breaking or BEC, or long-range order (and the relationship between the three) in the theory of superfluidity. While showing that there is no contradiction between Tisza's idea of superfluidity due to BEC and Landau's phenomenological theory, Bogoliubov's paper does not answer the central question of whether the presence of a BEC (containing almost all the particles) in a weakly interacting gas is only a fortuitous feature specific to his model or a crucial generic phenomenon required for superfluidity, irrespective of the system parameters. To illustrate the point, we note that Landau, when arguing in his 1948 paper [28] against Tisza's theory (by correctly pointing out certain misconceptions by Tisza), was referring to Bogoliubov's paper as supporting his (Landau's) picture, without any appreciation for the special role of the condensate.

---

* Bogoliubov did not address the energy gap for the vortex excitations. Correspondingly, there was no appreciation of the fact that superfluidity at finite temperature requires that this gap be infinite.

14.4   Off-Diagonal Long-Range Order, Vorticity Quantization, and Quantum-Statistical Insights

The fundamental relationship between BEC and long-range order and the direct relevance of the latter to superfluidity are revealed by Penrose in 1951 [29]. Penrose notices that if the single-particle density matrix $\rho(\mathbf{r}_1, \mathbf{r}_2)$ does not vanish at $|\mathbf{r}_1 - \mathbf{r}_2| \to \infty$, which is exactly the case in the presence of BEC, then the following asymptotic expression takes place

$$\rho(\mathbf{r}_1, \mathbf{r}_2) \to \psi(\mathbf{r}_1) \psi^*(\mathbf{r}_2) \qquad \text{at} \qquad |\mathbf{r}_1 - \mathbf{r}_2| \to \infty, \tag{14.1}$$

with ψ the wavefunction of the single-particle state into which the condensation takes place. He then gives a microscopic interpretation of the two-fluid hydrodynamics—thereby corroborating Tisza's conjecture—by observing that the slowly varying (in space) phase $\Phi(\mathbf{r})$ of the wavefunction $\psi(\mathbf{r})$ is a hydrodynamic parameter *independent* of the normal velocity $\mathbf{v}_n$ (the latter being introduced via a local Galilean transformation of ρ). The single-particle density matrix of the hydrodynamic flow thus has the form

$$\rho(\mathbf{r}_1, \mathbf{r}_2) = \rho_n(\mathbf{r}_1, \mathbf{r}_2) e^{i(\mathbf{r}_1 - \mathbf{r}_2) \cdot \mathbf{v}_n\left(\frac{\mathbf{r}_1 + \mathbf{r}_2}{2}\right)/\gamma} + \rho_s(\mathbf{r}_1, \mathbf{r}_2) e^{i\Phi(\mathbf{r}_1) - i\Phi(\mathbf{r}_2)}, \tag{14.2}$$

where $\rho_n(\mathbf{r}_1, \mathbf{r}_2)$ and $\rho_s(\mathbf{r}_1, \mathbf{r}_2)$ are certain real functions. Equation (14.2) explicitly yields the two-fluid picture of the flow:

$$\mathbf{j}(\mathbf{r}) = n_n(\mathbf{r})\mathbf{v}_n(\mathbf{r}) + n_s(\mathbf{r})\mathbf{v}_s(\mathbf{r}), \qquad \mathbf{v}_s(\mathbf{r}) = \gamma \nabla \Phi(\mathbf{r}), \tag{14.3}$$

where

$$n_n(\mathbf{r}) = \rho_n(\mathbf{r}, \mathbf{r}), \qquad n_s(\mathbf{r}) = \rho_s(\mathbf{r}, \mathbf{r}). \tag{14.4}$$

(Note that the difference between $n_s(\mathbf{r})$ and the condensate density $n_0(\mathbf{r}) = |\psi(\mathbf{r})|^2$ is seen here by inspection.) Our notation here is intentionally slightly different from that of Reference [29]. We replace $\hbar/m$ with γ to show that Penrose's analysis is *identical* for quantum and classical-field systems. We also use the phase Φ rather than the superfluid velocity potential ϕ [in our notation, $\phi(\mathbf{r}) = \gamma \Phi(\mathbf{r})$] to emphasize the topological aspect of superfluidity.

The latter circumstance is worth special discussion. The relationship between the off-diagonal long-range order [Equation (14.1)] and superfluidity revealed by Penrose corresponds to the most general picture of the phenomenon, provided we put Equations (14.2) through (14.4) in a broader context than the one considered in Reference [29]. First, recall that Equation (14.1) is a sufficient, but not necessary, condition for the picture (14.2) through (14.4) to take place; as we now know, the topological long-range order standing behind nonzero $\rho_s(\mathbf{r}_1, \mathbf{r}_2)$ does not necessarily imply the presence of the condensate.* Also, the system is not supposed to be

---

* The generalization of Equation (14.1) to the case with the topological long-range order without BEC is $\rho(\mathbf{r}_1, \mathbf{r}_2) \to \sum_\nu \psi_\nu(\mathbf{r}_1) \psi_\nu^*(\mathbf{r}_2)$ at $|\mathbf{r}_1 - \mathbf{r}_2| \to \infty$, where $\psi_\nu(\mathbf{r})$ is a complex field with significant spatial variations of the phase, but without macroscopically large (non-pinned) topological defects, and ν labels all relevant configurations of the field.

Galilean invariant; in the absence of Galilean invariance, one simply has $v_n \equiv 0$. But the most important generalization of the picture is allowing for the topological defects in the (otherwise potential) field of superfluid velocity. While substantiating Tisza's general idea of the condensate standing behind the two-fluid behavior of a superfluid, Penrose abandons Tisza's conjecture about "some sort of turbulent motion" that the superfluid component can support. We quote Penrose's conclusion about Equations (14.3) and (14.4): "These two equations are the basis of the two-fluid theory (Landau 1941, Tisza 1947); in Landau's version the superfluid velocity is irrotational." Penrose seems to rule out any turbulent motion of the superfluid component, and does not discuss the conjecture formulated in 1949 by Onsager in the celebrated footnote in Reference [30]: "Vortices in a superfluid are presumably quantized; the quantum of circulation is h/m, where m is the mass of a single molecule." However, Penrose might have noticed that the quantization of vorticity, including the value of the circulation quantum, follows immediately from Equations (14.3) and (14.4).

In the discussion following the paper by Gorter on the two-fluid model of liquid helium [Nuovo Cimento (Supplemento), 6, 249 (1949)], Onsager formulates the idea that quantized vorticity allows superfluid to mimic rotation: "If we admit the existence of quantized vortices, then a superfluid is able to rotate; but the distribution of vorticity is discrete rather than continuous." The alternative theories of the rotational response of a superfluid were advocated by Landau and Lifshitz in 1955 [31]. Noting that Andronikashvili's experiments on rapidly rotating superfluid helium are consistent with rigid-body rotation of the system, Landau and Lifshitz proposed that superfluid component can rotate by breaking into rotating coaxial cylindrical regions. Each of the regions, according to them, was characterized by a superfluid velocity b_i/r, where r is the distance to the center of the cylinder and b_i are different for each cylindrical region. Also, as early as 1946, H. London was proposing a similar "coaxial phase separation" model for rotating superfluid [32], explicitly discussing that boundaries of his coaxial regions are vortex sheets (in this work, he also discussed nonclassical rotational inertia of superfluid and contrasted it with the Meissner effect).

In 1955, Onsager's conjecture of vorticity quantization was developed by Feynman to a full-scale theory presented in his seminal paper [33]. Feynman argues that quantization of velocity circulation in units of h/m is implied by the single-valuedness of the many-body wavefunction around the vortex line. He then explores the implications of the existence of quantized vortex lines, predicting an array of vortex lines in a rotating superfluid, and he proposes a scenario for the critical velocity in a macroscopic channel based on the vortex generation and introduces the notion of superfluid turbulence as the tangle of quantized vortices. Furthermore, Feynman formulates a revolutionary* conjecture that the mechanism by which the superfluid state becomes normal is the appearance of macroscopic vortex lines. In Feynman's own words, "The superfluid is pierced through

---

* To the extent that the author writes "Of the following I am not sure, but it does seem to be an interesting possibility."

and through with vortex lines. We are describing the disorder of helium I." It is worth noting that a similar idea was mentioned earlier by Onsager,* but apparently it was not appreciated by the community.

As we understand now, quantization of the superfluid velocity circulation is the fundamental reason for having persistent supercurrents in the most general (e.g., disordered) case of toroid geometry. It is important to mention that before Feynman's paper, Reference [33], not only the reason but even the fact of universal existence of persistent supercurrents was not known. This point is perfectly illustrated by the footnote from Feynman's earlier paper [26] in which he incorrectly states that to support a persistent current, "the liquid must be completely confined with no free surface because the exchange of atoms between the rotating liquid and the stationary gas above it might cause a rapid damping of the angular momentum."

The experimental confirmation of the Onsager–Feynman quantization comes in two steps. In 1956, Hall and Vinen report [34] strong (though still indirect) evidence for the presence of a "vortex array" in rotating ^{4}He. They obtain this evidence by measuring the attenuation of second sound.† A few years later, Vinen performs an elegant experiment [35] revealing a single quantum of circulation in doubly connected rotating ^{4}He. Vinen uses a cylindrical cell pierced by a wire along the symmetry axis and employs the fact that the circulation of fluid around the wire removes the twofold degeneracy of its vibration spectrum; the resulting frequency separation between the two circularly polarized normal modes is proportional to the circulation.

Feynman's work on quantized vorticity is preceded by yet another fundamental contribution of his to quantum statistics in general and to the theory of superfluidity in particular. In 1953, Feynman introduces the path-integral representation of the partition function [36] and applies it to interacting bosons. In contrast to the evolution operator considered previously, he finds that the weights of path-integral configurations are positive definite. Thereby, he establishes an exact mapping between the Gibbs statistics of quantum bosons in d dimensions and classical statistics of closed paths (objects similar to ring polymers) in $(d + 1)$ dimensions. The size of the classical system along the extra dimension (often referred to as imaginary time) is proportional to the inverse temperature, and this is how the classical system knows about the physical temperature of its quantum counterpart. Nowadays, Feynman's path-integral representation is at the heart of first-principles Monte Carlo simulations of both continuous-space and lattice superfluids. What is less appreciated, however, is that the mapping onto the

---

* In the previously mentioned discussion [Nuovo Cimento (Supplemento), **6**, 249 (1949)] Onsager was saying, "We can have vortex rings in the liquid, and the thermal excitation of helium II, apart from the phonons, is presumably due to vortex rings of molecular size. As a possible interpretation of the λ-point we can understand that when the concentration of vortices reaches the point where they form a connected tangle throughout the liquid, then the liquid becomes normal." If not for the molecular size of vortex rings, that would be identical to Feynman's suggestion.

† The fact that vortices can form a lattice remained unrevealed until Abrikosov's work on the mixed state of type-2 superconductors (see Section 14.7).

$(d+1)$-dimensional "polymers" (or world lines) is a powerful tool for qualitative analysis. Qualitative insight provided by path integrals is central for Feynman's paper, which corroborates London's conjecture by showing that general features of BEC are not altered by interactions. Feynman achieves this goal by observing that the main qualitative feature associated with condensation is the appearance of macroscopically large exchange loops and that interactions, as long as they are not strong enough to result in solidification, do not alter this picture. Speaking in modern language, Feynman establishes the universality class* of the superfluid phase transition as the one corresponding to proliferation of closed oriented loops in d dimensions (the imaginary-time dimension is finite at the critical point). There remains, however, a missing connection: It is not at all obvious why the appearance of macroscopic permutation cycles necessarily means the onset of superfluidity. This connection was established years later in 1987 by Pollock and Ceperley [37], who derived a remarkable formula that related superfluid density to the variance of the world-line winding numbers.

A historical remark is in order here. While it is true that Feynman's path-integral representation is the most natural for introducing exchange cycles, these were envisioned for the first time without resorting to world lines. In his 1951 paper [38], Matsubara arrived at the idea of permutation cycles by studying the bosonic partition function in the coordinate representation. The essence of Matsubara's result is the relationship between the bosonic N-body density matrix, $R\left(\mathbf{r}_1,\ldots,\mathbf{r}_N;\mathbf{r}'_1,\ldots,\mathbf{r}'_N\right)$, and that of a system of distinguishable particles, $R_{\text{dist}}\left(\mathbf{r}_1,\ldots,\mathbf{r}_N;\mathbf{r}'_1,\ldots,\mathbf{r}'_N\right)$, of the same mass and interaction potential. The former is related to the latter by a symmetrization procedure that amounts to summing over all the permutations of the coordinates $(\mathbf{r}_1,\ldots,\mathbf{r}_N)$. Correspondingly, the partition function of bosons (the trace of the many-body density matrix) involves summation over the coordinate permutations. Matsubara observed that BEC in the ideal gas corresponds to the partition function being dominated by macroscopically long permutation cycles, and he conjectured that the same feature of the partition function corresponds to the superfluid state of interacting bosons.

In the years 1953–1954, after his work on the path-integral representation of quantum statistics and before his work on quantized vorticity, Feynman publishes two papers dealing with elementary excitations in helium-4 [26,39]. In these papers, he argues that the fact that phonons are the only low-lying excitations is nontrivial[†] and has to do with the symmetry of the wavefunction. By analyzing wavefunctions of low-energy elementary excitations [26], he shows that the only

---

* By using high-temperature expansions of the partition functions of XY/U(1)-type lattice models, one can show that proliferation of closed oriented loops corresponds to the classical XY [aka U(1)] universality class. We should also mention the difference between noninteracting (Gaussian) and generic XY universality classes at $d < 4$, crucial for the λ-phenomenon. Since Einstein's idea of the classical-field behavior of bosons was forgotten, it is not surprising that Feynman did not reveal the classical-field context of his exchange cycles.

† It goes without saying that here we are speaking of the quantum-particle perspective. The idea that superfluidity is the property directly inherited by bosonic quantum field from its classical counterpart did not exist at that time.

modes consistent with the permutation symmetry of the wavefunction are density waves (phonons), providing thus a general microscopic proof of Landau's picture. In the second paper in this series [39], Feynman addresses the question of the overall shape of the phonon spectrum* and, in particular, the roton minimum. He introduces the many-body wavefunction

$$\psi_{\mathbf{k}}(\mathbf{r}_1,\ldots,\mathbf{r}_N) = \phi(\mathbf{r}_1,\ldots,\mathbf{r}_N) \sum_{j=1}^{N} e^{i\mathbf{k}\cdot\mathbf{r}_j}, \tag{14.5}$$

with $\phi(\mathbf{r}_1,\ldots,\mathbf{r}_N)$ the ground state of N bosons, and argues that this wavefunction is not only an accurate wavefunction of a long-wave phonon with the wavevector $\mathbf{k}$—the latter is readily seen from the fact that (14.5) is the result of acting on the ground state with the kth harmonic of the number-density operator[†]—but also an accurate variational function of an elementary excitation with essentially finite k. He then derives an elegant formula for the dispersion relation of elementary excitations:

$$E_k = \hbar^2 k^2 / 2mS(k), \tag{14.6}$$

where $S(k)$ is the static structure factor in the ground state.[‡] In a strongly correlated liquid, $S(k)$ has a pronounced maximum at some $k = k_0$ on the order of interparticle distance. Equation (14.6) then translates this maximum into a local minimum of E_k.

For historical reasons, the minimum in E_k is called a roton excitation. Recall that originally Landau was thinking of rotons as an independent branch of excitations somehow associated with vorticity (thus "roton"). Landau later had to conclude that there is only one branch of elementary excitations with a minimum, but the term "roton" survived. Clearly, the wavefunction (14.5) does not imply that the finite-k excitations have anything to do with rotation; in fact, by continuity, it suggests that these excitations are phonon-like. In Feynman's own words: "States of low k will be called phonons, and the states of momentum near k_0 will be called rotons, ... in accordance with the terminology of Landau, although we do not necessarily mean to imply that rotons carry intrinsic angular momentum or represent vortex motion." Nevertheless, in his 1955 paper, Feynman arrives at the conclusion that rotons are small vortex loops. The origin of this misconception is as follows: In the same Reference [39], Feynman had introduced an improved variational wavefunction featuring a dipole backflow produced by the elementary excitation. Feynman had mistakenly thought of this backflow as indicative of a vortex ring. While it is true that the asymptotic behavior of the velocity field

---

* Apart from the end point. The theory of the end point was developed later by Pitaevskii.

[†] In the long-wave (hydrodynamic) limit, the kth harmonic of the number-density operator is a linear combination of creation and annihilation operators of the phonon with the wavevector $\mathbf{k}$.

[‡] Behaving as $S(k) \to \hbar k/2mc$ at $k \to 0$, consistently with the fact that (14.6) yields exact phonon dispersion in this limit. The experimental proof of such a behavior of $S(k)$ was obtained later: R. B. Hallock, *Phys. Rev. Lett.* **23**, 830 (1969).

of a vortex ring (or vortex–antivortex pair in 2D) corresponds to the dipole flow pattern, this pattern is generic for *any* object moving in an (incompressible) ideal fluid.

In 1956, Penrose and Onsager publish a paper with the title "Bose–Einstein Condensation and Liquid Helium" [40], in which they attempt to unify different approaches to what they refer to as an "ideal-gas analogy." In the introduction, they deliberately use that euphemism to emphasize that there are numerous viewpoints on superfluidity and the λ-transition in interacting bosonic systems but that the connection between them, or even the existence/necessity of such a connection, is not clear. Indeed, Landau theory, based essentially on the spectrum of elementary excitations, seems to imply that BEC is not directly relevant to the phenomenon. Feynman's microscopic theory while insisting on the crucial role of the permutation symmetry of the wavefunction does not address the question of BEC either. Symptomatically, Feynman does not cite Penrose's paper [29], which shows that BEC implies superfluidity. This is reminiscent of Landau's not citing London's paper [8], along with criticizing Tisza's conjecture that BEC is relevant to superfluidity. Even when stating explicitly that "London's view is essentially correct," Feynman does not mean, or, at least, does not show, that the λ-transition in an interacting system implies appearance of a macroscopic number of particles with zero momentum. His conclusion that "in a quantum-mechanical Bose liquid the atoms behave in some respects like free particles" comes from the observation that in both cases, the phase transition can be described (within the path-integral formalism) as appearance of macroscopically large exchange loops.

A likely reason that Feynman ignored Penrose's long-range order is that, after Landau and Bogoliubov, the form of the dispersion relation for elementary excitations was regarded as the key ingredient. The parabolic dispersion of an ideal gas is inconsistent with the stable superflow, while Penrose's order is present in an ideal gas as well. Interestingly, neither Penrose in Reference [29] nor Penrose and Onsager in Reference [40] address this subtlety. Meanwhile, the resolution of the problem is instructive for understanding what the condition (14.1) implies. If, instead of considering a superflow with finite velocity, we address the question of the system's response to a small gauge phase (or, equivalently, rotation with the angular velocity much smaller than the critical one), we will see that ideal and interacting systems respond similarly—the fact perfectly captured by the criterion (14.1). Hence, Equation (14.1) implies the superfluid response to the vanishingly small gauge field, but, speaking generally, does not guarantee the stability of a superflow with finite velocity.

According to Penrose and Onsager, "Bose–Einstein condensation is said to be present whenever the largest eigenvalue of the one-particle reduced density matrix is an extensive rather than intensive quantity." This largest eigenvalue, N_0, is identified with the number of condensate particles. The relationship between the formulated criterion and Penrose's original Equation (14.1) is as follows: With macroscopic accuracy, $\psi(\mathbf{r})$ is the eigenfunction—termed condensate wavefunction—corresponding to the eigenvalue N_0, implying that $\int |\psi(\mathbf{r})|^2 d^d r = N_0$. The authors present a first-principles argument that a generic ground-state wavefunction of a bosonic liquid satisfies their criterion. They also explain how the Matsubara–Feynman picture of proliferation of permutation cycles leads to their criterion. In addition to that, they argue that BEC is absent in a solid phase. Against the background of modern theory, all three arguments are not perfect. In 1D Bose fluids, genuine BEC is absent even in the ground state. In 2D superfluids, there is no genuine BEC at any finite temperature. Hence, the proliferation of permutational cycles, while implying superfluidity, does not necessarily mean BEC. And it is also known that a solid ground state can, in principle, have BEC and support a superflow. Nevertheless, all these counterexamples are the exceptions that prove the rule. In low-dimensional superfluids, the genuine long-range order implied by Penrose–Onsager criterion is destroyed only by the long-range fluctuations of the superfluid phase. As a result, the largest eigenvalue of the one-particle reduced density matrix behaves marginally. As opposed to the normal system, N_0 diverges with increasing the system size, but, in contrast to the case with genuine BEC, the ratio N_0/N vanishes in the thermodynamic limit. The argument that there is no BEC in the bosonic solid ground state is based on the assumption about the form of the solid wavefunction, which excludes vacancies and interstitials even at the microscopic scale. This argument may be considered as a precursor to the theorem (see Chapter 9) forbidding superfluidity in solids without vacancies and/or interstitials.

It is worth noting that, in their discussion section, Penrose and Onsager say that "condensation is present whenever a finite fraction of particles occupies one single-particle quantum state." One cannot object to the definition but has to keep in mind that if all particles are occupying the same state $\psi(\mathbf{r})$ that contains topological defects, or is turbulent, then the corresponding "condensate" is not superfluid! In this sense, not all condensates are BECs.

14.5 Meissner–Ochsenfeld Effect and London Phenomenology

We now return to the story of "supraconductivity." We left behind the remarkable achievements of the 1930s: the experimental discovery of perfect diamagnetic response of superconductors in 1933, the so-called Meissner–Ochsenfeld effect [41], and the development of the corresponding phenomenological macroscopic theory, the London equation, by the London brothers in 1935 [42].

The London brothers realize that the fact that superconductors completely cancel magnetic field from the bulk means that they are something more than just perfect conductors.* According to the Londons, the Meissner–Ochsenfeld effect is a consequence of some fundamental law, which they then establish under certain simplified assumptions and which is now called the London equation.

The starting point of the London theory is an observation that the absence of dissipation implies an "ideal acceleration equation" (time derivative of momentum equals force) for the supercurrent density $\mathbf{j}_s$:

---

* Note that this was done 3 years before the discovery of superfluidity and the realization, on the basis of unusual turbulent behavior, that superfluids are something more than just ideal fluids.

$$\Lambda \frac{\partial \mathbf{j}_s}{\partial t} = \mathbf{E}. \tag{14.7}$$

Here, $\mathbf{E}$ is the strength of the electric field, $\Lambda = m/n_s e^2$ with e the electron charge, m the electron mass, and n_s the superfluid density* of the superconducting electrons. The simplified assumption adopted by the Londons is that n_s can be treated as a constant with respect to spatial derivatives (incompressibility of superfluid component). Taking the curl of (14.7) and using Maxwell equations (neglecting the displacement current) leads to

$$\Lambda \nabla^2 \dot{\mathbf{H}} = \dot{\mathbf{H}}. \tag{14.8}$$

(Here and in the following, we adopt units such that the speed of light $c = 1$.) A straightforward time integration then yields

$$\Lambda \nabla^2 (\mathbf{H} - \mathbf{H}_0) = \mathbf{H} - \mathbf{H}_0, \tag{14.9}$$

where $\mathbf{H}_0$ is the magnetic field at time zero. Equation (14.9) implies that $\mathbf{H}$ approaches $\mathbf{H}_0$ inside the superconducting sample [42]: "The general solution means, therefore, that practically the original field persists forever in the supraconductor." But the Meissner and Ochsenfeld experiment demonstrates that the magnetic field inside the superconductor always vanishes, meaning that the only acceptable value of $\mathbf{H}_0$ is zero. This fact leads the Londons to the crucial conclusion: "If in reality $\mathbf{H}_0$ is always confined to the value zero, then this means that $\Lambda \nabla^2 \mathbf{H} = \mathbf{H}$ is to be considered as a fundamental law and not to be treated as a particular integral of a differential equation." They proceed further by rewriting this law of nature in the form (taking into account that $\nabla \times \mathbf{H} = \mathbf{j}_s$)

$$\nabla \times \Lambda \mathbf{j}_s = -\mathbf{H}, \tag{14.10}$$

which they postulate as a fundamental equation that "replaces Ohm's law in supraconductors." It is instructive to observe that at $H \equiv 0$, the London Equation (14.10), rewritten in terms of the supercurrent velocity $\mathbf{v}_s = \mathbf{j}_s/(n_s e)$, yields the $\nabla \times \mathbf{v}_s = 0$ condition proposed by Landau for superfluids 6 years after Londons' paper. At the most fundamental level, and for both superconductors and superfluids, the new law of nature envisioned by the London brothers has its origin in the complex-valued order parameter field. The general concept of the order parameter was proposed by Landau in 1937, as the key notion in his theory of second-order phase transitions. Retrospectively, it appears rather surprising that this series of three discoveries made within so short a time frame—the Londons' concept of the new law of nature associated with superconductivity, the Landau order parameter, and the discovery of superfluidity below the point of the second-order phase transition in ^{4}He—did not immediately result in raising a natural question of the order parameter responsible for superphenomena. This might be attributed in part to

---

* More precisely, the notion of superfluid density did not exist at that time. The Londons were using the total electron density instead of n_s.

the tremendous success of the Landau theory of superfluidity, which seemingly did not require any "extra" notion, be it BEC or some order parameter.

14.6   Magnetic Flux Quantization

In 1948, London publishes a paper [43] in which he arrives at the quantization of magnetic flux in multiply connected superconductors. This work is extremely interesting in the historical perspective. Nowadays, it appears hardly possible to derive the quantization of magnetic flux without simultaneously arriving at the quantization of the superfluid velocity circulation since they have the same topological origin. Meanwhile, London's paper is concerned exclusively with the quantization of magnetic flux. It seems equally impossible to speak of the magnetic flux quantization without touching upon the so-called Aharonov–Bohm effect [44,45]. However, London's paper while dealing with essentially the same physics does not discuss it.

It is characteristic of London's 1948 work that the superconducting long-range order is introduced without explicitly specifying the order parameter. Having noticed that Equation (14.10) can be written as $\nabla \times \mathbf{p} = 0$ in terms of the canonical momentum $\mathbf{p} = m\mathbf{v} + e\mathbf{A}$, with $\mathbf{A}$ the vector potential, London interprets it as a "kind of condensation in momentum space" leading to the "rigidity" of the ground state. This "rigidity" is thus the central conjecture of the London theory. The precise meaning of the postulate is that upon switching on the magnetic field, the wavefunction of the superconducting ground state *preserves its functional form*[*] (up to a gauge freedom). London then considers a massive toroid superconductor with the trapped magnetic flux. In the presence of magnetic field, the "rigid" ground state corresponds to a state with the supercurrent density

$$\mathbf{j} = (1/\Lambda)(\nabla\nu - \mathbf{A}), \tag{14.11}$$

where $\nu(\mathbf{r})$ is (up to the coefficient $e/\hbar$) the phase field that should be used to modify the wavefunction, ψ_0, to compensate for the gauge freedom:

$$\psi_0 \rightarrow \psi = \psi_0 \exp\left[i(e/\hbar) \sum_a \nu(\mathbf{r}_a)\right], \tag{14.12}$$

with $\mathbf{r}_a$ the coordinate of the a-th particle. The single-valuedness of ψ requires that[†]

$$(e/h) \oint \nabla\nu \cdot d\mathbf{l} = \text{integer}. \tag{14.13}$$

---

[*] Note that the conjecture is not always correct quantitatively. This is readily seen, for example, in the case of a disordered superconductor. Nevertheless, it does capture the essence of the magnetic response.

[†] In the case of a ground state with *paired* electrons not considered by London, the integer in Equation (14.13) should be replaced with a half-integer.

In accordance with (14.11), the requirement (14.13) is compatible with vanishing of the current density inside the superconducting toroid only if

$$\text{Magnetic flux} = \oint \mathbf{A} \cdot d\mathbf{l} = \oint \nabla v \cdot d\mathbf{l} = (h/e) \times \text{integer}. \tag{14.14}$$

Looking at Equation (14.11), a modern reader immediately identifies $v(\mathbf{r})$ with the superconducting phase (times $e/\hbar$) and, in particular, clearly sees that Equation (14.13), in the absence of $\mathbf{A}$, is nothing but the law of quantization of the superfluid velocity circulation. It is thus important to stress here that in 1948, this connection was not at all obvious and there was still a long way to go to understand the unifying role of the phase variable in superconductors and superfluids. London focuses on the electrically charged system and does not introduce $v(\mathbf{r})$ as an emergent collective variable that exists *without* $\mathbf{A}$. Rather, he focuses on the key aspect of local gauge invariance* where $v(\mathbf{r})$ is a mathematical entity, a gauge counterpart of $\mathbf{A}$, such that only their combination represents a physical observable, unless one considers integrals over a closed contour around a hole. London deliberately confines himself to the situation of finite particle charge and vanishing supercurrent density in the bulk of a massive superconductor. Apparently, it is this focus that prevented London not only from discovering quantization of the superfluid velocity circulation in a neutral superfluid but also from pinpointing the Aharonov–Bohm effect. London does realize that Equation (14.12) applies not only to superconductors. He writes that this equation is "evidently, quite generally valid for any ring-shaped quantum-mechanical system which embraces a magnetic flux in such a way as not to touch the flux itself." What was left out is the claim that if the embraced flux is different from the flux quantum h/e times integer, then it cannot be gauged out and thus has to be *observable from outside the flux tube.* Interestingly enough, in 1949, Ehrenberg and Siday in a paper published almost simultaneously with the London's work did point out the missing relationship to the superconducting ring: "The effect has ... a certain analogy in the existence of a permanent current in a superconducting ring due to a magnetic flux through it" [44]; this observation was made a decade before it was rediscovered by Aharonov and Bohm.

The expression for flux quantization in a superconducting cavity predicted by London to be $\Phi = hc/e$ had not been universally accepted until the 1960s. To appreciate the difficulties in understanding this phenomenon, we recall that in 1958, Onsager incorrectly conjectured that the phenomenon of flux quantization in a superconducting cavity is an intrinsic property of the electromagnetic field [46], despite Onsager being the author of the first example of topological quantization in classical fields (the superfluid velocity quantization). Flux quantization was first observed experimentally in 1961, the flux quantum to be half of the value predicted by London [47,48]. After that experiment, Onsager publishes a one-page paper explaining that London's picture of magnetic flux quantization is correct

---

* In a broader context, London was one of the first (along with Weyl and Fock) to discuss local gauge invariance concepts in quantum systems.

but the flux quantum should have an extra 1/2 factor as a consequence of electron pairing [49]. He also states there that his conjecture from 1958 was incorrect.

14.7   Ginzburg–Landau Phenomenology, Shubnikov Phase/Abrikosov Lattice

In 1950, Ginzburg and Landau publish their famous phenomenological description of superconductors [25] based on Landau theory of second-order phase transitions. They introduce the complex-field order parameter Ψ and write the free-energy functional for it by assuming that it can be expanded in powers of Ψ near the critical temperature.[*] The bilinear part of the functional is reminiscent of the Hamiltonian of a charged quantum particle in the magnetic field while the interaction term is the standard $|\Psi|^4$ coupling. The resulting equations of the theory read (in the gauge $\mathrm{div}\mathbf{A} = 0$)

$$\frac{1}{2m_*}(i\hbar\nabla + e_*\mathbf{A})^2\Psi + \alpha\Psi + \beta|\Psi|^2\Psi = 0, \tag{14.15}$$

$$\Delta\mathbf{A} = -4\pi\mathbf{j}_s, \tag{14.16}$$

$$\mathbf{j}_s = -\frac{e_*}{m_*}\mathrm{Re}\left[\Psi^*(i\hbar\nabla + e_*\mathbf{A})\Psi\right]. \tag{14.17}$$

The value of m_* is essentially arbitrary, depending on the normalization of the order parameter. The value of the effective charge e_* is, strictly speaking, not fixed, but it has to be linked to the particle charge in a (semi-)universal way to respect the gauge invariance. In view of the latter circumstance, Ginzburg and Landau conclude that "there are no grounds to believe that the charge is different from the electron charge."

The Ginzburg–Landau theory was a tremendous breakthrough in many respects. For the first time, it was proposed that supertransport can be associated with a complex order parameter, that is, with broken U(1) symmetry and essentially classical-field behavior of quantum particles. The theory covered the physics of the London equation (and thus the Meissner–Ochsenfeld effect) and magnetic flux quantization. Furthermore, it provided a convenient tool to deal with one apparent shortcoming of Londons' theory, which was predicting negative surface energy for the boundary between the normal and superconducting phases [51] while the experimental evidence was pointing to the opposite. The first-order phase transition to the normal state at a certain critical value of magnetic field indicated that the surface energy is positive. In the Ginzburg–Landau theory, the sign of the normal-superconductor surface energy was arbitrary and controlled by the parameter

$$\kappa = \frac{m_*}{e_*\hbar}\sqrt{\frac{\beta}{\pi}}. \tag{14.18}$$

---

[*] An earlier unsuccessful attempt to formulate a simple model of superconductivity was made by Landau in 1933, where, instead of the order parameter expansion near T_c, he was proposing to expand the free energy in powers of the supercurrent [50].

At $\kappa < \kappa_c = 1$, the surface energy is positive; at $\kappa > \kappa_c$, it is negative, the large κ limit corresponding to the Londons' theory.

For some reason, Ginzburg and Landau do not project their theory on neutral superfluids by taking the limit $e_* \to 0$.[*] Ginzburg comments on that in his Nobel lectures [53]: "After the development of the Ψ-theory of superconductivity, the transfer of something similar to the superfluidity case appeared to be rather obvious. ... By the way, Landau, the originator of the theory of phase transitions and superfluidity, for some reason was never concerned with this range of questions, as far as I know." Possibly, back in 1950, Landau was not interested in putting forward the Ψ-theory of superfluidity because he thought that BEC is not required to explain it, while in superconductors, the challenge was to understand the highly nontrivial magnetic response. We can only speculate why the irrotational flow condition for superfluids (which, in a way, is a superfluid counterpart of the Meissner–Ochsenfeld effect) appeared natural to Landau. A personal reason for Landau not to be excited about the Ψ-theory of superfluidity would be his hesitation to admit that his previous criticism of London–Tisza ideas had gone too far.

The hallmark of the Ginzburg–Landau description is the theory of type-2 superconductors in a magnetic field. Type-2 superconductors can be in the "mixed state" (often called the Shubnikov phase and the Abrikosov vortex lattice). Although the main motivation for Ginzburg and Landau is to correct the negative interface energy result, which at that time was considered to be an artifact of Londons' theory, they do acknowledge the possible existence of a new state in magnetic field for superconductors with $\kappa > 1$ in [25]. Namely, after establishing that the superfluid-normal surface energy changes sign at $\kappa = 1$, they conclude that there should be "some kind of instability of the normal phase." They write, however, that the character of the resulting state was not investigated because experiments suggest that in the existing superconductors, $\kappa \ll 1$.[†] In fact, substantial experimental evidence for type-2 superconductivity did exist (but was not trusted) for more than a decade before the Ginzburg–Landau works. It is reviewed in Ginzburg's and Abrikosov's Nobel lectures [53,54]. Magnetization curves characteristic for type-2 superconductors with two critical fields were first seen experimentally in 1935 by Casimir-Jonker and De Haas [55]. However, they dismissed the possibility of a new kind of superconductivity and explained it by the inhomogeneity of the sample.

The experimental discovery of type-2 superconductivity is attributed to Shubnikov, Chotkewitsch, Schepelew, and Rjabinin [56].[‡] Shubnikov's group carries out experiments with higher-quality samples. The conclusion is that the existence of two critical magnetic fields H_{c1} and H_{c2} is an intrinsic property of homogeneous systems—indicative of a new phase (the Shubnikov phase).

---

[*] This was done in 1958 by Ginzburg and Pitaevskii [52].

[†] Ginzburg and Landau start their paper by arguing that Londons' theory is unsatisfactory (it emerges only in the limit of large κ). Shortly afterward, Landau admitted that perhaps the $\kappa > 1$ case is also physical and applies to superconducting alloys [53].

[‡] Shubnikov worked at the same institution in Kharkov as Landau.

The discovery of type-2 superconductivity was dismissed as a "dirt effect" and was not widely accepted until the mid-1960s. Because Shubnikov was executed in 1937, he did not have a chance to carry out further studies of the mixed phase. Further experimental evidence of the existence of a phase with negative surface energy was obtained by Zavaritskii in experiments on thin superconducting films in the early 1950s. This prompted Abrikosov to investigate the regime between H_{c1} and H_{c2} with $\kappa > 1$ following Ginzburg and Landau's suggestion for the possible existence of a new phase [54].

In his first 1952 paper [57], Abrikosov investigates the linearized Ginzburg–Landau theory and the value of the second critical magnetic field H_{c2}, and does not discuss the vortex state. In his subsequent 1957 work [58], he explicitly focuses on the vortex lattice state motivating it by analogy with the vortex state in helium suggested by Onsager and Feynman (see Section 14.4). He finds that in the Shubnikov phase, the magnetic field penetrates a superconductor in the form of a vortex lattice, similar to the vortex array in a rotating superfluid.*

The Abrikosov lattice appears via a second-order phase transition at some critical value of the magnetic field $H_{c1}(T)$, called the lower critical field, while at the upper critical field $H_{c2}(T)$, the state becomes normal, thus providing the microscopic explanation for Shubnikov's phase. The vortex state in superconductors is introduced in modern textbooks as a straightforward counterpart of Onsager's quantum vortex picture, on the one hand, and as London quantization of (frozen) magnetic flux in a massive multiply connected superconductor, on the other hand. However, this was not generally understood in 1957. In particular, although it is well known today that the Abrikosov lattice features a universal flux quantization property—one magnetic flux quantum per vortex—this flux quantization is not explicitly mentioned in Abrikosov's paper [58], nor is London's work cited.† As discussed earlier, the whole idea of flux quantization was considered highly controversial at that time and, until 1961, arguments were still advanced (e.g., by Pippard [59]) that in a superconducting cavity, the flux will not be quantized.

Experimentally, the existence of Abrikosov lattices in type-2 superconductors was confirmed in the mid-1960s [60,61], thus proving that the observation of H_{c1} and H_{c2} by Shubnikov was indeed about a new kind of superconductivity rather than a "dirt effect."

14.8 BCS Theory

After the Ginzburg–Landau phenomenology was put forward, the main outstanding problem in the theory of superconductivity was to understand how fermions can develop long-range order. It was clear (and explicitly mentioned in

---

* According to Abrikosov's account [54], he obtained the vortex lattice state in 1953 but the publication was postponed because Landau disagreed with the whole idea.

† Also, at that time, the coupling constant to the vector potential **A** in the Ginzburg–Landau theory was not yet established, while this knowledge was required to get the flux quantization correctly.

numerous publications) that if electrons could form bound states—pairs or, more generally, bound complexes with an even number of particles—then the BEC of those complexes would solve the problem (speculations about electron pairing in the literature go back to at least the late 1940s [53]). Moreover, the discovery of the isotope effect in 1950 [62] pointed directly to the electron–phonon interaction as the crucial part of the binding mechanism. However, the details of the mechanism remained obscure until 1956 when Cooper constructed the ground state of a pair of fermions excited above the Fermi surface [63]. Cooper demonstrated that in the case of an attractive interaction, no matter how weak, the ground state of the pair is always bound (the Cooper effect).

In 1957, Bardeen, Cooper, and Schrieffer construct a many-body superconducting ground state in the form of a condensate of Cooper pairs [64], thus putting forward a microscopic theory of superconductivity—the so-called BCS theory—the accuracy of which is controlled by the small value of the attractive effective coupling constant between electrons at the Fermi surface.

While based on a small parameter, the BCS theory (as well as the Cooper effect it is based on) is essentially nonperturbative and thus quite nontrivial. According to the BCS theory, the ground state of an interacting system is a superposition of weakly excited states ("configurations") of the following structure: If a single-particle state $\mathbf{k} \uparrow$ is occupied in any configuration, then $-\mathbf{k} \downarrow$ is occupied as well. A fermionic elementary excitation with the momentum $\mathbf{p}$ is constructed by starting with the single-particle mode $\mathbf{p}$ occupied by either spin-up or spin-down particle (hole, if below the Fermi surface) and then using the rest of the single-particle modes to form the ground-state-type superposition of configurations. Thus, constructed single-particle excitations feature a gap* allowing the three authors to rely on the Landau theory of superfluidity.†

The BCS theory based on the phonon-mediated mechanism of attraction between fermions at the Fermi surface did not address the problem of strong Coulomb repulsion between electrons in metals. Meanwhile, it was the Coulomb repulsion that for many years had been a major obstacle in understanding the phenomenon of electron pairing. This puzzle was solved by Tolmachev and Tiablikov [66] who found that strong repulsive interaction at high excitation energy is suppressed by the logarithm of the ratio between the Fermi and Debye temperatures as the excitation energy is decreased. As a result, phonon-mediated attraction can overcome *renormalized* Coulomb repulsion near the Fermi surface. Retardation effects are crucial in the physical interpretation of this result: lattice deformation produced by one electron stays in place and attracts another electron long after the first electron is gone and traveled some distance away.

---

* In 1960, Giaever starts experimental studies of tunneling effects in superconductors and confirms the existence of the energy gap predicted by BCS theory [65].

† Similar to Landau (and, later, Bogoliubov), Bardeen, Cooper, and Schrieffer did not realize that for the superproperties to survive at nonzero temperature, the gap for vorticity should be infinite.

14.9   Advent of Field-Theoretical Methods

14.9.1   Green's Functions and Diagrammatics

The year of 1958 evidences a shock wave of glorious applications of Feynman's diagrammatic technique to problems of many-body quantum statistics. Within 1 year—and in one single volume of the *Journal of Experimental and Theoretical Physics*—three crucial developments are published. In one, Galitskii and Migdal [67] construct the diagrammatic formalism for normal fermionic systems. In another, Beliaev [68] develops the diagrammatic approach to bosonic ground state and applies it to a weakly interacting gas. Similarly to the Bogoliubov method, the superfluid order parameter ψ_0 plays the central part in Beliaev's approach by explicitly entering the theory in the form of anomalous average of the field operator, $\langle \hat{\psi} \rangle = \psi_0 \neq 0$, and generating an anomalous propagator $\langle \hat{\psi}'(1)\, \hat{\psi}'(2) \rangle \neq 0$ (with $\hat{\psi}' = \hat{\psi} - \psi_0$). The nonperturbative character of the effect of interaction—which Bogoliubov captures by his famous transformation—is readily accounted for by Dyson's resummation leading to a pair of coupled equations for normal and anomalous propagators (Dyson–Beliaev equations). Analogous to Bogoliubov theory, the phase Φ of ψ_0 in Beliaev's technique has an explicit time dependence; a new circumstance (the importance of which will be appreciated later, after the works of Josephson and Anderson) is that the time dependence is universally given by the chemical potential: $\Phi(t) = \Phi(0) + (\mu/\hbar)t$. In the third paper, following the steps of Beliaev's approach, Gor'kov [69] formulates Green's function equations for superfluid fermions. Gor'kov's theory deals with the explicit (Ginzburg–Landau) order parameter $\langle \hat{\psi}_\uparrow \hat{\psi}_\downarrow \rangle = \Psi \neq 0$ that generates anomalous propagators. Diagrammatically, the nonperturbative character of the Cooper effect is captured by Dyson's resummation leading to a coupled system of equations for normal and anomalous correlators (Gor'kov equations). The most important advantage brought by the diagrammatic technique to the microscopic theory of weakly interacting systems was a systematic way of going from microscopic to effective interactions and estimating higher-order corrections.

After publication of the BCS theory, its relationship to Ginzburg–Landau theory remained unclear. In 1959, Gor'kov [70] derives Ginzburg–Landau equations microscopically as an effective free-energy functional emerging from BCS theory near the critical temperature. This derivation also allows him to establish that the coupling constant to the gauge field in the Ginzburg–Landau theory is twice the electric charge: $2e$.

14.9.2   Gross–Pitaevskii Equation

Significant progress in the microscopic classical-field description of a weakly interacting Bose gas is made from 1957 through 1961. In 1957–1958, Gross formulates—in a semi heuristic way—the "classical theory of Boson wave fields" in the form of what today is called the time-dependent Gross–Pitaevskii equation [71]. (It took two decades after that to fully understand the scope and applicability of the classical-field description). In 1961, Gross [72] and Pitaevskii [73] publish two independent but very closely related works in which they argue that the time-dependent Gross–Pitaevskii equation can be used to accurately describe a vortex line in a weakly interacting Bose gas at zero temperature. Recall that Bogoliubov's paper [27] contained the time-dependent equation for the complex-valued condensate function, $\psi_0(t) = \psi_0(0)\exp(-iE_0 t/\hbar)$. To generalize Bogoliubov's approach to the case with a (straight) vortex line, Gross and Pitaevskii add the spatial dependence, $\psi_0 \equiv \psi_0(\mathbf{r}, t)$, and consider an *inhomogeneous* stationary solution, $\psi_0(\mathbf{r}, t) = \psi_0(\mathbf{r}, 0)\exp(-iE_0 t/\hbar)$. The stationary Gross–Pitaevskii equation for $\psi_0(\mathbf{r}, 0)$ has the mathematical form of the Ginzburg–Landau equation previously used by Ginzburg and Pitaevskii for a phenomenological description of a vortex line in a superfluid [52]. Since the works by Gross and Pitaevskii, the field $\psi_0(\mathbf{r}, t)$ of a weakly interacting gas acquires the precise meaning of the time-dependent microscopic field of the order parameter.

It is very characteristic that Gross and Pitaevskii do not go far beyond the stationary solution. An extension that they consider unquestionably applicable to a quantum system is the case of the normal modes found from the linearized equation. Gross writes: "We use special solutions of the semiclassical theory to suggest an approximate single particle basis, in terms of which the quantized field is expanded. We use the small oscillation analysis of the classical field to suggest a suitable quasiparticle transformation" [72]. Similarly, Pitaevskii comments (cf. previously discussed footnote in Bogoliubov's paper) that the reason one can rely on a purely classical-field analysis for finding the spectrum of elementary excitations is that the quantized harmonic modes have the same frequencies as their classical counterparts. Hence, as opposed to Einstein's general conjecture of classical-field behavior associated with large occupation numbers, the idea of Gross and Pitaevskii is fundamentally based on the presence of a BEC, whose shape is meant to be described by the Gross–Pitaevskii equation. The broader context of the time-dependent Gross–Pitaevskii equation, and especially the physical meaning of its weak-turbulence regime, was realized substantially later. Incidentally, it was in Pitaevskii's work [73] that the system of two coupled differential equations for the normal modes—nowadays, in a more general context that includes fermions and quantized rather than classical fields, known as "Bogoliubov–de Gennes equations"—appeared for the first time. Pitaevskii used these equations to derive the spectrum of normal modes of the vortex filament, the Kelvin waves.

14.10 Josephson Effect, the Phase, Popov Hydrodynamic Action

In 1962, Josephson predicts the *Josephson effect* [74]—a set of phenomena associated with the existence and time evolution of a current between two superconductors (or superfluids) with different chemical potentials connected by a weak link (tunneling barrier or the Josephson junction). As long as each of the two superconductors can be ascribed a global phase (of its order parameter), the Josephson current is given by $J = J_0 \sin(\Delta\Phi)$, where $\Delta\Phi$ is the phase difference between the two superconductors evolving in time as $\Delta\Phi(t) = \Delta\Phi(0) + (\Delta\mu/\hbar)t$ (here, J_0 is a constant characterizing the strength of the Josephson junction). To appreciate the role of Josephson's discovery as a conceptual turning point in the theory of super-fluidity, recall that by 1962, the time dependence of the condensate phase had already been known for 15 (!) years since Bogoliubov's work [27].* Clearly, it could hardly have been all about the missing idea of a weak link that the effect remained undiscovered for so long. Then what could have been the actual reason? Interesting evidence has been provided by Pitaevskii [75] and Anderson [76]. According to Pitaevskii, the idea that the phase of the classical-field order parameter is "real" was not adopted by the community at that time: only its gradient (i.e., the super-fluid velocity) was believed to be "real."† As to our question of what Anderson told Josephson during their discussions (acknowledged in Reference [74]), Anderson replied *that the phase is real.*

Even after the discovery of flux quantization and the Josephson effect, it takes a few years to understand certain subtleties of the description of superfluids in terms of the phase. The main theoretical objection (rather a prejudice) remains the broken gauge symmetry. In his 1966 paper [77], Anderson argues that "it is as legitimate to treat the quantum field amplitude as a macroscopic dynamical variable as it is the position of a solid body; both represent a broken symme-try which, however, cannot be conveniently repaired until one gets to the stage of quantizing." In the $T \to 0$ limit, Anderson explicitly constructs the operator of phase for a mesoscopic region of a superfluid, showing that the phase oper-ator is canonically conjugated to the operator of the total number of particles in the same region. Anderson arrives at the *local* (in the hydrodynamical sense) equation

$$\hbar(d\Phi/dt) = \partial E/\partial N = \mu \qquad (14.19)$$

(here, E is the ground-state energy) that generalizes Beliaev's result and empha-sizes the utmost importance of this relation for the theory of superfluidity: the uniformity of the chemical potential in a steady-state and frictionless acceleration in the case when there is a gradient of the chemical potential.

In the same year, Leggett [78] publishes a paper where he introduces phases for components of the two-band superconductor and proposes a physically observ-able mode associated with their phase difference (Leggett's mode).

In the early 1970s, Popov introduces his hydrodynamic action approach [79] that he derives from the functional-integral representation of the partition func-tion for bosons. The treatment, which Popov himself views as further development of Landau superfluid hydrodynamics, as well as the collective variables approach

---

* This happened simultaneously with the experimental discovery of magnetic flux quantization due to phase winding. These concepts were so controversial at the time that the effect was vigorously denied by Bardeen, who, for example, wrote in Phys. Rev. Lett. **9**, 147 (1962) "Pairing does not extend into the barrier, so that there can be no such superfluid flow."

† At that time, Gross and Pitaevskii believed that their classical-field description of a superfluid was limited to the Bogoliubov-type treatment. It also appears to be quite characteristic that Josephson himself was relying on a quantum, rather than classical-field, approach.

of Bogoliubov and Zubarev, proves very important from both technical and physical perspectives. By generalizing Bogoliubov's 3D finite-temperature result, Popov finds that asymptotic properties of off-diagonal correlations in any dimension are dictated by hydrodynamic fluctuations. In particular, for the single-particle density matrix in 2D at $T \neq 0$, Popov obtains [80]

$$\rho(r) \propto r^{-\alpha}, \qquad \alpha = \frac{mT}{2\pi\hbar n_s} \qquad (d = 2, \quad r \to \infty). \qquad (14.20)$$

Originally, Popov believed that the hydrodynamic treatment and, correspondingly, the result (14.20) is valid only in the limit of $\alpha \ll 1$, because at $\alpha \sim 1$, it becomes energetically favorable to create vortices, which formally requires to modifying the theory. Only significantly later, it was understood—note a remarkable similarity with the story of Equation (14.21)—that the role of vortices reduces to renormalizing n_s while preserving the structure of the final relation. Popov argues that, despite the absence of genuine long-range order, superfluidity is possible in 1D at $T = 0$ and in 2D at finite temperatures [80]. His argument is based on the notion of *seed condensate* essentially equivalent to the notion of algebraic/topological order. Popov's seed condensate is a condensate with a well-defined (no topological defects) but strongly fluctuating phase. The density of the seed condensate, $n_0(k_0)$, depends on the (low enough and arbitrary otherwise) momentum cutoff k_0 separating low-momentum harmonics, $n_0(k_0) \propto k_0^\alpha$, and thus is not a physical quantity. One should distinguish Popov's seed condensate from the *quasicondensate*. The latter is a more specific notion, meaningful only in the limit of small α (low temperatures or/and weak interactions), when the decay of $\rho(r)$ proceeds in two stages and allows one to controllably introduce the (approximate) notion of quasicondensate density, n_*, at short enough scales. Formally, n_* is given by $n_* = n_0(k_0 \sim k_*)$, with k_* the inverse healing length; at $\alpha \ll 1$, there is no need in defining the momentum scale more accurately because of slow dependence of n_0 on k_0. The seed condensate thus plays a purely auxiliary role of carrier of topological order, the only exact observable quantity being n_s.

The hydrodynamic approach allowed Popov to generalize Beliaev's diagrammatic technique to lower dimensions as well as to deal with the infrared divergence problems in 3D. Later on, Popov's hydrodynamic action proved very efficient for treating superfluid–insulator quantum phase transitions in 1D.

14.11   Superfluid Phase Transition

Now we turn to historical aspects of understanding superfluidity and superconductivity in the vicinity of transition to the normal state. The transition was known to be continuous and as such was naturally assumed to obey the general Landau theory of second-order phase transitions [24]; the latter was believed to be correct until the late 1950s.

As already mentioned, the Ginzburg–Landau theory of superconductivity was viewed as a specific case of Landau theory, asymptotically exact on approach to the critical point.* However, in the 1959–1960 works by Levanyuk [81] and Ginzburg [82], it was revealed that fluctuations of the order parameter inevitably render Landau theory inconsistent close enough to the critical point.[†] According to the *Levanyuk–Ginzburg criterion*, only weakly interacting superfluids/superconductors can be controllably described by Ginzburg–Landau theory; even in this case, there is a (small) fluctuation region in the vicinity of the critical point where the theory breaks down. By this criterion, all superconductors known at that time were extremely weakly interacting and thus, for all practical purposes, perfectly describable by Ginzburg–Landau theory in the vicinity of the critical point. Only for ^{4}He, the criterion was not satisfied. Within a decade from the works of Levanyuk and Ginzburg, the fluctuation theory of second-order phase transitions was developed.

The *Ginzburg–Landau–Wilson* theory associates finite-temperature phase transitions with ordering of an emergent *classical field*—the fluctuating order parameter—that gradually develops in the normal phase, becomes well-defined in the vicinity of the critical point, and describes the universal criticality of the transition. This theory became a distinct subfield of theoretical physics. The superfluid phase transition has been identified with the XY universality class, the name coming from the XY model of an easy-plane ferromagnet (in connection with superfluidity, the XY universality class is also known as $U(1)$ universality class). In the theory of second-order phase transitions, it plays the role of the "hydrogen atom" as the simplest universality class associated with breaking a continuous symmetry and is next to the Ising universality class in terms of the number of real components of the order parameter: one for Ising and two for $XY/U(1)$. The theory of continuous phase transitions deserves a separate historical review that we cannot afford here. We will thus confine ourselves to the discussion of the Berezinskii–Kosterlitz–Thouless (BKT) transition in 2D and the vortex–world-line duality in 3D, both topics being of prime conceptual importance to the theory of superfluidity/superconductivity.

14.11.1 BKT Transition: Topological Order

By the beginning of the 1970s, it had been well established due to the works of Peierls (1935), Mermin and Wagner (1966), and Hohenberg (1967) that in finite-temperature 2D systems with continuous symmetry, genuine long-range order is destroyed by thermal fluctuations of the corresponding Goldstone modes. By standard thinking, this implied the absence of an order parameter defining the phase and thus no phase transition. Nevertheless, theoretical arguments suggested that there might be some kind of a transition to a state with infinite susceptibility.

---

* For example, Ginzburg and Pitaevskii, when applying Ginzburg–Landau theory to neutral superfluids (in 1958) [52], conjectured that a deviation of the critical behavior of ^{4}He from that predicted by Landau theory could be due to a very small coefficient in front of the $|\psi|^4$ term in the free-energy expansion—the natural fix simply taking into account the $|\psi|^6$ term.

† Recall that, for typical second-order phase transitions, this statement is valid only in dimensions $d \leq 3$; in $d \geq 4$, the Landau theory is asymptotically exact on approach to the critical point.

The solution to the apparent paradox was found by Berezinskii [83] and Kosterlitz and Thouless [84] (and also by Popov; see Section 14.10) who predicted a new type of order known as *algebraic* or *topological* order. The term *algebraic order* emphasizes the power-law decay of the order-parameter correlation function as opposed to saturation to a constant in the case of genuine long-range order, while the term *topological order* emphasizes the existence of topological invariants (such as Burgers vectors in solids and phase winding numbers in superfluids) in the ordered phase. In 2D systems belonging to the XY/U(1) universality class, the destruction of algebraic/topological order happens only when there appear free point topological defects—vortices. The corresponding phase transition is now called the BKT transition.

Below the BKT transition temperature, T_{BKT}, the vortices are bound into topologically neutral vortex–antivortex pairs. Kosterlitz and Thouless developed an asymptotically accurate renormalization-group description of the BKT criticality, utilizing the fact that the relative concentration of macroscopically large vortex pairs vanishes at $T \to T_{\mathrm{BKT}}$, so that the theory has to deal with a rather simple regime of arbitrarily dilute gas of vortex pairs. Berezinskii did not present the details of his calculations, but it is clear that they contained some technical error because he claimed vanishing superfluid density at the critical point. The latter is known to be finite, obeying the universal Nelson–Kosterlitz relation [85]

$$\frac{mk_B T_{\mathrm{BKT}}}{\hbar^2 n_s} = \frac{\pi}{2}. \tag{14.21}$$

The history of this relation reveals conceptual difficulties that were not fully understood back in 1972, despite the asymptotic accuracy of the Kosterlitz–Thouless theory. Interestingly, the relation is already present in the short 1972 letter by Kosterlitz and Thouless [84]. Nevertheless, in their longer paper of 1973, they view it only as an approximate estimate, becoming exact only in the limit of infinitesimally small fugacity of vortex pairs. This sounds rather odd since the whole treatment is based on the vanishing relative fugacity of large vortex pairs—the number of pairs of size $\sim R$ per area $\sim R^2$—in the $R \to \infty$ limit, which automatically leads to (14.21) for the macroscopic (i.e., renormalized by all the vortex pairs) value of n_s from the zero vortex free-energy condition. Even more intriguingly, Nelson and Kosterlitz, when arguing in 1977 that (14.21) is an exact relation, were relying on the explicit analysis of the Kosterlitz–Thouless renormalization-group equations, without mentioning the simple macroscopic physics behind (14.21).

We believe that the difficulty was in the proper identification of fields in the original Kosterlitz–Thouless formalism that was employing the mathematically correct, but physically rather subtle, procedure of separating the phase field into two statistically independent parts—the one containing the vortices (it was required to satisfy the Laplace equation to guarantee its decoupling from the rest of the field) and the smooth part interpreted as "phonons." The price for this mathematical simplicity is that the theory becomes explicitly dependent on the "microscopic" superfluid density $n_s^{(0)}$ and "microscopic fugacity" of vortex pairs;

both quantities depend on the length scale used to define them and thus are rather ambiguous. In 1974, Kosterlitz notices [86] an intriguing conspiracy between the seemingly independent "phonons" and vortex pairs in terms of their contributions to the exponent describing the algebraic decay of the single-particle density matrix. While partial contributions of the two subsystems were explicitly dependent on $n_s^{(0)}$, the net result was a function of the genuine macroscopic n_s. Clearly, there was some qualitative similarity between the "vortex-phonon conspiracy" and the relation (14.21), pointing to the physically natural fact that the procedure of separation of the phase field into "phonons" and "vortices" can be done at an arbitrarily large scale of distance, with appropriately renormalized values of $n_s^{(0)}$ and vortex pair fugacity. Full understanding of the situation came later, when it was explicitly shown that the smooth field that decouples from the vortex part statistically *does not* correspond to phonons as soon as there is a finite concentration of vortex pairs (see Chapter 4). The long-range part of the "vortex-pair field," as defined originally in the Kosterlitz–Thouless theory, is indistinguishable from any nonsingular configuration of the field with the dipole asymptotic behavior and as such belongs to real phonons. The theory can then be rewritten in terms of phonons that *are* coupled to vortex pairs, and this naturally leads to the superfluid stiffness being renormalized by this coupling. Correspondingly, the vortex pairs of sizes smaller than some cutoff scale—appropriately large, but otherwise arbitrary—are absorbed into the renormalized phonon modes.

Compelling experimental observation of the BKT transition was performed (in 1978) by Rudnick [87] and Bishop and Reppy [88]. Direct evidence of topological order in a finite-temperature ^{4}He film was provided by Telschow and Hallock (1983) [89], who demonstrated the conservation of velocity circulation despite radical changes of the film thickness by evaporating/condensing ^{4}He vapor.

14.11.2   Vortex–World-Line Duality in 3D

In 1974, Halperin, Lubensky, and Ma argue that fluctuations of vector potential will render a superconducting transition to be first order, in contrast to superfluids [90]. However, in a subsequent 1981 work, Dasgupta and Halperin show that the statement is correct only for strongly type-1 superconductors. For type-2 superconductors in the London regime, the transition should be in the "inverted-XY" universality class [91], which is dual to the conventional XY-transition. The derivation of Dasgupta and Halperin is based on exact mapping of the statistical sum of a lattice London model of superconductor to that of an ordinary XY model with reversed temperature axis.

14.12   Multicomponent Order Parameters/Effective Theories
14.12.1   Helium-3

In 1971, Lee, Richardson, and Osheroff [92] discover long-anticipated superfluidity in helium-3 at 0.0026K (it is misidentified first as "a new phase of solid ^{3}He" before finding that both A- and B-transitions occur in a liquid [93]). The previous decade had seen an explosion of theoretical work proposing and analyzing various pairing mechanisms and order parameters for this unique Fermi liquid. Due to hard-core repulsion between the atoms, it was concluded that the pairing will most likely occur with finite angular momentum $L = 1$ and spin $S = 1$, and thus, helium-3 ought to be an unconventional anisotropic superfluid. In particular, the A-phase was part of the Anderson–Morel classification [94], and the Vdovin–Balian–Werthamer phase [95,96], or simply B-phase, was describing pairing with zero total momentum $J = L + S = 0$. (Vdovin's work was, and remains, unknown to most people to the extent that even comprehensive reviews and books miss it.) Both phases were introduced almost a decade before the actual experimental finding. Thanks to this theoretical activity, the correct identification of superfluid phases within the manifold of various anisotropic states—in this case, we are talking about an order parameter with 9 complex degrees of freedom—was very quick.

In 1972–1973, Leggett [97,98] shows how experimental nuclear magnetic resonance (NMR) frequency shifts can be produced by a p-wave state and matched to the calculated NMR susceptibility of the B-phase. He also predicts an entirely new "longitudinal" NMR mode that should be independent of the applied magnetic field. The discovery of superfluidity in ^{3}He, the first unconventional superfluid with finite angular momentum and spin of Cooper pairs, was a remarkable triumph of the BCS theory and understanding of pairing mechanisms in strongly correlated systems. It also revealed new effects for anisotropic states, such as textures, which were not paid much attention until then.

14.12.2 Multiband Superconductors

Almost immediately after the publication of the BCS theory, there appeared the first generalizations of superconductivity to the multicomponent case. In two independent works in 1959, first by Moskalenko [99] and shortly afterwards by Suhl, Mathias, and Walker [100], BCS superconductivity was generalized to the multiband case. In 1964, Tilley [101] generalizes Gor'kov's microscopic derivation of Ginzburg–Landau theory to multiband cases, arriving at the multicomponent classical field theory where classical-field components are not associated with order parameters.* It was recently proposed that multiple coherence lengths ξ_i—natural for multicomponent superconductors—should lead to the breakdown of the standard type-1/type-2 dichotomy in the classification of superconductors (see Chapter 6). A new regime that arises in these systems is where one of the coherence lengths is larger and another is smaller than the magnetic field penetration length (in traditional units $\xi_1 < \lambda < \xi_2$).

---

* In two-band superconductors, the superconducting components are not independently conserved; thus, the order parameter should be just a single complex field. Nonetheless, the system can indeed have a classical-field theory description where classical fields are not associated with order parameters.

14.13   Related Developments

14.13.1   Applications in High-Energy Physics/The Anderson Mechanism

Many ideas that were developed first in an attempt to understand superconductivity and superfluidity were successfully used in other branches of physics, ranging from liquid crystals to particle physics. To discuss all interdisciplinary connections is beyond the scope of this book. We briefly mention here two developments that became the cornerstone concepts in particle physics. One is the application of BCS ideas to particle physics by Nambu and Jona-Lasinio [102], who used the analogy with fermionic pairing to introduce the concepts of spontaneous symmetry breaking into elementary particle physics and suggested a mechanism for mass generation of nucleons. The other development deals with the mechanism of mass generation for gauge bosons. It is frequently attributed to Brout, Englert, and Higgs [103,104] in the high-energy community even though it was first introduced and explained by Anderson [105] in the context of superconductors and thus should be called the Anderson mechanism. More specifically, the phase field of a neutral superfluid (i.e., in the absence of coupling to the vector potential) is massless. In 1963, Anderson demonstrates—building on an earlier Schwinger's conjecture, with the superconductor as a prototypical model—that coupling to vector potential replaces massless excitations with gapped ones and insists that the same mechanism should be relevant in high-energy physics context: "It is likely, then, considering the superconducting analog, that the way is now open for a degenerate vacuum theory of the Nambu type without any difficulties involving either zero-mass Yang-Mills gauge bosons or zero-mass Goldstone bosons. These two types of bosons seem capable of 'canceling each other out' and leaving finite mass bosons only." In the subsequent paper, the analogy is acknowledged by Higgs [104]: "This phenomenon is just the relativistic analog of the plasmon phenomenon to which Anderson has drawn attention." Simultaneously, Higgs emphasizes the importance of the massive *scalar* collective excitation—known from Goldstone's work as a massive counterpart of the massless Goldstone mode in the superfluid relativistic vacuum [106].

14.13.2   Quantum Monte Carlo

In the mid-1980s, Ceperley and collaborators develop path-integral Monte Carlo techniques for the equilibrium statistics of interacting bosons. Their approach employs the generic Metropolis–Rosenbluth–Teller algorithm (1953)—originally introduced for simulating Gibbs distributions of classical particles—for sampling the statistics of Feynman paths in $(d + 1)$-dimensional space–imaginary time. This breakthrough opens the door for quantitative description of strongly correlated bosonic superfluids such as ^{4}He. Driven by the need to compute the superfluid density, Pollock and Ceperley discover an elegant formula relating the superfluid density to the variance of the winding number counting how many times a path goes around a periodic cell [37]. This development establishes a precise translation from the language of paths (or "polymers") into the language of quantum statistics of bosons. In particular, with the Pollock–Ceperley relation, it is seen that only macroscopic bosonic permutation cycles render a ground-state superfluid; this fact is very important for understanding why insulating bosonic crystals exist at $T = 0$.

14.13.3   Localization of Interacting Bosons, Supersolid

During the mid- to late 1980s, an important progress had been achieved in understanding the phenomenon of localization of interacting bosonic systems—in the presence of disorder and/or commensurate external potential. As opposed to fermions where Anderson's localization in a macroscopic system can be introduced and discussed at the single-particle noninteracting level, the ground state of disordered bosons in the limit of vanishing interactions is pathological: all particles occupy the same lowest-energy localized state. Hence, theoretical treatment of the quantum phase transition from the superfluid to an insulating state in a system of disordered bosons requires essentially nonperturbative many-body techniques. In 1987, Giamarchi and Schulz [107] develop an asymptotically exact renormalization-group approach for a 1D system that allows them to describe the superfluid–insulator transition. Building on this result, Fisher, Weichman, Grinstein, and Fisher formulate the concept of a *Bose glass* phase in any dimensions [108]. The signature property of this phase is the structure of its low-lying elementary excitations: these are spatially localized particles and holes with Pauli exclusion principle originating from interactions. In this sense, despite a very different microscopic picture, especially in the limit of small interactions, at the macroscopic low-energy level, the Bose glass phase is similar to localized fermions.

The history of understanding localization of interacting bosons in the absence of disorder also features interesting twists. The first, and so far the most reliable, theoretical model demonstrating that solid and superfluid orders can coexist was envisioned by Gross [71], who found that his classical wave equation may have a density wave state for certain interaction potentials and field densities. Unfortunately, his observation went unnoticed until it was reinvented several decades later. In 1969–1970, Andreev and Lifshitz [109], Thouless [110], and Chester [111] put forward arguments, based on BEC of equilibrium vacancies, that there is nothing fundamentally wrong with having coexisting crystalline and superfluid orders, as opposed to the earlier conclusion by Penrose and Onsager. The corresponding state was termed a "supersolid." In 1970, Leggett argues [112] that if the solid phase of helium-4 were a supersolid without vacancies, its ground-state superfluid density and, correspondingly, critical temperature would be orders of magnitude smaller than in the liquid phase. This started a prejudice—surviving to the present day [113]—that every pure solid ground state of bosons is a supersolid. Using the language of path integrals, one can prove rigorously (see Chapter 9) that continuous-space supersolids ought to have equilibrium vacancies/interstitials and that insulating bosonic ground states do exist (it is sufficient to examine the cost of macroscopic world-line cycles in the limit of large mass and/or strong interactions).

The theory of the superfluid–Mott-insulator quantum phase transition in a regular system in a commensurate periodic external potential was formulated in the previously cited paper by Fisher et al. [108]. The authors showed that in a d-dimensional system, the transition is in the universality class of the $(d + 1)$-dimensional XY model.

14.14 BEC in Ultracold Gases

Despite being a textbook subject for decades, the phenomenon of BEC in an almost ideal gas had not been observed experimentally until the mid-1990s. During the mid- to late 1970s, the idea of achieving BEC in a metastable gas of spin-polarized atomic hydrogen emerged [114], giving birth to what later on evolved into the interdisciplinary subfield of ultracold atoms. Many years of attempts to condense spin-polarized atomic hydrogen by cryogenic methods in a closed geometry [114] led to the realization that only a very dilute spin-polarized atomic system, in an open geometry (magnetic or optical trap), would be protected against recombination into molecules on approach to BEC. An evaporative cooling of a trapped gas was proposed for the final stage of the experimental protocol, rendering large amplitude of the two-particle s-scattering the most important optimization parameter (thermalization should occur on time scales much shorter than system decay). In this regard, alkaline atoms proved to be the best option. In 1995, three groups—that of Cornell and Wieman, that of Hulet, and that of Ketterle—within a few months achieve BEC of three different systems: rubidium [115], lithium [116], and sodium [117], respectively. After these experiments, the field of ultracold atoms explodes with results.

Given that, from the very beginning, Einstein was putting BEC in a broader context of classical-field behavior of quantum particles, it is curious to note that even in the early 1990s—about 65 years after Einstein's work—the physical importance of this context was not fully appreciated. Textbooks were treating BEC as a purely quantum-statistical phenomenon, and the vast majority of physicists working in the field of ultracold atoms were still under the impression that the dynamical description in terms of the classical field (by the time-dependent Gross–Pitaevskii equation) becomes possible only after the long-range order sets in. Even the equilibrium-statistical classical-field description of a weakly interacting Bose gas, natural from the general theory of second-order phase transitions, had not been developed until the early 2000s. However, in 1978, Levich and Yakhot argued [118] that the classical-field description of bosonic fields derived by Langer a decade earlier for the superfluid state [10] (see Section 14.1) also works for kinetics of BEC formation. They observed that, at large occupation numbers, the leading contribution to the collision term of quantum kinetic equation for bosons is nothing but the collision term of the weak-turbulence kinetic equation for the corresponding classical field. Levich and Yakhot then conjectured that classical-field dynamics will remain valid even after strong nonlinearity of the field renders the kinetic equation inapplicable. Unfortunately, until the early

1990s, the ideas of Levich and Yakhot were not fully adopted by the ultracold atomic community.

Full theoretical understanding of the scope and implications—both qualitative and quantitative—of the classical-field description was achieved in the period from the early 1990s until the mid-2000s. The scenario of strongly nonequilibrium BEC kinetics was developed (and later checked by experiments and numerical simulations). The universal parameters and scaling functions describing weakly interacting U(1) systems in the fluctuation region were calculated. In the early 2010s, experimental evidence of the presence of strongly turbulent classical field in both normal and superfluid states of a weakly interacting Bose gas in the vicinity of the critical point was obtained. This was achieved by measuring the equations of state of 2D and 3D systems and by comparing the non-mean-field contributions to those of purely classical-field systems.

We stop our brief historical overview here and do not consider more recent developments in various subfields. Additional details can be found in the Nature and Laboratory chapter.

References

1. H. Kamerlingh Onnes, Further experiments with liquid helium, *Commun. Phys. Lab. Univ. Leiden*; No. 120b (1911).

2. D. van Delft and P. Kes, The discovery of superconductivity, *Phys. Today*, September 2010, p. 38.

3. A. Einstein, Quantentheorie des einatomigen idealen Gases, *Ber. Berl. Akad.* 261 (1924); ibid. 3 (1925).

4. S. N. Bose, Plancks Gesetz und Lichtquantenhypothese, *Zeitschrift für Physik* **26**, 178 (1924). (The German translation of Bose's paper on Planck's law.)

5. R. J. Donnelly, A. W. Francis, and L. I. Dana, *Cryogenic Science and Technology: Contributions by Leo I. Dana*, Union Carbide Corporation, Danbury, CT (1985).

6. E. Schrödinger, An undulatory theory of the mechanics of atoms and molecules, *Phys. Rev.* **28**, 1049 (1926).

7. G. E. Uhelenbeck, Dissertation, Leiden, the Netherlands (1927).

8. F. London, The λ-phenomenon of liquid helium and the Bose–Einstein degeneracy, *Nature* **141**, 643 (1938); On the Bose–Einstein condensation, *Phys. Rev.* **54**, 947 (1938).

9. R. J. Glauber, Coherent and incoherent states of the radiation field, *Phys. Rev.* **131**, 2766 (1963).

10. J. S. Langer, Coherent states in the theory of superfluidity, *Phys. Rev.* **167**, 183 (1968).

11. W. H. Keesom and M. Wolfke, Two different liquid states of helium, *Leiden. Commun.* 190b (1927); W. H. Keesom and M. Wolfke, On the change of the dielectric constant of liquid helium with the temperature. Provisional measurements, *Proc. Roy. Acad. Sci. Amst.* **31**, 81 (1928).

12. W. H. Keesom and K. Clusius, Communication No. 219e from the Kamerlingh Onnes Laboratory at Leiden, *Proc. Roy. Acad. Sci. Amst.* **35**, 307 (1932); W. H. Keesom and A. P. Keesom, Communication No. 221d from the Kamerlingh Onnes Laboratory at Leiden, *Proc. Roy. Acad. Sci. Amst.* **35**, 736 (1932).

13. E. F. Burton, Viscosity of helium I and helium II, *Nature* **135**, 265 (1935).

14. A. Griffin, A brief history of our understanding of BEC: From Bose to Beliaev, Varenna Summer School (1999); New light on the intriguing history of superfluidity in liquid ^{4}He, *J. Phys. Condens. Matter* **21**, 164220 (2009).

15. W. H. Keesom and A. P. Keesom, New measurements on the specific heat of liquid helium, *Phys.* **2**, 557 (1935).

16. W. H. Keesom and A. P. Keesom, On the heat conductivity of liquid helium, *Physica* **3**, 359 (1936).

17. J. F. Allen, R. Peierls, and M. Zaki Uddin, Heat conduction in liquid helium, *Nature* **140**, 62 (1937).

18. P. Kapitza, Viscosity of liquid helium below the λ-point, *Nature* **141**, 74 (1938).

19. J. F. Allen and A. D. Misener, Flow of liquid helium II, *Nature* **141**, 75 (1938).

20. S. Balibar, The discovery of superfluidity, *J. Low Temp. Phys.* **146**, 441 (2007).

21. J. F. Allen and H. Jones, New phenomena connected with heat flow in helium II, *Nature* **141**, 243 (1938).

22. L. Tisza, Transport phenomena in helium II, *Nature* **141**, 913 (1938).

23. L. Landau, Theory of superfluidity of helium II, *Phys. Rev.* **60**, 356 (1941).

24. L. D. Landau, Theory of phase transitions, *Phys. Z. Sowjetunion* **11**, 26, 545 (1937).

25. V. L. Ginzburg and L. D. Landau, To the theory of superconductivity, *Zh. Eksp. Teor. Fiz.* **20**, 1064 (1950).

26. R. P. Feynman, Atomic theory of liquid helium near absolute zero, *Phys. Rev.* **91**, 1301 (1953).

27. N. Bogoliubov, On the theory of superfluidity, *J. Phys. USSR* **11**, 23 (1947).

28. L. Landau, On the theory of superfluidity, *Phys. Rev.* **75**, 884 (1948).

29. O. Penrose, On the quantum mechanics of helium II, *Phil. Mag.* **42**, 1373 (1951).

30. L. Onsager, Statistical hydrodynamics, *Nuovo Cimento* **6**, 279 (1949).

31. L. D. Landau and E. M. Lifshitz, On rotation of liquid helium, *Dokl. Akad. Nauk.* **100**, 669 (1955).

32. H. London, The superfluid flow of liquid helium II, Report of an international conference on fundamental particles and low temperatures held at the Cavendish Laboratory, Cambridge, U.K., July 22–27 1946. Vol. 2: Low temperatures (1946).

33. R. P. Feynman, Application of quantum mechanics to liquid helium, In: *Progress in Low Temperature Physics* (ed. C. J. Gorter), North-Holland Pub. Co., Amsterdam, the Netherlands, Vol. **1**, p. 17 (1955).

34. H. E. Hall and W. F. Vinen, The rotation of liquid helium II, *Proc. Roy. Soc. (Lond.)* **A238**, 204, 215 (1956).

35. W. F. Vinen, The detection of single quanta of circulation in liquid helium II, *Proc. Roy. Soc. (Lond.)* **A260**, 218 (1961).

36. R. P. Feynman, Atomic theory of the λ-transition in helium, *Phys. Rev.* **91**, 1291 (1953).

37. E. L. Pollock and D. M. Ceperley, Path-integral computation of superfluid densities, *Phys. Rev. B* **36**, 8343 (1987).

38. T. Matsubara, Quantum-statistical theory of liquid helium, *Progr. Theor. Phys. Jpn.* **6**, 714 (1951).

39. R. P. Feynman, Atomic theory of the two-fluid model of liquid helium, *Phys. Rev.* **94**, 262 (1954).

40. O. Penrose and L. Onsager, Bose–Einstein condensation and liquid helium, *Phys. Rev.* **104**, 576 (1956).

41. W. Meissner and R. Ochsenfeld, Ein neuer Eekt bei Eintritt der Supraleitfähigkeit, *Naturwissenschaften* **21**, 787 (1933).

42. F. London and H. London, The electromagnetic equations of the supraconductor, *Proc. Roy. Soc. (Lond.)* **A149**, 71 (1935).

43. F. London, On the problem of the molecular theory of superconductivity, *Phys. Rev.* **74**, 562 (1948).

44. W. Ehrenberg and R. E. Siday, The refractive index in electron optics and the principles of dynamics, *Proc. Phys. Soc. B* **62**, 8 (1949).

45. Y. Aharonov and D. Bohm, Significance of electromagnetic potentials in quantum theory, *Phys. Rev.* **115**, 485 (1959).

46. L. Onsager, Discussion, *Proceedings of the International Conference on Theoretical Physics*, Kyoto and Tokyo, Japan, September 1953 (Science Council of Japan, Tokyo, Japan, 1954), p. 935.

47. B. S. Deaver, Jr. and W. M. Fairbank, Experimental evidence for quantized flux in superconducting cylinders, *Phys. Rev. Lett.* **7**, 43 (1961).

48. R. Doll and M. Nabauer, Experimental proof of magnetic flux quantization in a superconducting ring, *Phys. Rev. Lett.* **7**, 51 (1961).

49. L. Onsager, Magnetic flux through a superconducting ring, *Phys. Rev. Lett.* **7**, 50 (1961).

50. L. D. Landau, Zur Theorier de Supraleitheitfahigkeit. I, *Physikalische Zeitschrift der Sowjetunion* **4**, 43 (1933).

51. F. London, *Superfluids*, John Wiley, New York, 1950.

52. V. L. Ginzburg and L. P. Pitaevskii, On the theory of superfluidity, *Zh. Eksp. Teor. Fiz.* **34**, 1240 (1958) [*Sov. Phys. JETP* **7**, 858 (1958)].

53. V. L. Ginzburg, Nobel lecture: On superconductivity and superfluidity (what I have and have not managed to do) as well as on the "physical minimum" at the beginning of the XXI century, *Rev. Mod. Phys.* **76**, 981 (2004).

54. A. A. Abrikosov, Nobel lecture: Type-II superconductors and the vortex lattice, *Rev. Mod. Phys.* **76**, 975 (2004).

55. J. M. Casimir-Jonker and W. J. De Haas, Some experiments on a supraconductive alloy in a magnetic field, *Physica* **2**, 935 (1935).

56. L. W. Schubnikow, W. I. Chotkewitsch, J. D. Schepelew, and J. N. Rjabinin, Magnetische Eigenschaften supraleitender Metalle und Legierungen, *Sondernummer Phys. Z. Sowjet. Arbeiten auf dem Gebiete tiefer Temperaturen*, Juni **1936**, 39 (1936); L. W. Schubnikow, W. I. Chotkewitsch, J. D. Schepelew, and J. N. Rjabinin, Magnetische Eigenschaften supraleitender Metalle und Legierungen, *Phys. Z. Sowjet.* **Bd.10, H.2**, 165 (1936); L. V. Shubnikov, V. I. Khotkevich, Yu. D. Shepelev, and Yu. N. Riabinin, Magnetic properties of superconducting metals and alloys, *Zh. Eksp. Teor. Fiz.* **7**, 221 (1937).

57. A. A. Abrikosov, Influence of the size on critical field of superconductors of the second group, *Dokl. Akad. Nauk. SSSR* **86**, 489 (1952).

58. A. A. Abrikosov, On the magnetic properties of superconductors of the second group, *Zh. Eksp. Teor. Fiz.* **32**, 1442 (1957) [*Sov. Phys. JETP* **5**, 1174 (1957)].

59. D. Einzel, 50 years of fluxoid quantization: 2e or not 2e, *J. Low Temp. Phys.* **163**, 215 (2011).

60. D. Cribier, B. Jacrot, L. M. Rao, and B. Farnoux, Mise en evidence par diraction de neutrons d'une structure periodique du champ magnetique dans le niobium supraconducteur, *Phys. Lett.* **9**, 106 (1964).

61. U. Essmann and H. Traeuble, The direct observation of individual flux lines in type II superconductors, *Phys. Lett. A* **24**, 526 (1967).

62. E. Maxwell, Isotope effect in the superconductivity of mercury, *Phys. Rev.* **78**, 477 (1950); C. A. Reynolds, B. Serin, W. H. Wright, and L. B. Nesbitt, Superconductivity of isotopes of mercury, *Phys. Rev.* **78**, 487, (1950).

63. L. N. Cooper, Bound electron pairs in a degenerate Fermi gas, *Phys. Rev.* **104**, 1189 (1956).

64. J. Bardeen, L. N. Cooper, and J. R. Schrieffer, Microscopic theory of superconductivity, *Phys. Rev.* **106**, 162 (1957); Theory of Superconductivity, *Phys. Rev.* **108**, 1175 (1957).

65. I. Giaever, Energy gap in superconductors measured by electron tunneling, *Phys. Rev. Lett.* **5**, 147 (1960); Electron tunneling between two superconductors, *Phys. Rev. Lett.* **5**, 464 (1960).

66. V. V. Tolmachev and S. V. Tiablikov, A new method in the theory of superconductivity, *Zh. Eksp. Teor. Fiz.* **34**, 66 (1958) [*Sov. Phys. JETP* **7**, 46 (1958)].

67. V. M. Galitskii and A. B. Migdal, Application of quantum field theory methods to the many body problem, *Zh. Eksp. Teor. Fiz.* **34** 139 (1958) [*Sov. Phys. JETP* **7**, 96 (1958)].

68. S. T. Beliaev, Application of the methods of quantum field theory to a system of bosons, *Zh. Eksp. Teor. Fiz.* **34**, 417 (1958) [*Sov. Phys. JETP* **7**, 289 (1958)]; Energy spectrum of a non-ideal Bose gas, *Zh. Eksp. Teor. Fiz.* **34**, 433, (1958) [*Sov. Phys. JETP* **7**, 299 (1958)].

69. L. P. Gor'kov, On the energy spectrum of superconductors, *Zh. Eksp. Teor. Fiz.* **34**, 735 (1958) [*Sov. Phys. JETP* **7**, 505 (1958)].

70. L. P. Gor'kov, Microscopic derivation of the Ginzburg–Landau equations in the theory of superconductivity, *Zh. Eksp. Teor. Fiz.* **36**, 1918 (1959) [*Sov. Phys. JETP* **9**, 1364 (1959)].

71. E. P. Gross, Unified theory of interacting bosons, *Phys. Rev.* **106**, 161 (1957); Classical theory of boson wave fields, *Ann. Phys.* **4**, 57 (1958).

72. E. P. Gross, Structure of a quantized vortex in boson systems, *Nuovo Cimento* **20**, 454 (1961).

73. L. P. Pitaevskii, Vortex lines in an imperfect Bose gas, *Zh. Eksp. Teor. Fiz.* **40**, 646 (1961) [*Sov. Phys. JETP* **13**, 451 (1961)].

74. B. D. Josephson, Possible new effects in superconductive tunneling, *Phys. Lett.* **1**, 251 (1962).

75. L. P. Pitaevskii, Private communication to the authors, August 2010.

76. P. W. Anderson, Private communication to the authors, June 2011.

77. P. W. Anderson, Considerations on the flow of superfluid helium, *Rev. Mod. Phys.* **38**, 298 (1966).

78. A. Leggett, Number-phase fluctuations in two-band superconductors, *Prog. Theor. Phys.* **36**, 901 (1966).

79. V. N. Popov, *Functional Integrals in Quantum Field Theory and Statistical Physics*, Reidel, Dordrecht, the Netherlands (1983).

80. V. N. Popov, On the theory of superfluidity of two-dimensional and one-dimensional Bose systems, *Theor. Math. Phys.* **11**, 354 (1971).

81. A. P. Levanyuk, Contribution on the theory of light scattering near the second-order phase-transition points, *Zh. Eksp. Teor. Fiz.* **36**, 810 (1959) [*Sov. Phys. JETP* **9**, 571 (1959)].

82. V. L. Ginzburg, Some remarks on phase transitions of second kind and the microscopic theory of ferroelectric materials, *Fiz. Tverd. Tela (Leningrad)* **2**, 2031 (1960) [*Sov. Phys. Solid State* **2**, 1824 (1961)].

83. V. L. Berezinskii, Destruction of long-range order in one-dimensional and two-dimensional systems possessing a continuous symmetry group. II. Quantum systems, *Zh. Eksp. Teor. Fiz.* **61**, 1144 (1971) [*Sov. Phys. JETP*, **34**, 610 (1972)].

84. J. M. Kosterlitz and D. J. Thouless, Long-range order and metastability in two-dimensional solids and superfluids, *J. Phys. C Solid State Phys.* **5**, L124 (1972); Ordering, metastability and phase transitions in two-dimensional systems, *J. Phys. C Solid State Phys.* **6**, 1181 (1973).

85. D. R. Nelson and J. M. Kosterlitz, Universal jump in the superfluid density of two-dimensional Superfluids, *Phys. Rev. Lett.* **39**, 1201 (1977).

86. J. M. Kosterlitz, The critical properties of the two-dimensional XY model, *J. Phys. C Solid State Phys.* **7**, 1046 (1974).

87. I. Rudnick, Critical surface density of the superfluid component in ^{4}He films, *Phys. Rev. Lett.* **40**, 1454 (1978).

88. D. J. Bishop and J. D. Reppy, Study of the superfluid transition in two-dimensional ^{4}He films, *Phys. Rev. Lett.* **40**, 1727 (1978).

89. K. L. Telschow and R. B. Hallock, Observations on ^{4}He persistent flow, *Phys. Rev. B* **27**, 3068 (1983).

90. B. I. Halperin, T. C. Lubensky, and S. K. Ma, First-order phase transitions in superconductors and smectic–A liquid crystals, *Phys. Rev. Lett.* **32**, 292 (1974).

91. C. Dasgupta and B. I. Halperin, Phase transition in a lattice model of superconductivity, *Phys. Rev. Lett.* **47**, 1556 (1981).

92. D. D. Osheroff, R. C. Richardson, and D. M. Lee, Evidence for a new phase of solid ^{3}He, *Phys. Rev. Lett.* **28**, 885 (1972).

93. D. D. Osheroff, W. J. Gully, R. C. Richardson, and D. M. Lee, New magnetic phenomena in liquid ^{3}He below 3 mK, *Phys. Rev. Lett.* **29**, 920 (1972).

94. P. W. Anderson and P. Morel, Generalized Bardeen-Cooper-Schrieffer states and the proposed low-temperature phase of liquid ^{3}He, *Phys. Rev.* **123**, 1911 (1961).

95. Y. A. Vdovin, In: *Application of Methods of Quantum Field Theory to Problems of Many Particles*, [in Russian] (ed. A. I. Alekseyeva), Gos Atom Izdat, Moscow, Russia (1963).

96. R. Balian and N. R. Werthamer, Superconductivity with pairs in a relative p-wave, *Phys. Rev.* **131**, 1553 (1963).

97. A. J. Leggett, Interpretation of recent results on ^{3}He below 3 mK: A new liquid phase?, *Phys. Rev. Lett.* **29**, 1227 (1972).

98. A. J. Leggett, Microscopic theory of NMR in an anisotropic superfluid ($_3$HeA), *Phys. Rev. Lett.* **31**, 352 (1973).

99. V. A. Moskalenko, Superconductivity in metals with overlapping energy bands, *Fiz. Metallov i Metallovedenie* **8**, 503 (1959).

100. H. Suhl, B. T. Matthias, and L. R. Walker, Bardeen-Cooper-Schrieffer theory of superconductivity in the case of overlapping bands, *Phys. Rev. Lett.* **3**, 552 (1959).

101. D. R. Tilley, The Ginzburg–Landau equations for pure two-band superconductors, *Proc. Phys. Soc.* **84**, 573 (1964).

102. Y. Nambu, A "superconductor" model of elementary particles and its consequencies, Talk given at a conference at Purdue (1960), reprinted in Broken symmetries, selected papers by Y. Nambu, (eds. T. Eguchi and K. Nishijima), World Scientific (1995); Y. Nambu and G. Jona-Lasinio, Dynamical model

of elementary particles based on an analogy with superconductivity. I, *Phys. Rev.* **122**, 345 (1961); Dynamical model of elementary particles based on an analogy with superconductivity. II, *Phys. Rev.* **124**, 246 (1961).

103. F. Englert and R. Brout, Broken symmetry and the mass of gauge vector mesons, *Phys. Rev. Lett.* **13**, 321 (1964).

104. P. W. Higgs, Broken symmetries and the masses of gauge bosons, *Phys. Rev. Lett.* **13**, 508 (1964).

105. P. W. Anderson, Plasmons, gauge invariance, and mass, *Phys. Rev.* **130**, 439 (1963).

106. J. Goldstone, Field theories with "superconductor" solutions, *Nuovo Cimento* **19**, 154 (1961).

107. T. Giamarchi and H. J. Schulz, Localization and interaction in one-dimensional quantum fluids, *Europhys. Lett.* **3**, 1287 (1987); Anderson localization and interactions in one-dimensional metals, *Phys. Rev. B* **37**, 325 (1988).

108. M. P. A. Fisher, P. B. Weichman, G. Grinstein, and D. S. Fisher, Boson localization and the superfluid-insulator transition, *Phys. Rev. B* **40**, 546 (1989).

109. A. F. Andreev and I. M. Lifshitz, Quantum theory of defects in crystals, *Zh. Eksp. Teor. Fiz.* **56**, 2057 (1969) [*Sov. Phys. JETP* **29**, 1107 (1969)].

110. D. J. Thouless, The flow of a dense superfluid, *Ann. Phys.* **52**, 403 (1969).

111. G. V. Chester, Speculations on Bose–Einstein condensation and quantum crystals, *Phys. Rev. A* **2**, 256 (1970).

112. A. J. Leggett, Can a solid be "superfluid"?, *Phys. Rev. Lett.* **25**, 1543 (1970).

113. P. W. Anderson, A Gross–Pitaevskii treatment for supersolid helium, *Science* **324**, 631 (2009).

114. T. J. Greytak and D. Kleppner, Lectures on Spin-Polarized Hydrogen. In: *New Trends in Atomic Physics, Les Houches Summer School, Session XXXVIII, 1982* (eds. G. Grynberg and R. Stora), North-Holland, Amsterdam, the Netherlands, Vol. 2, p. 1125 (1984); I. F. Silvera and J. T. M. Walraven, Spin-polarized atomic hydrogen. In: *Progress in Low Temperature Physics* (ed. D. F. Brewer), North-Holland, Amsterdam, the Netherlands, Vol. 10, p. 139 (1986); T. J. Greytak, Prospects for Bose–Einstein condensation in magnetically trapped atomic hydrogen. In: *Bose–Einstein Condensation* (eds. A. Griffen, D. W. Snoke, and S. Stringari), Cambridge University Press, Cambridge, U.K., p. 131 (1995).

115. M. H. Anderson, J. R. Ensher, M. R. Matthews, C. E. Wieman, and E. A. Cornell, Observation of Bose–Einstein condensation in a dilute atomic vapor, *Science* **269**, 198 (1995).

116. C. C. Bradley, C. A. Sackett, J. J. Tollett, and R. G. Hulet, Evidence of Bose–Einstein condensation in an atomic gas with attractive interactions, *Phys. Rev. Lett.* **75**, 1687 (1995).

117. K. B. Davis, M.-O. Mewes, M. R. Andrews, N. J. van Druten, D. S. Durfee, D. M. Kurn, and W. Ketterle, Bose–Einstein condensation in a gas of sodium atoms, *Phys. Rev. Lett.* **75**, 3969 (1995).

118. E. Levich and V. Yakhot, Time development of coherent and superfluid properties in the course of a λ-transition, *J. Phys. A Math. Gen.* **11**, 2237 (1978).

CHAPTER 15

Superfluid States in Nature and the Laboratory

15.1   Helium-4

After the term "superfluid liquid" was introduced in 1938 (covered in detail in Chapter 14), it remained synonymous, at least in the experimental sense, with helium. And before the discovery of superfluidity in the fermionic isotope ^{3}He, the words "superfluid helium" referred to the bosonic isotope ^{4}He.

Due to its unique combination of light atomic mass and weak interatomic interaction, helium happens to be the only substance forming a liquid at zero temperature and pressure; the ground states of all other substances are solids.* The effective strength of the interatomic interaction increases with increasing density, causing crystallization of the system at high enough pressures (see the phase diagram in Figure 15.1). Speaking of the phase diagram of ^{4}He, one has to appreciate the role of Bose statistics, which proves to be crucial not only for the existence of the superfluid phase but also for the shape and even topology of the phase diagram. First-principles simulations of "distinguishable helium-4" (distinguishable particles equivalent to ^{4}He atoms in terms of their mass and pairwise interaction) show that, at finite temperatures $\sim 0.5\,\mathrm{K}$, the system forms a crystal even at the saturated vapor pressure, implying the phase diagram with two disconnected liquid regions, one at low and the other at high temperature [1].

The spectrum of elementary excitations in ^{4}He at low temperature is characterized by the phonon dispersion curve, $\epsilon(k)$ (see Figure 15.2). The linear long-wave part of the curve corresponds to hydrodynamic sound, $\epsilon(k) \to ck$, with c the sound velocity. The pronounced minimum at the wavevector k_0 of the order of the inverse interparticle distance† reflects the nonlocal character of interatomic interactions. In this connection, it is worth reminding that for a classical matter field, the phonon dispersion with a minimum at some wavevector k_0 can take place for extended density–density interaction potential with negative Fourier harmonics [see the discussion after Equation (1.159)]. What distinguishes the phonon spectrum of ^{4}He from any classical matter-field counterpart is the end point k_c beyond which phonons are no longer well defined. End points in the spectra of normal modes is an essentially quantum phenomenon that takes its origin in spontaneous decay of quasiparticles into two (or more) other elementary excitations with large density of states. The end point and a

---

* The molecular hydrogen crystal is a perfect illustration of the fact that, for the ground state to be a liquid, the weakness of interatomic/intermolecular interaction is as important as the smallness of the mass.

† Same as the characteristic range of the pair interaction potential.

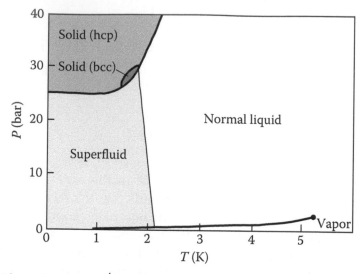

Figure 15.1   Phase diagram of ^{4}He. There are four qualitatively different phases: normal liquid/vapor (aka helium I), superfluid liquid (aka helium II), hcp solid, and bcc solid. Note that bcc solid occupies only in a small and essentially finite-temperature region of the phase diagram. The critical point of the superfluid transition (the so-called λ-point) at the saturated vapor pressure is $T_\lambda \approx 2.177\,$K. (Adapted from Reference [2].)

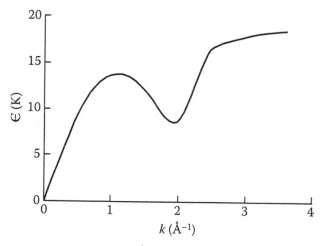

Figure 15.2   Phonon dispersion curve of ^{4}He. (Adapted from Reference [3].)

well-defined spectrum at $k < k_c$ are related features implying that the decay of elementary excitations is kinematically* suppressed. The shape of the dispersion curve in Figure 15.2 is consistent with the kinematic suppression of phonon decay (into a larger number of phonons), as well as with Pitaevskii's general theory [4] of end points. Physically, the end point corresponds to the threshold beyond

---

* In a strongly correlated system, a dynamic suppression at the momenta of the order of inverse interparticle distance is virtually improbable.

which the decay of an elementary excitation becomes kinematically possible and proceeds at a large rate.* On approach to the end point, the shape of the dispersion curve is strongly modified by virtual decay processes leading to vanishing group velocity at k_c; the decay width of the elementary excitation beyond the end point is on the order of its energy, so that one cannot talk of good elementary excitations, even approximately.

Obviously, phonons would decay into some other excitations, provided the latter existed within the kinematically appropriate domain of the energy-momentum space. And such a decay might seem possible in the previously mentioned "distinguishable ^{4}He," where there is a continuum of low-lying modes associated with ballistic motion of each individual atom. However, the permutation symmetry of the Hamiltonian excludes such a decay. Indeed, any permutationally symmetric state of "distinguishable ^{4}He"—and the phonon excitation is one of them—evolves *precisely* as the corresponding state of genuine ^{4}He, since the Hamiltonian preserves permutation symmetry of the state.† To avoid this loophole in a general argument, one can think of a system of distinguishable particles with different masses but the same average mass as that of ^{4}He. In this case, the phonons with $k \sim k_0$ acquire a decay width of the order of their energy and are no longer good elementary excitations. We conclude that the existence of stable phonons at momenta $k \sim k_0$ is deeply related to the indistinguishability of ^{4}He atoms—at least through the permutational symmetry of the Hamiltonian.

It is impossible to overestimate the role of ^{4}He in understanding the phenomenon of superfluidity. Experimentally, superfluidity itself and the vast majority of related effects were discovered with ^{4}He (see Chapter 14 for interesting details). Theoretically, ^{4}He is a strongly correlated bosonic system, the equilibrium properties of which are amenable to controllable *ab initio* analysis on the basis of Feynman path integral representation and efficient Monte Carlo simulations. The space-based experiment by Lipa et al. [5] on the superfluid transition in ^{4}He set the record in accurately measuring critical exponents, rendering the system important in a broader context of critical phenomena. In the area of superfluid turbulence, the issue of large inertial range characterized by the ratio of the system size to the vortex core radius is crucial: ^{4}He with this ratio at $>10^8$ has no analogs.

Apart from the bulk superfluid liquid phase, very important are ^{4}He-based lower-dimensional and/or locally inhomogeneous superfluid structures: films of ^{4}He atoms adsorbed on various surfaces,‡ helium in porous media (e.g., Vycor glass), superfluid dislocation cores in ^{4}He crystals, etc.

---

* In ^{4}He, the end point corresponds to the threshold at which the quasiparticle with momentum k_c can decay into two rotons with momentum $k_0 = k_c/2$ and energy $\epsilon(k_0) = \epsilon(k_c)/2$, which is consistent with the shape of the $\epsilon(k)$ curve in Figure 15.2.

† For that reason, the system is non-ergodic.

‡ Helium wets almost all known solid surfaces, cesium being a rare exception to the rule.

15.2   Helium-3

Helium-3 atoms are substantially lighter than helium-4 atoms. Given that the interatomic potential is essentially identical between the two isotopes, this ensures that ^{3}He remains liquid at the saturated vapor pressure (SVP) in the limit of $T \to 0$. The liquid-solid phase diagram of ^{3}He is similar to that of ^{4}He with one notable exception: In the low-temperature region, the first-order solid–liquid transition line noticeably bends down, favoring solid over the liquid phase as temperature is increased (see Figure 15.3). This is the manifestation of the Pomeranchuk effect: Atoms in the solid phase are tightly localized and, at temperatures above the nuclear spin ordering temperature, an entropy per particle in the solid is close to $s_S = \ln 2$. The entropy of the normal Fermi liquid in the limit $T \ll \epsilon_F$, where $\epsilon_F \sim 1\mathrm{K}$, is much smaller, $s_L = \left(\pi^2/2\right) T/\epsilon_F$. As a result, the free energy difference $F_S - F_L = (E_S - E_L) - NT (s_S - s_L)$ becomes negative with temperature even if we start from the liquid ground state.

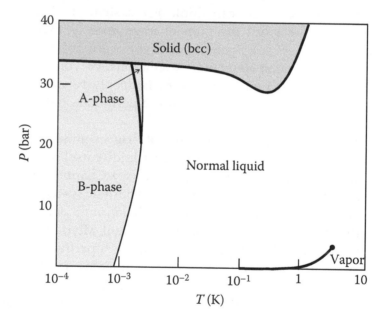

Figure 15.3   Phase diagram of ^{3}He in zero magnetic field with logarithmic temperature scale (see also Wheatley [6]). Notice the negative slope for the solid–liquid phase boundary at low temperature. We do not show here low-temperature magnetic transition lines within the solid phase. In zero magnetic field and low temperature, the transition between the normal and superfluid B-phase is second order. At pressures roughly above 20 bar, the second-order normal-superfluid transition is to the A-phases with subsequent first-order A–B transition (with large hysteresis under typical experimental conditions). In finite magnetic field, the size of the B-phase shrinks (until it disappears completely for fields above 0.5 T), while the A-phase occupies the rest of the superfluid domain and splits into the polarized A_1 and A_2 phases. The normal-superfluid transition is now to the A_1 phase, which forms a thin sliver near T_c. The value of the superfluid transition temperature to the B-phase at SVP is $T_B \approx 0.93\,\mathrm{mK}$ and to the A-phase at the melting curve $T_A = 2.491\,\mathrm{mK}$.

Even today, as hundreds of new superfluid/superconducting materials are discovered, ^{3}He remains the most prominent realization of the anisotropic superfluid featuring pairing correlations with nonzero angular momentum $L=1$ and spin $S=1$ (and the only one known that is not charged!). Its phase diagram is presented and discussed in Figure 15.3. The spectrum of elementary fermionic excitations in the B-phase characterized by zero total momentum of pairing correlations, $J = L + S = 0$, is fully gapped, and the value of the gap is the same over the Fermi surface. In the A-phase, pairing occurs with parallel orientation of spins and orbital momenta. The gap in the spectrum now has two nodes at the "poles" of the Fermi surface. As magnetic field is applied, S and L are polarized in the field direction leading to the A_1-phase. The physics of anisotropic states in a neutral superfluid is extremely rich, and even brief enumeration of exciting topics would be outside of the scope of our book. There are dozens of collective excitations, numerous topological defects with exotic excitations leaving on them, textures resulting from the order parameter orientation by vessel boundaries and external fields, interplay between the superfluid order parameter and nuclear magnetism, intricate kinetic phenomena within the phases and boundaries between them, etc. (see Reference [7]).

15.3   Dilute Trapped Ultracold Gases

A classical complex-valued matter field—the minimalistic model for superfluidity—can be experimentally realized with metastable dilute ultracold gases. Due to the weakness of effective interaction, highly occupied single-particle modes—and, in particular, Bose–Einstein condensations (BECs)—of dilute bosonic gases behave like harmonics of a classical matter field. Dynamically, they obey the Gross–Pitaevskii equation in the most typical case of short-ranged interatomic potentials.

Ultracold atomic gases are not stable. They decay by elementary processes of recombination into molecules. Kinematically, the recombination in the bulk involves at least three atoms, while the relaxation toward thermal equilibrium is due to pair collisions. Hence, in a dilute gas—when the interparticle distance is much larger than the characteristic size of interaction potential—the recombination is very slow compared to thermalization, which is a key circumstance behind cooling the system below the condensation temperature by evaporative protocols. Apart from being dilute, the ultracold gases must be detached from the wall to prevent adsorption-assisted recombination. That is the reason Bose–Einstein condensation of ultracold gases is achieved in (magnetic and/or optical) traps.

The behavior of single-particle modes with occupation numbers smaller than, or on the order of, unity is essentially quantum. The theory of Chapter 12 yields an accurate quantum-statistical description of equilibrium properties of dilute gases. If one neglects finite lifetime effects,[*] then answers obtained for dilute Bose gas and expressed in terms of 3D s-scattering length directly apply to metastable

---

[*] Their analysis goes beyond the scope of the book.

Figure 15.4   Interference fringes produced by two expanding condensates. (Adapted from Reference [8].)

ultracold gases. In particular, this is true for quasi-2D and quasi-1D systems, in which the effective 2D and 1D interaction depends on the s-scattering length of the underlying 3D system.

A remark is in order concerning the "ultracoldness" of those systems. The quantum-statistical degeneracy of a 3D gas of bosons or fermions develops at temperatures $\sim\hbar^2 n^{2/3}/m$ (with n the number density and m the particle mass) that are many orders of magnitude smaller—because of very small n—than, say, the temperature of the superfluid transition in ^{4}He. Hence, on an absolute scale, the degenerate dilute systems are indeed very cold. However, comparing ultracold gases with ^{4}He (or any other natural quantum liquid for that matter) in terms of the absolute value of temperature makes little physical sense. As a crucial condition for their existence, ultracold gases cannot have a direct contact with "condensed-matter world," and for that reason, their absolute temperatures are of academic interest only.

A simple direct manifestation of the classical-field behavior of ultracold gases is the interference pattern formed by two expanding atomic clouds produced by releasing two BECs from their traps (see Figure 15.4). The interference visualizes the fundamental fact that expanding condensate is a coherent classical matter wave.*

The crucial phenomenon of vorticity quantization in superfluids is also nicely visualized with BECs of ultracold atoms where one can create and observe (with absorption imaging techniques) one or a few vortices [9,10], as well as a lattice of vortices in a rotating condensate (like the one shown in Figure 1.2).

A more complicated classical-field regime takes place in the fluctuation region of the superfluid transition. As discussed in Chapter 12, the classical component of the quantum field is here strongly turbulent and must be described by nonperturbative methods. The experiments with ultracold atoms in two [11,12] and three

---

* For that reason, the expanding condensate setup is often referred to as *atom laser*.

[13] dimensions demonstrate excellent quantitative agreement with the classical-field theory of the fluctuation region.

A sophisticated multistage turbulent classical-field behavior is at the heart of strongly nonequilibrium kinetics of BEC formation in a dilute gas (as discussed in Chapter 13). The experiment with ultracold atoms reported in Reference [14], along with its direct numeric simulation by Gross–Pitaevskii equation with an appropriate turbulent initial condition, yields a strong evidence for the validity of the kinetic scenario presented in Chapter 13.

Often, trapped ultracold gases are in the so-called Knudsen regime: The mean free path for elementary excitations is significantly larger than the system size (this is not true for the strongly turbulent classical-field part in the fluctuation region where the mean free path is on the order of the healing length). Under these conditions, releasing the system from a trap results in a characteristic bimodal structure of the expanding atomic cloud (see Figure 15.5). The noncondensate particles expand in a semiclassical regime, as if they were *classical particles* distributed over momenta in accordance with the (spatially averaged) Bose–Einstein statistics. The parameter controlling the accuracy of the semiclassical particle regime for the noncondensate fraction is $\lambda_T/L \ll 1$, where λ_T is the thermal de Broglie wavelength and L is the system size. The very same parameter guarantees the applicability of local density approximation for noncondensate fraction in the trap. The local

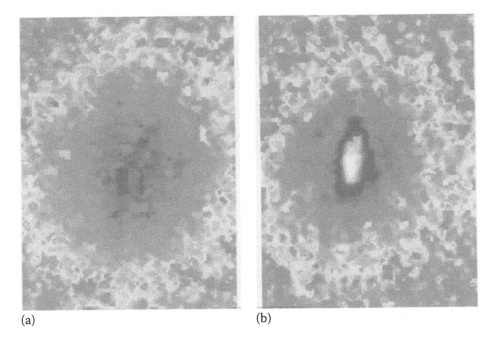

(a)   (b)

Figure 15.5 Absorption image of expanding atomic cloud. (a) Slightly above the critical temperature. (b) Slightly below the critical temperature. The condensed and noncondensed fractions result in a bimodal distribution of density. Adapted from the pioneering Reference [15]; note that the resolution of absorption images has increased significantly since then.

density approximation implies the isotropy of momentum distribution of noncondensate particles and thus the asymptotic isotropy of the shape of the expanding cloud of noncondensate particles.

The condensate released from a trap expands coherently, following the Gross–Pitaevskii equation. The expansion of the condensate is much slower than that of the noncondensate part because of the dramatic difference in characteristic momenta of the condensate and noncondensate particles (at temperatures high enough to guarantee an appreciable fraction of noncondensate particles). This explains the bimodality of the density distribution.

The expanding condensate fraction is isotropic only if the initial state is isotropic: that is, if the condensate is released from an isotropic trap. Along with the bimodality of the absorption image, the anisotropy of the central part of the image is a very characteristic feature indicative of the presence of condensate.

An "instructive confusion" may occur with regard to the classification (condensate or noncondensate) of particles forming highly occupied excited single-particle modes in the trap. In view of large occupation numbers, these particles can be attributed (at least at the initial stages of expansion) to the classical-field part of the system. In this connection, it is worth noting that the semiclassical particle picture does not exclude the classical-field description (mathematically, cf. the semiclassical regime of a single-particle Schrödinger equation). The necessary and sufficient condition for the (weakly interacting) classical field to demonstrate asymptotic semiclassical particle expansion is $kL \gg 1$, where k is the typical wavevector of the mode.

In accordance with the relation $U = 4\pi\hbar^2 a/m$, the sign of the coupling constant U in the Gross–Pitaevskii equation describing the classical-field regime of a dilute degenerate bosonic gas is dictated by the sign of the s-scattering length a. If the scattering length is negative (as, e.g., in the case of lithium-7), the coupling is attractive, and the uniform ground state is unstable. As is well known in nonlinear physics, the self-evolution of nonlinear Schröedinger equation with attractive coupling ultimately leads to singularities (collapses). Those collapses were observed for lithium-7 [16].

15.4   Resonant Fermions: From Ultracold Atoms to Neutron Stars

As we discussed previously, taking a dilute gas limit (large interparticle distance compared to the range of interaction) typically implies a weak effective interaction. An exception occurs in the case of strong *resonant interaction*. In a many-body system with the number density n, the interaction is called resonant if the following two conditions are satisfied: First,

$$|a| \gg R_0, \qquad n^{-1/3} \gg R_0 \qquad \text{(resonant interaction)}, \qquad (15.1)$$

where a is the s-scattering length and R_0 is the typical range of pairwise interaction. The second condition is a standard one for dilute gases with respect to the interaction range. The first condition is very specific since it requires fine-tuning of the interaction potential.[*] The following parameter (below k_F is the Fermi momentum,[†] and $\hbar = 1$)

$$\xi = ak_F \sim an^{1/3}, \tag{15.2}$$

characterizes now the strength of effective interaction (cf. the theory of weakly interacting Bose gas)—at $|\xi| \ll 1$, we are dealing with the weak-coupling case. At $|\xi| \gtrsim 1$, the effective interaction is strong, and the many-body correlations play a crucial part. The condition $n^{-1/3} \gg R_0$ guarantees that the details of scattering potential are not important and, up to subleading corrections, the state of the system depends only on four dimensionless parameters: the parameter ξ, the ratios of masses and number densities of the two components, and the temperature in units of Fermi energy, T/ϵ_F. Neglecting nonuniversal subleading terms, the behavior of the two-component resonant fermions is captured by a concise theoretical model in which the role of resonant interaction is played by a boundary condition on the many-body wavefunction when coordinates of any two particles of the opposite spin[‡] approach each other:

$$\Psi\left(\mathbf{r}_{\uparrow 1}, \ldots, \mathbf{r}_{\uparrow N_\uparrow}; \mathbf{r}_{\downarrow 1}, \ldots, \mathbf{r}_{\downarrow N_\downarrow}\right) \to \frac{A}{|\mathbf{r}_{\uparrow i} - \mathbf{r}_{\downarrow j}|} + B \quad \text{at} \quad |\mathbf{r}_{\uparrow i} - \mathbf{r}_{\downarrow j}| \to 0, \tag{15.3}$$

$$A/B = -a. \tag{15.4}$$

Here, Ψ is the wave function of the system containing $N_\uparrow$ "spin-up" and $N_\downarrow$ "spin-down" particles; $\mathbf{r}_{\uparrow i}$ and $\mathbf{r}_{\downarrow j}$ are, respectively, coordinates of the ith spin-up and jth spin-down particles; A and B depend on all coordinates but $\mathbf{r}_{\uparrow i}$ and $\mathbf{r}_{\downarrow j}$ (their ratio, however, being coordinate independent and equal to the negative scattering length).

The phase diagram of the model (15.3) and (15.4)—in the space of its four control parameters—is predicted to be rather rich (for a comprehensive discussion, see [17–19]), containing, in particular, Fulde–Ferell–Larkin–Ovchinnikov superfluid phase with broken translational symmetry.

In the case of equal masses and component densities, of fundamental interest is the Bardeen–Cooper–Schrieffer (BCS)–BEC crossover problem—the crossover from the weakly interacting BCS regime corresponding to small negative ξ, to the regime of BEC of weakly interacting molecules taking place at small positive ξ.[§] The $a = \infty$ case corresponds to the unitarity regime. In a two-body problem, the divergence of a takes place at the threshold corresponding to the appearance of a new bound state. In a macroscopic system, there are no singularities at the unitarity point, and all the thermodynamic functions are analytic if parameterized in terms of the variable $1/\xi$. This is understandable from Equations (15.3)

---

[*] In a generic case, we would have $|a| \lesssim R_0$.

[†] Since we focus on fermions, it is convenient to associate the typical momentum $(\sim n^{1/3})$ and energy $(\sim n^{2/3}/m)$ scales with Fermi momentum and Fermi energy, respectively.

[‡] Our pseudospin notation does not necessarily imply equality of the masses of the two components.

[§] The s-scattering length for molecules in the BEC regime is $\approx 0.6a$ [20].

and (15.4). What happens at unitarity is that the (subleading) term B in the small-distance asymptotic expansion vanishes. This leads to special properties of thermodynamic functions at $a \rightarrow \infty$ because they depend only on the ratio T/ϵ_F similarly to an ideal Fermi gas, while the physics itself is dramatically different due to strong correlations and the superfluid transition.

Of special interest is the problem of stability of model (15.3) and (15.4). Note that at $|\xi| \sim 1$ (i.e., in the most intriguing strongly correlated regime), the instability also implies the lack of *meta* stability, because at $|\xi| \sim 1$, there is no small parameter by which the decay time could be much smaller than the equilibration time. It is known that the bosonic version of (15.3) and (15.4) is unstable due to the Efimov effect—the existence of unlimited-from-below spectrum of three-body bound states, implying, in particular, the absence of the ground state in a system of more than two bosons (without u-v regularization). Similarly, if a system of three or more fermions features the spectrum of Efimov states, the corresponding macroscopic system is unstable, and vice versa. Thus, the necessary condition for stability of a macroscopic system is the absence of Efimov effect for any finite number of particles. In practice, the stability of the resonant Fermi gas is addressed experimentally or by controllable numerical solutions of few-body problems.

Experimentally,[*] the system of resonant fermions is realized with ultracold gases using the technique of Feshbach resonance.[†] Thanks to high the controllability of the effective model and accuracy of measurements, the system of resonant ultracold fermions can be viewed as a quantum emulator of the theoretical model (15.3) and (15.4). The emulator can, in particular, be used to test *ab initio* numeric techniques.[‡] The collection of all numeric and experimental results for the critical temperature at unitarity can be summarized (with a conservative error bar) as $T_c/\epsilon_F = 0.160(2)$.

Interestingly enough, the unitarity regime is approximately realized in the inner crust of neutron stars, where the neutron–neutron scattering length is nearly an order of magnitude larger than the mean interparticle separation. The key circumstance here is that the nuclear force is fine-tuned by nature in such a way that the s-scattering amplitude is about an order of magnitude larger than the interaction range.

15.5   Electronic Superconductors

Mercury was the first discovered superconducting material [23] with transition temperature around 4 K. Subsequently, many pure elements such as lead and tin also were found to have a superconducting state. Most of them are type-1 superconductors with some notable exceptions: for example, niobium and vanadium

---

[*] See, for example, Ku et al. [21] who report the first observation of superfluid transition in the system of resonant fermions.

[†] Feshbach resonance is achieved by tuning the external magnetic field in such a way that the energy of one of the compact (tightly bound) molecular states is close to zero.

[‡] See, for example, Van Houcke et al. [22] where the technique of diagrammatic Monte Carlo was experimentally validated.

are type-2 superconductors. Over the years, the list of superconducting systems has greatly expanded and now includes numerous alloy materials that became the first well-studied type-2 superconductors.

The first superconductors, for example, simple elements, had relatively low critical temperatures (below several Kelvin) and their basic properties were well described at the microscopic level by the BCS theory [24]. The key parameter in this mean-field-type theory is the gap Δ in the spectrum of elementary excitations at the Fermi surface, which defines the energy gain due to formation of electronic Cooper pairs. The BCS theory relates Δ to a number of experimentally observed phenomena such as the specific heat jump at T_c and the exponential low-temperature law $C(T) \propto \exp(-\Delta/T)$, signaling that it takes an energy Δ to excite the superconducting ground state. Similar gap-related characteristics appear in many other measurements such as tunneling, acoustic attenuation, and magnetic susceptibility. Also, for simple low-T_c superconductors, the critical temperature is directly related to the energy of breaking Cooper pairs,

$$r \equiv \frac{\Delta(T = 0)}{T_c} = 1.76. \tag{15.5}$$

This ratio is in excellent agreement with experimental results for Al ($r = 1.7$), Sn ($r = 1.8$), Zn, Ta, In ($r = 1.6$). As discussed previously, the BCS theory describes the weak-coupling limit of the superconducting state, and this allows one to neglect beyond-the-mean-field fluctuation effects. In a more general situation, this is no longer the case. Deviations from the BCS theory become visible in superconductors with stronger coupling or multiple gaps.

For three-quarters of a century, the discovered superconductors were weak-coupling single-gap materials with transition temperatures below 30K despite numerous theoretical ideas and enormous experimental efforts to increase T_c. Finally, in 1986, Bednorz and Müller discovered what is called now "high-T_c superconductivity" in cuprates [25]. The fist cuprate compound was La-Ba-Cu-O with $T_c = 30$K. Subsequently, other cuprate materials were found to have significantly higher transition temperatures, above 100K. These materials can be broadly classified as d-wave type-2 superconductors. The microscopic physics of these compounds is distinct from that of the BCS theory for simple elements and is still not fully understood.

The first clear-cut example of a superconductor that is described by a generalized BCS theory with multiple energy gaps $\Delta_{1,2}$ (i.e., a two-band superconductor) is MgB_2, which was discovered in 2001. It has two well-defined superconducting components with the interband Josephson coupling setting the preferential value for their phase difference: $\theta_1 - \theta_2 = 0$. Experimental evidence points to a non-trivial phase difference dynamics [26] in this material. And the set of discovered multiband materials keeps growing. There is evidence that some of the recently discovered iron-based superconductors with $T_c > 50$K are also best described as multiband superconductors but in this case, the interband Josephson coupling sets the preferential phase difference to π: $\theta_1 - \theta_2 = \pi$. Some of the recently discovered superconductors, such as Sr_2RuO_4 or some of the iron-based materials [27], have more than two superconducting bands. It seems that multiband superconductivity is more of a typical phenomenon than an exception.

Numerous theoretical studies over the years explored superconductors with higher broken symmetries that support a plethora of topological defects other than vortices. Let us mention here a few examples: As discussed in Chapter 6, additional discrete symmetry can be broken in superconductors with more than three bands and frustrated interband interactions. Candidates for such states are iron-based superconductors. Other candidates are superconductors with non-s-wave pairing. The best studied candidate for a superconductor with spin-triplet Cooper pairs is currently Sr_2RuO_4. Support for this point of view comes from a variety of measurements that include the complete suppression of the superconducting transition temperature with nonmagnetic impurities, nuclear magnetic resonance (NMR) Knight shift measurements that show no change in the spin susceptibility with temperature in the superconducting phase, muon spin measurements that suggest broken time-reversal symmetry in the superconducting state, polar Kerr effect showing the broken time-reversal symmetry and the presence of chirality in the superconducting phase, and phase-sensitive measurements consistent with a chiral spin-triplet state. However, the case for a chiral spin-triplet superconducting phase in Sr_2RuO_4 is not ironclad yet. In particular, detectable edge currents are expected to exist at sample boundaries and at domain walls between domains of opposite chirality. A search for these currents has been carried out and none have not been observed. In addition, there is active search for superconductors that would exhibit the so-called Fulde–Ferrell–Larkin–Ovchinnikov state that breaks both $U(1)$ and translation symmetry.

15.6 Superconductivity of Nucleons: Finite Nuclei, Neutron Stars, and Liquid Metallic Hydrogen

15.6.1 Finite Nuclei

In 1958, soon after the publication of the BCS theory, it was suggested that, in finite nuclei, nucleons (i.e., neutrons and protons) also form Cooper pairs and their superfluidity may be revealed through the reduced moment of inertia (see the works by Bohr et al. [28], Belyaev [29], and Migdal [30]). The main experimental evidence for this proposal is found in spectra of even–even neutron/protons nuclei that show a gap absent in odd systems. The moment of inertia of open-shell nuclei is nonclassical, which is consistent with the existence of the irrotational superfluid component that does not participate in rotation (Belyaev [29], and Migdal [30]). There are also neutron-pair transfer reactions similar to the Josephson effect.

15.6.2 Neutron Stars

Another example of superfluid nuclear matter is found in neutron stars, where a large part of the interior consists of a neutron liquid with a small admixture of protons and electrons. One then expects that neutrons form Cooper pairs, as discussed in previous sections of this chapter (this was first proposed by Migdal in 1959 [30,31]), but the same might be also true for protons. There are also proposals for the existence of other kinds of Cooper pairs such as those made out of Σ^- hyperons. If true, the interior of a neutron star features a mixture of superconducting components based on different particles.

It was observed by Ginzburg and Kirzhnits in 1964 [32] that superfluid neutrons should form vortex lattices. Superconducting properties of protons were discussed in 1969 by Baym et al. [33], who found that magnetic fields of neutron stars are substantially smaller than critical magnetic fields. However, after a star cools below the critical temperature of the superconducting transition, an estimate of a time scale necessary to expel the magnetic fields shows that it is extremely long. This suggests the presence of a metastable tangle of superconducting vortices. Initial estimates predicted that protons should form a type-2 superconductor, but more recent arguments advance an idea that various components in neutron stars form type-1 and type-1.5 superconductors. There are theories postulating that observable dynamics of pulsars could be explained through the dynamics of quantum vortex matter in their interior.

15.6.3 Liquid Metallic Hydrogen

For nearly a century, it has been expected that, under high-pressure conditions, hydrogen will become a metal. In 1935, Wigner and Huntington put forward a semiquantitative prediction that at zero temperature, hydrogen should become a metal at 25 GPa [34]. Since then, this pressure estimate has been revised several times. The latest estimates suggest very high, but in principle experimentally accessible, metalization pressure at around 400 GPa (in the low-temperature limit). With new breakthroughs in high-pressure experimental techniques, the metallic state of hydrogen is currently a subject of vigorous experimental pursuit [35]. In 1968, Ashcroft predicted that solid metallic hydrogen should become a superconductor with an exceptionally high T_c [36]. Moreover, due to a low mass of protons, it has also been suggested that quantum fluctuations could favor a liquid ground state [37]. Similar arguments were applied to other hydrogen isotopes such as deuterium, which is a spin-1 boson, as well as to the proton sublattice melting in hydrogen-rich alloys [38].

It has been discussed that liquid metallic states of hydrogen, its isotopes, their mixtures, as well as hydrogen-rich alloys may allow coexistent Cooper pairs of protons and electrons or Bose condensates of charged spin-1 deuteron coexistent with Cooper pairs of electrons. In contrast to multiband superconductors, the coexistent condensates here will not feature intercomponent Josephson coupling, because one cannot convert electrons to, say, protons. Yet the components should necessarily be coupled electromagnetically via common coupling to the vector potential **A**. As a result, such a system allows new aggregate "super" states: metallic superfluid and superconducting superfluid. For example, if we consider coexistence of electronic and protonic superconductivity, the superflow will be associated with co flow of electronic $|\Psi_e|\exp(i\theta_e)$, protonic $|\Psi_p|\exp(i\theta_p)$, and deuteronic $|\Psi_d|\exp(i\theta_d)$ components, without net charge transfer, while superconductivity will correspond to their counterflow. As discussed in Chapter 6, such flow will be endowed with a number of exotic properties. The metallic superfluid state arises due to the fact that long-range electromagnetic coupling between components can cause a phase transition to a state where individually the condensate's phases are disordered, while their phase sum or phase difference is ordered. For example, for a mixture of electronic and protonic superconducting components, this state corresponds to the situation where $\langle \Psi_e \rangle = 0$ and $\langle \Psi_p \rangle = 0$, while $\langle \Psi_p \Psi_e \rangle \neq 0$. In that case, only the coflow of electronic and protonic components is dissipationless. On the other hand, for a mixture of protonic Cooper condensate and deuteronic BEC,[*] the metallic superfluid state corresponds to the situation where $\langle \Psi_d \rangle = 0$ and $\langle \Psi_p \rangle = 0$, while $\langle \Psi_p^* \Psi_d^2 \rangle \neq 0$ (i.e., the order is only in the phase difference $\theta_p - 2\theta_d$). In that case, the only dissipationless flow corresponds to counter propagation of protonic and deuteronic components [39,40].

15.7   Color Superconductivity of Quarks

In 1965, an idea has been put forward that matter at densities higher than that of a "neutron–proton" superfluid away from the core of a neutron star will be in yet another superfluid state. That is, at extreme pressure, which may correspond to that in the core of a neutron star, and low temperature, a new state is possible: a degenerate liquid of quarks (Ivanenko and Kurdgelaidze [41]). In such a liquid, quarks should form Cooper pairs near the Fermi surface (Ivanenko and Kurdgelaidze [42,43]), resulting in the superconducting quark state that breaks the color gauge symmetry. The phase was termed "color superconductivity." Color superconductivity has recently been intensively studied theoretically (for a review, see Reference [44]).

15.8   Stable, Metastable, and Unstable Elementary Excitations

Under certain conditions, excited normal modes of a classical nonlinear field, as well as their direct quantum-field counterparts—bosonic elementary excitations, such as magnons, excitons, and exciton-polaritons [45–49]—can form a BEC. Speaking academically, we could exclude from the discussion the case of genuine thermodynamic equilibrium. Indeed, genuine equilibrium BEC of elementary excitations requires that their total number[†] be a constant of motion associated with certain exact U(1) symmetry of the system's Hamiltonian, which is broken by the formation of the condensate. In this case, however, it seems more natural to associate the constant of motion with the system as a whole

---

[*] The charges of both condensates have the same sign; the values of the charges differ by a factor of 2.

[†] Or its classical-field equivalent if we are talking of the normal modes of a classical field.

and speak of the broken symmetry of the ground state rather than the BEC of elementary excitations. Consider, for example, the supercounterfluidity (counter propagating superfluid currents) in a two-component system and its equivalent—supertransport of magnetic polarization in a system with conserved magnetization. Here, magnetic and supercounterfluid languages are totally equivalent, and neither of them requires invoking the notion of BEC of elementary excitations, magnons. The (conserved) amount of magnons in this case is nothing else but the difference between the particle numbers of two components, each of which is a constant of motion on its own. Rather than invoking the notion of elementary excitations, one can speak of the ground state with a broken symmetry, the latter being expressed by anomalous averages [Equation (15.7)]. An exception worth making would be only for the dilute limit of stable excitations (see, e.g., [50,51]), since this case implies extra universality and simplicity of description.

For excitations with a finite lifetime, conceptually, the simplest is the metastable setup when the relaxation of the total number of quasiparticles—initially created by pumping—is much slower than the thermalization. This is a direct analog of BEC of metastable ultracold atomic gases. Recalling that BEC is a classical-field phenomenon, it is clear that it has to emerge from the relaxation of classical normal modes (due to nonlinearities) provided there exists a well-defined (i.e., almost conserved) quantity equivalent to the total number of elementary excitations. If the classical field features the UV catastrophe, then the evolution of the metastable subsystem of its excited normal modes will always result in the formation of BEC since the evolution leads to vanishing temperature.

As a characteristic example, consider a classical ferromagnetic model dealing with a local magnetization vector $\mathbf{M}(\mathbf{r}, t)$ of a fixed length: $|\mathbf{M}(\mathbf{r}, t)| = M_0 = \text{const.}$ In the presence of a uniformly directed—but not necessarily constant—magnetic field, the ground state is $\mathbf{M}(\mathbf{r}, t) \equiv \mathbf{M}_0$, where $\mathbf{M}_0$ is a constant vector pointing along the magnetic field direction. With this ground state, the normal modes of the system, magnons, are the precessing waves of $\mathbf{M}(\mathbf{r}, t)$ about its ground-state value $\mathbf{M}_0$. Since the absolute value of $\mathbf{M}$ is fixed, its precession can be unambiguously parameterized with a complex field $\psi(\mathbf{r}, t)$, the real and imaginary parts of which represent Cartesian components of $\mathbf{M}(\mathbf{r}, t)$ in the plane perpendicular to $\mathbf{M}_0$. In this parameterization, the rotation symmetry of the problem with respect to the direction of magnetic field corresponds to the global U(1) symmetry in terms of the field ψ. In many practical cases, the relaxation of the "field amount"

$$N = \int |\psi|^2 \, d\mathbf{r} \tag{15.6}$$

is much slower than any other relaxation process in the system, so that N can be interpreted as an approximately conserved number of magnons. Given this (approximate) conservation law, the self-evolution of the field ψ should lead to the formation of BEC of magnons. In terms of the field ψ, this BEC represents a finite (in the macroscopic limit) fraction of the field amount, which has a uniformly evolving phase. Translating this back into the language of $\mathbf{M}(\mathbf{r}, t)$, the magnon condensate describes a finite fraction of magnetization directed perpendicular to $\mathbf{M}_0$.

The uniform evolution of the condensate phase then translates into the uniform rotation of the condensed magnetization fraction.

Let the initial magnon number N be produced by pumping. If pumping is spatially uniform, then the magnon condensate is directly created by pumping. This trivial option is a generic way to create BECs based on normal modes and/or elementary excitations. If pumping is incoherent—that is, a macroscopic number of magnon modes with different wavevectors is being excited simultaneously—while the amplitude of the zero-momentum mode remaining macroscopically small, the problem of subsequent relaxation becomes physically interesting: The system can develop BEC spontaneously before N relaxes. Complementary to the case of incoherent pumping of a uniform system is the case of uniform pumping of a system with quenched local disorder. For our purposes, pumping can be reduced to a quench-tilting of the magnetization field, which initially creates a uniform BEC of magnons. However, since the system is inhomogeneous, this uniform condensate has little to do with the spatially nonuniform quasi-equilibrium condensate that is supposed to minimize the energy (at a fixed amount of magnons) in a given disordered sample. The evolution can first lead to a destruction of the initial uniform condensate and subsequent formation of the inhomogeneous one. This situation is typical for experiments with magnon BECs [52].

By continuity, the existence of conditions under which BEC forms in a metastable (quasi-equilibrium) setup implies the possibility of achieving BEC in a strongly nonequilibrium steady-state regime, when statistical properties of the system have little to do with the Gibbs ensemble. To achieve this (rather instructive from a general point of view) regime, the system should be permanently and strongly pumped with elementary excitations, while the energy is simultaneously extracted out by some mechanism. To form a BEC, the pumping-to-extracting ratio must exceed a certain critical value, an analog of critical temperature in the equilibrium case.

For an essentially dissipative nonequilibrium steady-state system with BEC (or topological order in 2D), a question arises as to whether the system can support a superflow or persistent current. A positive answer requires a definition of superflow more subtle than a flow free of energy dissipation, because dissipation is inherent in a steady-state regime. One might thus try to associate superflow with the existence of persistent currents. However, in a dissipative system, a persistent current can occur as a result of dynamical symmetry breaking not associated with any preformed long-range order.* Nevertheless, recalling the theory of equilibrium supercurrent states, a precise definition of superflow in a dissipative system can be given based on the concept of an emergent constant of motion. We define a supercurrent state (in a torus) as a state characterized by conserved winding number of the phase of the complex field associated with BEC. With this definition, the term supercurrent reflects the stability of the state with respect to changes in external conditions (such as the pumping rate, the rate of energy drain, and amplitudes of external fields). In this connection, we would like to mention an experiment

---

* Benar cells are a good example of a macroscopic steady-state flow pattern.

with ^{4}He film [53] that demonstrated conservation of velocity circulation despite radical changes of the film thickness by evaporating/condensing ^{4}He vapor.

We want to emphasize here again that the most fundamental and universal property of superfluids and superconductors is the topological long-range order: that is, the absence of topological defects in the phase of the (appropriately coarse-grained) complex order parameter. *Any* other related property, such as the quantization of velocity circulation, quantization of magnetic flux, frictionless flow, or a genuine long-range order (i.e., genuine BEC), can be absent under certain specific conditions.

15.9 Superfluid States in Optical-Lattice Emulators

In free space, trapped ultracold atoms are dilute and for that reason are almost ideal (with an exception of long-range or resonant interaction). The physical situation changes dramatically if there is a sufficiently strong periodic external potential, with the period on the order of the interparticle distance. In the experiment with ultracold atoms, such a potential is created by standing waves of near-resonant light [54], and for that reason, it is referred to as *optical lattice*. A standard setup in three dimensions is a simple cubic lattice formed by three mutually perpendicular standing waves, the lattice period being equal to one half of the wavelength. In 2D, lattices can be created by design with the holographic techniques.

In an appropriately strong optical-lattice field, the motion of a particle reduces to weak intersite tunneling, with a single-band Hubbard model as an accurate effective Hamiltonian [55]. The intersite tunneling (hopping) amplitude naturally sets the low energy scale for the problem. In the units of hopping amplitude, the effective interaction (typically on-site in nature with notable exceptions for dipole, resonant, and Rydberg atom systems) can be rendered arbitrarily large by simply changing the depth of the optical potential. This simple control over the interaction strength in combination with other techniques allows one to create a wide class of generalized Hubbard-type Hamiltonians,* which, by design, render the ultracold optical-lattice systems unique for studying different phases of strongly correlated quantum matter. In this context, one often uses the term *optical-lattice emulator* (OLE), which emphasizes the following strategy: Based on existing theoretical understanding—even at the level of a hypothesis—of a certain phenomenon, one identifies a lattice model capturing the physics of interest. The behavior of the model is then emulated with the corresponding ultracold optical-lattice system having the same effective Hamiltonian. With OLEs, one can, in principle, get accurate answers for fundamental theoretical models that are not amenable to reliable analytic or numeric analysis. Clearly, the potential scope of OLEs goes far beyond the problems of superfluidity. Nevertheless, one of the most anticipated OLE tasks is the search for high-temperature superfluid states of the fermionic Hubbard model that, in accordance with numerous theoretical

---

* For example, with an artificial static vector potential.

expectations, is relevant to the problem of high-temperature superconductivity. The key question is whether the fermionic Hubbard model has a region in the interaction-density parameter plane with high temperature for the superfluid transition. The subtlety here is that the Cooper instability in the Hubbard model turns out to be rather weak. This implies exponentially small critical temperatures. Hence, the highest critical temperature in the fermionic Hubbard model is expected to be still much smaller than the hopping amplitude, in which case it cannot be assessed by dimensional analysis. Only *ab initio* methods—controllable numeric approaches or OLEs—can yield an accurate answer.

Another interesting class of superfluid states that can be studied with OLE are the counterflow states (discussed in Chapter 6) arising in a multicomponent bosonic or fermionic lattice systems with integer net filling factors [56–59] and characterized by the following order parameters:

$$\left\langle a_{s\alpha}^{\dagger} a_{s\beta} \right\rangle \neq 0 \qquad (\alpha \neq \beta), \qquad (15.7)$$

where $a_{s\alpha}^{\dagger} \left(a_{s\beta} \right)$ creates (annihilates) a particle of the sort $\alpha(\beta)$ on the site s. Along with the integer net filling, the conditions for the supercounterfluidity imply strong enough interaction to suppress the single-particle supertransport* via Mott physics, leaving exchanges of neighboring particles of two different kinds as the leading mechanism of supertransport. In a two-component case, the supercounterfluid state is equivalent to an easy-plane (anti-) ferromagnet [56–58]. In a generic multicomponent case, the best way of thinking of a supercounterfluid, suggested by Equation (15.7), is that the state is a condensate of pairs formed by particles of one component and holes of the other component.[†]

References

1. M. Boninsegni, L. Pollet, N. Prokof'ev, and B. Svistunov, Role of Bose statistics in crystallization and quantum jamming, *Phys. Rev. Lett.* **109**, 025302 (2012).

2. J. S. Brooks and R. J. Donnelly, The calculated thermodynamic properties of superfluid helium-4, *J. Phys. Chem. Ref. Data* **6**, 51 (1977).

3. R. J. Donnelly, J. A. Donnelly and R. N. Hills, Specific heat and dispersion curve for helium II, *J. Low Temp. Phys.* **44**, 471 (1981).

4. L. P. Pitaevskii, Properties of the spectrum of elementary excitations near the disintegration threshold of the excitations, *Sov. Phys. JETP* **9**, 830 (1959).

5. J. A. Lipa, J. A. Nissen, D. A. Stricker, D. R. Swanson, and T. C. P. Chui, Specific heat of liquid helium in zero gravity very near the lambda point, *Phys. Rev. B* **68**, 174518 (2003).

---

* That is to guarantee that $\langle a_{s\alpha} \rangle \equiv 0$.

† Note that Mott localization in the net particle channel renders the notion of a hole accurate.

6. J. C. Wheatley, Experimental properties of superfluid $_3$He, *Rev. Mod. Phys.* **47**, 415 (1975).

7. G. E. Volovik, *Exotic Properties of Superfluid Helium 3*, World Scientific, Singapore (1992).

8. M. R. Andrews, C. G. Townsend, H.-J. Miesner, D. S. Durfee, D. M. Kurn, and W. Ketterle, Observation of interference between two Bose condensates, *Science* **275**, 637 (1997).

9. M. R. Matthews, B. P. Anderson, P. C. Haljan, D. S. Hall, C. E. Wieman, and E. A. Cornell, Vortices in a Bose–Einstein condensate, *Phys. Rev. Lett.* **83**, 2498 (1999).

10. K. W. Madison, F. Chevy, W. Wohlleben, and J. Dalibard, Vortex formation in a stirred Bose–Einstein condensate, *Phys. Rev. Lett.* **84**, 806 (2000).

11. C.-L. Hung, X. Zhang, N. Gemelke, and C. Chin, Observation of scale invariance and universality in two-dimensional Bose gases, *Nature* **470**, 236 (2011).

12. T. Yefsah, R. Desbuquois, L. Chomaz, K. J. Günter, and J. Dalibard, Exploring the thermodynamics of a two-dimensional Bose gas, *Phys. Rev. Lett.* **107**, 130401 (2011).

13. R. P. Smith, R. L. D. Campbell, N. Tammuz, and Z. Hadzibabic, Effects of interactions on the critical temperature of a trapped Bose gas, *Phys. Rev. Lett.* **106**, 250403 (2011); R. P. Smith, N. Tammuz, R. L. D. Campbell, M. Holzmann, and Z. Hadzibabic, Condensed fraction of an atomic Bose gas induced by critical correlations, *Phys. Rev. Lett.* **107**, 190403 (2011); A. L. Gaunt, T. F. Schmidutz, I. Gotlibovych, R. P. Smith, and Z. Hadzibabic, Bose–Einstein condensation of atoms in a uniform potential, *Phys. Rev. Lett.* **110**, 200406 (2013).

14. C. N. Weiler, T. W. Neely, D. R. Scherer, A. S. Bradley, M. J. Davis, and B. P. Anderson, Spontaneous vortices in the formation of Bose–Einstein condensates, *Nature* **455**, 948 (2008).

15. M. H. Anderson, J. R. Ensher, M. R. Matthews, C. E. Wieman, and E. A. Cornell, Observation of Bose–Einstein condensation in a dilute atomic vapor, *Science* **269**, 198 (1995).

16. C. A. Sackett, J. M. Gerton, M. Welling, and R. G. Hulet, Measurements of collective collapse in a Bose–Einstein condensate with attractive interactions, *Phys. Rev. Lett.* **82**, 876 (1999).

17. L. Radzihovsky and D. E. Sheehy, Imbalanced Feshbach-resonant Fermi gases, *Rep. Prog. Phys.* **73**, 076501 (2010).

18. F. Chevy and C. Mora, Ultra-cold polarized Fermi gases, *Rep. Prog. Phys.* **73**, 112401 (2010).

19. W. Zwerger (ed.), *BCS-BEC Crossover and the Unitary Fermi Gas*, Lecture Notes in Physics, Springer, Heidelberg, Germany (2011).

20. D. S. Petrov, C. Salomon, and G. V. Shlyapnikov, Weakly bound dimers of fermionic atoms, *Phys. Rev. Lett.* **93**, 090404 (2004).

21. M. J. H. Ku, A. T. Sommer, L. W. Cheuk, and M. W. Zwierlein, Revealing the superfluid lambda transition in the universal thermodynamics of a unitary Fermi gas, *Science* **335**, 563 (2012).

22. K. Van Houcke, F. Werner, E. Kozik, N. Prokof'ev, B. Svistunov, M. J. H. Ku, A. T. Sommer, L. W. Cheuk, A. Schirotzek, and M. W. Zwierlein, Feynman diagrams versus Fermi-gas Feyman emulator, *Nat. Phys.* **8**, 366 (2012).

23. H. Kamerlingh Onnes, Further experiments with liquid helium. C. On the change of electric resistance of pure metals at very low temperatures, etc. IV. The resistance of pure mercury at helium temperatures, *Comm. Phys. Lab. Univ. Leiden*; No. 120b (1911); Further experiments with liquid helium. D. On the change of electric resistance of pure metals at very low temperatures, etc. V. The disappearance of the resistance of mercury, *Comm. Phys. Lab. Univ. Leiden*; No. 122b (1911); Further experiments with liquid helium. G. On the electrical resistance of pure metals, etc. VI. On the sudden change in the rate at which the resistance of mercury disappears, *Comm. Phys. Lab. Univ. Leiden*; No. 124c (1911).

24. J. Bardeen, L. N. Cooper, and J. R. Schrieffer, Microscopic theory of superconductivity, *Phys. Rev.* **106**, 162 (1957); Theory of superconductivity, *Phys. Rev.* **108**, 1175 (1957).

25. J. G. Bednorz and K. A. Müller, Possible high T_c superconductivity in the Ba-La-Cu-O system, *Z. Physik B* **64**, 189 (1986).

26. G. Blumberg, A. Mialitsin, B. S. Dennis, M. V. Klein, N. D. Zhigadlo, and J. Karpinski, Observation of Leggett's collective mode in a multi-band MgB_2 superconductor, *Phys. Rev. Lett.* **99**, 227002 (2007).

27. Y. Kamihara, T. Watanabe, M. Hirano, and H. Hosono, Iron-based layered superconductor La[O1-xFx]FeAs (x = 0.05–0.12) with T_c = 26 K, *J. Amer. Chem. Soc.* **130**, 3296 (2008).

28. A. Bohr, B. R. Mottelson, and D. Pines, Possible analogy between the excitation spectra of nuclei and those of the superconducting metallic state, *Phys. Rev.* **110**, 936 (1958).

29. S. T. Belyaev, Effect of pairing correlations on nuclear properties, *Mat. Fys. Medd. Dan. Vid. Selsk.* **31**, No. 11 (1959).

30. A. B. Migdal, Superfluidity and the moments of inertia of nuclei, *Nucl. Phys.* **13**, 655 (1959).

31. A. B. Migdal, Superfluidity and the moments of inertia of nuclei, *Zh. Eksp. Teor. Fiz.* **37**, 249 (1959) [*Sov. Phys. JETP* **10**, 176 (1960)].

32. V. L. Ginzburg and D. A. Kirzhnits, On the superfluidity of neutron stars, *Sov. Phys. JETP* **20**, 1346 (1965).

33. G. Baym, C. Pethick, and D. Pines, Superfluidity in neutron stars, *Nature* **224**, 673 (1969).

34. E. Wigner and H. B. Huntington, On the possibility of a metallic modification of hydrogen, *J. Chem. Phys.* **3**, 764 (1935).

35. M. I. Eremets and I. A. Troyan, Conductive dense hydrogen, *Nat. Mater.* **10**, 927 (2011).

36. N. W. Ashcroft, Metallic hydrogen: A high-temperature superconductor? *Phys. Rev. Lett.* **21**, 1748 (1968).

37. N. W. Ashcroft, The hydrogen liquids, *J. Phys. Condens. Matter* **12**, A129 (2000).

38. N. W. Ashcroft, Hydrogen dominant metallic alloys: High temperature superconductors? *Phys. Rev. Lett.* **92**, 187002 (2004).

39. E. Babaev, Phase diagram of planar U(1) x U(1) superconductors: Condensation of vortices with fractional flux and a superfluid state, *Nucl. Phys. B* **686**, 397 (2004).

40. E. Babaev, A. Sudbo, and N. W. Ashcroft, A superconductor to superfluid phase transition in liquid metallic hydrogen, *Nature* **431**, 666 (2004).

41. D. D. Ivanenko and D. F. Kurdgelaidze, Hypothesis concerning quark stars, *Astrophysics* **1**, 479 (1965).

42. D. D. Ivanenko and D. F. Kurdgelaidze, Remarks on quark stars, *Nuovo Cimento* **2**, 13 (1969).

43. D. D. Ivanenko and D. F. Kurdgelaidze, Quark stars, *Sov. Phys. J.* **13**, 1015 (1970).

44. M. G. Alford, A. Schmitt, K. Rajagopal, and T. Schafer, Color superconductivity in dense quark matter, *Rev. Mod. Phys.* **80**, 1455 (2008).

45. E. B. Sonin, Analogs of superfluid currents for spins and electron-hole pairs, *Sov. Phys. JETP* **47**, 1091 (1978); Spin currents and spin superfluidity, *Adv. Phys.* **59**, 181 (2010).

46. A. S. Borovik-Romanov, Yu. M. Bunkov, V. V. Dmitriev, and Yu. M. Mukharskiy, Long-lived induction signal in superfluid $_3$He-B, *JETP Lett.* **40**, 1033 (1984).

47. J. P. Eisenstein and A. H. MacDonald, Bose–Einstein condensation of excitons in bilayer electron systems, *Nature* **432**, 691 (2004).

48. J. Kasprzak, M. Richard, S. Kundermann, A. Baas, P. Jeambrun, J. M. J. Keeling, F. M. Marchetti et al., Bose–Einstein condensation of exciton polaritons, *Nature* **443**, 409 (2006).

49. S. O. Demokritov, V. E. Demidov, O. Dzyapko, G. A. Melkov, A. A. Serga, B. Hillebrands, and A. N. Slavin, Bose–Einstein condensation of quasi-equilibrium magnons at room temperature under pumping, *Nature* **443**, 430 (2006).

50. T. Nikuni, M. Oshikawa, A. Oosawa, and H. Tanaka, Bose–Einstein condensation of diluted magnons in $TlCuCl_3$, *Phys. Rev. Lett.* **84**, 5868 (2000).

51. Ch. Rüegg, N. Cavadini, A. Furrer, H.-U. Güdel, K. Krämer, H. Mutka, A. Wildes, K. Habicht, and P. Vorderwisch, Bose–Einstein condensation of the triplet states in the magnetic insulator $TlCuCl_3$, *Nature* **423**, 62 (2003).

52. Yu. M. Bunkov and G. E. Volovik, Bose–Einstein condensation of magnons in superfluid ^3He, *J. Low Temp. Phys.* **150**, 135 (2008).

53. K. L. Telschow and R. B. Hallock, Observations on ^4He persistent flow, *Phys. Rev. B* **27**, 3068 (1983).

54. M. Greiner, O. Mandel, T. Esslinger, T. W. Hänsch, I. Bloch, Quantum phase transition from a superfluid to a Mott insulator in a gas of ultracold atoms, *Nature* **415**, 6867 (2002).

55. D. Jaksch, C. Bruder, J. I. Cirac, C. W. Gardiner, and P. Zoller, Cold bosonic atoms in optical lattices, *Phys. Rev. Lett.* **81**, 3108 (1998).

56. M. Boninsegni, Phase separation in mixtures of hard core bosons, *Phys. Rev. Lett.* **87**, 087201 (2001).

57. A. B. Kuklov and B. V. Svistunov, Counterflow superfluidity of two-species ultracold atoms in a commensurate optical lattice, *Phys. Rev. Lett.* **90**, 100401 (2003).

58. E. Altman, W. Hofstetter, E. Demler, and M. Lukin, Phase diagram of two-component bosons on an optical lattice, *New J. Phys.* **5**, 113 (2003).

59. S. G. Söyler, B. Capogrosso-Sansone, N. V. Prokof'ev, and B. V. Svistunov, Sign-alternating interaction mediated by strongly-correlated lattice bosons, *New J. Phys.* **11**, 073036 (2009).

Index